Optoelectronic Technology and Lightwave Communications Systems

Optoelectronic Technology and Lightwave Communications Systems

EDITED BY

Chinlon Lin

Bellcore (Bell Communications Research)
Red Bank, New Jersey

VNR VAN NOSTRAND REINHOLD
New York

Project Supervision by The Total Book

Van Nostrand Reinhold
115 Fifth Avenue
New York, New York 10003

Van Nostrand Reinhold (International) Limited
11 New Fetter Lane
London EC4P 4EE, England

Van Nostrand Reinhold
480 La Trobe Street
Melbourne, Victoria 3000, Australia

Macmillan of Canada
Division of Canada Publishing Corporation
164 Commander Boulevard
Agincourt, Ontario MIS 3C7, Canada

16 15 14 13 12 11 10 9 8 7 6 5 4 3 2 1

Library of Congress Cataloging-in-Publication Data

Optoelectronic technology and lightwave communications
  systems.

  Bibliography: p.
    1. Optical communications.   2. Optoelectronic
devices.   I. Lin, Chinlon.
TK5103.59.O695   1989   621.38′0414   88-27724
ISBN 0-442-26050-4

Contents

Contributors

P. J. Anthony, Ph.D.
Department Head, Optoelectronic Devices Department
AT&T Bell Laboratories
Murray Hill, NJ 07974

J. E. Bowers, Ph.D.
Professor, Department of Electrical and Computer Engineering
University of California
Santa Barbara, CA 93106

J. A. Bucaro, Ph.D.
Head, Physical Acoustics Branch
Naval Research Laboratory
Washington, DC 20375

W. K. Burns, Ph.D.
Research Analyst
Naval Research Laboratory
Washington, DC 20375

J. C. Campbell, Ph.D.
Professor of Electrical and Computer Engineering
Cockrell Family Regents Chair in Engineering
University of Texas at Austin
Austin, TX 78712

A. R. Chraplyvy, Ph.D.
Distinguished Member of Technical Staff
AT&T Bell Laboratories
Holmdel, NJ 07733

L. G. Cohen, Ph.D.
Supervisor
AT&T Bell Laboratories
Murray Hill, NJ 07974

J. H. Cole, M.S.
Vice President, Dylor Corporation
Arlington, VA 22202

A. Dandridge, Ph.D.
Head, Optical Sensor Section
Naval Research Laboratory
Washington, DC 20375

N. K. Dutta, Ph.D.
Supervisor
AT&T Bell Laboratories
Murray Hill, NJ 07974

Luis Figueroa, Ph.D.
Manager, Photonics Laboratory
Boeing High Technology Center
Seattle, WA 98124

T. G. Giallorenzi, Ph.D.
Director, Optical Sciences Division
Naval Research Laboratory
Washington, DC 20375

Elmer H. Hara, Ph.D.
Professor, SaskTel-NSERC Industrial Research Chair
Electronic Information Systems Technology
Faculty of Engineering
University of Regina
Regina, Saskatchewan, Canada

T. Ikegami, Dr.Eng.
Executive Manager, Photonic Functional Laboratory
NTT Opto-electronics Laboratories
Kanagawa, Japan

Charles K. Kao, Ph.D.
Vice-Chancellor
The Chinese University of Hong Kong
New Territories, Hong Kong

Felix P. Kapron, Ph.D.
District Manager, Optical Systems Technology
Bellcore (Bell Communications Research)
Morristown, NJ 07960

Joseph Katz, Ph.D.
Director, Research and Development
Symbol Technologies, Inc.
Bohemia, NY 11716

Giok-Djan Khoe, Professor
Technical University Eindhoven, and Senior Scientist, Philips Research
    Laboratory
Eindhoven, The Netherlands

Kohroh Kobayashi, Dr.Eng.
Manager, Optical Devices Development Department
Fiber Optic Communications Development Division
NEC Corporation
Kanagawa, Japan

U. Koren, Ph.D.
Member of the Technical Staff
AT&T Bell Laboratories
Holmdel, NJ 07733

Masaki Koyama, Ph.D.
Executive Manager
NTT Transmission Systems Laboratories
Kanagawa, Japan

Chinlon Lin, Ph.D.
District Manager, Lightwave Technology Applications Research
Bellcore (Bell Communications Research)
Red Bank, NJ 07701

W. L. Mammel, M.Sc.
Former Member of Technical Staff
AT&T Bell Laboratories
Holmdel, NJ 07733

Tran Van Muoi, Ph.D.
Director, Research, Development & Engineering
PCO Inc.
Chartsworth, CA 91311

Nobuo Nishida, Dr.Eng.
Manager
NEC Corporation
Kanagawa, Japan

T. Okoshi, Dr.Eng.
Director and Professor, Research Center for Advanced Science and
    Technology
University of Tokyo
Tokyo, Japan

Yuzo Ono, Dr.Eng.
Research Manager
NEC Corporation
Kanagawa, Japan

Dan L. Philen, Ph.D.
Member of Technical Staff
AT&T Bell Laboratories
Norcross, GA 30071

W. A. Reed, Ph.D.
Distinguished Member of Technical Staff
AT&T Bell Laboratories
Murray Hill, NJ 07974

P. W. Shumate, Ph.D.
Division Manager, Optical Networks Research Division
Bellcore (Bell Communications Research)
Morristown, NJ 07960

D. W. Smith, B.Tech
Head of Section,
British Telecommunications Research Laboratories
Ipswick, Suffolk, England

I. W. Stanley, Ph.D.
Senior Engineering Advisor
British Telecommunications Research Laboratories
Ipswich, Suffolk, England

Nobuyuki Tokura, M.S.
Supervisor
NTT Transmission System Laboratories
Kanagawa, Japan

W. C. Young, M.S.
District Manager, Lightguide Technology Research
Bellcore (Bell Communications Research)
Red Bank, NJ 07701

Preface

Ever since the invention of the transistor, semiconductor-based microelectronics has made a revolutionary impact on the information society, as evident from the widespread application of microprocessor-based technology in our modern society.

The next wave of modern information technology, after transistors and microelectronics, is that of lasers and micro-optoelectronics. Optoelectronics, or optical electronics, based on lasers and related modern optical technology, has also become a very important field of science and technology in the past 20 years.

Electronics or microelectronics deals with (micro)electronic devices and components for generation, transmission, and processing of electronic signals. In contrast, in optoelectronics we deal with optoelectronic devices and components for the generation, transmission, and processing of lightwave signals. It is the interaction of lightwaves (photons) with matter that shows the uniqueness of optoelectronic technology; optical absorption and scattering, optical gain and amplification, material and waveguide dispersion, nonlinear optical effects, etc., are very much dependent on the material's intrinsic properties and the lightwave propagation effects.

Historically, the invention of lasers in 1960 has led to a wide range of very significant scientific and technological progresses. Laser science and technology constitutes a very important foundation of optoelectronics. In the last 20 years or so, the impact of laser and modern optoelectronic technology in basic scientific research, in industrial, medical, and biological applications has been spectacular. However, the impact of optoelectronic technology on telecommunications and information systems was not felt until after the 1970s. In 1970, the demonstration of both the low loss optical fiber waveguide transmission and the room-temperature continuous operation of the miniature semiconductor diode laser excited the telecommunications research laboratories worldwide and started a very important chapter in the history of optoelectronics for telecommunications.

Since then, the advances in optoelectronics and lightwave technology have been phenomenal; the practical realization of optical fiber transmission systems and broadband optical network for telecommunications is making the information age closer to reality. Starting in 1980, the technology of single-mode optical fibers, microelectronics, and semiconductor diode lasers combined is making a truly significant impact on lightwave communication systems. In addition, diode-laser-based optical information storage and processing systems (compact audio disks, video disks, optical data disks, laser printers, laser bar-code scanners, etc.) also see significant advances after 1980; these are important parts of advanced information systems.

As we move toward the 1990s, there is no doubt that the new era of information age will see the widespread impact of optoelectronics-based information technology and systems in every aspect of our life, including office, factory, school, hospital, home, etc.

As we move toward this new era of information age based on optoelectronic information technology, it is time to have a book which reviews the fundamentals of these key optoelectronic technologies and the essential aspects of some key lightwave system applications. This book is intended to serve that purpose. This book hopes to be one which provides the reader with an in-depth background and working knowledge of:

1. the foundations of key optoelectronic technologies used in information system applications;
2. the important features and design parameters of some important lightwave systems;
3. where the future is going in optoelectronic technology and systems applications; and
4. the significance of the impacts that optoelectronic technology and lightwave systems will have on the future information society.

The readers are assumed to have a physics or engineering background (B.S. degree or above), to have taken an introductory course in lasers and optical electronics, and to have an interest in optoelectronic technology and lightwave system applications such as optical communications and optoelectronic information transfer, processing, and storage. Most of fiber-optic engineers, communication engineers, optoelectronic scientists and engineers, and graduate students in engineering and physics can benefit from such a broad coverage of the key technologies and system applications. This is not an introductory book—there are at least ten such books in print in that category—but is an intermediate-to-advanced level book on lightwave technology and systems. For technology such as optoelectronics and lightwave systems which are found in more and more advanced applications, introductory books may not be enough to meet the needs of many in the field. In this book readers can choose the level at which they want to utilize this book, because the fundamentals are reviewed first; the more advanced topics which bring the readers close to the state of the art are then discussed.

This book focuses on the "micro-optoelectronic" technology of optical fibers and semiconductor lasers, as well as the applications in lightwave systems based on these technologies. The organization of the book should allow the reader to either go through the book sequentially or to go to a particular subject of interest, read it, and then come back to the other chapters for a broader coverage of the related topics.

The book is divided into seven parts:

1. **Optical Fiber Waveguides** *(Chapters 1–5)*
   - fiber waveguides and transmission properties
   - characterization and measurement of optical fibers
   - advanced single-mode fibers including dispersion-shifted and dispersion-flattened fibers
   - polarization-maintaining and single-polarization fibers
   - fiber transmission limitations due to nonlinear effects

2. **Fiber-Joining Technology and Passive Optical Components** *(Chapters 6–7)*
   - connectors, splices, wavelength-division-multiplexers (WDM), couplers, device-fiber coupling technology, etc.

3. **Semiconductor Laser Sources and Photodetectors** *(Chapters 8–14)*
   - basic physics of semiconductor lasers
   - semiconductor laser diode fabrication and characterization
   - transverse-mode control and spectral control, single-longitudinal-mode DFB laser characteristics
   - high-speed semiconductor lasers and direct modulation
   - high-power semiconductor laser diodes and diode arrays
   - photodetectors (PINs and APDs)

4. **Optical Transmitters and Receivers** *(Chapters 15–16)*
   - semiconductor laser transmitters and reliability considerations
   - optical receiver design considerations

5. **Applications of Optoelectronics in Lightwave Systems** *(Chapters 17–22)*
   - optical communications based on single-mode fiber transmission systems for long-haul and subscriber loop (Fiber-to-the-Home) applications
   - optical communications in local area networks between computers and workstations, etc., within buildings or campuses
   - broadband optical fiber network for information-age services such as broadcasting of NTSC TV and HDTV, video-on-demand, video library browsing, two-way video phone, interactive hypermedia (video/graphics/data), etc., based on optical laser disks for video/audio/data storage and optical fiber networks for communications

- free-space (e.g., intersatellite) optical communication system
- optical fiber sensor technologies for defense and industrial sensing applications
- optoelectronic information processing technology of laser printers, laser bar-code scanners, and related systems

6. **Future Optoelectronic Technology and Transmission Systems** *(Chapters 23–24)*
   - OEIC, optoelectronic integrated circuit technology
   - coherent optical fiber transmission technology and systems

7. **Impacts on the Information Society** *(Chapter 25)*

The contributors of these chapters are experts in their field of specialty; they are from various research laboratories around the world. The style of writing is necessarily different, but the unique flavor of each chapter is what makes this book interesting. While not all the chapters cover the state of the art, the most essential topics and principles have been discussed. Readers will find these chapters provide useful background knowledge and insights. For those interested in exploring more advanced topics and latest information, the references at the end of the chapters as well as the latest publications (for example, papers in *IEEE/OSA Journal of Lightwave Technology, Electronics Letters*, and the Technical Digests of major conferences on optical fiber communications, semiconductor lasers, communication systems, etc.) should be consulted.

* * *

I would like to express my sincere thanks to all the authors for their excellent contributions to this book. Many researchers in this field of optoelectronics and lightwave technology contributed to the spectacular advances of the last two decades or so, and they all deserve our thanks. Let us hope that together we scientists and engineers not only make spectacular advances in the optoelectronic technology and lightwave information systems, but also help to bring about true communication and human understanding in this small world of planet Earth.

I wish to thank Professor N. Holonyak, Jr., of the University of Illinois, Professors T. K. Gustafson, A. Dienes, and J. R. Whinnery of the University of California, Berkeley, and Mr. and Mrs. Jesse Cox for their encouragement over the years. I also would like to thank my wife, Helen (Huan-huan), for her assistance and encouragement. In addition, I would like to thank the editing staff of Van Nostrand Reinhold and Annette Bodzin of The Total Book for their expert assistance.

May the Light be with you always.

CHINLON LIN

PART 1

Optical Fiber Waveguides

CHAPTER 1

Transmission Properties of Optical Fibers

Felix P. Kapron

1.1 INTRODUCTION

Since the first low-loss fiber in 1970, fiber optics has emerged from being a laboratory curiosity to constituting a significant portion of the communications and sensors business. Throughout the world, optical cables are carrying plain old telephone-service across land and sea, along with data and video, at rates exceeding several gegabytes per second. Compared with radiowave and microwave communications, lightwave transmission has some distinguishing features:

Short wavelengths ($\lambda \sim 0.8-1.7$ μm), so that source, fiber, detector, and interconnection dimensions are small (~ 1 mm to 1 cm)
High frequencies ($v \sim 175-375$ THz), so that enormous modulation rates (to 1 Tb/s?) are potentially possible
Little electrical interference

However, lightwave transmission through the atmosphere has problems with

Scattering loss
Diffraction, requiring large receiving antennae
Negotiating bends around buildings and other obstacles
Signal security

Hence, *guided* transmission is more desirable, and optical fiber cable wins on many counts:

No suitable alternatives. Lens and prism trains, gas tube lenses, and hollow reflecting tubes are all too lossy and complex.

Low attenuation. 0.15–5 dB/km, uniform over a wide range of modulation frequencies.

High bandwidth. 10 MHz-km to over 1 THz-km.

Small size. Cables and components less than a few centimeters in diameter.

Light weight. Under 3 g per meter of cabled fiber for higher counts.

Flexibility. Can be bent to radii of a few centimeters without deleterious mechanical or optical effects.

Electrical immunity. All-dielectric construction, avoiding problems in radio frequency interference (RFI), electromagnetic pulses (EMP), grounding and lightning.

Preservation of coherence and polarization for coherent transmission.

Security. Negligible crosstalk, difficult to tap.

Falling costs, approaching $100 per kilometer of cabled fiber. Sources and detectors, splices and connectors, couplers, and multiplexers are all dropping in cost as well.

Long repeater spacing and high bit rates. Current and near-term installations are at 90 Mb/s to 2.5 Gb/s, at unrepeated distances up to 100 km; in excess of hundreds of kilometers is possible in the laboratory with optical amplification, where bit rates have exceeded 10 Gb/s.

Future inline optical amplification using semiconductor diodes or active fibers and fewer regenerative repeaters.

Future solution propagation for femtosecond pulses and terabit-per-second bit rates.

1.2 FIBER LIGHT GUIDANCE

Let us examine just why fibers are capable of guiding light the way they do.

1.2.1 The Basic Principle

An optical fiber or lightguide is a cylindrical low-loss dielectric structure consisting of a *cladding* region of uniform refractive index $n(\lambda)$ surrounding a *core* region of radius a and higher varying index,

$$n(r, \lambda) = n(\lambda)[1 + \Delta(r, \lambda)] \qquad r \leq a \qquad (1)$$

Here r is the radial coordinate and Δ is a relative index difference; the profile also depends upon wavelength λ. Figure 1.1 shows some simple profiles; for some of these the distinction between core and varying cladding is not clear.

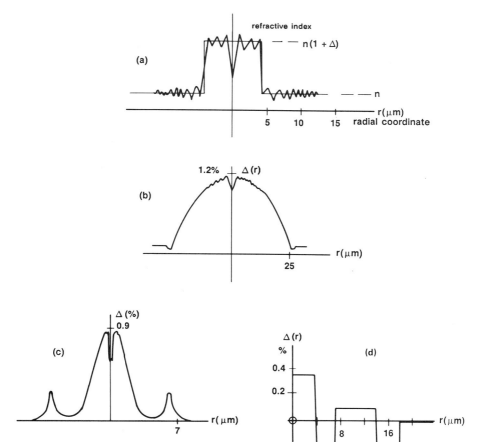

Figure 1.1 Some refractive index profiles. (a) Step-index single-mode (schematic and actual by MCVD); (b) parabolic index multimode (actual by MCVD); (c) triangular index single-mode (actual by OVD); (d) quadruply clad or segmented core single-mode (schematic).

Recall that in a homogeneous medium, light rays travel in straight lines. Any variation in the refractive index $n(\bar{\rho})$, where $\bar{\rho}$ is the position vector of any typical point along the ray, will cause a ray deviation rate satisfying [1]

$$\frac{d}{ds}\left[n(\bar{\rho})\hat{T}\right] = \text{grad } n(\bar{\rho}) \tag{2a}$$

Here s is the distance along the ray of the typical point from any fixed point on the rath path, and $\hat{T} = d\bar{\rho}/ds$ is a unit vector tangent to the ray at the typical point.

Differentiating through Eq. (2a) gives the curvature vector

$$\frac{d\hat{T}}{ds} = \bar{G} - \hat{T}(\hat{T} \cdot \bar{G}) \tag{2b}$$

directed toward the local center of curvature of the ray. The component $\bar{G} = (\text{grad } n)/n$ is in the direction of the steepest increasing gradient of log n. The second component is an "inertial" component antiparallel to the ray tangent $\hat{T}$. It reduces the tendency of the ray to head in the direction of $\bar{G}$ toward the higher-index region.

In a circularly symmetrical fiber, the vector of magnitude $G = dn/n \, dr$ is directed along the radial coordinate r, so that a ray in the outer core region near the cladding tends toward the fiber's central z-axis. However, the $\hat{T} \cdot \bar{G}$ magnitude of the opposing tangential component thereby increases. This means that the ray will go past the fiber axis and then experience a "turn-around" near the cladding when the ray and gradient are perpendicular. In this fashion, a guided ray periodic along the fiber z axis is established. The ray is periodic with a precessional movement in the transverse cross section as well. Both of these are shown in the side view and end view of Figure 1.2.

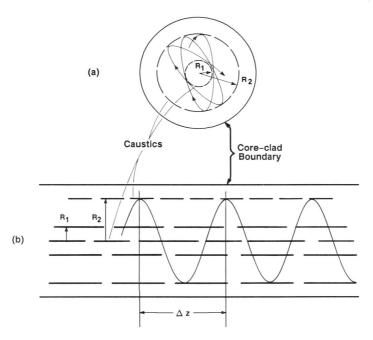

Figure 1.2 Ray picture of a mode contained within a tube bounded by inner and outer caustics at $r = R_1, R_2$, respectively. (a) Cross-sectional view showing precessional movement (b) longitudinal side view showing the period Δz. The latter may be quite complex depending upon the relative number of radial and axial oscillations.

1.2.2 Modes

An accurate description of fiber transmission requires the fundamental Maxwell equations with boundary conditions. Fortunately, these simplify when the electric ($\bar{E}$) and magnetic ($\bar{M}$) fields are taken to be harmonic in time t, in the angular cross-sectional coordinate ϕ, and along the longitudinal z axis:

$$\bar{E}, \bar{M} \propto \exp i(\omega t - \beta z \pm L\phi) \qquad L = 0, 1, 2, \ldots \tag{3}$$

Here $\omega = 2\pi \nu$ is determined by the light frequency ν or wavelength $\lambda = c/\nu$; equivalently $\omega = ck$, where $k = 2\pi/\lambda$ is the vacuum wavenumber. Then $\beta = n_p k$ is the propagation wavenumber, where n_p is a phase index; ω/β is the phase velocity c/n_p in the fiber.

The resulting simplified waveguide equations can be easily solved algebraically for only the ideal step index and parabolic index profiles of Figure 1.1. However, if the profile gradient $\bar{G}$ in Eq. (2b) (i.e., $d\Delta/\Delta\ dr$) is small enough, an approximate simple scalar waveguide equation results. Since the largest relative index difference Δ in common fiber ranges from about 0.25 to 2.5% over radial distances from 3 to 50 μm, respectively, this is valid for most cases.

Then

$$\frac{d^2 E}{dr^2} + \frac{dE}{r\ dr} + q^2(r)E = 0 \tag{4a}$$

where the transverse Cartesian component of the electric field $E(r)$ and its derivative are finite and continuous everywhere. The longitudinal fields are small in this approximation, so the mode groups are labeled LP_{LM} for linearly polarized. Here

$$q^2(r) = k^2 n^2(r, \lambda) - \beta^2 - (L^2 - 1/4)r^2 \tag{4b}$$

using the index profile from Eq. (1). The WKBJ-type solutions are approximately of the form

$$E(r) \propto r^{-1/2} \exp\left[\pm irq(r)\right] \tag{5}$$

Wherever $q(r)$ is real, the fields are oscillatory in the radial coordinate r, possessing M zero nodes such that

$$(M + 1/2)\pi = \int_{R_1}^{R_2} q(r)\ dr \qquad M = 0, 1, 2, \ldots \tag{6}$$

This *characteristic equation* determines discrete values of the propagation wavenumber $\beta(\lambda)$ for various *modes* specified by the (L, M) pair of integers. Each pair is in turn fourfold degenerate, corresponding to the plus/minus choice of the angular coordinate L in Eq. (3) and to the x, y polarizations of the transverse E field in Eq. (4a). The positions $r = R_1$, R_2 are the inner and outer *caustics*, shown in Figure 1.2, where $q(R_1$ or $R_2) = 0$. The light paths, or *rays*, exist inside this tube, whereas outside it (i.e., $r < R_1$, $r > R_2$) the fields are evanescently decaying beyond the ray turnarounds at the caustics.

In general, the propagation wavenumber β is multivalued for a fixed wavelength; that is, a number of modes travel through the *multimode* fiber. However, some fiber designs allow only one mode at wavelengths of interest; such *single-mode* fiber has particular advantages and disadvantages.

1.3 FIBER MATERIALS AND WAVELENGTHS

Let us next examine how certain materials are associated with wavelengths of interest for fiber optics.

1.3.1 Materials and Processes

One consideration is low attenuation, which requires a nonconducting dielectric. Performance depends strongly upon the material type and the operating wavelength. Dielectric materials such as plastics have only narrow wavelength ranges within which the attenuation coefficient is under several hundred decibels per kilometer. Low melting point "soft" glasses, usually made by drawing from multiple concentric crucibles, can get down to several decibels per kilometer over a somewhat wider wavelength window. The best performance to date, however, is reserved for high-silica (SiO_2) glasses.

Silica fibers are made by several vapor deposition processes. The modified chemical vapor deposition (MCVD) [2] or (inside vapor-phase oxidation (IVPO) is shown schematically in Figure 1.3. It begins with a clean, uniform, hollow substrate tube that is mounted in the concentric and synchronously rotating chucks of a glass-working lathe. Heating is commonly accomplished with a traversing oxyhydrogen torch to the 1300–1600°C range, while gases are continuously passed into (and exhausted out of) the tube. These are chlorides of silicon, germanium, phosphorus, and others, along with possible fluorides, all within inert carrier gases and oxygen. Although the chemical processes are very complex, the net result is of the form $SiCl_4(g) + O_2(g) \rightarrow SiO_2(s) + 2Cl_2(g)$. The oxides and silica fluorides form a glassy film deposit on the tube inner wall. The gas recipe is changed with each pass as first clad material and then core material is deposited during several tens of passes. Deposition rates of a few grams per minute are common.

In the laboratory, a plasma-assisted chemical vapor deposition (PA-CVD) process has been used [2]. Traversing water cooling outside the tube and an RF plasma inside the tube increase the deposition rate, while the flame now consolidates the soot into a clear glass layer. In the plasma-activated (PCVD) production process, the flame is eliminated [3]. Although deposition rates are not as high, a substrate tube temperature below 1200°C

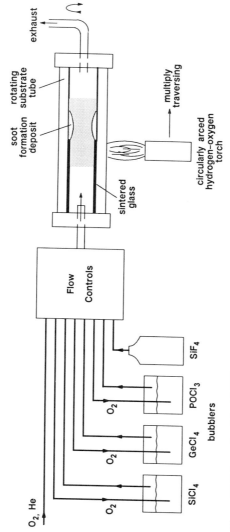

Figure 1.3 Schematic of the MCVD process.

reduces any tube deformation problems and possible OH ion migration from the tube to the deposit. Deposition efficiency may be higher, but the rates are lower; several hundred passes may be needed for careful profile control.

For all three variants of MCVD, the tube is heated to about 2000°C and then collapsed into a rod. The rod may be jacketed within another (undeposited) collapsing tube to obtain the final rod *preform* representing tens of kilometers of fiber. Alternatively, the collapsing and/or overjacketing may be done during fiber draw.

As compared with the inside deposition process just described, there are two types of outside deposition processes shown in Figure 1.4. In outside vapor deposition (OVD) [4] or outside vapor-phase oxidation (OVPO),

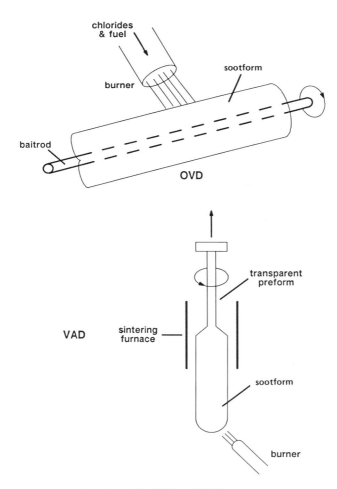

Figure 1.4 Schematic of the OVD and VAD processes.

halide gases are fed through a central orifice surrounded by a concentric orifice of shielding gas and then an outer orifice of methane fuel gas and oxygen. At ambient pressure, the heat of the reaction causes oxide particles of about 0.1 μm diameter to be formed and deposited onto the outside of a horizontally rotating refractory target rod. After hundreds of traversals, the diameter of the resulting *sootform* can exceed 15 cm, and such a large target area allows a deposition rate several times that of MCVD (though without the contribution of a starting tube). The bait rod is then removed or, in some processes, retained as a silica rod.

With vapor-phase axial deposition (VAD) [5] or axial vapor-phase deposition (AVD) there is no traversing. Rather, one or more torches are situated to deposit the full refractive index profile to grow a sootform longitudinally downward from a starting vertical bait rod. Torch angles and shape (e.g., rectangular with eccentric design) are critical.

All forms of vapor deposition have the advantage that a very high level of purity can be attained in the starting chemicals and in the delivery system, and impurities with a low vapor pressure will never volatilize. *Dopant* levels of several mole percent and higher can modify the refractive index. Germania (GeO_2, which is relatively expensive) and phosphorus pentoxide (P_2O_5) are common index enhancers. Alumina (Al_2O_3) and titania (TiO_2) are also under study. Index depressants include fluorine (F) and borosilica (B_2O_3), although the latter is used in only some fiber designs since it has attenuation problems beyond 1100 nm.

Unlike MCVD, the outside processes must undergo dehydration and consolidation (or drying and sintering). The first step occurs by placing the sootform at an elevated temperature in an $SiOCl_2$ or Cl_2 atmosphere to remove the OH ion in a reaction such as $2SiOH + Cl_2 \rightarrow SiOSi + 2HCl + \frac{1}{2}O_2$. The bore left by OVD is advantageous in allowing efficient drying in the region where light will propagate. In an appropriate mixture of inert and drying gases at even higher temperatures, the sootform is zone-refined into a rod of shrunken, densified, bubble-free glass. With VAD, this may be done online during the deposition as indicated in Figure 1.4.

There are a number of variations that can occur. The deposition/consolidation process may be repeated for complex profiles. The rod may be directly pulled into a fiber or jacketed with a tube as with MCVD. Alternatively, more outside deposition may be done to the rod, or to the rod after elongation and cutting. Doping may be done during drying and/or sintering. Combinations of VAD and OVD may be used, horizontally or vertically. Preforms representing several hundred kilometers may be produced.

In addition to index profile modification, silica dopants have the effect of lowering the deposition temperature and increasing the deposition rate. Doped silicas also have controllable thermal expansion coefficients. Through the stress-optic effect, this inherently modifies the index profile such that there is a small difference between preform profile prior to draw and fiber profile after draw. Moreover, built-in internal stresses are commonly used for fibers of specific polarization properties. These dopants can affect the non-linear behavior that occurs when very high optical power densities are maintained in the fiber. Others, such as neodymium (Nd^{3+}) or erbium (Er^{3+}) ions, can be used for fiber filters lasers, and amplifiers. Finally, cerium (Ce) can be added to improve resistance to loss damage due to nuclear radiation.

If its refractive index profile is measured to be suitable, the glass rod *preform* is heated to about 2200°C in a furnace at the top of a drawtower, as indicated in Figure 1.5 [6]. Usually a graphite electrical resistance furnace or a zirconia microwave induction furnace is used. The latter can operate in an ambient oxidating atmosphere, but the susceptor can suffer from thermal shock, and RF shielding is required at the generator. With a laser scanner and feedback control, the fiber outside diameter is maintained to typically 125 ± 1 μm. The fiber must cool to about 50°C before being coated. To maintain drawspeeds of several meters per seconds, the drawtower height often exceeds 8 m, or else some forced cold air cooling is used. So as not

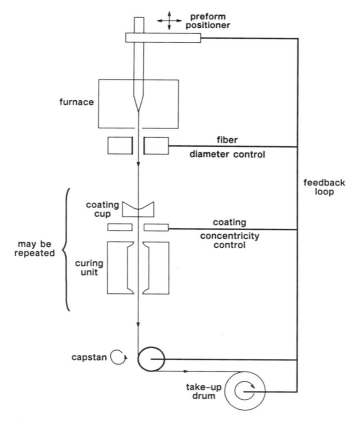

Figure 1.5 Schematic of a drawtower.

to weaken the fiber substantially, the region between the furnace and coater must be kept to a very low air dust particle count.

The fiber now passes into a coating cup and receives an application of a hot-melt thermoplastic elastomer, or silicone rubber, or nylon, or Teflon (which are all thermally cured) or an epoxy or urethane acrylate (increasingly common and cured by ultraviolet light). The coater may be pressurized to inhibit bubble formation, and it is followed by the thermal or UV curing unit. Often a low-modulus inner (primary) buffer cushions the fiber, while a high-modulus outer (secondary) jacket shields against lateral outside forces. Abrasion resistance and colorability are also important for cabling. Such a dual layer coating may be applied with another coater/curer offline, or on line with two in series, or with a concentric coater. A common outside diameter is 250 μm, but some tightly bound cable structures require 500–950-μm jacket diameters.

The stability of fiber attenuation with temperature relates closely to coating and jacket design. Because the expansion coefficient of glass is much smaller than that of the coating, fiber buckling can occur at low temperature, and fiber tension at high temperature. The former can produce microbend attenuation. To counteract this, the primary should be stiff and with a small diameter. However, outside forces argue for a soft, thick primary, so a compromise must be struck.

A take-up spool is set at a tension compatible with drawspeed and furnace temperature. The resulting viscosity is related to fiber attenuation and strength. The fiber is then *prooftested*; usually this occurs offline, although it may be done prior to take-up. A typical proofstrain is 0.5% (corresponding to a stress of about 50 kpsi or 0.345 GN/m²) for less than a half second. This helps ensure some level of service lifetime for the surviving fiber, as during cable fabrication or installation. More rugged submarine or military applications may require prooftesting above 2%.

As a liquid or vapor, water can eventually penetrate through a polymer; this can weaken a fiber if it is under permanent residual stress. To prevent this, especially at high temperatures, a hermetic coating, a few micrometers or less thick, may be applied prior to jacketing. Both metallic and dielectric coatings have been used. Other applications of such coatings are the prevention of excess hydrogen attenuation (discussed later) and the use of some fiber acoustic or electromagnetic sensors.

1.3.2 Attenuation

Spectral attenuation is the loss of useful signal power along a fiber. The attenuation coefficient can be written in the simplified form [7]

$$\alpha(\lambda) = A\lambda^{-4} + B + C \exp\frac{D}{\lambda} + E(\lambda) + F \exp\frac{-G}{\lambda} \tag{7}$$

in dB/km, and typical examples are plotted in Figure 1.6.

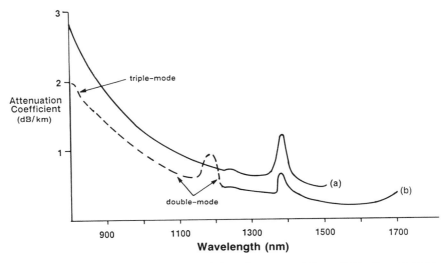

Figure 1.6 Typical spectral attenuation coefficients for commercially available fiber. (a) Multimode fiber; (b) single-mode fiber.

Absorption [the last three wavelength-dependent terms in Eq. (7)] is the conversion of light into heat or of photons into phonons. *Intrinsic absorption* in silica lies below 1 dB/km within the wavelength window 650–1800 nm. At smaller wavelengths, some draw-induced coloration can occur, but the largest UV absorption is due to electronic transitions between the valence and conduction bands that are distributed around 140 nm. However, according to Urbach's rule, the attenuation coefficient decreases exponentially as the third term of Eq. (7) into the visible and near-IR regions. At longer wavelengths above the attenuation window, IR absorption [the last term of Eq. (7)] is due to molecular vibration states having a primary peak around 8 μm but with higher-order combination and overtone bands at a number of shorter wavelengths.

Impurity absorption occurs primarily due to the OH^- ion in silica, and this has a fundamental vibration at 2725 nm. Typical levels of 10 parts per billion (ppb) (although under 1 ppb is achievable) result in the following overtone effects:

nm	880	945	1240	1383
dB/km	0.001	0.01	0.015	0.35

Only the last of these is of concern in todays's fibers, and the "water peak" shows up quite clearly in Figure 1.6. Dopants can interact both with SiO_2 and with OH^-; the germania peak is near 1410 nm. Boron should be avoided if the fiber is to be used past 1100 nm, while phosphorus must be held to low levels if the fiber is used past 1500 nm.

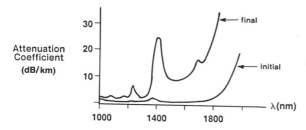

Figure 1.7 Effect of hydrogen on fiber attenuation. Conditions: 100°C for 200 hr in 1 atm H_2.

The tiny molecules of gaseous hydrogen can effectively penetrate into the silica matrix [8], causing multipeak absorption as shown in Figure 1.7; the fundamental is near 2420 nm. The diffusion naturally increases with pressure and temperature until saturation occurs. If this is done at room temperature and then the hydrogen source is removed, out-diffusion renders the effect reversible. However, with increasingly higher temperature there is more residual permanent OH loss around 1.4 μm and a (less important) UV tail. This is due to the formation of Si—OH, Ge—OH, and P—OH chemical bonds (in order of increasing effect); any fluorine will tend to anneal out some defects.

Using modern fibers and cables, hydrogen is of little concern for most applications. Except in undersea applications, it can usually escape. More important hydrogen generators such as polymer fiber jackets and cable materials, metallic outgassing, or undersea galvanic corrosion (zinc/iron, copper/aluminum) have been identified and avoided. Filler compounds and copper tubing serve as water and hydrogen barriers. Consequently, tests and theoretical extrapolations show that extra attenuation can be made negligible over the cable lifetime.

A final form of impurity absorption is due to transition metal ions such as iron, copper, chromium, vanadium, cobalt, and manganese. Modern fabrication methods reduce these levels to below 0.1 ppb, currently resulting in no apparent effects. The general impurity absorptions and defect absorptions (below) are lumped into $E(\lambda)$ of Eq. (7).

Defect absorption can be affected by the thermal history of the glass and can result, for example, in 630-nm absorption due to nonoptimized fiber draw viscosity and temperature conditions. Defects can also be caused by UV light below 400 nm such as occurs during coating cure, but such phenomena are not yet well understood. Still another cause is ionizing radiation such as gamma rays, X-rays, cosmic rays, beta particles, and neutrons due to nuclear reactions around power facilities or in warfare or the use of radioactive isotopes and sterilization technique in medicine, or occurring around X-ray equipment and particle accelerators, in both outer space and terrestrial applications. The Si unit of radiation is the rad, which is equal to the

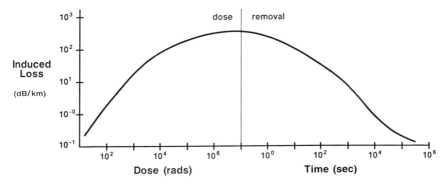

Figure 1.8 Effect of nuclear radiation on fiber attenuation, schematically showing the effects of dosage and recovery.

radiation-applied dose of 100 ergs/g absorbed by silicon. Dose rates in these applications can range from 10^{-4} to 10^4 rad/s.

Fiber behavior is extremely variable [9], but Figure 1.8 schematically shows the effects. On a log-log plot of attenuation increase versus irradiation time, the damage increases linearly to form a color center due to unpaired electrons or holes at a defect (vacancy or interstitial) or at an ion (dopant or impurity). Large pulses may cause luminescence, while very large doses can damage even the fiber coating and possibly weaken the fiber. Silica annealing, especially at higher temperatures, acts to reduce the net rate of increase; saturation may occur. This can cause partial recovery after the irradiation is removed. Photobleaching, due to transmitted or ambient light, accelerates the process. Recovery competes with dose darkening, so higher dose rates would have a steeper slope in the figure. Generally, induced attenuation decreases at wavelengths approaching 1550 nm.

Impurities such as OH and Cl are to be avoided to minimize radiation effects. Overall, pure silica is affected least, but this severely limits fiber design since a fluorine-doped cladding is usually necessary. This is easier to do for single-mode fibers than for multimode fibers. Although germanium doping gives high initial damage, recovery is fairly rapid with minimal permanent damage. The latter is not true with phosphorus doping, which may be helpful only at the lowest levels. Doping with cerium throughout has been shown to be beneficial.

Absorption can be measured directly by microcalorimetry. Over most wavelengths, however, small effects dictate very long fibers and extremely small temperature rises. Few such measurements have been made on modern fibers. The C, D, E, F, G coefficients in Eq. (7) will vary greatly with fiber design.

Scattering [the A and B terms of Eq. (7)] is the conversion of light that is bound to and propagating within the fiber into unbound or radiative or backscattered light. The λ^{-4} dependence is termed Rayleigh scattering; wavelength-independent scattering is given by the B term. *Intrinsic scattering* is due to refractive index fluctuations on a submicrometer molecular level. There are two contributions that can be expressed as

$$A = A_1 + A_2 \Delta^n \tag{8}$$

The term $A_1 = 0.7$ dB-μm^4/km is intrinsic to thermal fluctuations that occur at the glass formation temperature of fused silica. The second term is due to compositional fluctuations of silica dopants for which $A_2 \simeq 0.4$–1 dB-μm^4/km for Δ [see Eq. (1)] expressed in percent; the exponent n ranges from 0.7 to 1. To determine A for a fiber, one needs to know A_2 and n for each dopant, $\Delta(r)$ for the profile due to that dopant, and the effective overlap of the profile with the cross-sectional optical power density.

With modern single-mode fiber, overall $A \simeq 0.8$ dB-μm^4/km typically, and B is below a few hundredths of a decibel per kilometer. At 1300 nm, the UV absorption is only 0.02 dB/km, so overall attenuation coefficients below 0.35 dB/km are typical; 0.27 dB/km has been reported for undoped silica. With multimode fibers, higher doping levels [affecting numerical aperture (NA)] and higher-order mode losses can double these values, the lowest being 0.44–0.6 dB/km, depending upon NA. InGaAsP-based emitters and detectors (or Ge detectors) are used in this long-wavelength region. Before 1980, only GaAlAs emitters and silicon detectors were more readily available for the 850-nm region. Note from Eq. (7) that scattering is higher and UV absorption is about 0.14 dB/km; although the best fiber was not truly single-mode, 1.82 dB/km was achieved. Multimode fiber has shown lowest values of 2.2–2.9 dB/km with increasing NA.

This λ^{-4} behavior explains one of the dramatic improvements of long-wavelength transmission; dispersion, to be discussed later, is another. Note also that scattering and UV absorption both halve in going from 1300 to 1550 nm, the "third window" of operation. Despite slight IR absorption, single-mode attenuation below 0.2 dB/km is typical, and best values of 0.15 and 0.24 dB/km have been measured for single-mode and multimode fiber, respectively.

Rayleigh scattering out of the fiber is detrimental to optical attenuation, but the backscattered bound portion also has effects. First, the feedback of modulated signals back to the transmitter can cause laser instability. Although it can narrow a laser spectral line to below a megahertz frequency width, the line position will fluctuate several times a second over a gigahertz interval. Second, bidirectional transmission along the same fiber can be used to conserve on fiber costs, but with the price of having an input/output coupler and emitter/detector pair at each end. Backscatter can cause crosstalk between the oppositely directed channels. (There may be crosstalk at each coupler as well.) One solution is wavelength-division multiplexing

(WDM) at two wavelengths so that the demultiplexer at each end rejects the out-of-band backscattered signal.

There are effects in addition to intrinsic material scattering. *Waveguide* or *index profile scattering* can be due to profile (core, numerical aperture, mode diameter, etc.) fluctuations along the fiber length, or to a defective profile such as too thin a cladding. These may be only weakly wavelength-dependent and are lumped into the B term of Eq. (7). Leaky modes, which will be discussed later under multimode fibers, also contribute a length-transient radiative loss. *Environmental scattering* occurs in the form of macrobending to a radius of a few centimeters, as in a splice housing. Another form is microbending of micrometer amplitudes and millimeter periods, as in a cable or on a bobbin. These are included in the $E(\lambda)$ term, and we will have more to say on this later when modes are better understood.

Scattering can be measured directly by use of light-integrating spheres or cubes. As with absorption, low levels make this difficult.

Finally, *nonlinear scattering* occurs at high optical power levels in fibers and with very coherent sources. A later chapter will show that there are many beneficial effects, such as multiwavelength generation for versatile sources in various instruments and for fiber lasers (e.g., erbium-doped at 1536 nm, pumped with 514-nm argon). Fibers can be used for optical amplification for nonregenerative repeaters and preamplifiers, and for subpicosecond pulse compression. Such effects will extend repeater spacings beyond hundreds of kilometers and transmission rates beyond hundreds of gigabits per second. However, these same effects can limit the amount of power that can be coupled into a fiber (the remaining being backscattered) and can cause crosstalk between WDM or optical frequency division multiplexing (OFDM) channels.

Total attenuation is generally measured by the cutback method, wherein the power exiting a long fiber is compared to power from a 2-m length cut off near the source end. If the source-to-fiber coupling is not disturbed, the result can be very precise since the fiber-to-detector coupling is generally made quite repeatable. Insertion loss or attenuation by substitution avoids the fiber cut at the expense of greater uncertainty at the source end. With multimode fibers, selective mode excitation has been used to obtain differential mode attenuation (DMA) results especially interesting for central and near-cutoff modes. From a practical viewpoint, DMA means that for a light source (or splice) that injects a number of lossy fiber modes, the attenuation coefficient will decrease with length. Such behavior is important for shorter-haul local network applications.

Finally, fiber backscatter can have a beneficial effect for attenuation probing with an optical time domain reflectometer (OTDR). An OTDR sends an input pulse of 1-ns to 10-μs duration into a fiber; resolution is 1 m per 10 ns. The backscatter power, shown in Figure 1.9, starts off at a level proportional to the fiber backscatter coefficient and the pulse energy (for greater range) [10]. (A multimode fiber has larger backscatter than a single-mode

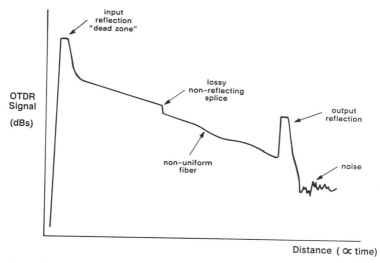

Figure 1.9 Example of a trace from an optical time-domain reflectometer (OTDR).

fiber has, and so a narrower pulse can be used.) The signal decays in time, where time is proportional to twice the distance along the fiber, to include the round trip from the input to the scattering section and back. Similarly, the signal has been doubly attenuated, so the display halves the dBs. Hence the logarithmic slope of the curve equals the attenuation coefficient (at the OTDR wavelength) plus a term proportional to the logarithmic derivative (usually small) of the backscatter coefficient. The instrument can do a nondestructive estimate of fiber or cable attenuation uniformity with length, typically at 1300 and 1550 nm, the latter being sensitive to bend conditions. The locations and magnitudes of discrete losses and reflections (termed "optical return loss") can be measured, for example, as due to splice installations in the field. Variation of the backscatter coefficient can sometimes lead to unreliable attenuation values, even splice "gains." Averaging bidirectional values gives better attenuation estimates, while the difference gives the length variation of the backscatter coefficient.

1.3.3 Dispersion

Recall that in moving from the 850-nm region to the 1550-nm region, and from multimode fiber to single-mode fiber, there was more than a ten-fold decrease in the attenuation coefficient. In this wavelength shift there is some slight penalty to be paid in power launched at the transmitter end, and receiver sensitivity is more than 5 dB lower at the longer wavelength. Nevertheless, the overwhelming attenuation coefficient decrease leads to a much longer power-limited repeater span. The other limiting factor is dispersion, and it is the second reason for the shift toward longer wavelengths.

To understand why, begin with phase velocity, defined by ω/β or c/n_p, where the phase index n_p equals the refractive index n of Eq. (1) in a material and equals β/k [see after Eq. (3)] in a fiber waveguide. However, this is the velocity of the wave phase fronts and not the velocity of optical power signals. Pulses that are narrow in frequency ω, that is, narrow in the wavelength spectrum, and not too short in time (i.e., not too broad in the Fourier frequency) actually travel at a group velocity. To see this, consider two equal-amplitude nearby signal frequencies ω and $\omega + \Delta\omega$, respectively, corresponding to two nearby propagation wavenumbers β and $\beta + \Delta\beta$. Then

$$\text{Total field} = \exp i(\omega t - \beta z) + \exp i[\beta + \Delta\beta]z - (\omega + \Delta\omega)t] \qquad (9a)$$

$$\text{Power} \propto |\text{field}|^2 \propto \cos^2\left[\tfrac{1}{2}(z\,\Delta\beta - t\,\Delta\omega)\right] \qquad (9b)$$

The group velocity in the limit is

$$\frac{c}{N} = \frac{d\omega}{d\beta} \qquad (10)$$

where

$$N = \frac{d\beta}{dk} = n_p - \frac{\lambda\,dn_p}{d\lambda}$$

is the group index. (As one can see from the notation, there are several ways of writing the same thing.) Figure 1.10 shows the phase and group indices for undoped silica over a broad range of wavelengths.

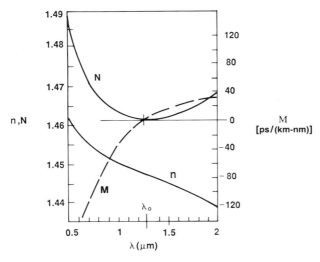

Figure 1.10 Phase index $n(\lambda)$, group index $N(\lambda)$, and material dispersion coefficient $M(\lambda)$ for fused silica.

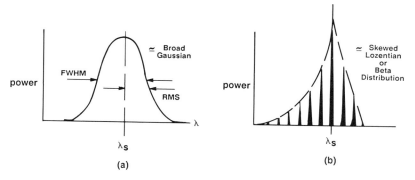

Figure 1.11 Schematic source spectra defining the mean wavelength λ_s and width $\Delta\lambda$ (FWHM or rms). (a) Continuous LED spectrum; (b) multilongitudinal mode laser diode spectrum.

The group delay per unit fiber length is $\tau(\lambda) = N(\lambda)/c$ (about 4.95 μs/km). Consider, as in Figure 1.11, a source spectrum with mean wavelength λ_s and full width at half-maximum (FWHM) or root-mean-square (rms) width $\Delta\lambda$ (in nm). Down a fiber length L (in km), two wavelengths separated by the width $\Delta\lambda$ will experience a differential time delay or link pulse broadening (in ps)

$$\Delta t(\lambda_s, \Delta\lambda) \simeq M(\lambda_s)L\,\Delta\lambda \tag{11}$$

Here

$$M(\lambda) = \frac{d\tau}{d\lambda} = \frac{dN}{c\,d\lambda}$$

is the *material dispersion coefficient* [in ps/(nm-km)], shown in Figure 1.10.
The spectral group delay can be given the empirical Sellemeier fit

$$\tau(\lambda) = A\lambda^2 + B\lambda^{-2} + C \qquad A, B, C, \text{ constants}$$

or [11]

$$\tau(\lambda) = \tau(\lambda_0) + \frac{S(\lambda_0)}{8}\left(\lambda - \frac{\lambda_0^2}{\lambda}\right)^2 \tag{12}$$

from which

$$M(\lambda) = S_0\,\frac{\lambda}{4}\left(1 - \frac{\lambda_0^4}{\lambda^4}\right) \tag{13}$$

Here λ_0 is the *zero-dispersion wavelength* where $M(\lambda_0) = 0$; $S_0 = S(\lambda_0)$ is the *zero-dispersion slope*, that is, the value of dispersion slope $S = dM/d\lambda$ at λ_0.
For silica, $\lambda_0 = 1273$ nm and $S_0 = 0.10$ ps/(nm²-km), so at λ_0 the pulse spreading in Eq. (11) is very small. It never goes to zero because, as we shall

see later, higher-order derivatives contribute there. Morever, both silica dopant and waveguide effects move λ_0 out past 1300 and modify S_0 (also later). At 850 nm, $M(\lambda) \simeq 0.1$ ns/km, so that from Eq. (11) a light-emitting diode (LED) 50 nm wide will experience a pulse spreading of about 5 ns/km down the fiber; lasers will be more than 10 times better. Around 1300 nm, a larger $\Delta\lambda$ of 80 nm will result in only about 0.2 ns/km second-order dispersion. With lasers this is smaller still, but for maximum benefit the source wavelength λ_s must be positioned quite close to λ_0. Around the 1550-nm lower attenuation region, special fiber designs are required to shift λ_0.

Thus we have seen that there are two materials factors favoring longer wavelengths. One is attenuation, with greatly reduced Rayleigh scattering and UV absorption. The other is material dispersion.

1.3.4 Far Infrared

The quest for even lower attenuation has prompted the search for materials with larger ions and weaker field strengths, resulting in IR absorption edges beyond 1800 nm. Dispersion properties must be kept in mind as well. Three classes have emerged.

One is *heavy-metal oxides*. These are oxides of Al, Ba, Bi, Ca, Ga, Ge, K, La, Pb, Sb, Ta, Te, W, and Zn that can be fabricated by the conventional crucible or vapor techniques used for silicate glasses. Intrinsic attenuation is below 0.1 dB/km between 2 μm and 3 μm, not much under that of silica; but only 4 dB/km at 2.0 μm for a Sb_2O_3-doped GeO_2 fiber has been reported to date. Minimum dispersion occurs at $\lambda_0 \simeq 1.7$ μm, so that waveguide dispersion would be important in shifting that wavelength. The cost of germania and the marginal attenuation advantage of these fibers may preclude extensive development.

Another class is *heavy-metal halides*. Theoretical minima as low as 3×10^{-4} dB/km occur at selected wavelengths within 2–12 μm, but scattering sites and impurities are still a problem. At the CO_2 laser wavelength of 10.6 μm, extruded polycrystalline halides such as thallium bromoiodide (KRS-5) and silver iodide have yielded only 190 dB/km, and zone-refined single-crystal fibers only 300 dB/km. The best results below 6 μm have been obtained on fluoride multicomponent glasses fabricated by casting, crucible, or vapor-phase oxide-to-fluoride techniques. A ZrF_4-BaF_2-LaF_3-AlF_3-Na 100-μm core composition with HfF_4 added in the cladding to a 140-μm outside diameter has yielded under 1 dB/km at 2.55 μm [12]. Only 0.12 dB/km was due to scattering; the remainder was impurity absorption. The minimum dispersion wavelength lies below 2 μm typically, but the zero-dispersion slope S_0 is one-third to one-fourth that of silica, so that low link dispersion will be easier over a broad spectral range. Other fluorides contain Cd, Li, Al, and Pb (CLAP), but simple two- or three-component glasses such as AlF_3-BeF_2 look promising. Glasses based on chlorides, bromides, and iodides can transmit much beyond 8 μm but are more difficult to form.

Finally there are *chalcogenide* (sulfide, selenide, telluride) glasses. These also contain arsenic, germanium, and phosphorus and are prepared by zone-refining or crucible techniques. Original estimates below 10^{-2} dB/km at 3–35 μm have been tempered by the discovery of a broad intrinsic absorption effect. To date, 35 dB/km at 2.44 μm, with an even smaller dispersion slope than above, has been attained for an As-S glass.

Initial applications for the above fibers (and for hollow fibers) have been mainly in short-length power delivery, for example, pyrometry, pollution spectroscopy, and laser surgery. Communications usage will require the solution of devitrification, hydroscopicity, aging, and toxicity problems as well as weak thermal and mechanical strength properties. Active devices have already been developed for spectroscopic applications. Around 2.55 μm, lasers are numerous: hydrogen fluoride (HF), deturium fluoride (DF), color center, and even He-Ne. Wavelength-tunable lead salts (Pb ternaries of Cd, S, Ge, Te, Se, Sn) have been made into laser diodes, and alloys of HgCdTe and structures of GaInAsSb and of InAsPSb have been made into emitter and detector diodes. Josephson devices and InAsSb diodes may also be used as detectors. Cryogenic cooling has been sometimes required for these small-bandgap materials, and several are too slow; but the technology learned with GaAlAs and InGaAsP room-temperature devices is being applied to these materials as well; emitters and detectors may be ahead of mid-IR fibers at this time. If fiber attenuation and dispersion properties can be improved, link repeater spacings of thousands of kilometers will be possible [13].

1.3.5 Strength

As-drawn fibers are said to have intrinsic strengths in excess of a million pounds per square inch (6.9 GN/m^2), corresponding to a breaking strain of about 10%. However, this depends upon the fiber length and test duration. Strength degradation is associated mainly with microscopic crack growth on the fiber surface due to environmental moisture and temperature conditions and due to the application of short-term or long-term stress.

In *prooftesting*, a constant stress σ_p is applied for a time t_p (<1 s usually, including the effects of rise time and fall time) along the whole fiber length. A particular crack of initial strength S_1 will grow and weaken to a strength S_2 given by [14]

$$S_2^{N-2} = S_1^{N-2} - \hat{S}^{N-2} \tag{14a}$$

where

$$\hat{S}^{N-2} = \sigma_P^N \frac{t_p}{B} \tag{14b}$$

The fatigue parameter N (15 or higher) and B value (10^5 kpsi^2-s within an order of magnitude) are related to crack dynamics. They are assumed to be time- and length-independent fiber constants but are affected by pH and humidity. Prooftesting guarantees that any flaws of initial strength $S_1 < \hat{S}$ will rupture, but it also severely weakens those within a few percent of $\hat{S}$ [14].

The parameters N and B can be obtained from *fatigue testing* to failure a fiber of gauge length L [14]. In static fatigue, a constant applied stress σ_a via tension or bending will produce a time to failure of

$$t_f = (BS_1^{N-2})\sigma_a^{-N} \qquad (15a)$$

A plot of log t_f versus log σ_a for a number of fiber samples should yield a straight line of negative slope $-N$ and statistically varying intercept BS_1^{N-2}. In dynamic fatigue, a constant applied stress rate σ_a will produce a failure stress $\sigma_f = \dot{\sigma}_a t_f$, where

$$t_f = (BS_1^{N-2})(N+1)\sigma_f^{-N} \qquad (15b)$$

$$\sigma_f^{N+1} = (BS_1^{N-2})(N+1)\dot{\sigma}_a \qquad (15c)$$

A plot of log t_f versus log σ_f should yield a straight line of negative slope $-n$ and a similar intercept. Alternatively, a plot of log σ_f versus log $\dot{\sigma}_a$ should yield a straight line of positive slope $(N+1)^{-1}$. Quantities B and initial strength S_1 are coupled in both cases. Note that strength depends upon stress rate.

Weibull testing is fatigue testing for many fiber samples. The survival probability is

$$P(L, \sigma, t) = \exp\left[-AL(\sigma_v^N t)^{M/(N-2)}\right] \qquad (16)$$

where A is a static or dynamic constant. M is the Weibull parameter obtainable from the slope of a linearized plot of cumulative failure $[\ln \ln (1/P)]$ versus the logarithm of failure time t for a constant applied stress (or strain) σ or versus the logarithm of failure stress σ or strain for a constant applied stress rate σ/t. Figure 1.12 shows a Weibull plot of the latter on two groups of samples, with and without prooftesting,

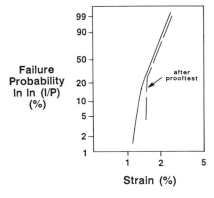

Figure 1.12 Weibull plot for stress-to-failure testing. Note the two regions of breakage, one of which is eliminated after prooftesting removes the weakest flaws. (Gauge length = 1 m.)

Note that the median fiber strength (at $P = 0.5$) decreases with increasing fiber length L. An important practical use of such testing is to estimate fiber survivability under repetitive or permanent stressing. Examples include the laying and repair of submarine cable and storage on a tight fiber bobbin at various temperatures. Relations such as Eqs. (15) and (16) can be used to give the probability distribution of times to failure for any stress history.

1.4 SINGLE-MODE FIBERS

The attenuation advantages of single-mode fibers (SMF) have already been discussed as due to the very low dopant levels. Here we will elaborate on a number of other properties.

1.4.1 Cutoff Wavelength

When the scalar waveguide equation, Eq. (4), is solved for fibers with only a few modes, the solutions must be more accurate than approximations (5) and (6). The results for the propagation wavenumbers or modal phase indices are schematically indicated in Figure 1.13 as a function of the normalized frequency

$$V = 2\pi \frac{a}{\lambda} n\sqrt{2\Delta} \tag{17}$$

Using the notation of Eq. (1), $n(1 + \Delta)$ is the maximum value of refractive index within the core of diameter $2a$ surrounded by an outer cladding of uniform index n. (For more complex profiles these terms may be interpreted a little differently.)

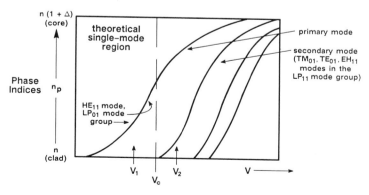

Figure 1.13 Phase indices, relative to the core and cladding indices, for a few low-order mode groups as a function of the V value.

Note that for larger V values there are a number of modes with various phase indices, and these modes are "cut off" as V decreases. Our interest will be around the cutoff value V_c of the second-order linearly polarized LP_{11} mode group. That group is actually quadruply degenerate in polarization modes that have very nearly the same modal phase indices. There are the transverse electric TE_{01}, transverse magnetic TM_{01}, and hybrid HE_{21} modes; the last has two polarization states. The mode power profile in the fiber cross section is annular (doughnut-shaped).

Corresponding to V_c is the *theoretical cutoff wavelength* from Eq. (17):

$$\lambda_{ct} = 2a \frac{\pi}{V_c} n\sqrt{2\Delta} \tag{18a}$$

or

$$\lambda_{ct} \text{ [nm]} \simeq 267 \cdot 2a \text{ } [\mu m] \cdot \sqrt{\Delta \text{ [\%]}} \tag{18b}$$

The second equation is in practical units for the step-index fiber of Figure 1.1. Typically, the core diameter $2a \simeq 8.5$ μm and index $\Delta \simeq 0.35\%$, so $\lambda_{ct} \simeq$ 1345 nm. The coefficient 267 is replaced by 183 for parabolic profiles (not often used in single mode) and by 147 for triangular profiles (detailed later).

For wavelengths $\lambda > \lambda_{ct}$ (i.e., at $V_1 < V_c$ in Fig. 1.13), only the primary LP_{01} mode group or the HE_{11} mode, which is doubly degenerate in polarization, can propagate. This "above-cutoff" situation is shown in Figure 1.14a; the secondary LP_{11} phase index is imaginary, and that mode does not

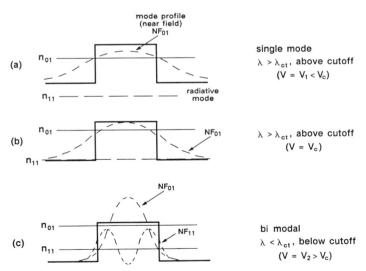

Figure 1.14 Mode profiles superimposed upon index profiles at various wavelengths. Also shown are the varying positions of the mode indices at wavelengths (a) above cutoff, (b) at cutoff; (c) below cutoff.

propagate. At cutoff, that index equals the clad index in Figure 1.14b, while the LP_{01} index rises slightly and the mode profile narrows and becomes tightly bound. Just below cutoff (Fig. 1.14c), both mode indices are real and the doughnut-shaped LP_{11} mode becomes more tightly bound for shorter wavelengths. The limit of this is reached at the cutoff value of the third-order mode group ($V = 3.832$ for step index). The ratio of first to second cutoffs is $2.405/3.832 \simeq 62.8\%$, or $\lambda_{ct} \simeq 845$ nm for the sample in the previous paragraph.

The theoretical cutoff wavelength for more complex profiles may be discussed in reference to Figure 1.15. Beginning with two mode groups in the *depressed clad* profile, movement toward longer λ causes the LP_{11} mode to become a *leaky mode* having outer regions of the refractive index at n above the mode index. The effect is accentuated because of the ring-shaped mode power profile. Note that at longer λ there is a cutoff for the primary LP_{01} mode as well, though the effect is more gradual because of the tighter mode profile. An advantage of the depressed clad design is that the primary mode experiences a high core/innerclad relative index difference $\Delta = \Delta^- + \Delta^+$ good for tight primary mode confinement (a small mode diameter as discussed later). On the other hand, λ_c is determined by Δ^+ in Eq. (18) and will not be as high as with a larger matched clad Δ. Finally, in the MCVD process, fluorine doping of the depression adds little attenuation, while germanium doping of Δ^+ causes less scattering loss than it would for Δ. (Phosphorus may be used throughout for lower-temperature processing.) Nevertheless, issues of matched clad versus depressed clad are not clear-cut, and both find favor commercially.

The *raised ring* profile is sometimes used with triangular core fiber that is dispersion-shifted (we'll learn the meaning later) to the 1550-nm region. This requires a core diameter of about 7 μm at the triangle base and relative

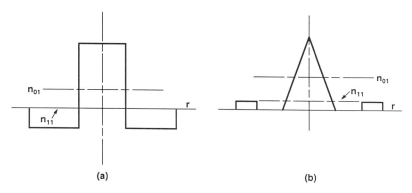

(a)   (b)

Figure 1.15 Cutoff condition for two more complex refractive index profiles. (a) Step-index depressed clad; (b) triangular core, raised ring.

index delta of about 0.9%. By Eq. (18), the cutoff wavelength is about 980 nm; the ring can raise this to the 1400-nm region.

The above are called the "theoretical" cutoff wavelengths λ_{ct} calculated from the index profiles. In practice one can operate at a wavelength somewhat below this because there the second-order mode is loosely bound and will disappear depending upon fiber length and bend conditions. This is best quantified by determining the spectral attenuation coefficient of the secondary LP_{11} mode. To do this, one can measure the spectral power throughput $P_1(\lambda)$ of a short length; $L = 2$ m with a gentle $R = 14$-cm bend radius has become a standard for measuring cutoff wavelength. This contains both LP_{01} (effectively unattenuated) and LP_{11}. Get rid of the latter by wrapping the fiber around several tight turns, $R < 3$ cm, and measure $P_2(\lambda)$. Finally, with the 14-cm bend again, cut back to a shorter length $L_0 < 0.5$ m to attenuate LP_{11} less and obtain $P_3(\lambda)$. Then the spectral attenuation coefficient of the secondary mode is

$$\alpha_{11}(\lambda)\,[dB/m] \simeq \frac{10}{(L - L_0)\,[m]} \log \frac{P_3 - P_2}{P_1 - P_2} \tag{19}$$

Figure 1.16 shows a representative result [15] that is independent of length but not curvature. Empirically, the curve may be fit to

$$\alpha_{11}(\lambda) = a \exp{(b\lambda)} \tag{20a}$$

where a, b depend upon curvature only. The *effective* or functional (as opposed to theoretical) cutoff wavelength λ_{ce} may be defined by a requirement that at any length L the secondary mode power loss for any length L is B dB, that is,

$$\alpha_{11}[\lambda_{ce}(L)]L = B \tag{20b}$$

From Eq. (20) one has a logarithmic decrease of effective cutoff wavelength with length

$$\lambda_{ce}(L) = \lambda_{ce}(2) - A \log \frac{L}{2} \tag{21}$$

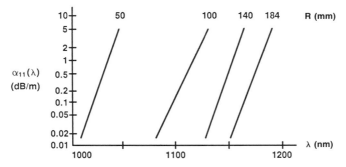

Figure 1.16 Spectral attenuation coefficient of the secondary LP_{11} mode for several fiber curvature conditions.

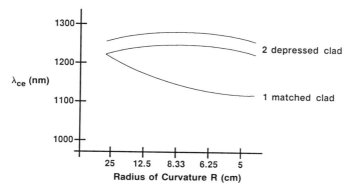

Figure 1.17 Curvature dependence of effective cutoff λ_{ce} for matched clad and depressed clad fibers.

normalized to a 2-m standard length. This occurs because longer lengths require a smaller attenuation coefficient, but this decrease will saturate after several kilometers due to coupling between the two mode groups.

The values of A may range from 20–50 nm/decade matched clad to 35–75 nm/decade depressed clad. The latter is somewhat more length-sensitive. For a fixed length, the curvature sensitivity may be fit to

$$\lambda_{ce}(R) = \lambda_{ce}(\infty) + BR^{-1} + CR^{-2} \qquad (22)$$

where B, C are fit parameters. The results of Figure 1.17 show that for cutoff behavior, depressed clad fibers for short lengths are less bend-sensitive than matched clad fibers.

The practical consequence of the above is that for a transmitter-to-receiver link to be truly single-mode the source wavelength λ_s should exceed the largest "cable" cutoff wavelength λ_{cc}. The latter is usually determined by the shortest cable repair section, perhaps 20 m with a couple of 1-m fiber ends with single loops of 7.5 cm diameter at both ends in the splice housings. If bimodal transmission does occur, *differential mode delays* of $\Delta T = 3$ ps/m, along with possible intermodal crosstalk, will severely limit dispersion. This effect can be eliminated, even at 850 nm, by use of loop filters near the receiver [16]. Second, if there are lossy connections between fiber lengths, *modal noise* is likely if a very coherent source has a coherence time $\lambda_s^2/c\,\Delta\lambda$ exceeding the pulse separation ΔTL. In this case the joint can experience loss fluctuations due to the bimodal speckle pattern there. For this coherence to be lost, such a joint may need to be several meters to several hundreds of meters, depending upon source characteristics, away from the source or earlier joint.

Cutoff wavelength can be measured in a number of ways, but the standardized methods call for the spectral power $P_1(\lambda)$ to be referenced to either the power $P_2(\lambda)$ through the more tightly bent fiber [as in Eq. (19)] or to a

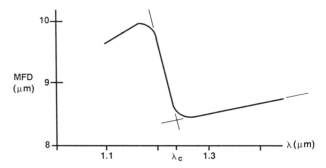

Figure 1.18 Behavior of spectral mode-field diameter (MFD), including the bimodal region. The intersection of the straight lines can be used to define the cutoff wavelength λ_c.

multimode fiber power $P_m(\lambda)$. The referencing eliminates effects of the source, detector, monochromator, etc. One then computes

$$A_j(\lambda) = 10 \log \frac{P_s(\lambda)}{P_j(\lambda)} \qquad j = 2, m \tag{23}$$

where A_2 is the small bend attenuation or A_m is relative attenuation; a 0.1-dB offset determines the fiber cutoff wavelength λ_{cf}. Values are typically below 1300 nm. Cable cutoff λ_{cc} replaces the short fiber with the arbitrary cable in the multimode-reference technique; values may be below 1200 nm.

We will now examine a second fundamental property of single-mode fibers, the mode diameter. Recall from Figure 1.14 how the width of the mode decreased with wavelength. This is shown in Figure 1.18 in which the apparent bimodal diameter increases below cutoff—another method of determining λ_{cf}.

1.4.2 Mode-Field Diameter

For several types of single-mode fibers around 1300 nm, the modal power profile or *near-field* intensity pattern $p(r)$ approximates a Gaussian

$$p(r) = p(0) \exp\left[-2\left(\frac{r}{w_0}\right)^2 \right] \tag{24}$$

where $2w_0$ is the mode-field diameter (MFD). [The field $f(r)$ is related by $p(r) \propto |f(r)|^2$.] This is shown for the step-index case in Figure 1.19, where it is apparent that the intensity at the mode radius $r = w_0$ has reduced to the $p(w_0)/p(0) = e^{-2} \simeq 13.5\%$ level; moreoever, a fraction e^{-2} of power resides beyond the MFD.

For step-index fibers, one has theoretically [17]

$$\frac{w_0}{a} \approx 0.799 + 0.290 \left(\frac{\lambda}{\lambda_{ct}}\right)^{2.505} \qquad \text{for } 0.7 < \frac{\lambda}{\lambda_{ct}} < 1.28 \tag{25}$$

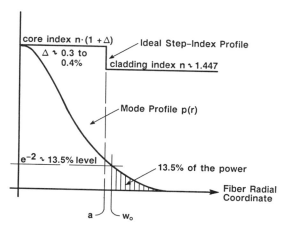

Figure 1.19 A step-index profile and a Gaussian mode field profile. Core diameter $2a \sim 8$–9 μm; mode diameter $2w_0$.

Note that for the single-mode regime the MFD has its minimum value at the theoretical cutoff wavelength λ_{ct}. Because of the lower measured λ_{cf} or λ_{cc}, for an operating wavelength $\lambda = 1305$ nm $= 0.97 \lambda_{ct}$ the MFD is about 6% larger than the core diameter; at 1550 nm, it is 20% larger, as indicated by Figure 1.19. The relative intensity at the core–cladding boundary and the fraction of total power lying outside the core are both

$$F = \exp\left[-2\left(\frac{a}{w_0}\right)^2\right] \tag{26}$$

At 1310 and 1550 nm, this is about 17 and 25%, respectively. A larger mode diameter with more power in the cladding is one reason single-mode fibers are increasingly bend-sensitive at longer wavelengths.

Also at longer wavelengths, or for more complex index profiles, mode profiles are quite non-Gaussian [18], with features such as central dips or very long tails beyond $r = w_0$. This has led to an rms form (relative to microbend loss or joint tilt loss)

$$w_{01}^2 = 2 \frac{\int_0^\infty r^2 f^2(r) r \, dr}{\int_0^\infty f^2(r) r \, dr} \tag{27a}$$

or to a derivative version (relevant to transverse offset loss or waveguide dispersion)

$$w_{02}^2 = 2 \frac{\int_0^\infty f^2(r) r \, dr}{\int_0^\infty (df/dr)^2 r \, dr} \tag{27b}$$

All fits to Eqs. (24) and (27) can be applied to measured *near-field* intensity patterns observed at the fiber end face. Another measurement method examines transmittance through a butt joint of the fiber ends as a function of transverse offset d:

$$T(d) = T(0) \exp\left[-\left(\frac{d}{w_0}\right)^2\right] \tag{28}$$

At a distance z from the fiber end face, the Gaussian nature of the mode field is preserved, but with a larger mode-field radius,

$$w_0(z) = \left[w_0^2 + \left(\frac{2z}{kw_0}\right)^2\right]^{1/2} \qquad k = 2\pi/\lambda \tag{29}$$

For large enough z, the second term in the square brackets dominates and the *far-field* intensity pattern (as would be cast on a screen) is

$$P(q) = P(0) \exp\left[-2\left(\frac{q}{W_0}\right)^2\right] \tag{30}$$

where

$$q = k \sin\theta \qquad \text{and} \qquad w_0 W_0 = 2$$

Here θ is the angle with the z axis, and W_0 is the angular mode diameter.

For more general profiles, one needs to do the Hankel transform [17]:

$$F(q) = \int_0^\infty f(r) J_0(qr) r \, dr \tag{31}$$

where

$$P(q) \propto |F(q)|^2$$

The inverse, going from far field to near field, also holds. This is useful for far-field measurements of MFD, which are conveniently done without major lensing optics by using detector scanning or variable aperture and knife-edge transmittances.

Relationship (31) does not apply to multimode fibers. The incoherent nature of multimode fibers means that the core diameter from the near field and the numerical aperture obtained from the far field are independent of each other. Both can be varied and maximized, for example, to affect splice loss and enhance source-to-fiber coupling. With single-mode fiber, enlarging the (spatial) mode diameter causes the angular mode diameter to shrink. Coupling with coherent sources and losses at splices depends mainly on MFD mismatch.

Nominal values of MFD tend to cluster around $9\frac{1}{2}$ and 8 μm for matched clad and depressed clad fibers, respectively, at wavelengths of use. Tolerances are typically $\pm 8\%$ to assist in low-loss splicing and connectorizing.

Practically, the MFD is quite useful. The fiber manufacturer can correlate it to a dopant index profile for minimum attenuation, and can monitor it for minimum bending losses (further discussion later). The MFD relates to cutoff wavelength (already mentioned) and to waveguide dispersion, which adds to material dispersion $M(\lambda)$ as [19]

$$G(\lambda) = \frac{\lambda}{2\pi^2 cn} \frac{d(\lambda w_0^{-2})}{d\lambda} \tag{32}$$

It implicitly modifies the fraction of backscattered light captured within the fiber, $\sim 1.8(\lambda/w_0)^2 \%$, which affects the OTDR signal; passage from a large to small MFD can produce an apparent splice gain.

From a user viewpoint, the greatest importance of the MFD is to source coupling or fiber-to-fiber coupling (replacing the core diameter and numerical aperture of multimode fibers). Losses are bidirectionally equal, except possibly within fusion splices that can deform the fiber structure. Intrinsic to the fiber, a mode-diameter mismatch $2\,\Delta w_0$ leads to

$$\text{dB loss} \simeq 4.343 \left(\frac{\Delta w_0}{w_0}\right)^2 \tag{33a}$$

A core offset or transverse alignment error x gives

$$\text{dB loss} \simeq 4.343 \left(\frac{x}{w_0}\right)^2 \tag{33b}$$

These are the most important errors. For a 9.5-μm mode diameter at a wavelength of 1310 nm, a 0.05-dB butt splice loss corresponds to a 10.7% MFD mismatch (not likely) or a 0.51-μm transverse offset. Other alignment errors include fiber longitudinal separation z,

$$\text{dB loss} \simeq 5.3 \left(\frac{\lambda z}{10 w_0^2}\right)^2 \tag{33c}$$

or relative tilt angle $\theta°$,

$$\text{dB loss} \simeq 2.7 \left(\frac{w_0 \theta}{10 \lambda}\right)^2 \tag{33d}$$

giving 0.05 dB at a large 17-μm separation (not likely) and 0.37° tilt (of concern), respectively.

Since fiber can be manufactured to keep the MFD mismatch loss of Eq. (33a) small, jointing effort has concentrated on alignment. Geometrically this can be done on the fiber outside surface (assuming good core–cladding concentricity) or upon cores made visible by special optics. Power monitoring is often used, either transmitted (with distant or local injection and detection) or local scattering. Good splices give below 0.05-dB loss, good connectors below 0.2 dB.

1.4.3 Chromatic Dispersion

In Section 1.3 we saw that the spectral group delay $\tau(\lambda)$ had a minimum and that the material dispersion coefficient $M(\lambda)$ vanished at the zero-dispersion wavelength, $\lambda_0 = 1273$ nm for silica. The dopant and index profile defining the single-mode fiber waveguide serves to give the variation of phase index n_p with wavelength as shown by the V value in Figure 1.13. Moreover the core-to-wavelength ratio a/λ and the index relative difference Δ both vary with wavelength; these contribute to *profile dispersion*. The sum of these interactive effects on $n_p(\lambda)$ leads to (after double differentiation) *waveguide dispersion* $G(\lambda)$, which is related to the spectral mode diameter via Eq. (32).

When the core and cladding refractive indices in Figure 1.13 are allowed to vary with wavelength, this material dispersion approximately adds to the waveguide dispersion to give the total chromatic dispersion

$$D(\lambda) = M(\lambda) + G(\lambda) \tag{34}$$

For the most common dispersion-unshifted fiber, the fit Eq. (13) for material dispersion still applies for $D(\lambda)$, but λ_0 increases to about 1310 nm and the zero-dispersion slope S_0 decreases to about 0.085 ps/(nm²-km). This is shown in Figure 1.20, along with a curve for dispersion-shifted fiber having λ_0 about 1550 nm, the region of minimum attenuation, with a smaller S_0 about 0.07 ps/(nm²-km):

$$D(\lambda) = S_0(\lambda - \lambda_0) \tag{35}$$

Shifting is obtained with index profiles similar to that of Figure 1.1c; control of this larger waveguide dispersion $G(\lambda)$ requires tighter profile control.

The ultimate case shown is dispersion-flattened (EIA class IVc) fiber that may have several dispersion zeros [20]. The design and fabrication of such fibers is quite complex, but as the OH⁻ water absorption peak around 1385 nm is progressively reduced, they will be increasingly important for operation over a 1200–1700-nm window. Wavelength-division multiplexing (WDM) and eventually coherent optical frequently division multiplexing (OFDM) should allow for thousands of channels to be placed within the window.

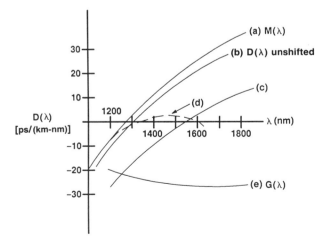

Figure 1.20 Dispersion coefficients for (a) pure silica, (b) dispersion-unshifted fiber, (c) dispersion-shifted fiber, and (d) dispersion-flattened fiber. Waveguide dispersion (e) is shown for (c).

Material dispersion has been measured by measuring $n(\lambda)$ for bulk materials (for example, using minimum deviation by a prism) and then differentiating. With fiber, the most common method is to use a multiwavelength source such as a Raman fiber laser, multiple laser diodes, a scanning diode (via temperature or drive electronics), or spectrally filtered light-emitting diodes (LEDs). Usually the normalized spectral group delay $\tau(\lambda)$ is measured as a time domain delay or as a phase shift $\phi(\lambda) = 2\pi f L\tau(\lambda)$. Alternatively, "wave-length wobbulation" of the source or monochromator about each of several wavelengths leads to a more direct measure of $D = \delta\tau/\delta\lambda$ at each wavelength. Finally, interferometic methods have been applied to 1–2-m lengths to extract the dispersion coefficient. This is particularly useful as a check on length uniformity but is not widely used.

Especially near a dispersion zero, a more accurate form of Eq. (11) is required. If T is the rms input Gaussian pulse width, then after a fiber length L the rms output non-Gaussian pulse width σ is determined [21] to be

$$\sigma^2 = T^2 + (LW_s)^2 \left[D_s^2 + \left(\frac{W_s^2}{2} \right) \left(S_s + 2D_s \frac{D_s}{\lambda_s} \right)^2 \right] \tag{36}$$

where

$$W_s^2 = (\Delta\lambda)^2 + \left(\frac{\lambda_s^2}{4\pi c T} \right)^2$$

and the subscripts denote evaluation at the source mean wavelength λ_s. Here W is an effective source spectral width, of which one component is the rms wavelength Gaussian spread $\Delta\lambda$, which usually dominates. The other (modulation) term has the value in nanometers of $0.45/T$ at 1300 nm and $0.64/T$ at 1550 nm for T in picoseconds. This component will dominate in fast (small T) coherent (small $\Delta\lambda$) systems. The ultimate bit rate is about 3 Tb-km-nm^2/s for a single-wavelength channel.

For dispersion-unshifted and some dispersion-shifted fibers, respectively, the Sellmeier form Eq. (13) and linear form Eq. (35) are adequate for D and $S = dD/d\lambda$ in Eq. (36). For other dispersion-modified fibers, other curve fits can be used. Broad non-Gaussian sources that may additionally be spectrally filtered by the fiber attenuation coefficient require alternative, more accurate descriptions.

1.4.4 Bending

Consider a single-mode fiber of curvature R. Note that in Figure 1.21 the mode power profile skews slightly away from the bend. Moreover, some refractive index profile distortion may occur due to bending stresses. More important, the mode phase index n_p in the plane of the bend decreases away from the fiber center, because in that area the traveling field must go faster to preserve the cross-sectional phase. This requires that

$$n_p(r) = \frac{n_p}{1 + r/R} \tag{37a}$$

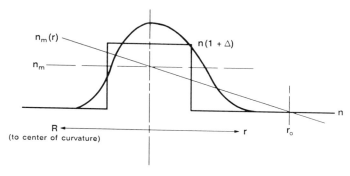

Figure 1.21 Mode field and mode index distortion for a bent step-index fiber.

The field will attempt to exceed the cladding phase velocity beyond a distance

$$r_0 = (n_p - n) \frac{R}{n} \tag{37b}$$

That part of the field radiates away. This effect has been used in local injection and detection (LID) splice sets, where radiated power is related to splice alignment loss.

Good bend resistance is achieved with fiber designs of small mode diameter ($w_0 < r_0$) and a large phase index difference $n_p - n$ between the mode and cladding. This is achieved with a small core radius a and a large index Δ, but with care not to dispersion-shift the fiber or increase scattering loss. Fibers are increasingly sensitive at longer wavelengths because w_0 grows while r_0 shrinks. Depressed clad fibers must be designed with an inner cladding radius large enough that cutoff of the primary mode does not occur below 1600 nm.

Constant-curvature macrobend *loss* is uniform along the fiber length with an attenuation coefficient [22]

$$\alpha = A R^{-1/2} \exp(-BR) \tag{38}$$

where A, B are complicated functions of the fiber V value. This is very sensitive to the curvature radius R dependence for a fixed fiber and wavelength. Comparisons between fibers and wavelengths are somewhat difficult to calculate. As shown by the measurement of Figure 1.22, long wavelengths and short cutoff wavelengths are detrimental.

Transition losses occur with bend changes. They are due to coupling between the primary mode and radiation modes and are oscillatory in fiber distance after the curvature change. The loss between curvature radii R_1 and R_2 is [23]

$$T = 2(nk)^4 w_0^6 \left(\frac{1}{R_1} - \frac{1}{R_2} \right)^2 \qquad k = \frac{2\pi}{\lambda} \tag{39a}$$

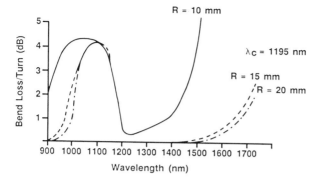

Figure 1.22 Dependence of bending loss upon fiber curvature and operating wavelength.

Between a straight and bent condition,

$$T = A(R \text{ [cm]})^{-2} \tag{39b}$$

so that for dispersion-unshifted fiber at 1300 nm, a 9.5-μm mode diameter gives $A = 0.39$ cm^2 (which is almost wavelength-independent).

Measurement conditions are being standardized gradually. A condition intended to simulate windings within splice housings between repeaters consists of 100 loose turns around a 7.5-cm-diameter mandrel. (In an actual situation there will be only a few turns every few kilometers. Transition losses and noncircular bend effects are also neglected here.) The bending loss should not exceed 1 dB at 1550 nm or about a tenth of that at 1300 nm.

Microbend losses are similar to numerous very closely spaced transition losses. They are very sensitive to the power spectrum of the curvature distribution, in terms of the spatial frequency Ω, often described as [18]

$$\Phi(\Omega) = A\Omega^{-2p} \tag{40}$$

Here A and p are constants that differ greatly depending upon fiber environmental conditions: jacketing, cabling, temperature, etc. One simple result for the microbend attenuation coefficient is

$$\alpha = A w_{02}^{-2} \qquad \text{for } p = -1 \tag{41a}$$

using the mode diameter in Eq. (27b). Another exact result is the lower limit

$$\alpha \geq A(\tfrac{1}{2} n k w_{01})^2 (\tfrac{1}{2} n w_{01} w_{02})^{2p} \qquad p \geq 0 \tag{41b}$$

using the mode diameter in Eq. (27a).

The above results are difficult to verify in practice because of uncertainty in the knowledge of A and p. They do underscore the importance of small mode diameter in microbend performance. Experimental work is usually aimed at comparing the performance of several fiber types by means of some

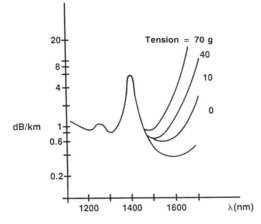

Figure 1.23 Spectral attenuation in the basket-weave test on a 15-cm-diameter drum.

test. One type of test passes a short length of fiber between rough sandpaper surfaces pressed together or through a container of ball bearings, wires, etc., thereby inducing an excessively large microbend loss per unit length. Another test is a basket-weave type, subjecting longer lengths to smaller incremental losses. Both macrobending and microbending are generated in winding the fiber onto a spool under tension with several layers. Figure 1.23 shows a typical result.

1.5 MULTIMODE FIBERS

Historically, high-loss step-index multimode fiber was first used for several short-distance light transfer applications. As we shall see, its bandwidth limitations were partially overcome by graded-index designs. These were applied commercially beginning in 1975 for short-wavelength links, and then in 1980 to long-wavelength links, allowing both lower attenuation and chromatic dispersion. Although the market today is dominated by single-mode fiber, special multimode fibers do have enduring applications in medium-bandwidth short-length distribution architectures.

1.5.1 Modes and Rays

In Section 1.2.2, we saw that an LP_{LM} mode could be characterized by periodicity integers L and M for the angular and radial cross-sectional coordinates, respectively. The propagation wavenumber β is a function of these, so any two of L, M, β specify the mode.

From a ray viewpoint, one begins with (2a) applied to a cylindrical geometry. Figure 1.24 shows a ray located at (r, ψ) with respect to the x, y axes

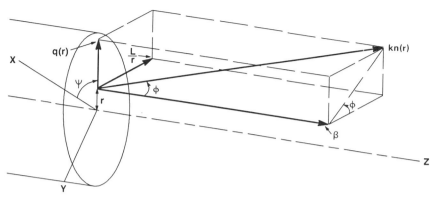

Figure 1.24 Ray components at an internal cross section of multimode fiber.

on an arbitrary cross section at z. Relative to a local (r, ϕ, θ) cylindrical coordinate system at that point, the ray has a local wavenumber $kn(r)$, where $n(r)$ is the refractive index of Eq. (1). (Recall that in a vacuum the photon wavenumber is k.) The ray components there are

$$q(r) = kn(r) \sin \theta(r) \sin \phi(r)$$

$$\frac{L}{r} = kn(r) \sin \theta(r) \cos \phi(r) \qquad (42)$$

$$\beta = kn(r) \cos \theta(r)$$

After some trigonometric identities, the radial wavenumber $q(r)$ is seen to be equivalent to Eq. (4b) in the wave/mode picture; L and the axial wavenumber β are ray constants, the latter being a generalization of Snell's law.

Modes and rays are thus connected by the L, β constants; M enters via Eq. (6). Figure 1.2 shows where a particular ray or mode is localized between caustics at $r = R_1$ and R_2 in the fiber cross section. Along the longitudinal direction the ray follows an oscillatory path with an axial period

$$\Delta z = 2\beta \int_{R_1}^{R_2} q^{-1}(r) \, dr \qquad (43)$$

Relations can be given for the other coordinates as well, but of more practical concern are the issues of light coupling and pulse broadening.

1.5.2 Mode Types

The ray components of Eq. (42) inside the fiber are launched from the outside world (of refractive index n_0) through the fiber input surface. That surface may be tapered or have a spherical or other shape to optimize coupling from an LED (light-emitting diode) or ILD (injection laser diode). For the case of a planar fiber end face, Snell's law for ray entry at r_0 gives

$$\beta = k[n^2(r_0) - n_0^2 \sin^2 \theta_0]^{1/2}$$
$$= kr_0 n_0 \sin \theta_0 \cos \phi_0 \tag{44}$$

where angular directions θ_0, ϕ_0 are just outside the fiber at the end face. Hence the ray/mode constants β, L are determined by the launch conditions.

The range of *bound modes* or *rays* are defined by β/k being between the minimum and maximum values of the fiber refractive index as we saw in Figure 1.13. The *local numerical aperture* at r is defined by the smallest β (near cutoff) where

$$n_0 \sin \theta_0(r) = [n^2(r) - n^2(a)]^{1/2} \tag{45a}$$

Note that this angular acceptance generally decreases near the core–cladding interface. Since the refractive index is usually largest at the fiber center, the *maximum theoretical numerical aperture* is

$$NA_t \equiv n_0 \sin \theta_0 = [n^2(0) - n^2(a)]^{1/2} \simeq n\sqrt{2\Delta} \tag{45b}$$

using Eq. (1).

The total number of bound modes is obtained by summing over allowed mode numbers with the result

$$M = k^2 \int_0^a [n^2(r) - n^2(a)]r \, dr \tag{46}$$

This is just the local numerical aperture squared and integrated over the core area.

In fact, even larger values of incidence angles θ can be accepted into *leaky modes* or *rays*. To understand mode types, return to Figure 1.2. For angles within $\theta_0(r)$, there are always (β, L) values such that the radial wavenumber $q(r)$ of Eq. (46) is real and the rays are radially oscillatory in the annular "doughnut" formed by the caustics R_1, R_2 within the core. Within the annulus $(0 < r < R_1)$ and outside of it $(r > R_2)$, $q(r)$ is imaginary, so by Eq. (5) the power is evanescent as the fields decay away from the caustics.

However, larger incidence angles $\theta_0(r)$, $\phi_0(r)$ reduce β such that a third caustic R_3 appears in the cladding. Beyond this region $(r > R_3)$, the field is again oscillatory, that is, rays again classically exist there. Light leaks or tunnels through the evanescent barrier between R_2 and R_3; leaky modes are sometimes called *tunneling modes* or *rays*. The corresponding attenuation coefficient is

$$\alpha = (\Delta z)^{-1} \exp\left[-2 \int_{R_2}^{R_3} |q(r)| \, dr \right] \tag{47}$$

These rays each the glass–jacket interface and are lost.

Unbound rays or *radiative* or *cladding modes* occur when $R_2 = R_3$ and the tunneling region vanishes. Apart from a central evanescent region, the

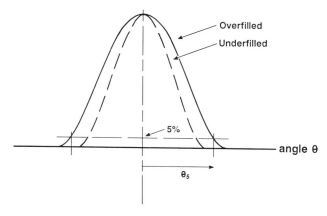

Figure 1.25 Typical multimode far-field intensity and measured numerical aperture.

power from such rays exists everywhere throughout the fiber and they are so lossy that they do not propagate.

Measurement of the core diameter and theoretical numerical aperture can be done from the fiber refractive index profile obtained from the refracted near-field method that applies to multimode and single-mode fibers.

A transmission measurement of NA is usually done at the exit end rather than the input end used in the definitions, Eqs. (45a) and (45b). An overfill launch is applied to a 2-m fiber length; a higher refractive index cladding mode stripper, such as the fiber jacket, is used. The angle corresponding to the 5% level of the far-field intensity in Figure 1.25 gives the measured NA. This value may be 5–10% below the theoretical NA_t because some of the highest-order bound and leaky modes past θ_5 are not included. Moreover, because of the higher attenuation coefficients of such modes, the measured NA will decrease along the fiber length [see Eq. (6)].

The *mode volume* (MV) is defined as the square of the product of core diameter and numerical aperture:

$$MV = (CD \cdot NA)^2 \tag{48}$$

The ratio of measured MV to theoretical MV should approximately equal the ratio of the actual number of modes to the total number of modes. The MV relates to the light-gathering power of a fiber from an optical emitter.

1.5.3 Fiber Types

In terms of materials, multimode fibers may be all-plastic, glass core and plastic clad, or all-glass. The first two categories are invariably step-index, with a wide variety of larger core diameters, numerical apertures, and outside diameters. Their high attenuations and low bandwidths for many applications are offset by easier source coupling and connectorization.

TABLE 1.1. Standard types of graded-index multimode fibers

Core diameter, 2a (μm)	Outside (clad) diameter (μm)	Nominal numerical aperature NA
50	125	0.19–0.25
62.5	125	0.27–0.31
85	125	0.25–0.30
100	140	0.25–0.30
		Δ = 0.85–2.1%

The all-glass category may be either step-index or graded-index; as we shall see later, the latter has a higher bandwidth. There are four EIA-recognized standard multimode types in the United States, which are shown in Table 1.1. Note that the nominal mode volumes range from about 90 to 900 μm^2, a tenfold range.

1.5.4 Power-Law Profiles

Some analytical simplicity and practical results come from using a special case of Eq. (1):

$$n(r, \lambda) \simeq n\left[1 + \Delta\left(\frac{r}{a}\right)^g\right] \qquad (49)$$

where g is the profile gradient parameter. Values $g = \infty$, 2, 1 correspond to the step, parabolic, and triangular core profiles, respectively. Then sets of modes coalesce or degenerate to the same phase index,

$$n_p = \frac{\beta}{k} = n\left[1 + \Delta\left(\frac{m}{\tilde{M}}\right)^{2g/(g+2)}\right] \qquad (50)$$

These *mode groups* have the compound mode number

$$m = 2L + M + 1 \qquad (51a)$$

with a total number of mode groups

$$\tilde{M} = \left[\frac{g}{2(g+2)}\right]^{1/2} V \qquad (51b)$$

The total number of modes is $\tilde{M}^2$, for example, $V^2/2$ for step index, $V^2/4$ for graded index.

A multitude of simplified relations result. A useful one relates the incident ray coordinates (r_0, θ_0) to the bound-mode group number:

$$\frac{m}{\tilde{M}} = \left[\left(\frac{r_0}{a}\right)^g + \left(\frac{\sin\theta_0}{NA}\right)^2\right]^{(g+2)/2g} \qquad (52)$$

For the step-index case, this depends on angle only. Relations can be written

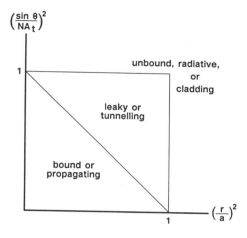

Figure 1.26 Ray coordinate phase-space diagram for parabolic index fiber, showing the three mode/ray regions.

for excitation conditions of leaky and radiative modes; these are useful for selective mode (group) excitation to measure differential mode attenuation and differential mode delay for index profile optimization. Moreover, modal multiplexing has been used to transmit several separate channels, but cross-talk is a problem. A pictorial representation of the ray coordinate "phase space" for the three mode types on parabolic-index fiber is given in Figure 1.26.

The modal power distribution (MPD) can be measured experimentally by determining the near-field pattern $N_F(r)$ for full injection and $N_P(r)$ for the partial fill. The MPD is then [24]

$$P\left(\frac{m}{\tilde{M}}\right) = \frac{dN_P/dr}{dN_F/dr} \qquad \text{at} \left(\frac{r}{a}\right)^2 = \frac{m}{M} \qquad (53)$$

Knowledge of the MPD is useful for characterizing multimode fibers after sources, splices, connectors, etc. These effects are significant in short-haul LAN-type applications.

1.5.5 Multimode Distortion

Whether derived by ray or wave methods, each mode has its own phase velocity and hence group velocity via Eq. (10). These mode delay differences vary with wavelength but exist even if a monochromatic source is used; they contribute to pulse distortion after passage through a length of fiber.

Many theoretical approaches have been applied to designing refractive index profiles that minimize the spread in these delays. The problem is complicated by the fact that in general the launch conditions determining the initial relative modal intensities are not known. Moreover, differential mode attenuation down the fiber length will change these intensities, and mixing among modes can occur (see Section 1.5.6). With the use of single-mode fiber for long-distance high-bit-rate applications, there is now less practical interest in maximizing multimode fiber bandwidth and minimizing bandwidth reduction with increasing length.

The power-law profile approach provides compact analytical expressions for the group index from Eq. (50) and for the rms spread σ in intermodal delays when all modes are taken to be equally excited. The spreading is minimized at a value of profile parameter [25]

$$g_0 = 2 + \varepsilon - \frac{(3 + \varepsilon)(4 + \varepsilon)\Delta}{5 + 2\varepsilon} \tag{54}$$

where

$$\varepsilon = -\frac{2n\lambda}{N\Delta} \frac{d\Delta}{d\lambda}$$

is the *profile dispersion parameter* (measurable from the spectral variation of NA). For fibers doped with phosphorus or germanium, g_0 has a value around 1.9 in the 1300-nm wavelength region but above 2 in the 850-nm region.

For a step-index fiber the rms intermodal spreading is

$$\sigma_s = \frac{LN\Delta}{2c\sqrt{3}} \simeq 14L \text{ [km] } \Delta[\%] \qquad \text{in ns} \tag{55a}$$

Multimode fibers typically have relative index differences of 1–2%. The optimum profile results in a greatly reduced pulse spread value

$$\sigma_0 \simeq 0.1\sigma_s\Delta \simeq 14L \text{ [km]}(\Delta \text{ [\%]})^2 \qquad \text{in ps} \tag{55b}$$

Because of practical fabrication tolerances minute deviations from an optimized profile result in distortion values 5–10 times that in Eq. (55b).

If a graded fiber has too small a value of profile parameter ($g < g_0$), then the higher-order modes generally travel faster than the lower-order modes. This is the opposite of the situation for $g > g_0$ (e.g., step-index fibers), and the profile is said to be *overcompensated*. The effect can be probed with selective mode excitation and by measuring the resultant differential mode delays (DMD). If an undercompensated fiber is connected to an overcompensated fiber, the resultant pulse broadening will be less than would be expected by simple addition. Some scenarios have been demonstrated for joining fibers of opposite compensations along a link to minimize pulse distortion. Again, the advent of single-mode transmission has reduced interest in such studies.

Measurement of multimode distortion can take place in the time domain where typically laser diodes or a Nd:YAG multiwavelength laser is used along with PIN or APD detectors. A cutback method similar to that used in attenuation measures the duration T_2 of the pulse shape $s_2(t)$ out of a long fiber length and $s_1(t)$ from a 2-m length. The impulse response duration is then

$$\Delta T = (T_2^2 - T_1^2)^{1/2} \tag{56}$$

The fiber is presumed to act as a linear system, so the result is "exact" for rms values, but it is often used for Gaussian-like FWHM values. Results may be length-normalized to nanoseconds per kilometer, but this is not always justified, as will be seen later. The pulses are often Fourier-transformed into the frequency f domain:

$$S_j(f) = \int_{-\infty}^{\infty} s_j(t) \exp(2\pi i f t) \, dr \qquad j = 1, 2 \tag{57}$$

from which the transfer function

$$H(f) = \frac{S_2(f)}{S_1(f)} \tag{58}$$

is computed. Occasionally the inverse to this is computed to yield the impulse response $h(t)$, which often shows some structure due to various mode group arrivals. In principle it has a width ΔT and convolutes with input $s_1(t)$ to give the output $s_2(t)$.

A single-modulation frequency signal experiences an amplitude reduction via the baseband response $|H(f)|$ of Figure 1.27 and a phase shift arg $H(f)$; usually the latter is neglected. Measurement can also use a scan in the frequency domain, yielding the baseband response directly from Eq. (58).

The fiber acts as a low-pass filter, and both the amplitude and phase show some structure related to the impulse response. Sometimes a weighted Gaussian fitting is done to smooth the curves, an effect that takes place naturally in concatenation. In either event, a common specification is the 3-dB optical (equal to the 6-dB electrical) bandwidth B. It is often length-normalized to megahertz-kilometers. The bandwidth varies with wavelength (since the sensitive profile parameter g does also), as typically shown in Figure 1.28.

Chromatic dispersion adds to multimode distortion. The former is similar to that encountered for single-mode fiber except that the zero-dispersion

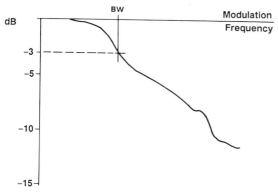

Figure 1.27 Multimode fiber baseband response and the definition of bandwidth. A weighted Gaussian fit is sometimes used.

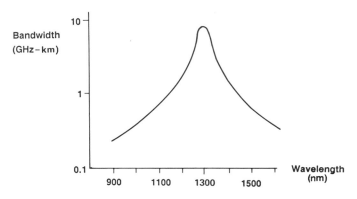

Figure 1.28 Typical variation of multimode bandwidth with wavelength for 1300-nm-optimized fiber.

wavelength can exceed $\lambda_0 = 1340$ nm, with a zero-dispersion slope above $S_0 = 0.095$ ps/(nm²-km). The associated pulse broadening, Eq. (35), adds quadratically to the multimode effect as indicated by Eq. (56) or multiples in transfer functions similar to Eq. (58). For LED sources of broad spectral width, chromatic bandwidth is usually comparable with multimode bandwidth, and more detailed computation is necessary.

Convenient Gaussian rules of thumb relating pulse broadening Δt and 3-dB optical bandwidth B are

$$B \, \Delta t_{\text{rms}} \simeq 187 \qquad B \, \Delta t_{\text{FWHM}} \simeq 441 \tag{59}$$

Units are B in megahertz or gigahertz, Δt in nanoseconds or picoseconds, respectively.

1.5.6 Length-Dependent Effects

Even with a minimum of environmental perturbations and with a uniform length of fiber, optical properties are not constant down the fiber. In the single-mode case there is a length-dependent shift of cutoff wavelength, although this may be accounted for by a uniform LP_{11} attenuation coefficient. Multimode behavior is somewhat more complex.

Differential mode attenuation of bound and leaky modes and coupling among modes can occur. The latter is due to Rayleigh scattering, small systematic or random longitudinal fluctuations in refractive index, macrobend or microbend effects, or the presence of inline splices, connectors, couplers, etc. A number of complex theories have been introduced to explain these phenomena, but none have been totally successful in practical simplicity or accuracy.

An example of a transient power loss due to source coupling at the fiber input end is shown in Figure 1.29. This was done by a destructive multiple

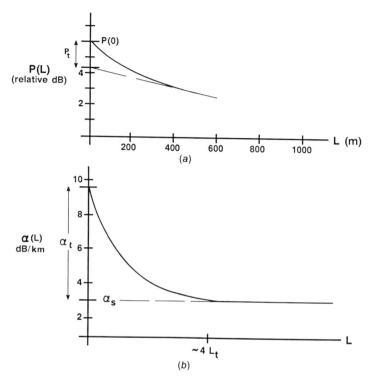

Figure 1.29 (a) Measured transient power loss due to LED-to-fiber coupling at 850 nm. (b) Resultant transient attenuation coefficient.

cutback measurement. The decay of power in decibels as a function of length L is empirically [26]

$$P(L) = P(0) - \alpha_s L - P_t\left[1 - \exp\left(\frac{-L}{L_t}\right)\right]$$

(60a)

where $P(0)$ is the power coupled in and α_s is the usual steady-state attenuation coefficient. The transient power P_t can range from below 0.2 dB for an ILD to almost 2 dB for an LED. This power is lost over four or five transient lengths L_t, where L_t ranges from below 100 m to several hundred meters. This power decay is equivalent to an exponentially decaying attenuation coefficient

$$\frac{dP}{dL} \equiv \alpha(L) = \alpha_s + \frac{P_t}{L_t}\exp\frac{-L}{L_t}$$

(60b)

Similar length decays are experienced by the effective core diameter, effective numerical aperture, and mode power distribution. Further down the fiber link, each splice or connector will act as a new source having two components of loss: local and transient. The latter can be described as in Eqs. (60a) and (60b).

Fiber macrobends can be described analogously to the single-mode case, except that the steady-state bend loss, Eq. (37), and transition loss, Eq. (38), must be derived for each mode. Ray arguments are commonly used, but because of the complicated MPD situation, few useful formulas have been derived. The transition loss is the main contributor; going into or out of a bend, the fraction of light lost is approximately

$$f = \frac{4\,CD}{R\,NA^2} \qquad \text{for } R > 4\frac{CD}{NA^2} \tag{61}$$

For a radius of curvature $R = 5$ cm, this corresponds to a loss of about 0.4 dB or less for the fibers of Table 1.1. As wavelength λ increases, the number of modes decreases as λ^{-2}, but both CD and NA vary little. The fractional loss is nearly constant with λ, unlike the single-mode case.

Microbending is sensitive to the power spectrum of the curvature distribution as with Eq. (39) for single-mode fibers. This curvature is as experienced by the fiber axis; there is some shielding effect of the fiber jacket from the environment. In the absence of a definitive model, the wavelength-independent microbend attenuation coefficient is written as [27]

$$\alpha_\mu = A\,CD^4\,OD^{-6}\,NA^{-6} \tag{62}$$

where the A term relates to the fiber profile and the curvature distribution. This result is useful for interfiber comparisons. For example, the second fiber of Table 1.1 is expected to have only one-third the microbend loss of the first fiber. Such loss effects are particularly important at low operating temperatures, where microbending may be induced by expansion coefficient mismatch between the fiber and jacket and cable.

Length-dependent effects on bandwidth are perhaps more dramatic than attenuation effects. The causes are similar to the above, but the fact that a particular light photon skips from mode to mode means that pulse spreading is generally sublinear with length. In contrast to Eq. (55), it is found empirically that [28]

$$\sigma \propto L^\lambda \qquad 0.5 \leq \gamma \leq 1 \tag{63a}$$

where γ depends on the separation of the operating and optimum wavelengths and decreases with mode mixing conditions. Then γ may depend weakly upon length. Recall that $\gamma = 1$ for the chromatic dispersion contribution, Eq. (11) or (35), a strong effect near 850 nm or for spectrally broad sources even around 1300 nm.

The associated bandwidth is sometimes written as

$$B(L) = B(l)L^{-\gamma} \tag{63b}$$

where the length-normalized bandwidth $B(l)$ is in units of MHz-km$^\gamma$. Typical specified gamma values range from 0.7 to 0.9, and this can lead to trade-offs of the form shown in Figure 1.30.

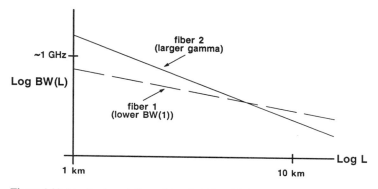

Figure 1.30 Idealized variation of bandwidth with length. Note that fiber 1, with lower initial bandwidth but with a smaller bandwidth decrease along the fiber length, can be better for longer lengths than fiber 2 of opposite characteristics (which is better for shorter length applications).

An additional complication arises when lengths of otherwise homogeneous fiber are concatenated. In that case, two fibers of bandwidths B_1 and B_2 (independent of length) provide a resultant bandwidth B, where

$$B^{-1/\gamma} = B_1^{-1/\gamma} + B_2^{-1/\gamma} \qquad (64)$$

Independent fibers would have $\gamma = \frac{1}{2}$ analogous to Eq. (56). Another interpretation is

$$B^{-2} = B_1^{-2} + B_2^{-2} + 2r_{12}B_1^{-1}B_2^{-1} \qquad (65)$$

where r_{12} is a correlation coefficient ranging from -1 to $+1$, depending both upon mode mixing ($r \rightarrow 0$) and upon the over- or undercompensation characteristics of the two fibers ($r \rightarrow -1$ for opposite compensations). Both Eqs. (64) and (65) can be generalized to a larger number of fibers.

The empirical fits of Eqs. (63b), (64), and (65) have met with very limited success in predicting end-to-end field bandwidth. Perhaps this helped accelerate the acceptance of single-mode fiber, which has a more predictable link concatenation behavior.

REFERENCES

Hundreds of references would be necessary to thoroughly cover the topics of this chapter alone. No attempt has been made to be representative or complete. Only some of the topics treated at length here are referenced. Emphasis is on more recent and review-type articles.

1. M. Born and E. Wolf, *Principles of Optics*, 4th ed., Pergamon, New York, 1970.
2. D. P. Jablonowski, "Fiber manufacture at AT&T with the MCVD process," *J. Lightwave Technol.*, **LT-4**: 1016 (1986).

3. H. Lydtin, "PCVD: a technique suitable for large-scale fabrication of optical fibers," *J. Lightwave Technol.*, **LT-4**: 1034 (1986).
4. R. V. VanDewoestine and A. J. Morrow, "Developments in optical waveguide fabrication by the outside vapor deposition process," *J. Lightwave Technol.*, **LT-4**: 1020 (1986).
5. H. Murata, "Recent developments in vapor phase axial deposition," *J. Lightwave Technol.*, **LT-4**: 1026 (1986).
6. U. C. Park, "High-speed high-strength fiber drawing," *J. Lightwave Technol.*, **LT-4**: 1048 (1986).
7. S. S. Walker, "Rapid modeling and estimation of total spectral loss in optical fibers," *J. Lightwave Technol.*, **LT-4**: 1125 (1986).
8. U. Naoya and N. Uesugi, "Infrared optical loss increase in silica fibers due to hydrogen," *J. Lightwave Technol.*, **LT-4**: 1132 (1986).
9. E. J. Friebele et al., "Overview of radiation effects in fiber optics," *Proc. SPIE*, **541**: 70 (1985).
10. A. Hartog and M. Gold, "On the theory of backscattering in single-mode optical fibers," *J. Lightwave Technol.*, **LT-2**: 76 (1984).
11. F. P. Kapron and T. C. Olson, "Accurate specification of single-mode dispersion measurements," *Tech. Digest*, Symp. Opt. Fiber Measurements, NBS Special Publication 683 (Boulder, CO, October 1984), p. 111.
12. O. Eknoyan, R. P. Moeller, W. K. Burns, D. C. Tran, K. H. Levin, "Transmission properties on 2.55 μm fluoride glass fibre," *Electron. Lett.*, **22**: 752 (1986).
13. M. M. Broer and L. G. Cohen, "Heavy metal halide glass fiber lightwave system," *J. Lightwave Technol.*, **LT-4**: 1509 (1986).
14. F. P. Kapron, "Standards issues in fiber environmental performance and reliability," *Proc. SPIE*, **842**: 14 (1987).
15. Corning Glass Works, "Applicability of cutoff wavelength measurements to system design for matched-clad fibers," CCITT (July, 1986).
16. M. Stern W. I. Way, V. Shah, M. B. Romeiser, W. C. Young, "800-nm digital transmission in 1300-nm optimized single-mode fiber," Conf. Opt. Fiber Commun./Integrated Opt. and Opt. Commun. (OFC/IOOC'87), Reno, NV, January 1987, paper MD2.
17. W. T. Anderson, "Consistency of measurement methods for the mode field radius in a single-mode fiber," *J. Lightwave Technol.*, **LT-2**: 191 (1984).
18. K. Petermann and R. Kuhn, "Upper and lower limits for the microbending loss in arbitrary single-mode fibers," *J. Lightwave Technol.*, **LT-4**: 2 (1986).
19. C. Pask, "Physical interpretation of Petermann's strange spot size for single-mode fibres," *Electron. Lett.*, **20**: 144 (1984).
20. B. J. Ainslie and C. R. Day, "A review of single-mode fibers with modified dispersion characteristics," *J. Lightwave Technol.*, **LT-2**: 967 (1986).
21. D. Marcuse and C. Lin, "Low dispersion single-mode fiber transmission—the question of practical versus theoretical maximum transmission bandwidth," *J. Quantum Electron.*, **QE-17**: 869 (1981).
22. D. Marcuse, "Curvature loss formula for optical fibers," *J. Opt. Soc. Am.*, **66**: 216 (1976).
23. W. Gambling et al., "Curvature and microbending losses in single-mode optical fibres," *Opt. Quantum Electron.*, **11**: 43 (1979).
24. G. Granand and O. Leminger, "Relations between near-field and far-field intensities, radiance, and model power distribution of multimode graded-index fibers," *Applied. Opt.*, **20**: 457 (1981).
25. G. Einarsson, "Pulse broadening in graded index optical fibers: correction," *Appl. Opt.*, **25**: 1030 (1986).
26. P. Vella et al., "Precise measurement of steady-state fiber attenuation," NBS Symp. Opt. Fiber Measurements (Boulder, CO, October 1982), p. 55.
27. R. Olshansky, "Distortion losses in cabled optical fibers," *Appl Opt.*, **14**: 20 (1975).
28. D. A. Nolan et al., "Multimode concatenation modal group analysis," 11th Eur. Conf. Opt. Commun. (ECOC '85), Venice, October 1985.

CHAPTER 2

Measurement and Characterization of Optical Fibers

Dan L. Philen

2.1 INTRODUCTION

The intense development and implementation of optical fibers in the past few years have been accompanied by a similar growth and development in fiber measurements. The data generated from fiber measurements are used by systems designers for planning transmission systems, by manufacturers to control product quality, and by researchers to explore new fiber properties. Thus, accurate measurements that can be easily used and implemented at a reasonable cost are needed. The development of optical fiber technology also shows that measurement techniques that were the best in the industry a few years ago are no longer in widespread use. As new measurement methods have been developed, the older ones have been replaced. Likewise, techniques that were in the research stage only a few years ago are beginning to be found in commercial instruments suitable for both the research laboratory and the field.

It is not the purpose of this chapter to describe every fiber measurement method in detail. The traditional measurements will be mentioned for completeness, but the emphasis will be on those newer techniques that are mostly applicable to single-mode fibers. The references in each section of this chapter are intended not only as documentation, but also to provide a suitable base for further reading and research.

2.2 GEOMETRIC MEASUREMENTS

Geometric measurements are used to determine the physical properties of the fiber. These are cladding diameter, core diameter, refractive index profile, and numerical aperture. The outside diameter of typical fibers is about 125 μm, or about the thickness of a piece of paper. The core diameter of typical multimode fibers is 50 μm, or half the paper thickness. For comparison, a typical human hair has a diameter of about 50 μm. Single-mode fibers have core diameters of about 10 μm, or one-tenth the paper thickness, and the accuracy required for measurements is about 0.1 μm, or one-thousandth of the paper thickness. Microscopes can be used for some measurements, but where more accuracy is required, optical methods are used.

2.2.1 Refractive Index Profiling

a. Refracted Near-Field-Method. A refractive index profiling method that has gained widespread use is the refracted near-field method. This method, originally proposed by Stewart [1], uses light that is refracted out of the fiber core rather than transmitted through it. The profile shape is measured by moving a small spot (relative to the core size) across the fiber core and detecting the refracted light that escapes from the core–cladding boundary. Light from a He-Ne laser is focused onto the fiber core with a cone of light that is larger than the acceptance angle of the fiber. Some of the light remains in the fiber core, some is lost to radiation modes, and some is lost to leaky modes. To avoid having to correct for leaky modes, an opaque mask is used to prevent the leaky modes, which escape at angles below a certain minimum (θ_{min}), from reaching the detector. The minimum escape angle, θ_{min}, corresponds to a certain input angle, θ'_{min}, and all light of $\theta' > \theta'_{min}$ must reach the detector. Input apertures are used to limit the numerical aperture of the input beam to the proper $\theta' = \theta'_{max}$, as shown in Figure 2.1.

The fiber is immersed in index-matching oil to prevent reflection at the outer cladding boundary, and measurements of the detected light power as a function of input beam position $P(r)$ are made. The angle θ_{min} is known by the size of the mask in front of the detector, and θ'_{max} is known from the incident light numerical aperture. The refractive index of the core is obtained from

$$n(r) = n_2 + n_2 \cos \theta_{min} \left(\cos \theta_{min} - \cos \theta'_{max} \right) \left[\frac{P(a) - P(r)}{P(a)} \right]$$

$P(r)$ is the detected light power, $P(a)$ is the detected power as the beam is scanned into the cladding, and n_2 is the refractive index of the cladding. This equation can be simplified to

$$n(r) = k_1 - k_2 P(r)$$

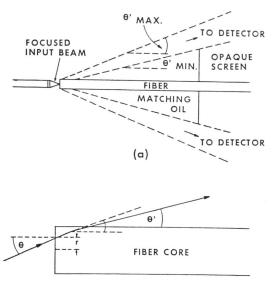

Figure 2.1 Refractive index measurement using the refracted near-field technique, showing the ray paths and detector location. (a) Experimental setup; (b) ray trajectories.

where k_1 and k_2 are constants that can be determined by calibration, since $n(r)$ is directly proportional to $P(r)$ [2]. The refracted near-field method is recognized as a definitive method for determining the index profile by standards organizations. However, to realize its inherent accuracy, clean fiber ends and care are required in the measurements. Thus, it is not usually used in manufacturing; but because sample preparation is easy and accuracy is high, it is often used in research and as a check of other methods.

b. Interferometric Methods. The interferometric or slab method was the first refractive index measurement method that was reliable and could give refractive indices of about 1 part in 10^4 [3]. However, it does require careful sample preparation in that the faces of the slab must be parallel and the thickness must be known accurately to determine the refractive index. A reference slab of known refractive index and thickness is placed in one arm of a specially modified interference microscope while the fiber slab sample (the unknown refractive index) is placed in the other. The optical paths of the two light beams are compared in the eyepiece. The phase delay corresponding to the change in optical path length is

$$L = n(r)d$$

and

$$\theta = kL$$

where θ is the phase retardation and $k = 2\pi/\lambda$. The phase shift S of the fringe depends on its position in the fiber core. The refractive index is calculated from the fringe shift and the fringe spacing D. The fringe shift $S(r)$ is the relative phase difference ψ between the phase retardation of the reference slab and the unknown (fiber) slab.

$$\psi = k[n(r) - n]d$$

Therefore

$$\frac{2\pi}{D} = \frac{\psi}{S(r)}$$

The refractive index of the core is

$$n(r) = n + \frac{\lambda S(r)}{Dd}$$

This method is seldom used today because of the difficulty in sample preparation. A variation of the technique is to use the fiber as the sample and look transverse to the fiber axis. The interferogram is obtained directly from the fiber, so no grinding and polishing of the sample is needed. Core diameter, core concentricity error, core noncircularity, refractive index profile, and delta (Δ) are easily obtained with one measurement technique.

2.2.2 Outside Diameter Measurement and Control

The outside diameter of the fiber is usually monitored during fiber drawing to maintain uniform diameter control and may also be checked by offline methods for verification. The most common method is to use a microscope to determine the outside diameter in several orientations. However, this is time-consuming and involves some operator interpretation on the placement of a cursor, but the procedure can be automated to make it faster and less subject to operator error [4].

Diameter control for online fiber drawing is a fiber diameter measurement, and several novel techniques have been developed. Probably the most reliable and most widely used is the forward-scattering technique developed by Watkins [5]. This method uses interference from a He-Ne laser beam striking the fiber at a right angle to the fiber axis. Interference is observed from a ray striking the cladding–air boundary and a ray passing through the fiber. Forward scattering between 6° and 68° contains the interference pattern, and the ability to link the fringe pattern to a feedback loop has made this method particularly suited to online diameter control of fiber drawing.

2.2.3 *Numerical Aperture*

Three parameters that are used to describe the difference in refractive index between the core and cladding in multimode fibers are the normalized refractive index difference, delta (Δ), the maximum theoretical numerical aperture calculated from Δ, and the measured numerical aperture from the far-field measurement. In principle, any multimode fiber could be described by any one of these three parameters and the other two would be known. In practice, there is some uncertainty in making the transition from Δ to theoretical NA to measured NA. The term *numerical aperture* is also used imprecisely to describe an optical waveguide. The precise terms *acceptance angle* and *radiation angle* are preferable [6]; however, the term numerical aperture has gained widespread use.

The theoretical NA is calculated from Δ:

$$NA_{th} = \sqrt{n_1^2 - n_2^2}$$

where n_1 is the core refractive index and n_2 is the cladding refractive index. The normalized refractive index difference Δ is

$$\Delta = \frac{n_1 - n_2}{n_2}$$

Thus,

$$NA_{th} = n_2\sqrt{(1 + \Delta)^2 - 1} \approx 1.45\sqrt{(1 + \Delta)^2 - 1}$$

where Δ is measured at 0.63 μm with the refracted near-field technique or at 0.54 μm by transverse interferometry.

The measured NA is determined at 0.85 μm according to EIA test method RS-455-47, using the far-field radiation pattern of the fiber, and is defined by

$$NA_{meas} = \sin \theta_5$$

where θ_5 is the half-angle of the radiation pattern scanned in the far field at the 5% intensity point. The value of NA measured will be less than the theoretical NA because of the different wavelengths and the way the measurements are defined. The theoretical value is defined at the baseline of the refractive index profile, while the measured value is arbitrarily defined at the 5% point of the radiation pattern above the baseline. There may also be some question in determining the peak value of the refractive index for the theoretical method.

Currently, there is about a 5% difference between NA calculated from Δ as measured by refracted near field and the lower measured NA value from the far-field data. There is a correction to the wavelength from 0.63 μm, but the largest factor in making the two measurements agree is the difference in the definitions. Currently the relationship between the two is $NA_{meas} = (0.95) NA_{th}$.

2.3 ATTENUATION

The optical attenuation in fibers consists of absorption and scattering and can be characterized by an equation of the form

$$\gamma = A\lambda^{-4} + B + C\lambda$$

where A is the Rayleigh scattering coefficient, B is the wavelength-independent loss, and C is the wavelength-dependent absorption from OH, etc.

   Attenuation is usually measured using a technique known as two-point loss measurement and also known as the cut-back method [7]. The same technique is used for both single-mode and multimode fibers, but minor differences exist in the implementation of the measurement. The attenuation of any fiber is the ratio of the power transmitted through the fiber, P_1, to the power launched into the fiber, P_0. The attenuation, expressed in decibels, is

$$\alpha = -10 \log \frac{P_1}{P_0}$$

The attenuation coefficient α is usually used to mean the optical loss of the fiber in decibels per kilometer. To determine the attenuation as a function of wavelength, interference filters or a monochromator can be used to vary the wavelength of the launched light. The entire wavelength region of interest is scanned, the fiber is broken for a short reference strap, and an identical wavelength scan is made. The short-strap power at every wavelength is compared to that of the long length at the same wavelength, and the loss is computed as above.

2.3.1 Multimode Attenuation

Multimode fibers contain hundreds of propagating modes, with each mode having different attenuation characteristics. Because of differential mode attenuation and mode coupling, the measured attenuation is dependent on the power distribution between the modes [8]. The measured attenuation can therefore depend on how the power is launched into the modes. After a certain length, called the coupling length, there exists a steady-state or equilibrium mode distribution, and the measured loss is a linear function of length, as shown in Figure 2.2. Unfortunately, the coupling length is not constant from fiber to fiber and may vary from a few hundred meters to many kilometers. Launching a steady-state mode distribution will result in an attenuation coefficient that is independent of length, but this is not always easy to achieve, and several methods exist for trying to obtain this distribution. The input beam may be shaped [8] to attempt to launch a steady-state mode distribution, or an overfilled distribution may be used that fully populates all the fiber modes. If an overfilled condition is used, mode filters are then used to remove the high-order modes and give the steady state [9].

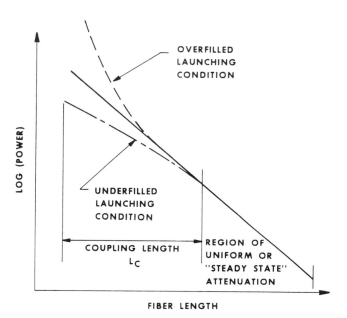

Figure 2.2 Effect of mode distribution with fiber length. At lengths greater than the coupling length, the attenuation is independent of fiber length.

2.3.2 Single-Mode Attenuation

The measurement of single-mode attenuation is not complicated by the presence of hundreds of modes, as in the multimode case, but there are some other considerations. Single-mode fibers are not single-mode across the entire wavelength spectrum. They are only single-mode above a certain wavelength, known as the cutoff wavelength. For accurate attenuation measurements the test fiber should be measured in the single-mode region. The addition of a small loop that causes abnormally high loss of the second-order mode, LP_{11}, but negligible loss of the fundamental mode, LP_{01}, is usually enough to ensure single-mode operation. The attenuation is measured with the loop in the portion of the fiber that is to become the short strap.

2.3.3 Reference Fiber Method

The two-point method, while being the preferred method because of its high accuracy, is a destructive method because the fiber must be broken to make the short-strap measurement. Frequently the attenuation of an installed or connectorized cable must be determined, and the cable cannot be broken for the short-strap measurement. If a short length of fiber is used as a reference, then inserting the cable between the short fiber and the detector and measuring the power through the two gives the loss of the cable. Measurements using this method are generally less accurate than those from the two-point method because of uncertainty in the connector loss and the short-fiber characteristics. In addition, these measurements are frequently made under harsh conditions, such as inside a manhole, where cleanliness and the availability of stable electric power are problems. Fortunately, it is usually only necessary to make measurements at one or a few selected wavelengths using laser diodes or light-emitting diodes (LEDs).

2.3.4 Optical Time Domain Reflectometry

Optical time domain reflectometry (OTDR) uses the scattering properties of the fiber to determine the total attenuation. A pulse of light with a short time duration is launched into the fiber, and a portion of the light traveling down the fiber in the forward direction is scattered and captured by the fiber in the backward direction. The incident pulse is attenuated as it travels down the fiber in the forward direction. Likewise, the scattered pulse traveling in the backward direction is attenuated the same amount. The resulting pulse is attenuated twice over any given length of fiber because the pulse has traversed the same length of fiber twice [10, 11]. An experimental arrangement and backscattering trace are shown in Figure 2.3.

This technique has found widespread use as a diagnostic tool for fiber attenuation measurements, splice loss measurements, and fault or break location and for length measurement. An OTDR trace is unique to the fiber and connectors because it shows the attenuation at every point along the fiber. The basic difference between OTDR and the two-point spectral measurement is that OTDR gives the loss at every point in the fiber for one selected wavelength. The spectral loss measurement gives the spectral loss for the composite structure, with no length information. The two methods complement each other in the data they provide, since each measurement provides important, but different, information.

Measurements using OTDR are one-ended measurements; access to both ends of a fiber or cable is not necessary, making it useful as a field measurement technique.

The development of OTDR for single-mode fibers has been rapid. The first research papers on single-mode measurements appeared in 1980 [12], and since that time substantial progress has been made in extending the length range of measurements [13–15]. Commercial OTDR instruments using laser diodes are available that operate at 0.85, 1.31, and 1.55 μm, and improvements in laser sources and detectors have made this type of measurement commonplace.

Attenuation measurements are made by examining the power relationships in the transmitted and reflected pulse. The power incident on a scatterer

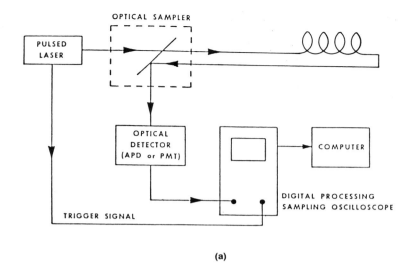

(a)

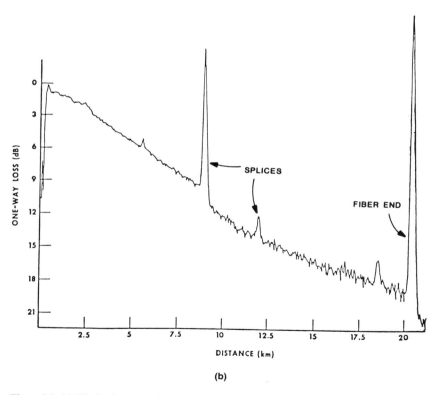

(b)

Figure 2.3 (a) Block diagram of optical time domain reflectometer (OTDR), illustrating the basic principle of OTDR measurements. (b) Backscatter trace of six fiber sections totaling about 20 km; the splices and fiber sections are easily visible.

at distance z into the fiber is

$$P(z) = P_i \exp\left[-\int_0^z 2\alpha_i(z)\, dz\right]$$

where $P(z)$ is the backscattered power, P_i is the input pulse power, α_i is the attenuation of the incident pulse, and z is the length coordinate. The attenuation of the incident (i) and scattered (s) pulses is not the same because the reflected and incident pulse do not have to occupy the same mode volume. In practice, α_i and α_s cannot be separated and what is measured is the average attenuation α_t. The loss coefficient α_t is obtained from the slope of the logarithm of the backscattered signal.

$$\alpha_t(z) = \frac{1}{2}\left(\frac{d}{dz}\right)\ln\frac{P_i}{P_s}$$

Care in controlling the mode group through launch conditions and filters aids in minimizing variations in α_s for multimode fibers. If accurate attenuation or splice loss measurements are to be made on single-mode fibers, the OTDR should be operated at the wavelength where the fiber will be used. For splice loss measurements, a measurement from each end of the fiber span is usually required to minimize the difference in scattering coefficients between nonidentical fibers on either side of a splice.

2.4 BANDWIDTH MEASUREMENTS

The bandwidth of an optical fiber is a measure of the information-carrying capacity of the fiber. The type of information required for both single-mode and multimode fibers is similar; but because of the different inherent bandwidths, the techniques used will be different.

Bandwidth is proportional to the inverse of the pulse broadening, and pulse broadening is caused by dispersion. Dispersion in multimode fibers is caused by waveguide, material, and intermodal dispersion. The largest of the three is usually intermodal dispersion where the time delay among the modes is not constant. In a fiber having hundreds of modes it is impossible to equalize the transit time of all the modes, and this limits the bandwidth of multimode fibers to a few gigahertz per kilometer. Single-mode fibers, having only one mode, have higher bandwidths. Dispersion is from material and waveguide effects, and these can be tailored to position the wavelength of minimum dispersion.

2.4.1 Multimode Bandwidth

Bandwidth measurements on multimode fibers are complicated by the same problems as are found in attenuation measurements. The intermodal dispersion depends on the power distribution among the various modes, the distribution of power that is launched into the modes, the differential mode attenuation, and the mode coupling [16–18]. Many attempts have been made to extract a length scaling factor that will give the appropriate bandwidth versus length. However, no universal constant has been forthcoming because of these problems. Tight control of the launching conditions is necessary for good accuracy and precision on multimode bandwidth [19, 20].

The measurement of bandwidth in multimode fibers is a measure of the impulse response of the fiber. To obtain the impulse response, the input pulse must be deconvolved from the output pulse, where the input pulse is measured through a short length of fiber and the output pulse through the long length. The detector and other electronics responses are canceled out since these effects are superimposed on both the input and output pulses, but the input pulse must also be attenuated to avoid any nonlinear effect in the detector. An experimental apparatus is shown in Figure 2.4.

To deconvolve the input pulse from the output pulse, it is easier to convert the time information to the frequency domain by the Fourier transform and determine the transfer function of the fiber [21]. The input and output pulse Fourier transforms are related by

$$P_2(f) = H(f)P_1(f)$$

where $P_1(f)$ and $P_2(f)$ are the Fourier transforms of the input $P_1(t)$ and output $P_2(t)$ in the time domain. $H(f)$ is the transfer function, also known as the baseband frequency response, of the fiber. The expressions in the above equation are complex, and several methods have been used for describing their mathematical representations [22, 23]. Probably the easiest to under-

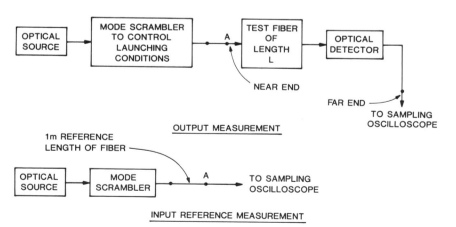

Figure 2.4 Schematic diagram of a bandwidth measurement in the time domain using the fast Fourier transform to obtain the baseband frequency response.

stand, and to relate to physical concepts, is the formulation in polar coordinates [21]. The transfer function can be written as

$$H(f) = M(f)^{e^{i\theta(f)}}$$

where $M(f)$ and $\theta(f)$ are real functions of f.

Since the impulse response is the inverse of the transfer function, the bandwidth is determined from the ratios of the fast Fourier transforms of the input and output waveforms and is defined as the point where the baseband frequency response has decreased by 3 dB.

Bandwidth or frequency response and rms pulse width (2σ) can be related in simple terms. If the output pulse is represented by a pure Gaussian, $f(t)$,

$$f(t) = \frac{1}{\sigma\sqrt{2\pi}} e^{r^2/2\sigma^2}$$

then the transform is

$$F(\omega) = e^{-\sigma^2\omega^2/2}$$

The relationship between the rms pulse width 2σ and the full width at half-maximum bandwidth (BW) is obtained from the Fourier transform of the pulse. The bandwidth is defined as the frequency where the frequency response has decreased by 3 dB; therefore,

$$F(\omega) = \frac{1}{2} = \exp\left(\frac{\sigma^2(2\pi\,\mathrm{BW})^2}{2}\right)$$

$$\ln\left(\frac{1}{2}\right) = -\frac{1}{2}(2\pi\,\mathrm{BW}\,\sigma)^2$$

The bandwidth is related to the rms pulse width by

$$\mathrm{BW} = \frac{\sqrt{2\ln 2}}{2\pi\sigma} = \frac{0.19}{\sigma_T}$$

where σ_T is the rms pulse broadening. The relationship between the impulse response in the time domain and the frequency response is shown in Figure 2.5.

2.4.2 Frequency Domain Measurements

The main advantage to frequency domain measurements is that the inherent precision is better because no complex deconvolution is involved. The bandwidth is a simple ratio of the output to the input frequency response [24]. Time domain measurements are generally used where fiber lengths are short and the total bandwidth may be in the gigahertz range, whereas frequency domain measurements are more widely used where the total fiber length is

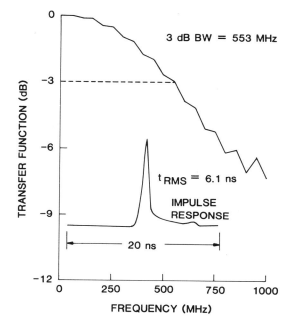

Figure 2.5 Bandwidth measurement. An impulse response (time domain) is shown with its corresponding fast Fourier transform in the frequency domain.

long and the bandwidth may be in the megahertz range. The ability to modulate a diode laser at megahertz rates and the compact and robust nature of the electronics have made this technique a widely used method for field measurements.

The simplest approach to measuring the bandwidth in the frequency domain is to modulate a laser with an RF sweep generator and detect the output with a wideband, linear RF detector. The transfer function is obtained by dividing the output frequency response with the fiber in place by the input response with the fiber removed. This method is sensitive to the signal-to-noise ratio and harmonic distortion introduced by the optical source.

Spectrum analyzers used in radio-frequency analysis have choices of gain, averaging, sweep rate, and detection bandwidth. These narrowband devices make good detectors for frequency domain measurements. However, maximum use of a spectrum analyzer is made by locking the laser transmitter source to the local oscillator of the spectrum analyzer. The output frequency then varies as the sweep of the spectrum analyzer. Such devices are known as tracking generators because the output tracks the input sweep. These are probably the most widely used instruments for frequency domain measurements because of their simplicity of construction and their commercial availability. They are limited only by how accurately the lock between the tracking generator and spectrum analyzer can be maintained. As long as only the magnitude of the transfer function is needed, these methods are adequate. Other, more sophisticated, methods have been used to extract phase information [24–26].

2.5 SINGLE-MODE DISPERSION

Pulse broadening in single-mode fibers is caused by material and waveguide effects. Material dispersion occurs because the group velocity of a pulse is a function of the refractive index variation with wavelength. Waveguide effects depend on the refractive index profile of the core and the refractive index profile of the cladding. These profile shapes can usually be changed slightly to give the desired dispersion minimum at some particular wavelength. The total dispersion is the parameter of interest and is referred to as intramodal dispersion, since there is only one mode, or chromatic dispersion. The term *dispersion* will be used in this discussion to mean the total dispersion.

The dispersion is the change in pulse broadening with wavelength as is expressed by

$$D(\lambda) = \frac{1}{L} \frac{\partial \tau}{\partial \lambda}$$

The dispersion is expressed in units of picoseconds per nanometer per kilometer. The pulse broadening is given by

$$\Delta \tau = D(\lambda) L \, \Delta \lambda$$

where D is the dispersion, L is the fiber length, and $\Delta \lambda$ is the wavelength spread of the source.

In a single-mode fiber, the pulse broadening is too small to measure directly by the methods used for multimode fibers. The method that is usually used is that of measuring the time delay of a series of pulses at different wavelengths. At the wavelength of zero dispersion, the time delay is a minimum. Thus, it is called the pulse delay method. An equation of the Sellmeier type is fit to the delay data, and the dispersion is calculated from the fitted coefficients of the equation. Such an equation can be represented by

$$r = A + B\lambda^2 + C\lambda^{-2}$$

The derivative of τ with respect to wavelength, $\partial \tau / \partial \lambda$, is the dispersion coefficient and is

$$\frac{\partial \tau}{\partial \lambda} = 2B\lambda - 2C\lambda^{-3}$$

A plot of the pulse delay curve and the calculated dispersion coefficient is shown in Figure 2.6. For some systems design applications, it is necessary

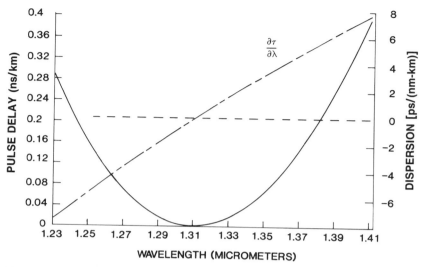

Figure 2.6 Pulse delay (solid curve) versus wavelength is shown for a fiber having its zero dispersion wavelength at 1.31 μm. The dispersion (dashed curve) is shown as the derivative of the pulse delay. Its zero crossing point is 1.31 μm.

to know the change in dispersion with wavelength. This is necessary when operating with multimode spectral sources that are not at the zero-dispersion wavelength. Setting the above equation to zero and solving for λ, we get the expression for the zero-dispersion wavelength:

$$\lambda_0 = \left(\frac{C}{B}\right)^{1/4}$$

Inserting this value for the zero-dispersion wavelength back into the expression for the dispersion coefficient and differentiating to get the change of dispersion with wavelength at the zero-dispersion wavelength gives

$$S_0 = 8B$$

where S_0 is the dispersion slope at the zero-dispersion wavelength in picoseconds per square nanometer per kilometer.

2.5.1 Dispersion vs. Bit Rate

Once the dispersion is determined, the bit rate (information-carrying capacity) of the fiber can be calculated. Unlike multimode bandwidth, the bit rate is a function of the input source spectral width and the dispersion of the fiber. The calculated bit rate of a single-mode fiber would seem to go to infinity at the zero-dispersion wavelength, except that higher-order terms, namely the dispersion slope, limit the bit rate.

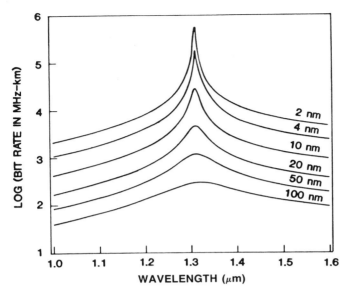

Figure 2.7 Bit rate versus wavelength is plotted for several different source spectral widths for a fiber like that shown in Figure 2.6. The bit rate is sharply peaked at the zero-dispersion wavelength for sources with narrow spectral widths but falls off rapidly as the spectral width increases. (For simplicity, the logarithm of the bit rate rather than the actual bit rate is plotted.)

The limiting bit rate is given by [27]

$$\text{BR} \leq \tfrac{1}{4}\{2(DL\,\Delta\lambda)^2 + [SL(\Delta\lambda)^2]^2\}^{-1/2}$$

where $\Delta\lambda$ is the spectral width of the source and S is the dispersion slope at any wavelength given by

$$S = 2B + 6C\lambda^{-4}$$

Near the zero-dispersion wavelength, the bit rate is limited by the dispersion slope, and at wavelengths far away from λ_0 it is limited by the dispersion. Also, the bit rate is a function of the source spectral width. This relation and the dependence on source width is plotted in Figure 2.7.

2.5.2 Experimental Methods

a. The Fiber-Raman Laser. The fiber-Raman laser was the first technique to give accurate dispersion data, and it is the method by which all other techniques are compared. This method uses narrow pulses of different wavelengths to measure the transit time through the fiber and give the pulse delay curve. The difficulty is in generating the narrow pulses over a wide spectral range in the 1.3–1.7 μm region [28, 29].

When light strikes any material, some is absorbed and some is scattered. Most of the scattered light is of the same frequency as the incident light, but a small percentage is shifted in frequency. This shift arises from the interaction of the incident light with the vibrating molecules of the medium and is known as Raman scattering. Normally, the intensity of Raman scattering is many times less than the incident intensity. If the incident light is focused into an optical fiber, the scattered light and incident light travel together through the fiber. The scattered light component increases in intensity because it is being "pumped" by the incident light. The phase relationship between the pump and scattered light can be maintained for distances of a few hundred meters, but they eventually get out of phase because of dispersion. As the Raman-scattered light increases in intensity, another Raman-scattered component begins to grow, and this process may be repeated several times to give five or six Raman components. Because the Raman-shifted light is broader in spectral width than the initial pump light, each successive shifted component is broader than the one before. Eventually, they overlap to form a continuum. This arrangement is sometimes referred to as a fiber-Raman laser. Figure 2.8 shows the spectrum from a fiber with Raman-shifted output.

High-power lasers of the Nd:YAG type make good pump lasers for this purpose, because with their strong 1.064-μm line, the fiber output is a con-

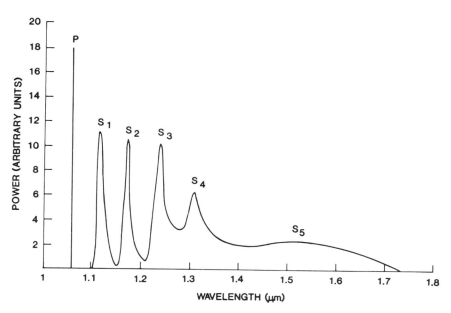

Figure 2.8 Spectral output of a Raman-shifted fiber. The pump laser is at 1.064 μm and the Stokes components are indicated as S_1, S_2, etc. The successive broadening effect of the Stokes scattering is to blend into a continuous spectrum over the wavelength range of interest for single-mode fibers.

tinuum from 1.2 to over 1.7 μm. Unfortunately, a Q-switched pulse of even 10 ns is too long for accurate delay measurements. A much shorter pulse is needed, which should be about 0.5 ns or less in duration. Modelocking is the preferred way to get short pulses of a few hundred picoseconds, but the repetition rate is about 100 MHz. The development that solved this problem was to run the Q-switch from the modelocker timer [28]. A modelocked pulse starts a countdown circuit that divides the modelock frequency by a preset factor and fires the Q-switch. The modelocker and Q-switch are now tied together through the timing circuit, and this results in the laser operating like a simple Q-switched laser, but under each Q-switch envelope are several modelocked pulses. Further, all the modelocked pulses have the same time relationship under the Q-switched pulses. Using normal sampling electronics, one modelocked pulse can be selected from the pulse train.

The Raman fiber's output versus wavelength is like that shown in Figure 2.8, and the output versus time is a series of modelocked pulses modulated at the Q-switched rate. The experimental arrangement for such an apparatus is shown in Figure 2.9. Different wavelengths are selected by the monochromator, the coefficients of the Sellmeier equation are determined from the fit to the delay data, and the zero-dispersion wavelength is calculated.

b. Other Transit Time Techniques. Instead of having to use a high-power Nd:YAG laser, several pulsed laser diodes have been used. Each laser provides a different wavelength, and this method has been called the "Gatling gun" approach because each different wavelength requires a different laser. The pulse delay of each wavelength is measured in the same way as in the fiber-Raman approach [30].

Rather than using pulsed lasers and timing the pulse, multiple phase detection techniques allow an apparent length measurement without having to produce a narrow pulse. A continuous-wave (cw) diode laser is sinusoidally

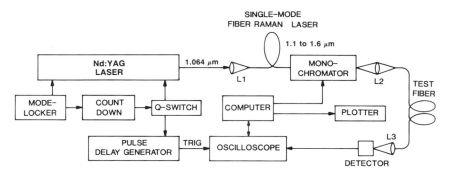

Figure 2.9 Experimental schematic for a Nd:YAG pumped Raman laser dispersion measurement. The output of the Raman fiber is like that shown in Figure 2.8, and this setup is used to generate the curves shown in Figure 2.6.

modulated, and the phases of the transmitted and received signals are compared to determine the length [31–34]. Length measurements are made at the various wavelengths and then converted to the corresponding transit time, and the delay curve is plotted as before.

Light-emitting diodes have also been used instead of lasers with the phase method. They have the advantage that their spectral output is broad enough that a wide wavelength range can be covered using one LED.

A novel LED phase technique is to modulate the monochromator at two closely spaced wavelengths. The phase of each wavelength is measured, and the dispersion is calculated by interpolating between the measured wavelengths. This has been called a differential phase technique and has the advantage of measuring the dispersion directly so that no assumption about a fitting function or differentiation of a fitted curve is needed.

These laser diodes and LED methods have the advantage of being compact, low-power, and rugged. Thus, they have the advantage of being useful for field measurements. They show good agreement with the fiber-Raman technique and are finding use in both manufacturing and field measurements.

c. Interferometric Techniques. The fiber-Raman technique requires the generation of narrow pulses and uses lengths of fiber about 1 km long. As an alternative, several researchers [35–37] have investigated making dispersion measurements using interferometry.

The phase delay of a cw light beam is measured as the beam traverses a short length of fiber. This technique is able to use short (0.5-m) lengths of fiber since the time delay corresponding to the measured phase shift is so

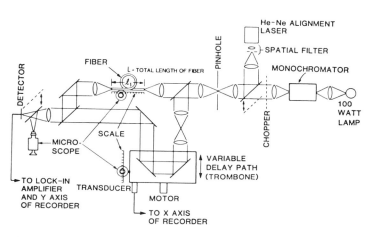

Figure 2.10 Experimental diagram of an interferometer used to measure dispersion. Much shorter lengths of fiber are used than are required in Figure 2.9, and comparable delay curves are generated.

much shorter than in the pulse delay method. Both Mach-Zehner [35, 36] and Michelson [37] interferometers have been used with good results. A typical measurement is made by scanning the interferometer with the fiber in one arm and with the fiber removed. A diagram of such an experiment is shown in Figure 2.10.

Each of these methods has a particular advantage. The fiber-Raman is the most versatile because any wavelength range or wavelength spacing can be chosen. The diode lasers and LEDs are useful for routine factory and field measurements. The interferometer is useful for investigating the length dependence of dispersion since it uses short lengths of fiber.

2.6 CUTOFF WAVELENGTH

Cutoff wavelength measurements on single-mode fibers probably cause more difficulty in interpretation than any other measurement. This is not because the measurement itself is ambiguous or difficult, but because the end users are often unfamiliar with the meaning of "cutoff wavelength." A requirement is placed on the cutoff wavelength because of concern about adequate attenuation of high-order modes to avoid interference effects at splices and connectors (modal noise). The theoretical cutoff wavelength is defined as the wavelength above which the high-order modes, LP_{11} and above, will not propagate. This is of little more than academic interest for the purposes of this discussion, because it is impossible to measure. The practical issue is the effective cutoff, or the wavelength where the high-order modes cannot propagate significant distances. The effective cutoff wavelength is some arbitrarily large value, typically about 20 dB, total attenuation of the high-order modes, in some arbitrary length (2, 5, or 20 m), with some arbitrary bending condition (none, 28 cm, inside cable, etc.). Thus, the measured cutoff wavelength depends on the length and configuration of the measured fiber and is chosen for convenience and reproducibility of the measurement, not to represent any typical system configuration [38]. The cutoff wavelength is not a well-defined number as is attenuation, dispersion, core diameter, etc. The cutoff wavelength has a strong length dependence that is different for different fiber designs. The difficulty in interpretation arises when a cutoff measurement is made on short lengths of fiber but installed fiber in a cable is kilometers long. End users often confuse the meaning of cutoff by using the short-length measurement to represent the cutoff of the long length.

There are two commonly used cutoff measurement techniques: the bend [38–41] and the power step [42]. The bend method uses a short (2–5 m) length of fiber, and the attenuation is measured with the strap straight (minimum bend radius of 28 cm) and with a 2.5-cm bend placed in it. The bend is intended to attenuate completely the higher-order modes while causing negligible attenuation of the fundamental mode. The wavelength is stepped

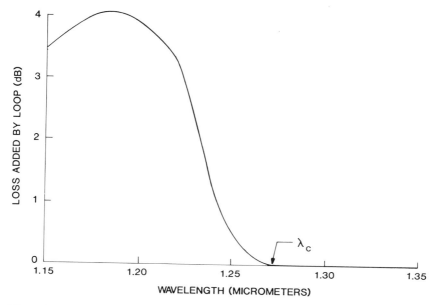

Figure 2.11 Determination of cutoff wavelength by the attenuation of an added loop. The loop causes the next higher order mode to be highly attenuated (cut off), and the attenuation caused by adding the loop is plotted versus wavelength. An arbitrary point (usually 0.1 dB) where the addition of the loop causes minimal increase in loss is taken as the cutoff wavelength.

in short wavelength intervals of about 10 nm, and the difference in the two strap measurements is plotted. The cutoff is the wavelength where the introduction of the small 2.5-cm loop causes the attenuation to approach a linear region. Figure 2.11 shows this effect.

In the power-step method, the power transmitted through the single-mode fiber is normalized to the power transmitted through a multimode fiber. The cutoff wavelength is determined as the point where the power in the single-mode fiber increases as indicated by the appearance of the LP_{11} mode. Figure 2.12 shows this effect. This method can be used for any fiber length and has been applied to cables. Both of these methods are easy to execute since they use the same equipment as that used in measuring cutback attenuation. The difficulty arises in the meaning and interpretation of the results.

2.6.1 Length Dependence

The explanation for the cutoff phenomena is that modes near cutoff become leaky. This leakage loss depends on the cladding profile, cladding thickness, core dimensions, core refractive index, and bend radius. Because it is an attenuation measurement, the value quoted for the cutoff wavelength is de-

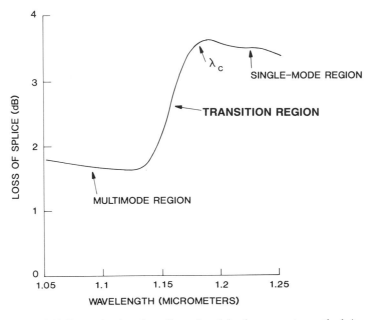

Figure 2.12 Determination of cutoff wavelength by the power-step method. A multimode fiber is spliced to a single-mode fiber and the transmitted power is measured against wavelength. Where the single-mode fiber is below cutoff, the effect is of two multimode fibers spliced together. Above cutoff, only one mode exists in the fiber and the transmitted power decreases; that is, the apparent splice loss increases.

pendent on how the fiber is measured and on the particular fiber design. The attenuation rate, the rate at which the measured cutoff wavelength decreases with length, is also a function of the fiber design. Different fiber designs have shown a wavelength dependence of 20 nm [43] to 70 nm [38] per decade of fiber length. An empirical relation has been proposed to predict the effective cutoff wavelength for any fiber design [44]:

$$\lambda_{ce} = \lambda_e - A \log L - \frac{B}{R}$$

where λ_e is the cutoff of 1 m of stretched fiber, L is the length, R is the bend radius, and A and B are the length dependence and bending sensitivity, respectively, of the cutoff wavelength.

The above empirical relation shows that the decrease in cutoff wavelength with length is a logarithmic function of the length. While systems have been demonstrated with transmitters at 1.26 μm and fiber cutoff wavelengths at 1.46 μm [45], a more conservative approach is usually taken. For a system operating at 1.31 μm, cutoff wavelengths of 1.35 μm measured on a 5-m length have been used. Allowing the cutoff as measured on a 2-m strap to be as large as possible, but avoiding modal noise, has several advantages from the point of view of systems design. It results in the fundamental mode being more tightly bound, making the fiber less bend-sensitive and giving a more robust fiber design. As an additional consideration, if the fiber is designed to operate at two widely spaced wavelengths, such as 1.31 and 1.55 μm, the bending loss at 1.55 μm due to macrobends is reduced because the fiber is less bend-sensitive.

2.7 MODE-FIELD DIAMETER

The fields in the core of a single-mode fiber can be approximated by a Gaussian function of the form

$$g(r) = g_0 \exp \frac{-r^2}{\omega_0^2}$$

where $g(r)$ is a Gaussian function of width ω_0. Since the field has a Gaussian distribution, there is a significant portion of it that propagates in the cladding. The degree of mode confinement is wavelength-dependent and decreases with increasing wavelength. Therefore, the mode-field diameter increases with wavelength, and splice loss and mode-filed diameter mismatch have been directly related. The most important reason for measuring the mode-field diameter is to determine the splice loss compatibility between fibers. Added loss from mode-field diameter mismatch is a slowly varying function of mode-field diameter, and fibers having different mode-field diameters will incur some added splice loss in both splices and connectors. The added splice loss between two fibers with different mode-field diameters $2\omega_1$ and $2\omega_2$ is predicted by

$$\alpha_{\text{splice}} = 20 \log_{10} \left[\frac{\omega_1^2 + \omega_2^2}{2\omega_1\omega_2} \right]$$

Several methods of measuring the mode-field diameter have been examined [46]. Realizing that the Gaussian model is an approximation of the exact fields, the mode-field radius is defined as the value of ω_0 that maximizes the launch efficiency integral:

$$I = \frac{\left[\int_0^\infty E_g(r)E_m(r)r \, dr \right]^2}{\int_0^\infty E_g^2(r)r \, dr \int_0^\infty E_m^2(r)r \, dr}$$

where $E_g(r) = E_0 \exp\left[-r^2/\omega_0^2\right]$ and represents the fields that are approximated in the fiber by a Gaussian with radial dependence r. $E_m(r)$ is the measured field distribution. This approach has physical significance since a Gaussian beam of this width would maximize the power launched into the fiber.

2.7.1 Measurement Methods

1. *Near field*. The near field of the fiber is imaged onto an IR vidicon, and the measured intensity is fit by the above equation. This method is probably the simplest to execute, but it is limited by the nonlinearity of the vidicon in both scan rate and intensity.
2. *Transverse offset*. The sensitivity of splice loss to transverse offset is also a good indicator of mode-field radius. A fiber is cleaved so that the ends are perpendicular and is then spliced together using precision translation stages. Since the fibers are identical, they have the same mode-field radius, and the transmission coefficient as a function of transverse offset d is

$$T(d) = T_0 \exp \frac{-d^2}{\omega_0^2}$$

The constant T_0 is not necessary since only the change in splice loss versus offset needs to be measured. This is easier than measuring the absolute splice loss. The disadvantages to this technique are the precision and stability required in the stages and the high quality needed for two fiber ends. The advantages are that many wavelengths may be used, the equipment is simple, and the measurement is rapid.
3. *Far field*. The far-field technique relies on the Gaussian approximation being made directly in the far field. If the near field is nearly Gaussian, then the far field is also nearly Gaussian. The mode-field diameter can be defined as

$$2\omega_0 = \frac{\lambda}{2\pi \tan \theta_\omega}$$

where θ_ω is the far-field beam angle, since the far-field data are measured as a function of angle rather than distance. An overlap integral is used to fit the angular far-field data, where the radial dependence is replaced with θ.
4. *Variable aperture*. The power transmitted through various size apertures is measured in the far field as a function of aperture size.
5. *Knife edge*. The power transmitted in the far field is measured as a knife edge is moved laterally through the field.

The agreement between these methods depends on the choice of definition of mode-field diameter and the type of fiber measured. The Gaussian definition shows small differences for near-step-index fibers at longer wavelengths [47], and differences near 10% have been measured for dispersion-shifted fibers having non-Gaussian fields [48]. These differences exist because the fields deviate from a perfect Gaussian, and thus the computational method used to fit the measured data to a Gaussian is also important. The fitting method and the range of data used in the fit should be known when using the Gaussian definition. While the Gaussian function gives a good indicator of splice loss for near-step-index fibers, it is not a universal fitting function.

An alternative to the Gaussian fitting technique has been proposed by Petermann [49, 50], and it has the potential of being a general fitting function for all methods. It is based on the second moment of the far field, and is defined by

$$2\omega = \frac{2}{\pi W}$$

W is given by

$$W = \left[2 \frac{\int_0^\infty F^2(q)q^3 \, dq}{\int_0^\infty F^2(q)q \, dq} \right]^{1/2}$$

where $q = \sin \theta / \lambda$ and F^2 is the far-field power. The use of this definition, frequently called the Petermann 2 definition, allows the five measurement methods above to be related from one measurement domain to another using functions such as the Hankel, Abel, and Fourier transforms.

2.8 SUMMARY

Numerous fiber measurement techniques have been discussed in this chapter, and an attempt has been made to suggest the appropriate uses for each, as well as to provide a theoretical basis for their use. Although traditional methods were mentioned, the newer ones that are primarily applicable to single-mode fibers have been emphasized.

Geometric measurements are useful for characterizing the physical properties of the fiber such as core diameter and refractive index profile. Optical attenuation measurements help to minimize absorption and scattering to produce the lowest possible fiber loss. In this area, optical time domain reflectometry has been particularly useful in field measurements on installed cables. Bandwidth measurements are used to determine the information-carrying capacity of multimode fibers and are usually obtained by time domain techniques in the laboratory or frequency domain methods on installed cables. The accurate measurement of single-mode dispersion is important for realizing the inherent capacity of single fibers for high-bit-rate systems and is usually done by pulse delay or phase methods. New developments in instrumentation are making dispersion measurements possible for field applications. Cutoff wavelength measurements can be source of confusion, but problems can be avoided by an understanding of the meaning of the term "cutoff wavelength." Finally, the mode-field diameter is important in determining the splice loss compatibility between fibers, and five measurement techniques were discussed.

The continued rapid change in optical fiber technology will quickly outdate some of these techniques, while others will see further enhancement and expansion in the years to come. The continued variety and depth of applications of optical fibers will dictate the direction and development of future fiber measurement techniques.

REFERENCES

1. W. J. Stewart, "A new technique for measuring the refractive index profiles of graded index optical fibers," *Tech. Digest*, Int. Conf. Integrated Optics and Opt. Fiber Commun., Tokyo, Japan, July 1977, pp. 395 Paper C2-2.
2. K. I. White, "Practical application of the refracted near-field technique for the measurement of optical fiber refractive index profiles," *Opt. Quantum Electron.*, **11**: 185 (1979).
3. H. M. Presby, "Profile characterization of optical fibers—a comparative study," *Bell Syst. Tech. J.*, **60**: 1335 (1981).
4. D. Marcuse and H. M. Presby, "Automatic geometric measurements of single mode and multimode optical fibers," *Appl. Opt.*, **18**: 402 (1979).
5. L. S. Watkins, "Scattering from side-illuminated clad glass fibers for determination of fiber parameters," *J. Opt. Soc. Am.*, **64**: 767 (1974).
6. *Optical Waveguide Communications Glossary*, U.S. Department of Commerce, Washington, DC., September 1979.
7. D. B. Keck and A. R. Tynes, "Spectral response of low loss optical waveguides," *Appl. Opt.*, **11**: 1502 (1972).
8. G. T. Holmes, "Propagation parameter measurement of optical waveguides," *SPIE Conf. Proc.*, Washington, DC, April 1980, p. 138.
9. A. H. Cherin and W. B. Gardner, "Measurement standards for multimode telecommunication fibers," *SPIE Conf. Proc.*, Washington, DC, April 1980, pp. 144.
10. M. K. Barnoski and S. M. Jensen, "Fiber waveguides, a novel technique for investigating attenuation characteristics," *Appl. Opt.*, **15**: 2112 (1976).
11. M. K. Barnoski, M. D. Rourke, S. M. Jensen, and R. T. Melville, "Optical time domain reflectometer," *Appl. Opt.*, **16**: 2375 (1977).
12. D. L. Philen, "Optical time domain reflectometry on single mode fibers using a Q-switched Nd:YAG laser," *Tech. Digest*, Symp. Opt. Fiber Measurements, *NBS Publ.* No. 597, Boulder, CO, October 1980.
13. D. L. Philen, I. A. White, J. F. Kuhl, and S. C. Mettler, "Single mode fiber OTDR: experiment and theory," *IEEE J. Quantum Electron.*, **18**: 1499 (1982).
14. M. Nakazawa, T. Tanifuji, K. Washio, and Y. Morishige, "Marked extension of diagnosis length in optical time domain reflectometry using a 1.32 μm YAG laser," *Electron. Lett.*, **17**: 783 (1981).
15. J. Stone, A. R. Chraplyvy, and B. L. Kasper, "Long range 1.5 micron OTDR in a single mode fiber using a D_2 gas-in-glass laser (100 km) or a semiconductor laser (60 km)," *Electron. Lett.* (1985), **21**(12): 541.
16. L. G. Cohen, H. W. Astle, and I. P. Kaminow, "Wavelength dependence of frequency response measurements in multimode optical fibers," *Bell Syst. Tech. J.*, **55**: 1509 (1976).
17. L. G. Cohen, I. P. Kaminow, H. W. Astle, and L. W. Stulz, "Profile dispersion effects on transmission bandwidth in graded index optical fibers," *IEEE J. Quantum Electron.*, **QE-14**: 37 (1978).
18. M. J. Buckler, "Differential mode delay measurement using single mode fiber selective excitation," Proc. Conf. Precision Electromagnetic Measurements, Braunschweig, Germany, June 1980, p. 224.

19. M. Eve, A. M. Hill, D. J. Malyon, J. E. Midwinter, B. P. Nelson, J. R. Stern, and J. V. Wright, "Launching independent measurements of multimode fibers," 2nd Eur. Conf. Opt. Commun., Paris, 1976.

20. P. R. Reitz, "Characterization of concatenated multimode optical fibers with time domain measurement techniques," Conf. Precision Electromagnetic Measurements, Boulder, CO, 1982, paper L-14.

21. D. L. Franzen and G. W. Day, "Measurement of optical fiber bandwidth in the time domain," *NBS Spec. Publ.* No. 637, **1**: 47 (1982).

22. L. G. Cohen, H. W. Astle, and I. P. Kaminow, "Frequency domain measurements of dispersion in multimode optical fibers," *Appl. Phys. Lett.*, **30**: 17 (1977).

23. Vassallo, "Linear power responses on an optical fiber," *IEEE. Trans. Microwave Theory and Technol.*, **MTT-25**: 572 (1977).

24. G. W. Day, "Measurement of optical fiber bandwidth in the frequency domain," *NBS Spec, Publ.* No. 637, **2**: 57 (1983).

25. G. Cancellieri and U. Ravaioli, *Measurements of Optical Fibers and Devices: Theory and Experiments*, Artech House, Dedham, MA, 1984.

26. D. Marcuse, *Principles of Optical Fiber Measurement*, Academic, New York, 1981.

27. L. B. Jeunhomme, *Single Mode Fiber Optics: Principles and Applications*, Marcel Dekker, New York, 1983.

28. C. Lin, L. G. Cohen, W. G. French, and V. A. Foertmeyer, "Pulse delay measurements in the zero material dispersion region for germanium and phosphorus doped silica fibers," *Electron. Lett.*, **14**: 170 (1978).

29. L. G. Cohen and C. Lin, "Pulse delay measurements in the zero material dispersion wavelength region for optical fibers," *Appl. Opt.*, **16**: 3136 (1977).

30. C. Lin, A. R. Tynes, A. Tomita, P. L. Liu, and D. L. Philen, "Chromatic dispersion measurements in single mode fibers using picosecond lnGaAsP injection lasers," *Bell Syst. Tech. J.*, **62**: 457 (1983).

31. D. L. Philen and P. D. Patel, "Measurements of strain in optical fiber cables using a commercial distance meter," 8th Eur, Conf. Opt. Commun., Cannes, France, September 1982, paper AVII-3.

32. K. Tatekura, H. Nishikawa, M. Fujise, and H. Wakabayshi, "High accurate measurement equipment for chromatic dispersion making use of the phase shift technique with LDs," *Tech. Digest*, Symp. Opt. Fiber Measurements, *NBS Spec.*, *Publ.* No. 683, Boulder, CO, October 1984.

33. P. J. Vella, P. M. Garel-Jones, and R. S. Lowe, "Measuring the chromatic dispersion of single mode fibers: a compliance test method," Conf. Opt. Fiber Commun, New Orleans, January 1984, paper TUN15.

34. R. Rao, "Field dispersion measurements—a swept frequency technique," *Tech. Digest*, Symp. Opt. Fiber Measurements, *NBS Spec. Publ.* No. 683, Boulder, CO, October 1984.

35. M. J. Saunders and W. B. Gardner, "Precision interferometric measurement of dispersion in short single mode fibers," *Tech. Digest*, Symp. Opt. Fiber Measurements, *NBS Spec. Publ.* No. 683, Boulder, CO, October 1984.

36. F. M. Sears, L. G. Cohen, and J. Stone, "Interferometric measurements of dispersion spectra variations in a single mode fiber," *J. Lightwave Technol.*, **LT-2**: 181 (1984).

37. W. D. Bomberger and J. J. Burke, "Interferometric measurement of dispersion of a single mode optical fiber," *Electron. Lett.*, **17**: 495 (1981).

38. W. T. Anderson, "Measuring the effective cut-off wavelength of a single mode fiber—the effect of fiber length," Proc. Conf. Precision Electromagnetic Measurements, Boulder, CO, 1982. Paper L-4.

39. K. Kitayama, M. Ohashi, and Y. Ishida, "Length dependence of effective cutoff wavelength in single mode fibers," *J. Lightwave Technol.*, **LT-2**: 629 (1984).

40. W. T. Anderson and T. A. Lenahan, "Length dependence of the effective cutoff wavelength in single mode fibers," *J. Lightwave Technol.*, **LT-2**: 238 (1984).

41. V. S. Shah, "Effective cutoff wavelength for single mode fibers: the combined effect of curvature and index profile," *Tech. Digest*, Symp. Opt. Fiber Measurements, *NBS Special Publ.* No. 683, Boulder, CO, October 1984.

42. C. C. Wang, C. A. Villarruel, and W. K. Burns, "Comparison of cutoff wavelength measurements for single mode waveguides," *Tech. Digest*, Symp. Opt. Fiber Measurements, *NBS Spec. Publ.* No. 641 Boulder, CO, October 1982.

43. N. K. Cheung and P. Kaiser, "Cutoff wavelength and modal noise in single mode fiber systems," *Tech. Digest*, Symp. Opt. Fiber Measurements, *NBS Spec. Publ.* No. 683, Boulder, CO, October 1984.

44. H. T. Nijnuis and K. A. Leeuwen, "Length and curvature dependence of effective cutoff wavelength and LP_{11} mode attenuation in single mode fibers," *Tech. Digest*, Symp. Opt. Fiber Measurements, *NBS Spec. Publ.* No. 683, Boulder, CO, October 1984.

45. N. K. Cheung and P. Kaiser, "Cutoff wavelength and modal noise in single-mode fiber systems," *Tech. Digest*, Symp. Opt. Fiber Measurements, *NBS Spec. Publ.* No. 683, Boulder, CO October 1984, p. 15.

46. W. T. Anderson and D. L. Philen, "Spot size measurements for single mode fibers—a comparison of four techniques," *J. Lightwave Technol.*, **LT-1**: 20 (1983).

47. W. T. Anderson, "Consistency of measurement methods for the field radius in a single mode fiber," *J. Lightwave Technol.*, **LT-2**: 191 (1984).

48. J. Augu and L. Jeunhomme, *Tech. Digest*, SPIE Symp., August 1985. Paper 559-03.

49. K. Petermann, "Constraints for fundamental mode spot size for broadband dispersion compensated single mode fibers," *Electron. Lett.*, **19**: 712 (1983).

50. C. Pask, "Physical interpretation of Petermann's strange spot size for single mode fibers," *Electron. Lett.*, **20**: 144 (1984).

3

Advanced Single-Mode Fiber Designs for Lightwave Systems Applications

L. G. Cohen, W. L. Mammel, and W. A. Reed

3.1 INTRODUCTION

Single-mode fibers serve as transmission highways for lightwave systems that transmit data at high rates over long distances. Fibers are currently being deployed with low losses and near-zero dispersion for lightwave systems applications at wavelengths near 1.3 μm. However, systems of the next generation will undoubtedly include 1.55-μm wavelength applications because intrinsic scattering losses in fused silica glasses are minimized there.

Two general system strategies exist. One is to use conventional step-index fibers, which have minimum dispersion near 1.3 μm but rather high dispersion near 1.55 μm. High-bandwidth systems applications near 1.55 μm would therefore require the use of single-longitudinal-mode (SLM) lasers. The alternative strategy is to use conventional multimode lasers in conjunction with high-bandwidth fibers that have low dispersion at one or many wavelengths between 1.3 and 1.55 μm. This paper will describe three advanced single-mode fiber designs that address these strategies.

Silica core fibers [1, 2] have little dopant in the core and therefore cause the the lowest losses. However, their minimum dispersion region is near 1.29 μm.

Dispersion-flattened fibers [3–5] belong to another class of lightguides that can broaden the near-zero-dispersion region over a wide wavelength range that, in the limit, can simultaneously encompass 1.3 and 1.55 μm. Such fibers could be useful for wavelength-division multiplexing many high-bit-rate information channels onto the same fiber.

Dispersion-shifted fibers [6–9] belong to a class of lightguides that can be tailored to simultaneously provide very low losses and zero dispersion at any wavelength near 1.55 μm.

3.2 TRANSMISSION MEDIA LIMITATIONS

Figure 3.1 illustrates the principal fiber media limitations that can degrade the performance of high-bit-rate lightwave systems. Input pulses are confined within time slots indicated by the tick marks. Fiber attenuation (Fig. 3.1a) causes a loss of light photons as the pulse propagates. That decreases the area within a pulse but does not distort its shape. Single-mode fibers have lower transmission losses than multimode fibers because the lower dopant concentration required in the core leads to lower intrinsic losses due to Rayleigh scattering effects. Figure 3.2 shows typical loss [10] and dispersion [11] characteristics plotted versus wavelength λ. Minimum losses are determined by the combination of Rayleigh scattering (which decreases according

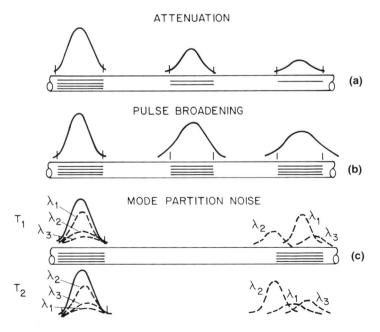

Figure 3.1 Fiber media limitations. (a) Attenuation; (b) pulse broadening; (c) mode partition noise.

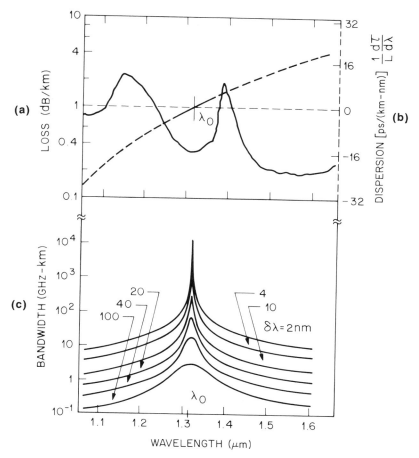

Figure 3.2 Single-mode fiber transmission characteristics. (a) Loss spectra (dB/km vs. λ in μm) with a 0.2 dB/km minimum value near $\lambda = 1.55$ μm. Loss peaks are due to cutoff of the second propagating mode near $\lambda = 1.25$ μm and OH ion absorption near $\lambda = 1.39$ μm. (b) Dispersion spectra [ps/(km-nm) vs. λ in μm] with $\lambda_0 = 1.3$ μm zero-dispersion crossover wavelength. (c) Bandwidth–distance spectra (GHz-km vs. λ in μm) are drawn for source spectral widths ranging from $\delta\lambda = 2$ nm to $\delta\lambda = 100$ nm. The bandwidths scale with $\delta\lambda^2$ at the peaks and $\delta\lambda$ elsewhere.

to λ^{-4}) and the tail of infrared absorption effects due to multiphonon molecular vibrations. The absorption of water in the form of OH ions causes an excess loss peak of 1 dB/km at $\lambda = 1.39$ μm if the OH concentration is 25 ppb.

Other limitations occur because semiconductor laser sources are not monochromatic and radiate a spectrum of wavelengths that will affect transmission due to the dispersion properties of fibers. Pulse broadening (Fig. 3.1b) occurs because fiber dispersion causes each laser wavelength to propagate with a different group velocity, which makes them arrive at different times at the end of the fiber. That broadens the output pulse width σ, beyond the initial time slot and causes intersymbol interference with pulses in adjacent time slots reaching the receiver.

$$\sigma = DL\,\delta\lambda \tag{1}$$

where D is the fiber dispersion, L is the fiber length, and $\delta\lambda$ is the spectral width of the pulse. The ultimate dispersion limitation of a system can be characterized by convolving the dispersion spectrum $[D(\lambda)$ in ps/(km-nm)$]$ of the fiber (Fig. 3.2b) with the spectrum of an input pulse with linewidth $\delta\lambda$. Minimum pulse broadening (σ_{min}) occurs for a source centered about λ_0, the minimum dispersion point, and is proportional to the slope of the dispersion spectrum multiplied by the square of the source linewidth [12],

$$\sigma_{min} \approx \frac{1}{8}\frac{dD}{d\lambda}\,L\,\delta\lambda^2 \tag{2}$$

It is assumed that the dispersion power penalty is small provided that pulse broadening σ is smaller than half a bit period ($B \approx 1.82B_w < L/2\sigma$). The bit-rate capacity B and bandwidth B_w spectra of fibers are inversely related to dispersion, as shown in Figure 3.2c. The peak values of BL (and B_wL) increase as $\delta\lambda^2$ decreases and are on the order of 1000 Gb/(s-km) at λ_0. The bit rate (and bandwidth)–distance product decreases rapidly with wavelength and is only about 50 Gb/(s-km) near $\lambda = 1.55$ μm for conventional fiber and $\delta\lambda = 1$ nm.

Another dispersion limitation is called mode partition noise. It is illustrated in Figure 3.1c for a constant-power laser in which the sum of the powers in individual wavelengths add up to the constant-power solid envelope. However, at different times the total power can be partitioned differently among the individual modes. For the example in the figure, most of the laser power is in the mode at λ_1 at T_1 while at time T_2 most of the power is in the mode at λ_2. Again, fiber dispersion makes the power at different wavelengths separate in time and the received signal becomes noisy because a fraction of the received power is randomly partitioned among the wavelengths. The quality of the laser source is parameterized by the rms width, $2\delta\lambda$, of the laser spectrum and the fraction K of the fluctuation power not in its dominant mode. System performance limits can then be represented by a bit rate–distance product (BL) that depends upon the $\delta\lambda$ and K parameters of the laser and the D and $dD/d\lambda$ chromatic dispersion properties of the fiber [13].

The equation

$$BL < \frac{341}{\delta\lambda\,D}\frac{\text{Gb}}{\text{s}}\text{-km} \tag{3}$$

represents pulse broadening effects on BL, and

$$BL < \frac{130}{\delta\lambda\,D\sqrt{K}}\frac{\text{Gb}}{\text{s}}\text{-km} \tag{4}$$

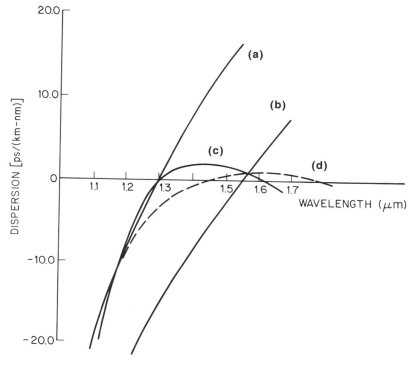

Figure 3.3 Alternative dispersion spectra for (a) conventional fiber with $\lambda_0 = 1.3\ \mu$m, (b) dispersion-shifted fiber with $\lambda_0 \approx 1.55\ \mu$m, (c) dispersion-flattened fiber with low dispersion for $1.3\ \mu$m $< \lambda < 1.55\ \mu$m, and (d) dispersion-flattened fiber with low dispersion for $1.45\ \mu$m $< \lambda < 1.7\ \mu$m.

shows that mode partition noise can impose even more stringent performance limitations on BL when $K > 0.1$ and λ is away from the wavelength of minimum dispersion. The BL limitation for the system is the lesser of Eqs. (3) or (4).

Much higher BL products are possible in fibers that have minimum dispersion close to the system's operating wavelength. Then $D \approx 0$ and $\sigma \approx (dD/d\lambda)L\ \delta\lambda^2$ (instead of $DL\ \delta\lambda$), which lead to dispersion constraints given by the lesser of Eqs. (5) or (6).

$$BL < \frac{11{,}207}{\delta\lambda^2(dD/d\lambda)}\ \frac{\text{Gb}}{\text{s}}\text{-km} \qquad (5)$$

$$BL < \frac{1173}{\delta\lambda^2(dD/d\lambda)\sqrt{K}}\ \frac{\text{Gb}}{\text{s}}\text{-km} \qquad (6)$$

Figures 3.3 and 3.4 plot dispersion spectra versus wavelength for the three generic lightguides that will be discussed. The total dispersion in single-mode

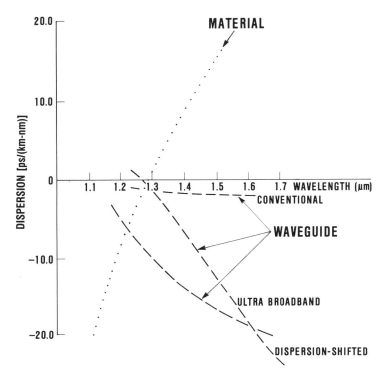

Figure 3.4 Material and waveguide dispersion spectra for conventional, dispersion-shifted, and dispersion-flattened fiber designs.

fibers is actually the sum of two effects. Material dispersion is a bulk property that occurs because the group velocity of a lightguide mode is a function of frequency through the nonlinear dependence of refractive index on wavelength. Zero material dispersion occurs very close to 1.3 μm because current single-mode fibers are made from lightly doped silica glasses. However, waveguide dispersion is a property whose value depends on the group velocity also being a function of frequency through its dependence on the structure of the waveguide (i.e., core diameter and shape of the refractive index profile). Thus, the key to controlling dispersion is to use the lightguide structure to tailor waveguide dispersion spectra so that its shape cancels the material dispersion spectrum at one or more required wavelengths [14]. The curves in Figure 3.4 illustrate the design flexibility for making waveguide dispersion large or small at particular wavelengths. Figure 3.3 shows the total dispersion resulting from the addition of material and waveguide dispersion components for the three generic fiber design alternatives that will be described in ensuing sections. Spectrum (a) is typical of step-index fibers that have zero dispersion near 1.3 μm but have about 20 ps/(km-nm) dispersion near 1.6 μm. Spectrum (b) is essentially obtained by shifting (a) to longer wavelengths. Its crossover wavelength is at 1.55 μm, but high dispersion occurs near 1.3 μm. This is achieved in dispersion-shifted fibers by raising the index across a smaller diameter core or equivalently by grading the core index with a profile shape like a triangle. The remaining spectra are dispersion-flattened between widely separated zero crossings. For example, the low-dispersion region can encompass the 1.3–1.55 μm window [as in (c)] or, as in (d), it can cover the lowest-loss region between 1.45 and 1.7 μm.

3.3 STEP-INDEX FIBERS

Figure 3.5a to e illustrates refractive index profile shapes for nominally step-index lightguides that have zero dispersion near 1.3 μm. The type in Figure 3.5a typically has a germania-silicate deposited core and fluorophosphosilicate deposited cladding whose refractive index difference Δ is tailored to match the index of an outer silica support tube. The profile in Figure 3.5e is characteristic of depressed-index fibers in which an inner cladding is doped with fluorine so that Δ^- is below that of the outer cladding [15, 16].

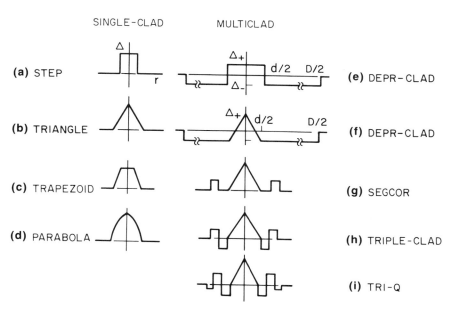

Figure 3.5 Refractive-index profiles for single-mode fiber media that can be tailored to control dispersion within the 1.3–1.6 μm spectral region. They include step- and graded-index core profiles and claddings with multiple-index wells.

Typically, the cladding (deposited cladding plus tube material) makes up 99.6% of the total fiber volume (i.e., core diameter $d \approx 8.5$ μm, fiber diameter $= 125$ μm). Therefore, preform processing times can be significantly reduced by minimizing the volume of cladding glass deposited within a thick support tube of a lower optical grade. However, if D/d is too small, then fiber losses may be increased by two mechanisms: light power penetration into the lossy substrate and leaky-mode effects in depressed-cladding light guides [17].

The first criterion for D/d is to ensure that the mode power profile P is not truncated and that propagating light power does not overlap the lossy material of the substrate tube. For example, if the substrate loss is approximately 100 dB/km, then the loss added to the fiber's propagating mode will be less than 0.01 dB/km if $P_s = P(\text{substrate})/P(\text{total}) < 10^{-4}$. The solid curve in Figure 3.6 illustrates the minimum D/d (required to maintain $P_s < 10^{-4}$) plotted versus V/V_{co}, which is the fiber V number normalized to the value for LP_{01} mode cutoff. Note that $D/d \approx 4$ if Δ and d are chosen to make $V/V_{co} \approx 1$. However, the required D/d increases rapidly if Δ is reduced to make V/V_{co} smaller [18].

The other criterion for D/d is that it be sufficient to prevent the fundamental LP_{01} mode from radiating into the outer cladding of lightguides with depressed-index claddings. Losses occur if a significant fraction of the total power leaks away from the core as it propagates along the lightguide axis. Such effects do not occur at wavelengths where the effective index N_e of the mode is greater than the index of the outer cladding (i.e., $N_e = \beta K_0 > N_0$, where β is the propagation constant of the LP_{01} mode and K_0 is the propagation constant in free space). However, as λ increases, the mode field penetrates further into the cladding and thereby reduces $N_e = \beta/K_0$. If $N_e < N_0$ for $\lambda > \lambda_c(01)$, then the LP_{01} is said to be cut off because power can radiate through the outer cladding. Note that if $\Delta^+ = 0$ as in Figure 3.5c, then $N_e < N_0$ and the LP_{01} mode is cut off at all wavelengths. However, that does not preclude using the fiber for long-distance communications because the fundamental mode leakage loss can be made arbitrarily small if the depressed cladding is made sufficiently thick. For example, if the cladding is made wide enough, then the mode power will decay to a negligibly small value at the outer diameter D of the index well from which the mode power leaks. The radiation leakage loss effects, described above, can become even larger when the fiber is wound on reels, which create constant-curvature macrobends. The procedure described in Ref. 17 can be used to calculate macrobending losses as a function of curvature for depressed-cladding lightguides. Figure 3.6 plots D/d versus V/V_{co} requirements that are necessary to overcome the effects of macrobending and lossy substrate tubes by keeping added losses to less than 0.01 dB/km.

The fiber represented in Figure 3.5e has been extensively tested for single-mode terrestrial and undersea applications [16]. It provides low losses and

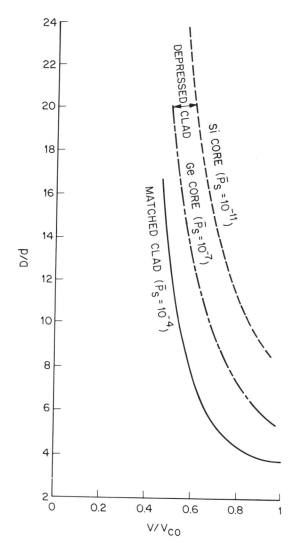

Figure 3.6 Deposited cladding thickness requirements plotted versus normalized V number (D/d vs. V/V_{co}) for three types of single-mode lightguides. P_s represents the power in the substrate normalized relative to the total power for matched cladding designs, GeO_2-SiO_2 depressed-index designs, and totally depressed designs with SiO_2 cores.

is relatively insensitive to macrobending effects that occur when the fiber is cabled. The lightguide structure is characterized by $\Delta^- \approx -0.12\%$, $\Delta^+ \approx 0.28\%$, $d \approx 8.3$ μm, and $D/d \approx 6.5$. Lower losses are achievable in silica core fibers, which minimize Rayleigh scattering by lowering the concentration of germania dopant in the core while depressing the index of the cladding with

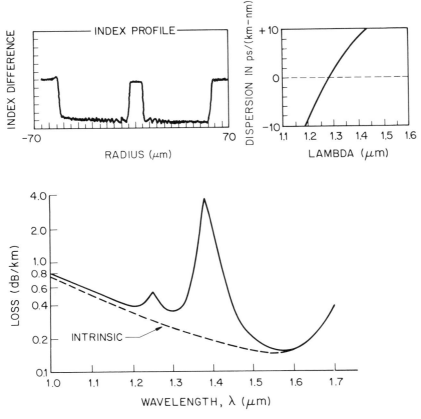

Figure 3.7 Refractive index profile and loss and dispersion spectra for a silica-core fiber with a minimum loss ≈ 0.16 dB/km.

fluorine in order to achieve a lightguiding profile. The fiber core has virtually the same refractive index as that of pure silica ($\Delta^+ < 0.05\%$). Lower Rayleigh scattering losses result in the reduction of optical losses by about 10% compared to those of conventional single-mode fibers. Losses as low as 0.16 dB/km [19] have been achieved, with median values of about 0.19 dB/km. Figure 3.7 compares the actual minimum loss obtained to the theoretical minimum achievable loss. It is representative of 300 km of fiber. The curve follows an A/λ^4 Rayleigh scattering dependence with the coefficient $A = 0.87$ dB/km chosen to fit data at $\lambda = 1$ μm. Minimum losses are approximately 0.19 dB/km and occur near $\lambda = 1.58$ μm. The loss value is about 0.04 dB/km higher than the Rayleigh scattering limit. Individual fibers were concatenated into long lengths. Average losses of spliced fibers were approximately 0.01 dB/km higher than the unspliced values. In order to provide good mode confinement and minimal spot size, the LP_{11} mode cutoff wavelength ranged from 1.33 to 1.55 μm and was generally within 10% of potential system wavelengths near 1.55 μm.

The trace amount of germanium that is added to the core is necessary to prevent the anomalous losses that have been observed with pure silica-core single-mode fibers. In addition to minimizing optical losses, the lower level of germanium in the core results in a decreased sensitivity to environmental effects such as hydrogen and radiation.

Losses due to microdeformations such as microbending and profile fluctuations depend on power confinement and can be greatly reduced by decreasing the mode-field radius (MFR), which is defined to be the half-width between $1/e$ points of the LP_{01} mode-field profile. This was accomplished by increasing Δ and decreasing the core diameter. The cladding was therefore made deeply depressed ($\Delta^- = 0.39\%$). MFR is plotted versus wavelength in Figure 3.8 for three different types of single-mode fibers with depressed-index claddings (i.e., type I: $\Delta^+ \approx 0.25\%$, $\Delta^- \approx 0.05\%$, $d \approx 9$ μm; type 2: $\Delta^+ \approx 0.28\%$, $\Delta^- \approx 0.12\%$, $d \approx 8.3$ μm; type 3: $\Delta^+ \approx 0$, $\Delta^- \approx 0.39\%$, $d \approx 8.5$ μm). Such mode-field radius spectra are good comparative indicators of the relative microbending loss effects in various types of fibers. Type 1 is essentially

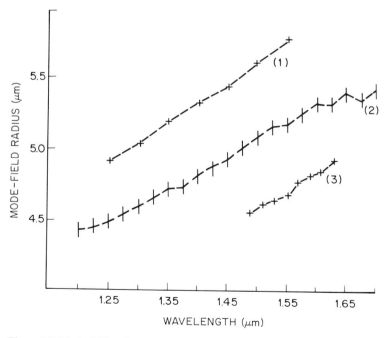

Figure 3.8 Mode-field radius versus wavelength for three different types of fiber with depressed-index claddings. (See text.)

a matched-cladding design and has the largest mode-field radius and the poorest microbending performance when cabled. The type 2 lightguide [16] has been extensively tested in cables and has been found to cause minimal added losses at $\lambda = 1.3$ μm. Type 3, the silica-core lightguide, has been specifically tailored to provide very low losses near $\lambda = 1.55$ μm. To this end, its mode-field radius at $\lambda \approx 1.55$ μm is 4.6 μm, which is nearly identical to the field radius for the type 2 fiber at $\lambda \approx 1.3$ μm. As a result, the cabling performance of the silica core fiber at $\lambda \approx 1.55$ μm is as good as that for the type 2 fiber at $\lambda \approx 1.3$ μm.

3.4 DISPERSION-FLATTENED FIBERS

Double-clad fibers with wide depressed claddings approximate step-index fibers with infinite claddings. However, if the depressed-cladding width is reduced as in Figure 3.9a, then single-mode lightguiding properties become

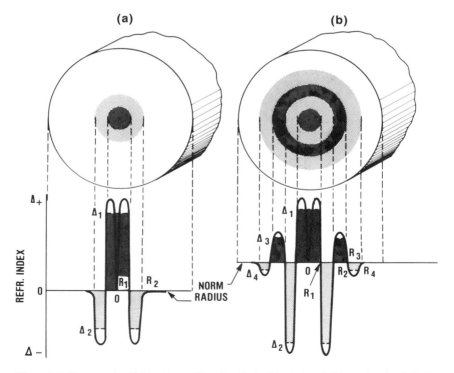

Figure 3.9 Comparative lightguide profiles for (a) double-clad and (b) quadruple-clad ultra-broadband fibers.

infuenced by the structure of the index well between the core and second cladding [14, 20]. One of the most important properties of this type of fiber is that its fundamental mode can be cut off because power spreads into the index well as the wavelength gets longer. If the well is deep enough, then the phase index of the mode, which is an average over the core and low-index inner cladding, can drop below the index of the outer cladding. Cutoff will occur if that happens, because the mode field will radiate through the cladding as it propagates along the axis of the fiber. As a result of this cutoff, waveguide dispersion effects can be large enough, even at wavelengths shorter than cutoff, to cancel material dispersion effects over an ultrabroad wavelength range.

Figure 3.10 qualitatively compares wavelength-dependent spectral properties of conventional and dispersion-flattened lightguides. Figure 3.10a illustrates group index or group delay plotted as a function of wavelength. At short wavelengths, the LP_{01} mode is well confined within the fiber core and its group index follows the group index of the core material. For single-clad

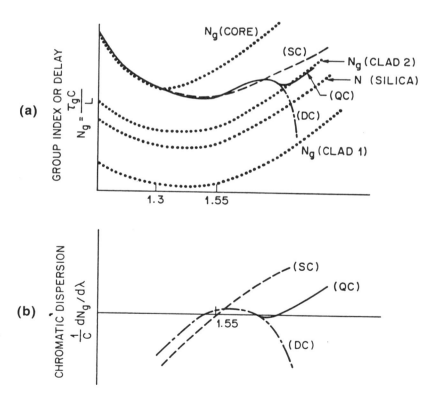

Figure 3.10 Evolution of group index N_g and dispersion for a quadruple-clad (QC) index profile. Comparisons are made with respect to single-clad (SC) and double-clad (DC) profiles.

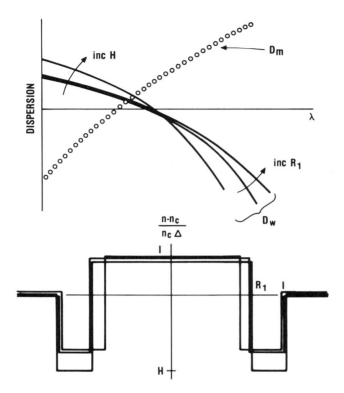

Figure 3.11 Parametric studies of double-clad lightguides illustrate how their waveguide dispersion spectra depend on the width, $1 - R_1$, of the depressed-index cladding and H, the index between the first and second cladding divided by the index of the core.

fibers, there is a transition to the long-wavelength region where the mode is no longer well confined and the group index of the mode follows the group index of the cladding material. By comparison, in a double-clad fiber, a transition region connects the core index with the inner cladding index, which is lower than the group index of the outer cladding. However, the LP_{01} mode becomes cut off when its group index goes below that of the outer cladding. Figure 3.10b shows chromatic dispersion spectra obtained from the first derivative or slope of group delay spectra. The double-clad fiber delay spectrum has one minimum and one maximum. Therefore the corresponding dispersion has two zero crossings with locations that can be controlled through the proper choice of fiber parameters.

Figure 3.11 summarizes results of computer-aided modeling studies [21–23] to determine how lightguide parameters affect the shape of waveguide dispersion spectra. Increasing the width of the first cladding $(1 - R_1)$ relative to the core radius R_1 increases the effect of its low index and makes the

LP_{01} mode become cut off at shorter wavelengths. The effect is to increase the curvature of the waveguide dispersion spectra for increasing cladding widths. The second important shaping effect is to be able to translate the waveguide dispersion characteristic up or down. The controlling waveguide parameter is H, the index difference between the first and second cladding divided by the index of the core. Increasing H translates the waveguide dispersion curve upward because the total index change between the core and first cladding becomes larger. For example, if low dispersion is required over the entire spectral region between 1.3 and 1.55 μm, then $D/d = 1.5$ and $H = 2$ have been found to be practical lightguide parameters.

Figure 3.12 shows the effects of changing core diameter on dispersion spectra. The dash-dot curve outlines the material dispersion spectrum for fibers with a germania-doped core and fluorosilicate claddings. The dashed curves outline waveguide dispersion spectra for lightguides with constant index parameters and different normalized radii. As the core radius becomes smaller, the V number of the lightguide becomes smaller. The mode becomes less well confined and therefore is cut off at a shorter wavelength. The effect

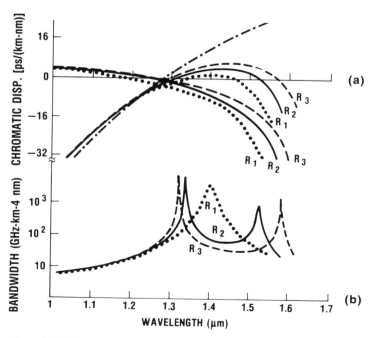

Figure 3.12 (a) Dispersion spectra calculated for double-clad fibers with different radii ($R_1 < R_2 < R_3$). Material, waveguide, and total dispersion curves are illustrated separately. The dot-dash curve applies to single-clad fibers. (b) Bandwidth spectra corresponding to total dispersion curves in (a). They are calculated for sources with $\delta\lambda = 4$ nm linewidths.

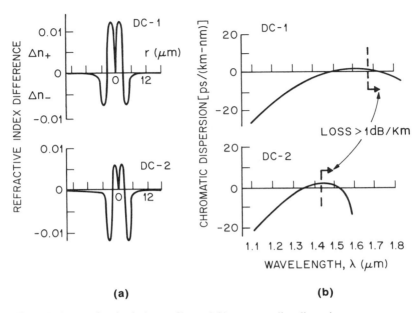

Figure 3.13 (a) Refractive index profiles and (b) corresponding dispersion spectra are compared for double-clad lightguides with various ratios of Δ^-/Δ^+.

is to bend the long-wavelength edge of waveguide dispersion spectra around more sharply. When these effects are added to the constant material dispersion characteristic, the overall effect is to displace the chromatic dispersion spectrum downward and move the two minimum dispersion wavelengths closer together as the radius becomes smaller. Similar behavior occurs if the core index is decreased while the fiber radius is kept constant. Corresponding bandwidth spectra are shown in the lower figure and are plotted in gigahertz-kilometers for laser sources with 4-nm linewidths. They are inversely related to the dispersion spectra, so bandwidth peaks occur when dispersion spectra cross the zero axis. In particular, note how the two corresponding bandwidth peaks coalesce into a single broad peak if the chromatic dispersion curve becomes tangent to the zero dispersion axis. The calculated bandwidths can remain larger than 25 GHz-km over the ultrabroadband wavelength region spanning 1.3–1.55 μm. The first published experimental confirmation of dispersion-flattened behavior in double-clad fibers was reported by Miya and others [24] from NTT. Figure 3.13 shows an evolution of refractive index profile shapes corresponding to an evolution of dispersion spectra. The top set are examples of dispersion-shifted fibers (DC-1) that have broadband characteristics within the 1.5–1.7-μm wavelength region [25]. The bottom set (DC-2) have dispersion-flattened spectra throughout the 1.35–1.55 μm region. The trend of these data clearly shows that the broadband region can be moved to shorter wavelengths by increasing Δ^-, the depth of the index well, relative to the raised core index Δ^+. The effect is to bend the dispersion spectra around more sharply at longer wavelengths by moving the fundamental mode cutoff wavelengths closer to the region of minimum dispersion. However, a small change in wavelength can cause the fundamental mode to change from a guided wave into a leaky wave that radiates out of the index well and through the second cladding. Since these leaky-mode effects occur at even shorter wavelengths in bent fibers, a useful guideline is to keep the fundamental mode cutoff at least 0.1 μm longer than the longest wavelength of interest.

Lightguides with multiple claddings (as shown in Fig. 3.5g, h, i) were conceived to alleviate the problem of fundamental mode leakage losses. The quadruple-clad lightguide structure (in Fig. 3.9b) contains one double-clad profile within another [7]. It has two raised-index guiding regions and two index wells. The leakage loss problem is addressed because light leaking out of the inner index well can be retrapped within the lightguide formed by the annular ring guide between the two index wells. For that reason, the area of the raised-index second cladding should be made as large as possible without moving the cutoff of the second propagating mode into the system wavelength window.

Quadruple-clad fibers have nine independent radial and index parameters that can be manipulated to achieve optimal results. An advantage is that the core and first cladding region can be optimized to control dispersion between 1.3 and 1.55 μm, while the third and fourth claddings can be optimized to control leakage loss characteristics. In other words, the area within the first index well should be larger than the area within the core in order to achieve dispersion-flattened spectra between 1.3 and 1.55 μm. However, the total area within the index wells should be less than the total area within the raised-index guiding regions in order to minimize leakage losses within the wavelength range of interest.

Figure 3.10 qualitatively illustrates the spectral behavior that can be expected for fundamental mode propagation in quadruple-clad (QC) lightguides. The solid (QC) curve follows the dashed (DC) curve to the wavelength at which cutoff would normally occur due to leaky-mode effects. However, the raised-index annular ring cladding retraps this light and raises the group-index spectra so that it bends through a third turning point. That moves the fundamental mode cutoff to longer wavelengths or prevents it entirely. The resultant chromatic dispersion curves in Figure 3.10 show that single-clad (SC) lightguides have one zero crossing, double-clad (DC) lightguides can have two zero crossings, and QC lightguides can have three zero crossings. One additional zero-dispersion wavelength can be added for each two additional raised and depressed-index cladding regions.

Figure 3.14 demonstrates the improved mode-confinement properties of QC fibers by plotting computer-generated modal power profiles for several

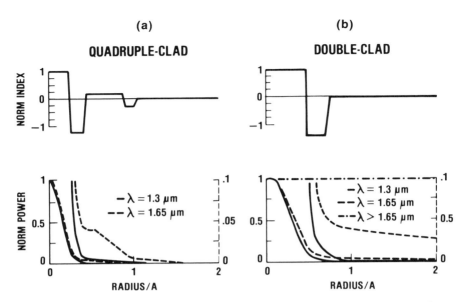

Figure 3.14 The advantage of (a) quadruple-clad relative to (b) double-clad lightguides is illustrated by comparing mode power profile versus normalized radius for several different wavelengths. The dashed curves refer to the expanded power scale on the right vertical axis.

different wavelengths. The dashed curves refer to the expanded power scale on the right vertical axis. Figure 3.14b applies to double-clad fiber. At short wavelengths, the solid curves show that the mode is well confined within the lightguide core. However, for wavelengths longer than 1.65 μm, the mode becomes cut off, and the dashed curve shows that the mode power becomes distributed throughout the cladding instead of being confined within the core. By comparison, the normalized QC profile in Figure 3.14a has an additional secondary guiding region that retraps the power that would otherwise leak through the cladding. The design philosophy is to make the index, or the width of the second cladding, as large as possible without moving the second mode cutoff into the operating wavelength window.

Figure 3.15 compares loss spectra for fibers with dispersion-flattened spectra. The dashed curves apply to double-clad lightguides. As the short-wavelength zero crossing decreases going from DC-1 to DC-2, the advantage of QC lightguides is clearly evident in enabling the long-wavelength loss edge to increase by more than 200 nm.

As one might expect, the attractive properties of dispersion-flattened fibers require rather tight tolerances. However, gradual variations of diameter and index can be tolerated within about 5% of the desired mean value because a dispersion spectrum is the length-weighted average of local values

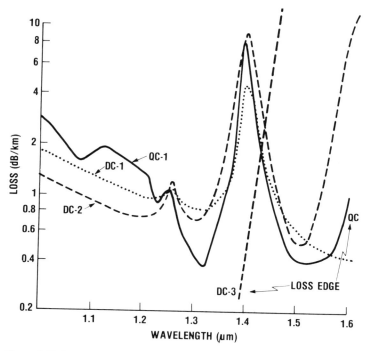

Figure 3.15 Loss spectra for dispersion-flattened fibers. The dashed curves apply to double-clad lightguides and the others to quadruple-clad lightguides. The advantage of lightguides is clearly evident in enabling the long-wavelength loss edge to increase by more than 200 nm.

along the fiber. Figure 3.16 illustrates the effect by comparing the dispersion of a long fiber with corresponding spectra obtained from interferometric dispersion measurements on short segments from each end [26]. Example (a) shows very little end-to-end variation. However, the fiber for (b) had a relatively large (3 μm) diameter taper between ends that were separated by 1 km. Although dispersion effects are significantly different for each end of the fiber, the overall spectrum is dispersion-flattened between 1.3 and 1.6 μm. It is simply the length-weighted average of the end segments. Table 3.1 lists loss and dispersion characteristics for four fibers making up a 20.3-km link at four wavelengths, $\lambda = 1.3$, 1.33, 1.51, and 1.54 μm. Although some variation is evident between individual fibers, the overall dispersion of the link would be less than 2.5 ps/(km-nm) at wavelengths between 1.3 and 1.54 μm, and the overall link loss is approximately 0.5 dB/km. Further improvements toward reducing losses of dispersion-flattened fibers have been made by using lightguides with silica cores [27] or graded-index germania-silicate cores

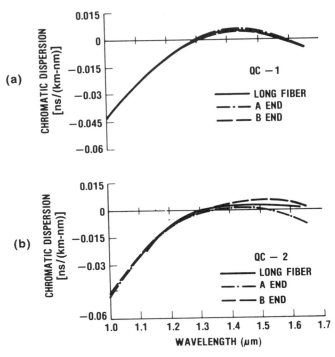

Figure 3.16 Chromatic dispersion spectra from interferometric measurements of quadruple-clad fiber for 4-m lengths from each end of the test fiber compared with chromatic dispersion of the 1-km test fiber measured by a conventional pulse delay method. Dispersion of the long length is approximately the average of the spectra taken from each end.

[28]. As an example, Figure 3.17 shows that lower losses can be achieved in dispersion-flattened fiber with a triangular core profile. The dispersion-flattened spectrum in Figure 3.17b has zero crossings at $\lambda = 1.33$ and $1.6 \ \mu m$ and low dispersion everywhere in between. The improved loss spectrum (Fig. 3.17c) has values of 0.5 dB/km near 1.3 μm and a minimum loss value of 0.3 dB/km near $\lambda = 1.54 \ \mu m$.

3.5 DISPERSION-SHIFTED FIBERS

Dispersion-shifted lightguides require large waveguide dispersion effects to counterbalance material dispersion near $\lambda = 1.55 \ \mu m$. The waveguide effects can be written in the form [29]

$$d_w = \frac{L}{c\lambda} \, n \, \Delta D_w(V) \tag{7}$$

TABLE 3.1. Loss and dispersion characteristics for four fibers in an ultrabroadband 20.3-km concatenated link

Fiber	Length (km)	λ = 1.305 μm		λ = 1.335 μm		λ = 1.51 μm		λ = 1.54 μm	
		Loss (dB/km)	Dispersion [ps/(km-nm)]	Loss (dB/km)	Dispersion [ps/(km-nm)]	Loss (dB/km)	Dispersion [ps/(km-nm)]	Loss (dB/km)	Dispersion [ps/(km-nm)]
1	5.4	0.62	−3	0.7	−1.5	0.54	0.8	0.55	0.6
2	5.4	0.71	−2	0.76	−1	0.56	0.1	0.57	−0.1
3	4.5	0.54	0	0.58	2	0.44	6.5	0.53	0.55
4	5	0.64	−1.5	0.7	0.8	0.65	3.5	0.77	3
	20.3	0.63	−1.8	0.69	0	0.55	2.5	0.6	0.4

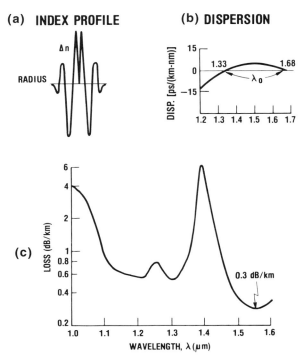

(a) INDEX PROFILE

(b) DISPERSION

(c)

Figure 3.17 Quadruple-clad graded-index profile and loss and dispersion curves for a dispersion-flattened fiber with a minimum loss of 0.3 dB/km near $\lambda = 1.55\ \mu$m.

where $D_w(V)$ is a dimensionless dispersion coefficient that depends on the V parameter of the fiber (Fig. 3.18). Its qualitative shape depends on the way power is distributed between the core and cladding as a function of V (or λ). For large V (short λ), $D_w \approx 0$, because many modes propagate and the fundamental mode is confined by the core independently of V (or λ). For small V (long λ), $D_w \approx 0$, because mode power is primarily in the cladding. For intermediate values of V ($0.25 < V/V_{co} < 0.5$), $D_w(V)$ goes through an extremum, because the power distribution between the core and cladding changes rapidly as a function of V (or λ). Therefore, large waveguide effects occur at V numbers well below the single-mode cutoff. The effects [Eq. (7)] can be exaggerated in small-diameter lightguides with a relatively large core index.

Computer-aided modeling studies have facilitated choices for such lightguide parameters as the diameter and shape of the refractive index profile separating a core from a cladding. As an example, Figure 3.19b compares numerical results for computer models represented by the solid curves for

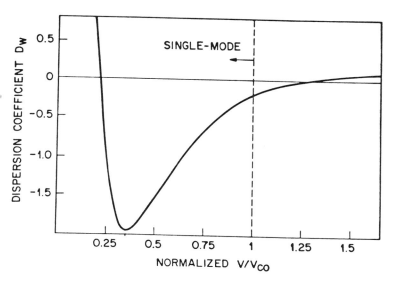

Figure 3.18 Waveguide dispersion coefficient [29] D_w versus V/V_{co}. Dispersion-shifted fibers require relatively large D_w and small V/V_{co}.

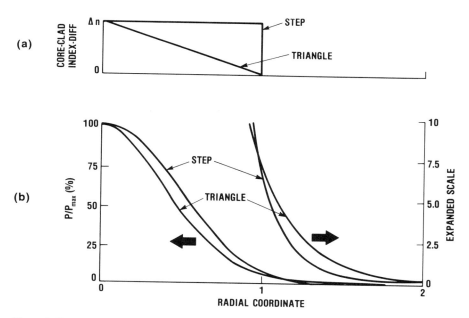

Figure 3.19 Light intensity versus radius are compared for step-index and triangular graded-index profiles. (a) Index profiles; (b) mode power profiles.

step-index profiles and dashed curves for triangular profiles plotted versus radius. The linear gradient forms a self-focusing profile that confines the propagating beam to a smaller spot inside the core than the corresponding step-index profile does. Since dispersion-shifted fibers require large dispersion at long wavelengths, and since waveguide dispersion becomes larger as the spot size becomes smaller, a triangular graded-index fiber can have a larger core diameter than a step-index fiber and still produce the same magnitude of waveguide dispersion. The triangular profile shape also provides the added benefit of having a smoother and less abrupt interface between the cladding and the higher-index core region, which has resulted in lower losses than for equivalent step-index profiles [4, 30].

Single-mode lightguide properties can be compared for various graded-index core profile shapes (step, trapezoidal, parabolic, and triangular) and SiO_2 cladding (Fig. 3.5a–d) by plotting the wavelength of zero dispersion, λ_0, as a function of (1) core diameter (Fig. 3.20), (2) V number at λ_0 normalized relative to the cutoff V value, V_0/V_{co} (Fig. 3.21), and (3) the effective core–cladding index difference, $\Delta_{eff} = (n_{eff} - n_0)/n_0$, of the fundamental mode (Fig. 3.22) (where n_{eff} is the ratio of the modal propagation constant β to the propagation constant $k = 2\pi/\lambda$ for a plane wave in free space). The variable parameter associated with each curve in Figures 3.20–3.22 is the maximum index difference between the core and the cladding (Δ, in %).

For the core profiles considered, the properties of the step and triangular profiles represent the extremes of the variations, with the properties of the parabolic and trapezoidal profiles lying in between. Since the curves go through a maximum, the choice for the peak core–cladding index difference can be chosen to minimize dimensional tolerance requirements by making the curve just tangent to the $\lambda_0 = 1.55$ μm line for dispersion-shifting applications. That defines a minimum $\Delta = \Delta_{min}$ necessary to achieve $\lambda_0 = 1.55$ μm. Note that of all the options studied the triangular profile allows the largest core diameter and the broadest variation in diameter for 1.55-μm applications. For $\lambda_0 = 1.55$ μm, $\Delta_{min} = 0.55\%$ and $d \approx 5.5$ μm for a step profile, whereas $\Delta_{min} = 0.82\%$ and $d \approx 5.5$ μm for a triangular profile. From Figure 3.21 we can then calculate that $\lambda_c^{step} \approx 0.77$ μm and $\lambda_c^{tri} \approx 0.72$ μm, and from Figure 3.22, $\Delta_{eff}^{step} \approx 0.06\%$ and $\Delta_{eff}^{tri} \approx 0.04\%$.

Figure 3.23 is an example of data that show that low losses and zero dispersion can be achieved near 1.55 μm in a dispersion-shifted fiber with a triangular graded-index profile shape between a germania-silicate core and a fluorophosphosilicate cladding [31]. The dispersion spectrum shown in Figure 3.23b has a zero crossing near 1.55 μm, and the remaining curves show loss data indicating that minimum loss also occurs near 1.55 μm. The intrinsic curve is an estimate for the fiber losses due to Rayleigh scattering and long-wavelength absorption due to intrinsic molecular vibrations. It is reassuring to see how closely the minimum measured loss of 0.21 dB/km for the fabricated fiber compares with the intrinsic level of 0.18 dB/km.

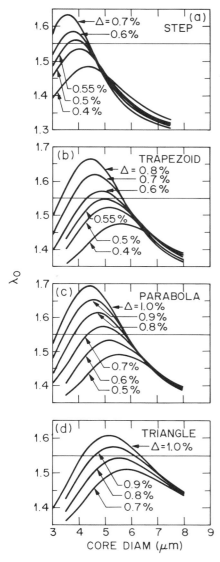

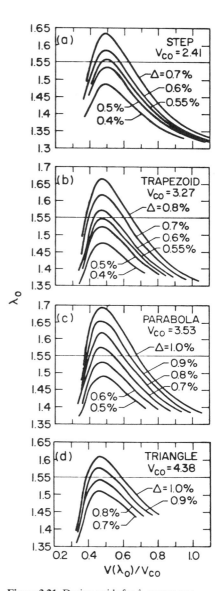

Figure 3.20 Design guide for λ_0 versus core diameter for step, trapezoidal, parabolic, and triangular profile geometries.

Figure 3.21 Design guide for λ_0 versus normalized V value $[V(\lambda_0)V_{co}]$ for step, trapezoidal parabolic, and triangular profile geometries.

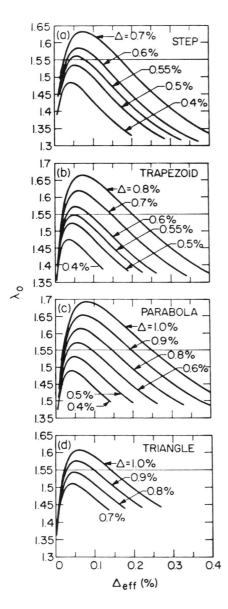

Figure 3.22 Design guide for λ_0 versus effective index (Δ_{eff}) for step, trapezoidal, parabolic, and triangular profile geometries.

One disadvantage of single-clad dispersion-shifted fibers is that they are usually low V-value lightguides with cutoff wavelengths that are considerably shorter than 1.55 µm. Because a significant fraction of the propagating modal power is distributed within the cladding, fiber losses may increase due to macro- and microbending effects that occur if the fiber is wrapped around

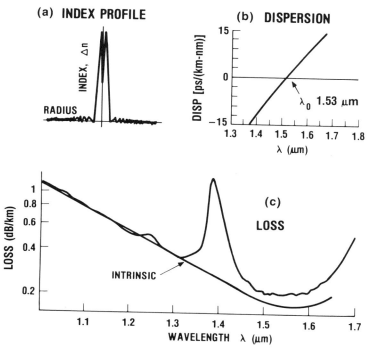

Figure 3.23 Triangular graded-index profile and loss and dispersion spectra for a dispersion-shifted fiber with a minimum loss of approximately 0.21 dB/km near $\lambda_0 = 1.53$ μm.

a drum or packaged within a cable. An attractive feature of a multiclad structure is that its core and innermost cladding can be designed to control dispersion spectra and the remaining claddings can be independently tailored to improve mode confinement and reduce bending-related losses. For example, the quadruple-clad lightguide has been described in the section on dispersion-flattened fibers. The triple-clad profile structure (Fig. 3.5h) is typical of a multiclad dispersion-shifted fiber. The triangular core is used to shift the dispersion zero to 1.55 μm, and the raised outer ring increases the cutoff wavelength for better mode confinement [6, 28]. Fibers with this design have been used to achieve $\lambda_0 = 1.55$ μm and low losses (0.21 dB/km).

Figure 3.24 shows an example of how microbending sensitivity decreases for multiclad lightguides [32, 33]. Qualitative comparisons are made based on basket-weave tests that are performed by winding alternating fiber layers across one another under a high (70 g) tension onto a small-diameter (6-in.) reel. Loss spectra are compared for two different winding arrangements. Lowest losses occur when the fiber is wound under zero tension. Loss increases due to basket weaving can then be compared for different fiber de-

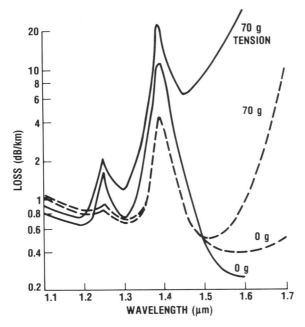

Figure 3.24 Comparison of microbending sensitivity based on basket-weave tests that are performed by winding alternating fiber layers across one another under various degrees of tension. Solid curves apply to dispersion-shifted fibers with matched silica claddings, and dashed curves apply to designs with multiple claddings.

signs. The figure applies to dispersion-shifted fiber with triangular-index core profiles. Note that loss increases between zero tension and high tension are much smaller for the multiple-cladding design (dashed curve) than for the matched-cladding design (solid curve).

3.6 CONCLUSION

This chapter has described the performance of three advanced single-mode fiber lightguides.

Silica core fibers are designed to minimize loss without any constraint on dispersion. Minimum losses of 0.16 dB/km have been achieved, with mean values of approximately 0.19 dB/km, at $\lambda = 1.57$ μm for long span lengths. Zero dispersion occurs near $\lambda = 1.29$ μm, and the magnitude of dispersion is approximately 17.5 ps/(km-nm) near $\lambda = 1.57$ μm. Therefore, high-bandwidth systems applications near 1.55 μm require the use of single-frequency lasers. These have been used to do milestone experiments including 420-Mb/s transmission through 203 km [34]. The link loss was 0.204 dB/km including 26 splices. Other experiments included 2-Gb/s transmission through 130 km, 1.37 Tb-km/s using 10 wavelength-division-multiplexed lasers [35], and 4 Gb/s through 117 km using a Ti:LiNbO$_3$ external modulator [36].

An alternative strategy is to use conventional multimode lasers in conjunction with high-bandwidth fibers that have low dispersion at one or more wavelengths between 1.3 and 1.55 μm. Such dispersion spectra are more sensitive to variations of fiber parameters than conventional spectra are. However, over- and undercompensated characteristics can be used since the spectrum of long span lengths is the length-weighted average of spectra from the individual sections.

Dispersion-shifted fibers generally contain graded-index cores. They belong to a class of lightguides that can be tailored to simultaneously provide low losses of about 0.25 dB/km and zero dispersion at wavelengths near 1.55 μm. Systems experiments have demonstrated that high-bit-rate spans can be longer than 100 km even with multilongitudinal mode lasers ($2\delta\lambda = 2.8$ nm). One experiment demonstrated a laser-sparing strategy by using a passive polarizing beam splitter to couple two transmitters into the long transmission fiber [37]. The span loss of a 150-km link including splices was 0.267 dB/km at $\lambda = 1.526$ μm and 0.258 dB/km at $\lambda = 1.542$ μm. The resultant dispersion penalty was negligible.

Dispersion-flattened fibers have more complex lightguide profiles that can contain four or more claddings with different index levels. They can provide moderately low losses of about 0.35 dB/km, but they can be used to extend the near-zero dispersion [<2 ps/(km-nm)] region over a very wide wavelength range that can encompass the lowest-loss region near 1.55 μm or the entire 1.3–1.6-μm spectrum. Such fibers could become useful for wavelength division multiplexing many high-bit-rate information channels onto the same fiber.

REFERENCES

1. K. Ciemiecki Nelson, D. L. Brownlow, L. G. Cohen, F. V. DiMarcello, R. G. Huff, J. T. Krause, P. J. Lemaire, W. A. Reed, D. S. Shenk, E. A. Sigety, J. R. Simpson, A. Tomita, and K. L. Walker, "The fabrication and performance of long lengths of silica core fiber," *J. Lightwave Technol.*, **LT-3**: 935–941 (1986).
2. H. Yokota, H. Kanamori, Y. Ishiguro, G. Tanaka, S. Tanaka, H. Takada, M. Watanam, S. Suzuki, K. Yano, M. Hoshikawa, and H. Shimba, "Ultra-low-loss pure-silica-core single-mode fibers and transmission experiment," *Tech. Digest*, OFC'86, Atlanta, GA, February 1986, Postdeadline paper.
3. L. G. Cohen, W. L. Mammel, and S. J. Jang, "Low-loss quadruple-clad single-mode lightguides with dispersion below 2 ps/km-nm over the 1.28 μm–1.65 μm wavelength range," *Electron. Lett.*, **18**: 1023–1024 (1982).
4. V. A. Bhagavatula, M. S. Spotz, W. F. Love, and D. B. Keck, "Segmented-core single-mode fibres with low loss and low dispersion," *Electron. Lett.*, **19**: 317–318 (1983).

5. P. K. Bachmann, D. Leers, H. Wehr, and D. U. Wiechert, "Ultrabroadband single-mode fibers prepared with PCVD," *Tech. Digest,* OFC'85, San Diego, CA, February 1985, p. 78.

6. L. G. Cohen, C. Lin, and W. G. French, "Tailoring zero chromatic dispersion into the 1.5–1.6 μm low-loss spectral region of single-mode fibers," *Electron. Lett.,* **15**: 334–335 (1979).

7. B. J. Ainslie, K. J. Beales, D. M. Cooper, D. R. Day, and J. D. Rush, "Monomode fibre with ultra-low loss and minimum dispersion at 1.55 μm," *Electron. Lett.* **18**: 842–843 (1982).

8. H. T. Shang, T. A. Lenahan, P. F. Glodis, and D. Kalish, "Dispersion-shifted depressed-clad triangular-profile single-mode fiber," *Tech. Digest,* OFC'85, San Diego, CA, February 1985, pp. 92–93.

9. T. D. Croft, J. E. Ritter, and V. A. Bhagavatula, "Low-loss dispersion-shifted single-mode fiber manufactured by the OVD process," *J. Lightwave Technol.,* LT-3: 931–934 (1986).

10. S. R. Nagel, J. B. MacChesney, and K. L. Walker, "An overview of the modified chemical vapor deposition (MCVD) process and performance." *IEEE J. Quantum Electron.,* **QE-18**: 459–476 (1982).

11. L. G. Cohen, W. L. Mammel, and S. Lumish, "Dispersion and bandwidth spectra in single-mode fibers," *IEEE J. Quantum Electron.,* **QE-18**: 49–53 (1982).

12. F. P. Kapron, "Maximum information capacity of fibre-optic waveguides," *Electron. Lett.,* **13**: 96–97 (1977).

13. K. Ogawa, "Considerations for single-mode fiber systems," *Bell Syst. Tech. J.,* **61**: 1919–1931 (1982).

14. L. G. Cohen, W. L. Mammel, and S. Lumish, "Tailoring the shapes of dispersion spectra to control bandwidths in single-mode fibers," *Opt. Lett.,* **7**: 183–185 (1982).

15. P. D. Lazay and A. D. Pearson, "Developments in single-mode fiber design, materials, and performance at Bell Laboratories," *IEEE J. Quantum Electron.,* **QE-18**: 504–510 (1982).

16. W. M. Flegal, E. A. Haney, R. S. Elliott, J. T. Kamino, and D. N. Ernst, "Mass production of depressed clad single-mode fiber," *Tech. Digest,* OFC'84, New Orleans, LA, January 1984, pp. 56–57.

17. L. G. Cohen, D. Marcuse, and W. L. Mammel, "Radiating leaky-mode losses in single-mode lightguides with depressed-index claddings," *IEEE J. Quantum Electron.,* **QE-18**: 1467–1472 (1982).

18. J. B. MacChesney, D. W. Johnson, P. J. Lemaire, L. G. Cohen, and E. M. Rabinovich, "Depressed index substrate tubes to eliminate leaky-mode losses in single-mode fibers," *J. Lightwave Technol.,* **LT-3**: 942–945 (1985).

19. R. Csencits, P. J. Lemaire, W. A. Reed, D. S. Shenk, and K. L. Walker, "Fabrication of low loss single-mode fibers," *Tech. Digest,* OFC'84, New Orleans, LA, February 1984, pp. 54–55.

20. S. Kawakami and S. Nishida, "Characteristics of a doubly clad optical fiber with a low-index inner cladding," *IEEE J. Quantum Electron.,* **QE-10**: 879–887 (1974).

21. L. G. Cohen, W. L. Mammel, and H. M. Presby, "Correlations between numerical predictions and measurements of single-mode fiber dispersion characteristics," *Appl. Opt.,* **19**: 2007–2010 (1980).

22. W. L. Mammel and L. G. Cohen, "Numerical prediction of fiber transmission characteristics from arbitrary refractive-index profiles," *Appl. Opt.,* **21**: 699–703 (1982).

23. T. A. Lenahan, "Calculation of modes in an optical fiber using the finite element method and Eispack," *Bell Syst. Tech. J.,* **62**: 2663–2694 (1983).

24. T. Miya, K. Okamoto, Y. Ohmori, and Y. Sasaki, "Fabrication of low dispersion single-mode fibers over a wide spectral range," *IEEE J. Quantum Electron.,* **15**: 858–861 (1981).

25. S. J. Jang, L. G. Cohen, W. L. Mammel, and M. A. Saifi, "Experimental verification of ultra-wide bandwidth spectra in double-clad single-mode fiber." *Bell Syst. Tech. J.,* **61**: 385–390 (1982).

26. F. M. Sears, L. G. Cohen, and J. Stone, "Interferometric measurements of dispersion-spectra variations in a single-mode fiber," *J. Lightwave Technol.,* **LT-2**: 181–184 (1984).

27. P. K. Bachmann, D. Leers, H. Wehr, D. U. Wiechert, J. A. v. Steenwijk, D. L. A. Tjaden, and E. Wehrhahn, "Dispersion-flattened single-mode fibers prepared with PCVD: performance, limitations, design optimization," *J. Lightwave Technol.*, **LT-4**: 858–863 (1986).

28. S. J. Jang, J. Sanchez, K. D. Pohl, and L. D. L'Esperance, "Graded-index single-mode fibers with multiple claddings," *Tech. Digest*, 4th Int. Conf. Integrated Opt. and Opt. Fiber Commun. (IOOC'83), Tokyo, Japan, June 1983, pp. 396–397.

29. S. E. Miller and A. G. Chynoweth (Eds.), *Optical Fiber Telecommunications*, Academic, New York, 1979, p. 105.

30. M. A. Saifi, S. J. Jang, L. G. Cohen, and J. Stone, "Triangular-profile single-mode fiber," *Opt. Lett.*, **17**: 43–45 (1982).

31. A. D. Pearson, L. G. Cohen, W. A. Reed, J. T. Krause, E. A. Sigety, F. V. DiMarcello, and A. G. Richardson, "Optical transmission in dispersion-shifted single-mode spliced fiber and cable," *J. Lightwave Technol.*, **LT-2**: 346–349 (1984).

32. S. J. Jang, J. Sanchez, K. D. Pohl, and L. D. L'Esperance, "Fundamental mode size and bend sensitivity of graded and step-index single-mode fibers with zero-dispersion near 1.55 μm," *J. Lightwave Technol.*, **3**: 312–316 (1984).

33. B. J. Ainslie, D. M. Cooper, S. P. Craig, and C. R. Day, "Low loss single mode fibres with 'W' profiles which are insensitive to bending losses," *Br. Telecom Technol. J.*, **4**: 65–70 (1986).

34. V. J. Mazurczyk, N. S. Bergano, R. E. Wagner, K. L. Walker, N. A. Olsson, L. G. Cohen, and J. C. Campbell, "420 Mbs Transmission through 203 km using very low loss fiber and a DFB laser," 10th Eur. Conf. Opt. Commun. (ECOC'84), Postdeadline paper, Stuttgart, W. Germany, 1984.

35. S. K. Korotky, G. Eisenstein, A. H. Gnauck, B. L. Kasper, J. J. Veselka, R. C. Alferness, L. L. Buhl, C. A. Burrus, T. C. D. Huo, L. W. Stulz, K. Ciemiecki Nelson, L. G. Cohen, R. W. Dawson, and J. C. Campbell, "4 Gbit/s transmission experiment over 117 km of optical fiber using a Ti-LiNbO₃ external modulator," *J. Lightwave Technol.*, **LT-3**: 1027–1031 (1985).

36. N. A. Olsson, J. Hegarty, R. A. Logan, L. F. Johnson, K. L. Walker, L. G. Cohen, B. L. Kasper, and J. C. Campbell, "68.3 km transmission with 1.37 Tbit km/s capacity using wavelength division multiplexing of ten single-frequency lasers at 1.5 μm," *Electron. Lett.*, **21**: 106–107 (1985).

37. N. S. Bergano, R. E. Wagner, H. T. Shang, and P. F. Glodis, "150-km 296-Mbit/sec dispersion-shifted fiber experiment," *Tech. Digest*, OFC'85, February 1985, San Diego, CA, pp. 50–51.

CHAPTER 4

Polarization-Maintaining Optical Fibers

T. Okoshi

4.1 INTRODUCTION

In ordinary axially symmetrical single-mode fibers, two mutually independent orthogonal HE_{11} modes can propagate. In the framework of Cartesian coordinates, these are the HE_{11x} mode, which has its principal electric field component in the x direction, and the HE_{11y} mode, which has principally the y component of the electric field. We hereafter omit the suffix 11 for simplicity and call these modes the HE_x and HE_y modes.

On the other hand, any linearly polarized HE_{11} mode can be expressed as a combination of two circularly polarized, clockwise and counterclockwise rotating HE_{11} modes. Hence, we may equivalently say that in an axially symmetrical fiber, two mutually independent, circularly polarized, clockwise and counterclockwise rotating modes can propagate. We hereafter call these clockwise and counterclockwise HE_{11} modes the HE^+ and HE^- modes, respectively.

If the fiber is completely round, all the HE_x, HE_y, HE^+, and HE^- modes are degenerate; that is, no difference exists among their propagation constants. Moreover, if the fiber is completely straight, no coupling exists between these modes. This means that if one of these modes is launched at the input, the same mode will appear at the exit end.

However, an actual optical fiber is neither completely axially symmetrical nor completely straight. As a result, the polarization state of the propagated light is subject to unstable fluctuation when ambient conditions change. In a multimode fiber, such instability usually causes little trouble except for its possible effect on modal noise. However, in a single-mode fiber, the following problems arise [1, 2]:

1. In an optical communication system, the received signal level fluctuates when the receiver is sensitive to the polarization. This situation takes place in heterodyne-type or homodyne-type optical communications in which a matching of the polarization state is required between the received signal and the local oscillator [3].
2. Even when a fiber is designed to be axially symmetric, slight elliptical deformation exists. This residual ellipticity separates the propagation constants of two orthogonal HE_{11} modes which are otherwise degenerate with each other and causes the so-called polarization mode dispersion in the group delay.
3. In measurements using a single-mode fiber, such as in magnetooptic current-sensing or laser gyroscopes, or in coherent optical communications such as those employing PCM-PSK (phase-shift keying) or PCM-FSK (frequency-shift keying), the polarization instability causes the measurement error or the bit error rate to deteriorate.

When the polarization-mode dispersion (the group-delay difference, item 2) is the only problem, a direct solution is to fabricate the fiber to be as circular as possible. Various efforts toward this target have been reported.

Typical achievements are the realization of very small propagation constant differences between the HE_x and HE_y modes ($\Delta\beta \triangleq |\beta_x - \beta_y| = 10$ deg/m) by the improvement of physical circularity [4] and the realization of $\Delta\beta = 2.6$ deg/m by equalizing the thermal expansion coefficients of core and cladding to reduce the residual stress that often deteriorates the circularity [5]. In the latter, SiO_2/GeO_2 and SiO_2/B_2O_3 are used as core and cladding materials, respectively.

Furthermore, to achieve a dramatic improvement in the circularity, Barlow et al. proposed to rotate the preform rod during the drawing process to smooth out its circular irregularity; they called the fiber thus produced the "spun fiber" [6, 7]. A propagation constant difference $\Delta\beta = 0.4$–0.6 deg/m and polarization-mode dispersion (group-delay difference) of 0.02 ps/km were obtained with this scheme.

However, in heterodyne or coherent optical fiber communications and in coherent optical fiber measurements, items 1 and 3 described above are also important. To solve these problems, various methods for avoiding the adverse effects of the polarization-state fluctuation at the input end of the receiver have been devised. These methods are basically classified into active control of the state of polarization and polarization diversity receiving schemes.

The active polarization controllers proposed so far are electromagnetic fiber squeezers [8], electrooptic crystals [9], "bend-and-twist" type fiber devices [10], and Faraday rotators [11]. In the second approach, the polarization diversity receiver [12], signals in two orthogonal polarizations are heterodyne-detected separately at the receiving end, and the two intermediate frequency (IF) signals are added electronically in the IF amplifier after appropriate phase compensation, or added after demodulation.

However, the more essential and complete solution to the problem of the polarization-state fluctuation is the use of polarization-maintaining or single-polarization fibers [1, 2], which is the subject of this chapter.

4.2 CLASSIFICATION OF POLARIZATION-MAINTAINING OPTICAL FIBERS

To begin with, a comment should be made about the confusion in presently existing terminology. Various terms such as polarization-maintaining fibers, polarization-holding fibers, single-polarization fibers, and single-polarization single-mode (SPSM) fibers are now used in this area, in a somewhat confusing manner, without widely accepted definitions. In the following, we will tentatively use "single-polarization fiber" to denote the so-called "truly" single-polarization fiber, and "polarization-maintaining fiber" to denote generically single-polarization fibers and "birefringent" fibers.

Various polarization-maintaining fiber schemes are classified as in Table 4.1 [13] using the above terminology. Short comments on each type follow.

A. *Linear-polarization-maintaining fibers (LPM fibers)*. Fibers belonging to this category are designed so that only one of the two linearly polarized modes (HE_x and HE_y modes) can propagate. These fibers can further be classified as A.1 and A.2.

   A.1. *Truly single-polarization fibers (differential attenuation fibers)*. The fiber is designed so that the transmission losses for the HE_x and HE_y modes are largely different, hopefully so that one propagation mode is cut off at a specific frequency (i.e., wavelength) range. Such a fiber was first called an "absolutely single-polarization fiber" [2] and later also referred to as "truly single-polarization fiber" or simply "single-polarization fiber" or "differential attenuation fiber" in contrast to "birefringent fiber." This scheme can further be classified:

      A.1.1. *Geometry-induced single-polarization fibers*, examples of which are side-pit and side-tunnel fibers.

      A.1.2. *Stress-induced single-polarization fibers*, an example of which is the differential-attenuation-type bow-tie fibers.

   A.2. *Linearly birefringent fibers*. In this case the fiber is designed so that the propagation constants of the HE_x and HE_y modes are different, as illustrated in Figure 4.1. As will be seen in the following sections,

TABLE 4.1. Classification of various polarization-maintaining optical fibers

A. Linear-polarization maintaining (LPM) fibers
    A.1. (Truly) single-polarization fibers (differential attenuation fibers)
        A.1.1. Geometry type
            (1) Side-pit
            (2) Side-tunnel
        A.1.2. Stress type
            (1) Bow-tie
            (2) Flattened depressed-cladding
            (3) Stress-guiding
    A.2. Linearly birefringent fibers
        A.2.1. Geometry type
            (1) Elliptical core
            (2) Dumbbell core
            (3) Side-pit
            (4) Side-tunnel
        A.2.2. Stress type
            (1) Elliptical cladding
            (2) Elliptical jacket
            (3) PANDA
            (4) Four-sector core
            (5) Bow-tie
B. Circular-polarization-maintaining (CPM) fibers
    Circularly birefringent fibers
        Stress type
            Twisted round fiber

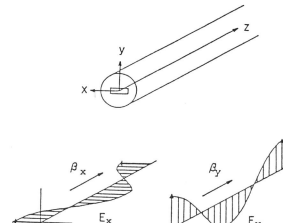

Figure 4.1 Principle of a linearly birefringent fiber. (After Okoshi [3].)

when the propagation constant difference increases, the two polarization modes are "decoupled," and the launched one will continue to travel with very little power transfer to the other over an appreciable length of the fiber. This scheme can further be classified.

A.2.1. *Geometry-induced birefringent fibers*, in which the propagation constant difference is produced by an axially nonsymmetrical refractive index distribution. Elliptical and dumbbell core fibers belong to this category. However, we should note that stress-induced birefringence (see the next item) also exists in such fibers and that the truly single-polarization fibers such as side-pit and side-tunnel fibers can also be used as birefringent fibers at frequencies outside the truly single-polarization region.

A.2.2. *Stress-induced linearly birefringent fibers*, in which the propagation constant difference is produced by an axially nonsymmetrical internal stress. The elliptical cladding fiber, elliptical jacket fiber, PANDA fiber, four-sector-core fiber, and birefringence-type bow-tie fiber are included in this category.

B. *Circular-polarization-maintaining (CPM) fibers.* A round (axially symmetrical) fiber is twisted as shown in Figure 4.2 to produce a difference between the propagation constants of the clockwise and counterclockwise circularly polarized HE_{11} modes. Thus, these two circular polarization modes—the HE^+ and HE^- modes—are decoupled. These fibers can

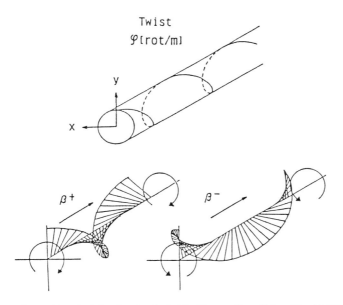

Figure 4.2 Principle of a circularly birefringent fiber. (After Okoshi [3].)

therefore also be called "circularly birefringent" fibers. (Note that a "truly single circular-polarization fiber" has not been invented.)

4.3 HISTORY OF RESEARCH AND DEVELOPMENT OF POLARIZATION-MAINTAINING FIBERS

As early as the mid-1970s, some of the optical fiber specialists who originally came from microwave research believed that axially nonsymmetrical single-mode fibers would become common, like rectangular waveguides at microwave frequencies. Vigorous research to produce such fibers, however, started in the late 1970s. The short history of the research and development of polarization-maintaining fibers can conveniently be divided into four periods.

1976–1977. In the first period, the modal birefringence caused by an unintentional elliptical core deformation was the principal subject of research. Intentional polarization-maintaining schemes were not investigated.

1978–1979. Various kinds of linear-polarization-maintaining (LPM) fiber schemes, such as elliptical core fibers [14, 15], dumbbell core fibers [16], and stress-induced (elliptical cladding) fibers [17–20], were proposed, mainly by the investigators of Bell Laboratories. The side-pit fiber was proposed in Japan [21]; this was the first proposal of the concept of the truly single-polarization (differential attenuation) fiber.

1980–1981. The center of research seems to have moved from the United States to Japan and Europe in this period. Researchers of CNET (France) proposed an entirely new scheme, the circular-polarization-maintaining (CPM) fiber, in the fall of 1980 [22]. Various novel LPM fibers, such as a new type of elliptical cladding fiber [23], elliptical jacket fiber [24], side-pit fiber [25, 26], PANDA fiber [27], and four-sector-core fiber [28] were proposed and/or experimented with in Japan.

1982–1984. The development of various fibers continued. On the other hand, British Telecom people reported in the fall of 1982 that the polarization-state fluctuation was not so drastic when the fiber was cabled and installed [29]. This report led to a new discussion on the polarization-maintaining capability that would be required in practical communications systems.

Three new proposals for the polarization-maintaining fiber structure appeared in 1982. The first is the side-tunnel fiber [30] in which the refractive index pits in side-pit fiber are replaced by vacant tunnels. The second is bow-tie fiber [31, 32], which is a version of the PANDA fiber in which the stress-producing part is shaped like a bow tie to maximize the birefringence.

Third, the possibility of realizing a truly single-polarization fiber taking advantage of stress-induced birefringence was discussed by Eickhoff [33].

In 1983, experiments of stress-induced truly single-polarization (differential attenuation) fibers were reported; these were the measurement of the truly single linear-polarization characteristics of "bent" bow-tie fibers [34, 35] and that of "flattened" depressed-cladding fiber by Simpson et al. [36, 37]. The PANDA fiber was improved further [38, 39]. Snyder and Rühl proposed a new concept of the anisotropic differential attenuation fiber [40–42].

Two other significant achievements were reported in 1983 by Birch et al. [43] and Okamoto et al. [44]. The first paper demonstrated experimentally that a truly single-polarization transmission was possible regardless of the wavelength (see the next section). The second showed theoretically that a low dispersion (several picoseconds per kilometer per nanometer) was obtainable over a wide wavelength range (1.5–1.7 μm) with the bow-tie or PANDA structure.

4.4 PRINCIPLES, CHARACTERISTICS, AND COMPARISON OF VARIOUS SINGLE-POLARIZATION FIBERS

4.4.1 Truly Single-Polarization Fibers (Differential Attenuation Fibers)

The difference between the dipersion characteristics of a (truly) single-polarization fiber and that of a linearly birefringent fiber is depicted in Figure 4.3 [13, 21]. This figure shows mode curves for the HE_x and HE_y modes, with the normalized frequency v on the abscissa and normalized propagation constant β/k on the ordinate, where k denotes the propagation constant in vacuum, that is, $k = \omega/c$. When the fiber has an axially nonsymmetrical structure, the curves for the two polarization modes will be separated. (In most cases the separation is very small; it can hardly be noticed on a graph of this size.)

In a truly single-polarization fiber, the two polarization modes must have different cutoff frequencies, as shown by the solid curves in Figure 4.3. However, in an ordinary optical fiber, no cutoff exists for HE_{11} modes, as shown by broken curves in the same figure. A hint for realizing a truly single-polarization fiber can be found in the following fact: In an axially *symmetrical* fiber having an "index valley" with appreciable width and depth around the core as shown in Figure 4.4, so that

$$\int [n(r) - n_2] 2\pi r \, dr < 0 \qquad (1)$$

where n_2 denotes the cladding index, a cutoff region appears for the HE_{11} mode [25].

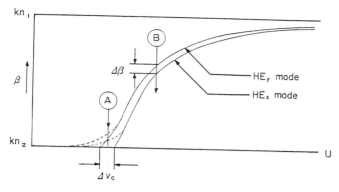

Figure 4.3 Comparison of the operating frequency for a truly single-polarization (differential attenuation) fiber (arrow A) and that for a linearly birefringent fiber (arrow B) (after Okoshi [13].)

We may then predict that, in index distributions such as those shown in Figures 4.5 and 4.6, where, for example (Fig. 4.5), $n_2 = 1.46$, $n_1 = 1.47$, and n_p (refractive index pit) = 1.45, or (Fig. 4.6) $n_2 = 1.46$, $n_1 = 1.47$, and n_p (tunnel) = 1.00, the HE_x and HE_y modes will have different cutoff frequencies. The structures corresponding to Figures 4.5 and 4.6 are called side-pit fiber [25, 26] and side-tunnel fiber [30], respectively.

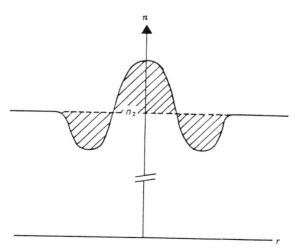

Figure 4.4 The condition for realizing a cutoff region for the HE_{11} mode in an axially symmetrical optical fiber. (After Okoshi [13].)

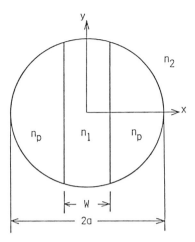

Figure 4.5 An example of the refractive index profile for side-pit or side-tunnel fibers. (After Okoshi and Oyamada [25].)

The realization of the truly single-polarization (differential attenuation) characteristics was reported first with a side-tunnel fiber [30]. However, the difference in attenuation was not satisfactorily large, typically 0.1 dB/m versus 3 dB/m, when the fiber was wound on a reel of 53-cm diameter.

The truly single-polarization fiber can also be realized by taking advantage of stress-induced birefringence. Two trends exist in the research toward such a scheme. The first one has to do with the ideas proposed by Eickhoff in late 1982 [33]. His two basic ideas are illustrated in Figure 4.7. In Figure 4.7a, the dopant profile and stress profile cancel each other in the y direction, making the y-polarized mode leaky. In Figure 4.7b the cancellation takes place in the x direction. Later, such an idea was investigated more thoroughly by Snyder and Rühl, who formulated the general waveguide theory, taking into account the material birefringence as well as the effect of higher-order

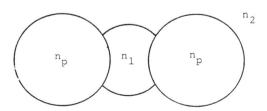

Figure 4.6 Another example of a refractive index profile for side-pit or side-tunnel fibers. (After Okoshi et al. [51].)

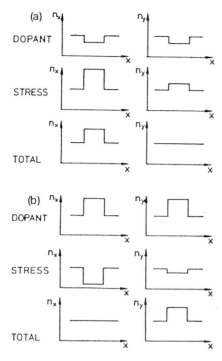

Figure 4.7 Schematic illustrations of refractive index profiles for realizing a truly single-polarization fiber. For details, refer to the text. (After Eickhoff [33].)

modes [40–42]. However, such a scheme, as it stands, has not been experimentally proved to be effective, because the stress-induced birefringence is actually not high enough to produce a significant difference in attenuation.

The second approach combines stress-induced birefringence and the idea of producing the cutoff for the HE_{11} mode by a refractive index valley [25]. This was first demonstrated experimentally almost simultaneously by Varnham et al. [34, 35] and Simpson et al. [36, 37], but at that time only for bent fibers.

The basic concept depicted by the latter is shown in Figure 4.8 [37], where the core index is higher than the cladding index for one polarization mode to propagate, but lower for the other orthogonal mode to be leaky. Figure 4.9 shows the refractive index profiles along the x and y axes measured on the preform. Figure 4.10 shows the truly single-polarization (differential attenuation) characteristics obtained when the fiber was wound four turns around a mandrel with 1.55-cm diameter.

On the other hand, in the experiment reported by Varnham et al. [34], the fiber structure itself was essentially the bow-tie fiber originally designed

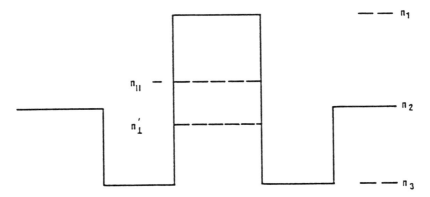

Figure 4.8 Schematic illustration showing the condition for producing truly single-polarization characteristics. (After Simpson et al. [37].)

as a strongly birefringent fiber. However, probably the effect of the axially nonsymmetrical index valleys (the side-pit effect) was combined with the strong stress-induced birefringence to produce truly single-polarization characteristics. Figure 4.11 shows the single-polarization characteristics of the bent bow-tie fibers for bending radii of 6 cm (dotted curves) and 15 cm (solid curves).

In the two experiments of Figures 4.10 and 4.11, significant single-polarization wavelength regions could be obtained only when an appropriate amount of bending was applied to the fiber. However, if the side-pit effect and a strong stress-induced birefringence are combined effectively, it should be possible to obtain a truly single-polarization property regardless of the wavelength. Such a fiber was realized by Birch et al. [43], who fabricated a fiber having strongly birefringent index profiles in the x and y directions.

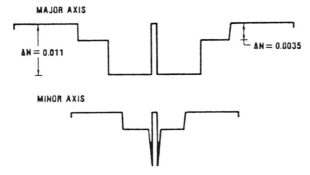

Figure 4.9 Refractive index profile in the x and y directions measured on the preform of the "flattened depressed cladding" fiber. (After Simpson et al. [37].)

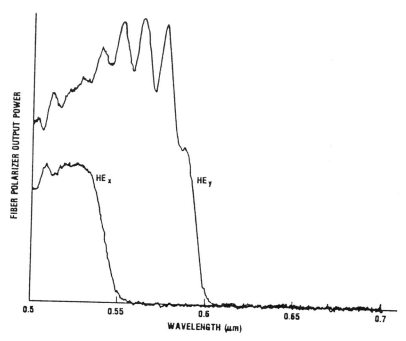

Figure 4.10 Wavelength dependence of the output power for two orthogonal polarizations from a 4.7-m long "flattened depressed-cladding" fiber, wound four turns around a mandrel with 1.55 cm diameter. (After Simpson et al. [37].)

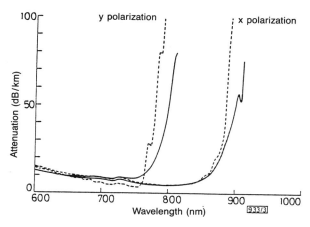

Figure 4.11 Wavelength dependence of the attenuation for two orthogonal polarizations in bow-tie fibers, for bending radius of 6 cm (dotted curves) and 15 cm (solid curves). (After Varnham et al. [34].)

They called it a "stress-guiding optical fiber" and found that it could propagate only *x*-polarized light at any wavelength when drawn as a single-mode fiber (fiber diameter 80 μm) as well as when drawn as a multimode fiber (fiber diameter 600 μm).

4.4.2 Characteristic Parameters of Truly Single-Polarization Fibers

An important parameter is the extinction ratio for a given length of fiber, that is, the ratio of the propagation-mode and leaky-mode attenuations. This parameter is particularly significant when the fiber is used as a "fiber polarizer."

The other parameter is the relative bandwidth of the truly single-polarization frequency region, defined as [25]

$$K = \frac{|v_{cx} - v_{cy}|}{(v_{cx} + v_{cy})/2} \tag{2}$$

where v_{cx} and v_{cy} denote normalized cutoff frequencies for the HE_x and HE_y modes, respectively [30].

However, in the really wideband scheme such as for stress-guiding optical fibers [43], the K parameter is less significant. In such a case a more significant parameter is the numerical aperture (NA), which gives the acceptance angle for the launched beam, because the index difference cannot be easily increased. The maximum NA so far obtained is 0.059 at a wavelength of 633 nm [43].

4.4.3 Linearly Birefringent Fibers

In a linearly birefringent fiber, both of the two polarization modes can propagate, but the propagation constant difference $\Delta\beta$ is made as large as possible. When a structural fluctuation is present along the axis, the power transfer between the two modes per unit length of fiber is known to be proportional to $\Delta\beta^{-4}$–$\Delta\beta^{-8}$ [45]. Therefore, we can reduce the coupling between the two modes effectively by increasing $\Delta\beta$.

The modal birefringence can be generated by an axially nonsymmetrical refractive index distribution or an axially nonsymmetrical stress distribution. Usually, the latter produces a much stronger birefringence.

One of the strongest birefringences has been reported by Katsuyama and others of Hitachi, Ltd., with a stress-induced birefringent fiber having a circular core, circular inner cladding, elliptical outer cladding, and circular jacketing [23]. A beat length between the two polarization modes (defined in the following subsection) shorter than 0.88 mm was obtained with this scheme.

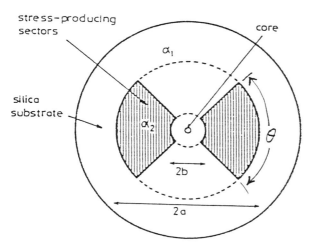

Figure 4.12 A cross section of the bow-tie fiber used in the optimum-design theory. The stress-producing sectors are highly doped to give an expansion coefficient α_2, which is much larger than α_1. After Birch et al. [31].)

There are many other types of stress-induced birefringent fibers; examples are the PANDA (polarization-maintaining and absorption-reducing) fiber [27, 38, 39] and bow-tie fiber [31]. In both of these schemes, the internal stress is produced by two "stress-applying regions" heavily doped with B_2O_3. These regions are made on both sides of the core but separated from it by an appropriate distance. This distance is particularly important for preventing greater loss due to the presence of lossy B_2O_3 regions.

An example of the normalized birefringence obtained is $B = 6 \times 10^{-4}$ for a refractive index difference between core and cladding of 0.6% at a wavelength of 1.06 μm [39]. [For the definition of B, see Eq. (3).]

The birefringence has been much enhanced by the bow-tie fiber [31] by optimizing the cross-sectional shape of the stress-applying regions, as shown in Figure 4.12. It was shown by stress analysis that the parameters in Figure 4.12 should be [31]

$2a$: 75% of the outer diameter

$2b$: as small as possible unless the transmission loss increases (actually about twice the core diameter)

θ: nearlfy equal to 90°

The highest birefringence obtained with this structure was $B = 1.16 \times 10^{-3}$, which gave $L = 0.55$ mm at a wavelength of 633 nm. The four steps in the fabrication of such a structure is illustrated in Figure 4.13.

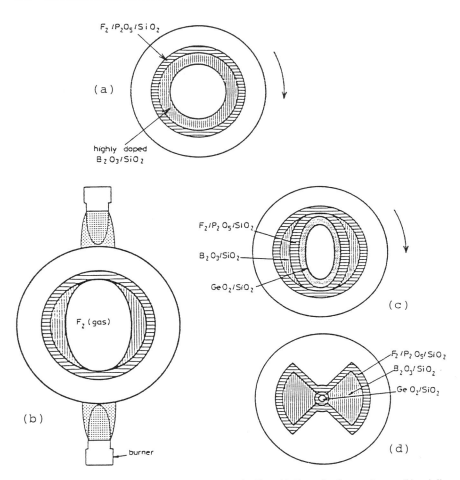

Figure 4.13 Four stages of fabrication of a bow-tie fiber: (a) Deposit of outer layers, (b) axially nonsymmetrical etch of B_2O_3-doped layer, (c) deposit of inner layers, and (d) collapse. (After Birch et al. [31].)

4.4.4 Characteristic Parameters of Linearly Birefringent Fibers

The most significant parameter is the difference between the propagation constants of the HE_x and HE_y modes, $\Delta\beta \triangleq |\beta_x - \beta_y|$. This is called the polarization birefringence or modal birefringence. Practically, a normalized quantity called the normalized birefringence,

$$B = \frac{\Delta\beta}{k} = 2n\frac{|\beta_x - \beta_y|}{\beta_x + \beta_y} \tag{3}$$

where n denotes the refractive index of the fiber material, and the beat length between the HE_x and HE_y modes,

$$L = \frac{2\pi}{\Delta\beta} = \frac{\lambda}{B} \tag{4}$$

where λ denotes the wavelength of light, are often used to express the magnitude of the birefringence.

When the beat length L is much shorter than the typical spatial period of geometrical perturbations in the fiber (bends, for example), the energy transfer between the HE_x and HE_y modes becomes negligible.

The magnitude of the polarization-mode dispersion (or polarization dispersion, for short) is usually expressed as the group-delay difference t_p between the two modes. Using the effective refractive indexes n_x and n_y defined as

$$n_x = \frac{c\beta_x}{\omega} \tag{5}$$

and

$$n_y = \frac{c\beta_y}{\omega} \tag{6}$$

where ω and c denote angular frequency and light velocity, we can express t_p as [46]

$$t_p = \left| \frac{d(\Delta\beta)}{d\omega} \right| = \left| \frac{n_x - n_y}{c} + \frac{\omega}{c}\left(\frac{dn_x}{d\omega} - \frac{dn_y}{d\omega} \right) \right| \tag{7}$$

In ordinary silica glass, the dispersion in the near-infrared region is weak, that is, $dn/d\omega \ll n/\omega$. Therefore, the approximate, simple relation

$$t_p \doteq \frac{|n_x - n_y|}{c} = \frac{B}{c} \tag{8}$$

holds. This equation tells us that when we increase B to obtain a high polarization-maintaining capability, the polarization dispersion t_p must increase.

4.4.5 Circularly Birefringent Fibers

The circularly birefringent fiber was proposed first by Jeunhomme and Monerie of CNET, France, in 1980 [22]. In this scheme, a twist with a rate of φ [rotations/m] is given to an axially symmetrical fiber, as shown in Figure 4.2. It is known both theoretically and experimentally that if a linearly polarized light is launched onto such a fiber, a "drag" of the polarization plane takes place in the same direction as the twist but with a rate $\alpha\varphi$ [rotations/m], where $\alpha = 0.07$. This means that a difference is generated between the propagation constants of the clockwise circular polarization (β^+) and that of the counterclockwise circular polarization (β^-), which is given as

$$\delta\beta \triangleq |\beta^+ - \beta^-| = 2\alpha\varphi \tag{9}$$

The circular birefringence thus obtained is usually much weaker than that obtained with stress-induced linearly birefringent fibers, as shown in Section 4.4.6.

4.4.6 Comparison of Linear-Polarization-Maintaining (LPM) Fibers and Circular-Polarization-Maintaining (CPM) Fibers

The features of LPM and CPM fibers are compared in Table 4.2 [47]. It is seen that the latter has many attractive features: ease of fabrication, connection/splicing, and adjustment of the polarization between the signal and local oscillator beams and small polarization dispersion. We should note, however, that a large connection loss would be generated if a lateral displacement were present between two CPM fibers to be connected; such a displacement is in a sense equivalent to an azimuthal displacement in the connection of two LPM fibers.

A comment follows on the fourth feature of the CPM fiber—its small polarization dispersion. According to the analysis by Monerie and Jeunhomme [48], the group-delay difference between the HE^+ and HE^- modes t_{pc} is given as

$$t_{pc} = \frac{\Delta\beta}{2(1 - \alpha)\varphi} \, t_{pL} \tag{10}$$

where $\Delta\beta = |\beta_x - \beta_y|$, α is the optical activity of the fiber material (0.07 for silica glass), φ is the twist rate in radians per meter, and t_{pL} denotes the group-delay difference between the HE_x and HE_y modes in the absence of the twist. For example, when $\Delta\beta = 2\pi \times 0.2$ rad/m and $\varphi = 2\pi \times 10$ rad/m, we obtain $t_{pc} \doteqdot t_{pL}/100$. This example suggests that the circularly birefringent fiber is much superior to linearly birefringent fibers with respect to polarization dispersion.

TABLE 4.2. Comparison of linear-polarization-maintaining (LPM) fibers and circular-polarization-maintaining (CPM) fibers

	LPM fibers	CPM fibers
Fabrication	Difficult	Easy
Connection and splicing	Difficult	Easy
Polarization dispersion	Large	Small
Polarization matching with LO injection	Tedious	Not difficult, if not unnecessary
Mode decoupling in birefringent fibers	Large	Small
Possibility of truly single-polarization scheme	Yes	No

Equation (10) might appear to be somewhat contradictory to our intuition. That is, readers might wonder why the group-velocity difference decreases when the fiber is twisted [i.e., $t_{pc} \propto \varphi^{-1}$; see Eq. (10)], despite the fact that the phase-velocity difference increases due to the twist [i.e., $\delta\beta \propto \varphi$; see Eq. (9)]. The answer to this question can best be given by computing t_{pc} for the case when no linear birefringence exists, that is,

$$t_{pc} = \frac{d}{d\omega}(\delta\beta) = \frac{d}{d\omega}|\beta^+ - \beta^-| = \frac{d}{d\omega}(2\alpha\varphi) \tag{11}$$

which tends to zero in the first-order approximation (note that α is almost independent of ω). Equation (11) suggests that in a circularly birefringent fiber the group-delay difference is essentially zero because $\delta\beta$ is independent of the light frequency in the first-order approximation. The group-delay difference given in Eq. (10) expresses a minor term originating from the presence of residual linear birefringence [49].

Finally, to compare the degree of decoupling between polarization modes in linearly birefringent and circularly birefringent fibers, we investigate the minimum beat length (in other words, the maximum B) for both types of birefringent fibers. For linearly birefringent fibers, the shortest beat length ever reported is 0.55 mm [31]. On the other hand, in a circularly birefringent fiber, we can write $L = 2\pi/\delta\beta = \pi/\alpha\varphi$, which gives $L_{min} = 10$ cm if we assume $\varphi_{max} = 2\pi \times 70$ rad/m (i.e., 70 rotations per meter, which is a practical limit from fiber elasticity) and $\alpha = 0.07$ (the value for SiO_2). Thus, we find that the beat length is about two orders of magnitude greater in circularly birefringent fibers than in linearly birefringent fibers; in other words, the polarization decoupling is much weaker. (We should note, however, that the types of structural perturbations that will cause intermodal coupling are different in the two kinds of birefringent fibers.)

On the other hand, there is still a debate over what magnitude of birefringence (i.e., polarization-maintaining capability) is really needed when a birefringent fiber is used in actual optical communication channels. A paper from British Telecom Research Laboratories reported that the polarization state in a cabled fiber is not so unstable: the time constant of the fluctuation is typically more than several hours [29]. If so, the use of the circularly birefringent fiber might become attractive in some cases because it can be manufactured more easily and makes connection and splicing easier at the sacrifice of a weaker birefringence.

4.5 SUMMARY AND CONCLUSIONS

The technical background, classification of various types, history of research and development, principles, and characteristics of polarization-maintaining optical fibers have been reviewed.

It is rather difficult to answer one possible question: What type of polarization-maintaining scheme will become dominant in the future? The difficulty is partly because of the wide variety of applications of polarization-maintaining transmission and partly because of the indefinite range of the "future." However, if we restrict ourselves to applications in very long-haul heterodyne/coherent optical fiber communications in the somewhat distant future (20–30 years from now), we will probably use one of the truly single-polarization schemes. The "stress-guidance" scheme, when assisted by the side-pit effect, can offer excellent characteristics. The side-tunnel scheme can also offer excellent differential-attenuation characteristics [50].

In this paper, polarization-maintained transmission of light for communication over an appreciable distance was considered. Another relevant topic is the application of truly single-polarization fiber techniques to the "fiber polarizers" to be used as optical circuit elements in communications and measurement. This technique is becoming important and making rapid progress, because the fiber polarizer, when spliced between two fibers, will offer the only means of low-loss polarization in fiber-oriented optical circuits. This is discussed in another review paper [51].

REFERENCES

1. L. P. Kaminow, "Polarization in optical fibers," *IEEE J. Quantum Electron.*, **QE-17** (1): 15–17 (1981).
2. T. Okoshi, "Single-polarization single-mode optical fibers," *IEEE J. Quantum Electron.*, **QE-17** (6): 879–884 (1981).
3. T. Okoshi, "Heterodyne and coherent optical fiber communications: recent progress" (Invited), *IEEE Trans. Microwave Theory Tech.*, **MTT-30** (8): 1138–1149 (1982).
4. H. Schneider, H. Harms, A. Rapp, and H. Aulich, "Low-birefringence single-mode fibers: preparation and polarization characteristics," *Appl. Opt.*, **17** (19): 3035–3037 (1978).
5. S. R. Normann, D. N. Payne, M. J. Adams, and A. M. Smith, "Fabrication of single-mode fibers exhibiting extremely low polarization birefringence," *Electron. Lett.*, **15** (11): 309–311 (1979).
6. A. J. Barlow, D. N. Payne, M. R. Hadley, and R. J. Mansfield, "Production of single-mode fibers with negligible intrinsic birefringence and polarization mode dispersion," Eur.Conf. Opt. Commun. (ECOC'81), Paper No. 2-3, September 1981, Copenhagen.
7. A. J. Barlow, J. J. Ramskov-Hansen, and D. N. Payne, "Birefringence and polarization-mode dispersion in spun single-mode fibers," *Appl. Opt.*, **20** (17): 2962–2968 (1981).
8. R. Ulrich, "Polarization stabilization on single-mode fiber," *Appl. Phys. Lett.*, **35** (11): 840–842 (1979).
9. M. Kubota, T. Oohara, K. Furuya, and Y. Suematsu, "Electro-optical polarization control on single-mode optical fibers," *Electron. Lett.*, **16** (15): 573 (1980).
10. H. C. Lefevre, "Single-mode fiber fractional wave devices and polarization controllers," *Electron. Lett.*, **16** (20): 778–780 (1980).
11. T. Okoshi, K. Kikuchi, and Y. Cheng, "A new polarization-control scheme for optical heterodyne receiver," *Tech. Digest. Eur. Conf. Common.* (ECOC'85), October 1985, Venice, Italy, pp. 405–408.

12. T. Okoshi, S. Ryu, and K. Kikuchi, "Polarization-diversity receiver for heterodyne/coherent optical fiber communications," 4th Int. Conf. Integral Opt. and Opt. Fiber Commun. (IOOC'83), paper 30C3-2, June 1983, Tokyo.

13. T. Okoshi, "Review of polarization-maintaining single-mode fiber" (Invited), 4th Int. Conf. Integral Opt. and Opt. Fiber Commun. (IOOC'83), paper 28A4-1, June 1983, Tokyo.

14. V. Ramaswamy, W. G. French, and R. D. Standley, "Polarization characteristics of noncircular core single-mode fibers," *Appl. Opt.*, **17**: 3014–3017 (1978).

15. R. B. Dyott, J. R. Cozens, and D. G. Morris, "Preservation of polarization in optical-fiber waveguides with elliptical cores," *Electron. Lett.*, **15**: 380–382, (1979).

16. V. Ramaswamy and W. G. French, "Influence of noncircular core on the polarization performance of single-mode fibers," *Electron. Lett.*, **14**: 143–144 (1978).

17. V. Ramaswamy, I. P. Kaminow, P. Kaiser, and W. G. French, "Single polarization optical fibers: exposed cladding technique," *Appl. Phys. Lett.*, **33**: 814–816 (1978).

18. R. H. Stolen, V. Ramaswamy, P. Kaiser, and W. Pleibel, "Linear polarization in birefringent single-mode fibers," *Appl. Phys. Lett.*, **33**: 699–701 (1978).

19. V. Ramaswamy, R. H. Stolen, M. D. Divino, and W. Pleibel, "Birefringence in elliptically clad borosilicate single-mode fibers," *Appl. Opt.*, **18**: 4080–4084 (1979).

20. J. P. Kaminow, J. R. Simpson, H. M. Presby, and J. B. MacChesney, "Strain birefringence in single-polarization germanosilicate optical fibers," *Electron. Lett.*, **15**: 677–679 (1979).

21. T. Okoshi, "Feasibility study of frequency division multiplexing optical communication system using optical heterodyne or homodyne schemes" (in Japanese), Tech. Group, IECE Japan, Paper No. OQE78-139, February 1979.

22. L. Jeunhomme and N. Monerie, "Polarization-maintaining single-mode fiber cable design," *Electron. Lett.*, **16** (24): 921–922 (1980).

23. H. Matsumura, T. Katsuyama, and T. Suganuma, "Fundamental study of single polarization fibers," Proc. 6th Eur. Conf. Opt. Commun., September 1980, York, England. pp. 49–52.

24. T. Katsuyama, H. Matsumura, and T. Suganuma, "Low-loss single-polarization fibers," *Electron. Lett.*, **17** (13): 473–474 (1981).

25. T. Okoshi and K. Oyamada, "Single-polarization single-mode optical fiber with refractive-index pits on both sides of the core," *Electron. Lett.*, **16** (18): 712–713 (1980).

26. T. Hosaka, K. Okamoto, Y. Sasaki, and T. Edahiro, "Single-mode fibers with asymmetrical refractive-index pits on both sides of core," *Electron. Lett.*, **17** (15): 191–193 (1981).

27. T. Hosaka, K. Okamoto, T. Miya, Y. Sasaki, and T. Edahiro, "Low-loss single-polarization fibers with asymmetrical strain birefringence," *Electron. Lett.*, **17** (15): 530–531 (1981).

28. K. Kitayama et al., "Polarization-maintaining single-mode fiber with azimuthally inhomogeneous index profile," *Electron. Lett.*, **17** (12): 419–420 (1981).

29. T. G. Hodgkinson et al., "Experimental 1.5 μm coherent optical fiber transmission system," Eur. Conf. Opt. Commun. (ECOC' 82), Paper No. AXII-5, September 1982, Cannes.

30. T. Okoshi, K. Oyamada, M. Nishimura, and H. Yokota, "Side-tunnel fiber: an approach to polarization-maintaining optical waveguiding scheme," *Electron. Lett.*, **18** (19): 824–826 (1982).

31. R. D. Birch, M. P. Varnham, D. N. Payne, and E. J. Tarbox, "Fabrication of polarization-maintaining fibers using gas-phase etching," *Electron. Lett.*, **18** (24): 1036–1038 (1982).

32. A. Ourmazd, R. D. Birch, M. P. Varnham, D. N. Payne, and E. L. Tarbox, "Enhancement of birefringence in polarization-maintaining fibers by thermal annealing," *Electron. Lett.*, **19** (4): 143–144 (1983).

33. W. Eickhoff, "Stress-induced single-polarization single-mode fiber," *Opt. Lett.*, **7** (12): 629–631 (1982).

34. M. P. Varnham, D. N. Payne, R. D. Birch, and E. J. Tarbox, "Single-polarization operation of highly birefringent bow-tie optical fibers," *Electron. Lett.*, **19** (7): 246–247 (1983).

35. M. P. Varnham, D. N. Payne, R. D. Birch, and E. J. Tarbox, "Bend behavior of polarizing optical fibers," *Electron. Lett.*, **19** (17): 679–680 (1983).

36. J. R. Simpson, F. M. Sears, and J. B. MacChesney, "Single-polarization fiber," Topical Meeting on Opt. Fiber Commun. (OFC'83), paper No. TuA2, February–March 1983, New Orleans.

37. J. R. Simpson, R. H. Stolen, F. M. Sears, W. Pleibel, J. B. MacChesney, and R. E. Howard, "A single-polarization fiber," *J. Lightwave Technol.*, **LT-1** (2): 370–374 (1983).

38. N. Shibata, Y. Sasaki, K. Okamoto, and T. Hosaka, "Fabrication of polarization-maintaining and absorption-reducing fibers," *J. Lightwave Technol.*, **LT-1** (1): 38–43 (1983).

39. Y. Sasaki, T. Hosaka, K. Takada, and J. Noda, "8 km-long polarization-maintaining fiber with highly stable polarization state," *Electron. Lett.*, **19** (19): 792–794 (1983).

40. A. W. Snyder and F. Rühl, "New single-mode single-polarization fiber," *Electron. Lett.*, **19** (5): 185–186 (1983).

41. A. W. Snyder and F. Rühl, "Novel polarization phenomena on anisotropic multimode fibers," *Electron Lett.*, **19** (11): 401–402 (1983).

42. A. W. Snyder and F. Rühl, "Practical single-polarization anisotropic fibers," *Electron. Lett.*, **19** (17): 687–688 (1983).

43. R. D. Birch, M. P. Varnham, D. N. Payne, and K. Okamoto, "Fabrication of a stress-guiding optical fibre," *Electron. Lett.*, **19** (21): 866–867 (1983).

44. K. Okamoto, M. P. Varnham, and D. N. Payne, "Polarization-maintaining optical fibers with low dispersion over a wide spectral range," *App. Opt.*, **22** (15): 2370–2373 (1983).

45. R. Olshansky, "Mode coupling effects in graded-index optical fibers," *Appl. Opt.*, **14** (4): 935–945 (1975).

46. S. C. Rashleigh and R. Ulrich, "Polarization mode dispersion in single-mode fibers," *Opt. Lett.*, **3** (8): 60–62 (1978).

47. T. Okoshi and K. Kikuchi, "Heterodyne-type optical fiber communications," *J. Opt. Commun.*, **2** (2): 82–88 (1981).

48. M. Monerie and L. Jeunhomme, "Polarization mode coupling in long single-mode fibers," *Opt. Quantum Electron.*, **12** (6): 449–461 (1980).

49. M. J. Adams, D. N. Payne, and C. M. Ragdale, "Birefringence in optical fibers with elliptical cross section," *Electron. Lett.*, **15**: 298–299 (1979).

50. T. Okoshi, T. Aihara, and K. Kikuchi, "Prediction of the ultimate performance of side-tunnel single-polarization fiber," *Electron. Lett.*, **19** (25/26): 1080–1082 (1983).

51. T. Okoshi, "Polarization-state control schemes for heterodyne or homodyne optical fiber communications," Special Joint Issue of *J. Lightwave Technol.* and *IEEE Trans. Electron Devices* (Special Issue on Lightwave Devices and Subsystems), **LT-3**: 1232–1237 (1985).

CHAPTER 5

Transmission Limitations in Fibers due to Nonlinear Optical Effects

A. R. Chraplyvy

5.1 INTRODUCTION

The trend in long-haul lightwave communications is toward higher information capacity and longer repeaterless spans. The former entails high bit rates and wavelength multiplexing, while the latter requires maximum transmitter power and higher receiver sensitivity. In both cases, the ultimate limitations in lightwave transmission will be nonlinear optical interactions between the optical waves and the transmission medium, the optical fiber. These optical nonlinearities can lead to severe signal distortion and fading. Until recently, the whole area of optical nonlinearities in fibers was viewed by systems engineers as a curiosity that requires large solid-state or gas lasers to generate the nonlinear effects. However, a number of recent experiments have generated nonlinearities in silica fibers using injection lasers or lasers with outputs of only a few milliwatts. As lightwave systems become more sophisticated and injection laser powers increase to over 10mW, various optical nonlinearities will become important.

There exists a rich collection of optical nonlinearities in silica fibers. Some nonlinearities affect the amplitude or wavelength of the carrier. Clearly such nonlinearities are potentially detrimental to both direct-detection and coherent systems. Other nonlinearities affect only the phase of the propagating optical wave and consequently should not affect direct-detection systems. Examples of nonlinearities that alter the amplitude and wavelength of the lightwave are stimulated Raman scattering and stimulated

Brillouin scattering. Cross-phase modulation is an example of a nonlinear interaction that gives rise to phase shifts and instabilities in coherent transmission systems.

This chapter describes optical nonlinearities in the context of lightwave system transmission limitations. Only nonlinearities occuring in silica fibers at reasonably low powers will be discussed. The nature and severity of system degradation caused by each nonlinearity will be described. (Most nonlinearities have an associated threshold transmitter power above which system degradation is unacceptable.) In addition, techniques to minimize effects of the nonlinearities will be described.

5.2 STIMULATED RAMAN SCATTERING

Stimulated Raman scattering (SRS) is perhaps the most interesting of the optical nonlinearities occurring in silica fibers. SRS can be a source of signal degradation in some lightwave systems, while on the other hand there is the possibility that it can provide optical amplification of weak signals in other configurations. In addition, SRS in silica optical fibers can be exploited to construct sources of infrared radiation that are useful in fiber diagnostics. The deleterious effects of SRS will be discussed first. This will be followed by a summary of the various amplification experiments relevant to lightwave systems. Finally, several examples of infrared sources based on SRS in optical fibers and their applications will be presented.

There are several ways that SRS can be described. Classically, Raman scattering can be viewed as modulation of light by the molecular vibrations in the silica matrix. This modulation produces sidebands spaced by a frequency equal to that of the vibrating molecules. The lower-frequency sideband is called the Stokes line, and the upper-frequency sideband is the anti-Stokes line. The Stokes line is typically significantly stronger than the anti-Stokes line. Quantum mechanically, in Raman scattering a pump photon is annihilated and a Stokes photon is created along with a quantum of vibrational energy in the scattering molecule. The detailed description of Raman scattering is not important for our purposes and can be found in numerous references [e.g., 1–4]. The important point is that if a probe signal at the Stokes frequency is coinjected into the Raman active medium along with the pump, the probe will be amplified at the expense of the pump, which will be depleted. The strength of this amplification is exponentially dependent on the pump intensity

$$P_s(L) = P_s(0)e^{gLI} \qquad (1)$$

where $P_s(0)$ and $P_s(L)$ are the power of the probe signal entering and exiting the Raman gain medium, respectively, L is the length of the gain

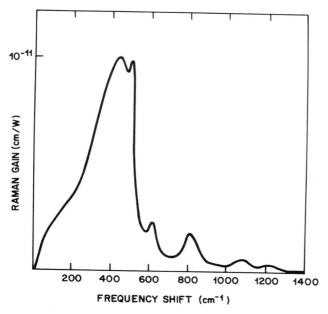

Figure 5.1 Raman gain versus frequency shift for fused silica at a pump wavelength of 1.0 μm. The gain coefficient scales inversely with pump wavelength [13].

medium, I is the pump intensity in W/cm², and g is the stimulated Raman gain coefficient in cm/W.

In a crystalline Raman gain medium, the molecular vibrations occur at specific well-defined frequencies. Consequently, no Raman amplification will occur unless the spacing between the pump and the probe is exactly equal to the vibration frequency. In silica fibers, the situation is quite different. Because silica is a glass, the various molecular vibration frequencies spread out into bands that overlap and create a continuum [5]. This gives rise to the gain coefficient g shown in Figure 5.1 [6]. Note that the gain coefficient increases approximately linearly with pump–probe frequency separation up to a separation of about 500 cm^{-1} (15,000 GHz). The magnitude of the gain coefficient shown in Figure 5.1 is for a pump wavelength of 1 μm. The gain coefficient scales inversely with wavelength [6] so that at 1.5 μm, which is the wavelength region of interest in the various examples that follow, the peak Raman gain coefficient is about 7×10^{-12} cm/W. Several minor modifications to Eq. (1) must be made for it to be applicable to single-mode optical fibers [7]. The correct expression is

$$P_s(L) = P_s(0) \exp \frac{gL_eP}{bA_e} \tag{2}$$

Here, P/A_e is the pump intensity, where P is the injected pump power and A_e is the effective area of the propagating waves. A_e can be calculated exactly by means of overlap integrals [8–10] and depends on the mode overlap between the pump and probe waves as well as on the fiber V number. However, in general if the pump and probe wavelengths are comparable and both are slightly longer than the fiber cutoff wavelength, then $A_e \approx A$, where A is the core area of the fiber [11]. L_e, called the effective fiber length, replaces the actual length L in order to account for the exponential decay of the pump power due to fiber loss as the pump propagates along the fiber. A simple integration shows that

$$L_e = \frac{1 - e^{-\alpha L}}{\alpha} \tag{3}$$

where α is the loss coefficient of the fiber. For long fiber lengths L and $\alpha = 0.25$ dB/km, $L_e = 17.4$ km. The factor b accounts for the relative polarizations of pump and probe and the polarization properties of the fiber. In a polarization-maintaining fiber with pump and probe polarization the same, $b = 1$. In a conventional fiber that does not maintain polarization, $b = 2$; this will be used in all the discussions.

In a single-channel system only one wavelength of light is injected into the fiber, and apparently no degradation due to SRS will occur because there is no probe signal to be amplified. However, the spontaneous Raman scattering from the injected light will generate probe photons in the fiber and these will then be amplified. It has been shown both theoretically [7, 12, 13] and experimentally [14–16] that amplification of these probe photons will cause severe degradation (50% pump depletion) when

$$\frac{gL_eP}{bA_e} = 16 \tag{4}$$

For long fibers with $\alpha = 0.25$ dB/km and $A_e = 5 \times 10^{-7}$ cm^2, the injected signal power required to produce system degradation is 1.3 W. It is clear that SRS will not be a factor in single-channel silica-fiber-based lightwave systems.

5.2.1 Two-Channel Wavelength-Division Multiplexed Systems

In wavelength-division multiplexed (WDM) systems, the situation is fundamentally different, because light at numerous wavelengths is injected into the fiber and the signals at longer wavelengths will be amplified by the short-wavelength signals. Initially, consider a two-channel system [17] in which the two channels are spaced such that SRS couples the two channels. This assumption will usually be satisfied in the 1.5-μm region because the broad stimulated-Raman gain profile of silica (Fig. 5.1) will couple channels that are separated in wavelength by up to 100 nm. Let channel 1 (pump) operate at a wavelength λ_1, which is shorter than λ_2, the wavelength of channel 2 (probe). Assume initially that both channels have equal optical power injected into the fiber. Suppose that in a return-to-zero (RZ) modulation format the bit pattern of the two channels is as shown in Figure 5.2a. Then, schematically, the effect of SRS is to produce bit patterns as shown in Figure 5.2b. Thus far we have ignored the effects of dispersion. Note that whenever there is a mark in both channels, the pump channel (λ_1) is depleted and the probe channel (λ_2) is intensified due to SRS. If a space (zero light intensity) appears in either channel, no intensity change occurs. (In conventional crosstalk a mark in channel 1 can produce a signal in channel 2 even if there is no mark in channel 2.) Furthermore, the effects of SRS on the two channels are not symmetric. Channel 1 experiences a partial closing of the eye pattern due to the depletion of individual bits, whereas the opening of the eye in channel 2 is unaffected because in the worst case some of the bits are amplified while the rest of the bits are unaltered. Therefore, the overall effect of SRS in ASK systems is a degradation in signal-to-noise ratio for the pump channel but no penalty for the longer-wavelength channel.

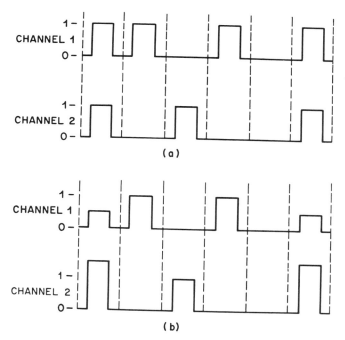

Figure 5.2 (a) Bit pattern in two-channel wavelength multiplexed system with no stimulated Raman interaction between channels. (b) Bit pattern with stimulated Raman interaction ($\lambda_1 < \lambda_2$) [17].

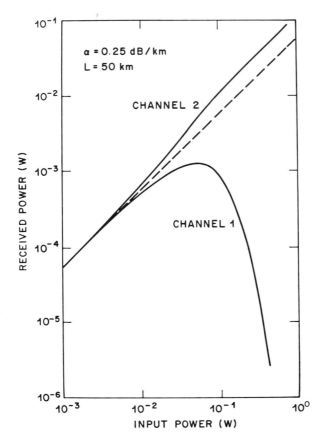

Figure 5.3 Received power versus input power of channel 1 (pump) and channel 2 (probe in presence of stimulated Raman interaction in a 50-km fiber with 0.25 dB/km loss [17].

The amplified probe power in the presence of pump depletion, and the depleted pump power, can be calculated numerically starting with Eq. (2). For a 50-km fiber with an effective core diameter of 8 μm, 0.25 dB/km loss, and a channel spacing corresponding to maximum Raman gain, the received pump (channel 1) and probe (channel 2) powers as a function of injected powers are shown in Figure 5.3. The broken line represents the received power with no Raman interaction between channels. Note that received power does not increase monotonically with input power. Increasing the transmitter power above 50 mW actually results in lower received power in the pump channel. Therefore, there is a maximum usable transmitter power that can be injected into the short-wavelength channel. This maximum power scales directly with fiber loss.

These results have been calculated assuming that the two channels have equal group velocities. In general, this assumption does not hold and a more detailed analysis is required. It has been shown [18] that in most instances the effect of group velocity dispersion on the nonlinear Raman interaction decreases the strength of the interaction by a factor between 1 and 2. For small dispersion and low bit rates, there is no reduction and the results in Figure 5.3 hold. For high bit rates and larger group velocity dispersion, the nonlinear interaction is reduced by a factor of 2. In this case the results of Figure 5.3 are valid if the values on the abscissa are increased twofold. Figure 5.3 can be generalized to a universal curve that describes the effects of SRS on two-channel systems for any core area, fiber loss, or gain coefficient. Figure 5.4 shows this universal curve (where the effective length is assumed to be $1/\alpha$). Once g, A, and α are specified, the abscissa corresponds to injected power, and plots such as Figure 5.3 result. With angle modulation the digital message to be conveyed is impressed on the phase

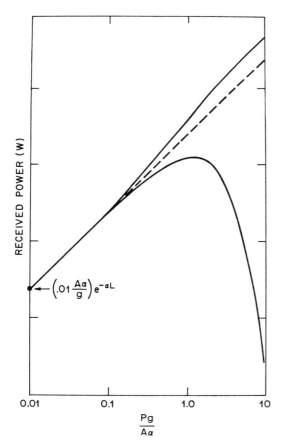

Figure 5.4 Universal curve of received power versus the ratio $Pg/\alpha A$ [17].

angle of the optical carrier rather than on the amplitude, which is nominally constant. In this case the gain in the probe channel is uniform rather than sporadic, because the optical intensities are unmodulated. Thus, one can obtain useful gain in the long-wavelength channel provided the concomitant loss in the short-wavelength channel can be tolerated.

The effects of SRS in two-channel configurations have been measured in several ways [19–21]. The crosstalk due to SRS was measured [20] using light from two injection lasers multiplexed on a fiber 21 km long. The power in the long-wavelength channel (1.34 μm) was deliberately reduced to 0.05 mW to avoid depletion of the short-wavelength channel (1.26 μm). Thus, the effects of SRS were monitored not by measuring the degradation of the short-wavelength channel but by measuring the amplification of the long-wavelength channel as a function of power in the short-wavelength channel. Both continuous wave (cw) and modulated (230-MHz square wave) experiments were performed. For 1 mW of power at 1.26 μm a crosstalk of -25 dB was measured, in good agreement with theoretical prediction.

The degradation of a short-wavelength channel due to depletion by a long-wavelength channel in a two-channel configuration was measured directly by determining power penalties from bit-error-rate (BER) curves [21]. Light from a DFB injection laser operating at 1.5 μm was transmitted through 43 km of fiber. The laser was modulated with a $2^{15} - 1$ pseudorandom bit stream at 1 Gb/s, and the BER was measured as a function of received power. Light from a color center laser (FCL) emitting up to 150 mW at 1.57 μm was then also injected into the fiber and the BER curves were measured for several FCL powers. The BER measurements displayed power penalties up to 2.5 dB. The measured power penalties corresponded to the observed depletion levels in the short-wavelength channels and agreed with the predicted degradations.

5.2.2 Multichannel Systems

Eventually, the complexity of WDM systems will undoubtedly increase, and there exists the prospect of very high levels of multiplexing in coherent systems. Larger numbers of channels can lead to SRS degradation at much lower power levels than in the two-channel case. In this section we estimate the degradation due to SRS in wavelength-multiplexed systems containing an arbitrary number of channels with arbitrary (but equal) channel spacings [22].

Assume N equally spaced channels with channel separation v (Hz). The degradation caused by SRS will be most severe for the shortest-wavelength channel (call this the zeroth channel). Therefore, system performance limits can be estimated by calculating the depletion of the zeroth channel. Maximum depletion will occur when there is a binary 1 (optical power "on") in each channel. In this case, assuming scrambled polarization and the Raman gain to be in the linear regime, the fractional power D lost by the zeroth channel is

$$D = \sum_{i=1}^{N-1} \frac{\lambda_i}{\lambda_o} P_i g_i \frac{L_e}{2A_e} \tag{5}$$

where P_i is the injected power (watts) in the ith channel, λ_i is the wavelength of the ith channel, and g_i is the Raman gain coefficient coupling the ith and zeroth channels. Equation (1) was derived assuming no Raman interactions among the other $N - 1$ channels; that is, the various P_i are assumed to remain constant over the length L_e. In the case of linear Raman gain (for $P_i < 10$ mW), the error arising from this assumption is less than 1%. To facilitate calculation assume that the λ_i and P_i are the same for all channels. In addition, let the actual Raman gain profile of silica (broken curve in Fig. 5.5) be approximated by a triangular function (solid curve), that is, the peak Raman gain coefficient occurs at 500 cm^{-1}, and assume there is no Raman interaction at larger channel separation. Consequently,

$$g_i = \frac{i\,\Delta v}{1.5 \times 10^{13}}\, g_p \qquad \text{for } i\,\Delta v < 1.5 \times 10^{13} \text{ Hz} \tag{6}$$

$$g_i = 0 \qquad \text{for } i\,\Delta v > 1.5 \times 10^{13} \text{ Hz} \tag{7}$$

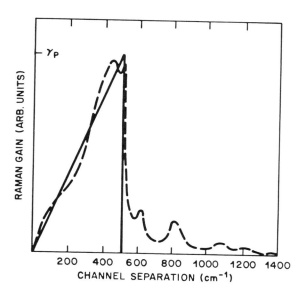

Figure 5.5 Actual Raman gain profile (broken curve) and approximation used in calculations (solid curve). 1 cm$^{-1} = 30$ GHz.

There are two possibilities to consider. First assume that all the channels fall within the Raman gain profile, that is, $(N - 1) \Delta v < 1.5 \times 10^{13}$ Hz; then

$$D_1 = \frac{\Delta v \, g_p PL_e}{3 \times 10^{13} \, A} \frac{N(N - 1)}{2} \tag{8}$$

For typical system parameters, that is, $\alpha = 0.25$ dB/km, $A = 5 \times 10^{-7}$ cm^2, $g_p = 6 \times 10^{-12}$ cm/W,

$$D_1 = 3.3 \times 10^{-13} \, P \, \Delta v \, N(N - 1) \tag{9}$$

For the second possibility assume $(N - 1) \Delta v > 1.5 \times 10^{13}$ Hz; only $M = 1.5 \times 10^{13}/\Delta v$ channels will contribute to depletion of the zeroth channel. In this case if $M \gg 1$,

$$D_2 = 7.4 \times 10^{13} \frac{P}{\Delta v} \tag{10}$$

and is independent of the total number of channels.

The power penalty X (dB) due to SRS is $X = -10 \log (1 - D)$. If we require $X < 0.5$, then the maximum allowable transmitter power per channel in the two cases is

$$P_1 < \frac{3.3 \times 10^{11}}{\Delta v \, N(N - 1)} \tag{11}$$

and

$$P_2 < 1.5 \times 10^{-15} \, \Delta v \tag{12}$$

These inequalities can now be used to estimate the transmitter power limits for a wide variety of wavelength-multiplexed systems. For example, in a 10-channel system with channel separation $\Delta \lambda = 10$ nm ($\Delta v = 1.3 \times 10^{12}$ Hz at 1.5 μm) the maximum allowable transmitter power per channel is [Eq. (11)] 3 mW. Note that when all channels fall within the Raman gain profile, the maximum allowable total optical injected into the fiber ($P_1 N$) scales inversely with the total occupied bandwidth $N \Delta v$. This can be summarized by

$$P_T \, (\text{BW}) < 3000 \text{ nm-mW} \tag{13}$$

where P_T is total optical power injected into all channels (mW) and BW is the total occupied optical bandwidth in nanometers.

In a recent wavelength-multiplexing experiment [23], 10 channels occupying a total optical bandwidth of 30 nm were multiplexed on a 68-km long fiber. The average total injected power in all channels was about 5 mW. Therefore, P_t BW = 150 nm-mW, well below the figure needed to produce a 0.5-dB penalty [Eq. (13)]. Indeed, no power penalty due to SRS was observed in the BER measurements [23].

5.2.3 Signal Amplification

There is considerable interest in exploiting SRS to amplify optical signals in fibers. Fiber-optic Raman amplification offers the possibility of replacing conventional regenerators with simple amplifiers consisting of a pump laser and a wavelength-selective coupler. This would lead to extremely large repeater spacings [24]. The transmission medium also serves as the amplification medium (this is referred to as active transmission lines), thereby eliminating various optical and electrical components normally required to regenerate or amplify a signal. Furthermore, fiber-optic Raman amplifiers have rather high saturation powers, which is important in amplifier chain configurations. A schematic diagram of a Raman amplifier is shown in Figure 5.6. The information-bearing signal is transmitted in the usual fashion. However, in addition to the light from the signal laser, shorter-wavelength pump light is injected into the fiber either copropagating with or counterpropagating to the signal wave. (In some applications, such as soliton amplification, it is desirable to inject both copropagating and counterpropagating pump light.) The pump and signal light is combined or separated by some wavelength-selective element such as a dichroic mirror, grating, or wavelength-selective coupler. The counterpropagating configuration probably is preferred because such a system is much less sensitive to pump power fluctuations. In effect, the signal channel sees an average pump intensity over the effective length of the fiber (~ 18 km) that equivalently corresponds to averaging the pump power for about 100 μs.

There are two major requirements on the pump used in an active transmission line. It must provide several hundred milliwatts of cw power, and it must have spectral properties that are not conducive to creation of stimulated Brillouin scattering (SBS). The latter point will be discussed in the

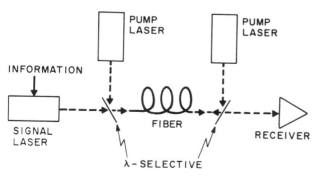

Figure 5.6 Schematic diagram of Raman amplification of propagating signal. Both forward and backward Raman amplification is shown.

next section. The cw power requirements initially limited demonstrations of active transmission lines to those regions of the spectrum near Nd:YAG laser lines, 1.06 μm [19, 25–27] and 1.32 μm [28–30]. The highest gains (up to 45 dB) were achieved with pulsed or Q-switched pump lasers [26]. However, pulsed amplification is not applicable in communication systems. Active transmission lines using cw Nd:YAG pump sources at 1.06 μm [25] and 1.32 μm [28] have provided optical gain of 3 dB at 1.118 μm and over 20 dB at 1.4 μm, respectively.

Recently the emphasis in amplification experiments has shifted to measuring not only optical gains but also bit error rates using a pseudorandom bit stream in the signal channel. A gain of 14 dB at 1.4 μm was achieved [29] using a cw modelocked Nd:YAG laser at 1.32 μm to amplify a 100-Mb/s NRZ signal. A small (1-dB) penalty was measured and attributed to fluctuations of the pump power. An active transmission line in the low-loss region of silica was demonstrated [31] using a tunable color center pump laser operating on two longitudinal modes at 1.48 μm. The 1.57-μm signal laser was modulated at 1 Gb/s with a pseudorandom bit stream. An optical gain of 5–8 dB was achieved, but a 1.4-dB power penalty attributed to back-scattered pump light reduced the net increase in receiver sensitivity to 4.4 dB.

The noise properties of Raman amplifiers have been measured [32] using a color center pump laser and a 1.55-μm DFB laser. The Raman amplifier was shown to have no excess noise for signal powers between -30 and -50 dBm using an APD receiver with an excess noise factor of 10. However, when signal levels are low, the Raman amplifier does add shot noise due to spontaneous Raman-scattered light. For low-bit-rate systems (< 500 Mb/s) optical filtering of the spontaneous Raman scattering is required [32]. The performance of Raman amplifiers, with respect to signal-to-noise ratio, is comparable to that of semiconductor amplifiers. The advantage Raman amplifiers possess over semiconductor laser amplifiers is their high saturation powers. Gain saturation in Raman amplifiers occurs when the amplified signal power becomes comparable to the pump power, which is typically a few hundred milliwatts. Semiconductor amplifiers saturate at powers between 0.1 and 1 mW [32]. The biggest problem confronting implementation of active transmission lines is the lack of suitable pump sources. Until a compact pump source (preferably an injection laser) is developed, Raman amplifiers are not practical for communications use.

Besides being potentially useful in active transmission lines, SRS in single-mode fibers has provided a convenient source of infrared light that can be used for diagnostic purposes in optical fibers. A few hundred meters of conventional single-mode fiber pumped by a Q-switched and/or modelocked 1.06-μm Nd:YAG laser will emit a wavelength continuum of light [33] between 1.1 and 1.7 μm (Fig. 5.7). After spectrometric filtering this light serves as a tunable source of infrared light. Infrared pulses as short as 100 ps or as

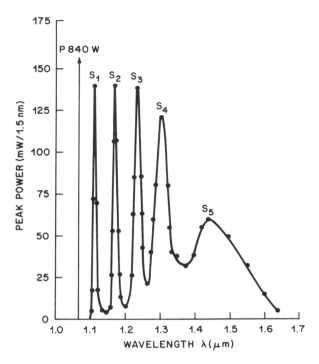

Figure 5.7 Spectral emission curve for a single-mode fiber Raman laser [33]. Power was measured through a monochromator with 1.5-nm resolution. The pump was a 1.064-μm wavelength Nd:YAG laser. Five orders of stimulated Stokes emission S_1 to S_5 are generated.

long as hundreds of nanoseconds are available from such a source. The short pulses can be used for measurements of group velocity dispersion [33], whereas the long pulses can be used as pump pulses in SRS amplification measurements [26, 27, 30]. Single-mode silica fibers can also be used as host matrices for diffused molecular species such as H_2 and D_2 [34, 35]. SRS from these molecules can provide extremely large single-step Raman shifts (3000–4000 cm^{-1}). For example, D_2 in silica-fiber Raman lasers converts 1.06-μm Nd:YAG light efficiently in a single-step process to 1.56-μm light [36, 37], which can be used for various measurement purposes. Pulses as short as 15 ps [36, 37] can be used to study nonlinear propagation effects, whereas pulses several hundred nanoseconds long [37] have been used for extremely long range optical time domain reflectometry [38]. Single-pass fiber Raman lasers all share the common quality that efficient conversion of pump light to Raman light can be realized at relatively low powers (hundreds of watts), orders of magnitude smaller than those required in bulk samples.

5.3 STIMULATED BRILLOUIN SCATTERING

Superficially, stimulated Brillouin scattering (SBS) [39] is similar to SRS except that SBS involves acoustic phonons rather than molecular vibrations. In this respect, both scattering processes are three-wave processes in which the incident (pump) light is converted into (Stokes) light of longer wavelength with a concomitant excitation of a molecular vibration (SRS) or acoustic phonon (SBS). However, there are a number of significant differences between SBS and SRS that lead to markedly different systems consequences.

First and foremost, the peak SBS gain coefficient in single-mode fibers is over two orders of magnitude larger ($g_B \approx 4 \times 10^{-9}$ cm/W) [9] than the gain coefficient for SRS. Consequently, under the proper conditions SBS will be the dominant nonlinear process. Second, the optical gain bandwidth Δv_R for SRS is on the order of 200 cm^{-1} FWHM (6×10^{12} Hz). Therefore, there is essentially no reduction in Raman gain for pump lasers with large line-widths. The optical bandwidth Δv_B for SBS, on the other hand, varies as λ^{-2} [40] and at 1.5 μm is about 20 MHz. Maximum SBS gain will occur for pump lasers with linewidths less than 20 MHz. For lasers with linewidths Δv_L larger than 20 MHz, SBS gain decreases as the ratio $\Delta v_B/\Delta v_L$, that is, $g = g_B \Delta v_B/\Delta v_L$ [9], where g_B is the peak steady-state Brillouin gain. Unlike SRS, which can occur in copropagating or counterpropagating geometries, SBS gain is zero in the forward direction and maximum in the backward direction, and therefore SBS occurs in the backward direction. This process obviously depletes the incident wave, and, in addition, a potentially strong scattered beam propagates back toward the transmitter [9, 13]. For 1.5-μm light, the backward beam is shifted by about 11 GHz from the pump beam and will probably affect the spectral properties and stability of the transmitter unless it is optically isolated.

Backward SBS was observed in experiments schematically depicted in Figure 5.8 [41, 42]. In one experiment [41], a dye laser with output wavelength of 0.7 μm was used. In the other [42], light from a narrow-linewidth single-frequency Nd:YAG laser was transmitted through a 13.6-km-long single-mode silica fiber. Both the backward-scattered light and the forward-transmitted light were monitored. Figure 5.9 shows the output power from each end of the fiber as a function of launched power. At low input power, the output power monitored in the backward direction was due only to the Fresnel reflection from the cleaved end of the fiber. At input powers above 5 mW the backward light increased rapidly in a nonlinear fashion due to SBS. An optical spectrum analyzer measured the frequency of the backscattered light to be shifted by 12.7 GHz from the pump frequency, in good agreement with the predicted value. The light transmitted through the fiber was related linearly to the input power at low incident powers. However, at input powers exceeding 6 mW the output power no longer increased linearly with input power, and at input powers above 10 mW the output

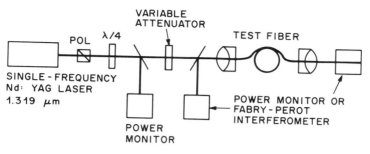

Figure 5.8 Experimental arrangement for observing stimulated Brillouin scattering [42].

reached a saturated maximum of about 2 mW. More recently [43] the onset of SBS at a wavelength of 1.52 μm was observed to occur in a 30-km fiber at input powers as low as 2 mW.

Although the onset of SBS has been shown to occur at very low incident laser powers, these experiments represent idealized situations in which the laser is operated in the cw mode with no information impressed on the optical wave. For systems considerations, the case of interest is light that is modulated (in amplitude, phase, or frequency) to transmit data. In this case the optical spectrum is significantly broadened and a reduction in stimulated Brillouin amplification can be expected. The analysis developed for SBS

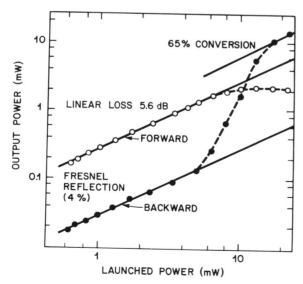

Figure 5.9 Power emitted from distant (forward) and near (backward) ends of 13.6-km single-mode fiber as a funcion of launched power [42].

generated by a narrow-linewidth source [44] can be extended [45, 46] to multimode sources and finally to sources with pseudorandom modulation [47] used in communications. The results depend on the particular encoding scheme used: amplitude shift keying (ASK), frequency-shift keying (FSK), or phase-shift keying (PSK), and on the ratio $B/\Delta v_B$, where B is the bit rate.

For ASK the launched field can be described by

$$E(t) = E_0\{1 - [1 - m(t)][1 - (1 - k_a)^{1/2}]\} \tag{15}$$

where the binary data stream is represented by the function $m(t)$ that can take values of 0 and 1 with equal probability, and k_a is the depth of intensity modulation ($0 < k_a \leq 1$). In this case the SBS gain is

$$g = g_B\left[\left(1 - \frac{a}{2}\right)^2 + \frac{a^2}{4}\left(1 - \frac{B}{\Delta v_B}(1 - e^{-\Delta v_B/B})\right)\right] \tag{16}$$

where $a = 1 - (1 - k_a)^{1/2}$.

The SBS gain for ASK can be minimized by using a 100% modulation depth ($k_a = 1$). For bit rates much smaller than the Brillouin linewidth $g \rightarrow g_B/2$. For high bit rates $g \rightarrow g_B/4$. The dependence of g on $B/\Delta v_B$ is summarized in Figure 5.10.

In PSK, the information is impressed on the phase of the electric field

$$E(t) = E_0 e^{i\phi(t)} \tag{17}$$

where $\phi(t) = k_p m(t)$ and k_p is the keyed phase shift. The SBS gain for PSK is

$$g = g_B\left[\frac{1}{2}(1 + \cos k_p) + \frac{1}{2}(1 - \cos k_p)\left[1 - \frac{B}{\Delta v_B}(1 - e^{\Delta v_B/B})\right]\right] \tag{18}$$

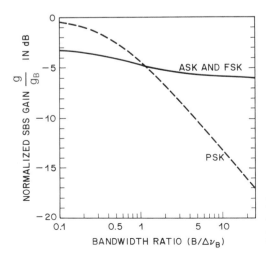

Figure 5.10 Normalized Brillouin gain as a function of $B/\Delta v_B$ [47]. For ASK $k_a = 1$ is assumed. For PSK $k_p = (2n + 1)\pi$ is assumed.

The gain for PSK is minimized for $k_p = \pi(2n + 1)$. For high bit rates the SBS gain decreases linearly with $B/\Delta v_B$ (Figure 5.10).

In wide-deviation FSK modulation the laser frequency is modulated between two relatively widely spaced frequencies ω_1 and ω_2, that is, the modulation depth $k_f = \omega_1 - \omega_2$ is at least several times the bit rate B. Consequently, the FSK spectrum is just the sum of two ASK spectra with $k_a = 1$ centered about ω_1 and ω_2. Therefore, the SBS gain for FSK is simply

$$g = g_B \left[\frac{1}{2} - \frac{B}{4\Delta v_B} (1 - e^{\Delta v_B/B}) \right] \tag{19}$$

As shown in Figure 5.10, the dependence of g on $B/\Delta v_B$ for FSK is the same as ASK. At high bit rates the SBS gain decreases to $\frac{1}{4}$ the cw value.

To summarize, SBS is a very strong nonlinear process that exhibits backward gain. This nonlinearity is most detrimental in systems employing narrow-linewidth lasers. In general, encoding pseudorandom data on the optical wave will reduce the effects of SBS. Maximum reduction occurs by using 100% modulation depth ($k_a = 1$) in ASK systems and $k_p = \pi(2n + 1)$ in PSK systems. The SBS gain decreases with increasing bit rates. In ASK and FSK systems the maximum reduction is a factor of 4. For high-bit-rate PSK systems the SBS gain decreases linearly with B.

Backward SBS has recently been exploited to amplify laser signals in single-mode silica fibers [48, 49]. The pump light was injected counterpropagating to the signal light at the receiver end of the fiber. Gains as high as 4.3 dB/per milliwatt of pump power were measured with a total gain of 18 dB. Bit-error-rate measurements at 100 Mb/s showed no penalty due to the amplification process. One advantage of SBS amplification, as with SRS amplification, is that the transmission medium is also the amplifier. However, this kind of amplifier, unlike an SRS amplifier, requires narrow-linewidth pump lasers and precise control of the difference frequency between the pump and transmitter laser.

5.4 CARRIER-INDUCED PHASE NOISE

An example of an optical nonlinearity that affects only the phase of the propagating signal is the optical Kerr effect. In most materials, the refractive index is not a constant but depends weakly on the intensity of the propagating light. Normally, extremely high optical intensities are required to induce significant changes in the refractive index. In silica optical fibers, however, even at small optical powers the minuscule change in refractive index is enhanced by the extremely long propagation lengths, and significant changes in the phase of the received optical fields can be observed. In angle-modulated systems, the digital message to be conveyed is impressed on the phase of the optical carrier rather than on the amplitude. Consequently, system performance will be degraded by any disturbances of the phase of the carrier caused by fluctuating laser power.

It can be shown [50, 51] that in single-channel systems the phase change in the received signal due to the nonlinear refractive index is linearly proportional to the optical power propagating in the fiber core. In particular, for long (i.e., >30 km) fibers the rms phase fluctuations σ_ϕ in the received signal is related to the rms power fluctuations σ_P:

$$\sigma_\phi = 0.02\sigma_P \tag{20}$$

where σ_ϕ is in radians and σ_P is in milliwatts.

Fluctuations measurements of InGaAsP injection lasers at power levels of a few milliwatts show $\sigma_P \approx 0.1$ mW [52]. These fluctuations increase roughly as the square root of the optical power [53], so that for transmitter powers up to 100 mW we expect the power fluctuations σ_P to be less than 1 mW.* The resultant phase noise [Eq. (20)] is less than 0.02 rad, which is negligibly small in angle-modulated systems [54] (0.15 rad of phase noise corresponds to a power penalty of roughly 0.5 dB).

Thus far, the phase effects of an optical beam acting on itself, that is, self-phase modulation, have been considered. In wavelength-multiplexed systems, in addition to self-phase modulation, there are cross-phase modulation effects due to power fluctuations in other optical channels. In a system with N channels, the rms phase fluctuations in a particular channel due to power fluctuations in the other channels is

$$\sigma_\phi = 0.04\sqrt{N}\sigma_P \tag{21}$$

were σ_ϕ is in radians and σ_P is in milliwatts. The power fluctuations σ_P in all the channels have been assumed to be the same.

For transmitter powers less than 25 mW, unacceptable levels of phase noise will occur when N is about 50, a rather sophisticated level of wavelength multiplexing.

Carrier-induced phase noise can be much greater, of course, in intensity-modulated systems because of large (intentional) changes occurring in the transmitter power. For incoherent demodulation these phase changes are unimportant, but they can cause serious impairment if the signal is coherently demodulated (e.g., ASK homodyne [55, 56]). Moreover, in wavelength-multiplexed systems, the worst-case phase shifts will grow linearly with the number of channels N unlike the $\sqrt{N}$ behavior associated with angle-modulated system (Eq. [R].) This phase shift is given by [57]

$$\Delta\phi = 0.04P(N - 1) \tag{22}$$

---

* For the approximate calculations we assume that the bandwidth of the power fluctuations is comparable to or less than the information bandwidth of the transmission system. This is a reasonable approximation for data rates in the gigabit-per-second range.

where P is the power per channel. To limit the degradation due to phase crosstalk to an acceptable level, the power in each channel is restricted to

$$P < \frac{3.5}{N-1} \, \text{mW} \tag{23}$$

This restriction on P may render coherent detection of ASK unattractive for wavelength-multiplexed systems applications. In frequency-modulated systems, the amount of residual amplitude modulation is restricted by similar arguments.

Cross-phase modulation has been experimentally observed using light from two conventional InGaAsP injection lasers multiplexed on a 15-km-long single-mode fiber [57]. In the experiment to measure the effects of cross-phase modulation, a novel self-reflexive interferometer was employed. The channel 1 source was a cw 1.5 μm InGaAsP single-frequency distributed-feedback laser with a coherence length of several meters, and channel 2 used a 1.3-μm InGaAsP V-groove buried-crescent multifrequency laser. The two beams were combined by a dichroic mirror and coupled into a 15-km-long depressed-step-index single-mode fiber. The effects of on–off modulation in channel 2 on the phase of the cw light in channel 1 were measured. A 1-mW change in the power of channel 2 produced a 0.024-rad (1.4°) phase shift in channel 1. The predicted value from Eq. (23) is 0.022 rad.

5.5 CONCLUSION

Optical nonlinearities will ultimately limit the optical power and information capacity of lightwave systems. These nonlinearities manifest themselves in silica fibers at modest powers because of the long interaction lengths and small beam diameters in optical fibers. Typical system degradations caused by nonlinearities are loss of power, excess phase noise, frequency conversion, and crosstalk in multiplexed systems. The onset of these nonlinearities occurs at powers between several and several hundred milliwatts depending on the nonlinearity and the particular system configuration. The detrimental effects of some nonlinearities can be reduced by judicious choice of system parameters. A fact that is not widely appreciated is that fused silica has relatively weak optical nonlinear characteristics. Other materials, for example, some of the new midinfrared fiber materials, have much stronger nonlinearities. Consequently, nonlinear effects will occur at lower optical powers. In addition, if the ultralow losses of the midinfrared materials are ever realized, the interaction lengths for nonlinear interactions will be much larger than in present silica fibers. This will also lead to lower threshold powers.

On the other hand, optical nonlinearities can be exploited to enhance system performance or can even be the basis for radically new types of systems. Examples of this are soliton systems [58] and nonlinear wavelength multiplexed networks [59–61].

REFERENCES

1. N. Bloembergen, "The stimulated Raman effect," *Am. J. Phys.*, **35**: 989(1967).
2. W. Kaiser and M. Maier, "Stimulated Rayleigh, Brillouin and Raman spectroscopy," in *Laser Handbook*, F. T. Arecchi and E. O. Schulz-Dubois (Eds.), North-Holland, Amsterdam, 1970, p. 1077.
3. A. Yariv, *Quantum Electronics*, 2d ed., Wiley, New York, 1975.
4. B. P. Stoicheff, "Characteristics of stimulated Raman radiation generated by coherent light," *Phys. Lett.*, **7**: 186 (1963).
5. R. Shuker and R. W. Gammon, "Raman-scattering selection-rule breaking and the density of states in amorphous materials," *Phys. Rev. Lett.*, **25**: 222 (1970).
6. R. H. Stolen and E. P. Ippen, "Raman gain in glass optical waveguides," *Appl. Phys. Lett.*, **22**: 276 (1973).
7. R. G. Smith, "Optical power handling capacity of low-loss optical fibers as determined by stimulated Raman and Brillouin scattering," *Appl. Opt.*, **11**: 2489 (1972).
8. K. O. Hill, D. C. Johnson, B. S. Kawasaki, and R. I. MacDonald, "CW three-wave mixing in single-mode optical fibers," *J. Appl. Phys.*, **49**: 5098 (1978).
9. R. H. Stolen, "Nonlinear properties of optical fibers," in *Optical Fiber Telecommunications*, S. E. Miller and A. G. Chynoweth (Eds.), Academic, New York, 1979, p. 130.
10. R. H. Stolen and J. E. Bjorkholm, "Parametric amplification and frequency conversion in optical fibers," *IEEE J. Quantum Electron.*, **QE-18**: 1062–1072 (1982).
11. D. C. Johnson, K. O. Hill, and B. S. Kawasaki, "Brillouin optical-fiber ring oscillator design," *Radio Sci.*, **12**: 519 (1977).
12. J. Auyeung and A. Yariv, "Spontaneous and stimulated Raman scattering in long low-loss fibers," *IEEE J. Quantum Electron.*, **QE-14**: 347 (1978).
13. R. H. Stolen, "Nonlinearity in fiber transmission," *Proc. IEEE*, **68**: 1232 (1980).
14. Y. Ohmori, Y. Sasaki, M. Kawachi, and T. Edahiro, "Fibre-length dependence of critical power for stimulated Raman scattering," *Electron. Lett.*, **17**: 593 (1981).
15. M. Ikeda, "Spectral power handling capability caused by stimulated Raman scattering effect in silica optical fibers," *Opt. Commun.*, **37**: 388 (1981).
16. R. H. Stolen, Clinton Lee, and R. K. Jain, "Development of the stimulated Raman spectrum in single-mode silica fibers," *J. Opt. Soc. Am. B*, **L**: 652 (1984).
17. A. R. Chraplyvy and P. S. Henry, "Performance degradation due to stimulated Raman scattering in wavelength-division-multiplexed optical-fibre systems," *Electron. Lett*, **19**: 641 (1983).
18. D. Cotter and A. M. Hill, "Stimulated Raman crosstalk in optical transmission: effects of group velocity dispersion," *Electron. Lett.*, **20**: 185 (1984).
19. M. Ikeda, "Stimulated Raman amplification characteristics in long span single-mode silica fibers," *Opt. Commun.*, **39**: 148 (1981).
20. A. Tomita, "Crosstalk caused by stimulated Raman scattering in single-mode wavelength-division multiplexed systems," *Opt. Lett.*, **8**: 412 (1983).
21. J. Hegarty, N. A. Olsson, and M. McGlashan-Powell, "Measurement of the Raman crosstalk at 1.5 μm in a wavelength-division-multiplexed transmission system," *Electron Lett.*, **21**: 395 (1985).
22. A. R. Chraplyvy, "Optical power limits in multichannel wavelength-division-multiplexed systems due to stimulated Raman scattering," *Electron. Lett.*, **20**: 58 (1984).
23. N. A. Olsson, J. Hegarty, R. A. Logan, L. F. Johnson, K. L. Walker, L. G. Cohen, B. L. Kasper, and J. C. Campbell, "68.3-km transmission with 1.37 Tbitkm/sec capacity using wavelength-division-muliplexing of ten single-frequency lasers at 1.5 μm," *Electron. Lett.*, **21**: 105 (1985).
24. K. Mochizuki, "Optical fiber transmission systems using stimulated Raman scattering: theory," *J. Lightwave Technol.*, **LT-3**: 688 (1985).

25. G. A. Koepf, D. M. Kalen, and K. H. Greene, "Raman amplification at 1.118 μm in single-mode fibre and its limitation by Brillouin scattering," *Electron. Lett.*, **18**: 942 (1982).
26. E. Desurvire, M. Papuchon, J. P. Pocholle, and J. Raffy, "High gain optical amplification of laser diode signal by Raman scattering in single-mode fibres," *Electron. Lett.*, **19**: 751 (1983).
27. M. Nakazawa, M. Tokuda, Y. Negishi, and N. Uchida, "Active transmission line: light amplification by backward-stimulated Raman scattering in polarization-maintaining optical fiber," *J. Opt. Soc. Am. B*, **1**: 80 (1984).
28. Y. Aoki, S. Kishida, H. Honmou, K. Washio, and M. Sugimoto, "Efficient backward and forward pumping cw Raman amplification for InGaAsP laser light in silica fibers," *Electron. Lett.*, **19**: 620 (1983).
29. Y. Aoki, S. Kishida, K. Washio, and K. Minemura, "Bit error rate evaluation of optical signals amplified via stimulated Raman process in an optical fiber," *Tech. Digest*, Conf. Opt, Fiber Commun. paper Tu04, San Diego, 1985.
30. M. Nakazawa, "Highly efficient Raman amplification in a polarization-preserving optical fiber," *Appl. Phys. Lett.*, **46**: 628 (1985).
31. J. Hegarty, N. A. Olsson, and L. Goldner, "CW pumped Raman preamplifier in a 45-km long fibre transmission system operating at 1.5 μm and 1 Gbit/s," *Electron Lett.*, **21**: 290 (1985).
32. N. A. Olsson and J. Hegarty, "Noise properties of a Raman amplifier," *J. Lightwave Technol.*, **LT-44**: 396 (1986).
33. L. G. Cohen and Chinlon Lin, "A universal fiber optic measurement system based on a near-IR fiber Raman laser," *IEEE J. Quantum Electron.*, **QE-14**: 855 (1978).
34. J. Stone, A. R. Chraplyvy, and C. A. Burrus, "Gas-in-glass—a new Raman-gain medium: molecular hydrogen in solid-silica optical fibers," *Opt. Lett.*, **7**: 297 (1982).
35. A. R. Chraplyvy, J. Stone, and C. A. Burrus, "Optical gain exceeding 35 dB at 1.56 μm due to stimulated Raman scattering by molecular D_2 in a solid silica optical fiber," *Opt. Lett.*, **8**: 415 (1983).
36. A. R. Chraplyvy and J. Stone, "Synchronously pumped D_2 gas-in-glass fiber Raman laser operating at 1.56 μm," *Opt. Lett.*, **9**: 241 (1984).
37. A. R. Chraplyvy and J. Stone, "Single-pass mode-locked or Q-switched pump operation of D_2 gas-in-glass fiber Raman lasers operating at 1.56 μm wavelength", *Opt. Lett.*, **10**: 344 (1985).
38. J. Stone, A. R. Chraplyvy, and B. L. Kasper, "Long-range 1.5 μm OTDR in a single-mode fibre using a D_2 gas-in-glass laser or a semiconductor laser," *Electron. Lett.*, **21**: 541 (1985).
39. E. P. Ippen and R. H. Stolen, "Stimulated Brillouin scattering in optical fibers," *Appl. Phys. Lett.*, **21**: 539 (1972).
40. D. Heinman D.S. Hamilton, and R. W. Hellwarth, "Brillouin scattering measurements on optical glasses," *Phys. Rev. B*, **19**: 6583 (1979).
41. N. Uesugi, M. Ikeda, and Y. Sasaki, "Maximum single-frequency input power in a long optical fiber determined by stimulated Brillouin scattering," *Electron. Lett.*, **17**: 379 (1981).
42. D. Cotter, "Observation of stimulated Brillouin scattering in low-loss silica fibre at 1.3 μm," *Electron. Lett.*, **18**: 495 (1982).
43. D. Cotter, "Optical nonlinearity in fibers: a new factor in systems designs," *Br. Telecom Technol. J.*, **1**: 17 (1983).
44. C. L. Tang, "Saturation and spectral characteristics of the Stokes emission in the stimulated Brillouin process," *J. Appl. Phys.*, **37**: 2945 (1966).
45. Y. Aoki and K. Tajima, "Dependence of the stimulated Brillouin scattering threshold in single-mode fibers on the number of longitudinal modes of a pump laser," *Tech Digest*, Conference on Lasers and Electrooptics, Paper TuHH5, Baltimore, (1987).
46. E. Lichtman and A. A. Friesem, "Stimulated Brillouin scattering excited by a multimode laser in single-mode optical fibers," *Opt. Comm.*, **64**: 544 (1987).
47. E. Lichtman, R. G. Waarts, and A. A. Friesem, "Stimulated Brillouin scattering excited by a modulated pump wave in single-mode fibers," *J. Lightwave Technol.*, **7**: 1 (1989).

48. N. A. Olsson and J. V. vanderZiel, "Cancellation of fiber loss by semiconductor laser pumped Brillouin amplification at 1.5 μm," *Appl. Phys. Lett.*, **48**: 1329 (1986).

49. N. A. Olsson and J. P. vanderZiel, "Fiber Brillouin amplifier with electronically controlled bandwidth," *Electron Lett.*, **22**: 488 (1986).

50. A. R. Chraplyvy, D. M. Marcuse, and P. S. Henry, "Carrier induced phase noise in angle-modulated optical-fiber systems," *J. Lightwave Technol.*, **LT-2**: 6 (1984).

51. R. H. Stolen and C. Lin, "Self-phase modulation in silica optical fibers," *Phys. Rev. A*, **17**: 1448 (1978).

52. P.-L. Liu, J.-S. Ko, I. P. Kaminow, T. P. Lee, and C. A. Burrus, "Steady-state intensity fluctuations, photon statistics and mode partitioning of injection lasers," *Tech. Digest*, 6th Topical Meeting, Opt. Fiber Commun., New Orleans, 1983, paper PD-3.

53. Y. Yamamoto, S. Saito, and T. Mukai, "AM and FM quantum noise in semiconductor lasers. Part II," *IEEE J. Quantum Electron.*, **QE-19**: 47 (1983).

54. V. K. Prabhu, "PSK performance with imperfect carrier phase recovery," *IEEE Trans. Aero Electron. Syst.*, **AES-12**: 275 (1976).

55. Y. Yamamoto, "Receiver performance and evaluation of various digital optical modulation-demodulation systems in the 0.5–10 μm wavelength region," *IEEE J. Quant. Electron.*, **QE-16**: 1251 (1980).

56. T. G. Hodgkinson, R. Wyatt, and D. W. Smith, "Experimental assessment of a 140 Mb/s coherent optical receiver at 1.52 μm, *Electron. Lett.*, **18**: 523 (1982).

57. A. R. Chraplyvy and J. Stone, "Measurement of crossphase modulation in coherent wavelength-division multiplexing using injection lasers," *Electron. Lett.*, **20**: 996 (1984).

58. L. F. Mollenauer and K. Smith, "Demonstration of soliton transmission over more than 4,000 km in fiber with loss periodically compensated by Raman gain," *Opt. Lett.*, **13**: 675 (1988).

59. A. R. Chraplyvy and R. W. Tkach, "Narrowband tunable optical filter for channel selection in densely-packed WDM systems," *Electron. Lett.*, **22**: 1084 (1986).

60. R. W. Tkach, A. R. Chraplyvy, R. M. Derosier, and H. T. Shang, "Optical demodulation and amplification of FSK signals using AlGaAs lasers," *Electron. Lett.*, **24**: 260 (1988).

61. R. W. Tkach, A. R. Chraplyvy, and R. M. Derosier, "Performance of a WDM network based on stimulated Brillouin scattering," to be published in *Electron. Lett.*, (1989).

PART 2

Fiber-Jointing Technology and Passive Optical Components

CHAPTER 6

Optical Fiber Connectors, Splices, and Jointing Technology

W. C. Young

6.1 INTRODUCTION

In recent years the state of the art of optical fiber technology has progressed to where the achievable attenuation levels for the fibers are very near the limitations due to Rayleigh scattering. As a result, optical fibers, and particularly single-mode fibers, can be routinely fabricated with attenuation levels of about 0.5 dB/km at 1300 nm and 0.25 dB/km at 1550 nm. Employing these fibers in lightwave systems requires precise jointing devices such as connectors and splices. Considering the small size of the fiber cores, less than 10 μm in diameter for single-mode fibers and less than 100 μm for multimode fibers, it is not surprising that these components can easily introduce high optical losses. Furthermore, since single-mode fibers have practically unlimited bandwidth, they have recently become the favorite choice for most of the lightwave systems presently being designed for telecommunication networks and in the future may be used in local area networks as well. To provide low-loss connectors and splices for these single-mode fibers, alignment accuracies in the submicrometer range are required, and these submicrometer alignments must be both reliable and cost-effective. Achieving these goals is presently the challenge facing the jointing technologist.

This chapter will review the fundamental technology presently used for both demountable connectors and splices. In particular, since single-mode fibers will more than likely dominate lightwave systems and require the greatest precision, most of our attention will be directed to the particular problems encountered in the jointing of these fibers.

We begin by defining the implied differences between a fiber connector and a fiber splice. The term *connector* is commonly used when referring to the jointing of two fibers in a manner that permits and anticipates their unjointing by its design intent. Connectors are usually used for terminating components, for system configuration, testing, and maintenance. Generally, the connectors are either factory-installed or field-installed depending on the particular application. In this sense, the term *field* usually refers to an environment outside the connector factory. However, in the case of single-mode fibers, the required submircometer alignment tolerance generally dictates that the connector installation be done in the connector factory. Presently, this limitation hampers single-mode fiber installation, although it is expected that recent results in this area by research laboratories will soon become common installation practices.

In contrast with the term connector, the term *splice* is commonly used when referring to the jointing of two fibers in a manner that does not lend itself to unjointing. Splices are usually used when the total span length can be realized only by the concatenation of shorter fiber sections. The splicing may be done either in the factory or during cable installation as required by practical fabrication and installation processes. Splices are also used for repairing broken or damaged fiber or cable lengths. In applications using single-mode fibers, splicing is also being used to attach preconnectorized short lengths of fibers (pigtails) to the ends of installed cables, fiber-terminated lasers, and other components terminated with single-mode fibers. As will be pointed out later, the practice of splicing preconnectorized single-mode fiber pigtails onto cable ends and component pigtails will probably be replaced by field-installing single-mode fiber connectors directly onto the fibers as is presently the practice in multimode fiber applications.

6.2 FACTORS CAUSING OPTICAL LOSSES

Factors causing optical losses (low coupling efficiency) in both connectors and splices can be conveniently divided into two groups (Table 6.1). Factors extrinsic to the optical fiber, both single-mode and multimode, such as lateral offset between fiber cores, longitudinal offset (end gap), angular misalignment (tilt), end-face quality, and reflections, are directly caused by the techniques used to join the fibers. The other group, intrinsic factors, are directly related to the particular properties of the two optical fibers that are joined. In particular, these intrinsic factors include mismatches in fiber core diameters, mismatches in index profiles, and the ellipticity of the cores.

TABLE 6.1. Factors causing coupling losses in fiber joints

Extrinsic factors
Lateral fiber-core offset (offset)
Longitudinal fiber-core offset (end gap)
Angular misalignments (tilt)
End-face quality
Reflections

Intrinsic factors
Mismatch in fiber core diameters
Mismatch in index profiles
Ellipticity of cores

In addition to these extrinsic and intrinsic factors, the coupling efficiency of a joint may also depend on the characteristics of the optical source, such as the center wavelength and coherence length, and, in the case of multimode fibers, on the relative locations of the optical source and the joint as well as the characteristics of the fibers, such as mode mixing and differential mode attenuation. For example, if the source is an incoherent LED and the length of continuous fiber before the joint is sufficient so that the modal power distribution has reached a steady state, the effect due to offsets is less than when the fiber is sufficiently short such that the steady-state mode distribution is not achieved and an overmoded state still exists. Therefore, when using multimode fibers and incoherent sources, the effects of overmoded or uniform excitation conditions and the steady-state conditions must be understood and considered when predicting the coupling efficiencies of connectors and splices. Because of these mode distribution effects, a much-practiced method of evaluating joints and other components is to use an overmoded launch followed by a mode filter that selectively filters the higher-order modes, thereby creating an approximate steady-state mode distribution. A further discussion on these launch conditions and their consequences will follow, but for now we will consider two distinct launch conditions for graded-index multimode fibers, namely, the uniform launch and steady-state launch conditions, and their effect on coupling efficiency.

6.2.1 Extrinsic Factors

Various experiments [1, 2] and analytical models [3, 4], for both single-mode and multimode fibers, have been used to quantify the effect that extrinsic factors have on the coupling losses of fiber joints. On the basis of these

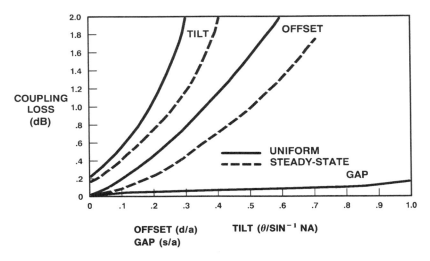

Figure 6.1 Coupling loss as a function of normalized lateral, longitudinal, and angular offset for graded-index multimode fibers.

studies we have plotted the coupling loss as a function of lateral, longitudinal, and angular offset for both graded-index multimode fiber (Fig. 6.1) and single-mode fiber (Fig. 6.2). In the multimode fiber's case, Figure 6.1 shows the effect of the misalignments for both uniform and steady-state launch conditions. From these curves it is clear that as the quality of the

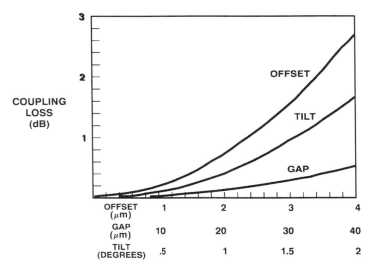

Figure 6.2 Coupling loss as a function of lateral, longitudinal, and angular offset for single-mode fiber having a mode field diameter of 10 μm.

alignment improves, the effect of the presence of higher-order modes diminishes. It must be reemphasized that it is very important to know the mode distribution that will be present in the particular application of multimode connectors and splices to correctly predict the performance of a particular connector design or splicing technique. Also, from these two figures it can be seen that the coupling loss is significantly more sensitive to lateral misalignments, for both single-mode and multimode fibers, than either longitudinal or angular misalignments. For example, for low-loss joints (0.5 dB), lateral offsets must be controlled to submicrometer accuracies for single-mode fibers and to micrometer accuracies for multimode fibers.

With regard to fiber end-face quality, flat mirrored finishes having end-face angles of less than 1° can be achieved with various fiber-cleaving tools. Although these cleaved ends are usually used with splicing procedures, lately there has been a trend to also employ cleaved fibers in connector installation procedures. In general, though, connectorization employs polishing procedures for end-face finishing. These polishing procedures usually require special tools for controlling end-face angles and sometimes for length control as well, and the tooling and procedures are usually designed to produce the desired end-face flatness and surface finish. In some cases the ends are polished slightly concave so that the fibers in the assembled connector cannot contact each other. In other cases, the tools are designed to create a flat and even sometimes a convex surface to guarantee fiber end-face contact. Finally, the end faces can also be polished at a slight angle to maximize the return loss of the connector. The surface finish that is generally acceptable for low-loss connectors is about 0.025 μm for non-index-matching connectors and can be as much as 0.18 μm for connectors employing index-matching fluids. The end face shown in Figure 6.3 is usually acceptable for a low-loss non-

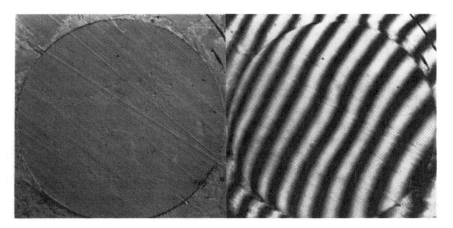

Figure 6.3 Microinterferogram of connector/fiber end face.

index-matching connector. The surface finish is about 0.025 μm, and the flatness shown in the microinterferogram is less than 0.25 μm.

The final extrinsic factor to be discussed is the effect of reflections on coupling efficiency. For joints having an air gap between the adjacent fibers, Fresnel reflections exist, and in this case for the two air–glass interfaces the loss amounts to about 0.31 dB for silica fibers. Furthermore, when these gaps are small (≲ 10 μm for LED sources and significantly larger for LD sources), interference effects can exist that may have an additional effect on the coupling efficiency [5]. While maximum coupling efficiency can be achieved when the fiber end faces .are in contact, small gaps can cause multiple reflections, and the resulting loss can be as high as 0.7 dB. Figure 6.4 shows the effect of small changes in the air gap between two single-mode fibers on the coupling efficiency of the joint. It should be noted that these reflections

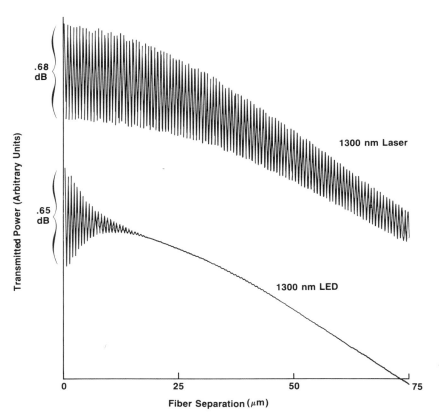

Figure 6.4 Transmission versus fiber end separation for single-mode fiber at a wavelength of 1300 nm.

can be minimized by using an index-matching medium (fluid, gel, etc.) between the fiber end faces. This approach is quite common in the splicing of fibers, but due to practical considerations like cleanliness and contamination, it is usually not followed in demountable connectors.

6.2.2 Intrinsic Loss Factors

Like extrinsic factors, intrinsic factors can have a large effect on the coupling efficiency of fiber joints. While extrinsic factors have a less significant effect on the loss of multimode fiber joints than on single-mode fiber joints, intrinsic factors have the opposite effect. That is, when evaluating the coupling efficiency of multimode fiber joints, one must consider the characteristics of the fibers on either side of the joint, and the direction of propagation of the optical power through the joint must also be taken into account. In the single-mode fiber's case, the differences in the characteristics of the fibers has a practically insignificant effect on the coupling efficiency, particularly when compared with the effect caused by submicrometer-type lateral misalignments. Also, with practical single-mode fiber joints, the law of reciprocity applies, and therefore the coupling efficiency is independent of the direction of propagation through the joint.

Again, various experiments [6] and analytical models have been used to quantify these effects. First, for the case of multimode fiber joints, the dependence of the coupling loss on mismatches in numerical apertures and core radii is summarized in Figure 6.5. It should be noted that this figure

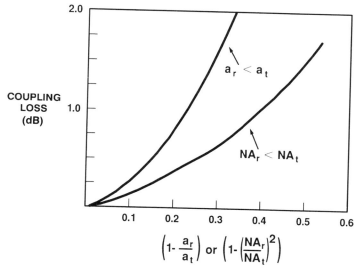

Figure 6.5 Coupling loss as a function of mismatch in numerical aperture and core diameters for graded-index multimode fibers.

shows the dependence of loss as the optical power propagates from larger values of numerical aperture (NA) to smaller values and from larger core radii to smaller core radii. When the direction of propagation is from smaller to larger NAs and core radii, there is no associated loss.

In the case of single-mode fiber joints, it has been shown [7] that the fields of most single-mode fibers being used today, those designed for use at 1300 nm, are nearly Gaussian, and therefore the coupling losses for the joints can be calculated by evaluating the coupling between two misaligned Gaussian beams. On the basis of this model the following general equation has been derived [8] for calculating the coupling loss between single-mode fibers (Fig. 6.6) that have unequal mode-field diameters (intrinsic factor) and lateral, longitudinal, and angular offsets as well as reflections (extrinsic factors):

$$\text{Coupling loss} = -10 \log \left(\frac{16n_1^2 n_2^2}{(n_1 + n_2)^4} \frac{4\sigma}{q} \exp \frac{-\rho u}{q} \right) \text{dB} \qquad (1)$$

where

$$\rho = (kw_1)^2 \qquad q = G^2 + (\sigma + 1)^2$$

$$u = (\sigma + 1)F^2 + 2\sigma FG \sin \theta + \sigma(G^2 + \sigma + 1) \sin^2 \theta$$

$$F = \frac{x}{kw_1^2} \qquad G = \frac{z}{kw_1^2}$$

$$\sigma = \left(\frac{w_2}{w_1} \right)^2 \qquad k = \frac{2\pi n_2}{\lambda}$$

n_1 and n_2 are the refractive index of the fibers and the medium between the fibers, respectively; λ is the wavelength of the source; x and z are the lateral and longitudinal offset, respectively; θ is the angular misalignment and w_1 and w_2 are the spot size of the transmitting and receiving fiber ($1/e$ power, radius), respectively.

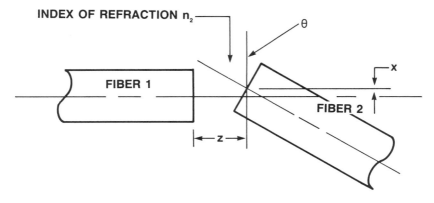

Figure 6.6 Typical offsets for butt-jointed fibers. n_1 is the index of refraction for the fiber; s_1 and s_2 are the spot sizes of fibers 1 and 2, respectively, $1/e$ power (radius).

It should be noted that Eq. (1) does not consider interference effects that may exist in joints with fiber end-face separation that do not employ index-matching fluids or gels.

Using this general equation, which has been found to have very good correlation with various experimental investigations, one can determine the coupling loss due to intrinsic factors. Assuming that no losses are present due to extrinsic factors, Eq. (1) reduces to

$$\text{Coupling loss} = -10 \log \left[4 \left(\frac{W_2}{W_1} + \frac{W_1}{W_2} \right)^{-2} \right] (\text{dB}) \qquad (2)$$

With this expression, the loss due to a 10% mismatch in mode-field diameters (intrinsic factor), a typical value for today's fibers, and with no other factors present is calculated to be 0.05 dB. It can also be seen in this expression that the loss is independent of the direction of propagation through the single-mode fiber joint.

6.3 INSERTION LOSSES

Unfortunately, for multimode fiber joints there exists no unique insertion loss value that is valid for all applications. The optical source, the number of joints and their location along the fiber, and the mode-mixing properties and differential mode attenuation of the particular fibers all play an important role in the performance, or, more correctly, the apparent performance, of the joint. Besides the previously mentioned effect that the actual incident mode distribution at the joint has on the insertion loss (due to extrinsic loss factors), for special cases additional effects can be realized. For example, consider the case when a long length of fiber exists after a joint and there is also a parameter mismatch between the transmitting and receiving fibers. In the case when the parameter mismatches cause the transmitting fiber to underfill the numerical aperture of the receiving fiber (i.e., the receiving fiber has a larger NA and/or core size), the actual loss of the receiving fiber may be much less than the steady-state loss of the fiber that is usually used in fiber characterization. In fact, when the extrinsic loss of the joint is very small, the decrease in fiber loss may be greater than the joint loss and result in an apparent negative loss for the joint. Furthermore, this redistribution of modal power may cause a substantial change in the incident mode distribution at the next joint and result in a different joint loss than the expected steady-state loss. For these reasons and others, it is not easy to assign a unique value of insertion loss for multimode fiber joints, and therefore when one evaluates joints, or plans a fiber transmission system, one must have a thorough understanding of the total system configuration.

Compared to multimode systems, with single-mode systems it is straightforward to evaluate and predict the insertion loss of fiber joints. Therefore, it is possible to include the measured value of insertion loss of single-mode fiber joints directly in the system loss budget. Since only one mode is propagating, the source and length dependence associated with multimode fibers is no longer a concern. The only condition that must be satisfied is that only the fundamental mode propagates both before and after the joint. However, with single-mode fibers there exists a wavelength at which the first higher-order mode is cut off or sufficiently attenuated such that above this wavelength the fiber guides only the fundamental mode [8] and below this wavelength two modes propagate. Theoretically, this cutoff wavelength is defined as the wavelength at which the mode index equals the cladding index. Unfortunately, the theoretical cutoff wavelength is very difficult to measure, since various loss mechanisms such as microbending, curvature, and stress that exist in practical systems strongly increase the loss of the first high-order mode. Consequently, the theoretical cutoff wavelength has little significance in practical fiber systems, and a more useful concept of defining an effective cutoff wavelength has been adopted. The *effective cutoff wavelength* is defined as the wavelength above which the power in the first higher-order mode is below a specified level when a specified length of fiber is deployed with a controlled radius of curvature. (CCITT G.652 recommends a 2-m length and a radius of curvature of 14 cm.) Therefore, it should be noted that for single-mode fiber joints having sufficiently large offsets, it is possible to launch power into the first higher-order mode, and if the length of fiber between the joints is not long enough or curved enough to attenuate the higher-order mode, then degradation in performance due to modal noise may result. For this to happen, the second joint would also have to have a substantial offset. Usually, with well-designed fibers (cutoff wavelength) and low-loss joints this type of modal-noise problem does not exist.

6.4 CONNECTORS

Low-loss connectors for lightwave systems, both multimode and single-mode, require precise alignments to control the extrinsic factors discussed in Section 6.2.1. These precise alignments must also be maintained under various operating conditions such as shock, vibration, and repetitive engagement and separation, as well as when changes occur in environmental conditions such as temperature and humidity. To be practical, these connectors must permit easy and simple installation outside the environment of the connector factory. To achieve these goals, considerable work has been directed at two types of connector designs—the butt-joint design and the expanded-beam design.

6.4.1 Butt-Joint Connectors

In the discussion of butt-joint connectors, we will concentrate on single-mode fiber requirements, since the goal, precise alignment, is the same as for multimode fiber except that the required alignment accuracy is extended to submicrometer accuracies instead of micrometer accuracies.

As previously explained, the most critical and also the most difficult alignment to achieve is the lateral offset between adjacent fiber cores. The resulting lateral offset in a connector assembly is influenced by the fiber's core/cladding offset and outside diameter variations, and in the connector by the hole diameter variation and the hole-to-reference surface offset. In evaluating and designing various connector parts, the combined effect of these inaccuracies must be completely understood and controlled. As an example, we can use the field-installable butt-jointed type of connector plug shown schematically in Figure 6.7. In this example, the reference surface may be the outside diameter of a cylindrical ferrule, the tapered alignment surface of a biconic connector plug, or any other type of alignment element. From this schematic, it is obvious that a simple linear summation of these offsets does not represent a true measure of the expected offset between the fiber-core axis and the reference-surface axis of any practical connector assembly. A more realistic expected offset can be calculated statistically, by using appropriate distributions for offsets of the hole-to-reference surface

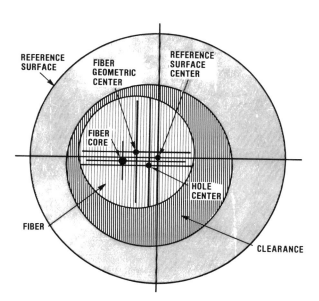

Figure 6.7 Accumulation of inaccuracies in a typical field-installable butt-jointed connector plug.

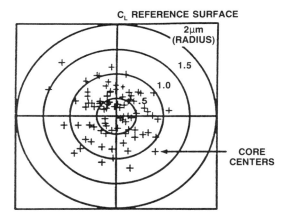

Figure 6.8 Scatter plot of the calculated locations of fiber core centers in a field-installable connector plug. Given: Number of connectors = 100; hole eccentricity, mean = 0.5 μm, $\sigma = 0.2$ μm; diametral clearance, mean = 0.7 μm, $\sigma = 0.2$ μm; fiber eccentricity, mean = 0.3 μm, $\sigma = 0.1$ μm. Results: core to reference surface C_L, mean = 0.6 μm, $\sigma = 0.3$ μm.

axis, and the fiber core-to-cladding axis, as well as the variations in the outside diameter of the fiber and the diameter of the hole in the ferrule or plug. The appropriate distributions for these factors are clearly dependent upon the fabrication processes as well as on any sorting processes that may be employed in selecting the fibers and connectors. With this in mind, a Monte Carlo simulation based on distributions representing good quality fibers and the connectors can be used to calculate the expected core-to-reference surface offset for a particular set of connector plugs.

Shown in Figure 6.8 is a scatter plot of the calculated locations of these core centers, assuming fiber core-to-cladding offsets having a mean of 0.3 and a standard deviation of 0.1 μm, hole offsets having a mean of 0.5 μm and a standard deviation of 0.2 μm, and diametral clearances between the fiber and hole having a mean of 0.7 μm and a standard deviation of 0.2 μm. As a result of this calculation, the mean and standard deviation for this particular connector model were calculated to be 0.6 and 0.3 μm, respectively. This type of distribution has been achieved routinely in laboratories by using special connector-trimming techniques, through using high-quality fibers, and by measuring and matching the fiber and hole diameters prior to installing the connectors [8].

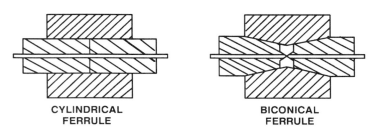

CYLINDRICAL FERRULE   BICONICAL FERRULE

Figure 6.9 Examples of two popular alignment mechanisms used in fiber-optic connectors.

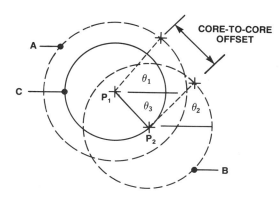

Figure 6.10 Schematic representation of the combined offsets in a typical butt-jointed-type connector assembly. Radius A = core/taper offset, plug 1; radius B = core/taper offset, plug 2; radius C = adapter offset.

After calculating the distribution of the fiber-core to reference-surface offsets statistically, we must next consider the ferrule-to-ferrule (plug-to-plug) alignment mechanism. The two types of connectors shown in Figure 6.9 are by far the most popular type of butt-joint connector designs presently being used in both single-mode and multimode fiber systems. The resulting lateral offsets between fibers in these types of connectors can be represented schematically, as shown in Figure 6.10. With the calculated distribution for the core-to-reference surface offset from above, and typical offset effects attributed to the connector alignment adapters (a mean of 0.5 μm and a standard deviation of 0.2 μm), the distribution of the expected offset for adjacent fiber cores in a connection can also be calculated. From these distributions and a Monte Carlo simulation the mean and standard deviation for the core-to-core offset were calculated to be 0.9 and 0.4 μm, respectively.

With application of the general loss equation of Section 6.2.2 to these results, the distribution of the expected loss for these connectors, employing single-mode fibers operating at 1300 nm, was calculated to have a mean of 0.3 dB and a standard deviation of 0.1 dB. This calculation assumed an angular offset of 0.8°, which is typical for many connector designs. Experiments [8] using fibers and connectors representing the assumptions used in this example have resulted in similar distributions for insertion loss measurements (see Fig. 6.11). It should also be mentioned that many effects have not been considered in this example such as fiber end-face quality, precise circularity and straightness of the reference surfaces (hole, fibers, periphery of the ferrules), centering effects of epoxies (if any are used), and static friction.

The above example was used to show the complexity that one must consider when designing and characterizing a connector. The same philosophy applies whether the connector will be used with multimode or single-mode fibers, the only difference being that the clearances and accuracies can be traded off for cost and ease of assembly.

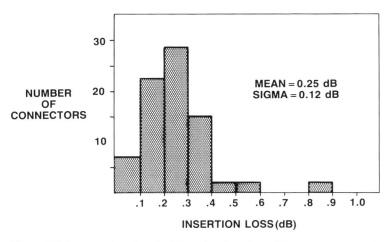

Figure 6.11 Insertion loss data for 75 low-loss butt-jointed single-mode connectors.

6.4.2 Expanded-Beam Connectors

A typical expanded-beam connector [10], schematically shown in Figure 6.12, consists of an optical element that collimates the beam radiating from the transmitting fiber and focuses the expanded beam onto the core of the receiving fiber. Two advantages over the butt-joint connector can be readily seen from this schematic. First, engagement and separation of the connector take place within the expanded beam, meaning that the connector should be much less dependent on lateral alignments and should also be more tolerant of dirt or any other contamination on the end faces. A second advantage is that optical processing elements, to perform functions such as beam-splitting and switching, can be easily inserted into the expanded beam.

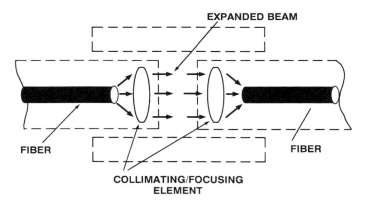

Figure 6.12 Schematic representation of a typical expanded-beam connector.

A further analysis of the expanded-beam design reveals that the demanding submicrometer and micrometer alignments for single-mode and multi-mode fibers, respectively, must still be satisfied. Although the "expanded beams" do not require alignments to these accuracies, the fibers must be positioned with respect to the axes of the collimating and focusing element to these same demanding tolerances. Also, angular alignment of the expanded beam is critical, since misalignments cause the focusing point to be displaced laterally from the fiber core. For these reasons, the major trade-off in butt-joint and expanded-beam connector designs is in the control of either lateral or angular alignments. Furthermore, with these designs, even when perfect alignment is achieved, losses may also be incurred due to aberrations of optical elements.

Even with the advantages that expanded-beam connectors apparently have, the majority of connectors used today are of the butt-joint type. However, interest in expanded-beam connectors is increasing, partially because of the advancement in the molding of glass aspheric lenses, and partially because of the increased anticipation in the optical processing of the signal, such as wavelength-division multiplexing and switching. Also, like butt-joint connectors, expanded-beam connectors must be field-installable and cost-effective, even though they will usually contain more parts and be more complex than butt-joint-type connectors.

6.5 SPLICES

Fiber splicing is used when it is not expected that it will be necessary to disconnect or separate the fibers in a joint. The major application for splices is therefore in the installation of fiber spans that exceed length limitations due to fiber/cable fabrication processes and/or cable deployment practices. Since there are generally many more splices in a span than connectors, perhaps more than one per kilometer, it is important that the splicing procedure yield lower losses than connectorizing (see Fig. 6.13). Also, since the splicing operation is usually done only once, procedures can be used to achieve low loss that may be impractical for other jointing operations. The splicing procedures presently being used can be characterized as (1) mechanical or (2) fusion procedures.

6.5.1 Mechanical Splices

Mechanical splicing techniques are similar to those used in butt-joint-type connectors. Since the factors that cause coupling losses are the same, similar techniques can be used to evaluate this type of splice. The ability of most mechanical splices to achieve lower losses than similar types of connectors is attributed to two major differences. First, it is possible to use a continuous

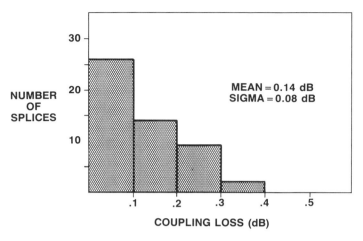

Figure 6.13 Loss data for 50 low-loss fusion splicing of single-mode fibers.

reference surface [10] that mates with both fibers, thereby minimizing the accumulation of inaccuracies in mating parts. A second difference is that since the operation is done only once, adjusting techniques can be used [11] to optimize the alignments while monitoring the throughput power. The first approach, a continuous reference surface for both fibers, has been used for low-loss splicing of multimode fibers, and the adjustment technique has recently been used for low-loss splicing of single-mode fibers.

One of the disadvantages with splicing in general, and particularly mechanical splicing, is the resulting buildup in the size of the fiber or cable being spliced. In some instances, this increase in size can be tolerated, but in other cases it cannot. For example, for factory splices prior to cabling the finished splice should not exceed the original coated-fiber size. Also, when cables have a large fiber count, the enclosure required to contain mechanical splices can become excessively large. As far as strength is concerned, the finished splices are usually isolated from external forces by using strain reliefs on the splice enclosure. Also, to ensure long-term stability the splice may employ epoxy or an equivalent bonding agent.

6.5.2 Fusion Splicing

In fusion splicing [12] the fibers are heated locally at the joint until they are softened and can be fused together (see Fig. 6.14). This technique is a widely practiced method for making low-loss splices with both multimode and single-mode fibers. The heat source most commonly used is an electric arc, but flame and laser sources are also used for fusion splicing. The first step in fusion splicing is preparation of the fiber end faces. The fibers should

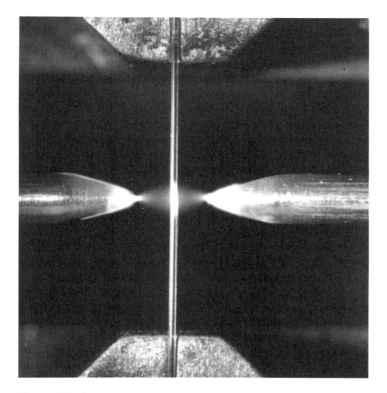

Figure 6.14 Photograph of arcfusion splicing of single-mode fibers.

have smooth, clean end faces with face angles less than 1°. Fiber-cleaving tools are available that achieve these requirements, but care must be taken in this step or high-loss splices can result due to the improper flow of molten silica during the heating operation. The next step is to load the uncoated fibers into holding chucks and align them either visually with a microscope or by maximizing the throughput power in the fiber. The fibers are then separated a fixed distance and a prefusion operation is used to slightly round their ends. This is usually done by applying heat with the fusion source. This important step is used to ensure core-to-core contact and proper flow during fusion. Finally, the aligned fibers are pressed together during fusion. The amount of overlap of fiber end faces, from the pressing stroke, is usually about 10–25 μm and depends largely on the end-face angle and prefusion process.

In setting up a fusion splicer, care must be taken to correctly optimize many interdependent parameters. Also, the parameters must be adjusted for environmental conditions, particularly with respect to humidity, and for variations in the melting or softening temperatures of different fibers. In general, for splicing single-mode fibers, the overlap distance must be optimized to within 1 μm, and the arc time to 0.1 sec.

With proper techniques, multimode and single-mode fibers can be spliced whereby coupling losses can be less than 0.1 dB. Also, in contrast to mechanical splices, the fused fibers can be recoated to nearly the original coated fiber's dimension, as is usually done for factory splicing of fiber prior to cabling. Although very high strength fusion splices have been demonstrated with flame fusion, it is not uncommon for fusion splices to have a tensile strength equal to about 50% of the original fiber's strength, and failure usually occurs slightly away from the joint, at the boundary of the heated zone. In general, like mechanical splices, the completed fusion splice is enclosed and protected from external forces by cable strain reliefs.

6.6 OTHER OPTICAL PROPERTIES

In the previous sections, the discussions concentrated mainly on the insertion loss of the joint. However, there are other optical properties such as return loss and modal noise [14] that should not go unmentioned in any discussion on jointing technology, particularly in the case of connectors.

6.6.1 Return Loss

Joints in fiber spans can sometimes cause reflections that result in the return of optical power along the input fiber (return loss). In laser systems, this reflected power can cause system degradation. The magnitude of this degradation is dependent on such factors as the location of the reflection with respect to the laser, the type of system (analog or digital), the laser structure, and, of course, the amount of power that reaches the laser. As previously mentioned, most nonfusion splicing techniques employ some form of index-matching material that nearly eliminates reflections from the fiber joint. No gap exists between fused fibers, and therefore there is no reflection. However, with connectors, reflections can become a concern, particularly in the case of the first connector in the span, which is usually located close to the source. Various approaches exist to minimize the power penalty resulting from the reflections. First, an index-matching material can be used. In this case the return loss is usually greater than 30 dB. A second approach is to have fiber end-face contact without index-matching material (a typical design goal for some connectors). With this approach return losses of 25–30 dB have been measured, compared to about 12 dB for connectors with air gaps. Finally, if required, return losses of 45 dB can be achieved by using end faces that have a face angle of about 5° for single-mode fibers and 10° for multimode fibers. It should be noted that the first two approaches also reduce the insertion loss of the connector, while the last approach usually increases the insertion loss while also increasing the complexity of the connector.

6.6.2 Modal Noise

As previously mentioned with respect to multimode fiber systems, some amount of redistribution of power among the propagating modes usually exists at all discontinuities in a fiber span such as connectors and splices. Interference effects between the propagating modes at these discontinuities can cause modal noise and, as a result, system degradation. When a significant amount of power is present in the first higher-order mode in a short section of fiber between two connectors or splices, similar interference effects can also occur in single-mode fiber systems. Consequently, modal noise in single-mode fiber systems may occur in short connectorized patch cords and laser pigtails. To avoid this problem, it is important to properly specify the cutoff wavelength of the fibers used in these applications, such that the operating wavelength is well above the effective cutoff wavelength for fibers deployed in these lengths and configurations. In long lengths of fiber, any power launched into the first higher-order mode is sufficiently attenuated and modal noise is not a problem.

6.7 FUTURE DIRECTIONS

The jointing of single fibers, both multimode and single-mode, has not yet become a routine and trouble-free process. Many developments, such as interface standardization, are required to improve the consistency of performance in jointing technologies, particularly in the area of single-mode fiber connectors. Finally, the state of the art in the packaging of optical components, with regard to connectorization, requires that the total connectorized component technology be addressed in order to achieve cost-effective components.

REFERENCES

1. R. B. Kummer and S. R. Fleming, "Monomode optical fiber splice loss," *Proc. Opt. Fiber Commun.*, April 1982, p. 44.
2. T. C. Chu and G. R. McCormick, "Measurement of loss due to offset, end separation and angular misalignment in graded-index fibers," *Bell Syst. Tech. J.*, **57** (3): 595–602 (1978).
3. C. M. Miller and S. C. Mettler, "A loss model for parabolic profile fiber splices," *Bell Syst. Tech. J.*, **57** (9): 3167–3180, (1978).
4. S. Nemota and T. Makimoto, *Opt. Quantum Electron.*, **11**: 447 (1979).
5. R. E. Wagner and C. R. Sandahl, "Interference effects in optical fiber connections," *J. Appl. Opt.*, **21** (8): 1381–1385 (1982).

6. P. Kaiser, W. C. Young, N. K. Cheung, and L. Curtis, "Loss characterization of biconic single-fiber connectors," *Tech. Digest*, Symp. Opt. Fiber Measurements, NBS Spec. Publ. No. 597, pp. 73–76, 1980.

7. D. Marcuse, "Loss analysis of single-mode fiber splices," *Bell Syst. Tech. J.*, **56**: 703–717 (1977).

8. W. C. Young, "Single-mode fiber connectors—design aspects," *Proc. 17th Ann. Connector Interconnector Tech. Symp.*, *Electronic Connector Study Group*, September 1984, pp. 295–301.

9. Y. Katsuyama et al., "New method for measuring V-value of a single-mode optical fiber," *Electron. Lett.*, **12**: 669 (1976).

10. J. C. Baker and D. N. Payne, "Expanded-beam connector design study." *J. Appli. Opt.*, **20** (16): 2861–2867 (1981).

11. A. J. Hoffman, "Mechanical splice for permanent single-mode and multimode fiber optic installation," *SPIE Proc.*, **574**: 72–77 (1985).

12. C. M. Miller and G. F. DeVeau, "Simple high-performance mechanical splice for single-mode fibers," *Proc. Opt. Fiber Commun.*, February 1985, p. 26.

13. D. H. Taylor and S. L. Saikkonen, "Factory splicing of optical waveguide fiber," *Proc. 32nd Int. Wire Cable Symp.*, November 1983, pp. 63–69.

14. N. K. Cheung, "Reflection and modal noise associated with connectors in single-mode fibers," *SPIE Proc.* **479**: 56–59 (1984).

CHAPTER 7

Passive Components for Optical Coupling and WDM Applications

Giok-Djan Khoe

7.1 INTRODUCTION

Optical fibers, semiconductor lasers, and photodetectors are the key components of optical communication systems. The realization of such systems requires the use of optical passive components and suitable means to interconnect and package the active and passive devices for practical use.

This chapter presents a short tutorial review of active to passive component interfacing, wavelength-division multiplexing, functional passive components, and key parts such as fiber end terminations. Emphasis is placed on single-mode technology, because there is already a vast literature concerning multimode versions. Exceptions will be made for cases where the multimode devices are manufactured with the aid of a new technology.

Theory is only briefly discussed, but an attempt is made to provide a clear view of where approximations fail, and the reader is referred to suitable theoretical papers. Design considerations are introduced, and an indication is given of the practical problems to be solved. Several important results are highlighted, and it is shown how new technology can influence the solutions in the future.

Section 7.2 deals with the preparation and packaging of optical fiber ends and aims at providing a basic physical understanding of some typical problems as well as an introduction to design concepts in the field of fabrication of passive components. Section 7.3 discusses laser-to-fiber coupling, the optical aspects as well as solutions to packaging and mechanical problems. Section 7.4 is devoted to wavelength-division multiplexing (WDM).

Section 7.5 contains a collection of important results in the field of some functional passive components such as taps and star couplers. The last two sections touch on the feasibility of some components in the context of system design and trends in electronics.

7.2 FIBER END PROCESSING

Suitable terminations must be prepared on fibers; the ends must be joined to the next fiber section, or the fiber must be terminated in a unit that includes passive components, sources, or detectors. Demountable or permanent joints between fiber ends fall into the category of fiber connectors and splices. These topics are treated in Chapter 6. We will focus on other applications.

Different types of fiber terminations are needed for different applications or for alternative solutions to the same coupling or launching problem. Most types can be found in the diagram shown in Figure 7.1. Essentially, the fiber end can be considered an aperture with an associated electromagnetic field. Cylindrical dielectric waveguide modes have been described in detail by Snitzer [1]. However, a reasonable approximation for the electromagnetic field in common single-mode fibers is the Gaussian field. The single-mode fiber end can be adequately described, within the framework of the approximation, by a Gaussian beam waist. Lenses can be used to make images of this beam waist with enlarged or reduced size, if required. Theoretical descriptions of the Gaussian beam approximation for single-mode fibers and the manipulations of Gaussian beams with the aid of lenses and other optical components have been published by Marcuse [2] and

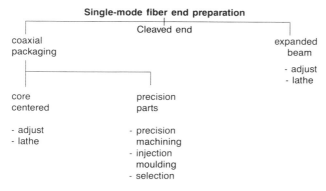

Figure 7.1 Different types of fiber terminations are needed for coupling to electrooptical devices, WDM, and other optical components.

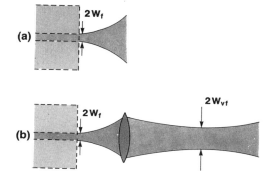

Figure 7.2 The end of a single-mode fiber can be represented by a Gaussian beam waist. An enlarged (or reduced) image of this beam waist can be made, if necessary, with the aid of a lens. The Gaussian beam is highly collimated when the waist is sufficiently enlarged.

Kogelnik and Li [3]. Single-mode fiber ends with associated Gaussian beam waists are shown schematically in Figure 7.2.

It is advantageous to know accurately the position of the beam waist in view of further handling of the fiber end. It is not possible to use the cylindrical surface of the fiber cladding as a reference for accurate positioning if the core-to-cladding concentricity of the bare fiber itself is not adequate. When the fiber end is packaged, it is important that the position of the waist be related to a reference on the outside of the package.

The following sections are devoted to fiber end preparation with emphasis on methods to provide practical mechanical references for the beam waist location for various types of applications. A comprehensive list of publications on this subject can be found in a review paper by Khoe et al. [4].

7.2.1 Coaxial Packaging

The bare fiber end is packaged and protected in a cylindrical body. The outer surface of the cylinder, when properly prepared, is concentric with the beam waist on the end surface of the fiber. This is a desirable situation for application in optoelectronic device packages and planar waveguide assemblies. The coaxial ferrules do not necessarily contain a simple fiber end. A tapered fiber end and a laser diode, to be coupled to each other, can each be provided with a coaxial ferrule prior to assembly in a cylindrical package. The only position to be adjusted during assembly is the distance between the components. A metal ferrule allows easy hermetic sealing of the package, for example, by laser welding. Coaxial glass ferrules can be desirable for coupling to glass-based planar waveguides or other components if the fiber itself if not coaxial. The glass-based component must be provided with suitable grooves or other references for the attachment of the ferrule.

Three methods of coaxial fiber end preparation are shown schematically in Figure 7.3. The first method is essentially different from the other two in that it relies on a coaxial bare fiber. Other differences are practical in nature,

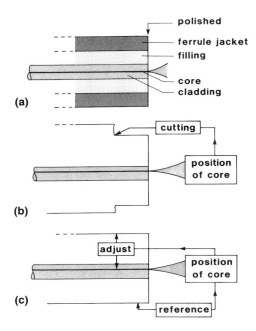

Figure 7.3 Coaxial packaging of single-mode fiber ends. The key aspect is the concentricity of the core. (a) Packaging made of coaxial parts. The fiber itself must have a concentric core. Tolerances between the parts must be very small. (b) The outer surface of the package can be made coaxial to the core by turning or grinding. Light from the core is used as a reference. The parts and fiber can have much larger tolerances. (c) The package is provided with parts to adjust the concentricity. The viewing system is a key instrument in methods b and c. The end surface is usually polished.

and the success of each method partly depends on the ability to develop ingenious tools and appliances for the preparation of the coaxial package. The second method is not necessarily a turning or cutting procedure; grinding and polishing can be used as well to correct the outer surface of the cylinder. The second and third methods rely heavily on the viewing system, which must be sufficiently accurate for determining the lateral position of the beam waist on the end surface of the fiber. One viewing system, described in detail by Khoe and van Leest [5], can be used for routine alignment to within 0.1 μm. Two methods can be used to obtain light from the core. One method is to couple light into the far end of the fiber; the other is to expose the fiber to ultraviolet light. Ultraviolet light will give rise to fluorescence in germanium-doped fiber cores. Part of this light, which is visible blue, will be guided in the core and is available for reference at the end.

7.2.2 Expanded Beam Termination

Passive components usually have fiber inputs and outputs. In between, the light beam may be intercepted and manipulated by filters, mirrors, or other optical elements. To obtain low loss it is necessary to have a parallel beam at these interception points. A Gaussian beam can be highly collimated with the aid of a single lens, as was shown in Figure 7.2. In many situations, it is desirable to combine the fiber end and a lens in one ferrule. Again, it is desirable to relate the position of the beam waist to the outer surface of the

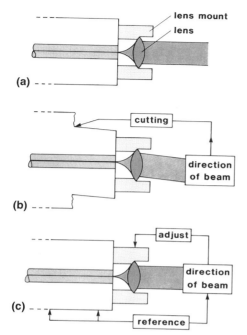

Figure 7.4 Expanded beam ferrules. Key parameter is the direction of the collimated expanded beam. The distance between lens and fiber is usually set by means of a spacer. (a) Assembly using precision parts. (b) The outer surface is turned or ground parallel to the beam. (c) The direction of the beam is adjusted to the surface of the cylinder by moving the lens laterally.

ferrule. Because the beam waist is far from the fiber end in this case, it is important to know accurately the direction of the expanded and collimated beam. Thus the axis of the ferrule must be parallel to the axis of the light beam.

Lens ferrules can be prepared in a number of ways, as shown in Figure 7.4. The viewing system is used to determine the direction of the beam. A properly designed viewing system can detect deviations of about 0.01°. A detailed analysis on lens ferrules was made by Nicia [6].

In principle, it is possible to use coaxially packaged fiber ends, as discussed in the preceding section. A coaxial ferrule can be used as a building block in the first example of Figure 7.4, which is based on precision parts. There is no need to use a coaxial bare fiber if a properly prepared ferrule is used as a subassembly.

Different types of lenses can be used for the expanded beam ferrules. Examples are shown in Figure 7.5. The ball and GRIN (graded-index) lenses are widely used. High-quality ball lenses are commercially available. The diameters can be chosen between 0.2 mm and a few millimeters, and the refractive index can be chosen between 1.5 and 1.9. GRIN lenses are also commercially available. An attractive property of these lenses is the possibility to attach them directly to other glass bodies without air spaces in between.

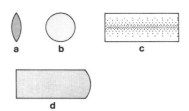

Figure 7.5 Examples of lenses for application in expanded-beam ferrules. (a) A common biconvex lens. (b) Homogeneous ball lens. (c) GRIN (graded-index) lens. This lens can be directly attached to other glass bodies without air spacings in between. (d) A homogeneous rod with rounded end. GRIN lenses can also be provided with rounded ends to increase the numerical aperture, but direct attachment to other glass bodies is not possible in that case.

7.2.3 Tapered Fiber End

The combinations of fiber end and lens do not necessarily consist of configurations with a separately aligned lens, as just discussed. A lens directly attached to the fiber end can be made by shaping the fiber end itself by etching or heating it or by dipping it in a suitable lens material. These configurations have been reviewed by Khoe et al. [4]. Among the various possibilities, the tapered end with lens seems to be the most widely used. Examples are shown schematically in Figure 7.6. A photograph of a taper with a high-index lens on its end is shown in Figure 7.7. The diameter of the end of the taper is commonly between 25 and 33 μm.

One of the consequences of the tapering is lateral enlargement of the extent of the electromagnetic field in the fiber. As a result, the influence of possible eccentricity between core and cladding is reduced. The absolute value of the eccentricity will remain the same, but the ratio between the error and the lateral size of the field will be smaller. It is therefore possible to align the lens or lens-shaped surface with the cladding of the fiber. Another advantage of tapered ends is that it is possible to choose a lens diameter smaller than the original diameter of the fiber cladding.

The Gaussian beam approximation fails to describe the field in the tapered region properly. The V-number in this region is much smaller than

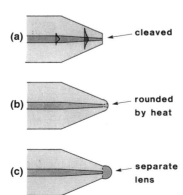

Figure 7.6 Tapered single-mode fiber ends for coupling to light sources. (a) The fiber end is tapered and cleaved. The diameter of the end is commonly between 25 and 30 μm. (b) The cleaved end is rounded by heating. (c) A separate lens is attached to the cleaved end. The index of the lens can be chosen.

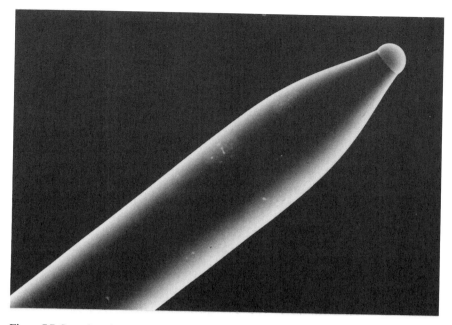

Figure 7.7 Scanning electron micrograph (SEM) of the tapered end with a separate lens. The refractive index of this lens is 1.9.

in common single-mode fibers, and this is the situation where the approximation deviates from the exact field, as pointed out by Marcuse [2]. Unfortunately, it is not possible either to describe analytically the field propagation in a weekly guiding taper with the aid of methods used earlier by Snitzer [1]. One possibility is to describe the electromagnetic field with the aid of the beam propagation method, described in detail by Fleck et al. [7]. One aspect that must not be neglected in the description of the taper is the curved phase front of the field. This curvature is small, but neglecting it will introduce large errors in calculating coupling efficiency.

7.2.4 Array Configurations

Array fiber terminations are needed for optical fiber coupling to multichannel WDM devices and to planar components that have many input or output ports in parallel. In principle, the beam waists at the fiber end surface must be arranged to form an accurate linear array. One possible method is to place the fibers in an array of precision grooves made on the surface of a silicon wafer by etching. Satake et al. [8] and Khoe et al. [19] have described array ferrules made with the aid of grooved silicon.

One unfortunate aspect of array configurations is that the fiber itself must be sufficiently coaxial, because the arrangements are usually based on the alignment of the fiber claddings. In practice, this could mean that selected fibers must be used for the pigtailing of array configurations. Another problem that can occur in array assemblies is that the fiber cladding is too thick. This problem may be encountered in grating-type WDM devices, where a close spacing between selected wavelengths usually means that the fiber cores in the array must also be very closely spaced. In this case, the cladding size may be reduced by etching, or a special single-mode fiber with thin cladding can be fabricated.

Expanded beam arrays may also be needed if the functional optical components are so arranged. In this case, the expanded beam array unit must be made by adjusting a fiber array to a separately made array of lenses. Again, the quality of the product will be strongly influenced by the concentricity of each fiber. The use of selected fibers may be advantageous. Expanded beam array configurations have been investigated by Iga et al. [9].

7.3 LASER-TO-FIBER COUPLING

Single-mode fiber communication systems require properly protected laser diodes, efficient launching of light from these sources into the fiber, and electrical connections in the package that allow high-speed modulation. High launching loss can severely limit the distance between repeaters. Other factors in the optical coupling scheme can also influence the system span length. Reflections from the surfaces of lenses in the launching optics may affect the mode spectrum and thereby cause system degradation.

The mechanical part of the coupling assembly must ensure that high launching efficiency is maintained under a variety of environmental conditions, such as vibration and temperature variations. These factors impose serious restrictions on the mechanical design of the assembly.

High bit rates are an inherent feature of single-mode trunk systems. For the time being, the requirements for high-speed modulation do not present severe problems in the design of the package, owing to the low currents through the laser diode. More attention must be paid to the reference signal used for monitoring and stabilization of the light output. The coupling between the monitoring photodiode and the laser must be reliable; otherwise errors may occur in the reference signal.

7.3.1 Optical Configurations

The light radiated from index-guided laser diodes of good quality can also be reasonably approximated by a Gaussian beam, as was confirmed by Lee

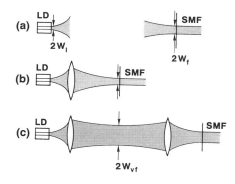

Figure 7.8 The relevant optical characteristics of laser and fiber can be represented by Gaussian beams with different waist sizes. (a) The difference can be bridged by using (b) one lens or (c) two lenses. In the latter case, final adjustment can be done between the enlarged images, thus relaxing the lateral precision.

and coworkers [10]. The optical coupling problem can therefore be adequately described, within the framework of the approximation, by a simple mismatch between the Gaussian beam waists of laser and fiber [2, 3]. Index-guided laser diodes usually have a waist radius of 1.5 μm, and the common value for single-mode fibers is 4.5 μm. In theory, it is possible to eliminate the mismatch by magnifying the spot of the laser to three times its original size by means of a suitable optical system.

Various optical systems have been reported on in the past. Khoe et al. [4] have reviewed these methods, and a list of the relevant publications can be found at the end of their paper. Generally, these designs can be grouped into either single-lensed or expanded-beam designs. The characteristics of each are illustrated in Figure 7.8. The expanded-beam design uses lenses to create enlarged images of the beam waists of both laser and fiber. In practice, the fiber with one lens is often prepared separately as an expanded-beam ferrule, as explained in Section 7.2.2. The effect of beam expansion is that the adjustment between the enlarged waists is far less sensitive. It should be noted that the alignment accuracy of the individual lenses and fiber is the same as that required for the single-lens case. Practical realizations of the first category use either a separately aligned lens or direct attachment to the fiber. An example of the last method is the tapered fiber end (Section 7.2.3).

The choice to be made depends on many considerations. The expanded-beam method allows the insertion of additional components between the lenses, such as an isolator. A disadvantage is the need to eliminate reflections from many surfaces in the assembly. It is not possible to insert an isolator between the tapered fiber end and the laser, but this configuration has a minimum number of reflecting boundaries, and the alignment procedure can be done in one step.

Coupling efficiencies of about 50% or slightly higher can be routinely obtained when index-guided lasers are used. The remaining loss is mainly caused by reflections at the surface of a lens or by imperfections in the laser beam, such as a slight astigmatism.

A monitoring photodiode is usually coupled to the rear of the laser by placing it directly in the light beam without any coupling optics. The photodiode is situated obliquely to prevent unwanted reflections into the laser.

7.3.2 Packaging

Different types of encapsulations and adjustment procedures are needed for different optical coupling configurations. Common types are illustrated in Figure 7.9. The first type is based on the expanded-beam method. The final adjustment step can be done between two separately prepared expanded-beam units. The other three methods are used in combination with specially prepared fiber ends, such as tapers. In both cases, a hermetic fiber seal is required.

The second and third methods differ in the manner of adjustment. A one-step procedure is employed in the second method. Part of the hermetic encapsulation is deformed in such a way that the fiber is adjusted in all

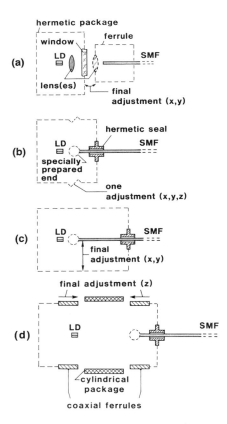

Figure 7.9 Methods for laser to single-mode fiber packaging and alignment. (a) Two expanded-beam units are adjusted. (b) One-step adjustment by deforming part of the package. (c) The fiber end is aligned inside the package. The axial position must be adjusted before the hermetic fiber seal is made. (d) Two coaxial ferrules are used, one with a fiber inside and the other with a laser. The outer surface of the ferrules can be centered to the fiber or laser by turning (see Fig. 7.3b). Both ferrules are inserted in a cylindrical package and sealed. Only the distance must be adjusted in the final step.

three directions. The part of the fiber that is protruding inside the encapsulation can be kept very short. A longer piece of fiber is needed inside the third package. In this case, the fiber end is adjusted in two lateral directions with the aid of an additional support inside the package. The distance between laser and fiber end can be varied by sliding the fiber inside the seal prior to soldering or glass melting. The sequence of adjustment depends on the particular construction of the encapsulation.

A fourth method is based on the coupling of two coaxial ferrules, one with a fiber inside and the second with a laser. The fiber can be provided with a tapered end, as discussed earlier. The ferrules can be prepared with the aid of the cutting process indicated in Figure 7.3b. Finally, the two ferrules can be inserted in a cylindrical package, adjusting the distance at the same time.

Stability against mechanical and temperature influences is a very important requirement for laser encapsulations. These influences could be either shocks and vibrations or slow, repeated temperature variations. Such requirements impose serious limitations on the design. The best packages are usually small and cylindrical. Common packages have a diameter of about 1 cm and a length of 2–3 cm.

7.3.3 Future Needs and Possibilities

A laser diode has many qualities that have not yet been fully exploited. Laboratory experiments have indicated that semiconductor laser structures are potentially suitable for use as switches, oscillators in coherent systems, external modulators, and active star points, among other things. In most of the applications just mentioned, it is necessary to couple more than one optical fiber to the diode. In principle, it is possible to couple two fibers to the laser structure simultaneously by using both ends of the diode, the front and the rear. However, such configurations give rise to serious problems in the design of a practical package. All fibers must therefore be coupled to the front end of the laser. This is possible because owing to the symmetry of the diode, the same end can be used as either input, output, or star point.

Experiments by Khoe et al. (11) have indicated that it is possible to couple double single-mode pigtails efficiently to the front end of laser diodes. It is possible to couple double pigtails to lasers with a total efficiency of 50% or more. More details can be found in Ref. 11. A photograph of the double pigtail is shown in Figure 7.10.

Array configurations will become important in the future, when optoelectronic integrated circuits (OEICs) can be made with sufficiently high yield. These devices will usually have multiple optical inputs and outputs. As a consequence, array pigtails will be needed. The first coupling methods will probably consist of the adjustment of an array fiber ferrule to the array of light sources or detectors. More advanced configurations will depend on

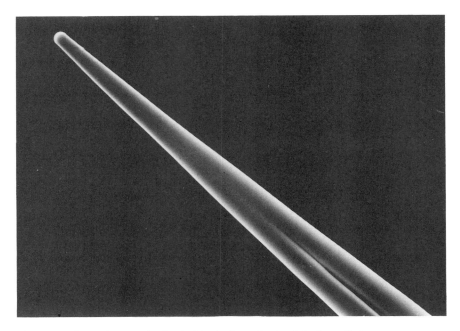

Figure 7.10 SEM photograph of a double single-mode pigtail for laser diodes.

the possibility of providing the semiconductor IC itself with waveguides, lenses, or references for fiber coupling.

7.4 OPTICAL MULTIPLEXING

In principle, optical multiplexing is quite similar to electrical multiplexing. However, much higher frequencies are employed in optical multiplexing, and the lack of oscillators of sufficient quality is the main reason for the delay in field application. Heterodyne or homodyne optical systems can be developed as soon as the oscillators are available, much in the same way as was done before for lower frequencies. We call this "true" optical multiplexing. This topic is treated in the chapter on coherent optical communication systems.

In contrast, wavelength-division multiplexing (WDM) is not a multiplexing technique that makes use of optical frequencies. In fact, WDM makes very lavish use of the bandwidth potential of optical frequencies. WDM divides a certain region of wavelengths into a number of coarse bands of many nanometers each. One band therefore represents a bandwidth of many terahertz, but this potential is not exploited because each band is only used to pass one signal with a relatively low information content. A disadvantage of present-day WDM components is that the assemblies are very expensive in comparison with electronic multiplexing equipment. As a consequence, WDM is not yet widely used, and the capacity of optical fiber systems has been gradually increased by means of electronic multiplexing instead.

The future of WDM depends on many aspects. Electronic multiplexing may become difficult or expensive at speeds far above 1 Gb/s. Planar WDM components may become available in the future at much reduced costs. On the other hand, heterodyne or homodyne systems may be preferred when stable single-frequency laser oscillators are available.

The rest of this section is devoted to the basic optical components needed for WDM. A detailed review, including a comprehensive list of references is given by Ishio et al. [12].

7.4.1 Basic Considerations for WDM

The diagram in Figure 7.11 shows a number of ways and means for optical multiplexing. "True" optical frequency division multiplexing (OFDM), with heterodyne or homodyne, will not be discussed in this section. Two WDM methods are commonly preferred because they offer the best possibilities for the realization of devices with many bands and close spacings in between. These are WDM devices that employ a grating or multilayer dielectric thin-film filters. A prism, offering a material-dispersion-based wavelength separation, can also be used for WDM, but practical realization is hampered by the low dispersion.

Key components in the filter types are, as shown schematically in figure 7.12, the filters and expanded-beam ferrules. The filters create a sequence of

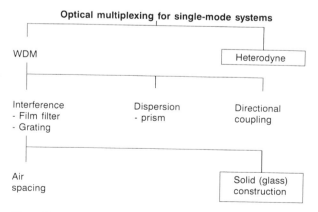

Figure 7.11 Optical multiplexing methods. The most promising methods seem to be the heterodyne type of optical frequency multiplexing and wavelength-division multiplex (WDM) using configurations without air space between the optical parts.

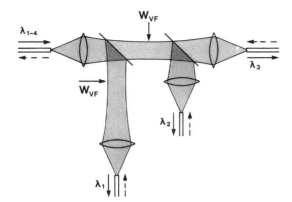

Figure 7.12 Basic configuration of a dielectric multilayer filter type WDM. Key components are the expanded-beam ferrules and filters. The expanded beams between the lenses are intercepted by filters with different characteristics. The mutiplexing and demultiplexing functions depend on a particular arrangement of filters.

wavelength bands, and the expanded-beam ferrules are needed to make the proper images of beam waists on the fiber ends and to provide highly collimated beams suitable for interception by the filters. In the configuration of Figure 7.12, the filters are arranged as wavelength-selective mirrors. The basic components of a grating type are one grating and an expanded-beam array ferrule, as shown in Figure 7.13. The grating reflects light of different wavelengths along different angles, and these directions are transferred into different positions on the array by the lens. Again, we can adequately describe the optical propagation characteristics with the aid of the Gaussian beam approximation. The wavelength separation aspects can be derived from the filter characteristics as provided by the manufacturer or from the standard formulas for the reflection grating.

Passive components are not the only critical part in a system with WDM. A serious difficulty up to now has been the lack of wavelength-selected light sources, both in quality and availability. It is not possible to simply purchase laser diodes with an exact sequence of given wavelengths. As a consequence, each band in the WDM must be provided with a wide window,

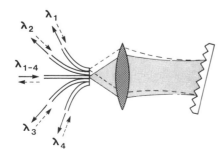

Figure 7.13 Principle of a grating-type WDM. The basic elements are an expanded-beam array ferrule and a reflection-type grating. The grating reflects light with different wavelengths along different angles. The different angles are transformed into an array of positions by the lens. The configuration can be used for multiplexing or demultiplexing.

commonly between 5 and 10 nm. The sensitivity of the laser diode wavelength to temperature variations is yet another reason why wide windows are required in each band. The variation is about 0.5 nm/°C for laser diodes with conventional cavities and about 0.1 nm/°C for distributed feedback (DFB) lasers. The spacing between the bands depends on the slope of the filters or on the spacing between the cores of the fibers in the array ferrule. The attenuation of the device depends on the quality of the imaging system, the adjustment procedure, and the losses of the filters or grating.

Directional couplers can also be used for WDM. We will discuss this in Section 7.5.

7.4.2 Practical Configurations

In principle, it is possible to fabricate WDM components in the way shown schematically in the figures. The filter type can be assembled from separately prepared expanded-beam ferrules and multilayer dielectric film filters. In addition, a suitable mechanical support and some parts for the adjustment and fixation must be made to complete the assembly. A grating type can be made by aligning and mounting the expanded-beam array ferrule and the grating on one support. This class of WDM components is indicated as the air spacing type in Figure 7.11. Such components are suitable for laboratory use, but they can easily fail under adverse conditions in the field, because their stability against mechanical shock is questionable.

WDM components without air spaces between the optical parts are more suitable for field use. The filter type can be made by assembling all parts on a solid glass block and by using GRIN lenses in the ferrules. An example is illustrated in Figure 7.14. The grating type can be made in a similar way; both sides of a GRIN lens provide suitable surfaces for direct attachment, the fiber array at one end and the grating at the other. The filter types can also be made without the use of expanded beams. In that case, a thin-film filter is embedded between two oblique fiber end surfaces and part of the light is reflected to a third fiber end [13]. The coupling of light to the third fiber is not as efficient as for the other two because it is not possible to place the third fiber end surface directly at the same filter.

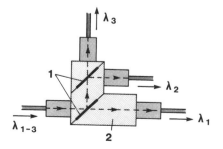

Figure 7.14 Example of a filter type demultiplexer without air spacing. The filters are surrounded by glass. One end of the GRIN lenses is directly attached to the solid glass body, and the other end is attached to the fiber.

A 10-channel grating-type device has been made with a channel spacing of about 2 nm and insertion loss of less than 3 db. Filter-type WDM components have been made with up to six channels, with spacings of about 30 nm and an insertion loss of less than 5 dB. More examples can be found in the paper of Ishio et al. [12].

7.4.3 Integrated Optics

Wavelength-division multiplex components are currently expensive because they consist of an assembly of separately fabricated parts. Research is now concentrated on the development of new technology more suitable for mass production. This means in general that the optical parts must be made on a single substrate, with the aid of methods such as photolithography, etching, deposition of layers, diffusion, and beam lithography. Geodesic optical principles might be interesting for application in this field. A variety of geodesic structures is discussed by Chang and Voges [14].

The first devices of this kind may be multimode types, because of less stringent fabrication tolerances. One of the most difficult parts to make seems to be a low-loss planar grating. Another difficult part, especially for single-mode devices, is the coupling between the substrate-based device and the fibers.

7.5 OTHER COMPONENTS

Optical fibers can be joined to each other by butt coupling or by coupling via lenses and other optical components. The latter method is employed in the passive components discussed in the preceding sections. A different method is to bring the cores of the fibers into close proximity and allow coupling to occur through optical tunneling. An explanation of this phenomenon can be derived from the theory of Marcuse [15].

A directional coupler can be based on parallel cores of single-mode fibers sufficiently closely spaced that energy is transferred by the process called optical tunneling. Devices of this kind are called directional couplers because, in principle, the direction of propagation is maintained. If the waveguides are identical, energy from one guide can be gradually coupled into the other and back again. The fraction of power coupled per unit length depends on the distance separating the parallel cores, the interaction length over which overlapping of the modes is present, the extent of mode penetration into the space between the cores, and the wavelength. To couple a given fraction of energy at a given wavelength, it is sufficient, in theory, to bend away the second core at a proper point. If the waveguiding is free of loss, this point can be chosen at a point where the alternate energy transfer has taken place many times.

This section deals with the fabrication of directional couplers and of other couplers for optical energy distribution, such as star couplers and power splitters based on straightforward branching methods. A list of publications on couplers and power splitters can be found in Ref. 4.

7.5.1 Fiber-Based Technology

The practical situation is far from the ideal one just described. A 100% energy transfer is possible only when the propagation constants of both waveguides are equal. In addition, it is difficult to fabricate a coupler with the coupled fraction of power given in advance, because the result is very sensitive to tolerances in the refractive index, the dimensions of the structure, and the point where the core is bent away.

One possible means of overcoming the severe fabrication tolerances is to monitor the optical properties during one of the final manufacturing steps. The final steps in manufacturing the fiber-based coupler are usually the minute adjustment of two specially prepared fibers and the fixation of the positions with the aid of glue or fusion. This is a cumbersome process because it is not easy to maintain the desired positions during the curing of the glue or freezing of the molten parts of the fiber. In a method reported by Digonnet and Shaw [16], two single-mode fiber cores are brought into close proximity by first grinding away part of the cladding. The two parts are then carefully adjusted to obtain the desired coupling fraction, and the positions are fixed by means of glue. Other methods differ in the way part of the cladding is removed and in the method of assembling and adjustment.

Fiber-based devices are commercially available. These devices are usually based on the use of locally constricted or etched fibers that are fused together. An interesting paper on this topic is that of Kawasaki et al. [17].

7.5.2 Substrate-Based Technology

Fiber-based devices are usually not well suited for mass production. However, the alternatives are also hampered by serious problems in fabrication.

In principle, it is not difficult to fabricate two single-mode waveguides on a glass substrate and to bring parts of the guides close enough to each other to allow optical tunneling. This device is shown schematically in Figure 7.15. However, it is again difficult to make a device with a given coupling ratio, because, in this case also, the exact value is very sensitive to many tolerances. Hence, a device with the desired value can be found only by selecting from a batch. Details on the fabrication of substrate-based single-mode waveguides can be found in the publication of Imoto et al. [18].

More research is needed in this field to have low-cost couplers with stable coupling characteristics for large-scale applications in fiber systems.

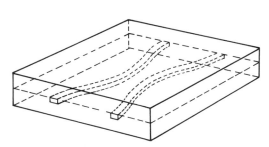

Figure 7.15 Typical configuration of a glass-substrate-based directional coupler. Parts of the waveguides are sufficiently close to each other to allow energy transfer. The exact amount of coupling is very sensitive to many tolerances and is therefore difficult to predict. The device can be made with the aid of local reactive ion etching and subsequent deposition of layers with different refractive indexes.

7.5.3 Examples and Applications

Directional couplers will be needed in systems employing heterodyne methods. The coupler can be used to combine optical signals on the transmission line, to couple laser diode devices with each other, and to couple lasers with optical frequency-selective fiber loops. In these examples, it is usually an important advantage that the coupling is present in only one direction.

Directional couplers are used in some measuring equipment. With the aid of the device, it is possible to couple a transmitter and receiver at the same end of a fiber simultaneously, with sufficient optical isolation between the light source and the detector.

A directional coupler is wavelength-sensitive, because the interaction length and propagation parameters are dependent on the wavelength. This phenomenon has motivated researchers to develop WDM components for multiplexing 1300-nm and 1550-nm lasers based on single-mode fiber directional couplers.

Directional couplers can be used as taps for the simple distribution of power, for example, from one fiber equally into two fibers. Another possibility is to branch off small portions of energy from an optical fiber loop to separate stations. For such applications, it is not necessary to use the principle of directional coupling. Power splitting can also be achieved with the aid of simple branching devices, as shown in Figure 7.16. Such devices are less sensitive to fabrication tolerances. To obtain low losses, it is important to

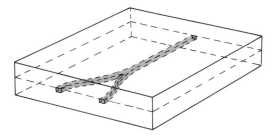

Figure 7.16 Optical power divider based on straightforward branching. This device is less sensitive to tolerances. The angle at the branching point must be kept small to minimize loss.

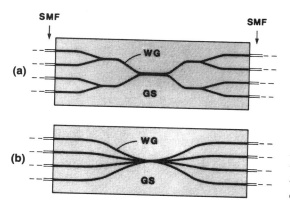

Figure 7.17 Examples of star couplers based on straightforward branching. For a device with n inputs and n outputs, the attenuation of each signal is proportional to $n \times n$, because it is only possible to combine two optical signals with the same wavelengths at the cost of 3-dB loss.

keep the aperture angle at the branching point as small as possible, preferably to less than 1°. Note that with this device it is only possible to combine optical signals with the same wavelength at the cost of 3-dB loss. The branching device can be analyzed with the aid of the beam propagation method [7]. Power splitting can also be achieved with the aid of filters. The construction can be similar to that shown in Figure 7.12. In this case the filters are not used to select optical wavelengths but serve as partially reflecting mirrors.

Star couplers can be made by employing a cascade of branching devices or directional couplers. The branching type, shown in Figure 7.17, have high loss because, as mentioned before, it is only possible to combine two signals at the cost of 3-dB loss. A disadvantage of the second type, based on directional couplers, is that it is not easy to fabricate the entire star on one substrate because of the existence of crossings. An example of this type with fiber interconnection is shown in Figure 7.18.

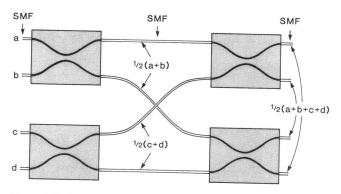

Figure 7.18 A star coupler based on directional couplers. In theory, this device can have zero excess loss. However, it is not easy to integrate the total device on one substrate because of the presence of crossings.

Multimode waveguides can easily be made in glass with the aid of ion migration. Star dividers and branching devices can be made with this method. Multimode star dividers can be used in local subscriber systems for distributive services. Star-based local area networks have also been considered. However, single-mode fibers are being seriously considered for use in both types of systems.

REFERENCES

1. E. Snitzer, "Cylindrical dielectric waveguide modes," *J. Opt. Soc. Am.*, **51**: 491–498 (1961).
2. D. Marcuse, "Loss analysis of single-mode fiber splices," *Bell Syst. Tech. J.*, **56**: 703–718 (1977).
3. H. Kogelnik and T. Li, "Laser beams and resonators," *Appl. Opt.*, **5**: 1550–1559 (1966).
4. G. D. Khoe, H. G. Kock, D. Kuppers, J. Poulissen, and H. M. deVrieze, "Progress in monomode optical-fiber interconnection devices," *J. Lightwave Technol.*, **LT-2**: 217–227 (1984).
5. G. D. Khoe and J. H. F. M. van Leest, "Single-mode fiber connector using core-centered ferrules" *IEEE J. Quantum Electron.*, **QE-18**: 1573–1580 (1982).
6. A. J. A. Nicia, "Lens coupling in fiber-optic devices," *Appl. Opt.*, **20**: 3136–3145 (1981).
7. J. A. Fleck, J. R. Morris, and M. D. Feit, "Time-dependent propagation of high energy laser beams through the atmosphere," *Appl. Phys.*, **10**: 129–160 (1976).
8. T. Satake, S. Nagasawa, and H. Murata, "Low-loss multifibre connectors with plug-guide-grooved silicon," *Electron. Lett.*, **17**: 828–830 (1981).
9. K. Iga, S. Misawa, and M. Oikawa, "Stacked planar optics by the use of planar microlens array,"*Proc. 10th Eur. Conf. Opt. Commun. Stuttgart*, September 1984, pp. 30–31.
10. T. P. Lee, C. A. Burrus, D. Marcuse, and A. G. Dentai, "Measurement of beam parameters of index guided and gain guided single frequency InGaAsP injection lasers," *Electron. Lett.*, **18**: 902–904 (1982).
11. G. D. Khoe, L. J. Meuleman, and J. Poulissen, "Laser diode devices for coherent fibre optics and special applications," *Proc. 10th Eur. Conf. Opt. Commun., Stuttgart*, September 1984, pp. 134–135.
12. H. Ishio, J. Minowa, and K. Nosu, "Review and status of wavelength division multiplexing technology and its application," *J. Lightwave Technol.*, **LT-2**: 448–463 (1984).
13. A. Reichelt, H. Michel, and W. Rauscher, "Wavelength division multi demultiplexers for two channel single-mode transmission systems," *J. Lightwave Technol.*, **LT-2**: 675–681 (1984).
14. W. L. Chang and E. Voges, "Geodesic components for guided wave optics," *Arch. Elektron. Ubertragungstech.*, **34**: 385–393 (1980).
15. D. Marcuse, "The coupling of degenerate modes in two parallel dielectric waveguides," *Bell Syst. Tech. J.*, **50**: 1791–1816 (1971).
16. M. J. F. Digonnet and H. J. Shaw, "Analysis of a tunable single mode optical fiber coupler," *IEEE J. Quantum Electron.*, **QE-18**: 746–754 (1982).
17. B. S. Kawasaki, M. Kawachi, K. O. Hill, and D. C. Johnson, "A single-mode fiber coupler with a variable coupling ratio," *J. Lightwave Technol.*, **LT-1**: 176–178 (1983).
18. N. Imoto, N. Suzuki, H. Mori, and M. Ikeda, "Sputtered silica waveguides with an embedded three dimensional structure," *J. Lightwave Technol.*, **LT-1**: 289–294 (1983).
19. G. D. Khoe, A. C. Jacobs, and L. J. C. Vroomen, "Single-mode ribbon connector exhibiting 0.2 dB loss without index matching during 300 cycles of repeated connection test," *Proc. 14th Eur. Cont. Opt. Commun.*, Brighton, September 1988, pp. 585–588.

PART 3

Semiconductor Laser Sources and Photodetectors

CHAPTER 8

Basic Physics of Semiconductor Lasers

N. K. Dutta

8.1 INTRODUCTION

8.1.1 Introductory Remarks

Since its invention in 1962 [1–4] the semiconductor injection laser has emerged as an important device in many optoelectronic systems such as optical recording, high-speed data transmission through fibers, optical sensors, high-speed printing, and guided wave signal processing. Perhaps the most important impact of semiconductor lasers is in the area of lightwave transmission systems where the information is sent through encoded light beams propagating in glass fibers. These lightwave transmission systems, which are currently being installed throughout the world, offer a much higher transmission capacity at a lower cost than coaxial copper cable transmission systems.

The advantages of semiconductor lasers over other types of lasers—gas lasers, dye lasers, and solid-state lasers—lie in their considerably smaller size and lower cost and their unique ability to be modulated at gigahertz speeds simply by modulation of the injection current. These properties make the laser diode an ideal device as a source in several optoelectronic systems, especially optical fiber transmission systems.

8.1.2 Basic Laser Diode

The concept of a semiconductor laser diode is a unique blend of semiconductor device physics and quantum electronics. The basic laser diode chip consists of two parallel cleaved facets, which form an optical cavity (Fig.

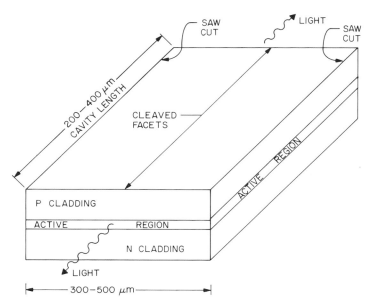

Figure 8.1 Schematic of the basic semiconductor laser diode.

8.1). The other two edges are saw cut. The device has a *pn* junction near the light-emitting region (active region) for effective current injection. The typical cavity length is about 200–400 μm.

The light current characteristics of a laser diode are shown in Figure 8.2. As the current injected is increased beyond a certain value (threshold current), the light output from the facet increases dramatically. Associated with this is a narrowing of the spectral emission from ~ 600 Å in the below-threshold region to ~ 30–40 Å in the above-threshold region. The light below threshold is emitted spontaneously as the injected electrons and holes recombine in the active region. Above threshold, stimulated emission into a low divergence beam characteristic of laser action is observed. For digital lightwave systems, the light signal is encoded by modulating the injection current, which generally switches the laser between the "on" state (stimulated emission regime) and the "off" state (spontaneous emission regime).

A close relative of the injection laser is the light-emitting diode (LED) [5]. The LED is a spontaneous emission device; its emission characteristics are similar to those of lasers below threshold. LEDs generally emit lower power and have a broader spectral emission and lower modulation speed than lasers.

The basic concepts and physics of semiconductor lasers are described in this chapter. The semiconductor laser is essentially an oscillator where the feedback is provided by the cleaved mirror facets and the amplification

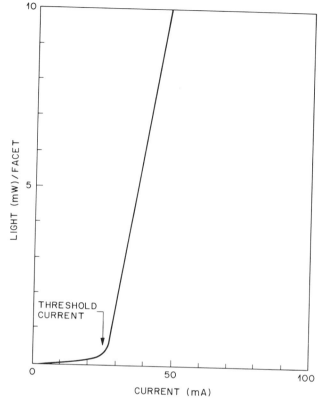

Figure 8.2 Light emitted from the facet versus the injected current for a semiconductor injection laser. The device is a proton stripe GaAs laser. (Courtesy R. L. Hartman.)

(or gain) is provided by electron–hole radiative recombination in the active region. The condition for sustained oscillation (laser action) is that gain equals loss.

8.1.3 *Materials*

The choice of materials for semiconductor lasers is principally determined by the requirement that the probability of radiative recombination should be sufficiently high that there is enough gain to overcome the cavity losses. This is usually satisfied for "direct gap" semiconductors. The various semiconductor material systems along with their range of emission wavelengths are shown in Figure 8.3. Many of these material systems are ternary (three-element) and quaternary (four-element) crystalline alloys that can be grown lattice-matched over a binary substrate.

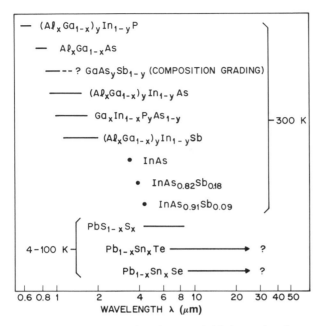

Figure 8.3 Direct-gap semiconductors suitable for semiconductor materials at different regions of the spectrum. (After Casey and Panish [12].)

8.1.4 Epitaxy

The lattice-matched crystalline growth of one semiconductor over another is called epitaxy. The development of sophisticated epitaxial growth techniques has been of major significance in the development of high-quality reliable semiconductor lasers. The commonly used techniques are liquid-phase epitaxy (LPE) [6], vapor-phase epitaxy (VPE) [7], and molecular beam epitaxy (MBE) [8].

In LPE, the epitaxial layer is grown by cooling a saturated solution of the components of the layer to be grown while that solution is in contact with the substrate.

In VPE, the epitaxial layer is grown by the reaction of gaseous elements or compounds at the surface of a heated substrate. The VPE technique has also been called chemical vapor deposition (CVD) depending on the constituents of the reactants. A variant of the same technique is metal-organic chemical vapor deposition (MOCVD) [9], which has been very successful for lasers in which metal alkyls are used as the compound source.

In MBE, the epitaxial layer growth is achieved by the reaction of atomic or molecular beams of the constituent elements (of the layer to be grown) with a crystalline substrate held at high temperature in ultrahigh vacuum.

8.1.5 Lightwave System Applications

As mentioned previously, one of the principal applications of semiconductor lasers is in the area of lightwave transmission systems. These systems have been made possible by two major technological advances: the development of low-loss silica (glass) fibers, which act as a transmission medium, and the development of high-performance reliable semiconductor lasers. Early lightwave systems used multimode fibers and an operating wavelength of about 0.85 μm. AlGaAs semiconductor lasers are used as sources for these systems [10]. Current systems using single-mode fibers and an operating wavelength near 1.3 μm offer longer repeater spacings because of the lower silica fiber loss near 1.3 μm than at $\sim$0.85 μm. Even larger repeater spacing is allowed for an operating wavelength near 1.55 μm, where the fiber loss is minimum. Semiconductor lasers fabricated using the InGaAsP material system are by and large the only sources for commercial high-data-rate long-laul lightwave systems operating near 1.3 and 1.55 μm. Because of its substantial impact on lightwave technology, a large portion of semiconductor laser research and development has been on AlGaAs ($\lambda \approx 0.85$ μm) and InGaAsP ($\lambda \approx 1.3$ and 1.55 μm) lasers. For the same reason, we shall use AlGaAs and InGaAsP material systems as examples to illustrate the concepts in semiconductor lasers.

The principles of semiconductor injection operation are described in Section 8.2. The radiative recombination mechanism that forms the basis of laser action is described in Section 8.3. Nonradiative recombination mechanisms, principally Auger recombination and its effect on the performance of long-wavelength lasers, are discussed in Section 8.4. The recombination mechanisms in a relatively new type of higher-performance lasers, that is, quantum well lasers, are described in Section 8.5. Early work in the area of semiconductor lasers has been extensively reviewed in three books [11–13]. Much of the material presented in this chapter concerns recent developments.

8.2 PRINCIPLES OF OPERATION

8.2.1 The Dielectric Waveguide

Conceptually, stimulated emission in a semiconductor laser arises from electron–hole radiative recombination in the active region, and the light generated is confined and guided by a dielectric waveguide (Fig. 8.4). The active region has a slightly higher index than the p- and n-type cladding layers, and the three layers form a dielectric waveguide. The energy distribution of the fundamental mode of the waveguide is also sketched in Figure 8.4. A fraction of the optical mode is confined in the active region. Two types of fundamental transverse modes can propagate in the waveguide: the transverse electric (TE) and the transverse magnetic (TM) modes. The

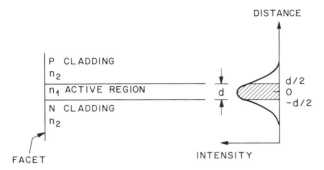

Figure 8.4 The dielectric waveguide of the semiconductor laser. n_2 is the refractive index of the cladding layers and n_1 that of the active region. $n_1 > n_2$. The cladding layers are of a higher bandgap material than the active region. The intensity distribution of the fundamental mode is shown. The cross-hatched region represents the fraction of the mode (Γ) within the active region.

confinement factor (Γ), the fraction of mode in the active region, has been previously calculated (Ref. 12, Part A, p. 34). Figure 8.5 shows the calculated Γ as a function of active layer thickness for the TE and TM modes for the $\lambda = 1.3\ \mu$m InGaAsP double heterostructure with p-InP and n-InP cladding layers.

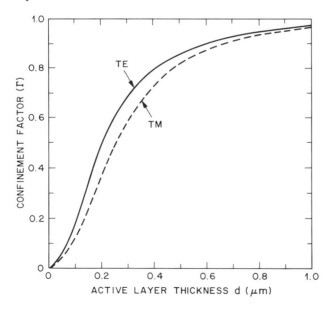

Figure 8.5 Confinement factor of the fundamental TE and TM modes for a waveguide with InGaAsP ($\lambda = 1.3\ \mu$m) active layer and InP cladding layers as a function of the thickness of the active region. The refractive indexes of the active and cladding region are 3.51 and 3.22, respectively.

8.2.2 Threshold Condition

At threshold, the optical gain equals the total optical loss in the laser cavity. The condition for threshold is

$$\Gamma g_{th} = \alpha_a \Gamma + (1 - \Gamma)\alpha_c + \frac{1}{L} \ln \frac{1}{R} \tag{1}$$

where g_{th} is the threshold gain in the active region; α_a, α_c are the absorption losses in the active and cladding regions, respectively; L is the cavity length; and R is the mirror facet reflectivity. Typically, $L \approx 300 \ \mu m$ and $R \approx 0.3$ For a 0.2-μm-thick active layer, Γ(TE) is ~ 0.47 (from Figure 8.5), and using $\alpha_a \approx 30 \ cm^{-1}$ the calculated $g_{th} \approx 150 \ cm^{-1}$. This compares with a value of $10^{-2} - 10^{-3} \ cm^{-1}$ for He-Ne lasers [14].

8.2.3 Condition for Stimulated Emission

Sufficient numbers of electrons and holes must be excited in the semiconductor for stimulated emission or net optical gain. The condition for net gain at photon energy E is given by [15]

$$E_{fc} + E_{fv} = E - E_g \tag{2}$$

where E_{fc}, E_{fv} are the quasi-Fermi levels of electrons and holes, respectively, measured from the respective band edges (positive into the band) and E_g is the band gap of the semiconductor. For zero net gain (transparency), the above condition becomes $E_{fc} + E_{fv} = 0$. For undoped material at a temperature T the quasi-Fermi energy E_{fc} is related to the injected carrier (electron or hole) density n by

$$n = N_c \frac{2}{\sqrt{\pi}} \int \frac{d\varepsilon}{1 + \exp(\varepsilon - \varepsilon_{fc})} \tag{3}$$

with

$$N_c = 2 \left(2\pi m_c \frac{kT}{h^2}\right)^{3/2} \quad \text{and} \quad \varepsilon_{fc} = E_{fc/kT}$$

where k is the Boltzmann constant, h is Planck's constant, T is the temperature, and m_c is the effective mass of the electrons in the conduction band. A similar equation holds for holes. Figure 8.6 shows the variation of the injected carrier density for transparency as a function of temperature for undoped $\lambda = 1.3 \ \mu m$ InGaAsP. The parameter values used in the calculation are $m_c = 0.061 \ m_0$, $m_{hh} = 0.45 \ m_0$, $m_{lh} = 0.08 \ m_0$, where m_0, m_{hh}, m_{lh} are the free electron, heavy hole, and light hole mass, respectively. Figure 8.6 shows that n_t is considerably smaller at low temperatures. The threshold carrier density is ~ 20–30% higher than n_t.

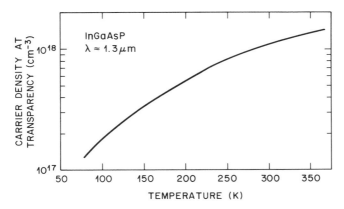

Figure 8.6 The calculated injected carrier density for transparency as a function of temperature for undoped $\lambda = 1.3$ μm InGaAsP.

The high carrier densities needed for gain can be generated in a semiconductor by optical excitation or current injection. The first injection laser operation was demonstrated in 1962 [1–3]. The injection laser utilized the unique characteristics of the *pn* junction to confine carriers near the depletion region under forward bias.

8.2.4 The PN Junction Laser

Under high current injection through a *pn* junction, a region near the depletion layer can have a high density of electrons and holes. These electrons and holes can recombine radiatively if the interfaces are free of traps. The device will lase if the threshold condition [Eq. (2)] is satisfied.

The energy-band diagram of a *pn* junction between two similar semiconductors (homojunction) at zero bias is shown in Figure 8.7. The dashed line represents the Fermi level. Under forward bias $V \approx E_g/e$, both electrons and holes are present near the depletion region. This region can have net gain if the electron and hole densities are sufficiently high. However, the thickness of the gain region is very small (~ 100 Å), which makes the confinement factor (Γ) for an optical mode very small. Hence from Eq. (1) it follows that the threshold gain, and hence the threshold current, is high for a homojunction laser.

The threshold current of an injection laser was historically reduced using a double heterostructure for carrier confinement, which increased the size of the region of optical gain [16, 17]. The double heterostructure laser utilizes a *pn* heterojunction for carrier injection. A heterojunction is a junction in a single crystal between two dissimilar semiconductors. Thus the fabrication of heterojunctions had to wait until the development of epitaxial growth techniques. The energy-band diagram of a double heterostructure laser at zero

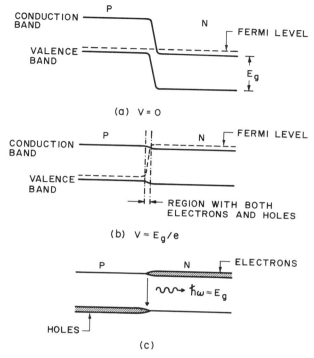

Figure 8.7 Energy band diagram of a *pn* junction at (a) zero bias, (b) forward bias $V \approx E_g/e$. (c) Schematic representation of electrons and hole densities under forward bias $V \approx E_g/e$.

bias is shown in Figure 8.8. The device consists of a narrow-gap semiconductor (*p*-type, *n*-type, or undoped) sandwiched between higher-gap *p*-type and *n*-type semiconductors. The narrow-gap semiconductor is the light-emitting region (active region) of the laser. The dashed line represents the Fermi level. For the purpose of illustration we have chosen the active region to be *p*-type in Figure 8.8. The band diagram under forward bias $V \approx E_g/e$ is shown in Figure 8.8b. Electrons and holes are injected at the heterojunction and are confined in the active region. Thus the region of optical gain is determined by the thickness of the active region in double heterostructure lasers. In addition, the refractive index of the lower-gap active region is higher than that for the *n*-type and *p*-type higher-gap confining (also known as cladding) layers. These layers form a waveguide for the lasing optical mode, as discussed in Section 8.2.1.

The active region thickness of the double heterostructure lasers is typically in the range 0.1–0.3 μm. Within the last few years, with the development of MBE and MOCVD growth techniques, it has been possible to fabricate very thin epitaxial layers (< 300 Å) bounded by higher-gap cladding

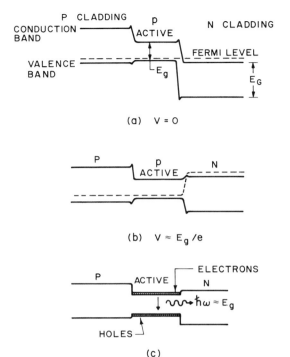

Figure 8.8 Energy band diagram of a double heterostructure laser. (a) Zero bias; (b) forward bias $V \approx E_g/e$.

layers. These double heterostructures are called quantum well double hetero-structures because the kinetic energy for carrier motion along the thickness of the active region is quantized, similar to that for a one-dimensional potential well [18]. The modification of the electron–hole recombination characteristics (which is the basis for laser action) in quantum wells is described in Section 8.5. For the purposes of this chapter, a laser device with an active layer thickness greater than 300 Å is a double heterostructure laser, and for smaller active layer thickness we shall call it a quantum well laser.

8.3 RADIATIVE RECOMBINATION

The basis of light emission in semiconductors is the recombination of an electron in the conduction band with a hole from the valence band; the ex-cess energy is emitted as a photon (light quantum). The process is called radiative recombination. The energy versus wavevector diagram of the elec-trons and holes in a cubic (zinc blende type) semiconductor is shown in Figure 8.9. For direct gap semiconductors, the bottom of the conduction band and the top of the valence band are at the same point in momentum space or $\vec{k}$ space ($\vec{k} = 0$ in Fig. 8.9). This allows both energy and momentum

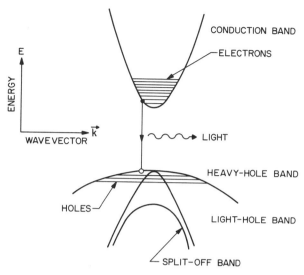

Figure 8.9 Energy versus wavevector of the four major energy bands for a zinc blende type of direct gap semiconductor.

conservation in the process of photon emission by electron–hole recombination. For indirect gap semiconductors (e.g., silicon), the momentum conservation can be achieved with the assistance of a phonon (lattice vibration), which significantly decreases the probability of radiative recombination

The valence band in many III–V semiconductors is represented by three major subbands. These are the heavy hole band, the light hole band, and the spin-splitoff band. The radiative transitions occurring near bandgap energies are due to the recombination of electrons with heavy holes and light holes. A heavy hole, as the name implies, has larger effective mass than a light hole, which makes the density of states (and also the available number of heavy holes for a given Fermi level) larger than that for the light holes.

8.3.1 Gain Calculation

The quantities associated with a radiative recombination are the absorption spectrum, emission spectrum, gain spectrum, and total radiative emission rate. The optical absorption or gain for a transition between the valence band and the conduction band at an energy E is given by (Ref. 12, Part A, p. 129)

$$\alpha(E) = \frac{e^2 h}{2\varepsilon_0 m_0^2 cnE} \int_{-\infty}^{\infty} \rho_c(E')\rho_v(E'')|M(E', E'')|^2$$
$$\times \left[f(E'' = E' - E) - f(E') \right] dE' \tag{4}$$

where m_0 is the free electron mass, e is the electron charge, ε_0 is the permittivity of free space, E is the photon energy, n is the refractive index at energy E, and ρ_c and ρ_v are the densities of state for unit volume per unit energy in the conduction and valence band, respectively. $f(E')$ is the probability that a state of energy E' is occupied by an electron, and M is the effective matrix element between the conduction band state of energy E' and the valence band state of energy E''. Hwang [19] has shown that the contribution of the band-tail impurity states can be significant for photon energies near the band edges. Several models for the density of states and matrix element for transition between band-tail states exist. (e.g., Ref. 11, Part A, Chap. 3). The latest of such models that take into account the contributions of the parabolic bands and impurities is a Gaussian fit of the Kane form to the Halperin-Lax model of band tails. This was first proposed by Stern [20] and used to calculate the gain and recombination rate in GaAs.

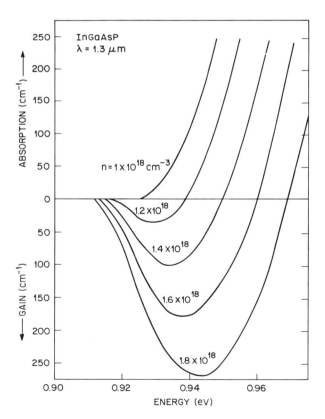

Figure 8.10 Calculated gain or absorption as a function of photon energy for $\lambda = 1.3$ μm ($E_g = 0.96$ eV) InGaAsP at various injected carrier densities. (After Dutta [22].)

The matrix element M may be expressed as a product of two terms: $M = M_b M_{env}$. The quantity M_b arises from the band-edge Bloch functions and M_{env} arises from the envelope wave functions. For III–V semiconductors using the Kane model [21],

$$|M_b|^2 = \frac{m_0^2 E_g}{12 m_c} \frac{E_g + \Delta}{E_g + (\frac{2}{3})\Delta} \tag{5}$$

where m_c is the conduction-band effective mass, E_g is the energy gap, and Δ is the spin–orbit coupling. The envelop matrix element for the band-tail states have been calculated by Stern.

The calculated spectral dependence of absorption or gain at various injected carrier densities is shown in Figure 8.10 for InGaAsP ($\lambda \approx 1.3~\mu$m) with acceptor and donor concentrations of 2×10^{17} cm^{-3}. The calculation was done using Gaussian Halperin-Lax band tails and Stern's matrix element. The material parameters used in the calculation are $m_c = 0.061 m_0$, $m_{hh} = 0.45 m_0$, $m_{lh} = 0.08 m_0$, $\Lambda = 0.26$ eV, $E_g = 0.96$ eV, and $\varepsilon = 11.5$. Figure 8.10 shows that the gain peak shifts to higher energies with increasing injection. Figure 8.11 shows the maximum gain as a function of injected carrier density at different temperatures. Note that considerably lower injected carrier density is needed at a lower temperature to achieve the same gain. This is the origin of the lower threshold current at low temperature. It is often convenient to use a linear relationship between the gain g and the injected carrier density n of the form

$$g = a(n - n_0) \tag{6}$$

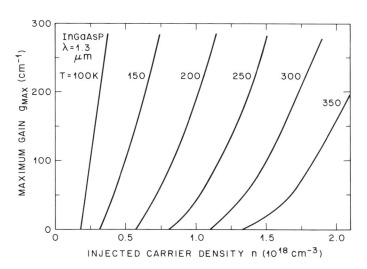

Figure 8.11 The maximum gain as a function of injected carrier density for undoped InGaAsP ($\lambda = 1.3~\mu$m) at different temperatures. (After Dutta and Nelson [27b].)

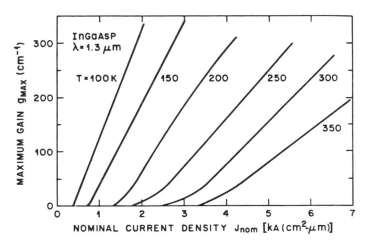

Figure 8.12 The maximum gain as a function of nominal current density for InGaAsP (λ = 1.3 μm) at different temperatures. (After Dutta and Nelson [27b].)

where a is the gain coefficient and n_0 is the injected carrier density at transparency.

8.3.2 Spontaneous Emission Rate

At unity quantum efficiency, the total spontaneous radiative recombination rate R equals the excitation rate. The latter is usually expressed in terms of the nominal current density J_n [20, 22]:

$$J_n \, [\mathrm{A/(cm^2\text{-}\mu m)}] = eR \tag{7}$$

with
$$R = \int r_{spon}(E) \, dE$$

where e is the electron charge and the thickness of the active region is assumed to be 1 μm. $r_{spon}(E)$ is the spontaneous emission rate at a photon energy E. It is given by [22]

$$r_{spon}(E) = \frac{4\pi n e^2 E}{m^2 \varepsilon_0 h^2 c^3} \int_{-\infty}^{\infty} \rho_c(E')\rho_v(E'')|M(E', E'')|^2 \, f(E')[1 - f(E'')] \, dE' \tag{8}$$

The calculated maximum optical gain g as a function of the nominal current density at various temperatures is shown in Figure 8.12 for InGaAsP ($\lambda \approx 1.3$ μm). The calculation is for an undoped lightly compensated material with 10^{17} cm^{-3} of acceptors and donors, respectively. Note that gain varies linearly with J_n above a certain gain.

The total spontaneous radiative recombination rate R can be approximated by

$$R = Bnp \tag{9}$$

where B is the radiative recombination coefficient and n, p are the electron and hole densities, respectively. For undoped semiconductors, Eq. (9) becomes $R = Bn^2$. For GaAs, the measured $B = 1 \times 10^{-10}$ cm^3/s. Calculation of the radiative recombination rate shows that B decreases with increasing carrier density [20]. This has been recently confirmed by Olshansky and coworkers [23] using carrier lifetime measurements.

8.3.3 Threshold Current Calculation

In double heterostructure lasers, the injected carriers can recombine by both radiative and nonradiative recombination. The injected current density J is simply the sum of the radiative R and nonradiative R_{nr} recombination rates in the absence of carrier leakage:

$$J = e(R + R_{nr})d = J_r + J_{nr} \qquad (J_r = J_n d) \tag{10}$$

where e is the electron charge, d is the active layer thickness, and J_r, J_{nr} are the radiative and nonradiative components of the current density, respectively.

For a GaAs double heterostructure laser where the quantum efficiency is believed to be close to unity, that is, most of the injected carriers recombine radiatively, $J \approx J_r$. The relation between the optical gain and the nominal current density for GaAs material system has been calculated by Stern [20]. It is given by

$$g = 0.045(J_{\text{nom}} - 4222) \text{ cm}^{-1} \tag{11}$$

where J_{nom} is expressed in amperes per square centimeter per micrometer. The threshold current of a 250-μm-long GaAs-AlGaAs double heterostructure laser with 2 μm $\times$ 0.2 μm active region can now be calculated using Eqs (1) and (11). The calculated $\Gamma \approx 0.6$ for 0.2-μm-thick GaAs-Al$_{0.36}$Ga$_{0.64}$As double heterostructure. Assuming $R = 0.3$ and $\alpha_a = \alpha_c = 20$ cm^{-1}, we get $g_{\text{th}} = 114$ cm^{-1}, $J_{\text{th}} = 1.35$ kA/cm^2, and a threshold current of 6.7 mA. This compares well with the experimentally observed threshold currents in the range 5–10 mA for GaAs-AlGaAs double heterostructure lasers with good current confinement. Figure 8.12 shows that the current density needed to achieve a certain optical gain increases with temperature; that is, the threshold current density of a laser is expected to increase with increasing temperature. It is experimentally observed that the threshold current density J_{th} of a laser varies with the temperature T as $J_{\text{th}} \approx J_0 \exp{(T/T_0)}$, where T_0 is a characteristic temperature determining the temperature sensitivity. For $g_{\text{th}} \approx 114$ cm^{-1}, the calculated $T_0 \approx 210$ K from Figure 8.12 in the temperature range 300–350 K. This agrees well with the measured values for GaAs lasers.

For InGaAsP lasers, a significant amount of nonradiative recombination is believed to be present at carrier densities comparable to lasing threshold near room temperature. The nonradiative recombination can increase the threshold current density, as discussed in the next section.

8.4 NONRADIATIVE RECOMBINATION

An electron–hole pair can recombine nonradiatively, meaning that the recombination can occur through any process that does not emit a photon. In many semiconductors—for example, pure germanium or silicon—the nonradiative recombination dominates radiative recombination. The measurable quantities associated with nonradiative recombination are internal quantum efficiency and carrier lifetime. The variation of these quantities with parameters such as temperature, pressure, and carrier concentration are by and large the only way to identify a particular nonradiative recombination process.

The effect of nonradiative recombination on the performance of injection lasers is to increase the threshold current. If τ_{nr} is the carrier lifetime associated with the nonradiative process, the increase in threshold current density is given approximately by

$$J_{nr} = \frac{e n_{th} d}{\tau_{nr}} \tag{12}$$

where n_{th} is the carrier density at threshold, d is the active layer thickness, and e is the electron charge.

The nonradiative recombination processes described in this section are the Auger effect, surface recombination, and recombination at defects or traps.

8.4.1 Auger Effect

Since the pioneering work by Beattie and Landsberg [24] it is generally accepted that Auger recombination can be a major nonradiative recombination mechanism in narrow-gap semiconductors. Recent attention to the Auger effect has been in connection with the observed greater higher temperature dependence of threshold current of long-wavelength InGaAsP lasers compared to short-wavelength AlGaAs lasers [25–29]. It is generally believed that the Auger effect plays a significant role in determining the observed high-temperature sensitivity of threshold current of InGaAsP lasers emitting near 1.3 and 1.55 μm.

There are several types of Auger recombination processes. The three major types are band-to-band processes, phonon-assisted Auger processes, and trap-assisted Auger processes.

The band-to-band Auger processes in direct gap semiconductors are shown in Figure 8.13. The three processes are labeled CCCH, CHHS, and CHHL, where C stands for the conduction band and H, L, S stand for heavy-hole, light-hole, and split-off valence band hole, respectively. The CCCH mechanism involves three electrons and a heavy hole and is dominant in n-type material. The process was first considered by Beattie and Landsberg

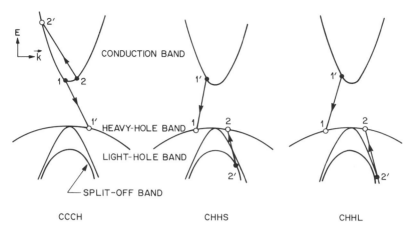

Figure 8.13 Band-to-band Auger recombination processes in a direct gap semiconductor.

[24]. The CHHS process involves one electron, two heavy holes, and a split-off band hole. The CHHL process is similar to the CHHS process except that it involves a light hole. The CHHS and CHHL mechanisms are dominant in p-type material. Under the high injection conditions present in lasers, all these mechanisms must be considered.

Band-to-band Auger processes are characterized by a strong temperature dependence and bandgap dependence, the Auger rate decreasing rapidly either for low temperature or for high-bandgap materials. These dependencies arise from the energy and momentum conservation that the four free particle states involved (1, 2, 1′, 2′ in Fig. 8.13) must satisfy. This may be seen in the following way for the CCCH process. The momentum and energy conservation laws give rise to a threshold energy E_T for each of the processes. For the CCCH process, if we assume $E_1 \approx E_2 = 0$, only holes with energies greater than $\sim(E_T - E_g) = \delta E_g$ can participate (δ is a constant that depends on effective masses). The number of such holes varies approximately as $\exp(-\delta E_g/kT)$ for nondegenerate statistics and thus the Auger rate varies as $\exp(-\delta E_g/kT)$ in the nondegenerate case. In the absence of momentum conservation, there is no threshold energy E_T. Thus the strong temperature dependence does not appear if the individual particle states are not states of definite momentum, for example, if they are trap states, or if momentum conservation is satisfied through phonon assistance. Examples of a CCCH-type phonon-assisted process and a shallow donor trap assisted process are shown in Figure 8.14.

The Auger rate R_a in n-type semiconductors with a carrier concentration n_0 varies as [24]

$$R_a = Cn_0^2 \delta p \tag{13}$$

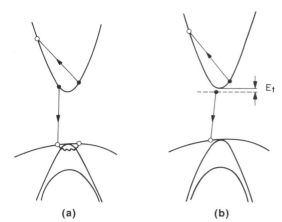

Figure 8.14 (a) Phonon-assisted CCCH-type Auger recombination process. (b) Auger recombination process using a donor trap.

where $\delta p \ll n_0$ is the injected minority carrier (hole) density and C is the Auger coefficient. Thus the minority carrier lifetime is given by

$$\tau_A = \frac{R_a}{\delta p} = \frac{1}{Cn_0^2} \tag{14}$$

The calculated τ_A for n-InGaAsP with a carrier concentration of 10^{18} cm^{-3} is shown in Figure 8.15. The phonon-assisted process dominates at high bandgap, and the band-to-band processes dominate for low-bandgap semiconductors.

The active region of an InGaAsP laser is nominally undoped. Under high injection the Auger rate R_a varies approximately as

$$R_a = Cn^3 \tag{15}$$

where n is the injected carrier density. Calculations of the Auger coefficient using the Kane band model [21] yield a value of 1×10^{-28} cm^6/s for $\lambda \approx 1.3$ μm InGaAsP. Haug [29] has calculated a value of $\sim 2.5 \times 10^{-29}$ cm^6/s for $\lambda \approx 1.3$ μm InGaAsP using a different model for the band structure. The experimental values of the Auger coefficient for this material are in the range $2 \times 10^{-29} - 7 \times 10^{-29}$ cm^6/s. For GaAs, $C \approx 10^{-31}$ cm^6/s. From Eqs. (10) and (15), the current lost to Auger recombination at threshold is given by

$$J_A = edCn^3 \tag{16}$$

Thus the effect of Auger recombination on the threshold current of GaAs lasers is small compared to that for InGaAsP lasers because the Auger coefficient in GaAs is smaller by two orders of magnitude. Since the threshold carrier density increases with increasing temperature, the carrier loss to the nonradiative Auger effect increases with increasing temperature, which results in a more rapid increase of threshold current with increasing temperature (low T_0) for long-wavelength InGaAsP lasers than for AlGaAs lasers. The

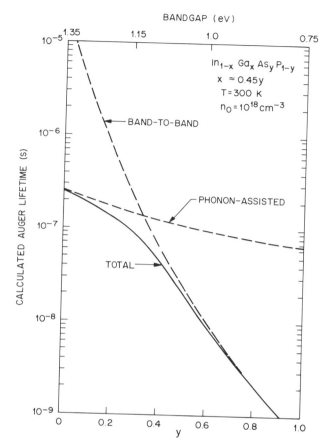

Figure 8.15 The calculated Auger lifetime for n-InGaAsP with a carrier concentration of 10^{18} cm^{-3}. (After Dutta and Nelson [27b].)

measured variation of threshold current density as a function of temperature for InGaAsP lasers can be expressed by the relaton $J_{th} \approx J_0 \exp (T/T_0)$ with $T_0 \approx 50$–70 K in the temperature range 300–350 K. Auger recombination plays a significant role in determining the smaller T_0 values of InGaAsP ($\lambda \approx 1.3$ μm, 1.55 μm) lasers compared to shorter wavelength ($\lambda \approx 0.85$ μm) AlGaAs lasers.

8.4.2 Surface Recombination

In an injection laser, the cleaved facets are surfaces exposed to the ambient. In addition, in many index-guided laser structures, the edges of the active region can be in contact with curved surfaces, which may not be a perfect lattice. A surface, in general, is a strong perturbation of the lattice, creating many dangling bonds that can absorb impurities from the ambient. Hence a high concentration of defects can occur that can act as nonradiative recombination centers. Such localized nonradiative centers, in addition to increasing the threshold current, can cause other performance problems (e.g., sustained oscillations) in lasers.

The recombination rate of carriers at the surface is expressed in terms of a surface recombination velocity S. If A is the surface area and n_{th} the threshold carrier density, then the increase in threshold current, ΔI_{th}, due to surface recombination is given by

$$\Delta I_{th} = e n_{th} S A \tag{17}$$

where e is the electron charge. The surface recombination velocity for InP surface exposed to air is about two orders of magnitude smaller than that for GaAs.

8.4.3 Recombination at Defects

Defects in the active region of an injection laser can be formed in several ways. In many cases, they are grown in during the epitaxial growth process. They can also be generated, multiply, or propagate during a stress aging test [30]. Defects can propagate along a specific crystal axis in a strained lattice. The well-known dark line defect (DLD; dark region of linear aspect) is generally believed to be responsible for the high degradation rate (short life span) of early AlGaAs lasers.

Defects in general produce a continuum of states in a localized region. Electrons or holes that are within a diffusion length from the edge of the defect may recombine nonradiatively via the continuum of states. The rate of recombination at a defect or trap is usually written as

$$R = \sigma v N_t \tag{18}$$

where σ is the capture cross section of the trap, N_t is the density of traps, and v is the velocity of the electrons or holes. A trap can preferentially capture electrons or holes. The study of recombination at defects in semiconductors is a vast subject, a detailed discussion of which is beyond the scope of this chapter.

8.5 QUANTUM WELL LASERS

As shown in Figure 8.1, a conventional double heterostructure laser consists of an active layer sandwiched between higher-gap cladding layers. The active layer thickness is typically 0.1–0.5 μm. In the last few years, double heterostructure lasers have been fabricated with an active layer thickness ~ 100 Å or less. The carrier (electron or hole) motion normal to the active layer in these structures is restricted. This may be viewed as carrier confinement in a one-dimensional potential well, and hence these lasers are called quantum well lasers [31, 32].

8.5.1 Energy Levels

When the thickness L_z of a narrow-gap semiconductor layer confined between two wide-gap semiconductors becomes comparable to the de Broglie wavelength ($\lambda = h/p \sim L_z$), quantum-mechanical effects are expected to occur. The energy levels of the carriers confined in the narrow-gap semiconductor can be determined by separating the Hamiltonian into a component normal to the layer (z component) and into the usual (unconfined) Bloch function components (x, y) in the plane of the layer. The resulting energy eigenvalues are

$$E(n, k_z, k_y) = E_n + \frac{\hbar^2}{2m_n^*} (k_z^2 + k_y^2) \tag{19}$$

where E_n is the nth confined-particle energy level for carrier motion normal to the well, m_n^* is the effective mass of the nth level, $\hbar$ is Planck's constant ($h/2\pi$), and k_x, k_y are the usual Bloch function wavevectors in the x and y directions.

Figure 8.16 shows schematically the energy levels E_n of the electrons and holes confined in a quantum well. The confined particle energy levels E_n are

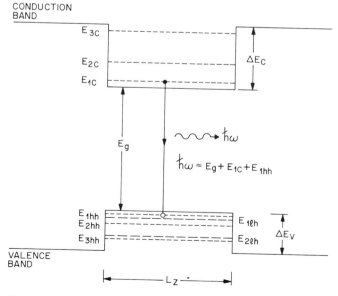

Figure 8.16 Energy levels in a quantum well structure. (After Holonyak et al. [32].)

denoted by E_{1c}, E_{2c}, E_{3c} for electrons; E_{1hh}, E_{2hh} for heavy holes; and E_{1lh}, E_{2lh} for light holes. The calculation of these quantities is a standard problem in quantum mechanics for a given potential barrier (ΔE_c, ΔE_v). For an infinite potential well, the following simple result is obtained:

$$E_n = \frac{h^2 n^2}{8L_z^2 m_n^*} \tag{20}$$

Since the separation between the lowest conduction band level and the highest valence band level is given by

$$E_q = E_g + E_{1c} + E_{1hh} \approx E_g + \frac{h^2}{8L_z^2}\left(\frac{1}{m_c} + \frac{1}{m_{hh}}\right) \tag{21}$$

it follows that in a quantum well structure the energy of the emitted photons can be varied simply by varying the well width L_z. Figure 8.17 shows the

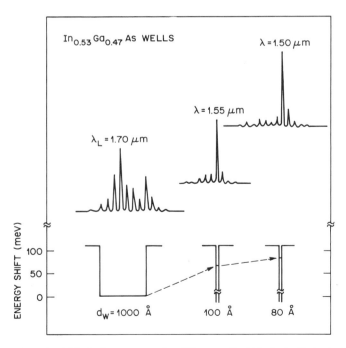

Figure 8.17 The lasing spectrum for different well widths for InGaAs single quantum well structures. (Courtesy H. Temkin.)

experimental results of Temkin et al. [31] for an InGaAs quantum well bounded by InP cladding layers.

8.5.2 Gain and Radiative Recombination

The optical gain and spontaneous emission rate in a quantum well structure can be calculated using Eqs. (4) and (8) with an appropriate modification of the density of the states. If the bands in the x, y directions are assumed parabolic, the density of states (per unit area) for a given subband (nth level) is given by

$$\rho = \frac{m_n^*}{\pi \hbar^2 L_z} \tag{22}$$

where m_n^* is the effective mass of the subband. Since the density of states is independent of energy, in quantum well structures a group of electrons with nearly the same energy (e.g., E_{1c}) can recombine with a group of holes also with nearly the same energy (e.g., E_{1hh}). In bulk semiconductors, $\rho \approx E^{1/2}$; that is, the recombining electrons and holes are distributed over a wide energy range with smaller densities near the band edges where the Fermi factors allow more occupancy. Thus in quantum well structures the optical gain at a given injected carrier density can be larger than that for bulk semiconductors.

Electron–hole recombinations in a quantum well follow the selection rule $\Delta n = 0$; that is, the electrons in states E_{1c} (E_{2c}, E_{3c}, etc.) can combine with the heavy-hole states E_{1hh} (E_{2hh}, E_{3hh}, etc.) and light-holes states E_{1lh} (E_{2lh}, E_{3lh}, etc.). Note, however, that since $E_{1lh} > E_{1hh}$, the light-hole transitions are at a higher energy than the heavy-hole transitions.

The optical gain calculation in quantum well structure has been calculated using both the $\dot{k}$ selection rule (for the x, y direction) and a constant matrix element approximation [33–35]. Figure 8.18 shows the calculated maximum gain for the $1c \rightarrow 1hh$ transition as a function of injected carrier density for a 200-Å thick GaAs quantum well. Note that very high gain ($\sim 10^3$ cm^{-3}) can be obtained at a lower injected carrier density than shown in Figure 8.11.

Figure 8.19 shows the maximum gain for the transition $1c \rightarrow 1hh$ plotted as a function of nominal current density [defined in Eq. (7)]. J_n is calculated using both light- and heavy-hole transitions. Sugimura [35] has theoretically investigated the effect of well thickness on the maximum optical gain. Although the theoretical investigations provide a methodology for analysis, the numerical accuracy of the calculations is limited because of the many assumptions made with respect to transition matrix elements, effective masses, and so on.

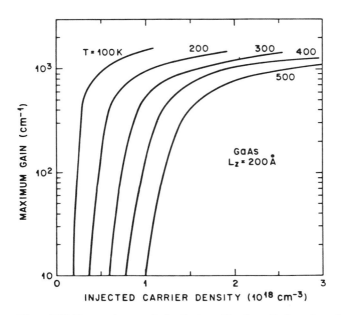

Figure 8.18 The maximum gain for the transition from the lowest conduction band level to the highest heavy-hole level ($1c \rightarrow 1hh$) plotted as a function of injected carrier density. (After Dutta [34b].)

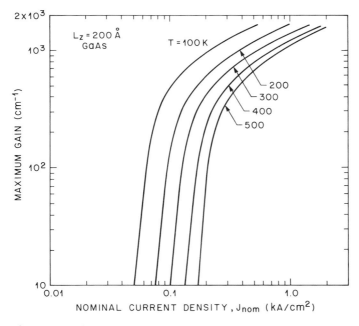

Figure 8.19 Calculated maximum gain for the $1c \rightarrow 1hh$ transition plotted as a function of nominal current density. (After Dutta [34b].)

8.5.3 Threshold Current Calculation

The threshold gain of a laser is given by Eq. (1). The quantum well laser has a small confinement factor Γ of the lasing mode; for example, for a 200-Å-wide GaAs-Al$_{0.52}$Ga$_{0.48}$As quantum well double heterostructure, calculated $\Gamma \approx 0.04$. Using $\alpha_a = \alpha_c = 10 \text{ cm}^{-1}$, $R = 0.3$, and $L = 380 \ \mu\text{m}$, we get $g_{\text{th}} \approx 10^3 \text{ cm}^{-1}$, which is an order of magnitude larger than the usual values for conventional double heterostructure lasers. From Figure 8.19, the threshold current density is $\sim 550 \text{ A/cm}^2$ at 300 K for $g_{\text{th}} = 10^3 \text{ cm}^{-1}$. This compares well with the results of Tsang and Ditzenberger [36] of 810 A/cm^2 for a 200-Å-wide GaAs single quantum well laser.

The confinement factor of a single quantum well (SQW) heterostructure can be significantly increased using a graded-index cladding layer. This allows the use of the intrinsic advantage of the quantum well (QW) structure (high gain at a low carrier density) without the penalty of a small mode confinement factor. Threshold current densities as low as 200 A/cm^2 have been reported for a GaAs graded-index (GRIN) quantum well laser [37].

Another device that allows the use of the intrinsic advantage of QW structure along with a high mode confinement factor is the multi-quantum well (MQW) laser. Threshold current densities as low as 250 A/cm^2 have been reported for AlGaAs MQW lasers [38]. Current injection in MQW lasers is an interesting problem. It has been shown that a proper choice of barrier height, barrier width, and active layer well thickness may be necessary in order to achieve low threshold operation [38, 39].

8.5.4 Nonradiative Recombination

The interest in nonradiative recombination for quantum well lasers arises from two important factors:

1. The ratio of the epitaxial interface area to the active volume is considerably larger for quantum well active layer lasers than that for conventional double heterostructure lasers.
2. The possibility of reduced importance of nonradiative Auger recombination rate in quantum well structures may allow the fabrication of InGaAsP lasers with low temperature dependence of threshold current.

Since both SQW and MQW lasers with low threshold current have been fabricated, it is possible to conclude that defect-free epitaxial interfaces for the AlGaAs material system have been grown by MBE or MOCVD. Furthermore, GaAs MQW lasers with a median lasing lifetime of about 5000 hr at 70°C have been reported. A more interesting observation is that of Petroff [40], who found that in contrast to a regular double heterostructure the dislocations in an MQW structure do not behave as nonradiative centers.

The importance of Auger recombination on the temperature sensitivity of threshold current of InGaAsP lasers has been theoretically investigated by several authors. These Auger rate calculations are similar to those for bulk semiconductors except that the density of states is modified. For band-to-band Auger processes the Auger rate increases with decreasing bandgap and with increasing temperature. The temperature dependence and bandgap dependence arise from the energy and momentum conservation (in the direction normal to the well) that the four particle states involved must satisfy.

Since the band structure of a quantum well system away from the band edge and the transition matrix elements are not well known, only an order-of-magnitude estimate of the Auger rate can be obtained. These estimates suggest somewhat lower temperature sensitivity of threshold current ($T_0 \approx$ 80–130 K) for InGaAsP ($\lambda \approx 1.3$ μm) quantum well lasers compared to regular InGaAsP ($\lambda \approx 1.3$ μm) double heterostructure lasers, which exhibit T_0 values in the range 55–65 K in the temperature range 10–60°C. Low threshold InGaAsP ($\lambda \approx 1.3$ μm) quantum well lasers with T_0 values higher than those of conventional double heterostructure lasers have been reported [41].

Acknowledgment

The author gratefully acknowledges discussions with of G. P. Agrawal, P. J. Anthony, R. W. Dixon, J. E. Geusic, R. L. Hartman, C. H. Henry, W. B. Joyce, F. R. Nash, T. M. Shen, and D. P. Wilt.

REFERENCES

1. R. N. Hall, G. E. Fenner, J. D. Kingley, T. J. Soltys, and R. O. Carlson, *Phys. Rev. Lett.*, **9**: (1962).
2. M. I. Nathan, W. P. Dumke, G. Burns, F. H. Dill, Jr., and G. Lasher, *Appl. Phys. Lett.*, **1**: 63 (1962).
3. T. M. Quist, R. H. Rediker, R. J. Keyes, W. E. Krag, B. Lax, A. L. McWhorter, and H. J. Ziegler, *Appl. Phys. Lett.*, **1**: 91 (1962).
4. N. Holonyak, Jr., and S. F. Bevacqua, *Appl. Phys. Lett.*, **1**: 82 (1962).
5. C. A. Burrus and R. W. Dawson, *Appl. Phys. Lett.*, **17**: 97 (1970).
6. H. Nelson, *RCA Rev.*, **24**: 603 (1963).
7. W. F. Finch and E. W. Mehal, *J. Electrochem. Soc.*, **111**: 814 (1964)).
8. A. Y. Cho, *J. Vac. Sci. Technol.*, **8**: 531 (1971).
9. R. D. Dupuis, *J. Crystal Growth*, **55**: 213 (1981).
10. R. W. Dixon. *Bell Syst. Tech. J.*, **59**: 669 (1980).
11. H. Kressel and J. K. Butler, *Semiconductor Lasers and Heterojunction LEDs*, Academic, New York, 1977.
12. H. C. Casey, Jr., and M. B. Panish, *Heterostructure Lasers*, Academic, New York, 1978.
13. G. H.B. Thompson, *Physics of Semiconductor Laser Devices*, Wiley, New York, 1981.
14. A. Yariv, *Quantum Electronics*, 2nd ed., Wiley, New York, 1976, p. 306.
15. M. G. A. Bernard and G. Duraffourg, *Phys. Stat. Solidi*, **1**: 699 (1961).

16. Zh. I. Alferov, V. M. Andreev, E. L. Portnoi, and M. K. Trukan, *Sov. Phys. Semiconductors,* **3**: 1107 (1970).
17. I. Hayashi, M. B. Panish, P. W. Foy, and S. Sumski, *Appl. Phys. Lett.,* **17**: 109 (1970).
18. R. Dingle, W. Wiegman, and C. H. Henry, *Phys. Rev. Lett.,* **33**: 827 (1974).
19. C. J. Hwang, *Phys. Rev., B,* **2**: 4117 (1970).
20. F. Stern, *J. Appl. Phys.,* **47**: 5382 (1976); *IEEE J. Quantum Electron.,* **QE-9**: 290 (1973).
21. E. O. Kane, *J. Phys. Chem. Solids,* **1**: 249 (1957).
22. N. K. Dutta, *J. Appl. Phys.,* **51**: 6095 (1980); **52**: 55 (1981).
23. R. Olshansky, C. B. Su, J. Manning, and W. Powaznik, *IEEE J. Quantum Electron.,* **QE-20**: 838 (1984).
24. A. R. Beattie and P. T. Landsberg, *Proc. R. Soc. London,* **249**: 16 (1959); *Ser. A.,* **258**: 486 (1960).
25. Y. Horikoshi and Y. Furukawa, *Jap. J. Appl. Phys.,* **18**: 809 (1979).
26. G. H. B. Thomson and G. D. Henshall, *Electron. Lett.,* **16**: 42 (1980).
27. N. K. Dutta and R. J. Nelson, (a) *Appl. Phys. Lett.,* **38**: 407 (1981); (b) *J. Appl. Phys.,* **53**: 74 (1982) and references therein.
28. A. Sugimura, *IEEE J. Quantum Electron.,* **QE-17**: 627 (1981).
29. A. Haug, *Appl. Phys. Lett.,* **42**: 512 (1983).
30. B. C. DeLoach, Jr., B. W. Hakki, R. L. Hartman, and L. A. D'Asaro, *Proc. IEEE,* **61**: 1042 (1973).
31. R. Dingle, *Festkorperprobleme* **XV**: 21–47 (1975).
32. N. Holonyak, Jr., R. M. Kolbas, R. D. Dupuis, and P. D. Dapkus, *IEEE J. Quantum Electron.,* **QE-16**: 170 (1980).
33. K. Hess, B. A. Vojak, N. Holouyak, Jr., R. Chin, and P. D. Dapkus, *Solid State Electron.,* **23**: 585 (1980).
34. N. K. Dutta, (a) *Electron. Lett.,* **18**: 451 (1982); (b) *J. Appl. Phys.,* **53**: 7211 (1982).
35. A. Sugimura, *IEEE J. Quantum Electron.,* **QE-20**: 336 (1984).
36. W. T. Tsang and J. A. Ditzenberger, *Appl. Phys. Lett.,* **39**: 193 (1981).
37. S. D. Hersee, B. de Cremoux, and J. O. Duchemin, *Appl. Phys. Lett.,* **44**: 476 (1984).
38. W. T. Tsang, *Appl. Phys. Lett.,* **39**: 786 (1981).
39. N. K. Dutta, *IEEE J. Quantum Electron.,* **QE-19**: 794 (1983).
40. P. Petroff, in *Defeats in Semiconductors,* J. Narayan and T. Tan (Eds.), North-Holland, Amsterdam, 1984, p. 457.
41. N. K. Dutta, S. G. Napholtz, R. Yen, R. L. Brown, T. M. Shen, N. A. Olsson, and D. C. Craft, *Electron. Lett.,* **20**: 727 (1984).

CHAPTER 9

Fabrication and Characterization of Semiconductor Lasers

P. J. Anthony

9.1 INTRODUCTION

The semiconductor laser is an amazing device that would merely be another scientific curiosity except that commercial use is growing very rapidly in several areas. These business applications drive the technology to continually find ways to improve the laser characteristics and capabilities, fabrication techniques, and testing procedures. Although the development of semiconductor laser products provides the resources that allow interesting science to be done, it can also create problems such as restrictions of proprietary information and specification gamesmanship. Despite these concerns, semiconductor lasers are readily available for investigation. More semiconductor lasers have been made than all other types of lasers put together, and some versions sell for less than $10.

It is the interplay between the electrons, photons, phonons, and defects in the material that makes the semiconductor laser extremely interesting to study. A device smaller than a grain of sand (Figure 9.1) can produce more power than the typical He-Ne gas laser, but with an electric to optical energy conversion efficiency greater than 10%. The actual volume of the device involved in lasing is smaller still, typically only 10^{-10} cm^{-3}. Although only about 10^4 photons reside in the lasing cavity at any time, that still produces a power density at the laser mirror 100 times that at the surface

Figure 9.1 Photograph of an InGaAsP laser mounted onto a ceramic heat sink.

of the sun. Some lasers can operate from liquid helium temperatures to over 250°C, and the output can be modulated faster than 20 GHz by simply modulating the supply current. The best of the devices are expected to operate as lasers continuously for decades, but the ones that do not provide a rich variety of defects for study.

When the possibility of lasing action in direct gap semiconductors was suggested and refined around 1960, the materials and technology were barely adequate to produce lasing in GaAs crystals at liquid nitrogen temperatures. In the following decade, much research and invention went into producing better lasers, primarily by making advances in the materials, in the hope of producing room-temperature lasing [1–3]. The enhanced optical and electrical confinement produced by the double heterostructure construction finally resulted in room-temperature continuous lasing action in AlGaAs devices in 1970. In a double heterostructure, a thin layer of semiconductor is sandwiched between two layers of another semiconductor with higher bandgap and lower index of refraction to confine electrons and photons in the thin layer.

Three fortunate properties of nature helped to drive the technology in the early 1970s. The first was that silica glass fiber (which still had some residual water absorption peaks) had its minimum optical absorption at wavelengths in the near-infrared (0.8–0.9 μm), matching the emission wavelength of GaAs. The second was that the lattice parameters of GaAs and AlAs are nearly the same, so that high-quality crystals can be grown over the entire (Al, Ga)As composition range to make double heterostructures. The third is that the lower-bandgap material required for electrical carrier confinement also has a higher optical index, producing optical confinement in the same region. These allowed the development, of a relatively simple laser structure in an easy-to-manipulate material system in order to meet the needs of optical fiber transmission.

That technology window provided the opportunity for the semiconductor laser to develop on steep learning curves in semiconductor physics, in materials growth and characterization, and in device design, processing, and characterization. The knowledge gained later allowed the rapid development of lasers in the more complex InGaAsP materials system to match the new absorption minima at 1.3 and 1.6 μm in higher purity glass fibers and the production of inexpensive devices for consumer applications.

The following sections on fabrication and testing will focus on the currently dominant laser technologies of InGaAsP and GaAlAs. Fabrication involves the growth of a double heterostructure on a substrate, defining of an optical cavity, imposition of current confinement, and mounting. Section 9.3 describes tests that are performed on an unmounted chip, a bonded device, or a packaged laser. Trends for the future are considered in the concluding section.

9.2 LASER FABRICATION

9.2.1 Group III–V Materials

The group IV elemental semiconductors such as Si and Ge have indirect bandgaps, and thus the probability of producing photons upon electron–hole recombination is negligible in these materials. However, many covalent compounds of group III and group V elements have direct bandgaps and high radiative efficiencies. Single-crystal wafers (substrates) with diameters of ≲10 cm are used as a base on which to grow several layers of the same lattice constant a but different compositions to provide the double heterostructure bandgap steps for carrier confinement. Large single crystals suitable for slicing into substrates have been produced in the III–V material system primarily for binary (two-element) crystals by the Czochralski and Bridgman techniques [4]. Consequently, commercial lasers at present are restricted to compositions that match the lattice constants of the several binary substrates available, for example, InP and GaAs (Fig. 9.2). As noted above, the ternary AlAs-GaAs system is in close match over the entire composition

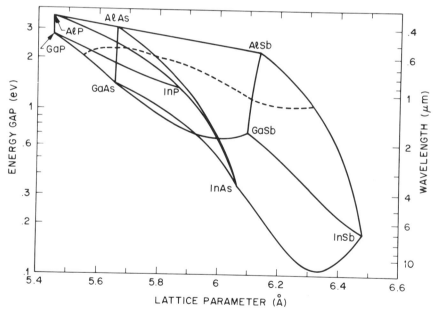

Figure 9.2 Plot of the direct bandgaps of group III–V semiconductor compounds versus lattice parameter. At energies approximately above the dashed line, the compounds are indirect semiconductors. The indirect bands change more slowly with composition than the direct bands; thus, above the dashed line, the indirect bandgaps are smaller than the direct gap values plotted (Based on data from Casey and Panish [2].)

range (the maximum $\Delta a/a$ is 1.3×10^{-3}), but lattice matching to InP requires the more complicated quaternary InP-In$_{0.53}$Ga$_{0.47}$As system.

For efficient radiative recombination, the density of nonradiative recombination sites such as dislocations and precipitates must be minimized in the active volume. Since dislocations can grow and impurities can diffuse from the substrate into subsequently grown layers, it is important to use high-quality substrates to grow laser structures. Commercial vendors of GaAs and InP substrates are continually making improvements in the purity level, doping control, and dislocation density of their products. However, after the $\sim\frac{1}{2}$-mm-thick slices are cut from the large single crystals and characterized for crystallographic orientation, doping level, and defect density, damage can be introduced into the substrate by improper handling. The III–V semiconductors are generally much softer than silicon, and wafer-processing techniques must be gentle (Fig. 9.3). Handling should be minimized and must be confined to unused portions of the wafer. Mechanical polishing introduces hidden damage beneath the surface that should be removed by chemical etching. Submonolayer coverages of impurities on the substrate surface can contaminate layers grown on top and must be removed by cleaning procedures developed with the help of surface analytical techniques. Fastidious attention to substrate quality is necessary for success in the even more difficult and subtle process of growing defect-free epitaxial layers.

9.2.2 Epitaxial Growth

Epitaxial growth is the trick of getting chemicals in either the liquid or gas phase to react, condense out, or precipitate onto a substrate, growing atomic layer by atomic layer a crystal that continues the subtrate's crystal lattice. The difficulty lies in growing a layer of crystal nearly defect-free, of a precise thickness, and of a composition such as In$_{.76}$Ga$_{.24}$As$_{.55}$P$_{.45}$. For the crystal to be latticed-matched to within $|\Delta a/a| \approx 10^{-4}$ so that misfit dislocations do not form requires control of the component proportions in the grown crystal to 0.1%. Since the incorporation rates into the crystal vary for different elements, the compositions in the gas or liquid phases do not correspond to those in the crystal. Because the thermodynamics and kinetics of the growth process are not precisely known, fine-tuning of the starting compositions to achieve acceptable lattice matching must be done empirically for each specific growth condition. Since the growth of most III–V crystals proceeds well only at temperatures $\gtrsim 600°$C, additional problems of differences in thermal expansion of the various compositions and evaporative loss of the more volatile species make epitaxial growth seem at times more of an art than a science. However, rigorous attention to all details and a fixation on cleanliness and reproducibility allow the growth of crystals that make excellent lasers.

Figure 9.3 X-ray topograph of an InP substrate revealing internal damage induced by handling that is not visible on the water surface. (W. R. Wagner.)

To grow a laser structure, lattice-matched layers are grown in sequence with the correct bandgap energy to form a double heterostructure with energy barriers on either side of a thin active layer. The barriers should be $\gtrsim 0.3$ eV high, roughly 10 times the thermal energy of electrons at room temperature. For efficient electrical injection of carriers, the p-type and n-type doping of the layers must be in the neighborhood of 10^{18} cm^{-3} with a pn junction very near the active volume.

The easiest laser structure to grow consists of a series of planar layers, shown in Figure 9.4. Additional optical and electrical confinement can then be achieved during processing. However, the highest performance laser structures incorporate either idiosyncrasies of a particular epitaxial growth technique or multiple growth runs with patterning of the wafer to produce confinement in the crystal around the active volume. Several of these buried heterostructure designs will be described in later chapters.

The growth technique first developed was liquid-phase epitaxy (LPE). In this process, a graphite assembly called a boat is used to hold several liquid melts in separate bins. The boat is placed in a furnace with temperatures controlled to better than 0.1°C, in a carefully controlled atmosphere such as H_2 passed through a palladium purifier. Each bin contains a melt primarily composed of one of the to-be-grown crystal elements such as indium or gallium. The other constituents of the melts are precisely determined to give the desired crystal compositions at the growth temperature. The melts can either be entirely liquid (single phase) and be on the border of supersaturation or they can be in equilibrium with another solid such as a piece of GaAs. In either case, as the temperature is lowered at a rate of about 0.1°C/min the melt components begin to crystallize onto a substrate, which is slid beneath each melt in turn. As the components become depleted near the substrate–melt interface, diffusion from the bulk of the melt tends to restore the depleted components, but at varying rates. Convection and mixing due

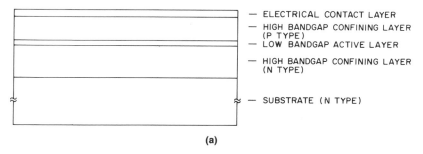

(a)

Figure 9.4 (a) Line drawing of the basic double heterostructure. (b) Photograph of the surface on an InGaAsP laser wafer grown by liquid-phase epitaxy. (c) Scanning electron micrograph of a cleaved section through the layers grown. The active layer is 0.15 μm thick.

(b)

(c)

Figure 9.4 (continued)

to the sliding of the substrate under the melt also contribute to the complexity of the growth process. With this technique, however, low-defect layers can be grown with precise crystal compositions, from several monolayers to several micrometers thick, and with uniformities within several percent over several square centimeters of substrate.

In spite of its success in growing good crystals, the liquid-phase epitaxy process has a multitude of interactions that must be precisely controlled. Several vapor-phase growth techniques have been used to grow good crystals with even better uniformity, and the interactions in the vapor phase are conceptually easier to control. One of these, molecular beam epitaxy (MBE), consists simply of heated crucibles of the crystal components in an ultrahigh vacuum system. The thermally evaporated molecules travel by line of sight to the substrate, where they condense to grow additional crystal layers. The crystal composition is controlled by the temperatures of the crucibles and thus by the relative evaporation rates. Other vapor-phase epitaxy processes pass streams of gases containing the desired components over a heated substrate, where the gases react. The crystal composition is controlled by regulating the gas flows. The several techniques use some gases in common but are differentiated by their use of chlorides, hydrides, or metal-organics to transport some of the components. (For more detailed discussions of the several growth technologies, see Refs. 5 and 6.)

All of the standard epitaxial growth techniques and their variants are able to produce the generic laser structure. The quality of the crystal grown, however, can vary widely and may need to be assessed using a number of destructive and nondestructive analysis techniques. The lattice parameters of the various layers can be measured using X-ray diffraction, and defects such as dislocations can be imaged using X-ray topography. A careful examination of the crystal surface by optical microscopy and scanning electron microscopy (SEM) can reveal much about how uniformly the crystal grew, and an SEM determination of the layer thicknesses on a portion cleaved from the wafer provides essential information. Photoluminescence evaluation of the material at both room temperature and low temperatures can provide information on the intentional and unintentional doping of the layers, the layer bandgaps and compositions, and the radiative efficiency of the material. Analysis techniques such as Auger spectroscopy, secondary ion mass spectroscopy, transmission electron microscopy, and transmission cathodoluminescence can also provide useful information on crystal quality, but the ultimate and most meaningful test is to make devices from the material and evaluate the lasers.

9.2.3 Device Processing

The goals of device processing are to provide optical and electrical confinement, electrical contacts, mounting surfaces, and mirrors for the lasers while doing no damage to the grown laser crystal. The elaborateness of the processing depends very much on the device structure, ranging all the way to the complexity of integrated circuits, as discussed in Chapter 23. Here we focus on the simpler laser structures.

To reduce the amount of processing required, it is always preferable to perform an operation on the entire wafer rather than after it is separated into individual lasers. Also, all of the processing procedures must be designed to minimize the amount of stress that the crystal experiences in order to limit the introduction and growth of defects in the active volume.

To confine the optical field and the carrier recombination to a small active volume, many different device structures (see Chapter 10) have been devised which have portions of the crystal etched away before, after, or between crystal growths. Thus the normally planar layers produced by the epitaxial growth processes may be broken along the plane, and material of different optical index and electrical properties substituted to make an active lasing region in a restricted volume.

As an example, the main procedures used to produce one type of InGaAsP channeled substrate buried heterostructure laser are shown in Figure 9.5. An InP substrate of n-type doping has a layer of p-type InP either grown on or diffused into the upper 2 μm or so of material. Dielectric material is applied to the surface, and stripes several micrometers wide along the length of the wafer are photolithographically opened in the dielectric. An acidic solution is used to etch away a vee in the InP.

After the remaining layers are epitaxially grown, ohmic contacts to the n and p material need to be made. To do this, metals are evaporated, sputtered, or plated $\gtrsim 1000$ Å thick onto the top and bottom surfaces and sometimes alloyed into the surface of the crystal by heat treatments in the 400°C range. Common p-type contacts are CrAu and AuZn, and n-type contacts include AuGe and AuSn [7]. These contacts may be restricted in area to aid in confining the current to the active volume and to improve the frequency response of the lasers by reducing parasitic capacitances. Additional metal layers may be added to facilitate mounting.

Since typical semiconductor lasers are only a few hundred micrometers wide and long, each wafer contains on the order of 1000 devices. The lasers may be separated by etching, saw cutting, or cleaving. Although both etching and cleaving can produce flat surfaces at the ends of the lasing volume that act as mirrors, cleaving thus far produces the better surfaces. To aid the cleaving operation, the wafer is generally thinned sometime after growth from an easy-to-handle thickness of ~ 250 μm to an easy-to-cleave thickness of ~ 10 μm. Some device structure designs then require the application of lower optical index half-wavelength protective coatings, quarter-wavelength antireflective coatings, or high-index reflective coatings to the mirror facets.

To produce lasers that operate in a single longitudinal mode (see Chapter 11), additional process procedures can be incorporated to make, for example, distributed feedback lasers with gratings etched into the crystal near the active volume.

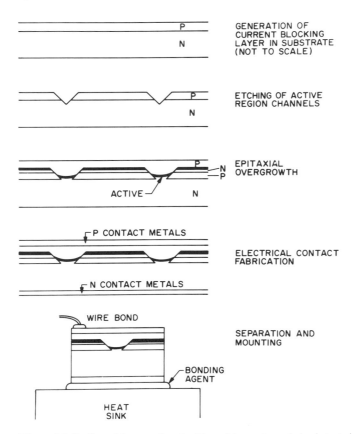

Figure 9.5 Outline of the steps involved in making a channeled substrate buried heterostructure laser.

9.2.4 Device Mounting

Since semiconductor lasers are not 100% efficient optical emitters, operating them continuously at the several-milliwatt output power level generates fractions of a watt of heat that must be removed. The laser must therefore be bonded to a heat sink, but with low strain to prevent degradation due to defects.

Materials such as diamond, SiC, and BeO are good choices for heat sinks, since they closely match the thermal expansion coefficients of III–V materials, although their thermal conductivities are not as high as those of some metals. Solders such as PbSn, AuSn, and In can be used to attach the laser to the heat sink, but care must be taken in the design so that the metallurgical interactions do not produce strain, whiskers, or voids. Epoxy may also be used in some applications. Electrical contact is generally made on one side via the heat sink and on the other side by a wire bond.

For most applications, the laser must then be placed in a package to provide convenient electrical and optical connections. Some packages provide additional features such as a detector to monitor the optical power level or a thermoelectric cooler and control circuitry to provide temperature stability for improved laser performance.

There are three common difficulties with laser packages. The first is that of maintaining coupling stability of the laser output into a fiber or lens. Motion with time and temperature on the order of 1 μm can cause factors of 2 changes in the coupling to a single-mode fiber. The second is minimization of optical reflections. Laser characteristics are very sensitive to external optical feedback, so all optical discontinuities inside and outside the package must be controlled. The third is that rapid changes in technology and the conflicting requirements for low costs and high performance have so far inhibited standardization of package designs.

9.3 LASER CHARACTERIZATION

9.3.1 General Principles

The complexity of the characterization of semiconductor lasers depends on the intended use for the device and also on whether the test engineer is a maker or user of lasers. The user is primarily interested in ensuring that the performance characteristics essential to the system are within specification, while the maker is generally concerned with a wider range of parameters and also seeks to understand the causes of abnormal laser behavior. Thus, tests for lasers are generally based on several routine procedures that can be extended to a more detailed analysis of the device.

The first principle of characterization is that the test procedures for lasers to be used in systems should not harm the device. This is not always easy to achieve, since anyone who begins to experiment with semiconductor lasers soon discovers that they are incredibly fast fuses. Transients generated by connecting the laser to a power supply, turning a power supply on, or electrostatic discharge can in a nanosecond literally melt and explode the mirror facets off lasers, as microscopic examination of failed devices too often shows. Power supplies should be regulated to deliver only the current (~ 100 mA) and voltage (~ 2 V) necessary to drive the laser, a current-limiting element such as a 10-Ω resistor should be connected in series with the laser, and the current should be brought up from zero after connection. Similar precautions and protection schemes are required of any system that incorporates semiconductor lasers.

Although rapid progress is being made in semiconductor laser technology, absolute uniformity of laser characteristics has not been achieved. The second principle is therefore that specifications need to be carefully considered, taking into account the range of characteristics actually produced wherever possible. If the initial design of the system allows a wide range on most of the laser characteristics, considerable expense can be saved in device testing as well as device yield. Standard statistical tests on sample lots can then be used on the less critical characteristics.

The tests that have been made on semiconductor lasers are quite diverse but are generally variations on measurements in time and space of the amplitudes of current, voltage, light, and temperature. Since the number of photons emitted depends on the number of electrons injected into the active volume, current is often the independent parameter. Derivatives of voltage or light versus current are very instructive in revealing the higher-order peculiarities of each laser. Operating the laser with current pulses less than about 1 μs at duty cycles under 1% give the characteristics of the laser roughly independent of the heat sink characteristics. Tests that use longer pulses and duty cycles approach the dc characteristics, which include the effects of device heating of $\sim 10°$C.

9.3.2 Standard Tests

Measurement standards for semiconductor lasers are only now beginning to be established for the most easily characterized laser parameters, and universal test procedures will be a long time in coming. More specialized characterization will always remain a matter of the good judgment and practice of the test engineer.

The most common characteristic of the semiconductor laser is shown in Figure 9.6a. There the measured voltage and light output from the laser are plotted versus the applied current. The voltage should show a standard diode characteristic with a turn-on voltage approximately equal to the bandgap of the laser active layer. The light characteristic should show a slow increase in spontaneous emission light up to a current called the threshold current, after which there is a rapid increase in stimulated laser light. The slope of the light versus current curve gives the external differential quantum efficiency of the laser.

Most semiconductor lasers have an approximate temperature dependence given by $I_{th}(T) = I_{th}(T') \exp\left[(T - T')/T_0\right]$, where T_0 is an empirical fitting parameter. Therefore plots of log I_{th} versus T for pulsed measurements provide a measure of T_0. Comparison of dc measurements of I_{th} with pulsed measurements gives an indication of the temperature rise of the active layer due to heating and thus a measure of the thermal impedance to the heat sink.

Derivatives of the characteristics in Figure 9.6a can graphically provide additional information about the laser. For an ideal laser, the current–voltage characteristic is given by $I = I_0(T)[\exp(eV_a/nkT) - 1]$ with the applied voltage V_a the sum of the series resistance term and the pn junction voltage: $V_a = V_J + IR$. At and above threshold, V_J ideally remains constant. The first

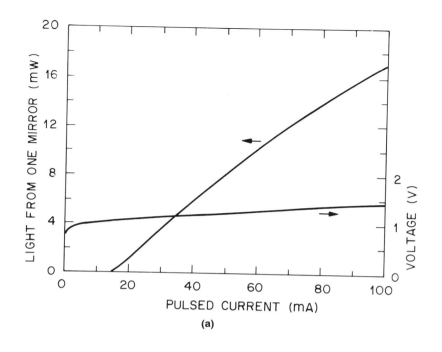

(a)

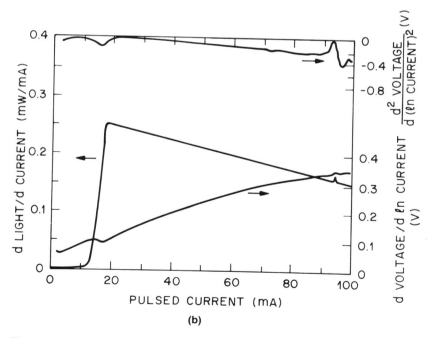

(b)

Figure 9.6 (a) Experimental plot of light and voltage versus current for a laser. (b) Derivatives of the above data that allow much more detailed analysis of the laser properties such as the "kink" at 95 mA.

and second derivatives of V_a with respect to $\ln I$ are simply

$$\frac{\partial V_a}{\partial \ln I} = \begin{cases} \dfrac{nkT}{e} + IR & \text{for } I < I_{th} \\[2mm] IR & \text{for } I > I_{th} \end{cases}$$

$$\frac{\partial^2 V_a}{\partial(\ln I)^2} = \begin{cases} -\dfrac{nkT}{e} & I < I_{th} \\[2mm] 0 & I > I_{th} \end{cases}$$

The graphical presentation of the voltage derivatives (Fig. 9.6b) provides information about the quality of the junction, the series resistance, and threshold. Deviations from ideality are indicative of deficiencies in the laser such as leakage currents around the active volume.

Plots of the derivatives of the light with respect to current versus current provide clear indications of the threshold current as well as the dependence of external differential quantum efficiency η on current. Again deviations from ideality are of interest. The onset of lasing in higher-order waveguide modes or in another polarization shows up as a change in η. The resulting "kink" in the light versus current curve is much more readily observed in dL/dI (Fig. 9.6b). For lasers operated dc, the quality of thermal heat sinking can be evaluated by the amount of decrease in dL/dI at high currents compared to pulsed measurements. Commercial test sets are available to make some of the above measurements, and computer-controlled current sources, light power meters, and voltage meters make the construction of more specialized equipment extremely practical. A calibration uncertainty of ~ 2 dB for optical power meters at 1.3 μm was resolved by the U.S. National Bureau of Standards in 1986, but optical power measurements still require great care to achieve reasonable accuracies.

The output spectrum of the laser is usually measured dc and shows the number of longitudinal modes that are present (Fig. 9.7). Pulsed operation broadens each individual line to ~ 1 Å from $\sim 10^{-3}$ Å due to the changing carrier density in the laser and also increases the amplitude of the weaker longitudinal modes. The measurement of this wavelength "chirp" and the suppression of unwanted side modes are critical for lasers designed to have a single longitudinal mode. Automated commercial test sets are available for routine spectral measurements.

The laser light is emitted from the laser facet only near the active region in a pattern called the laser near field, which reveals the distribution of the optical field in the laser cavity. As the laser beam propagates away from the confined active region, diffraction broadens the beam into the laser far-field pattern. The near field (Fig. 9.8) can be viewed indirectly with a camera (sensitive to the emission wavelength of the laser) attached to a good microscope. The nearly circular Gaussian far-field pattern is usually measured only as slices perpendicular and parallel to the *pn* junction (Fig. 9.9) and characterized by the full width at half-maximum (FWHM) amplitude. The

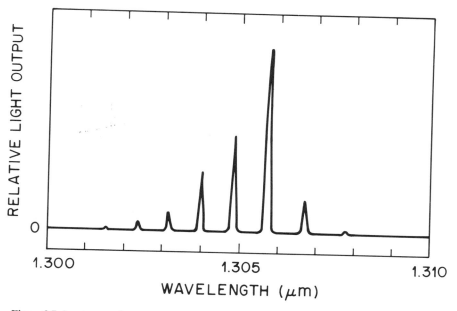

Figure 9.7 Spectrum of a multilongitudinal-mode laser, plotted as light intensity versus wavelength.

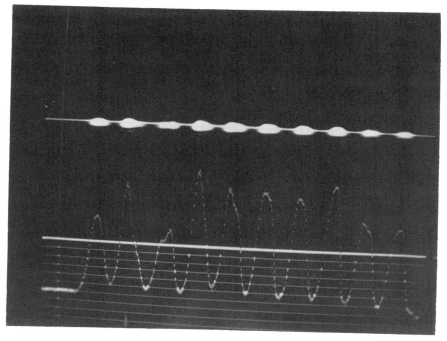

Figure 9.8 Photograph of the near field of a laser array.

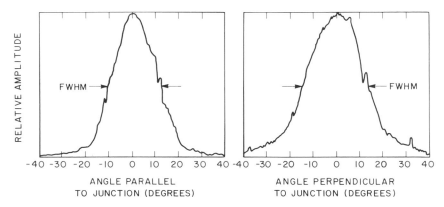

Figure 9.9 The far-field pattern of a laser, taken as slices perpendicular and parallel to the *pn* junction. The full width at half maximum is indicated by the arrows.

measurement obviously only partially characterizes the far-field pattern, but it provides a reasonable figure of merit from which relevant parameters such as fiber coupling can be estimated. Again, deviations from ideal behavior are of interest, such as the side lobes on the central Gaussian characteristics that appear when higher-order transverse modes are present. Laser arrays provide the most interesting near- and far-field patterns, since coupling between adjacent lasers can cause interference effects.

The frequency response characteristics are dominated by the natural resonance frequency of the laser and by parasitic leakage currents and capacitances. The resonance frequency, on the order of a few gigahertz, increases with power, but its value and the strength of the resonance can vary depending, for example, on the guiding mechanism in the particular laser structure. The parasitics are also dependent on the laser structure and on mounting details such as the inductance in wire bonds. These parameters can be measured in the small-signal limit with a fast detector together with a commercial sweep generator and a spectrum analyzer. The large-signal characteristics are often presented as eye diagrams, that is, repetitive oscilloscope traces of the laser output driven by a pseudorandom bit stream that modulates the laser between low- and high-power states. Commercial word generators, fast oscilloscopes, and bit-error-rate detectors are available for assembling test stations.

Since lasers operate with so few ($\sim 10^4$) photons in the cavity, small perturbations such as spontaneously generated photons or reflected light can cause appreciable noise in the output power. Measurements using spectrum analyzers and multichannel analyzers show intrinsic fluctuations that are in good agreement with theory. Additional noise problems such as pulsations due to saturable light absorbers in the laser cavity can be identified and removed from the tested population.

9.3.3 Reliability Tests

Semiconductor lasers have proved to be most different from other semiconductor components in the area of reliability. That is because the laser characteristics depend on optical as well as electrical effects, and the optical properties are nonlinear and sensitive to atomic level imperfections in the laser material. As in light-emitting diodes, lasers can be affected by impurities such as copper, which can reduce the radiative efficiencies of the active region. The recognition that stress plays a strong role in the development of dislocations (dark line defects) that cause rapid and catastrophic degradation in lasers led to substantial improvements in laser reliability.

Present devices are estimated to be able to emit laser light for periods in excess of tens of years under normal operating conditions. To establish such lifetime estimates and, sometimes more important, the expected number of near-term failures requires extrapolation of aging rates to times much longer than the test time. Most lasers appear to have a primary aging mechanism that is a function of temperature with an activation energy of somewhat less than 1 eV. Thus, accelerated aging of lasers at high temperatures produces results that can be extrapolated to normal operating temperatures to give an estimate of laser lifetimes. The precision and accuracy of measurements of laser degradation rates and assumptions about the functional form of the degradation with time also affect estimates of how long the lasers will operate. Several types of reliability tests are necessary for selecting lasers to be used in applications such as undersea transmission cables where failed lasers are difficult to replace [8].

Reliability test apparatus varies greatly in complexity, since a simple constant-current burn-in may suffice in some applications. More elaborate equipment provides feedback to the drive current of the laser to maintain a constant optical power while keeping the laser temperature constant within 0.1°C. Some test sets provide various overstress conditions to remove weaker devices, with the intention of eliminating infant failures from the population.

Often, however, the light-emitting lifetime of a laser represents only the upper bound to its useful life. Other parameters that affect the functionality of the laser in its intended use may change and go out of specification before the device fails to emit laser light. For instance, as the laser ages, its operating wavelength typically changes, and after a few years it may no longer match the dispersion or absorption minima in fiber. Or else the power from the laser may not stay coupled into the fiber if the mounting is not stable. Because of the complexity of monitoring all the relevant laser parameters for changes in time, the long time required for some tests, and the expense of test apparatus, the use of statistical sampling techniques is often the only practical approach laser manufacturers can use to certify their products.

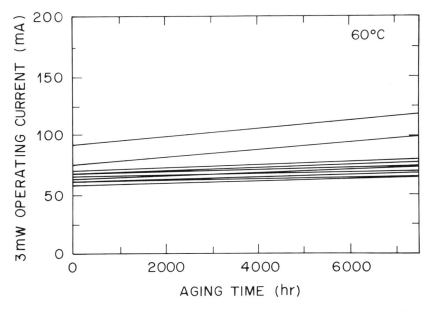

Figure 9.10 Aging characteristics for InGaAsP lasers operated at 60°C and 3 mW. The more rapidly degrading lasers can be eliminated from the population on the basis of their initial aging rates.

The state of the art in both the GaAlAs and InGaAsP material systems are lasers that last for years even at the higher temperatures used to accelerate the aging. The results of testing one type of device are shown in Figure 9.10.

9.3.4 Failure Mode Tests

As in many endeavors, the study of what went wrong can occupy more time than the study of the successful work. Failure mode testing provides critical feedback to the device designer, crystal grower, device fabricator, and test engineer. However, failure mode analysis in semiconductor lasers is especially difficult because of the active volume's small size. The time spent dissecting a failure again depends on the interests of the test engineer, with less time expended by a laser user on a random failure from a large population than by the device designer on a common failure mode in a new structure.

Laser failures can be put into two classes: the degradation of one or more operating characteristics of a laser that still lases and the absence of lasing under normal operating conditions. The first type are easier to study, since many of the routine tests can still be made on the devices. Even though many variables affect these tests, modeling of the abnormalities observed often leads to a correct assessment of the problem. The combination of analysis of the light-current-voltage characteristics of a laser and inspection of the emission patterns from the laser facet using infrared detection is particularly powerful in determining where the applied current is going if it is not producing laser light. Examination with SEM is useful to confirm that the epitaxial layers are as intended.

Optical inspection with a high-power microscope is usually a good starting point for a failed laser. Has the laser been physically damaged, and are the facets still as-cleaved? If mounted, is the device mounting correct? Push-off tests and wire pull tests can reveal problems in the device metallization and bonding. Transmission of below-bandgap light through the chip using crossed polarizing filters can show regions of unusual strain. Defects that grow in the active volume can be observed in the luminescence emitted normal to the junction if windows are opened in the contact metallizations. Dark line defects, dark spots, laser cavity fabrication problems, and blown mirror facets are all easily observed. To uncover sources of some of the problems, the lasers can be profiled and analyzed using techniques such as secondary ion mass spectroscopy, Auger spectroscopy, and transmission electron microscopy.

9.4 FUTURE TRENDS

The learning curve for device fabrication improvements in the future will be driven primarily not by fiber-optic communications but by consumer products such as the compact disk player. The millions of lasers required for these applications provide incentives for more automated manufacture and testing, higher yield processes, and low-cost mounting arrangements. In the past, much of the equipment and many procedures used for semiconductor lasers were adapted from silicon technology. Often considerable modification was required. For example, in device testing, the addition of optical measurements to standard electrical tests made home-built additions essential The continued growth of the semiconductor laser business as well as the expanding effort in GaAs integrated circuits is resulting in increased commercial availability of equipment intended specifically for group III–V material processing and laser characterization.

The improved control and throughput for materials grown by vapor-phase techniques as they are developed should mean a gradual deemphasis of liquid-phase growth technology. The increased uniformity provided by vapor-phase growth may even be required by the trend to small-scale integration of drive transistors and power monitors on the same chip as the laser. For fiber-optic communication systems, the past trend toward lasers with longer emission wavelengths will again depend on fiber technology. If fibers can be made with absorption minima beyond 2 μm, then semiconductor laser development will move past the antimonides in the III–V semiconductors and on to the group II–VI compounds. In the other direction, information storage technology is providing a need for shorter-wavelength devices, into the visible range.

Lasers will continue to be improved in their characteristics, with longer lifetimes, higher powers, higher modulation speeds, higher efficiencies, and narrower linewidths being the obvious ongoing trends. It is possible that the multiplicity of laser designs now made will evolve into several optimal structures for the various applications. Also, standardization in mounting and specifications will become more common, to the benefit of laser users.

In the years since the first room-temperature lasers, many obstacles to commercial devices have been overcome, continually improved characteristics have been produced, and new types of lasers have been developed. If the next twenty years bring as many developments, the semiconductor laser field will remain an important area of science and technology.

REFERENCES

1. H. Kressel and J. K. Butter, *Semiconductor Lasers and Heterojunction LEDs*, Academic, New York, 1977.
2. H. C. Casey, Jr. and M. B. Panish, *Heterostructure Lasers*, Academic, New York, 1978.
3. G. H. B. Thompson, *Physics of Semiconductor Laser Devices*, Wiley, New York, 1980.
4. H. C. Freyhardt (Ed.), *Crystals. Growth, Properties, and Applications. III–V Semiconductors*, Springer-Verlag, New York, 1980.
5. W. T. Tsang (Ed.), *Material Growth Technologies. (Semiconductors and Semimetals*. Vol. 22A) Academic, New York, 1985.
6. G. H. Olsen and T. J. Zamerowski, in *Progress in Crystal Growth and Characterization*, Vol. II, B. R. Pamplin (Ed.), Pergamon, London, 1980.
7. A. G. Milnes and D. C. Feucht, *Heterojunctions and Metal-Semiconductor Junctions*, Academic, New York, 1972.
8. R. L. Hartman (Ed.), *Assuring High Reliability of Lasers and Photodetectors for Submarine Lightwave Cable Systems*, special issue of *AT & T Tech. J.*, **64**(3) (1985).

CHAPTER 10

Transverse Mode Control in Semiconductor Lasers

Kohroh Kobayashi

10.1 INTRODUCTION

The transverse mode for semiconductor laser diodes can be discussed by considering two directions independently: the direction perpendicular to the junction and the direction parallel to the junction. In this chapter, only the horizontal or lateral transverse mode, which is the spatial mode in the junction plane, is considered, because the transverse mode in the direction perpendicular to the junction has been well established to be the fundamental mode when a double heterostructure with a thin active layer is adopted.

In the very early stage of GaAs laser diode development, a stripe contact [1] was introduced into the laser diode structure, where an SiO_2 insulating layer, with a stripe window about 10–50 μm wide, was inserted below the contact metal. Current flows into the active layer with a limited stripe shape in the laser diode crystal. At first, the stripe contact was set up to eliminate the filamentary lasing common in broad contact laser diodes. It was also used to reduce the threshold current and to increase the operating temperature under continuous wave (cw) oscillation. These results indicate that the stripe contact was not at first intentionally used for transverse mode control, although a beautiful Gaussian beam profile was observed in a stripe-contact GaAs homojunction laser diode [2].

The stripe geometry laser diodes exhibited some peculiar behavior, such as nonlinearity in light output versus current characteristics. At first, these peculiarities were thought to be due to some nonuniformities or defects in the laser diode crystal. With improvement in crystal quality, however, such unwanted phenomena did not disappear; instead, they were found to appear very reproducibly. Experimental and theoretical efforts have revealed that the strange phenomena are a result of horizontal mode instability. They were determined to be due to the fact that both the gain or loss and the refractive index depend on the carrier density.

Since then quite a large number of transverse mode stabilization structures have been devised. They can be divided into three categories: gain-induced waveguides, refractive-index-induced waveguides, and intermediate waveguides. In addition to transverse mode stabilizaton, current confinement structures have also been incorporated to realize efficient diode laser operation. Nowadays, laser diodes with a well-defined transverse mode are widely used in optical fiber communication systems as well as in optical disk systems. Furthermore, some of the transverse mode stabilized structures are now used as the base structures for single longitudinal mode laser diodes, which are essential for future long-distance, high-bit-rate optical communications or coherent optical communications.

10.2 TRANSVERSE MODE INSTABILITY

10.2.1 Unwanted Lasing Properties

Many unwanted or anomalous properties have been observed in which laser diodes are not transverse-mode-stabilized. The first universal and reliable evidence of these phenomena is kinking in the light output (L) versus current (I) characteristics in AlGaAs laser diodes with a planar stripe contact [3]. Examples are shown in Figure 10.1a, with $I–L$ curves of mode-stabilized LDs for comparison (Fig. 10.1b) [4]. The light output saturates strongly with the current at the point denoted by arrow A in Figure 10.1a. Another kink is indicated by arrow B. This second kink has been found to be caused by additional higher-order transverse mode lasing. In this section, discussions will be mainly concentrated on the phenomena related to kink A.

Figure 10.2 shows the transverse mode deformation observed in the vicinity of the $L–I$ kink [5, 6]. In these figures, peaks in the near-field and far-field patterns move when the excitation exceeds I_k, which is the current corresponding to the kink. The near-field pattern peak movement indicates a shift in the near-field peak position toward the stripe edge at the mirror facet. The far-field pattern peak movement indicates that the output beam direction changes with the current. The near-field intensity peak shift and the beam direction change are 2–3 μm and 2–3° for a current increase of ~ 30 mA. These transverse mode deformations may cause serious problems in practical application, because such deformations degrade, for example, coupling between a laser diode and an optical fiber or an optical waveguide. It should be noted that the near-field and far-field patterns have well-defined

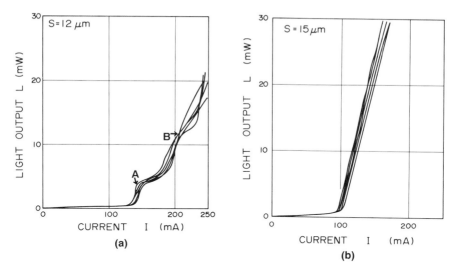

Figure 10.1 Light output versus current characteristics for (a) planar stripe laser diodes (transverse mode not stabilized) and (b) transverse-mode-stabilized deep zinc diffused planar stripe laser diodes.

profiles, which can mostly be described as a fundamental Gaussian function, for excitations above as well as below the kink current. This suggests that these deformations are not caused by higher-order mode oscillation. Spectrum measurements indicate that the number of lasing longitudinal modes increase abnormally above I_k.

Anomalies have also been observed in dynamic lasing characteristics, such as enhanced relaxation oscillation [6]. An anomalous transient response is observed for a current pulse with a peak current above I_k, as shown

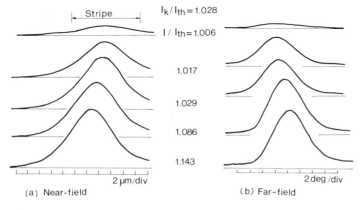

Figure 10.2 Anomalous transverse mode deformation that appears in connection with the $L–I$ kink. (a) Near-field patterns; (b) far-field patterns.

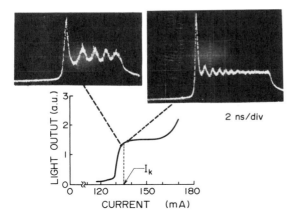

Figure 10.3 Anomalous dynamic lasing properties that appear in connection with the L–I kink.

in Figure 10.3. The relaxation oscillation in the transient response consists of a first spike of high peak intensity followed by strongly damped oscillation peaks. Near-field and far-field pattern measurements, using a time-resolved technique, revealed that the same kind of horizontal mode deformation as that described above occurs throughout the course of the relaxation oscillation.

Unwanted phenomena that occur in association with kink B are multiple horizontal mode and excess noise. For excitation above kink B current, near-field and far-field patterns have multiple peaks and their widths usually increase. The intensity fluctuation curve with respect to the excitation has an additional peak [7] corresponding to lasing of higher-order modes.

10.2.2 Cause of Transverse Mode Instability

Horizontal transverse mode instability described above has been observed mainly in laser diodes having a $10-20$-μm-wide current-confinement region. In these laser diodes, the transverse mode can be supported by the gain-induced guiding mechanism discussed by Cook and Nash [8]. Starting from the gain-induced waveguiding mechanism, the transverse mode deformation can be explained as follows.

Spatial gain and refractive index distributions are provided by the carrier density profile induced by the stripe-shape current flow in the active layer. Stimulated emission occurs strongly in the central stripe portion, where the gain has its maximum. Carrier consumption in the central portion is stronger than at the stripe edge, since the stimulated emission is stronger there, leading to spatial hole-burning. This results in a flatter gain and refractive index profile in the central portion. Therefore, the force that acts to hold the horizontal mode in the central portion is reduced. If there is a slight asymmetry in the gain or refractive-index profile in the horizontal direction, the

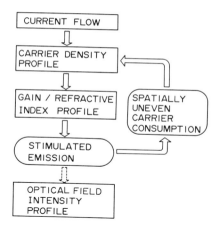

Figure 10.4 Transverse mode instability mechanism for gain-induced waveguiding laser diodes.

near-field intensity profile can easily move toward the stripe edge. Since the refractive index changes inversely with the carrier density, the feedback loop, as depicted in Figure 10.4, tends to accelerate the transverse mode movement.

The uneven spatial carrier consumption and the resulting flatter carrier profile shown in Figure 10.5, where a spontaneous emission profile is used as a measure for the carrier profile, have been clearly observed by Kirkby et al. [9]. The output beam deflection simultaneously observed with the near-field intensity shift can also be explained by taking into account an asymmetrical gain profile, which induces wavefront asymmetry. When the near-field intensity profile shifts toward the stripe edge, the optical field receives less gain than that which was received before the shift. Therefore, lasing efficiency decreases, resulting in light output power saturation, which causes the $I-L$ kinking (kink A).

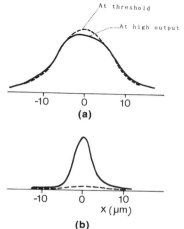

Figure 10.5 Spontaneous and lasing emission near-field profiles for a stripe contact laser diode with a 15-μm stripe width. Uneven carrier consumption and a resulting flatter carrier profile can be seen [9]. (a) Spontaneous profile at 8000 Å; (b) lasing profile at 8100 Å.

For laser diodes with a current stripe wider than about 20 μm, higher-order mode oscillation rather than mode deformation occurs with the current increase. In this case, because the lasing optical field width is less than the carrier density profile width, the carrier density outside the lasing mode continues to increase with increasing current, forming a dip in the carrier density profile. Due to the inverse relationship between refractive index and carrier density, the carrier density dip makes a convex refractive index profile (self-focusing effect). Therefore, no mode deformation occurs for the initially oscillating mode. However, an additional higher-order mode can oscillate, because the gain outside the dip increases with increasing current.

In the transient response for laser diodes without transverse mode stabilization, the deformation in the transverse mode and in the carrier density profile occurs during the relaxation oscillation. Thus, the efficiency saturates, resulting in anomalously damped relaxation oscillation.

10.3 TRANSVERSE MODE CONTROL

10.3.1 Fundamental Concepts

As discussed in the preceding section, the transverse mode instability originated from the fact that the lasing optical field profile changes with excitation because of the change in the carrier density profile, since the gain and the refractive index are related to the carrier density. Therefore, transverse mode stabilization can be achieved by making a fixed optical waveguide, which results in an unchanged optical field profile, even though spatially uneven carrier density consumption still exists. Fundamental to realizing the optical field profile independence is the establishment of a built-in optical waveguide inside the laser diode structure.

Transverse mode properties in an optical waveguide can be described by a scalar Maxwell wave equation, as follows.

$$\left[\frac{d^2}{dx^2} + k^2\varepsilon(x)\right]\psi(x) = \beta^2\psi(x) \tag{1}$$

where x denotes the spatial distance from the center of the waveguide in the junction plane, $\psi(x)$ is the field wave function, k is the wavenumber, and β is the eigenvalue. The complex dielectric constant $\varepsilon(x)$ can be written as

$$\varepsilon(x) = \left[\eta(x) + \frac{iG(x)}{2k}\right]^2 \tag{2}$$

where $\eta(x)$ is the refractive index and $G(x)$ is the gain.

The optical field can be supported in this waveguide when either the real part or the imaginary part of the complex refractive index has a suitable profile. It is called refractive-index-induced waveguiding when the real part takes an important role in the waveguide. In the same manner, the gain profile can support an optical field, which is called gain-induced waveguiding. When both the real and imaginary parts cooperate to confine the optical field in the waveguide, it is called intermediate waveguiding or gain/refractive-index-induced waveguiding.

In the following sections, typical examples of mode-stabilized laser diodes are described with their structures and characteristics.

10.3.2 Gain-Induced Waveguiding

Kogelnik showed theoretically that a cylindrical structure with a radial gain profile can support a Gaussian beam of constant diameter, even if there is no refractive index profile [10]. This is the fundamental concept for gain-induced waveguiding. The waveguiding mechanisms in a stripe geometry LD in a direction parallel to the junction is, in principle, based on this idea, because a gain or loss distribution is provided by the carrier distribution along that direction. The carrier density distribution, however, induces not only gain or loss distribution but also refractive index distribution. Schlosser analyzed the gain-induced mode in a planar structure with stepwise discontinuties of gain or loss [11]. Cook and Nash derived modes in a waveguide with a parabolic gain or loss and refractive index profile [8]. No mutual relationship was taken into account in their analysis between the gain or loss and the refractive index through the carrier density.

As described earlier, unstable horizontal mode behavior exists in stripe geometry LDs with a specific stripe width, since the optical field intensity profile, carrier density profile, and gain/loss and refractive index profiles are related to each other. Lang [12, 13], Kirkby et al. [9], Thompson et al. [14], Chinone [15], and Buss [16] have given theoretical consideration to the horizontal modes in stripe geometry LDs, where the mutual relationships between them have been taken into account. They predicted that by reducing the stripe width to well below 10 μm, the light output power at which the "*I–L* kink" appears can be increased. Mode-stability enhancement with stripe width reduction has also been confirmed experimentally [17, 18].

A carrier density profile and optical field intensity profiles are depicted in Figure 10.6 for narrow-stripe laser diodes. When the stripe width becomes narrow, the optical field spreads over the carrier density profile. The carrier density consumption by the stimulated recombination is not strongly localized at the central portion, as can be seen in the case of a wider stripe. Thus, the carrier density profile does not markedly change with the change in excitation, resulting in horizontal mode stabilization up to higher output power. Figure 10.7 shows calculated light output at the kink with respect to stripe width [13]. The kink light output power increases markedly when the stripe width is reduced to about twice the carrier diffusion length.

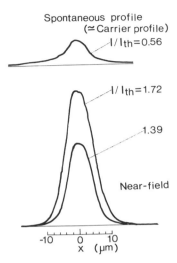

Figure 10.6 Carrier density profile and lasing optical field intensity profile for narrow-stripe laser diodes.

Representative laser diode structures with a narrow stripe geometry are shown in Figure 10.8 [1, 3, 19]. Experimental results obtained from AlGaAs-GaAs planar stripe laser diodes and proton-bombarded stripe laser diodes clearly support the theoretical predictions, as shown in Figure 10.9 [18].

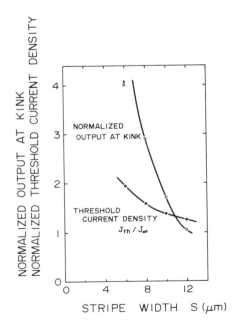

Figure 10.7 Effect of the stripe width on the light output power at the kink [12]. Calculated normalized light output power at the kink and normalized threshold current are shown as a function of stripe width.

(a) Oxide Defined Stripe

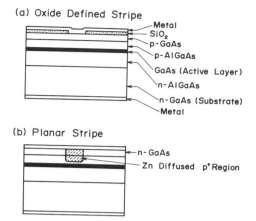

(b) Planar Stripe

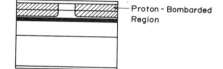

(c) Proton-Bombarded Stripe

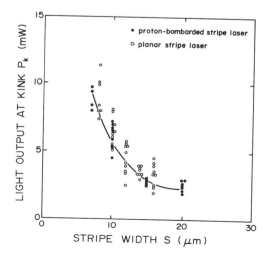

Proton - Bombarded
Region

Figure 10.8 Cross sections of transverse-mode-stabilized laser diodes using gain-induced waveguiding; narrow-stripe laser diodes. (a) Oxide stripe contact, (b) planar stripe, and (c) proton-bombarded stripe laser diodes.

One of the disadvantages of the narrow-stripe laser diodes is an excessive increase in the threshold current density, which is caused by the carrier density spreading. Another disadvantage is the light output beam astigmatism, that is, the light output beam waist does not coincide between directions parallel to and perpendicular to the junction. The beam astigmatism

Figure 10.9 Experimentally observed relation between the light output power at the kink and the stripe width for proton-bombarded stripe and planar stripe laser diodes [18].

originates from the fact that the wavefront in the horizontal direction has a finite curvature, while the double heterostructure gives a flat wavefront in the vertical direction, where the refractive index waveguide is formed. The astigmatism introduces an excessive difficulty in focusing a light beam to a tiny light spot. It could be a serious problem in some applications, such as optical disk reading and writing by laser diodes.

10.3.3 Refractive-Index-Induced Waveguiding

Another horizontal transverse mode stabilization scheme is refractive-index-induced waveguiding, where a built-in refractive index waveguide is formed in the laser diode structure. Requirements on the refractive index profile for transverse mode stabilization are (1) the refractive index step should be high enough to compensate the refractive index change induced by the carrier change, and (2) the waveguide width should be less than the carrier diffusion length. The latter is needed to suppress the carrier spatial hole burning.

The waveguide can be formed fundamentally in two ways: by real refractive index difference and by effective refractive index difference. Figure 10.10 illustrates typical waveguide structures. The real refractive index waveguide can be made by surrounding a core material (active region) with a cladding material having a refractive index lower than that of the core material (Fig. 10.10a). Doping concentration difference can also be used to induce the real refractive index difference (Fig 10.10b). An effective refractive index waveguide has been realized by introducing a thicker portion into the active layer (Fig. 10.10c) or into the waveguiding layer adjacent to the active layer

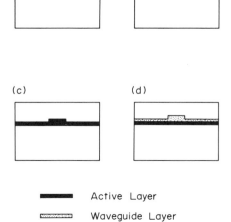

(a) (b)

(c) (d)

■■■ Active Layer

▨▨▨ Waveguide Layer

Figure 10.10 Fundamental concepts for refractive index waveguides. Real refractive index difference attained by use of different materials (a) and by different doping concentrations (b); effective refractive index difference attained by thickness variation in the active region (c) and in the waveguiding region (d).

(Fig. 10.10d) in the lateral direction. The optical field can be confined in the thicker portion.

Typical examples of the real refractive index waveguiding laser diodes are shown in Figure 10.11 [20–25], where different semiconductor material combinations are utilized to form the waveguide. The most basic structure devised with this scheme is a buried heterostructure (BH) laser diode for the 0.8-μm wavelength regions [20]. A narrow GaAs stripe is embedded in AlGaAs layers. For laser diodes in the long-wavelength region (1.2–1.6 μm), an InGaAsP stripe is fully embedded in an InP crystal [21]. Important factors that must be taken into account to realize stable and efficient BH laser diodes are suitable waveguide design, efficient current confinement, and ease of fabrication. Several devices have been developed based on the BH structure. Among them are the buried crescent (BC) [22], the V-grooved substrate buried heterostructure (VSB) [23], the double-channel planar buried heterostructure (DC-PBH) [24], and the current-confinement mesa substrate buried heterostructure (CCM) [25] laser diodes. Laser diodes shown in Figure 10.11a–e are fabricated by two-step epitaxy. In BH and DC-PBH laser diode fabrication procedures, first, narrow stripe mesas are made on a

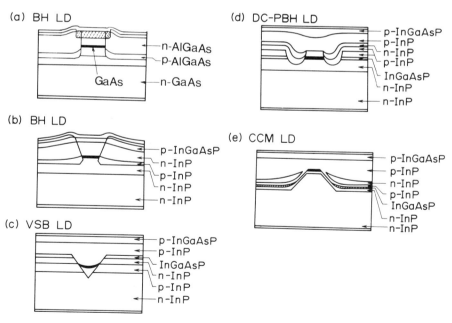

Figure 10.11 Cross sections of transverse-mode-stabilized laser diodes using real refractive index waveguiding with different material combinations. (a), (b) buried heterostructure, (c) V-grooved substrate buried heterostructure, (d) double-channel planar buried heterostructure, and (e) current-confinement mesa-substrate buried heterostructure laser diodes.

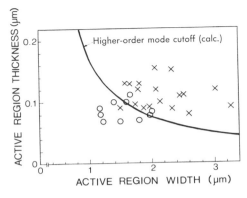

Figure 10.12 Observed transverse mode stability for real refractive-index waveguiding laser diodes in terms of the relation between the active region thickness and the width. The solid line indicates calculated higher-order mode cutoff condition for a passive waveguide with a rectangular cross section.

double heterostructure (DH) wafer. Then they are embedded by *pnp* or *pnpn* crystal layers in the second crystal growth. These layers form a current-blocking structure. In BC and VSB laser diode fabrication procedures, the first epitaxy makes the current-confinement structure, which consists of *pnp* or *pnpn* layers. An active region with a crescent shape is made in grooves in the second epitaxy. Suitable thickness and doping concentration of the current-blocking structure enable efficient current confinement in the narrow active region. The CCM laser diode is fabricated by one-step liquid-phase epitaxy. It improves the lasing properties by introducing a current-confinement structure into a mesa substrate buried heterostructure laser diode [26].

In realizing horizontal-mode-stabilized BH and BH-based laser diodes, the waveguide design is most important. Figure 10.12 shows observed transverse mode stability in terms of a relation between the active region width and the thickness for 1.3-μm InGaAsP DC-PBH laser diodes. A higher-order mode cutoff condition, calculated on the basis of a simple passive waveguide model, is also plotted in the figure. The open circles indicate a device whose transverse mode is stable up to at least five times the threshold. The crosses correspond to a device with an unstable transverse mode. The result indicates that the waveguide parameters should satisfy the higher-order mode cutoff condition to maintain the transverse mode stability up to a high excitation level. For 1.3-μm devices, typical optical waveguide dimensions are 1.5 μm × 0.1 μm. A current versus light output characteristic and far-field patterns for a transverse-mode-stabilized 1.3-μm DC-PBH laser diode are shown in Figure 10.13. Light output power as high as 140 mW has been emitted with a stable fundamental transverse mode in a 1.3-μm DC-PBH laser diode with optimized facet reflectivities [27]. Threshold current, differential quantum efficiency, and far-field radiation angles parallel and perpendicular to the junction for these BH and BH-based laser diodes are typically 10–30 mA, 40–60% (both facets), 10–20°, and 30–50°, respectively.

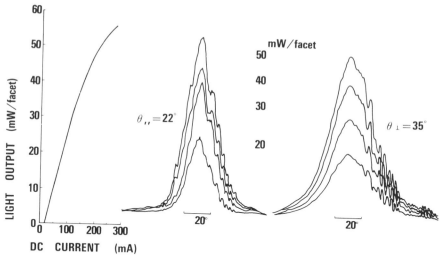

Figure 10.13 L–I curve and far-field patterns for a 1.3-μm DC-PBH laser diode.

Laser diodes with a built-in real refractive index waveguide formed by doping control are transverse junction stripe (TJS) laser diodes [28] and deep zinc diffusion stripe (DDS) laser diodes [29]. Their configurations are depicted in Figure 10.14. Zinc diffusion followed by driving-in through thermal treatment into an n-type GaAs-AlGaAs DH wafer provides a p^+pn

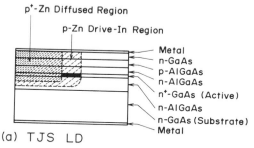

(a) TJS LD

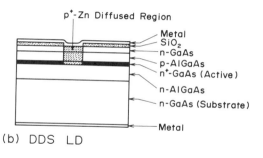

(b) DDS LD

Figure 10.14 Cross sections of transverse-mode-stabilized laser diodes using real refractive index waveguiding with a doping control. (a) Transverse junction stripe and (b) deep zinc diffusion stripe laser diodes.

doping profile along the active layer in TJS laser diodes. The p region refractive index is higher than the surrounding p^+, and n regions due to the reduced plasma effect, resulting in optical waveguide formation. In DDS laser diodes, a high doping concentration in the n-type active region is compensated for by heavy zinc diffusion. The reduced plasma effect again acts to raise the refractive index in the active region and forms a real built-in refractive index waveguide.

Thickness variation in the active layer itself, or in the adjacent waveguiding layer along the active layer, has also been utilized to confine the optical field, for example, in terraced substrate (TS) laser diodes [30], constricted double heterojunction, large optical cavity (CDH-LOC) laser diodes [31], plano-convex waveguide (PCW) laser diodes [32], and ridge waveguide (RW) laser diodes [33]. Their cross-sectional configurations are depicted in Figure 10.15. Current confinement is provided by a stripe contact or a planar stripe contact. Carrier diffusion along the active layer causes the threshold current to become higher than that for laser diodes with a tight

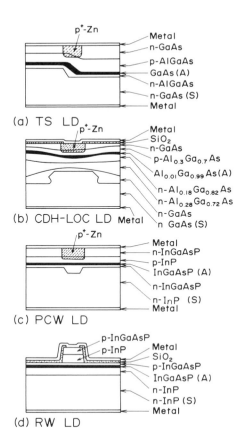

(a) TS LD

(b) CDH-LOC LD

(c) PCW LD

(d) RW LD

Figure 10.15 Cross sections of transverse-mode-stabilized laser diodes using effective refractive index waveguiding. (a) Terraced substrate, (b) constricted double heterojunction large optical cavity, (c) plano-convex waveguide, and (d) ridge waveguide laser diodes. A, active layer; S, substrate.

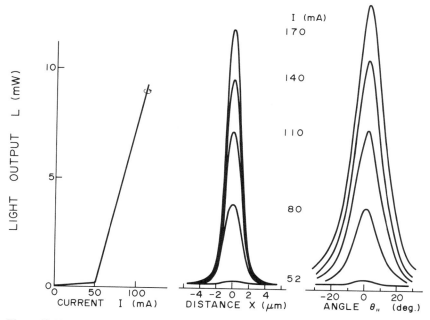

Figure 10.16 $L–I$ curve, near-field and far-field patterns of a 1.3-μm PCW laser diode.

current confinement structure, such as BH or BH-based laser diodes, as shown above. Typical characteristics include threshold currents of 30–60 mA at room temperature and differential quantum efficiencies of 40–70%. Since the effective refractive index difference is small and the optical waveguide boundary along the lateral direction is smooth, the observed far-field patterns are smooth, and interference fringes are rarely observed. These fringes are usually observed in BH or BH-based laser diodes. Near-field and far-field patterns for a 1.3-μm InGaAsP PCW laser diode are shown in Figure 10.16, with the $L–I$ curve.

10.3.4 Gain/Refractive-Index-Induced Waveguiding

In addition to gain-induced waveguiding and refractive-index-induced wave-guiding, a mixed mode is utilized to stabilize the horizontal transverse mode in laser diodes. This is gain/refractive-index-induced waveguiding, where both gain or loss and refractive index play an important role in confining the op-tical field along the horizontal direction. The fundamental model is depicted in Figure 10.17. There is a lossy region outside the active stripe region of the planar DH layers. The active layer is thin enough for the optical field to spread into the lossy region. Therefore, the optical field sees a larger loss outside the stripe region than in the stripe region, resulting in optical field

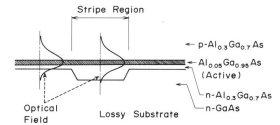

Figure 10.17 Fundamental model for gain/refractive-index-induced waveguiding.

confinement in the stripe region. Theoretical calculation indicates that an effective refractive index, which the optical field feels, can be higher in the stripe region than in the outside region, even though the lossy region material has a higher refractive index than the cladding material. The magnitudes of the effective refractive index difference and the gain/loss difference between the stripe region and the outside region depend on the active layer thickness, the stripe width, and the distance between the active layer and the lossy region. The contributions of refractive index waveguiding and gain/loss waveguiding depend on these parameters.

Cross-sectional views of representative examples for the mixed-mode waveguiding LDs are shown in Figure 10.18. In channeled substrate planar stripe geometry (CSP) laser diodes [34], DH layers are grown on an n-GaAs wafer with an etched channel. The evanescent wave tail reaches the lossy GaAs substrate in the region outside the channel; thus the optical field is

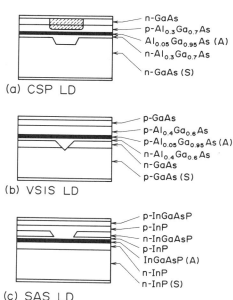

Figure 10.18 Cross sections of transverse-mode-stabilized laser diodes using mixed mode waveguiding. (a) Channeled substrate planar stripe, (b) V-grooved substrate inner stripe, and (c) self-aligned structure laser diodes.

confined to the channel region. The active layer thickness is 0.1 μm, and the adjacent AlGaAs layer thickness is 0.4 μm. With these parameters, a refractive index difference of 3×10^{-3} exists, and stable fundamental lateral mode oscillation has been demonstrated up to at least 2 times the threshold. The threshold current and the differential quantum efficiency at room temperature are typically 40–90 mA and 35%, respectively.

n-Type GaAs layers are formed on both sides of a V-grooved channel in a p-type GaAs substrate in V-grooved substrate inner stripe (VSIS) laser diodes [35]. The n-type GaAs acts as a lossy material to stabilize the lateral transverse mode as well as a current-confinement barrier. For long-wavelength regions, self-aligned structure (SAS) laser diodes [36] are included in this category, where InGaAsP layers constitute the lossy region.

Current confinement is provided by a zinc diffused stripe region in an n-type cap layer in CSP laser diodes or by lossy regions with a conductivity type different from the cladding material in VSIS and SAS laser diodes. The current profile and the lossy regions assist in stabilizing the lateral mode. This gain/loss-induced waveguiding causes output beam astigmatism because of the curved wavefront in a direction parallel to the junction. The ratio of the refractive-index-induced waveguiding contribution to the gain-induced waveguiding contribution can be controlled by the dimensional parameters, such as channel width, active layer thickness, and separation between active layer and lossy region. When the gain-induced waveguiding contribution is enhanced, the emission spectrum tends to consist of multilongitudinal modes. This behavior is utilized to reduce feedback-induced noise by decreasing the coherence in AlGaAs laser diodes for video disk applications.

10.4 CONCLUDING REMARKS

Horizontal transverse mode stabilization is most fundamental in realizing high-performance laser diodes. Horizontal transverse mode stabilization schemes for semiconductor laser diodes have been described in three categories: gain-induced waveguiding, refractive-index-induced waveguiding, and intermediate, or gain/refractive-index-induced, waveguiding. Although quite a few configurations have been devised, only major examples are described.

Because the transverse mode instability can induce longitudinal mode instability, such as two-mode lasing and mode jumping, it is important for single longitudinal mode laser diodes such as distributed feedback or distributed Bragg reflector laser diodes to be based on structures with good transverse mode stabilization characteristics. In addition, for stable high-power operation and a good spatial profile for ease of optical coupling and optical system design, it is essential that transverse-mode-stabilized laser diodes be designed and used.

REFERENCES

1. J. E. Ripper, J. C. Dyment, L. A. D'Asaro, and T. L. Paoli, "Stripe-geometry double-heterostructure junction lasers: mode structure and CW operation above room temperature," *Appl. Phys. Lett.*, **18**: 155–157 (1971).

2. T. H. Zachos, "Gaussian beams from GaAs junction lasers," *Appl. Phys. Lett.*, **12**: 318–320 (1968).

3. H. Yonezu, I. Sakuma, K. Kobayashi, T. Kamejima, M. Ueno, and Y. Nannichi, "A GaAs-AlGaAs double heterostructure planar stripe laser," *Jap. J. Appl. Phys.*, **12**: 1585–1592 (1973).

4. H. Yonezu, Y. Matsumoto, T. Shinohara, I. Sakuma, T. Suzuki, K. Kobayashi, R. Lang, Y. Nannich, and I. Hayashi, "New stripe geometry laser with high quality lasing characteristics by horizontal transverse mode stabilization—refractive index guiding with Zn doping," *Jap. J. Appl. Phys.*, **16**: 209–210 (1977).

5. K. Kobayashi, R. Lang, H. Yonezu, I. Sakuma, and I. Hayashi, "Horizontal mode deformation and anomalous lasing properties of stripe geometry injection lasers—theoretical model," *Jap. J. Appl. Phys.*, **16**: 207–208 (1977).

6. K. Kobayashi, R. Lang, H. Yonezu, Y. Matsumoto, T. Shinohara, I. Sakuma, T. Suzuki, and I. Hayashi, "Unstable horizontal transverse modes and their stabilization with a new stripe structure," *IEEE J. Quantum Electron.*, **QE-13**: 659–661 (1977).

7. R. Lang, K. Minemura, and K. Kobayashi, "Low-frequency intensity noise in C. W. (GaAl)As D. H. lasers with stripe geometry," *Electron. Lett.*, **13**: 228–230 (1977).

8. D. D. Cook, and F. R. Nash, "Gain-induced guiding and astigmatic output beam of GaAs lasers," *J. Appl. Phys.*, **46**: 1660–1672 (1975).

9. P. A. Kirkby, A. R. Goodwin, G. H. B. Thompson, and P. R. Selway, "Observations of self-focusing in stripe geometry semiconductor lasers and the development of a comprehensive model of their operation," *IEEE J. Quantum Electron.*, **QE-13**: 705–719 (1977).

10. H. Kogelnik, "On the propagation of Gaussian beams of light through lenslike media including those with a loss or gain variation," *Appl. Opt.*, **4**: 1562–1566 (1965).

11. W. O. Schlosser, "Gain-induced modes in a planar structures," *Bell Syst. Tech. J.*, **52**: 887–905 (1973).

12. R. Lang, "Horizontal mode deformation and anomalous lasing properties of stripe geometry injection lasers—theoretical model," *Jap. J. Appl. Phys.*, **16**: 205–206 (1977).

13. R. Lang, "Lateral trasverse mode instability and its stabilization in stripe geometry injection lasers," *IEEE J. Quantum Electron.*, **QE-15**: 718–726 (1979).

14. G. H. B. Thompson, D. F. Lovelace, and S. E. H. Turley, "Kinks in the light/current characteristics and near-field shifts in (GaAl)As heterostructure stripe lasers and their explanation by the effect of self-focusing on a built-in optical waveguide," *IEEE J. Solid-State Electron Devices*, **2**: 12–30 (1978).

15. N. Chinone, "Nonlinearity in power-output-current characteristics of stripe geometry injection lasers," *J. Appl. Phys.*, **48**: 3237–3243 (1977).

16. J. Buss, "Multimode field theory explanation of kinks in the characteristics of DH lasers," *Electron. Lett.*, **14**: 127–128 (1978).

17. T. Kobayashi, H. Kawaguchi, and Y. Furukawa, "Lasing characteristics of very narrow planar stripe lasers," *Jap. J. Appl. Phys.*, **16**: 601–607 (1977).

18. M. Ueno, "Linear light output power dependence on effective gain width in stripe geometry (Al, Ga)As DH lasers," *Jap. J. Appl. Phys.*, **16**: 1399–1402 (1977).

19. J. C. Dyment, L. A. D'Asaro, T. C. North, B. I. Miller, and J. E. Ripper, "Proton bombardment formation of stripe-geometry heterostructure lasers for 300 K cw operation," *Proc. IEEE.*, **60**: 726–728 (1972).

20. K. Saito, and R. Ito, "Buried-heterostructure AlGaAs lasers," *IEEE J. Quantum Electron.*, **QE-16**: 205–215 (1980).

21. M. Hirao, S. Tsuji, M. Mizuishi, A. Doi, and M. Nakamura, "Long wavelength InGaAsP/InP lasers for optical fiber communication systems," *J. Opt. Commun.*, **1**: 10–14 (1980).

22. R. Hirano, E. Oomura, H. Higuchi, Y. Sakakibara, and Y. Suzaki, "Low threshold current 1.3 μm InGaAsP buried crescent lasers," *Jap. J. Appl. Phys.*, **22**: 231–234 (1983).

23. H. Ishikawa, H. Imai, T. Tanahashi, K. Hori, and K. Takahei, "V-grooved substrate buried heterostructure InGaAsP/InP laser emitting at 1.3 μm wavelength," *IEEE J. Quantum Electron.*, **QE-18**: 1704–1711 (1982).

24. I. Mito, M. Kitamura, K. Kobayashi, S. Murata, M. Seki, Y. Odagiri, H. Nishimoto, M. Yamaguchi, and K. Kobayashi, "InGaAsP double-channel-planar-buried-heterostructure laser diodes (DC-PBH LD) with effective current confinement," *J. Lightwave Technol.*, **LT-1**: 195–202 (1983).

25. M. Sugimoto, A. Suzuki, H. Nomura, and R. Lang, "InGaAsP/InP current confinement mesa substrate buried heterostructure laser diode fabricated by one-step liquid-phase epitaxy," *J. Lightwave Technol.*, **LT-2**: 496–503 (1984).

26. K. Kishino, Y. Suematsu, Y. Takahashi, T. Tanbun-ek, and Y. Itaya, "Fabrication and lasing properties of mesa substrate buried heterostructure GaInAsP/InP lasers at 1.3 μm wavelength," *IEEE J. Quantum Electron.*, **QE-16**: 160–164 (1980).

27. M. Yamaguchi, H. Nishimoto, M. Kitamura, S. Yamazaki, I. Mito, and K. Kobayashi, "High-power cw operation over 100 mW at 1.3 μm in DC-PBH LD with reflectivity-optimized mirror facets," *CLEO'85*: paper no. THI1 (1985).

28. H. Namizaki, H. Kan, M. Ishi, and A. Ito, "Transverse-junction stripe-geometry double-heterostructure lasers with very low threshold current," *J. Appl. Phys.*, **21**: 2785–2786 (1974).

29. M. Ueno and H. Yonezu, "Stable transverse mode oscillation in planar stripe laser with deep Zn diffusion," *IEEE J. Quantum Electron.*, **QE-15**: 1189–1196 (1979).

30. T. Sugino, K. Itoh, H. Simizu, M. Wada, and I. Teramoto, "Reduction of threshold current in GaAlAs terraced substrate lasers," *IEEE J. Quantum Electron.*, **QE-17**: 745–750 (1981).

31. D. Botez, J. C. Connolly, M. Ettenberg, and D. B. Gilbert, "Very high cw output power and power conversion efficiency from current-confined CDH-LOC diode lasers," *Electron. Lett.*, **19**: 882–883 (1983).

32. M. Ueno, I. Sakuma, T. Furuse, Y. Matsumoto, H. Kawano, Y. Ide, and S. Matsumoto, "Transverse mode stabilized InGaAsP/InP (λ = 1.3 μm) plano-convex waveguide lasers," *IEEE J. Quantum Electron.*, **QE-17**: 1930–1940 (1981).

33. I. P. Kaminow, R. E. Nahory, M. A. Pollack, and J. C. Dewinter, "Single-mode cw ridge-waveguide laser emitting at 1.55 μm," *Electron. Lett.*, **15**: 764–765 (1979).

34. K. Aiki, M. Nakamura, T. Kuroda, J. Umeda, R. Ito, N. Chinone, and M. Maeda, "Transverse mode stabilized AlGaAs injection lasers with channeled-substrate planar structure," *IEEE J. Quantum Electron.*, **QE-14**: 89–94 (1978).

35. T. Hayakawa, N. Miyaguchi, S. Yamamoto, H. Hayashi, S. Yano, and T. Hijikata, "Highly reliable and mode-stabilized operation in V-channeled substrate inner stripe lasers on p-GaAs substrate emitting in the visible wavelength region," *J. Appl. Phys.*, **53**: 7224–7234 (1982).

36. H. Nishi, Y. Yano, Y. Nishitani, K. Akita, and M. Takusagawa, "Self-aligned structure InGaAsP/InP DH lasers," *Appl. Phys. Lett.*, **35**: 232–234 (1979).

CHAPTER 11

Longitudinal Mode Control in Laser Diodes

T. Ikegami

11.1 COHERENCE AND OPTICAL FIBER TRANSMISSION

The most interesting feature of light emitted from lasers is its coherence. Research on laser diodes (LDs) has been aimed at improving the quality of the light as well as improving electrical properties, for instance, reducing the threshold current.

There are three aspects of coherence. The first is coherence in space, which characterizes periodicity or the regularity of the phase relation in a plane normal to the direction of propagation, and which we must consider the transverse mode property or the near-field pattern on an optical port of a laser diode. In a laser diode, the transverse mode is governed by an optical waveguide structure formed with the active thin stripe layer surrounded by other materials, in contrast to a gas laser, the mode of which is usually determined by an arrangement of end mirrors. Transverse mode control was successfully attained around 1975 [1, 1a]. Reproducibility of the near-field and far-field patterns, lasing spectra, and so on, was dramatically improved in the device oscillating in a single stable transverse mode.

The second aspect of coherence is coherence in time, that is, a definite relation between the phases of the light observed at two separate points on the propagation axis. The distance over which this relation holds corresponds to the coherence length of the light. We often express this property in terms of the purity of the lasing wavelength or the lasing longitudinal mode. Control of the longitudinal mode in laser diodes is an important issue. Recent developments show stable single-longitudinal-mode (SLM) operation can be obtained in LDs incorporating longitudinal-mode-selection structures; this is the main topic of this chapter. The next issue is to reduce the linewidth of the lasing mode in wavelength or frequency in order to realize coherent wave transmission in which information is carried by the phase or the frequency of the optical field; at present, lightwave transmission is based on intensity modulation.

The third aspect of coherence is polarization of the emission. Conventional LDs have transverse electric polarization, since the cleaved facets have strong polarization selectivity for guided waves [2]. Since the threshold of the TE mode is lower than that of the TM mode except for special cases, even LDs with built-in gratings operate in the TE mode.

The outstanding features of laser diodes are small size, high efficiency, low voltage requirements, and long life. They are also robust and potentially inexpensive, mass-produced devices that meet the stringent requirements of optical fiber transmission systems. One other fantastic capability of LDs is direct modulation [3]. Information in an electric current can be transformed into information carried by a lightwave by applying the modulated current to the device without an external modulator.

In this chapter we will discuss how to utilize the coherence of LDs and what degree of coherence is required in fiber transmission systems.

Spatial coherence provides high launching efficiency of light from LDs into fibers with a core diameter of 50 μm for multimode fibers and 10 μm for single-mode fibers. More than -3-dB coupling into the single-mode fiber has already been reported. On the other hand, requirements for time coherence have changed over time or are dependent on the specific transmission systems. At the beginning of the multimode study, there was little concern about the spectral distribution of LDs; however, decreasing fiber loss and increasing transmission distance demanded a smaller number of lasing longitudinal modes, in particular for graded-index multimode fibers where the modal dispersion is small. The rapid development of single-mode fibers has pushed multimode fibers away from long-distance large-capacity transmission systems into local area network systems. However, the most inexpensive and convenient analog transmission system still uses multimode fibers. It is interesting that the primary issue with respect to LDs for analog systems is how to degrade their coherence, since the combination of coherence (strictly, depending on direct modulation), multiple paths of the ray in the multimode fiber, and effective spatial mode filters (connectors, splices, etc.) results in "modal noise" [4]. As a countermeasure, increasing the spectral width of the lasing longitudinal mode is more effective than increasing the number of longitudinal modes, and thus a method of RF-modulation superposition on LDs for enhancing chirping linewidth has been used.

As for the PCM single-mode fiber transmission systems, the transmission capacity and distance limitation governed by waveform distortion due to fiber dispersion and the spectral distribution of the optical source was analyzed in the 1960s. On the other hand, from actual measurements performed around 1975 it was found that the limit was much lower than the predicted value estimated under the assumption that spectral distribution was constant over time. At that time, it was already known that the number of lasing longitudinal modes of LDs under high-speed direct pulse modulation increased even though the devices operated in a single longitudinal mode under direct drive current. Later, temporal measurements indicated that the distribution of lasing longitudinal modes under direct modulation differed from pulse to pulse because the LD was in its transient state rather than its steady state [5]. This stochastic process will eventually become "noise" at the receiving end due to the dispersion of the single-mode fiber. For this reason both the transmission capacity and transmission distance are degraded [6, 6a]. We refer to this new limit as the *mode partition noise limit*. For example, a 400-Mb/s light signal from a conventional laser diode can travel through single-mode fiber up to only 20 km without error in the 1.5-μm wavelength range where the dispersion coefficient is ~ 17 ps/(nm-km). To overcome the mode partition noise limit, laser diodes that operate in a single longitudinal mode under direct modulation or strong perturbations, are required. One can eliminate the mode partition noise limit by utilizing longitudinal-mode-selective mechanisms in the laser cavity design. This type of LD is referred to as a single-longitudinal-mode (SLM), dynamic single-mode (DSM), or single-frequency laser diode; in this chapter SLM-LD will be used to avoid confusion. Presently, among the many kinds of SLM laser diode structures, the distributed feedback (DFB) laser diode seems most promising in terms of practical advantage.

Using SLM laser diodes, transmission capacity and distance can increase by more than one order of magnitude, for example, 400 Mb/s over 200 km at a laser wavelength of 1.5 μm. However, a new limitation exists even for SLM-laser diodes: the "chirping limit." Chirping is a wandering of lasing wavelength during transient operation of the laser due to the change in the refractive index in the cavity corresponding to the change in carrier density during high-speed direct modulation. The width of the swing is several angstroms and is large when the light intensity variation is large. Chirping causes optical signal transmission to deteriorate due to dispersion of the single-mode fiber [7].

We have discussed the relation between the coherence of LDs and fiber transmission on the basis of direct modulation. If an external modulator is used, we can easily reach the fiber loss limit because chirping is avoided. However, in this case we have to give up direct modulation, which is a highly desirable feature of LDs from a practical point of view.

Interest in coherent optical transmission where information is carried by the phase or frequency of light, as opposed to intensity modulation, has recently revived. The basic concept is very popular in conventional microwave systems, and many workers were drawn to the idea by the success of the ruby laser. Severe requirements are imposed on light sources; in fact, one main bottleneck of the coherent optical system is the laser diode source. Nevertheless, success in 400-Mb/s frequency-shift-keyed (FSK) coherent transmission over 290 km in the 1.5-μm wavelength range casts light on this new area of lightwave communications [8].

11.2 TRANSIENT BEHAVIOR OF LASING MODE

We often find reports or data sheets illustrating even conventional laser diodes lasing in single-longitudinal-mode (SLM) operation; however, this occurs only when the device is operated under direct drive current and at fixed temperature. The spectrum of the LD under dc operation has a multi-longitudinal-mode structure just above threshold, and with increasing drive current it changes toward SLM operation. This change is consistent with laser theory on the assumption that the gain distribution as a function of wavelength is associated with homogeneous gain broadening [9]. Moreover, this change is more definite in index-guided LDs than in gain-guided LDs. With increasing drive current and ambient temperature, the wavelength of the SLM lasing mode will shift to the next longitudinal mode or a longitudinal mode several mode numbers away; that is, "mode hopping" occurs [10].

If we apply a current pulse train instead of dc current to a conventional LD that shows SLM lasing under dc operation, the SLM lasing may turn into multi-longitudinal-mode lasing. Particularly, when we observe time-resolved spectra at the moment of switch-on in the pulse train with a monochromator, a fast photodetector, and a sampling oscilloscope, lasing consisting of 5–10 longitudinal modes within a bell-shaped envelope is obtained. Figure 11.1 shows measured half-width of the spectral envelope of longitudinal mode versus elapsed time after the initial peak of damping relaxation oscillation in the optical transient [11]. This transient phenomenon can be analyzed with the rate equations. Physically the suppression of longitudinal side modes is not sufficient during the transient even though the gain distribution is homogeneous. Consequently, time-averaged spectra of the light emitted from conventional LDs under high-speed direct modulation (more than several 100 Mb/s) show a multi-longitudinal-mode structure.

We have discussed the spectra averaged over many light pulses with the same shape that can be observed using a sampling oscilloscope. To know what is actually happening in an individual light pulse requires time-resolved measurement. Time-resolved observation of spectra of individual pulses indicates that the intensity distribution of the lasing longitudinal mode is not

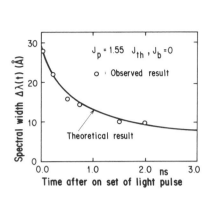
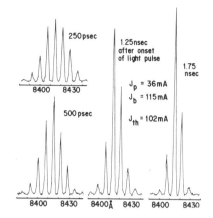

Figure 11.1 Transient behavior of spectral width (left) and spectrum (right) just after onset of light pulse. These results are averaged and smoothed over many pulses on a sampling oscilloscope. A quick change in light results in multi-longitudinal-mode lasing, and the number or lasing modes decreases with time toward the steady state [11].

uniform but random. Figure 11.2 shows the intensity of one longitudinal mode of a laser diode selected by the monochromator under direct modulation with a regular current pulse train [5]. Note that the intensity is not uniform but irregular. During the transient the longitudinal modes compete with each other, so the spectral energy distribution, or the energy partition among the modes, has statistical fluctuations and is not constant. Theoretically, the partition is governed mainly by the spontaneous emission just before the onset of the lasing, which corresponds to the initial conditions of the rate equations and is random by nature [6, 6a]. Note that since the total energy is predetermined, the averaged envelope is unique and the time-averaged gain distribution of LDs is bell-shaped (Fig. 11.1).

The temporal change in mode partition imposes severe limitations on single-mode fiber transmission systems due to the mode partition noise, as pointed out in the previous section.

11.3 SINGLE-LONGITUDINAL-MODE (SLM) LASER DIODE

The longitudinal modes of conventional LDs are spaced ~ 5–10 Å apart, whereas the optical gain distribution spreads over several hundred angstoms. The lasing mode selection is very sensitive to the gain difference; for example, even 1% is sufficient for SLM operation in the steady-state condition. To obtain a margin to hold SLM operation against strong perturbations such

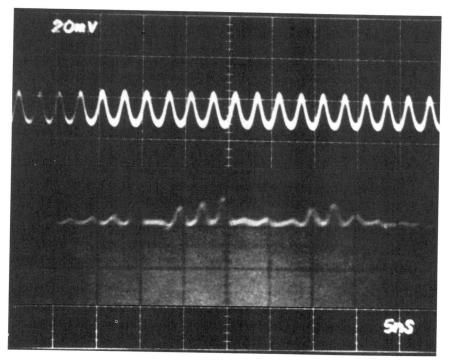

Figure 11.2 Many traces on high-speed oscilloscope of total light output pulse train (upper trace) and light output pulse train of one longitudinal lasing mode at 8464.5 Å (lower trace). The latter shows that the light waveforms are not uniform from pulse to pulse. Even though the total output and the averaged spectral distribution are uniform as shown in Figure 11.1, the temporal spectral distribution is not uniform; that is, mode partition in power is different from pulse to pulse [5].

as high-speed direct modulation, the actual gain roll-off is insufficient, as discussed in the previous section. A more stable SLM selection method is to introduce loss difference in the resonator modes by adding mechanisms that feed back part of the light into the original cavity. These can be arranged to give favorable interference conditions in which wavelength-dependent feedback adds constructively to the desired longitudinal mode. Figure 11.3 illustrates many examples that have been constructed [12]. An interfering reflection can be produced with an external mirror, by bends and other irregularities in waveguides, by an external grating, by a second passive or active optical cavity, by an external grating incorporated in the end section of the active waveguide, which is called a distributed Bragg reflector (DBR) [13], and by a grating incorporated in the active waveguide, which is called a distributed feedback (DFB) structure [14]. The simplest mechanism is a short cavity type, but it is not very effective unless the cavity is ultrashort.

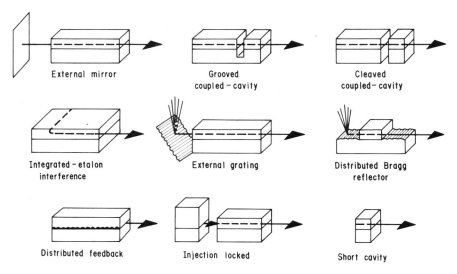

Figure 11.3 Various configurations to achieve single-longitudinal-mode (SLM) lasing operation [12] where only one longitudinal mode has the lowest loss by using a wavelength-selective mechanism. Injection locking is the other method in which the master laser oscillates in the single longitudinal mode in dc operation and the slave laser diode can be directly modulated. In shortcavity laser diodes, the adjustment of gain roll-off and mode spacing is essential.

Injection locking is also used in obtaining SLM laser diodes but is more complicated.

Note that to obtain stable SLM operation, single transverse mode control is essential.

Figure 11.4 indicates how much loss difference is required to provide a sufficient side-mode suppression ratio during the transient generation when a pulse current is applied. The ratio depends on β, the rate of the spontaneous emission coupled into the resonant mode, and the bias current J_b. Small β, large J_b ($>J_{th}$ desired), and a smooth optical waveform are beneficial for SLM operation. A loss difference of about $5-10$ cm^{-1} is sufficient for SLM operation in practical laser diodes.

11.4 DOUBLE-CAVITY SLM LASER; C^3 LASERS

As one basic structure of double-cavity LDs, the operation of an external mirror type LD is explained in Figure 11.5. The spectrum of the solitary LD usually has a multi-longitudinal-mode structure due to its gain profile and perturbations or fluctuations, as discussed in the previous section. When the external mirror is placed facing one facet of the solitary LD at a distance of

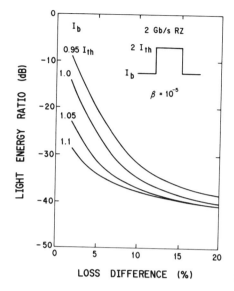

Figure 11.4 Light energy ratio of the side lasing mode to the main lasing mode for the given loss difference between the side mode and the main mode under 2-Gb/s return to zero (RZ) direct modulation. The absolute value of the ratio is often referred to as the side-mode suppression ratio or discrimination ratio. The ratio degrades with β, the rate of the spontaneous emission coupled into the resonant mode, assumed to be 10^{-5}.

about 0.05 of the laser length, the effective reflectivity $\hat{\rho}_e$ at the facet of the laser changes due to interference between the reflected lights from the facet and the external mirror. Figure 11.5 shows that the discrimination of $\hat{\rho}_e$ is greater than that of the gain profile itself. The outer peaks of $\hat{\rho}_e$ are of no concern when the gain peak coincides with the main peak, since the roll-off of the gain profile significantly decreases the gain at the outer peaks. However, the gain peak moves with temperature (5–8 Å/°C), resulting in a lasing mode jump to the neighboring longitudinal mode or the next peak of $\hat{\rho}_e$.

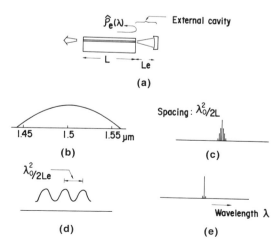

Figure 11.5 Longitudinal mode selection with (a) an external mirror configuration. (b) Gain profile of laser diode. (c) Lasing spectrum of solitary LD. (d) Reflectivity $\hat{\rho}_e(\lambda)$ by external mirror effect. (e) Lasing spectrum.

The performance of this kind of structure is expressed in terms of the temperature range, mechanical stability, and drive current range for stable SLM operation.

The concept of the composite or coupled laser cavity to obtain SLM operation was proposed in the 1960s for gas lasers and solid-state lasers. The application of the concept to laser diodes seems very attractive since the gain is very high (> 100 cm^{-1}) and, in addition to the gain or loss, the refractive index can be easily controlled by current. One device that makes the best use of these advantages is the C^3 LD.

The cleaved coupled cavity (C^3) laser was developed in 1983 for SLM operation [15]. The background was the success in cw operation of the distributed feedback laser in the 1.5-μm range in 1982 [16, 17], which had good performance but difficulties in fabrication at that time. The C^3 laser consists of two solitary LDs, which are fabricated by cleaving one LD into two parts and each with a separate contact, as shown in Figure 11.6. This configuration had been used in experiments for optical logic, and similar fabrication techniques had already been part of the basic technology for making integrated optics.

The three-terminal device in Figure 11.6 consists of a main laser 1 and a control laser 2. We will discuss tuning of the lasing mode, one of the possible functions of the C^3 laser. The spectrum of the main LD is assumed to be fixed ($I_1 > I_{1,\text{th}}$) while that of the control LD can be changed with the current I_2. Actually, the wavelength of LD 2 depends on the driving

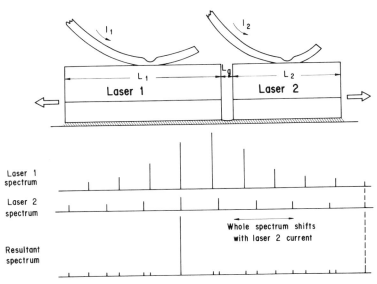

Figure 11.6 Single-longitudinal-mode (SLM) operation in a cleaved cavity (C^3) laser diode [15].

conditions of both LDs and on the light-coupling conditions. Since the mode spacing of the two laser diodes is different ($L_1 \neq L_2$), the configuration forms a vernier, which can be used as a measurement tool, and oscillation occurs at coincidences. Increasing the current into LD 2 shifts its resonant mode spectra toward a shorter wavelength as a whole and moves the coincidences more rapidly in the same direction if LD 2 is longer, and in the opposite direction if it is shorter, causing hopping of the lasing longitudinal mode. The mode hopping essentially ceases when LD 2 reaches its threshold ($I_2 = I_{2,th}$) because of the quenching of carrier density in LD 2 above the threshold. Consequently, the C^3 laser diode can be used either as a means of controlling intentional mode hopping if $I_2 < I_{2,th}$ or as a means of selecting a single mode if $I_2 > I_{2,th}$.

Figure 11.7 shows observed spectra of the C^3 LD operating in the 1.5-μm wavelength region, (a) under cw operation ($I_1 = 13$ mA, $I_2 = 59$ mA), and (b) and (c) under direct modulation at 2 Gb/s NRZ format with $I_1 = 43$–75 mA and $I_2 = 13$ mA [18]. Note that SLM operation is obtained even under high-speed direct modulation if the drive condition is properly adjusted. The C^3 LD was used as a light source in single-mode fiber transmission experiments, such as 420 Mb/s over 120 km (1983) and 1 Gb/s over 120 km (1983), in which the device eliminated the mode partition noise limit even though there is a large dispersion for the single-mode fiber at 1.5 μm.

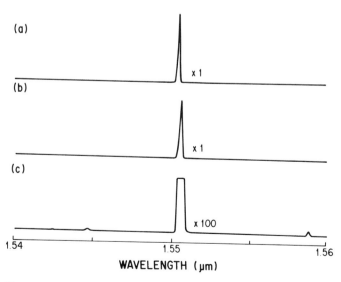

Figure 11.7 Observed spectra of C^3 laser diode [18]. (a) dc or cw operation ($I_1 = 13$ mA, $I_2 = 59$ mA); (b), (c) under direct modulation at 2 Gb/s (NRZ) ($I_1 = 13$ mA, $I_2 = 43$–75 mA for modulation).

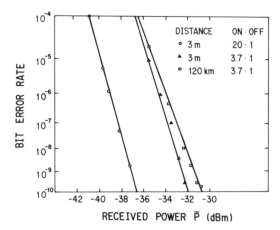

Figure 11.8 Bit-error-rate curves for 120-km single-mode fiber transmission experiment at 1 Gb/s using a C³ laser diode lasing at 1.55 μm. [Ref. 18, pp. 310–313]. When the on/off ratio is 20, the signal cannot be transmitted due to "chirping limit" (see Section 11.7). By increasing the bias level with the on/off ratio of 3.7 to reduce chirping, a low bit error rate is obtained.

[Ref. 18, pp. 310–313] Performance of the later experiment is shown in Figure 11.8, where the side-mode suppression ratio is 5000, the ratio of peak to bottom (or on to off) in the light pulse is 3.7, and the time-averaged spectral width of the lasing mode due to chirping is 1.4 Å. The large peak-to-bottom ratio is beneficial for low chirping. Analysis of the C³ LD is performed using an effective reflectivity at the interface facet facing the first LD, which takes the second LD cavity and the gap between the LDs into account [19]. Theory indicates that characteristics of the C³ LD are very sensitive to the gap separation, which is a common tendency in composite cavities. The performance of the C³ LD varies considerably between samples, and selected devices give good mode discrimination.

The three-terminal devices have many interesting application possibilities; however, distributed feedback LDs are superior to C³ LDs for eliminating the mode partition noise limit in the view of system people, who prefer a two-terminal device because of its simplicity in operation. Nevertheless, considering the next generation of coherent fiber transmission systems, we find that three-terminal DFB or C³-type laser diodes have interesting features that are essential for the frequency-tunable optical sources needed in such systems [20].

11.5 DISTRIBUTED FEEDBACK (DFB) LASERS

Lasers with a built-in grating were demonstrated in a distributed feedback dye laser by Kogelnik and Shank at Bell Laboratories in 1971 [14]. Room-temperature cw oscillation of a GaAlAs DFB laser diode with grating used in third-order was first reported by Nakamura et al. in 1975 [21]. As to the most interesting lasing wavelength range of 1.5 μm, room-temperature cw oscillation of InGaAsP DFB-laser diodes was demonstrated in September 1981 and was reported by Utaka et al. [16] and Matsuoka et al. [17]. Continuous wave operation of a distributed Bragg reflector (DBR) laser diode was also reported in 1981 by Koyama et al. [22]. The basic theory for lasers with a built-in grating was essentially completed in the 1970s by Kogelnik and Shank [23], Yariv [24], Wang [25], and Streifer et al. [26]. The most popular treatment for DFB laser diodes is that of the practical theory of Streifer et al., which appeared in 1975; it will be introduced later. The basic coupling equations are simple; however, numerical calculations are required to obtain the characteristics for each practical model.

Figure 11.9 illustrates the typical structure of laser diodes with a built-in grating. First of all, the necessary condition for single-longitudinal-mode

(a) Distributed Bragg Reflector (DBR) LDs

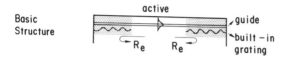

(b) Distributed Feedback (DFB) LDs

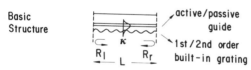

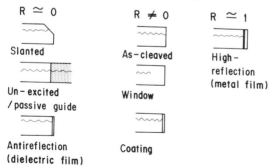

Figure 11.9 Structures of laser diodes with built-in grating. (a) In distributed Bragg reflector (DBR) laser diodes, the grating acts as a wavelength-selective mirror terminating the end of the active waveguide. (b) In distributed feedback (DFB) laser diodes, the grating is incorporated in parallel with the active waveguide. κ is the coupling coefficient between field and grating. The performance of DFB laser diodes depends on the end preparation, and various cases are shown in the figure.

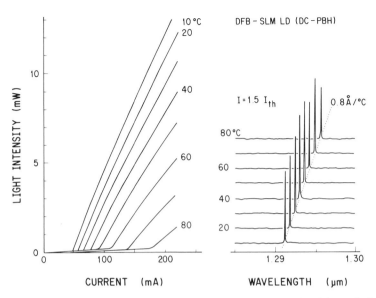

Figure 11.10 Lasing properties of well-behaved distributed feedback laser diodes.

(SLM) oscillation is the maintenance of stable fundamental transverse-mode operation because weak confinement of the transverse mode often disrupts SLM operation. Lasing properties of well-behaved DFB laser diodes in cw operation are presented in Figure 11.10. SLM operation is observed over the operating range, and the lasing wavelength changes by about 0.1 nm/deg, corresponding to an effective refractive index increase with temperature in InGaAsP lasers. The observed temperature dependency of the threshold current is almost the same as that of conventional Fabry-Perot InGaAsP lasers with T_0 of 50–60 K $[T_0 = \ln \{I_{th}(T_1)/I_{th}(T_2)\}/(T_1 - T_2)]$. Figure 11.11 shows the mode structure of the DFB-laser diode with two cleaved facets

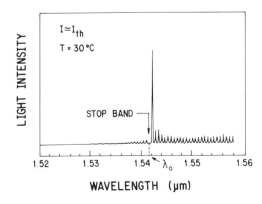

Figure 11.11 Spectrum of a DFB laser just below threshold. The device having two cleaved facets shows the stopband centered at Bragg wavelength λ_0, where no resonant mode exists. Periodic mode structure is determined by the length of a DFB laser diode.

just below threshold. The main mode, a longitudinal mode group determined mainly by a length L of the optical resonator, and a stopband at the center are observed. Since the waves cannot propagate at the Bragg wavelength λ_0, there is a range of lasing just outside the stopband where the waves may propagate but where they still interact strongly with the grating. Note that the width of the stopband is approximated to be 2κ only when $R \approx 0$ in Figure 11.9 [23].

The basic structure of a 1.5-μm wavelength DFB laser diode is shown in Figure 11.12. Fabrication of the device consists of four steps: (1) fabricating the grating on the substrate, (2) growing the crystal layer covering over the grating, (3) fabricating the fundamental transverse mode waveguide, and (4) forming the laser chip. The alternative to step 1 is to form the grating on the waveguide layer over the active layer, since an active layer grown on the grating seems to be less desirable owing to reliability concerns; however, presently there is no evidence of reliability problems.

The grating is formed by holographic photolithography or electron beam exposure and chemical etching. The pitch of the first- and second-order gratings is 2360 Å and 4520 Å, respectively, for 1.5-μm lasers. Crystal regrowth

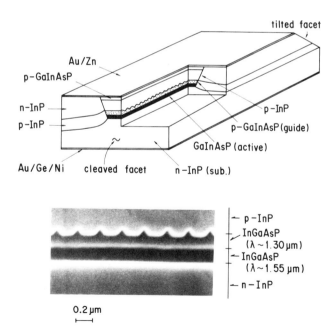

Figure 11.12 Cross-sectional view of a DFB laser with a cleaved facet and a slanted or tilted facet. Corrugation can be fabricated with holographic photolithography by using an argon or He-Cd laser or electron beam direct writing. The corrugation is formed over an active layer; however, it can also be made on a substrate.

on the corrugation can easily distort the grating shape. For example, special care should be paid to avoid thermal deformation during heating up of a furnace. In addition to liquid-phase epitaxy (LPE), metal-organic chemical vapor deposition (MOCVD), vapor-phase epitaxy (VPE) and hybrids of them have been used for the epitaxial crystal growth. Since the thickness of each layer is critical in determining the performance of DFB lasers, the MOCVD or VPE crystal growth method may be preferable to the LPE method for high-yield production. For step 3, buried heterostructures (BH), Schottky barrier-delineated stripes, and ridge waveguides, which have been used in conventional lasers, have been fabricated.

End preparations of the grating include slanted end by etching, antireflection (AR) coating, and lossy waveguide to reduce the reflected light at the end of the grating and as-cleaved and metal coating to provide relatively high reflection. The slanted end has the advantage of suppressing the Fabry-Perot mode oscillations; however, it is not suitable for an output port of the device. The alternative is the antireflection-coated end, which makes high output power available from the port.

Among these, DFB laser diodes with as-cleaved facets are expected to be used first because of their simplicity of fabrication and low threshold current, even though the margin for SLM operation is narrower than that for

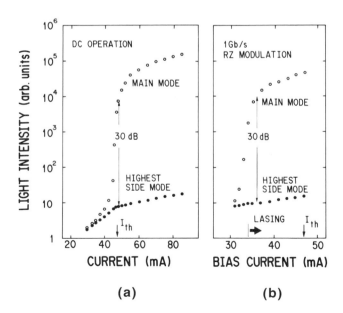

Figure 11.13 Observed light intensities of the main lasing mode and the highest side lasing mode of typical DFB laser diodes under (a) dc or cw operation and (b) direct modulation at 1 Gb/s with a pulse current of 15 mA. A side-mode suppression ratio of more than 30 dB can be obtained even under direct modulation.

the device with a low-reflectivity end. Some compromise with respect to performance will be made, and the device structure will be chosen according to applications.

Figure 11.13 shows intensities of the main and side modes under dc operation and under high-speed direct modulation at 1 Gb/s [27]. Note that even in the case of pulsed direct modulation with the bias current below threshold, the side-mode suppression ratio remains more than 30 dB in DFB laser diodes.

The SLM property of DFB laser diodes having reflection at the end or the facet ($R = 0$ in Fig. 11.9 or $\hat{\rho} \neq 0$) is sensitive to the location of the end or facet relative to the grating (refer to the next section). We denote the location of the facet relative to the grating periodicity as the *facet phase*. We can improve a poorly behaved DFB laser diode to make it an SLM device by etching one facet with an ion beam sputtering machine, as shown in Figure 11.14, or by coating the facet [27].

Theory indicates that a more reliable method to obtain SLM operation is to introduce a $\lambda/4$ (quarter-wavelength) shift in the center of the grating [28]. Several demonstrations [29] seem to prove the effect; for example, DFB laser diodes incorporated with a small dip in the middle, which effectively makes a $\lambda/4$ shift in the grating, oscillate at the Bragg wavelength with J_{th} of 40 mA and $\Delta\eta_{ex}$ of 40% per facet in the 1.5-μm wavelength range [30]. However, actual yield in laboratories tells us that even DFB laser diodes

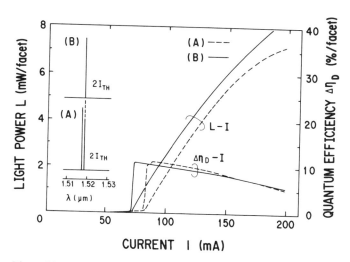

Figure 11.14 One example showing how the performance of DFB laser diodes is sensitive to facet location relative to the corrugation. (a) An as-fabricated device in which SLM operation breaks into two-mode operation at $2 \times I_{th}$. After etching one facet of the device with an ion beam by 45 nm, the device is cured to be a good SLM DFB laser diode as shown in (b).

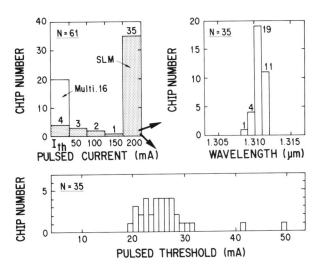

Figure 11.15 Yield of laboratory DFB lasers. Of the 61 chips, 35 show SLM operation with a pulsed current of 200 mA, more than 5 times the threshold current. The devices have two cleaved facets and a second-order grating.

having facets cleaved or cleaved and coated are quite good in SLM operation, as shown in Figure 11.15 [31].

Figure 11.16 illustrates the output power of some high-power DFB laser diodes reported so far, with the output power of conventional solitary LDs (indicated by FP) shown for comparison. CLV, LR, and HR denote as-cleaved facet, low-reflectivity (several %) coating end, and high-reflectivity (>80%) coating end. Note that the power output, threshold current, and

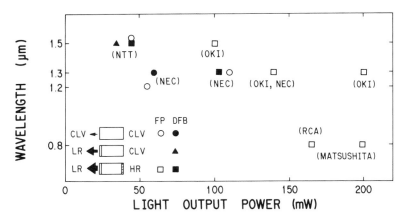

Figure 11.16 Output power of some high-power Fabry-Perot lasers (indicated as FP) and DFB lasers reported. These data are from solitary devices operating in cw and single transverse mode.

external differential quantum efficiency of DFB laser diodes are comparable with those of Fabry-Perot laser diodes in the 1-μm wavelength range.

11.6 ANALYSIS OF DISTRIBUTED FEEDBACK (DFB) LASERS

An analysis of laser action in a periodic structure was presented first by Kogelnik and Shank in 1972 [23]. The basic concept had been used in analyzing Bragg diffraction of X-rays, light diffraction by acoustics, etc. After the analysis, Wang [25] and Streifer et al. [26] presented a more generalized theory for semiconductor lasers with built-in gratings, that is, DFB lasers and DBR lasers.

Figure 11.17 illustrates the operation of a distributed feedback structure. Two waves are indicated, one of which travels to the left (S-wave) and the other to the right (R-wave). Each wave receives a small portion of the other wave at each of the scattering points that are distributed along the length of the periodic structure. This creates a distributed feedback mechanism, and spectral selection can be obtained because of the wavelength sensitivity of the Bragg effect. The Bragg wavelength is determined by the spacing of scattering points, that is, the pitch of the periodic structure. Figure 11.18 shows a waveguide with periodic scattering points. A guided wave with a propagation constant $\beta_{B,m}$ becomes in phase with the scattered wave under the following conditions.

$$2\beta_{B,m}\Lambda = 2\pi m \tag{1}$$

or

$$\Lambda = m\frac{\lambda_{B,m}}{2\hat{n}_{eg}} \qquad m = 1, 2, \ldots$$

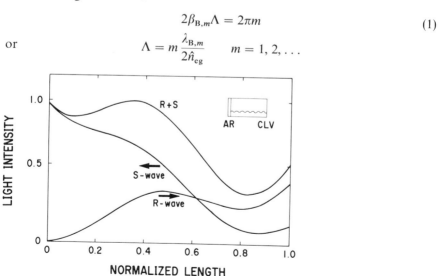

Figure 11.17 Intensity distribution of the field in a DFB laser diode with antireflection coated end and cleaved facet. S-wave traveling to the left and R-wave traveling to the right couple with each other through the periodic structure of the grating ($\kappa L = 1$).

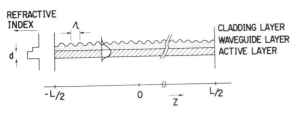

Figure 11.18 Waveguide with periodic structure. The periodic structure is terminated at $Z = \pm L/2$.

where m is the order of grating, $\hat{n}_{eq}$ is the equivalent index of the waveguide, and $\lambda_{B,m}$ is the Bragg wavelength. The third-order grating was used in a 0.8-μm DFB laser diode with $\Lambda \approx 400$ nm; however, the second-order ($\Lambda \approx 400$ nm) or first-order ($\Lambda \approx 200$ nm) grating or corrugation is formed in 1.3–1.5-μm DFB laser diodes.

The periodic structure can operate as a resonant cavity only if it has a termination at each end that breaks the periodicity. In a convenient method to solve this problem, we usually define a reflection coefficient $\hat{\rho}$ at the end. In this chapter, we follow an analysis based on that of Streifer et al. [26].

First we let the Bragg propagation constant with the order m in which we are interested be β_0.

$$\beta_0 = m \frac{\pi}{\Lambda} \qquad (2)$$

The total field is described by the sum of the two waves in Figure 11.17 as

$$E(Z) = R(Z) \exp(-j\beta_0 Z) + S(Z) \exp(j\beta_0 z) \qquad (3)$$

where $R(Z)$ and $S(Z)$ are complex amplitudes that vary slowly along the Z direction. On substitution in the wave equations including the perturbation term that stems from the grating, and after isolating expressions with similar Z dependence, we can obtain constant-coefficient linear differential equations:

$$\frac{dR(Z)}{dZ} = \left(\frac{\alpha}{2} - j\delta\right) R(Z) - j\kappa S(Z)$$

$$\frac{dS(Z)}{dZ} = -\left(\frac{\alpha}{2} - j\delta\right) S(Z) + j\kappa^* R(Z) \qquad (4)$$

where α is the gain coefficient in power ($\alpha/2$ is the gain coefficient in intensity) and

$$\delta = \beta - \beta_0 = 2\pi \hat{n}_{eq}(\lambda^{-1} - \lambda_0^{-1})$$

β is the propagation constant of the resonant mode, λ and λ_0 are the wavelengths of the resonant mode and Bragg resonance, respectively, and δ is the detuning parameter from the Bragg propagation constant β_0. κ in Eqs. (4) is the coupling coefficient between the S- and R-waves due to distributed scattering points, that is, the distributed feedback.

Since Eqs. (4) are linear coupling equations, they have exponential solutions of the form

$$R(Z) = r_1 e^{\gamma z} + r_2 e^{-\gamma z}$$
$$S(Z) = s_1 e^{\gamma z} + s_2 e^{-\gamma z} \tag{5}$$

The substitution of Eqs. (5) into Eqs. (4) requires the dispersion relation

$$\gamma^2 = \left(\frac{\alpha}{2} - j\delta\right)^2 + \kappa^2 \tag{6}$$

to obtain nontrivial solutions.

In order to determine the properties of the resonant mode, that is, α and δ, boundary conditions at each end ($z = \pm L/2$) should be imposed. Using field reflection coefficients $\hat{\rho}_l$ and $\hat{\rho}_r$, we can write down the boundary conditions as follows:

$$R\left(-\frac{L}{2}\right) e^{j\beta_0 L/2} = \hat{\rho}_l S\left(\frac{L}{2}\right) e^{-j\beta_0 L/2} \qquad \text{at } Z = -\frac{L}{2}$$

$$S\left(\frac{L}{2}\right) e^{j\beta_0 L/2} = \hat{\rho} R\left(\frac{L}{2}\right) e^{-j\beta_0 L/2} \qquad \text{at } Z = \frac{L}{2} \tag{7}$$

Equations (5)–(7) give the eigenvalue equation as

$$\gamma L = \frac{-j\kappa L \sinh(\gamma L)}{(1 + \rho^2)^2 - 4\rho^2 \cosh(\gamma L)}$$
$$\times \{(\rho_l + \rho_r)(1 - \rho^2) \cosh(\gamma L) \pm (1 + \rho^2)$$
$$\times [(\rho_l - \rho_r)^2 \sinh^2(\gamma L) + (1 - \rho^2)^2]^{1/2}\} \tag{8}$$

where

$$\rho_l = \hat{\rho}_l e^{j\theta_l} \qquad \rho_r = \hat{\rho}_r e^{j\theta_r} \qquad \rho^2 = \rho_l \rho_r = \hat{\rho}_l \hat{\rho}_r e^{-j2\beta_0 L} \tag{9}$$

In Eqs. (9), θ_l and θ_r indicate the locations of the grating at the ends of the cavity (Fig. 11.19), that is, at $z = \pm L/2$. If the reflection at the end is not zero, those values become important in Eq. (8). For example, since ρ of the DFB

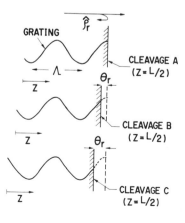

GRATING

CLEAVAGE A
($Z = L/2$)

CLEAVAGE B
($Z = L/2$)

Figure 11.19 The location of the grating termination and the facet phase. Cleavages A, B, and C have different values of the facet phase θ_r.

CLEAVAGE C
($Z = L/2$)

laser diode with a cleaved facet has $0.56 \exp(j\theta)$, α and β depend on θ even under the same κL. In that case, we call the phase θ that determines the cleavage location relative to the grating the *facet phase*. The reason the DFB resonator is sensitive to the facet phase θ is that even though the periodic structure has its own resonance mechanism by the distributed feedback, the reflected light at the facet acts as a strong perturbation to the resonator and the phase difference of the reflected light at the facet plays an important role in determining the properties of the DFB laser diode. We may regard the facet as an external mirror in a multicavity configuration.

α is the amount of power gain required to overcome the DFB resonator loss, which is equivalent to $\ln(1/R)/L$ in the conventional Fabry-Perot laser with facet reflectivity R, corresponding to the power escaping from the laser cavity. The threshold gain can be obtained using

$$\Gamma G(N_{th}) = \alpha + \alpha_f + \alpha_s \tag{10}$$

where Γ is the transverse confinement factor of the field in the active region, G is a gain function, α_f is absorption including free-carrier absorption, and α_s is loss due to scattering or diffraction. Assuming a linear relation $G(N) = A(N - N_0)$, where N_0 is the carrier density required for the active layer to be transparent, the threshold carrier density N_{th} is given by

$$N_{th} = N_0 + \frac{\alpha + \alpha_f + \alpha_s}{A\Gamma} \tag{11}$$

The threshold current density can be obtained using

$$J_{th} = qdr_c^{-1}\left(\frac{1}{\tau_{nr}} + BN_{th} + CN_{th}^2\right)N_{th} \tag{12}$$

where q is the amount of electron charge, d is the thickness of the active layer, and r_c is the ratio of the current flowing into the lasing region to the total current. $\tau_{nr}, (BN_{th})^{-1}$, and C are the nonradiative recombination lifetime, the spontaneous lifetime, and the Auger recombination coefficient, respectively. The amount in parentheses in Eq. (12) is estimated to be several nanoseconds in InGaAsP laser diodes.

External differential quantum efficiency $\Delta\eta_{ex}$ is given as

$$\Delta\eta_{ex} = \frac{\alpha}{\alpha + \alpha_f + \alpha_s}\eta_i r_c \tag{13}$$

where η_i is the internal quantum efficiency and r_c is the ratio of current flowing through the active region overlapping the lasing field to the total current. Note that r_c depends on the drive current level even above threshold in some devices.

From the result of the perturbation analysis, the coupling coefficient κ for the TE mode is given by [32]

$$\kappa = \frac{k_0^2}{2\beta}\frac{\iint\Delta(\varepsilon(x, y, z))E_y^2(x, y)\,dx\,dy}{\iint E_y^2(x, y)\,dx\,dy} \tag{14}$$

where k_0, is the free-space wavenumber, β is the TE mode propagation constant, and $\Delta(\varepsilon(x, y, z))$ is the grating-induced dielectric perturbation. Numerical results are presented in the paper by Streifer et al. [32]. Supposing the same tooth height of Δ, κ is largest in the rectangular grating and decreases in inverse proportion to the order of the grating, as $1/m$. In the sinusoidal grating, κ decreases faster than $1/m^2$ with increasing order. Figure 11.20 shows numerical results in the 1.5-μm wavelength region.

Numerical results are shown in Figure 11.21 for the required threshold gain α in power versus wavelength deviation from the Bragg wavelength $(\sim \delta\lambda_0/\beta_0)$ for the resonant modes of a DFB laser diode with two cleaved facets, $\rho_l = 0$ and $\rho_r = 0$. The coupling coefficient κ multiplied by cavity length L is a parameter, and n indicates the longitudinal modes determined by finite L in the DFB laser diode. The mode spacing among the longitudinal modes is not constant, and it becomes narrower in the vicinity of the Bragg wavelength ($\delta = 0$). In this case, the stopband centered with the Bragg wavelength where no resonant mode exists is shown.

Note that the symmetrically shaped gratings and symmetrical boundary conditions at the ends ($\rho_l = \rho_r$) result in the symmetrical resonant mode arrangement about the Bragg wavelength ($\delta = 0$) and there is degeneracy in modes having the lowest α; that is, there is threshold degeneracy in the $+1$

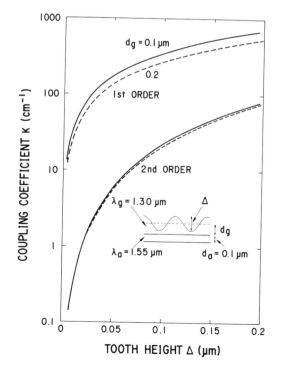

Figure 11.20 Coupling coefficient κ between S-wave and R-wave through the sinusoidal corrugation.

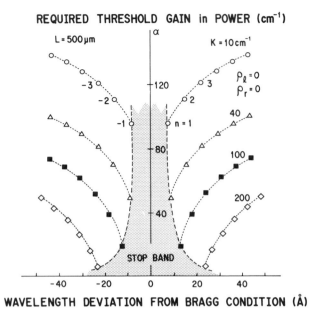

Figure 11.21 Required threshold gain α and resonant wavelength of the DFB modes. α is for power expression (occasionally, for field expression). The structure is assumed to be symmetric without reflection at the ends of the grating. In this case, $+n$ and $-n$ modes are symmetric in α and the resonant wavelength relations, so that $+1$ and -1 modes have the same threshold and two-mode operation may occur instead of SLM operation.

and -1 modes. This threshold degeneracy could cause the SLM oscillation to deteriorate and lead to two-mode operation. Theory points to three means of removing the threshold degeneracy:

1. Higher-order grating ($m \geq 2$)
2. Asymmetric structure
3. Quarter-wavelength ($\lambda/4$) shift or phase jump structure in the grating

In the second-order grating, diffraction emitted in the direction normal to the periodic structure surface is in phase, and this first-order diffraction loss [α_s in Eq. (10)] between the $+1$ and -1 modes differs, for example, by 10 cm^{-1} from the calculated value [33]. Experimental results show that a DFB-LD with a second-order grating seldom operates in two-mode lasing and the intensity of the radiation normal to the lasing direction is very weak [34]. These facts suggest that only one mode having the lower diffraction loss oscillates in the device.

Figure 11.22 shows the power gain α versus resonant mode wavelength for the asymmetric cavity structure [35], where $\rho_l = 0$ and $\rho_r = 0.56e^{j \cdot \pi}$. The

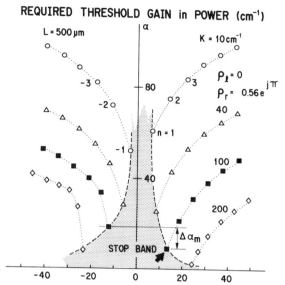

REQUIRED THRESHOLD GAIN in POWER (cm⁻¹)

WAVELENGTH DEVIATION FROM BRAGG CONDITION (Å)

Figure 11.22 Required threshold gain α and resonant wavelength of the DFB modes. The structure is assumed to be asymmetric with no-reflection end ($\rho_1 = 0$) and cleaved end with the facet phase of $\rho_r = 0.56 \exp j\pi$. The results are not symmetric (see Fig. 11.21), so that, in case of $\kappa = 100$ cm⁻¹, only the $+1$ mode oscillates in SLM operation with the loss difference of $\Delta\alpha_m$ against the submode (-1 mode).

threshold degeneracy is clearly eliminated, and this mode arrangement is very similar to the one in Figure 11.11. As illustrated in Figure 11.15, even DFB laser diodes with two cleaved facets demonstrated high yields of stable SLM operation. Even though we try to make symmetrical structures, asymmetric geometries are unintentionally formed in the grating and facet phases, and they can enable SLM operation to dominate. Theoretical yields, that is, statistics on the number of DFB lasers that can oscillate in SLM operation among DFB lasers having various facet-phase combinations, also show fairly good results [36].

A more reliable method for removing threshold degeneracy is to make a $\lambda/4$ phase jump at the middle of the corrugation with $R(\pm L/2) = 0$. In this arrangement, the stopband (Fig. 11.21) disappears and lasing occurs just at the Bragg wavelength [28]. One example is shown in Figure 11.23 where the phase jump of the grating is fabricated by using negative and positive resist films [37]. Electron beam lithography is also an appropriate method for making the phase jump. $\lambda/4$ phase shift can be regarded as part of asymmetric structures. For example, by changing the position of the phase jump, the ratio of output power emitted from one facet to that of another facet can

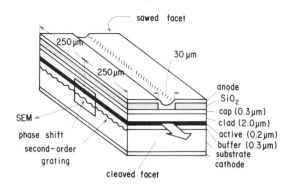

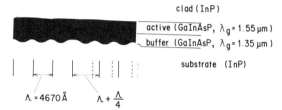

Figure 11.23 A quarter-wave-length-shifted or "phase jump" grating DFB laser diode. In this example, a $\lambda/4$-shifted grating is made by simultaneous holographic exposure of positive and negative photoresist [29].

be changed [38]. If we can precisely control the thickness of crystal layers and the shape of the corrugation, the phase shift method could be more easily controllable than facet phase control for obtaining SLM operation.

It is worthwhile to point out that κL has a reasonable value, around 1, for stable SLM laser diodes. Degradation of SLM operation above threshold due to spatial hole burning along the laser axis can be eliminated at $\kappa L = 1.25$ [39].

11.7 CHIRPING

In every laser diode, the time-averaged spectral width of its lasing longitudinal mode increases when the device is directly modulated at high speed regardless of SLM or non-SLM operation, as shown in Figure 11.24 [40]. This phenomenon was observed in AlGaAs and InGaAsP laser diodes showing self-pulsation or gain-switched high-speed pulses. Systems people pay much attention to it because it is a practical degradation factor, referred to as the "chirping limit." This chirping of the lasing wavelength occurs because of the modulation of refractive index in the active layer caused by the fluctuating injected carrier density during the damped relaxation oscillation in light intensity. The ratio of fractional change in the index to that of the free carrier density is measured to be -4×10^{-21} cm^3 for AlGaAs laser diodes and -7×10^{-21} cm^3 for InGaAsP laser diodes, which is mainly determined by

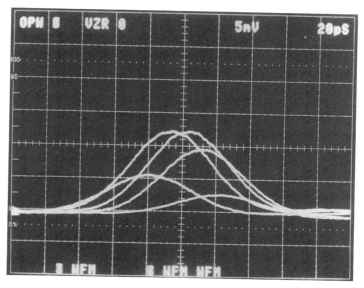

Figure 11.24 Dynamic lasing wavelength chirping of one longitudinal mode in an InGaAsP laser diode under pulse current modulation (the current pulse width is about 60 ps). Several traces correspond to the different spectral components with a separation of 4 Å for the five successive traces. Since the longitudinal mode spacing of this laser diode is about 40 Å, these traces show wandering of one lasing longitudinal mode in wavelength. The longer-wavelength component oscillates at a later time than the shorter-wavelength component, that is, there is a red shift. (20 ps/div.)

the plasma effect. The value of chirping is in the range of several tenths of a nanometer [41].

Figure 11.25 shows the wandering of peak emission wavelength corresponding to the relaxation oscillation of light output with a relative phase shift of $\pi/2$ [42]. The chirping can be analyzed with the rate equations, and the result tells us that in order to reduce it, a smooth light waveform without relaxation oscillation, if possible, is preferable. The observed data seem to depend on the device structure. However, chirping is an inherent feature of a directly modulated solitary LD, so we must reach a compromise on the degree of chirping. The easiest way to reduce chirping is to set the bias level above threshold [43] (Fig. 11.26). In this case there will be a penalty due to the small on/off ratio of the light pulses.

Chirping results in undesirable system penalties for single-mode fiber systems because of fiber dispersion. The bit rate–distance product is estimated to be about 20 (Gb/s)km$^{1/2}$ in the 1.5-μm wavelength range for a -1-dB power penalty in the received power [44, 44a]. The maximum distance decreases quadratically with increasing bit rate, because the analysis

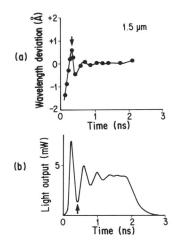

Figure 11.25 Temporal observation of spectrum from DFB laser diode [42]. Chirping in (a) follows the change of carriers in the active region of the device (arrows indicate the same time). A quick change in light output results in large chirping.

assumes a Gaussian light waveform. Note that reported high-capacity and long-distance transmission experiments are usually based on the optimization of drive conditions where the bias current is above threshold (for example, the case in Fig. 11.8). Thus, the results often exceed the above bit rate–distance product; for example, the product is 40 $(Gb/s)km^{1/2}$ when calculated, compared with the experimental data of 4 Gb/s over 103 km [45].

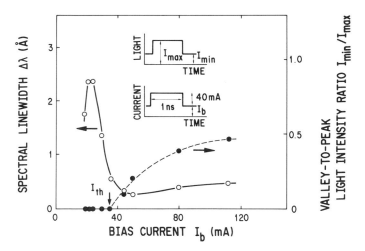

Figure 11.26 Chirping dependence of bias current. Time-averaged spectral linewidth of DFB laser diode under direct modulation can be reduced by increasing the bias current; however, the valley/peak light intensity ratio I_{min}/I_{max} would be degraded (on/off ratio would be degraded).

11.8 SINGLE-FREQUENCY LASERS

Since the great success in the development of optical fiber transmission, research based on intensity modulation is moving toward multigigabit transmission. On the other hand, optical fiber transmission utilizing an optical field instead of intensity as a carrier has been investigated for the next generation of systems. This method is very popular in microwave transmission, where information is carried by the frequency or phase of the electromagnetic field. Since the heterodyne or homodyne detection method is used in the system, it is referred to as optical heterodyne transmission or coherent transmission (see Chapter 24).

In order to realize a coherent transmission system, light oscillators having a narrow linewidth Δv and a well-controlled center frequency f_c are required. In this section, light sources for coherent transmission, in particular, "single-frequency" laser diodes, are discussed.

Typical linewidth Δv of laser diodes is measured to be several tens of megahertz, which is larger than that of gas lasers. The reasons for this are (1) the Q value of the resonator is low due to the short laser cavity length, low reflection coefficient of laser mirrors, etc., and (2) phase fluctuation is enhanced by the refractive index fluctuation due to appreciable changes in carrier density in the active region. The source of fluctuation of lasing frequency is spontaneous emission similar to that which occurs in common lasers. In laser diodes, the fluctuation of photons changes the carrier density through the change in gain. The refractive index is related to the carrier density, so the effective cavity length fluctuates, and the phase fluctuation of the lasing is enhanced to form a wider linewidth. That is, the indirect effect of carrier fluctuation is poor coherence in the laser diode. This indirect enhancement effect reflects the strong dependency of gain and refractive index profiles upon the carrier density in semiconductor lasers.

The linewidth Δv of solitary laser diodes can be expressed by the modified Schawlow-Townes formula including the enhancement effect mentioned above as [46]

$$\Delta v = v_g^2 h v n_{\rm sp}(\alpha_i + \alpha_m)\alpha_m \frac{1 + \alpha_e^2}{8\pi P} \tag{15}$$

where v_g is the group velocity, hv is the photon energy, $n_{\rm sp}$ is the spontaneous emission factor, α_i is the photon loss in the cavity due to absorption and scattering, α_m is the photon loss in the cavity due to the power output from the ends, α_e is the enhancement factor, and P is the output power.

The enhancement factor is expressed as

$$\alpha_e = \frac{d\hat{n}_1/dN}{d\hat{n}_2/dN} \tag{16}$$

where $\hat{n}_1$ and $\hat{n}_2$ are the real and imaginary parts, respectively, of the reflective index in the active layer and N is the carrier density. $d\hat{n}_2/dN$ corresponds to the derivative of the gain (or the loss) with respect to N. The observed values of α_e scatter between 2 and 5, and the power–linewidth product $P\,\Delta v$ is measured to be 20–200 MHz-mW in solitary lasers including Fabry-Perot and DFB laser diodes. Figure 11.27 shows linewidth Δv versus the inverse of optical power of DFB laser diodes lasing at 1.3 μm. Δv decreases with increases in cavity length, and the power–linewidth product is constant in the range of medium P [47].

Reduction of the linewidth Δv can be realized by a long laser cavity L, high output power operation, and reduction of the enhancement factor α_e. In Figure 11.27, 3 MHz is obtained with L of 780 μm at $P = 6$ mW. Presently linewidths of a few megahertz are reported in solitary DFB lasers in the 1.3–1.5-μm region; however, linewidth is never less than 1 MHz. Reduction of the enhancement factor α_e is another alternative to reduce the linewidth in DFB laser diodes, in which the lasing wavelength can be controlled or detuned from the peak of the gain profile of the active layer. $d\hat{n}_2/dN$ in Eq. (16) is larger at the shorter-wavelength side of the gain peak in semiconductor lasers, and detuning toward the shorter wavelengths reduces Δv, as shown in Figure 11.28 [48] as well as in Figure 11.27.

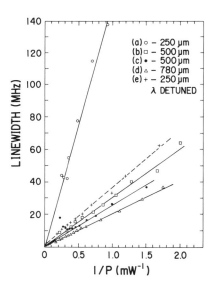

Figure 11.27 Measured linewidths versus inverse output power for DFB lasers with various lengths (solid lines). The dashed line is for a detuned DFB laser for which the Bragg wavelength is about 150 Å shorter than the peak wavelength of the gain profile [47].

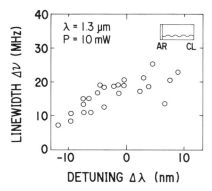

Figure 11.28 Linewidth control by detuning $\Delta\lambda$ of the Bragg wavelength of the gain profile. Each point corresponds to a device with a different grating pitch [48].

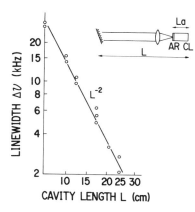

Figure 11.29 Measured 3-dB linewidth of an external cavity laser [49]. The results are well fit by $\Delta v \, [\text{kHz}] = 4.5 \times 10^3 / P[\text{mW}] L^2 \, [\text{cm}]$. The output power is 3 mW, and the minimum Δv is 2.5 kHz.

An external cavity configuration is a more effective means of reducing Δv. Figure 11.29 shows experimental results for an external cavity with a DFB laser diode with antireflection coating on the facet that faces an external grating [49]. A linewidth of 2 kHz is obtained with a total cavity length of 25 cm. In the external cavity configuration, linewidth behavior is very sensitive to power feedback into the laser diode. Figure 11.30 indicates that a strong coupling of less than -10 dB is necessary to get stable narrowing in linewidth. This can be realized by using optical fibers or a GRIN rod to improve stability; however, monolithic integration of a laser diode and an

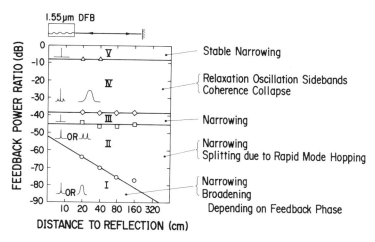

Figure 11.30 Regimes of feedback effect in 1.5-μm DFB lasers [50]. Plots showing the power levels at which the transition between regimes occur as a function of distance to the external mirror. Lines are drawn through the data points for clarity. Regime V is optimal for the operation of an extended cavity laser to obtain stable narrowing; however, it can be attained for antireflection-coated lasers.

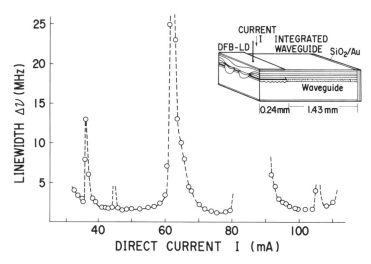

Figure 11.31 Monolithic extended cavity 1.3-μm DFB lasers [51]. The minimum linewidth is 1.2 MHz.

external cavity may be a practical approach to construction of a robust device. Figure 11.31 shows one example of a monolithic external cavity DFB laser diode [51] with a minimum linewidth of 1.2 MHz.

Direct modulation is advantageous even in a coherent transmission scheme. For example, a multiterminal DFB laser diode provides flat frequency modulation characteristics, as shown in Figure 11.32 [52]. Adjusting the two currents, we can eliminate a cross-frequency dividing thermal effect and a carrier-injection effect that exist in a solitary DFB laser diode. This three-terminal DFB laser diode can be modulated at high speeds since both sections operate in the lasing condition, which reduces carrier lifetime.

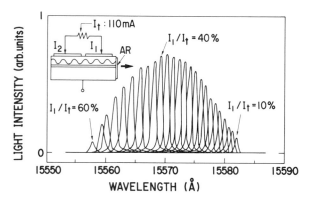

Figure 11.32 Three-terminal DFB lasers. Lasing wavelength and output power can be controlled by combining currents I_1 and I_2 [52].

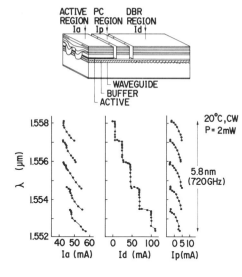

Figure 11.33 Four-terminal monolithic wavelength-tuning device. The lasing wavelength can be tuned by controlling three currents through an active section, a phase control section, and a distributed Bragg reflection section [53].

Frequency tunability is one of the interesting features of laser diodes that will be required in future systems including a circuit switching in local networks as well as coherent transmission. Multiterminal or multisection devices are also being investigated. Figure 11.33 illustrates an example of a tunable DBR laser diode [53] that consists of an active region for amplification, a phase control region, and a distributed-Bragg-reflection (DBR) region. The lasing wavelength can be tuned by the currents over 58 Å, or the equivalent of 720 GHz.

These devices are under investigation, and the grating structure is one of the key technologies for providing the stable operation that will be required in the next generation of optical communications.

11.9 SINGLE-LONGITUDINAL-MODE (SLM) LASERS FOR OPTICAL FIBER COMMUNICATIONS

In this chapter, we have discussed the role of coherence in laser diodes used in optical fiber transmission systems. The partition noise limit can be removed with SLM laser diodes, and we are now near the chirping limit, which may be regarded as the limit due to direct modulation. Note that evaluation of a transmission system should consider all components and the specific environment to be implemented. The practical fiber loss, dispersion-shifted fibers having zero dispersion at 1.55 μm, dispersion-controlled fibers, practical external modulators, etc. affect the specific requirements of the laser diode's coherence. Nevertheless, SLM operation of the light source will always be essential.

As for the DFB laser diodes, the first demonstration of a 400-Mb/s transmission experiment over 100 km at 1.55 μm in 1983 pushed the device into the realistic and practical domain [54]. The reliability of InGaAsP DFB laser diodes appears to be promising. Preliminary data indicate that light emission properties of the device are basically as reliable as those of conventional LDs [55]; however, special care should be paid to keep enough margin for SLM operation, which is still an issue. As an example of systems, 1.6-Gb/s systems using conventional single-mode fibers and DFB laser diodes at 1.3-μm lasing wavelength are being implemented for commercial service in NTT, as a forerunner of systems using SLM laser optical sources.

REFERENCES

1. T. Tsukada, "GaAs-Ga$_{1-x}$Al$_x$As buried heterostructure injection lasers," *J. Appl. Phys.*, **45**: 4899–4906 (1974).

1a. K. Aiki, M. Nakamura, T. Kuroda, J. Umeda, R. Ito, N. Chinone, and M. Maeda, "Transverse mode stabilized Al$_x$Ga$_{1-x}$As injection lasers with channel-substrate-planar structure, *IEEE J. Quantum Electron.*, **QE-14**: 89–94 (1978).

2. T. Ikegami, "Reflectivity of mode at facet and oscillation mode in double heterostructure injection lasers," *IEEE J. Quantum Electron.*, **QE-8**: 470–476 (1972).

3. T. L. Paoli and J. E. Ripper, "Direct modulation of semiconductor lasers," *Proc. IEEE*, **58**: 1457–1965 (1970).

4. R. E. Epworth, "The phenomenon of modal noise in analogue and digital optical fiber systems," *Tech. Digest*, 4th Eur. Conf. Opt. Commun., 1978, p. 492.

5. T. Ito, S. Machida, K. Nawata, and T. Ikegami, "Intensity fluctuations in each longitudinal mode of a multimode AlGaAs laser," *IEEE J. Quantum Electron.*, **QE-13**: 574–579 (1977).

6. Y. Okano, K. Nakagawa, and T. Ito, "Laser mode partition noise evaluation for optical fiber transmission," *IEEE Trans. Commu.*, **COM-28**: 238–243 (1980).

6a. K. Ogawa, "Analysis of mode partition noise in laser transmission system," *IEEE J. Quantum Electron.*, **QE-18**: 849–855 (1982).

7. R. A. Linke, B. L. Kasper, J.-S. Ko, I. P. Kaminow, and R. S Vodhanel, "1 Gbit/s transmission experiment over 101 km of single mode fiber using a 1.55 μm ridge guide C^3 laser," *Electron. Lett.*, **19**: 776–777 (1983).

8. K. Iwashita, T. Matsumoto, C. Tanaka, and G. Motosugi, "Linewidth requirement evaluation and 290 km transmission experiment for optical CPFSK differential detection," *Electron. Lett.*, **22**: 791–792 (1986).

9. L. W. Casperson, "Threshold characteristics of multimode laser oscillators," *J. Appl. Phys.*, **46**: 5194–5201 (1975).

10. N. Chinose, T. Kuroda, T. Ohtoshi, T. Takahashi, and T. Kajimura, "Mode-hopping noise in index-guided laser," *IEEE J. Quantum Electron.*, **QE-21**: 1264–1270 (1985).

11. T. Ikegami, "Spectrum broadening and tailing effect in direct modulated injection lasers," *Proc. 1st Eur. Conf. Opt. Commun.*, London, 1975, pp. 111–112.

12. T. E. Bell, "Single-frequency semiconductor lasers," *IEEE Spectrum*, **20**(12): 43 (1983).

13. H. Kogelnik and C. V. Shank, "Stimulated emission in a periodic structure," *Appl. Phys. Lett.*, **18**: 152–154 (1971).

14. I. P. Kaminow and H. P. Weber, "Poly(methyl methacrylate) dye laser with internal diffraction grating resonator," *Appl. Phys. Lett.*, **18**: 497–499 (1971).

15. W. T. Tsang, N. A. Olsson, R. A. Linke, and R. A. Logan, "1.5 μm wavelength GaInAsP C^3 lasers—single-frequency operation and wide band frequency tuning," *Electron. Lett.*, **19**: 415–416 (1983).

16. K. Utaka, S. Akiba, K. Sakai, and M. Matsushima, "Room-temperature CW operation of distributed-feedback buried-heterostructure InGaAsP/InP lasers emitting at 1.57 μm," *Electron. Lett.*, **17**: 961–963 (1981).
17. T. Matsuoka, H. Nagai, Y. Itaya, Y. Noguchi, U. Suzuki, and T. Ikegami, "CW operation of DFB-BH GaInAsP/InP lasers in 1.5 μm wavelength region," *Electron. Lett.*, **18**: 27–28 (1982).
18. W. T. Tsang, *Semiconductors and Semimetals*, Vol. 22, Part B, Academic, New York, 1985.
19. W. Streifer, D. Yevick, T. Paoli, and R. Burnham, "An analysis of cleaved coupled cavity lasers," *IEEE J. Quantum Electron.*, **QE-20**: 754–764 (1984).
20. S. W. Corzine and L. A. Coldren, "Continuous tunability in three-terminal coupled-cavity lasers," *Appl. Phys. Lett.*, **48**: 1190–1192 (1986).
21. M. Nakamura, K. Aiki, J. Umeda, and A. Yariv, "CW operation of distributed feedback GaAs-GaAlAs diode lasers at temperatures up to 300 K," *Appl. Phys. Lett.*, **27**: 403–405 (1975).
22. F. Koyama, S. Arai, Y. Suematsu, and K. Kishino, "Dynamic spectral width of rapidly modulated 1.58 μm GaInAsP/InP buried-heterostructure distributed Bragg reflector integrated twin guide lasers," *Electron. Lett.*, **17**: 938–939 (1981).
23. H. Kogelnik and C. V. Shank, "Coupled-wave theory of distributed feedback lasers," *J. Appl. Phys.*, **43**: 2327–2335 (1972).
24. A. Yariv, "Coupled-mode theory for guided-wave optics," *IEEE J. Quantum Electron.*, **QE-9**: 919–933 (1973).
25. S. Wang, "Principles of distributed feedback and distributed Bragg-reflector lasers," *IEEE J. Quantum Electron.*, **QE-10**: 413–427 (1974).
26. W. Streifer, R. D. Burnham, and D. R. Scifres, "Effect of external reflectors on longitudinal modes of distributed feedback lasers," *IEEE J. Quantum Electron.*, **QE-11**: 154–161 (1975).
27. T. Matsuoka, Y. Yoshikuni, and H. Nagai, "Verification of the light phase effect at the facet on DFB laser properties," *IEEE J. Quantum Electron.*, **QE-21**: 1880–1886 (1985).
28. H. A. Haus and C. V. Shank, "Antisymmetric taper of distributed feedback lasers," *IEEE J. Quantum Electron.*, **QE-12**: 532–539 (1976).
29. K. Utaka, S. Akiba, K. Sakai, and Y. Matsushima, "λ/4-shifted InGaAsP/InP DFB lasers by simultaneous holographic exposure of positive and negative photoresists," *Electron. Lett.*, **20**: 1008–1010 (1984).
30. B. Broberg, S. Koentjoro, F. Koyama, Y. Tohmori, and Y. Suematsu, "Mass transported 1.53 μm DFB lasers with improved longitudinal mode control," *Proc. 10th Eur. Conf. Opt. Commun.*, Stuttgart, 1984, Postdeadline Paper no. 2.
31. H. Nagai, T. Matsuoka, Y. Noguchi, Y. Suzuki, and Y. Yoshikuni, "InGaAsP/InP distributed feedback buried-heterostructure lasers with both facets cleaved structure," *IEEE J. Quantum Electron.*, **QE-22**: 450–457 (1986).
32. W. Streifer, D. R. Scifers, and R. D. Burnham, "Coupling coefficients for distributed feedback single- and double-heterostructure diode lasers," *IEEE J. Quantum Electron.*, **QE-11**: 867–873 (1975).
33. W. Streifer, R. D. Burnham, and D. R. Scifres, "Radiation losses in distributed feedback lasers and longitudinal mode," *IEEE J. Quantum Electron.*, **QE-12**: 737–739 (1976).
34. C. H. Henry, R. F. Kazarinov, R. A. Logan, and R. Yen, "Observation of destructive interference in the radiation loss of second-order DFB lasers," *IEEE J. Quantum Electron.*, **QE-21**: 151–154 (1985).
35. Y. Itaya, T. Matsuoka, K. Kuroiwa, and T. Ikegami, "Longitudinal mode behaviors of 1.5 μm range GaInAsP/InP distributed feedback lasers," *IEEE J. Quantum Electron.*, **QE-20**: 230–235 (1984).
36. J. Buus, "Mode selectivity in DFB lasers with cleaved facets," *Electron. Lett.*, **21**: 179–180 (1985).
37. M. Usami, S. Akiba, and K. Utaka, *Proc. Int. Semiconductor Laser Conf.*, Kanazawa, Japan, 1986, paper E-3.

38. M. Yamaguchi, Y. Koizumi, T. Numai, I. Mito, and K. Kobayashi, "High single longitudinal mode yield in 1.55 μm phase shifted DFB-DC-PBH LDs with a novel cavity end structure," *Proc. 13th Eur. Conf. Opt. Commun.* Helsinki, 1987, pp. 51–54.

39. H. Soda, H. Ishikawa, and H. Imai, "Design of DFB lasers for high-power single-mode operation," *Electron. Lett.*, **22**: 1047–1049 (1986).

40. C. Lin, T. P. Lee, and C. A. Burrus, "Picosecond frequency chirping and dynamic line broadening in InGaAsP injection lasers under fast excitation," *Appl. Phys. Lett.*, **42**: 141–143 (1983).

41. F. Koyama, S. Arai, Y. Suematsu, and K. Kishino, "Dynamic spectral width of rapidly modulated 1.58 μm GaInAsP/InP buried-heterostructure distributed Bragg reflector integrated twin guide lasers," *Electron. Lett.*, **17**: 938–940 (1981).

42. L. D. Westbrook, A. W. Nelson, P. J. Flddument, and I. D. Henning, "Performance of strongly-coupled distributed feedback lasers operating at λ = 1.5 μm," 9th IEEE Int. Semiconductor Laser Conf., Rio de Janeiro, 1984, pp. 14–15.

43. K. Kurumada and T. Ikegami, "Distributed feedback laser for optical transmission system," Conf. Opt. Fiber Commun., San Diego, February 1985, paper WC-1.

44. T. L. Koch and J. E. Bowers, "Nature of wavelength chirping in directly modulated semiconductor lasers," *Electron. Lett.*, **20**: 1038–1039 (1984).

44a. F. Koyama and Y. Suematsu, "Dynamic wavelength shift of dynamic-single-mode (DSM) lasers and its influence on the transmission bandwidth of single mode fibers," Nat. Conv. Rec. of IECE Japan, Matsuyama, S5-8, 1984.

45. A. H. Gnauk, B. L. Kasper, R. A. Linke, R. W. Dawson, T. L. Koch, T. J. Bridges, E. G. Burkhardt, R. T. Yen, D. P. Wilt, J. C. Campbell, K. Ciemiecki Nelson, and L. G. Cohen, "4 Gb/s transmission over 103 km of optical fiber using a novel electronic multiplexer/demultiplexer," Conf. Opt. Fiber Commun. San Diego, February 1985, Postdeadline Paper no. 2.

46. C. H. Henry, "Theory of the linewidth of semiconductor lasers," *IEEE J. Quantum. Electron.*, **QE-18**: 259–264 (1982).

47. K. -Y. Liou, N. K. Dutta, and C. A. Burrus, "Linewidth-narrowed distributed feedback injection lasers with long cavity length and detuned Bragg wavelength," *Appl. Phys. Lett.*, **50**: 489–491 (1987).

48. S. Ogita, M. Yano, H. Ishikawa, and H. Imai, "Linewidth reduction in DFB laser by detuning effect," *Electron. Lett.*, **23**: 393–394 (1987).

49. R. A. Linke and K. J. Pollack, "Linewidth vs. length dependence for external cavity laser," 10th IEEE Int. Semiconductor Laser Conf., Kanazawa 1987, pp. 118–119.

50. R. W. Tkatch and A. R. Chraplyvy, "Regimes of feedback effects in 1.5 μm distributed feedback lasers," *J. Lightwave Technol.*, **LT-4**: 1655–1661 (1986).

51. S. Murata, S. Yamazaki, I. Mito, and K. Kobayashi, "Spectral characteristics for 1.3 μm monolithic external cavity DFB lasers," *Electron. Lett.*, **22**: 1197–1198 (1986).

52. Y. Yoshikuni and G. Motosugi, "Multielectrode distributed feedback laser for pure frequency modulation and chirping suppressed amplitude modulation," *J. Lightwave Technol.*, **LT-5**: 516–522 (1987).

53. S. Murata, I. Mito, and K. Kobayashi, "Over 5.8-nm continuous wavelength tuning of 1.5 μm wavelength tunable DBR laser," OFC/IOOC'87, Reno, 1987, paper WC3.

54. T. Ikegami, K. Kuroiwa, Y. Itaya, S. Shinohara, K. Hagimoto, and N. Inagaki, "1.5 μm transmission experiment with distributed feedback laser," 8th Eur. Conf. Opt. Fiber Commun., Cannes, 1983, Postdeadline Paper no. 6.

55. G. Motosugi, M. Saruwatari, and M. Suzuki, "Distributed feedback lasers and laser modules for the F-1.6G system," *Rev. NTT Electrical Commun. Lab.*, **35**: 239–245 (1987).

CHAPTER 12

Modulation Properties of Semiconductor Lasers

J. E. Bowers

12.1 INTRODUCTION

Semiconductor lasers are critically important elements in optical communications systems because of their small size, high power, ease of modulation and, unlike other lasers, their potentially very low cost because the entire laser can be fabricated using planar processing. The intensity and frequency of semiconductor lasers can be conveniently modulated by modulating the current to the lasers. These lasers can be modulated at extraordinarily high frequencies because stimulated emission considerably shortens the carrier lifetime below the level of other nonlasing devices such as FETs, photodiodes, and light-emitting diodes.

The high-speed capability of semiconductor lasers was demonstrated early in their development [1–10]. Modulation rates up to a few gigahertz were the norm for the first 15 years as the effects of carrier diffusion, multiple transverse and longitudinal mode effects, and spectral hole burning effects became understood [11–22]. In the early 1980s the bandwidths of GaAs lasers were increased up to 12 GHz as the parasitic capacitance of lasers was reduced and the limitation of facet damage at high optical powers was reduced [23–29]. A number of groups around the world [30–54] developed high-frequency structures in the GaInAsP system, important for optical fiber communications systems with low loss and low dispersion. These lasers have low parasitics and high power and are not limited by the facet damage problems of GaAs lasers. With these lasers, an 8 Gb/s communication system was demonstrated using a single laser [55], and a 16 Gb/s time-division multiplexed system was demonstrated by combining the outputs of four lasers [56]. Direct modulation of a semiconductor laser at 16 Gb/s has also been demonstrated [57], and even higher bit rate modulation is possible with improvements in electronics. The prospect for high data rate communications and instrumentation using directly modulated semiconductor lasers looks very promising.

This chapter will concentrate on the modulation properties of semiconductor lasers with particular emphasis on the high-speed capabilities and limitations of semiconductor lasers. Several other excellent chapters on intensity modulation of GaAlAs lasers [58, 59] and GaInAsP lasers [60] have been written. This chapter concentrates on long-wavelength lasers and describes both the intensity and frequency modulation characteristics. The frequency modulation characteristics are important in system applications when fiber dispersion and laser spectral bandwidth may limit the transmission distance. Frequency modulation is also important in some coherent communication systems and other applications where a frequency-agile source is required.

The organization of this chapter is as follows. The basis for a theoretical understanding of laser modulation, the carrier and photon rate equations, are described, followed by a small-signal analysis of these equations. The effects of device parasitics are considered, and the measured modulation characteristics of a number of different structures are compared to the theoretical predictions. The effect of current modulation on the laser frequency is presented and compared to experimental results. Large-signal modulation characteristics are described, and ultimate limits to laser modulation are discussed.

12.2 RATE EQUATIONS

The dynamic characteristics of semiconductor lasers have been described by numerous authors to various levels of complexity. Following these papers, we make a number of assumptions to obtain simple, yet accurate and fairly general, expressions that are valid for most high-speed lasers. For example, we assume that only a single lateral mode is present and the waveguide is less than a diffusion length wide, so lateral variations in the carrier density can be neglected, carrier diffusion can be ignored, and the laterally varying optical mode can be represented as a uniform photon density S with a time- and space-independent confinement factor Γ_y. We make a similar assumption for the narrower vertical (transverse) direction. The final assumptions about spatial variations are that the cavity is uniform with just two reflecting facets (external and coupled cavities not considered here), the current injection is uniform, and the mirror loss is small enough compared to the internal loss that the carrier and photon variations along the laser length can be ignored. For lasers with an internal loss $\alpha_i \gtrsim 20 \text{ cm}^{-1}$ and a length of 300 μm, this is true for mirror reflectivities greater than 0.2 [59]. We further assume that the optical gain is a linear function of the carrier density N. We include a simple model for nonlinear suppression that is most applicable to a single-mode lasers [19]. An analysis for multimode lasers is given in Manning et al. [61].

With these assumptions the rate equations can be expressed as follows:

$$\frac{dN}{dt} = -\frac{g_0(N - N_t)S}{1 + \varepsilon S} + \frac{I}{qV} - \frac{N}{\tau_n} \tag{1a}$$

$$\frac{dS}{dt} = \frac{\Gamma g_0(N - N_t)S}{1 + \varepsilon S} - \frac{S}{\tau_p} + \frac{\beta \Gamma N}{\tau_n} \tag{1b}$$

where N and S are the electron and photon densities, g_0 is the differential gain, N_t is the carrier density for transparency, τ_n and τ_p are the spontaneous electron lifetime and photon lifetime, respectively, Γ is the optical confinement factor, ε is a parameter characterizing the nonlinear gain, β is the fraction of spontaneous emission coupled into the lasing mode, I is the current through the active layer, q is the electron charge, and V is the volume of the active layer.

12.3 SMALL-SIGNAL MODULATION

An analytic expression for the intensity modulation response can be obtained by assuming a sinusoidal variation for the three relevant variables: $I = I_0 + ie^{j\omega t}$, with similar expressions for N and S. We have explicitly ignored distortion terms at $2\omega, 3\omega, \ldots$, which give rise to second- and higher-order harmonic distortion. These terms clearly arise in Eq. (1) since both rate equations have products of N and S and thus source terms at 2ω. These effects as well as intermodulation distortion effects have been described by Lau and Yariv [62] and Darcie et al. [63]. Our approximation is valid in the small-signal regime ($n \ll N_0, s \ll S_0$). Substituting the small-signal expressions into Eqs. (1) gives expressions for the dc quantities N_0 and S_0 and for the rf quantities n and s. These can be combined and rewritten to give the transfer function for the photon density [5]:

$$\frac{s}{i} = \frac{A}{\omega_0^2 - \omega^2 + j\gamma\omega} \tag{2}$$

where A, γ, and ω_0^2 are real:

$$A = \frac{1}{qV}\left[\frac{\Gamma g_0 S_0}{1 + \varepsilon S_0} + \frac{\Gamma \beta}{\tau_n}\right]$$

$$\gamma = \frac{g_0 S_0}{1 + \varepsilon S_0} + \frac{1}{\tau_n} - \frac{\Gamma g_0 (N_0 - N_t)}{(1 + \varepsilon S_0)^2} + \frac{1}{\tau_p} \tag{3}$$

$$\omega_0^2 = -\left[\frac{g_0 S_0}{1 + \varepsilon S_0} + \frac{1}{\tau_n}\right]\left[\frac{\Gamma g_0 (N_0 - N_t)}{(1 + \varepsilon S_0)^2} - \frac{1}{\tau_p}\right]$$
$$+ \frac{g_0 (N_0 - N_t)}{(1 + \varepsilon S_0)^2}\left[\frac{\Gamma g_0 S_0}{1 + \varepsilon S_0} + \frac{\Gamma \beta}{\tau_n}\right]$$

$$N_0 - N_t = \frac{S_0/\tau_p - \Gamma \beta N_t/\tau_n}{\Gamma g_0 S_0/(1 + \varepsilon S_0) + \Gamma \beta/\tau_n}$$

This transfer function is plotted in Figure 12.1 for several values of the non-linear gain coefficient. Typical values for the variables in Eq. (3) for 1.3-μm

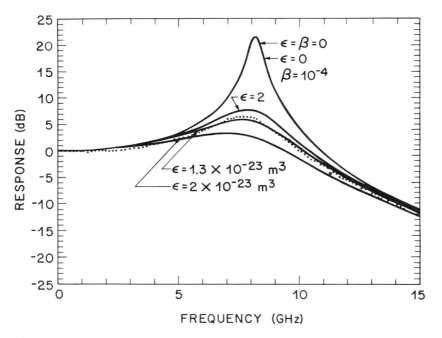

Figure 12.1 Transfer function given in Eq. (2) for no damping ($\beta = \varepsilon = 0$), spontaneous emission damping only ($\varepsilon = 0$), and several levels of nonlinear gain. A typical experimental curve (dotted line) is also shown for an output power of 4 mW. Values for the variables in Eq. (3) are given in Table 12.1. (After Bowers [50].)

TABLE 12.1. Parameters used in calculations for
1.3-μm InGaAsP lasers

Variable	Symbol	Unit	Value
Bandgap energy	E_g	eV	0.95
Waveguide thickness	t	μm	0.15
Waveguide width	w	μm	1.0
Confinement factor	Γ	—	0.34
Mirror reflectivity	R	—	0.35
Distributed loss	α_i	cm^{-1}	25
Gain coefficient	g_o	m^3/s	1.8×10^{-12}
Gain suppression	ε	m^{-3}	1.2×10^{-23}

InGaAsP lasers were used in this figure and are listed in Table 12.1. This resonance behavior has been extensively analyzed and compared to numerous experiments. The major controversy has to do with the source of damping of the resonance peak. Coupling of spontaneous emission into the lasing mode provides about 0.3 dB of damping, reducing the resonance peak height from 22.8 dB to 22.3 dB, which is clearly insufficient to explain experimental results (Fig. 12.1). Lateral carrier diffusion within the active layer damps the resonance peak and is important; however, at least in the case of long-wavelength lasers with narrow active layers, there is no value of diffusion coefficient that gives sufficient damping. An additional effect is needed such as a nonlinear gain or nonlinear absorption. Using a simple expression for nonlinear gain suppression yields the results in Figure 12.1, where a value of $\varepsilon = 1.3 \times 10^{-23}$ m^3 is indicated. This value agrees with calculated values [64–66].

The transfer function [Eq. (2)] has poles (zeros in the denominator) at

$$\omega = \frac{j\gamma}{2} \pm \left(\omega_0^2 - \frac{\gamma^2}{4} \right)^{1/2} \tag{4}$$

The damping of the response is given by the first term. The resonance frequency is approximately the second term in Eq. (4). The maximum in the response is at the frequency

$$\omega_p = \left(\omega_0^2 - \frac{\gamma^2}{2} \right)^{1/2} \tag{5}$$

The frequency corresponding to the maximum in the transfer function is usually called the resonance frequency or relaxation-oscillation frequency, since it is this peak and not the real part of the pole that is experimentally measurable.

The expression for the peak frequency, Eq. (5), can be simplified by noting that $N_0 - N_t \approx (g_0 \Gamma \tau_p)^{-1}$ and neglecting $g_0 S_0$ compared to $1/\tau_p$, so

$$\omega_p \approx \left[\frac{g_0 S_0}{\tau_p (1 + \varepsilon S_0)} \right]^{1/2} \tag{6}$$

Following Lau and Yariv [59], we have written this dependence as a function of internal photon density S_0 but have included nonlinear gain. A more convenient form for comparison with experimental data is to express this dependence in terms of the output optical power P,

$$P = \frac{SVhv\alpha_m}{2\Gamma\tau_p(\alpha_m + \alpha_i)} \tag{7}$$

where α_m is the mirror loss, $\alpha_m = (1/l) \ln (1/R)$, l is the cavity length, and R is the mirror reflectivity; α_i is the distributed loss; and hv is the photon energy. Thus,

$$f_p = D\sqrt{P} \tag{8}$$

where

$$D = \frac{1}{2\pi} \left(\frac{2g_0 \Gamma [\alpha_i + (1/l) \ln (1/R)]}{hvwt \ln (1/R)} \right)^{1/2} \tag{9}$$

w and t are the width and thickness of the waveguide and the active layer volume is $V = wtl$.

An experimental measurement of the dependence of resonance frequency on $P^{1/2}$ is shown in Figure 12.2 for a constricted-mesa laser. As expected from Eq. (8), a square root dependence on P is seen. We also show calculations based on Eq. (2) for several levels of damping. In general, there is a maximum bandwidth and a maximum resonance frequency given by $g_0/[2\pi\sqrt{2}(\varepsilon + g_0\tau_p)]$.

Depending on the application for direct modulation of lasers, the frequency of interest that characterizes the laser may be the frequency where the response has fallen from the resonance peak to its dc value ($f_{0\,dB}$), or 3 dB below its dc value ($f_{3\,dB}$), or one-half its dc value ($f_{6\,dB}$). Using the transfer function, Eq. (2), and evaluating these points, we find the following simple relations between these quantities [50]:

$$f_{0\,dB} = \sqrt{2}f_p \tag{10}$$

$$f_{3\,dB} \approx (1 + \sqrt{2})^{1/2}f_p) \tag{11}$$

$$f_{6\,dB} \approx \sqrt{3}f_p \tag{12}$$

The first expression is exact. The second and third expressions neglect terms of order $(\varepsilon S)^2$.

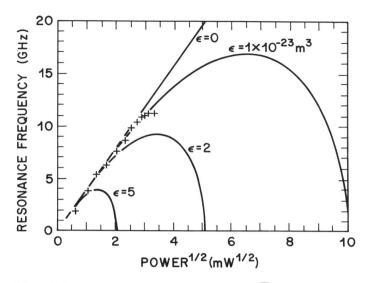

Figure 12.2 Dependence of resonance frequency on $\sqrt{P}$. The data points are measurements on a constricted-mesa laser. The solid lines are calculations assuming several values of damping. (After Bowers [50].)

The above results are valid if parasitics are insignificant. If this is not the case, then $f_{0\,dB}$, $f_{3\,dB}$, and $f_{6\,dB}$ will all be smaller than the values given above. In particular, the resonance frequency will still depend on the square root of P [Eq. (8)], but $f_{6\,dB}$, for example, will approach a frequency given by the details of the parasitics. Consequently [50],

$$\frac{f_{6\,dB}(P)}{f_p(P)} = \begin{cases} \sqrt{3} & P \ll (f_{par}/D)^2 & \text{inherent bandwidth} \\ f_{par}/D\sqrt{P} & P \gg (f_{par}/D)^2 & \text{parasitic bandwidth} \end{cases} \tag{13}$$

These two limits are plotted in Figure 12.3 along with some experimental points for a high-speed laser and a DCPBH laser. In practice, when a laser of unknown bandwidth is being characterized, one finds that the bandwidth increases linearly with $\sqrt{P}$ up to a point where the bandwidth saturates and may even decrease with increasing power [Fig. 12.2]. In GaAs lasers, this decrease is often catastrophic and nonreproducible and is usually due to facet damage [26]. In long-wavelength lasers, this saturation is very common [31, 38–41] and reproducible. Depending on the structure, it is due to parasitics in some lasers [31, 38], while in other lasers it is due to heating of the active layer and a resultant decrease in the gain coefficient [67]. In still other lasers [61] it may be due to optical nonlinearities, as indicated by the nonlinear dependence of f_p on $\sqrt{P}$ in Eq. (9). These three possibilities are indistinguishable in a plot of f_r or $f_{3\,dB}$ versus $\sqrt{P}$ but can be distinguished by a plot of the ratio $f_{6\,dB}/f_p$ versus $\sqrt{P}$. This is because heating or optical nonlinearities

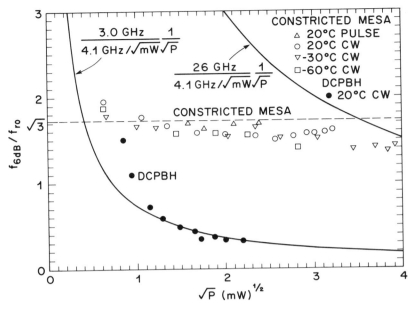

Figure 12.3 Power dependence of the ratio of 6-dB bandwidth to resonance frequency for a constricted-mesa laser and a DCPBH laser.

cause a saturation in f_p, $f_{3\,dB}$, and $f_{6\,dB}$, while parasitics affect only $f_{3\,dB}$ and not f_p (to first order).

12.4 FREQUENCY MODULATION

The above treatment has been concerned with intensity modulation of a semiconductor laser: the ratio s/i. Here, we follow the treatment of Koch and Bowers [68] and analyze the behavior of frequency modulation. This is an extremely important topic for long-distance transmission, where the repeater spacing is often limited by laser chirping and fiber dispersion.

We neglect thermal contributions to frequency modulation. Thermal contributions are important at low modulation frequencies and have been extensively analyzed by Kobayashi et al. [69]. We examine the laser frequency shift Δv due to a shift of the real part of the complex index of refraction $n = n' + jn''$. A change in the gain $\Delta g = 4\pi\,\Delta n''/\lambda$ is linked to a change in the frequency $\Delta v/v = \Delta n'/n$ by the wavelength-dependent material constant α defined as $\alpha = (\partial n'/\partial N)/(\partial n''/\partial N)$. α is commonly called the linewidth enhancement factor [70] because the linewidth of a semiconductor laser is increased over the Schawlow–Townes formula by the factor $1 + \alpha^2$. Westbrook [71] measured the wavelength dependence of α (Fig. 12.4) for 1.5-μm

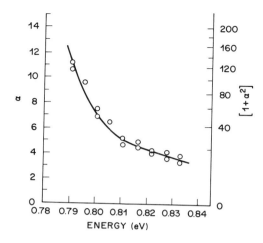

Figure 12.4 Measured variation in the linewidth enhancement factor α with photon energy for a GaInAsP laser that lased at 1.53 μm. (After Westbrook [71].)

GaInAsP lasers and found that it varies from 4 to 11 depending on the incident wavelength. The rapid increase in α near the band edge is due primarily to the wavelength dependence of the imaginary part of the index.

Using the photon rate equation [Eq. (1b)] and expanding about a dc operating point $g(t) = g_0 + \Delta g(t)$, $s(t) = S_0 + \Delta s(t)$, then we can relate $\Delta s(t)$ to $\Delta g(t)$. Using the expression for α, we can (to a good approximation except at very low powers) relate the frequency shift Δv to the output power [Eq. (7)]:

$$\Delta v(t) = -\frac{\alpha}{4\pi}\left[\frac{\partial P/\partial t}{P} + \kappa P\right] \tag{14}$$

where $\kappa = 2\Gamma\varepsilon(\alpha_m + \alpha_i)/Vhv\alpha_m$.

For sinusoidal modulation, $P = P_0 + \Delta P e^{j\omega t}$, and the rf component of Eq. (14) becomes

$$\frac{\Delta v}{\Delta P} = -\frac{j\alpha}{2}\frac{f}{P_0} + \kappa \tag{15}$$

where $\Delta v/\Delta P$ has been termed the chirp-to-modulated power ratio (CPR) [37, 68]. This is plotted in Figure 12.5 and compared to experimental measurements for a vapor-phase-transported (VPT) distributed feedback (DFB) laser. A predominantly linear dependence on f is observed, indicating the importance of the first term in Eq. (15). However, at low frequencies ($f <$ 1 GHz), where most communications systems operate, the second term dominates and may limit the transmission distance. Several lessons can be learned from Eq. (15) and Figure 12.5: First, chirping gets worse at higher frequencies (higher bit rates) since CPR is proportional to f. Second, rapid oscillations in the power due to relaxation oscillations produce a large dP/dt and thus a lot of chirping. Thus, anything that damps the excitation of relaxation oscillations such as laser parasitics or external filtering will reduce the

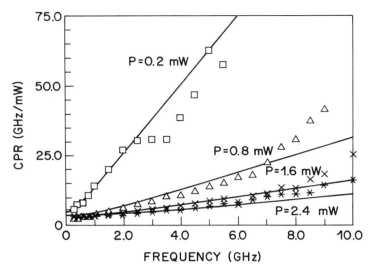

Figure 12.5 Frequency dependence of the chirp to modulated power ratio (CPR) for a VPT laser. Calculations based on Eq. (15) are also shown.

chirping. Third, the first term in Eq. (15) can be minimized by decreasing the active layer volume since this has the beneficial effect of increasing the slope D and thus pushes the resonance frequency out of the frequency band excited by the laser driver. Finally, it should be noted that chirping can be minimized by utilizing the wavelength dependence of α. Since fiber considerations or wavelength-division multiplexing considerations usually determine the desired operation wavelength and thus the grating period, the linewidth enhancement factor α can be decreased by adjusting the material composition for the smallest bandgap compatible with low-threshold DFB operation.

The important point of this discussion is that the intensity and frequency modulation characteristics are linked; a change in the structure, parasitics, or driver to optimize the intensity modulation through faster rise and fall times will also affect the chirping characteristics.

This linkage of the time dependence of the power and frequency of a semiconductor laser allows the calculation of system consequences of particular waveforms without restriction to a particular laser structure or design. An important example is the time–bandwidth product of a Gaussian pulse with power FWHM in the time domain of T [68, 72, 73]. The power spectrum FWHM [68] is

$$\Delta v_{1/2} = \frac{2 \ln 2 \sqrt{1 + \alpha^2}}{\pi T} \tag{16}$$

which is broadened over its transform-limited value by $\sqrt{1 + \alpha^2}$. Measurements have confirmed this limit, although one group [75] has reported transform-limited pulses from a semiconductor laser. This could be due to careful or fortuitous shaping of the pulse such that the two terms on the right-hand side of Eq. (14) cancel each other.

The above analysis of laser chirping can be combined with fiber dispersion to directly calculate transmission system characteristics without arbitrarily assuming a laser spectral width. Typical results of such an analysis are shown in Figure 12.6 for 4-Gb/s transmission over 100 km using VPT (Fig. 12.6a) and ridge-waveguide (Fig. 12.6b) lasers. For small extinction ratios, there is a significant extinction ratio penalty, since only a fraction of the received power is modulated. For larger extinction ratios, there is an additional system penalty, since the laser is driven close to threshold and larger relaxation oscillations develop, causing excessive chirping. The curves for the

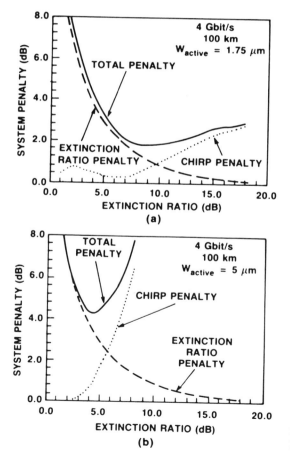

Figure 12.6 Extinction, chirp, and total system penalties for 4-Gb/s transmission over 100 km of 1.55-μm dispersive fiber [$\rho = 17$ ps/(nm-km)]. (a) 1.75-μm wide active layer device, (b) 5-μm wide active layer device. (After Corvini and Koch [76].)

two lasers are different because the second term in the chirping equation [Eq. (14)] is structure-dependent and, in particular, depends on the active layer volume.

12.5 STRUCTURE EFFECTS

12.5.1 Comparison of Measurement Techniques

In previous sections, our predominantly theoretical presentation emphasized the common characteristics of all semiconductor lasers. In this section we consider the differences of various structures and methods of characterizing the impedance and modulation response of a structure and describe ways to optimize a structure for high-speed operation.

Small-signal intensity modulation characteristics and laser impedances are usually measured with an arrangement similar to that shown in Figure 12.7. The laser is driven directly by the well-matched s-parameter test set using bias tees internal to the test set. Any reflections in the connecting cables or internal to the test set are removed mathematically by the network analyzer using standard 12-term error-correcting techniques. The optical isolator is optional, but the reduction in reflections back to the laser eliminates the small ripples in the modulation response that sometimes occur due to optical feedback. This measurement system measures the product of the

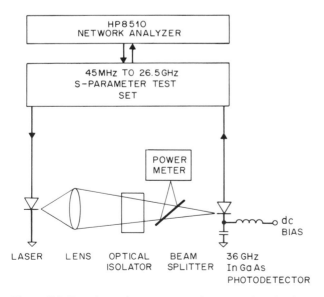

Figure 12.7 Experimental arrangement for measuring the laser modulation response and impedance.

modulation response of the laser and photodetector. The modulation response of the photodetector can be independently measured by illuminating the photodetector with a repetitive stream of picosecond pulses and measuring the response with a spectrum analyzer. Provided that the optical pulses are sufficiently short (FWHM < 5 ps for measurements up to 40 GHz), the spectrum analyzer output is directly the modulation response of the photodetector. Other techniques for measuring photodetector responses are compared in Ref. 77.

Typical modulation responses of many of the long-wavelength laser structures as of 1986 are compared in Figure 12.8. Some of the structures are optimized for high power such as dual-channel planar buried heterostructure (DCPBH) lasers, while others are optimized for speed (constricted-mesa and three-channel lasers), and still others are optimized for fabrication simplicity (ridge-waveguide lasers). The modulation characteristics in Figure 12.8 are representative of the state of the art in 1986. That is not to say that each of these structures cannot be improved; in particular, optimization for high speed is the subject of this section.

The differences in the various modulation responses in Figure 12.8 are almost entirely due to parasitics. The capacitance of each structure at threshold is shown in Figure 12.8, where it can be seen that the smaller the capacitance, the faster the laser. In particular, the parasitic frequency $1/2\pi RC$ is approximately the 3-dB bandwidth in each case except for the constricted-mesa laser. This is because the other structures are limited by parasitics while the bandwidth of the constricted-mesa laser is limited not by parasitics but by heating.

12.5.2 Parasitics

At this point we can be more specific about parasitics, circuit models, and their measurements. We shall use the simplest circuit model of any validity, namely, an inductance due to the bond wire in series with the parallel combination of the diode resistance and a parasitic capacitance (Fig. 12.9a). This model is quite valid for weakly guided structures such as ridge-waveguide and heteroepitaxial ridge overgrown (HRO) lasers as well as strongly guided structures that use dielectric layers as the predominant means of current confinement (constricted-mesa lasers). Strongly guided structures that use *pn* junctions as the predominant means of current confinement have more complicated circuit models because of the various current paths, some of which are resistively decoupled, which necessitates the use of distributed ladder networks. We will see the extent to which this model is valid for these two types of strongly guided lasers.

We are neglecting the impedance of the active-layer double heterostructure itself since its forward-bias active-layer resistance is usually less than an ohm. Since we typically measure the cw characteristics of the lasers after

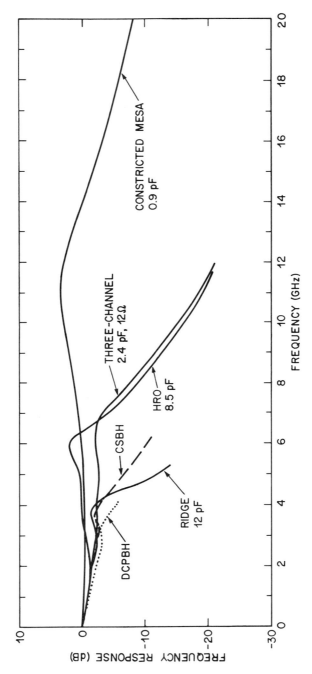

Figure 12.8 Modulation characteristics of several laser structures. The measured parasitic capacitance is listed for each device. (After Bowers [50].)

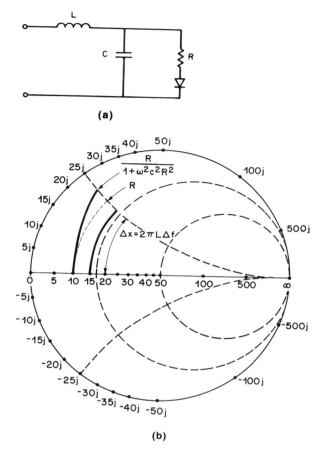

Figure 12.9 (a) Schematic of a simple model of laser parasitics. (b) Determination of laser parasitics from a Smith chart plot of impedance.

bonding to a stud, we include in our model one stud parameter, the bond-wire inductance, which properly is not a structural parameter of the laser.

A method for quickly looking at a Smith chart plot of the microwave impedance of a laser and determining its inductance, capacitance, and resistance is indicated in Figure 12.9b. At low frequencies, the impedance is real, $Z \approx R$, so the intercept on the horizontal axis gives R. The length of the arc along a constant-resistance line gives the inductance $L = \Delta X/(2\pi \, \Delta f)$. Obviously, this is most accurate when the capacitance is small. The capacitance under lasing conditions is somewhat difficult to measure because it is shunted by the low resistance of the forward-biased laser. It is most easily estimated from the zero-bias impedance at low frequencies where ωL can be neglected and $1/Z \approx 1/R_0 + j\omega C$ and R_0 is large, typically 1 kΩ. The capacitance under forward-bias conditions can be estimated by examining the change in the real part of the impedance at two frequencies, zero and ω:

$$\text{Re}(Z(\omega)) = \frac{R}{1 + \omega^2 R^2 C^2}$$

as indicated in Figure 12.9b.

The measured impedance of two laser structures is shown in Figure 12.10 along with the circuit values obtained by fitting the model predictions (also shown in Fig. 12.10) to the measured data.

Several features are evident from Figure 12.10. First, there is a great difference in the inductance of the two laser mounts, from 2 nH to 0.1 nH. There is certainly no need to endeavor to have a short bond wire if the laser is RC-limited or if the intended modulation frequencies are low; nonetheless, Figure 12.10 shows that many studs should not be expected to work at microwave frequencies. Second, almost all the laser resistances are the same (≈ 3–5 Ω). This is because most researchers have striven for a low resistance to increase the maximum cw optical power. The resistances of these two lasers are higher (7 and 9 Ω) because these lasers have a short cavity (170 μm versus typically 250 μm). The median resistance of 250-μm-long constricted-mesa lasers is 4–5 Ω.

How small do the laser parasitics have to be? The ratio of current flowing into the active layer to the current from an ideal current source with a

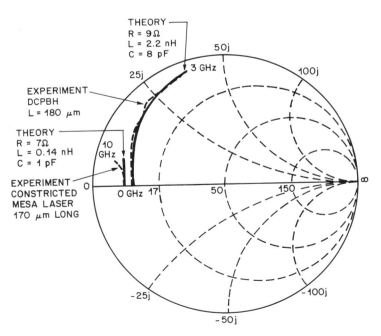

Figure 12.10 Measured impedance of DCPBH and constricted-mesa lasers.

source resistance R_s is [59]

$$\frac{I_L}{I_D} \sim \frac{1}{1 + j\omega/\omega_0 Q + (j\omega)^2/\omega_0^2} \tag{17}$$

where

$$\omega_0 = \left[\frac{R_s + R}{LRC}\right]^{1/2} \tag{18}$$

$$Q = \frac{[LRC(R_s + R)]^{1/2}}{L + RR_s C} \tag{19}$$

which has the form of a second-order low-pass filter. (The expression for Q is corrected from Refs. 23 and 59.) The -3-dB bandwidth of this circuit is

$$\omega_{3\,dB} = \omega_0\left[1 - \frac{1}{2Q^2} + \left(2 - \frac{1}{Q^2} + \frac{1}{4Q^4}\right)^{1/2}\right]^{1/2} \tag{20}$$

A laser's maximum bandwidth may be larger than this value because the resonance in the laser's inherent response can peak up the total response.

Suppose we wish to design a laser with a particular 3-dB parasitic bandwidth and with R and R_s equal to $4\,\Omega$ and $50\,\Omega$, respectively. Contours of constant parasitic bandwidth in the LC plane are shown in Figure 12.11.

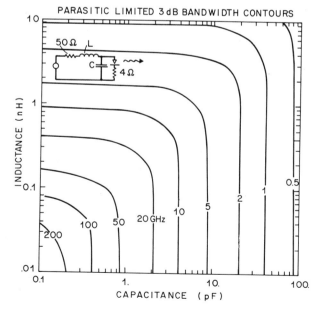

Figure 12.11 3-dB bandwidth contours in the plane of capacitance and inductance assuming $R_s = 50$ and $R = 4\,\Omega$. (After Bowers [50].)

Clearly a laser with subpicofarad capacitance is needed for response to millimeter-wave frequencies. Section 12.6 describes the constricted-mesa laser, which has demonstrated subpicofarad capacitance. First, in Section 12.5.3, we describe the optimization and qualities of this laser structure.

12.5.3 Optimization

To make a very high frequency laser, the parasitics must be small and the resonance frequency must be sufficiently high. The constricted-mesa laser has a very low capacitance because (1) the bonding pad capacitance is typically only 0.35 pF due to the thick polyimide layer under the bonding pad, and the bonding pad itself is only 50 or 100 μm wide (a viable alternative to achieve low bonding pad capacitance is to use semi-insulating substrates), and (2) the capacitance of the current confining structure, the constricted mesa, is only 0.18 pF and is not significantly bias-dependent. The laser resistance must also be low, which requires low-resistance contacts (a highly doped quaternary cap layer significantly reduces the p contact resistance over that obtained with a p-InP contact), a highly doped substrate, and a p-InP layer as highly doped and as thin as possible. We have demonstrated that this level of parasitics is possible, which should allow bandwidths well above 30 GHz.

After reducing parasitics, the second requirement for a high-frequency laser is to achieve a high resonance frequency. The maximum resonance frequency from Eq. (8) is $f_r^{max} = DP_{max}$, so we wish to maximize D and P_{max}, the maximum power from the laser. There have been six prominent suggestions to increase D: (1) decrease the waveguide width [41], increase the gain coefficient by (2) cooling the laser [67], (3) doping the active layer [35, 49], (4) confining the carriers in one or more dimensions [78], (5) "loading" the laser off its Fabry-Perot resonance [79], and (6) decrease the cavity length [26, 34, 45]. These options will be discussed in each of the following paragraphs.

Equation (9) predicts that D is proportional to Γ/w, where w is the lateral waveguide width and Γ is the optical power confinement factor. For weakly guided structures such as ridge-waveguide and HRO lasers, decreasing w from the 3–5-μm widths typically used down to 1 μm is actually detrimental, since Γ is proportional to w^2 for such a weak guide with $w \approx 1$ μm. However, for strongly guided structures, and particularly for constricted-mesa lasers that have a low-index dielectric layer (SiO$_2$ in our case) close to the active layer, the confinement factor Γ does not significantly change for $w = 1$–5 μm. The effect of waveguide width on the dependence of 3-dB bandwidth on $\sqrt{P}$ for lasers with $w = 2$ μm and $w = 1$ μm is shown in Figure 12.12. The ratio of the slopes is approximately $\sqrt{2}$ as expected from Eq. (9). Since the maximum power output is about the same in the two sets of lasers,

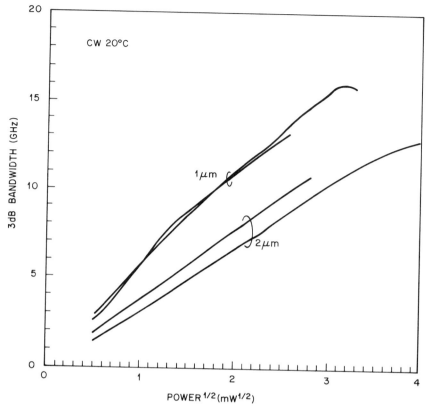

Figure 12.12 Dependence of 3-dB bandwidth on output optical power for two lasers with an active layer width w of 1 μm and two lasers with $w = 2$ μm.

it is the increase in D that allowed us to increase the 20°C cw bandwidths from 10–12 GHz [40] to 12–16 GHz [41].

Cooling the laser has two beneficial results: increasing the gain coefficient and increasing the maximum output power. Figure 12.13 shows that decreasing the temperature by 80°C increased the slope D from 5.3 to 8.7 GHz/mW$^{1/2}$ and increased the maximum power from 12 mW to 16 mW. Note that these bandwidth increases occur only because parasitics are not a significant limitation at room temperature. Cooling a DCPBH laser sometimes decreases, not increases, the bandwidth.

Su and Lanzisera [35] have published measurements on gain-guided lasers indicating that the gain coefficient increases with p doping in the active layer, with a doping level of 2.5×10^{18} cm^{-3} giving the highest gain coefficient. Similarly, Uomi et al. [49] have observed an increase in D by a factor of 2.5 with beryllium doping of multiquantum wells in GaAs lasers.

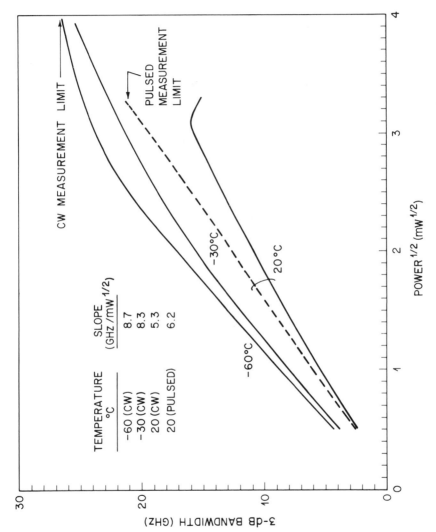

Figure 12.13 Dependence of 3-dB bandwidth on output optical power for three temperatures. The same measurement using pulsed rf and dc current is shown by a dashed line.

318

They note that it is important to decrease the growth temperature to minimize the diffusion of beryllium and mixing of the GaAs and AlGaAs interfaces. While these papers indicate an increase in D, it is important to note that it is the product DP_{max} that is important, and the effect of doping on the $L-I$ curve must also be considered. p doping of the active layer increases band tailing, intervalence band absorption, and Auger recombination, resulting in higher thresholds. Our measurements on constricted-mesa lasers with 3×10^{18} cm^{-3} zinc doping in the active layer [80] indicate that the median threshold is higher, the quantum efficiency is less, and the average peak cw output is significantly smaller than that for lasers without doping in the active layer. Consequently, the lasers with doped active layers were slower than the undoped lasers.

Carriers in a quantum well have a restricted density-of-states function and a high differential gain. Consequently, quantum well lasers should have a larger slope D than conventional lasers. Arakawa et al. [78] have calculated this increase (Fig. 12.14) for GaAs lasers, showing a factor of 2 improvement in D as the well width is decreased from > 200 Å to 50 Å. Even greater changes are predicted for quantum "wire" lasers (Fig 12.15) and quantum "dots" lasers where the electrons are confined in two and three dimensions, respectively. While similar increases are in principle possible for GaInAsP lasers, much higher carrier densities are required for wells, wires, and particularly dots than for conventional lasers, and the higher level of Auger recombination in GaInAsP lasers [81] may negate any overall increase in g_0.

Vahala and Yariv [79] have proposed and demonstrated the concept of detuned cavity loading and its effect on the modulation response. The idea

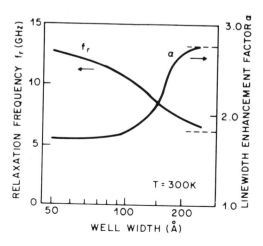

Figure 12.14 Resonance frequency and linewidth enhancement factor dependence on well width for a quantum well laser. (After Arakawa et al. [78].)

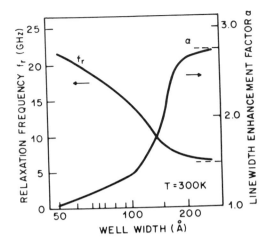

Figure 12.15 Same parameters as in Figure 12.14 for a quantum wire laser with width equal to thickness. (After Arakawa et al. [78].)

is to tune the laser to the short-wavelength side of the natural cavity resonance via an external mirror or grating, a coupled cavity, or an internal DFB or DBR grating. On the increasing current portion of an rf cycle, the laser chirps to longer wavelengths, closer to the cavity resonance, and effectively higher gain results. This effect on the gain coefficient has been demonstrated at modulation frequencies of a few gigahertz; it remains to actually demonstrate an increase in bandwidth with this technique.

The other approach to increasing the resonance frequency is to increase the maximum power. In obtaining a low-parasitic constricted-mesa laser, we

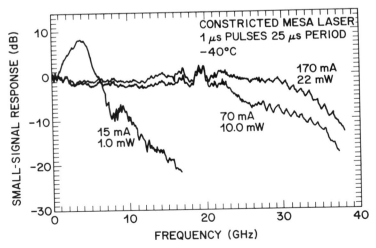

Figure 12.16 Small-signal pulsed modulation response of a constricted mesa laser at −40°C. (After Bowers [42].)

have increased the thermal resistance. Rather than using undercut mesas and thick polyimide layers, it is preferable from a thermal point of view to have a fully buried structure with a broad-area metal contact. 1.3-μm wavelength lasers of this type [82] have demonstrated 130 mW cw at 20°C but have modulation bandwidths of only 1 or 2 GHz. Many constricted-mesa lasers emit 40 mW pulsed at 20°C, and, with proper heat sinking, cw powers of this order should be possible. The other possibility is to just operate the lasers on a pulsed basis. In a 1985 experiment [42], microwave and millimeter-wave equipment was used to characterize the modulation properties of pulsed constricted-mesa lasers up to 40 GHz. One set of modulation responses at −40°C is shown in Figure 12.16. The power dependence of the bandwidth (Fig. 12.17) does not show significant saturation. These results indicate that the present cw limitations on constricted-mesa lasers are thermal, and much can be gained with proper attention to heat sinking.

The bandwidths of *planar* lasers have been rapidly increasing. Planar laser structures such as DCPBH lasers have been limited by the parasitic capacitance associated with the *pn* junction used for current confinement. By etching 5-μm wide mesas around a DCPBH active region laser, the parasitic capacitance was greatly reduced and the bandwidth increased from several gigahertz to 14 GHz. In truly planar structures, several research groups [51, 52, 83] have increased the bandwidth of buried crescent lasers by using

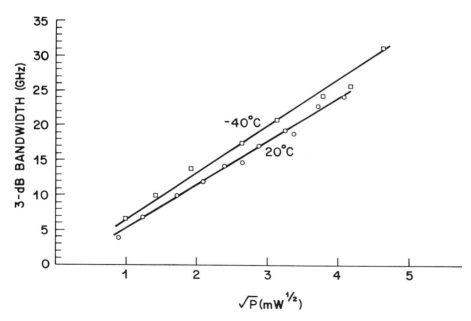

Figure 12.17 Dependence of 3-dB frequency on total output optical power per facet for pulsed operation at 20°C and −40°C. (After Bowers [42].)

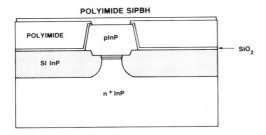

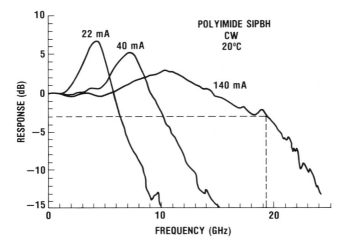

Figure 12.18 (a) Schematic diagram and (b) frequency response of an SIPBH laser. (After Bowers et al. [53].)

semi-insulating current blocking layers instead of reverse-biased *pn* junctions. Bandwidths as high as 13 GHz have been demonstrated.

Several groups have used semi-insulating layers regrown around etched mesas to achieve low-capacitance, high-speed lasers with high-power and high-temperature operation. Figure 12.18 shows a schematic diagram of one such structure, a polyimide semi-insulating planar buried heterostructure (SIPBH) laser [53]. This figure also shows the frequency response characteristics. The maximum bandwidth of 19 GHz indicates fairly low capacitance and high output power. The laser operated cw up to 105°C, although the bandwidth rapidly decreased with heating.

The above analysis and results have concentrated on small-signal modulation results, and we have seen that the modulation response depends strongly on the dc bias. We now discuss large-signal modulation and examine the expected pulse shapes and system response.

12.6 LARGE-SIGNAL MODULATION

12.6.1 Impulse Response

This section examines the large-signal laser rise and fall times. We are interested in the optical pulse shape in response to a current impulse or a step in current. Driving the laser from below threshold with a short current pulse is usually called gain switching. As the name implies, gain switching involves modulating the gain, which for semiconductor lasers usually means modulating the current drive. The goal is usually the emission of a short optical pulse, that is, a single relaxation oscillation. This optical pulse is often shorter than the electrical bandwidth would seem to allow, and the pulse shaping or shortening is due to the nonlinear aspects of semiconductor lasers. This behavior is different from that required in the next section where a non-return-to-zero (NRZ) digital code requires that for a 1-1 sequence the optical output does not drop to near zero. Here the applications are, for instance, time multiplexing a number of short-pulse laser outputs, each with its own external modulator.

To analyze the optical waveform, we again use the rate equations and make some simplifying assumptions. For a short, high current pulse applied to a laser biased below threshold, the carrier density rises to a high value and remains constant while the optical photon density builds up to a level where significant reduction in the electron density occurs. In this case, the shape of the current pulse is unimportant; only the charge, Q, in the pulse matters. We can rewrite Eq. (1b) as [87]

$$\frac{1}{S}\frac{dS}{dt} = \frac{1}{\tau_r} = g_0\Gamma(N(t) - N_{th}) = g_0\Gamma\left(\frac{Q}{qV} + N_0 - N_{th}\right) \tag{21}$$

where N_{th} is the threshold carrier density and N_0 is the dc carrier level. We see from Eq. (21) that for a constant carrier density (short electrical pulse), the photon density rises exponentially with a time constant that depends on how far the carrier density can be driven above threshold. Experimental measurements using a 17-ps electrical pulse and a high-speed laser confirm this behavior, as shown in Figure 12.19. The rise time in this figure is just 5 ps, which is only slightly longer than the round-trip transit time of 3 ps.

The fall time of a gain-switched pulse is also given by Eq. (21). The peak of the pulse occurs when the carrier density has been reduced to the threshold carrier density. The fall time is dependent on how far the carrier density can be reduced below threshold, either by recombination or by pulling charge out of the heterojunction electrically. The fall time is also approximately exponential with a fall time that is typically a factor of 2 or more times slower than the rise time. In the case shown here, the FWHM is 13 ps. Shorter pulses can be generated if short pulses with more charge can

GAIN SWITCHING

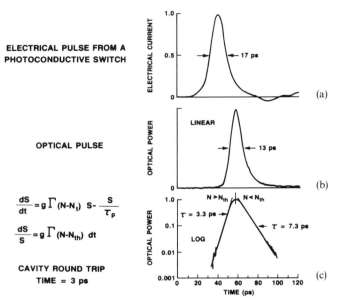

ELECTRICAL PULSE FROM A PHOTOCONDUCTIVE SWITCH

OPTICAL PULSE

$$\frac{dS}{dt} = g\,\Gamma\,(N-N_t)\;S - \frac{S}{\tau_p}$$

$$\frac{dS}{S} = g\,\Gamma\,(N-N_{th})\;dt$$

CAVITY ROUND TRIP TIME = 3 ps

Figure 12.19 Impulse response of a high-speed 1.3-μm GaInAsP constricted-mesa semiconductor laser biased at threshold. (a) Measured electrical pulse from the photoconductive pulse generator. (b) Measured optical output plotted on a linear scale showing an FWHM of 7.3 ps. (c) Measured optical output plotted on a log scale showing exponential rise and fall (After Downey et al. [87].)

be generated. An alternative technique for investigating the short-pulse gain-switching behavior of semiconductor lasers is to pump the lasers optically with a subpicosecond pulse. Arakawa et al. [84] have generated pulses as short as 2 ps in this way and found interesting, although unfortunate, differences in pulse-to-pulse behavior depending on the spontaneous emission conditions at the time of the input optical pump pulse.

We have just analyzed the impulse response of the laser. The large-signal response of a laser to a step in current can be analyzed similarly [85, 86], and the dependence of the laser rise and fall times on the step height and dc current can be calculated.

12.6.2 Pseudorandom Modulation

We now examine the effects of limited rise and fall times as well as laser ringing at the resonance frequency on large-signal pseudorandom modulation of a constricted-mesa laser at 16 Gb/s. The input electrical waveform is shown in the top pair of photos in Figure 12.20, the amplified electrical drive signal is shown in the middle pair of photos, and the measured optical waveforms are shown below in that figure. The horizontal trace in the bottom photo shows the zero light level. The electrical waveforms were generated by combining eight time-delayed versions of a 2-Gb/s pseudorandom signal of length $2^{15} - 1$. The drive signal was 1.5 V_{pp} into 50 Ω. The laser was a constricted-mesa laser with a threshold of 20 mA and a bandwidth of 15 GHz at an optical power of 10 mW per facet at a drive current of 120 mA. The output in Figure 12.20 was obtained at a bias current of 90 mA. The non-linearity of the laser and the finite bandwidth of the laser and detector cause some eye closure. Bit error rates below 10^{-9} were measured [88].

The full modulation capabilities of present lasers were demonstrated in a system experiment at 8 Gb/s over a 30-km length of fiber [55]. The system used a multi-longitudinal-mode constricted-mesa laser at 1.3 μm and a high-speed avalanche photodiode. The dependence of error rate on received power is shown in Figure 12.21 for transmission over short (3 m) and long (30 km) lengths of fiber. Slightly more than 1 dB of penalty is due to dispersion. There were predominantly four modes lasing in this laser under modulation, corresponding to a spectral width of 4.7 nm. Much longer distances could be transversed using a single-longitudinal-mode laser (to reduce the dispersion) at 1.55 μm, where the loss is less than half the loss (0.45 dB/km) at 1.30 μm for the fiber used in this experiment.

16 Gbit/s CONSTRICTED MESA LASER
DIRECT MODULATION

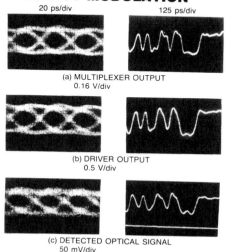

20 ps/div   125 ps/div

(a) MULTIPLEXER OUTPUT
0.16 V/div

(b) DRIVER OUTPUT
0.5 V/div

(c) DETECTED OPTICAL SIGNAL
50 mV/div

Figure 12.20 (a) Multiplexer output, (b) amplified electrical drive signal, and (c) detected optical signal for 16-Gb/s optical modulation experiment. (After Gnauck and Bowers [57].)

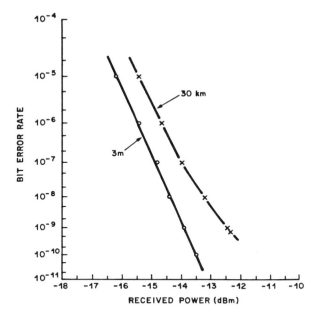

Figure 12.21 Dependence of bit error rate on received power for a constricted mesa laser and an avalanche photodiode. (After Gnauck et al. [55].)

The prospects for even higher modulation rates are good. Tucker et al. [86] calculate that the maximum bit rate is approximately 1.2 times the system bandwidth. For comparison, a linear system can transmit at a bit rate that is approximately 1.8 times the system bandwidth. We expect, therefore, that a laser with 30 GHz bandwidth should be capable of 30 Gb/s transmission. The present limitations are a lack of multiplexers and demultiplexers that operate at that rate as well as broadband amplifiers with a flat response from 1 MHz to 20 GHz or more.

12.6.3 Modelocking

In gain switching, the optical pulse builds up each time from spontaneous emission. Shorter, more stable pulses can be obtained if the optical pulse is seeded by a previous emission. The shortest pulses from a semiconductor laser have been obtained with active mode locking of high-frequency laser structures [89]. Figure 12.22 shows a schematic diagram of the cavity and the autocorrelation of the optical pulse. Short optical pulses are the result of resonantly pumping the gain region each time the optical pulse returns. Widths as short as 0.58 ps have been observed. One of the big problems is reducing the amplitude of trailing pulses, which occur as multiples of the semiconductor laser round trip time and are due to reflections off the

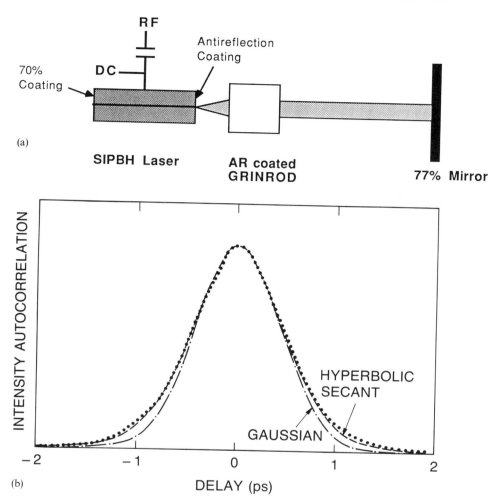

Figure 12.22 (a) Schematic diagram and (b) comparison (...) of the measured autocorrelation of the optical output of a high-frequency SIPBH laser actively modelocked at 16 GHz to autocorrelation of Gaussian and hyperbolic secant shaped pulses. The deconvolved pulse width is 0.58 ps. (After Corzine et al. [89].)

antireflection-coated semiconductor laser facet. All modelocking experiments with pulse widths under 2 ps published to date have shown this multiple pulse behavior, which significantly broadens the effective pulse width for applications such as time-division multiplexing or electrooptic sampling. The trailing pulses can be reduced with lower reflectivity coatings, shorter electrical pulses, and faster semiconductor lasers. This effects also produce shorter optical pulses. Several groups have investigated monolithic cavities [90] using long GaAs lasers and fiber optic cavities [91]. A drawing of the laser/fiber

COMPOSITE-CAVITY LASER

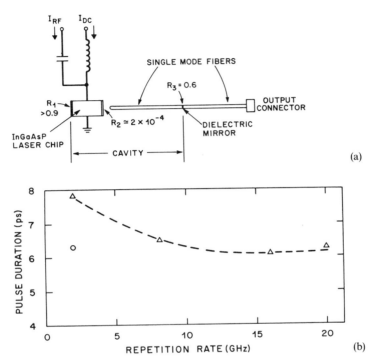

(a)

(b)

Figure 12.23 Schematic diagram of an external-fiber-cavity modelocked laser and (b) dependence of pulse width on repetition frequency. (After Eisenstein et al. [91].)

modelocked device is shown in Figure 12.23 along with a plot of the dependence of pulse width on modulation frequency. Even at repetition rates of 20 GHz, pulse widths as short as 7 ps have been obtained.

Another technique to generate short optical pulses is passive modelocking. In this case, the short pulse is obtained by a combination of a saturable gain, as in active modelocking, which shortens the trailing edge of the pulse, and a saturable absorber that acts to shorten the leading edge of the pulse. In this way, pulse widths of 0.8–5 ps have been obtained.

12.7 ULTIMATE LIMITS

The bandwidths of GaInAsP lasers are rapidly increasing, from 1 GHz in 1982 to more than 20 GHz in 1987. GaAs laser bandwidths have remained constant over the past few years due to the limitation on photon density imposed by the problem of facet damage. The threshold for facet damage in GaInAsP lasers is at least two orders of magnitude higher than for GaAs lasers, and consequently the bandwidths of GaInAsP lasers have continued to rise. What is the ultimate bandwidth obtainable from a semiconductor laser? We will discuss three limits to the bandwidth: parasitics, resonance frequency, and optical nonlinearities.

Figure 12.11 showed the limits that parasitic capacitance and inductance place on bandwidth, and it can be seen that a 100-GHz bandwidth can be obtained with $L \leq 0.05$ nH and $C \leq 0.2$ pF. Lower inductance than the 0.2 nH presently demonstrated can be achieved by using gold mesh instead of wire, or by beam-lead mounting the laser on microstrip or in waveguide. The 0.5–1.0-pF capacitance (depending on laser cavity length) presently demonstrated can be halved by (1) decreasing the bonding pad capacitance through thicker polyimide layers or semi-insulating substrates and (2) decreasing the mesa capacitance by making devices along the [011] direction instead of the [01$\bar{1}$] direction so that the mesa is rectangular instead of mushroom-shaped. The mesa capacitance is then smaller for the same contact opening. Consequently, present bandwidths are a long way from an inherent parasitic limit.

Resonance frequencies as high as 16 GHz at room temperature have been demonstrated. The present limitation to obtaining higher laser bandwidths is a mundane one, namely, obtaining more power with less heating of the active layer. The optimization that reduced the capacitance (constricted mesas and polyimide layers) also increased the thermal resistance. Thus what is needed is a higher slope D (six ways to do this were described in Section 12.5.3) or a higher output power. Since 130 mW cw in a low-speed structure at 1.3 μm has already been achieved, certainly better results should be obtained in a high-speed structure than the 10–20 mW presently demonstrated. Using a slope $D_{3 \text{ dB}}$ of 4.5 GHz/mW$^{1/2}$ and a power of 100 mW in Eqs. (8) and (11), a bandwidth of 45 GHz should be obtainable at room temperature. Perhaps one of the six ways to increase the slope D described in Section 12.5.3 will make possible bandwidths even higher than 45 GHz.

A potentially strong limitation is that the linear dependence of f_r and $f_{3 \text{ dB}}$ on the square root of optical power [Eq. (8)] is only approximate, and optical nonlinearities such as spectral hole burning, transient carrier heating or two-photon absorption cause a saturation in bandwidth [see the earlier, more exact, equations (3), (5) and (6)]. Thus there is a maximum 3-dB bandwidth for single-frequency lasers that depends strongly on the level of nonlinear gain ε. This maximum imposes a limit that is probably between 25 and 60 GHz. Higher-frequency operation is possible by increasing the optical gain coefficient g_0, perhaps by using quantum confinement. Thus further improvement is possible, and further experimentation is needed to determine the exact levels of optical nonlinearities and to increase the gain coefficient and photon densities of high-frequency lasers. Semiconductor laser bandwidths of 40 GHz and modulation at frequencies up to 60 GHz or more appears feasible.

REFERENCES

1. B. S. Goldstein and J. D. Welch, "Microwave modulation of a GaAs injection laser," *Proc. IEEE*, **52**: 715 (1964).

2. K. Konnerth and C. Lanza, "Delay between current pulse and light emission of a GaAs injection laser," *Appl. Phys. Lett.*, **24**: 120 (1964).

3. B. S. Goldstein and R. M. Weigand, "X band modulation of GaAs lasers," *Proc. IEEE*, **53**: 195 (1965).

4. S. Takamiya, F. Kitasawa, and J. Nishizawa, "Amplitude modulation of diode laser light in the millimeter-wave region," *Proc. IEEE*, **56**: 135 (1968).

5. T. Ikegami and Y. Suematsu, "Resonance-like characteristics of the direct modulation of a junction laser," *Proc. IEEE*, **55**: 122 (1967).

6. T. Ikegami and Y. Suematsu, "Direct modulation of semiconductor junction laser," *Electron. Commun. Japan*, **51-B**: 51 (1968).

7. T. L. Paoli and J. E. Ripper, "Direct modulation of semiconductor lasers," *Proc. IEEE*, **58**: 1457 (1970).

8. G. J. Lasher, "Analysis of a proposed bistable injection laser," *Solid State Electron.*, **7**: 707 (1964).

9. D. A. Kleinman, "Master rate equations and spiking," *Bell Syst. Tech. J.*, **43**: 1505 (1964).

10. H. Haug, "Quantum mechanical rate equations for semiconductor lasers," *Phys. Rev.*, **184**: 338 (1969).

11. M. Chowa, A. R. Goodwin, D. F. F. Lovelace, G. H. B. Thompson, and P. R. Selway, "Direction modulation of double heterostructure lasers at rates up to 1 Gbit/s," *Electron. Lett.*, **9**: 34 (1973).

12. P. M. Boers, M. T. Vlaardingerbroek, and M. Danielson, "Dynamic behavior of semiconductor lasers," *Electron. Lett.*, **11**: 206 (1975).

13. R. Lang and K. Kobayashi, "Suppression of the relaxation oscillation in the modulated output of semiconductor lasers," *"IEEE J. Quantum Electron.*, **QE-12**: 194 (1976).

14. M. Yano, K. Seki, T. Kamiya, and H. Yanai, "High speed direct modulation of DH semiconductor lasers—effect of multimode operation," *Electron. Commum. Japan*, **60-C**: 110 (1977).

15. M. J. Adams and B. Thomas, "Detailed calculations of transient effects in semiconductor injection lasers," *IEEE J. Quantum Electron.*, **QE-13**: 580 (1977).

16. Y. Suematsu and T. H. Hong, "Suppression of relaxation oscillation in light output of injection lasers by electrical resonance circuit," *IEEE J. Quantum Electron.*, **QE-13**: 756 (1977).

17. K. Furuya, Y. Suematsu, and T. H. Hong, "Reduction of resonance-like peak in direction modulation due to carrier diffusion in injection laser," *Appl. Opt.*, **17**: 1949 (1978).

18. K. Seki, M. Yano, T. Kamiya, and H. Yanai, "The phase shift of the light output in sinusoidally modulated semiconductor lasers," *IEEE J. Quantum Electron.*, **QE-15**: 791 (1979).

19. D. J. Channin, "Effect of gain saturation on injection laser switching," *J. Appl. Phys.*, **50**: 3858 (1979).

20. G. Arnold and P. Russer, "Modulation behavior of semiconductor injection lasers," *Appl. Phys.*, **14**: 255 (1977).

21. T. L. Paoli, "Optical response of a stripe geometry junction laser to sinusoidal current modulation at 1.2 GHz," *IEEE J. Quantum Electron.*, **QE-17**: 675 (1981).

22. M. Maeda, K. Nagans, M. Tanaka, and K. Chiba, "Buried heterostructure laser packaging for wideband optical transmission system," *IEEE Trans. Commun.*, **COM-26**: 1076 (1978).

23. K. Y. Lau and A. Yariv, "Ultra high speed semiconductor lasers," *IEEE J. Quantum Electron.*, **QE-21**: 121 (1985).

24. K. Y. Lau, N. Bar-Chaim, I. Ury, and A. Yariv, "An 11 GHz direct modulation bandwidth GaAlAs window laser on semi-insulating substrate operating at room temperature," *Appl. Phys. Lett.*, **45**: (1984).

25. K. Y. Lau, C. Harder, and A. Yariv, "Ultimate frequency response of GaAs injection lasers," *Opt. Commun.*, **36**: 472 (1981).

26. K. Y. Lau, N. Bar-Chiam, I. Ury, C. Harder, and A. Yariv, "Direct amplitude modulation of short cavity GaAs lasers up to X-band frequencies," *Appl. Phys. Lett.*, **43**: 1 (1983).

27. L. Figueroa, C. W. Slayman, and H. W. Yen, "High frequency characteristics of GaAlAs injection lasers," *IEEE J. Quantum Electron.*, **QE-18**: 1718 (1982).

28. R. S. Tucker and D. J. Pope, "Microwave circuit models of semiconductor injection lasers," *IEEE Trans. Microwave Theory Tech.*, **MTT-31**: 289 (1983).

29. R. S. Tucker and D. J. Pope, "Circuit modeling of the effect of diffusion on damping in a narrow stripe semiconductor laser," *IEEE J. Quantum Electron.*, **QE-19**: 1179 (1983).

30. I. P. Kaminow, L. W. Stulz, J. S. Ko, A. G. Dentai, R. E. Nahory, J. E. DeWinter, and R. L. Hartman, "Low-threshold InGaAsP ridge waveguide lasers at 1.3 μm," *IEEE J. Quantum Electron.*, **QE-19**: 1312–1318 (1983).

31. R. S. Tucker and I. P. Kaminow, "High frequency characteristics of directly modulated InGaAsP ridge waveguide and buried heterostructure lasers," *J. Lightwave Technol.*, **LT-2**: 385–393 (1984).

32. G. Eisenstein, U. Koren, R. S. Tucker, B. L. Kasper, A. H. Gnauck, and P. K. Tien "High-speed analog and digital modulation of 1.51-μm wavelength, three-channel buried crescent InGaAsP lasers," 9th IEEE Int. Semiconductor Laser Conf., Rio de Janiero, 1984, paper M-3.

33. T. L. Koch, L. A. Coldren, T. J. Bridges, E. G. Burkhardt, P. J. Corvini, and B. I. Miller, "Low-threshold high speed 1.55 μm vapor phase transported buried heterostructure lasers (VTPBH)," *Electron. Lett.*, **20**: 856 (1984).

34. R. S. Tucker, C. Lin, C. A. Burrus, P. Besomi, and R. J. Nelson, "High frequency small signal modulation characteristics of short cavity InGaAsP lasers," *Electron. Lett.*, **20**: 393 (1984).

35. C. B. Su and V. Lanzisera, "Effect of doping level on the gain constant and modulation bandwidth of InGaAsP semiconductor lasers," *Appl. Phys. Lett.*, **45**: 1302–1304 (1984).

36. J. E. Bowers, T. L. Koch, B. R. Hemenway, D. P. Wilt, T. J. Bridges, and E. G. Burkhardt, "High frequency modulation of 1.52 μm vapor-phase-transported InGaAsP lasers," *Electron. Lett.*, **21**: 392–393 (1985).

37. J. E. Bowers, W. T. Tsang, T. L. Koch, N. A. Olsson, and R. A. Logan, "Microwave intensity and frequency modulation of heteroepitaxial-ridge overgrown distributed feedback lasers," *Appl. Phys. Lett.*, **46**: 233 (1985).

38. R. Linke, "Direct gigabit modulation of injection lasers—structure dependent speed limitations," *J. Lightwave Technol.*, **LT-2**: 40–43 (1984).

39. U. Koren, G. Eisenstein, J. E. Bowers, A. H. Gnauck, and P. K. Tien, "Improved bandwidth of three channel buried crescent lasers," *Electron. Lett.*, **21**: 500 (1985).

40. J. E. Bowers, B. R. Hemenway, A. H. Gnauck, T. J. Bridges, and E. G. Burkhardt, "High frequency constricted mesa lasers," *Appl. Phys. Lett.*, **47**: 78–80 (1985).

41. J. E. Bowers, B. R. Hemenway, D. P. Wilt, T. J. Bridges, and E. G. Burkhardt, "26.5-GHz-bandwidth InGaAsP constricted mesa lasers with tight optical confinement," *Electron. Lett.*, **21**: 1090–1091 (1985).

42. J. E. Bowers, "Millimeter wave response of InGaAsP lasers," *Electron. Lett.*, **21**: 1195–1197 (1985).

43. J. E. Bowers, B. R. Hemenway, A. H. Gnauck, and D. P. Wilt, "High speed InGaAsP constricted mesa lasers," *J. Quantum Electron.*, **QE-22**: 833–844 (1986).

44. R. S. Tucker, "High speed modulation properties of semiconductor lasers," *J. Lightwave Technol.*, **LT-3**: 1180–1192 (1985).

45. C. B. Su, V. Lanzisera, W. Powazinik, E. Meland, R. Olshansky, and R. B. Lauer, "12.5-GHz direct modulation bandwidth of vapor phase regrown 1.3-μm InGaAsP buried heterostructure lasers," *Appl. Phys. Letts.*, **46**: 344–346 (1985).

46. H. Burkhardt and K. Kuphal, "Three-and-four-layer LPE InGaAs(P) mushroom stripe laser for $\lambda = 1.30$, 1.54, and 1.66 μm," *IEEE J. Quantum Electron.*, **QE-21**: 650 (1985).

47. H. Nishimoto, M. Yamaguchi, I. Mito, and K. Kobayashi, "High frequency response for DFB LD due to a wavelength detuning effect," *IEEE J. Quantum Electron.*, **QE-23** (1987).

48. R. Olshansky, W. Powazinik, P. Hill, V. Lanzisera, and R. B. Lauer, "InGaAsP buried heterostructure laser with 22 GHz bandwidth and high modulation efficiency," *Electron. Lett.*, **51**(2): 78–80 (1987).

49. K. Uomi, T. Mishima, and N. Chinone, "Ultrahigh relaxation oscillation frequency (up to 30 GHz) of highly p-doped GaAs/GaAlAs multiple quantum well lasers," *Appl. Phys. Lett.*, **51**(2): 78–80 (1987).

50. J. E. Bowers, "High speed semiconductor laser design and performance," *Solid State Electron.*, **30**: 1–11 (1987).

51. W. H. Cheng, K. D. Buehring, C. P. Chien, J. W. Ure, D. Perrachione, D. Renner, K. L Hess, and S. W. Zehr, "Low threshold 1.51 μm InGaAsP buried crescent lasers with semi-insulating Fe-doped InP current blocking layers," *Appl. Phys. Lett.*, **51**: 1783 (1987).

52. C. E. Zah, C. Caneau, S. G. Menocal, F. Favire, T. P. Lee, A. G. Dentai, and C. H. Joyner, "High speed 1.3 μm GaInAsP p-substrate buried crescent lasers with semi-insulating Fe/Ti-doped InP current blocking layers," *Electron. Lett.*, **24**: 695 (1988).

53. J. E. Bowers, U. Koren, B. I. Miller, C. Soccolich, and W. Y. Jan, "High speed polyimide based semi-insulating planar buried heterostructures," *Electron. Lett.*, **24**(24): 1263–1265 (1988).

54. U. Koren, B. I. Miller, G. Eisenstein, R. S. Tucker, G. Raybon, and R. J. Capik, "Semi-insulating blocked planar BH GaInAsP/InP laser with high power and high modulation bandwidth, "*Electron. Lett.*, **24**: 138 (1988).

55. A. H. Gnauck, J. E. Bowers, and J. C. Campbell, "8 Gb/s transmission over 30 km of optical fiber," *Electron. Lett.*, **22**: 600 (1986).

56. R. S. Tucker, G. Eisenstein, and S. K. Korotky, "Optical time-division multiplexing for very-high-bit-rate systems," CLEO 1988, paper TUK5, Anaheim, CA, April 1988.

57. A. H. Gnauck and J. E. Bowers, "16 Gbit/s direct modulation of an InGaAsP laser," *Electron. Lett.*, **23** (1987).

58. G. H. B. Thompson, *Physics of Semiconductor Laser Devices*, Wiley, New York, 1980, pp. 402–443.

59. K. Y. Lau and A. Yariv, "High-frequency current modulation of semiconductor injection lasers," in *Lightwave Communications Technology*, W. T. Tsang (Ed.), Academic, New York, 1985, pp. 70–152.

60. G. P. Agrawal and N. K. Dutta, *Long-Wavelength Semiconductor Lasers*, Van Nostrand, New York, 1986.

61. J. Manning, R. Olshansky, D. M. Fye, and W. Powazinik, "Strong influence of nonlinear gain on spectral and dynamic characteristics of InGaAsP lasers," *Electron. Lett.*, **21**: 496–497 (1985).

62. K. Y. Lau and A. Yariv, "Intermodulation distortion in directly modulated semiconductor lasers," *Appl. Phys. Lett.*, **45**: 1034–1037 (1985).

63. T. E. Darcie, R. S. Tucker, and G. J. Sullivan, "Intermodulation and harmonic distortion in InGaAsP lasers," *Electron. Lett.*, **21**: 665–666 (1985).

64. M. Asada and Y. Suematsu, "Density-matrix theory of semiconductor lasers with relaxa-

tion broadening model—gain and gain suppression in semiconductor laser," *IEEE J.,* **QE-21**: 434 (1985).

65. R. Olshansky, D. M. Fye, J. Manning, and C. B. Su, "Effect of nonlinear gain on the bandwidth of semiconductor lasers," *Electron. Lett.,* **21**: 721 (1985).
66. G. P. Agrawal, "Gain nonlinearities in semiconductor lasers: theory and application to distributed feedback lasers," *IEEE J.,* **QE-23**: 860 (1987).
67. F. Stern, "Calculated spectral dependence of gain in excited GaAs," *J. Appl. Phys.,* **47**: 5382–5386 (1976).
68. T. L. Koch and J. E. Bowers, "Nature of wavelength chirping in directly modulated semiconductor lasers,"*Electron. Lett.,* **20**: 1038 (1984); "Factors affecting wavelength chirping in directly modulated semiconductor lasers," CLEO, Baltimore, paper WB2, 1986.
69. S. Kobayashi, Y. Yamamoto, M. Ito. and T. Kimura, "Direct frequency modulation in AlGaAs semiconductor lasers," *IEEE J. Quantum Electron.,* **QE-18**: 582–595 (1982).
70. C. H. Henry, "Theory of the linewidth of semiconductor lasers," *J. Quantum Electron.,* **QE-18**: 259–264 (1982).
71. L. D. Westbrook, "Analysis of dynamic spectral width of dynamic-single-mode (DSM) lasers and related transmission width of single-mode fibers," *Electron. Lett.,* **22**: 1018 (1985).
72. F. Koyama and Y. Suematsu, "Analysis of dynamic spectral width of dynamic single mode (DSM) lasers and related transmission bandwidth of single mode fibers," *J. Quantum Electron.,* **QE-21**: 292 (1985).
73. J. Buus, "Dynamic line broadening of semiconductor lasers modulated at high frequencies," *Electron. Lett.,* **21**: 129–131 (1985).
74. C. Lin and J. E. Bowers, "High speed large-signal digital modulation of a 1.3 μm InGaAsP constricted mesa laser at a simulated bit rate of 16 Gbit/s," *Electron. Lett.,* **21**: 906–908 (1985).
75. N. Onodera, H. Ito, and H. Inaba, "Fourier-transform-limited, single-mode picosecond optical pulse generation by a distributed feedback InGaAsP diode laser," *Appl. Phys. Lett.,* **45**: 843–845 (1984).
76. P. J. Corvini and T. L. Koch, "Computer simulation of high bit rate optical fiber transmission using single-frequency lasers," *J. Lightwave Technol.,* **LT-5**: 1591 (1987).
77. J. E. Bowers and C. A. Burrus, "Ultrawide-band long-wavelength p-i-n photodetectors," *J. Lightwave Technol.,* **LT-5**(10): 1339–1350 (1987).
78. Y. Arakawa, K. Vahala, and A. Yariv, "Quantum noise and dynamics in quantum well and quantum wire lasers," *Appl. Phys. Lett.,* **45**: 950–952 (1984).
79. K. Vahala and A. Yariv, "Detuned loading in coupled cavity semiconductor lasers—effect on quantum noise and dynamics," *Appl. Phys. Lett.,* **45**: 501–503 (1984).
80. J. E. Bowers, B. R. Hemenway, and D. P. Wilt, unpublished, 1986.
81. N. K. Dutta and R. J. Nelson, "The case for Auger recombination in InGaAsP," *J. Appl. Phys.,* **53**: 74–92 (1982).
82. T. Asano and T. Okumura, "1.3 μm high power BH laser on p-InP substrates," *J. Quantum Electron.,* **QE-21**: 619–622 (1985).
83. D. P. Wilt, B. Schwartz, B. Tell, E. D. Beebe, and R. J. Nelson, "Channeled substrate buried heterostructure InGaAsP/InP laser employing a buried ion implant for current confinement," *Appl. Phys. Lett.,* **44**: 290 (1984).
84. Y. Arakawa, T. Sogawa, M. Nishioka, M. Tanaka, and H. Sakaki, "Picosecond pulse generation (< 1.8 ps) in a quantum well laser by a gain switching method," *Appl. Phys. Lett.,* **51**: 1295–1297 (1987).
85. R. S. Tucker, "Large signal switching transients in index-guided semiconductor lasers," *Electron. Lett.,* **20**: 802–803 (1984).
86. R. S. Tucker, J. M. Wiesenfeld, P. M. Downey, and J. E. Bowers, "Propagation delays and transition times in pulse modulated semiconductor lasers," *Appl. Phys. Lett.,* **48**: 1707–1709 (1986).

87. P. M. Downey, J. E. Bowers, R. S. Tucker, and E. Agyekum, "Picosecond dynamics of a gain-switched InGaAsP laser," *IEEE J. Quantum Electron.*, **QE-23**: 1039 (1987).

88. A. H. Gnauck and J. E. Bowers, unpublished, 1987.

89. S. W. Corzine, J. E. Bowers, G. Przybylek, U. Koren, B. I. Miller, and C. E. Soccolich, "Active mode locked GaInAsP laser with subpicosecond output," *Appl. Phys. Lett.*, **52**: 348 (1988).

90. K. Y. Lau and A. Yariv, "Direct modulation and active mode locking of ultrahigh speed GaAlAs lasers at frequencies up to 18 GHz," *Appl. Phys. Lett.*, **46**: 326–328 (1985).

91. G. Eisenstein, R. S. Tucker, U. Koren, and S. K. Korotky, "Active mode-locking characteristic of InGaAsP-single mode fiber composite cavity lasers," *IEEE J. Quantum Electron.*, **QE-22**: 142 (1986).

CHAPTER 13

High-Power Semiconductor Lasers

Luis Figueroa

13.1 INTRODUCTION

This chapter will summarize the principles that govern the operation of major types of high-power (cw) GaAlAs/GaAs and GaInAsP/InP single-mode lasers that are either commercially available or have demonstrated exceptional laboratory results. In addition, a summary of the operation of novel structures such as phaselocked arrays will be presented.

The chapter is divided into six sections. Section 13.2 describes the many applications of higher-power laser diodes now and in the future. Section 13.3 discusses the semiconductor laser device concepts that are applicable to high-power operation and provides a detailed summary of the basic concepts for high-power operation, important GaAlAs/GaAs commercial structures, and high-power 1.3-μm lasers. Section 13.4 describes semiconductor phase-locked laser arrays. A summary of the current device structures and methods for operating laser arrays in a single diffraction-limited beam are described. In Section 13.5 we describe the work on very high power two-dimensional laser arrays that can emit cw powers in excess of 100 W. Finally, Section 13.6 discusses promising areas of research.

13.2 APPLICATIONS FOR HIGH-POWER SEMICONDUCTOR LASERS

Up until recently, the need for mode-stabilized diode lasers was limited to optical communication applications requiring relatively low power levels (3–5 mW/facet). However, with the recent demonstration of very high power levels from group III–V compound semiconductor (GaAlAs/GaAs and GaInAsP/InP) lasers, a whole new branch of potential application areas has been opened. In particular, high-power semiconductor lasers are being considered for high-speed optical recording, high-speed printing, single- and multimode database distribution systems, long-distance transmission, submarine cable transmission, free-space communication, local area networks, Doppler optical radar, optical signal processing, high-speed optical microwave sources, pump sources for other solid-state lasers, and medical applications. There are two generic types of single-element lasers:

1. Gain-guided devices, whereby lateral mode control is obtained by the injected carrier profile
2. Index-guided devices whereby lateral mode control is obtained by a built-in dielectric waveguide

More recently, a third type of mode-stabilized laser—the phase locked multielement array—has been developed. In this structure, mode stabilization is accomplished by both the mutual coupling between the elements of the array and the tailoring of both the gain and lateral index profile across the array. Gain-guided lasers tend to operate multimode and have a highly astigmatic beam that is non-Gaussian in shape and can vary with drive conditions. On the other hand, index-guided lasers tend to operate in a single spectral mode, have little astigmatism, and have a beam that is more nearly Gaussian in shape. In general, the gain-guided devices are used with multimode fiber systems, while index-guided lasers are used with single-mode fiber systems. The phaselocked laser arrays are used for freespace communications and for pumping other solid-state lasers and could be used as surgical tools in the future.

Up until very recently all the major high-power laser diodes were fabricated using GaAlAs/GaAs heterostructures. However, there have been such significant advances in the GaInAsP/InP system that these lasers have performance levels similar to the GaAlAs/GaAs variety. Coupled with the inherently higher reliability, these types of lasers will experience a high demand in the future. One of the major applications for these high-power laser diodes is in local area networks. Such networks will be widely used in high-speed computer networks, avionic systems, satellite networks, and high-definition TV. These systems have a large number of couplers, switches, and other lossy interfaces that determine the total system loss. In order to maximize the number of terminals, a high-power laser diode will be required. At present, components operating in the 0.8–0.9-μm wavelength region have an edge due to their superior receiver technology, which utilizes silicon. In the final comparison, other factors such as system reliability, performance requirements, and cost need to be considered, and thus it is too early to say which technology will win out in the long run.

Long-haul fiber-optic systems operating at high bit rates require the use of single-mode fibers. It is generally accepted that index-guided lasers will be used with such systems due to their inherently high efficiency of coupling to the fiber, better mode stability, and spectral purity. In particular, it would be highly desirable to space repeaters as far apart as possible in submarine cable systems. The use of higher-power laser diodes will permit greater repeater spacing. In particular, it would be highly beneficial if repeaters could be located in more accessible areas and thus permit easier replacement.

For very high speed optical recording systems (> 100 Mb/s), the use of high-power diodes operating at relatively short wavelengths ($\lambda < 8000$ Å) is required. High-power operation is necessary since the pulse width must be reduced for high-bit-rate recording, and thus the peak power must be increased in order to keep the total energy constant.

Finally, with the advent of efficient high-power laser diodes, it has become practical to replace flash lamps for the pumping of solid-state lasers such as Nd:YAG. Such an approach has the advantages of compactness and high efficiency.

13.3 CONCEPTS FOR HIGH CW POWER OPERATION

13.3.1 Introduction

There are several useful methods for stabilizing the lateral modes of an injection laser [1–7]. In this section we will discuss techniques for the achievement of high-power operation in a single spatial and spectral mode. There are several physical mechanisms that limit the output power of the injection laser:

Spatial hole-burning effects lead to multi-spatial-mode operation and are intimately related to multi-spectral-mode operation.

Temperature increases in the active layer will eventually cause the output power to reach a maximum.

Catastrophic facet damage will limit the ultimate power of the laser diode (GaAlAs/GaAs).

Thus, the high-power laser designer must optimize these three physical mechanisms to achieve maximum power. In this section we discuss the design criteria for optimizing the laser power.

13.3.2 Catastrophic Facet Damage

The ultimate limit in output power for GaAlAs/GaAs semiconductor lasers is produced by catastrophic damage at the laser facets. This phenomenon occurs when the light intensity in the active layer is raised beyond a critical level. A schematic of the physical process involved is shown in Figure 13.1. It has been observed [8, 9] that the catastrophic damage is produced by a localized melting of the facets. The melting arises from significant absorption of light at the facet as a result of the high surface recombination velocity at the air–semiconductor interface, and thus the effect is strongly dependent on the type of semiconductor lasers used. For example, GaAlAs/GaAs lasers have a critical intensity of $\sim 2 \times 10^6$ W/cm^2 for an uncoated facet operating cw [10]. On the other hand, catastrophic damage has not been observed for GaInAsP/InP lasers (i.e., optical output power is limited by either heating or carrier leakage effects), but the catastrophic intensity is believed to be an order of magnitude larger.

Over the years, researchers have discovered various methods for increasing the catastrophic intensity. One of the most important methods involves the use of dielectric coatings on the laser facets. The use of dielectric coatings such as Al$_2$O$_3$ [11], Si$_3$N$_4$ [12], and SiO$_2$ [13] have several beneficial effects. First, the use of a protective coating with thickness $\lambda_o/2n_{die}$ (i.e., one that does not affect the facet reflectivity and n_{die} is the index of the dielectric) has been found to increase the intensity where catastrophic damage occurs. It is believed that the improvement (a factor of 2–3 for Al$_2$O$_3$) might be related to a reduction in the surface recombination velocity at the interface. A second reason might be related to the increased heat flow provided by a coating with a reasonably high thermal conductivity. Of the many

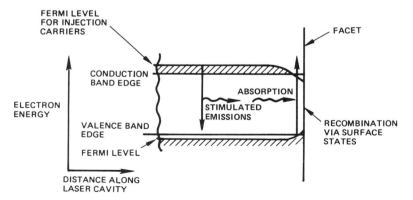

Figure 13.1 Schematic diagram of the energy-band structure near the laser facets. Note that there is significant band bending near the laser facets resulting from high surface recombination. Thus, light can actually be absorbed in a region approximately one diffusion length away from the facet. The absorption leads to significant heating at the facets, which eventually leads to catastrophic damage. (After Henry et al. [8].)

coatings that have been used, Al_2O_3 appears to be the most useful due to its close thermal expansion match to GaAs, high thermal conductivity, and good protection from moisture. A second helpful use of a dielectric coating is to reduce the facet reflectivity. It was shown early [14] that a decrease in facet reflectivity can increase the catastrophic power damage level P_c of the laser diode. It can be shown that [15]

$$\frac{P_c(\text{coating})}{P_c(\text{no coating})} = \frac{n(1 - R)}{(1 + R^{1/2})^2} \tag{1}$$

where n is the index of refraction of the semiconductor (3.6 for GaAs) and R is the power reflectivity of the facet. As $R \to 0$, P_c can be increased by a factor of n over an uncoated facet.

13.3.3 High-Power Mode-Stabilized Lasers with Reduced Facet Intensity

One of the most significant concerns for achieving high-power operation and high reliability is to reduce the facet intensity while at the same time providing a method for stabilizing the laser lateral mode. Over the years researchers have developed four approaches for performing this task:

1. Increase the lasing spot size both perpendicular to and in the plane of the junction, and at the same time introduce a mechanism for providing lateral mode-dependent absorption loss to discriminate against higher-order modes.
2. Modify the facet reflectivities by providing a combination of high-reflectivity and low-reflectivity dielectric coatings.
3. Eliminate or reduce the facet absorption by using structures with non-absorbing mirrors (NAM).
4. Use laser arrays and unstable resonator configurations to increase the mode volume.

Techniques 1 and 2 are the commonly used techniques and will be further discussed in this section. Techniques 3 and 4 (laser arrays) are discussed in Sections 13.3.5a and 13.4, respectively.

Given the proper heat sinking, in order to increase the output power of a semiconductor GaAlAs/GaAs laser we must increase the size of the beam and thus reduce the power density at the facets for a given power level. The first step in increasing the spot size involves the transverse direction (perpendicular to the junction). There are two approaches for accomplishing this, with the constraint of keeping threshold current low:

1. Thinning the active layer in a conventional double-heterostructure (DH) laser (Fig. 13.2a) below 1000 Å.
2. Creating a large optical cavity structure (Fig. 13.2b).

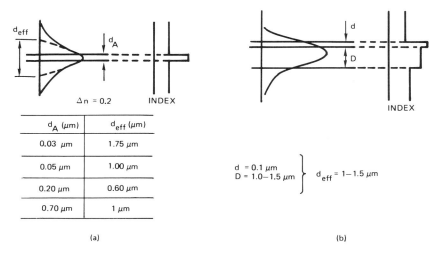

d_A (μm)	d_{eff} (μm)
0.03 μm	1.75 μm
0.05 μm	1.00 μm
0.20 μm	0.60 μm
0.70 μm	1 μm

(a)

(b)

Figure 13.2 Schematic diagram of the two most commonly used heterostructure configurations for fabricating high-power laser diodes. (a) DH structure; (b) layer large-optical-cavity structure. The d_{eff} calculations are after Botez [16].

Thinning the active layer from a conventional value of 0.2 μm to 0.03 μm causes the transverse-mode spot size to triple for a constant Δn [16]. The catastrophic power level is proportional to the effective beam width in the transverse direction, d_{eff}. The asymmetric large optical cavity (A-LOC) concept [17, 18] involves the epitaxial growth of an additional cladding layer referred to as the guide layer (d_G), with index of refraction intermediate between the n-AlGaAs cladding layer and the active layer. By using a relatively small index of refraction step ($\Delta n = 0.1$) versus 0.20–0.30 for DH lasers it is possible to force the optical mode to propagate with most of its energy in the guide layer. The effective beam width for the A-LOC can be approximately expressed as

$$d_{eff} \approx d_A + d_G \qquad (2)$$

Mode spot sizes in the transverse direction of approximately 1.5 μm can be achieved.

13.3.4 Important Commercial High-Power Diode Lasers

In the last few years several important high-power laser geometries have either become commercially available or have demonstrated impressive laboratory results. Table 13.1 summarizes the characteristics of the more important structures. It is evident that the structures that emit the highest cw

TABLE 13.1. Summary of mode-stabilized high-power laser characteristics (GaAlAs/GaAs)

Manuf. [Ref.]	Geometry	Const.	Rated power (mW)	Max. power (mW)	Spectral qual. (cw)	Spatial qual. (cw)	I_{th} (mA)	Slope EEF (mW/mA)	Far field	Substrate type
General Optronics [20a]	CNS-DCC	Two-step LPE	—	60	SLM(50)	SSM(50)	50	0.67	12° × 26°	n
Hitachi [5]	CSP	One-step LPE (TA)	30	100	SLM(40)	SSM(40)	75	0.5	(10–12°) × 27°	n
MATS. [19]	TRS	One-step LPE	25	115	SLM(50)	SSM(80)	90	0.43	6° × 20°	n
MATS. [20]	BTRS	Two-step LPE	40	200	SLM(50)	SSM(100)	50	0.8	6° × 16°	p
NEC [21]	BCM	Two-step LPE	—	80	SLM(80)	SSM(80)	40	0.78	7° × 20°	n
RCA [22]	CC-CDH	One-step LPE	—	165	SLM(50)	SSM(50)	50	0.77	6° × 30°	n
RCA [23]	CSP	One-step LPE	—	190	SLM(70)	SSM(70)	50	—	6.5 × 30°	n
Sharp [24]	VSIS	Two-step LPE	30	100	SLM(50)	SSM(50)	50	0.74	12° × 25°	p
Sharp [25]	BVSIS	Two-step LPE	—	100		SSM(70)	50	0.80	12° × 25°	p
HP [26]	TCSM	One-step MOVCD	—	65	SLM(65)	SSM(40)	50	0.4	—	p
TRW [27]	ICSP	Two-step MOCVD (AR/HR)	—	100CW	SLM(30)	150 (50% duty cycle)	60	0.86	(8–11°) × 35°	n
Ortel [28]	BH/LOC (NAM)	Two-step LPE (AR/HR)	30	90	—	90	75		—	n
							30–50	0.85		n

Approaches for achieving high-power GaAlAs lasers

Thin active (TA) or A-LOC layer to decrease facet power density
Tight current confinement to produce high current utilization
Combination of low/high reflectivity facet coatings (AR/HR) to produce high differential efficiency and lower facet intensity

SLM = single longitudinal mode; SSM = single spatial mode.

power (> 100 mW) (TRS, BTRS, CC-CDH-LOC, and the BVSIS) have several common features:

Large spot size (CDH-LOC, TRS, BTRS)
Low threshold current and high quantum efficiency
A combination of low and high reflectivity coatings to increase the power input

All the lasers with the highest powers, except for the CDH-LOC, use the thin active laser design.

Figure 13.3 contains schematic diagrams for three of the more common laser designs for high cw power operation, and Figure 13.4 shows plots

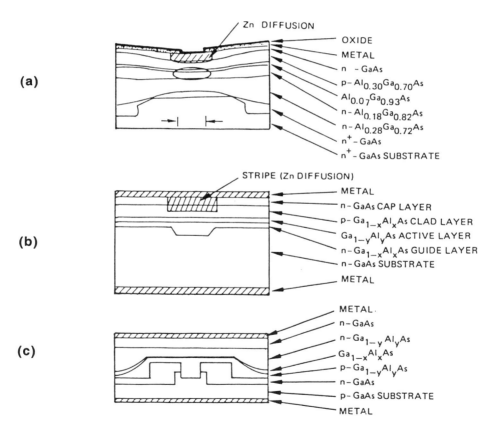

Figure 13.3 Geometries for several important high-power diode lasers. (a) Constricted double-heterostructure laser (CC-CDH-LOC) [22]; (b) channel substrate planar laser (CSP) [5]; (c) broad-area twin-ridge structure (BTRS) [20].

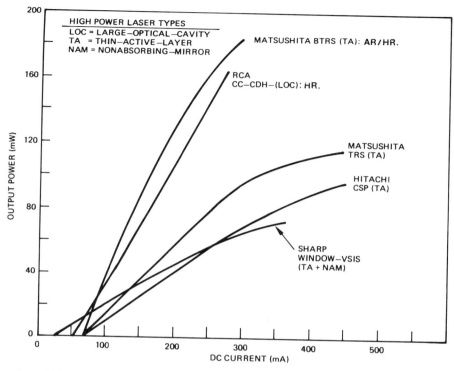

Figure 13.4 Plots showing output power versus cw current for the major high-power laser diodes. The maximum power in a single spatial mode is in the range of 100–150 mW and total cw power can approach 200 mW.

of output power versus current for various important geometries listed in Table 13.1.

The CC-CDH-LOC device with improved current confinement [22] (Fig. 13.3a) is fabricated by one-step liquid-phase epitaxy (LPE) above a mesa separating two dovetail channels. Current confinement is provided by a deep zinc diffusion and a narrow oxide stripe geometry. The final cross-sectional geometry of the device is very dependent on the exact substrate orientation and LPE growth conditions [29]. By properly choosing these conditions, it is possible to grow a convex-lens-shaped active layer ($Al_{0.07}Ga_{0.93}As$) on top of a concave-lens-shaped guide layer ($Al_{0.21}Ga_{0.79}As$). The combination of the two lead to a structure with antiwaveguiding properties and a large spot size. Discrimination against higher-order modes is provided by a leaky-mode waveguide. The cw threshold current is in the range of 50–70 mA, while T_0 is $\approx$ 150–185 K. Single-mode operation has been obtained to 60 mW under 50% duty cycle, and the maximum cw power from the device is 165 mW. The power conversion efficiency at this power level is 35% considering only the front facet. Due to the large near field, the far field is quite narrow ($6° \times 30°$).

The CSP laser [5] (Fig. 13.3b) is fabricated by one-step LPE above a substrate channel. The current stripe is purposely made larger than the channel to ensure uniform current flow across the channel. However, this leads to some waste of current and thus a lower differential efficiency than other similar high-power laser structures (BVSIS, BTRS). Lateral mode control is very effectively obtained by the large difference in the absorption coefficient α between the center and edges of the channel and by changes in the index of refraction that result from changes in the geometry. By proper control of the active and n-cladding layer thicknesses it is possible to obtain $\Delta\alpha \approx 1000 \text{ cm}^{-1}$ [30] and $\Delta n \approx 10^{-2}$. Threshold currents are in the range of 55–70 mA, and T_0 is in the range of 120–160 K. As shown in Table 13.1, typical beam widths are in the range of $10° \times 27°$. The transverse far field is relatively narrow due to the very thin active layer. Researchers from RCA have obtained power levels in excess of 150 mW (cw) with a CSP-type laser [23].

More recently, Matsushita [19] has developed a CSP-like structure that uses a 400-Å active layer thickness. The structure has demonstrated fundamental mode cw power to 80 mW and single-longitudinal-mode cw power to 50 mW. The maximum available power for the TRS laser is 115 mW, and threshold currents are in the range of 80–120 mA. The far-field pattern is $\sim 6° \times 16°$. It appears that even though their geometry is similar to that of the CSP, lasers with ultra thin and planar active layers have been fabricated. It should be further pointed out that one of the keys to achieving ultra high power from CSP-like structures is the achievement of ultra thin (< 1000 Å) active layers that are highly uniform in thickness. Small nonuniformities in the active layer thickness lead to a larger Δn difference, and thus a smaller lateral spot size, which will lead to lower power levels and reduced lateral mode stability.

In the past several years there have been significant efforts to fabricate high-power laser diodes on p-substrates in order to improve the current confinement and efficiency of the devices. The first laser structures on p-substrates were fabricated by workers from Sharp Corporation in 1981 [31]. However, prior to their work, researchers from Spectronix [32] and Motorola [33] had fabricated LEDs with superior performance by using p-substrates. Matsushita has developed an improved version of their TRS laser using p-substrates [20]. The structure (Fig. 13.3c) requires two steps of crystal growth. The first step involves the growth of an n-GaAs blocking layer. This is followed by a channel-etching step (analogous to the TRS) and the conventional DH laser growth, with the exception of reversing the sequence in which layers are grown. Typical threshold currents are 50–70 mA, and the structure has been able to produce 200 mW of total cw power and 100 mW of cw power in a single spatial mode by using a combination of low-reflectivity and high-reflectivity coatings. For an active layer of 300 Å, the far-field pattern is 6° × 16°.

The use of a p-substrate in the fabrication of many of the recent high-power semiconductor laser diodes appears to improve the laser performance for three reasons:

1. Improved current confinement is obtained as a result of the reduced current spreading in the first p-cladding layer.
2. The characteristics of the *pnpn* current-blocking structure are improved compared to a similar structure fabricated on an n-type substance.
3. According to the results in Section 13.3.5, the thermal properties of the laser are improved.

The reliability of high-power laser devices fabricated on p-type substrates has also been reported. Researchers from Sharp observed a degradation rate of 5 mA/khr for $P = 30$ mW (cw)$/T = 50°C$ for the BVSIS structure [25].

More recently metal-organic chemical vapor deposition (MOCVD) has been used to fabricate lasers with high optical power in a single spatial and longitudinal mode. This crystal growth technique, due to its inherent uniformity of layer growth and the possibility of using large-area wafers, promises to revolutionize laser diode production. The excellent layer uniformity leads to a reduced spectral width and more uniform threshold characteristics. Two MOCVD laser structures with demonstrated high-power capability are schematically shown in Figure 13.5, and their characteristics are summarized in Table 13.1. Figure 13.5a shows the twin-channel substrate mesa guide (TCSM) laser [26]. The fabrication consists of growing a DH laser structure over a chemically etched twin-channel structure using MOCVD. Optical guiding is provided by the curvature of the active layer. The TCSM laser has achieved cw powers of 40 mW in a single spatial mode and 65 mW in a single longitudinal mode. The latter number represents one of the best reported values for an MOCVD type laser. The inverted channel substrate planar (ICSP) laser [27] is schematically shown in Figure 13.5b. This structure is one MOCVD version of the very successful CSP structure (Fig. 13.5b) [34]. The ICSP laser has achieved powers in excess of 150 mW (50% duty cycle) in a single spatial mode and a 100 mW cw catastrophic power level. These power levels are some of the highest reported for a single-element MOCVD structure operating in a single spatial mode. A more recent trend is the use of Quantum well active layers for higher power operation.

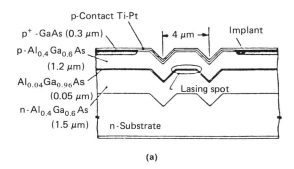

(a)

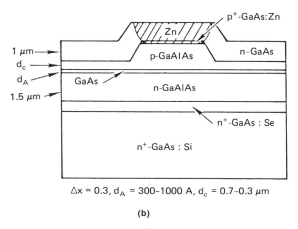

$\Delta x = 0.3, d_A = 300\text{--}1000 \text{ A}, d_c = 0.7\text{--}0.3 \ \mu m$

(b)

Figure 13.5 Geometries for two important high-power semiconductor lasers fabricated by MOCVD. (a) Twin-channel substrate mesa guide (TCSM) [26]; (b) inverted-channel substrate planar (ICSP) laser [27].

13.3.5 Future Directions for High-Power Lasers

a. Nonabsorbing Mirror Technology. We have pointed out in Section 13.3.2 that catastrophic facet damage is the ultimate limit to the power from a semiconductor laser. In order to prevent such effects, one has to create a region of higher-energy bandgap and low surface recombination at the laser facets. Thus, the concept of a laser with a nonabsorbing mirror (NAM) was developed. The first NAM structure was demonstrated by Yonezu et al. [35] by selectively diffusing zinc along the length of the stripe except near the facets. This created a bandgap difference between the facet and bulk regions and permitted a three- to fourfold increase in the cw facet damage threshold and a four- to fivefold increase in pulse power operation [36]. More recent structures have involved several steps of liquid-phase epitaxy [37, 38].

The incorporation of the NAM structure is strongly device-dependent. For example, in the diffused device structures, such as deep-diffused stripe (DDS) [35] and transverse junction stripe (TJS) lasers, NAM structures have been formed by selective diffusion of zinc in the cavity direction [36]. The n-type region will have a wider bandgap than the diffused region, and thus there will be little absorption near the facets. However, most index-guided structures require an additional growth step for forming the NAM region [37, 38]. The NAM structures in the past have suffered from several problems.

Due to their complex fabrication, they tend to have low yields. Furthermore, cw operation has been difficult to obtain.

Cleaving must be carefully controlled for NAM structures having no lateral confinement in order to avoid excessive radiation losses in the NAM region. The NAM length is a function of the spot size.

The effect of the NAM structure on lateral mode control has not been documented but could lead to excessive scattering and a rough far-field pattern.

It is now becoming clearer that the use of a NAM structure will be required for the reliable operation of high-power GaAlAs/GaAs laser diodes. Experimental results [39] appear to indicate that laser structures without a NAM region show a decrease in the catastrophic power level as the device degrades. However, most of the approaches currently being implemented require elaborate processing steps. A potentially more fundamental approach would involve the deposition of a coating that would reduce the surface recombination velocity and thus enhance the catastrophic intensity level.

Recently, the use of NAM technology has been appearing in commercial products. The crank TJS laser can operate reliably at $P = 15$ mW (cw), while the TJS laser without the NAM can only operate at $P = 3$ mW (cw) [39]. The Ortel Corporation has developed a buried heterostructure (BH) laser with significantly improved output power characteristics compared to conventional BH lasers [28]. The NAM BH laser is rated at 30 mW (cw) [28] compared to 3–10 mW for the conventional BH/LOC device.

Lastly, the use of alloy disordering, whereby the bandgap of a quantum well laser can be increased by diffusion of various types of impurities (for example, Zn and Si) [40], can lead to a very effective technique for the fabrication of a NAM structure. Such structures have produced an enhancement of the maximum pulsed power by a factor of 3–4.

b. High-Power 1.3-μm Lasers. Previous sections have discussed high cw power operation from (GaAl)As/GaAs laser devices. In the past several years there have been reports of the increasing power levels achieved with GaInAsP/InP lasers operating at $\lambda = 1.3$ μm. The physical mechanisms limiting high-power operation in this material system are quite different than those for GaAlAs/GaAs lasers. The surface recombination at the laser facets is significantly lower than in GaAlAs/GaAs, and thus catastrophic damage has not been observed. Maximum output power is limited by either heating or carrier leakage effects. With the advent of structures having low threshold current density and high quantum efficiency, it was just a matter of time before high-power results would become available. Furthermore, since facet damage is not a problem, the only real need for facet coatings is for improving the output power from one facet and sealing the device for improved reliability.

In Figure 13.6, we schematically show the two most common laser structures that have demonstrated high cw power operation. In Figure 13.6a, the double-channel planar buried heterostructure (DC-PBH) is shown [41, 42]. The structure requires a two-step LPE growth process. The first step is the growth of the first and top cladding layers in addition to the active layer. This is followed by the etching of the structure, which is followed by a regrowth to form the blocking and contact layers. LPE growth of this material

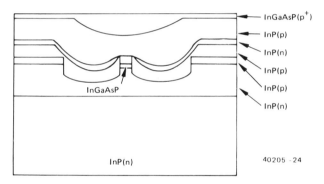

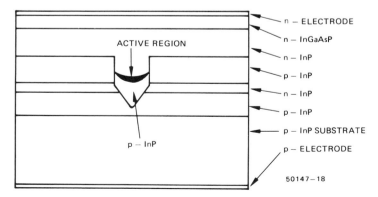

Figure 13.6 Schematic diagrams showing two representative high power InGaAsP/InP laser structures. (a) Double-channel planar buried heterostructure (DC-PBH) [41]; (b) buried crescent laser [43].

system is such that if the mesa region is narrow enough no growth occurs on top of it during the deposition of the blocking layer, and this occurs for mesa widths less than ~ 5 μm. Low threshold current is achieved due to the narrow mesa geometry and the good carrier and current confinement.

The DC-PBH has proved to be a laser structure with excellent output characteristics and high reliability. NEC has been able to obtain thresholds as low as 10 mA with 70% quantum efficiency. Degradation rates of the order of 10^{-6}/hr at $P = 5$ mW, $T = 70°C$ have also been obtained. More recently, NEC has obtained 140-mW power in a single spatial mode [41]. Lasers at $P = 50$ mW and $T = 25°C$ have been placed on lifetest and show relatively low degradation rates after several hundred hours. Degradation rates at $P = 20$ mW and 30 mW ($T = 50°C$) are 1.3×10^{-5}/hr and 2.22×10^{-5}/hr, respectively. TRW has also been working with DC-PBH/PBC-like laser diodes and has obtained 100 mW cw [42]. A summary of the various high-power $\lambda = 1.3$-μm laser diode structures and characteristics is given in Table 13.2.

The other structure that has demonstrated high cw power is the buried crescent laser first investigated by Mitsubishi (Fig. 13.6b) [43, 44]. The structure is grown using a two-step LPE process and a p-substrate. The final structure resembles a channel laser with an active layer that tapers to zero near the edges of the channel. The tapering provides good carrier and optical confinement. The structure has achieved maximum power levels of 200 and 140 mW in a single spatial mode [45]. Recent lifetest results [46] at high power levels by researchers from Oki with a structure similar to the Mitsubishi structure have demonstrated a mean time to failure of $\sim 7 \times 10^4$ hr ($T = 20°C$) at $P = 0.75P_{max}$, where P_{max} is the maximum cw power

TABLE 13.2. Summary of mode-stabilized high-power laser characteristics (GaInAsP/InP)

Manufacturer	Geometry	Construction	Max. power (mW)	Spatial quality	I_{th} (mA)
Mitsubishi	PBC [43, 44]	Two-step LPE (p-subst.)	140	SSM(70)	10–30
NEC	DC-PBH [40, 41]	Two-step LPE	140	SSM(140)	10–30
OKI	PEC [45]	Two-step LPE (p subst.)	200	SSM(200)	10–30
TRW/EORC	DC-PBH type [42]	Two-step LPE	100	SSM(70)	10–30
TRW/EORC	PBC [42]	Two-step LPE (p-substrate)	107	SSM(78)	10–30

Approaches for high-power GaInAsP/InP lasers

Tight current confinement to reduce the threshold current
Facet coatings (reflector/low reflecting front facet)
Diamond heat sinking
Long cavity length

SSM = single spatial mode.

from the laser ($P_{\text{max}} = 25-85$ mW). These results appear to indicate that
1.3-μm lasers are reliable for high-power applications.

An important parameter in the operation of high-power laser diodes is
the optimization of thermal properties of the device. In particular, optimizing
the laser geometry for achieving high-power operation is an important design
criterion. Arvind et al. [47] used a simple one-dimensional thermal mode
for estimating the maximum output power as a function of laser geometry
(cavity length, active layer thickness, substrate type, etc.). The results obtained for GaInAsP/InP narrow stripe lasers were as follows:

Maximum output power is achieved for an optimum active layer thickness
in the 0.15-μm region.

Significantly higher output powers (25–60%) are obtained for lasers fabricated on p-substrates compared to those on n-substrates. The result is
based on the lower electrical resistance of the top epitaxial layers in the
p-substrate compared to n-substrate.

Significantly higher output powers ($\sim 60\%$) are obtained for lasers mounted
on diamond rather than silicon heat sinks as a result of the higher thermal conductivity of diamond compared to silicon [22 vs. 1.3 W/°C-cm)].

Significantly higher output powers ($\sim 100\%$) are obtained for lasers having
a length of 700 μm compared to the conventional 300 μm. The higher
power results from the reduced threshold current density and thermal
resistance for the longer laser devices. A plot of the calculation and experimental data from Oki [45] is given in Figure 13.7. Note that there
are no adjustable parameters in the calculation.

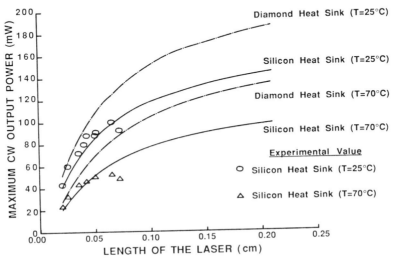

Figure 13.7 Calculated plot of maximum cw output power from a PBC type laser. The data
are taken from Kawahara et al. [45], and the calculation is from Arvind et al. [47].

An important conclusion from the thermal modeling is that longer cavity semiconductor lasers (700–1000 μm) will be able to operate at higher heat sink temperatures when the power level is nominal ($\sim$ 5 mW) compared to shorter cavity devices ($\sim$ 100–300 μm).

13.4 SEMICONDUCTOR LASER ARRAYS

One of the most common methods used for increasing the power from a semiconductor laser is to increase the width of the emitting region. However, as the width is increased, the occurrence of multilateral modes, filaments, and lateral mode instabilities become more significant. A far-field pattern is produced that is not diffraction-limited and has reduced brightness. The most practical method to overcome this problem is to use a monolithic array of phaselocked semiconductor lasers. Such lasers have been used to generate powers in excess of 8 (cw) [48] and over 50 W [90] from a single laser bar.

It was not until 1978 that Scifres and co-workers [49] first reported on the phaselocked operation of a monolithic array consisting of five closely coupled proton-bombarded lasers. The original coupling scheme involved branched waveguides, but this was quickly abandoned in favor of evanescent field coupling by placing the individual elements of the array in close proximity (Fig. 13.8a). More recently, arrays of index-guided lasers [50] have been fabricated; one example is shown in Figure 13.8b.

Recent emphasis has been on achieving higher cw power and controlling the output far-field distribution. Some of the more significant events in the development of practical semiconductor laser arrays are summarized in Table 13.3.

The subject of array mode stability has become of great interest. In a series of significant papers, Butler et al. [66] and Kapon et al. [67] recognized that to a first approximation an array can be modeled as a system of N weakly coupled waveguides. The results indicate that the general solution for the field amplitudes will consist of a superposition of these array modes. The analytic results permitted, for the first time, a simple explanation for the observed far-field patterns and provided a means for designing device structures that would operate in the fundamental array mode (i.e., all elements in phase). This particular mode will provide the greatest brightness [$B = (P/cm^2)/\Omega$, where Ω is the solid angle of the emitted radiation].

Many techniques have been used for improving array mode selection [56–65, 68–73] and thus achieving a well-controlled single-lobe beam. Two of the most successful techniques involve (1) incorporation of optical gain in the interelement regions (gain coupling) of the laser array [56–61, 69] and (2) use of interferometric techniques that involves Y-coupled junctions [63, 65, 68]. Figures 13.9 and 13.10 illustrate representative schematic diagrams of

SCHEMATIC OF THE TWO BASIC TYPES OF ARRAY STRUCTURES

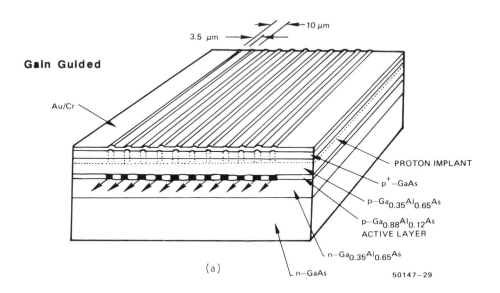

(a)

50147–29

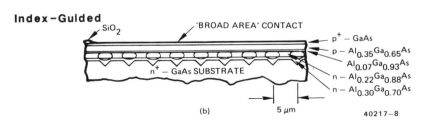

(b)

40217–8

Figure 13.8 Schematic diagram showing two types of laser array structures: (a) Gain-guided phase-arrayed using quantum well active layers and grown by MOCVD (After Scifres et al. [49].) (b) Index-guided phased array using CSP-LOC structures and grown by LPE. (After Botez and Connally [50].)

the two most prominent approaches. The gain-coupled arrays (Fig. 13.9) achieve mode selectivity by introducing optical gain in the interelement regions and thus increasing the gain of the fundamental array mode since this mode has a significant portion of its energy in the interelement regions. The first demonstration of this approach was the twin-channel laser (TCL) developed by researchers from TRW [56–58]; since then there have been other demonstrations [59–61, 69].

The Y-coupled junction (Fig. 13.10) is a more recent development, and the theoretical foundations were first described in a paper by Chen and Wang

TABLE 13.3. Summary of high-power phaselocked laser arrays

Laser group	No. of elements	Material system	Type of array*	Maximum power (controlled mode)† (mW)	maximum power† (mW)	Far field‡
Xerox, 1978 [49]	5	GaAlAs-GaAs	GG	60 (P)	130 (P)	SL (2°)
HP, 1981 [51]	10	GaAlAs-GaAs	IG	1 W (P)	1400 (P)	DL
Xerox, 1982 [52]	10	GaAlAs-GaAs	GG	200 (P)	270 (cw)	SL (1°)
Xerox, 1983 [53]	40	GaAlAs-GaAs	GG	800 (P)	2600 (cw)	DL
RCA, 1983 [50]	10	GaAlAs-GaAs	IG		1000 (P)	DL
Siemens, 1984 [54]	40	GaAlAs-GaAs	GG	400 (P)	1600 (cw)	DL
Bell Labs, 1984 [55]	10	GaAlAs-GaAs	IG		–	DL
TRW, 1984 [56–58]	2	GaAlAs-GaAs	IG		–	SL (4–6°)
UC Berkeley, 1984 [59]	10	GaAlAs-GaAs	IG	75 (cw)	115 (cw)	SL (2–7°)
Cal-Tech, 1984 [60]	5	GaAlAs-GaAs	IG		200	SL (3°)
Xerox/Spectra Diode, 1985 [61]	10	GaAlAs-GaAs	Offset stripe GG	575 (P)	–	SL (1.9°)
Bell Labs, 1985 [62]	10	InGaAsP-InP	GG	100 (cw)	600	SL (4°)
Sharp, 1985 [63]	2	GaAlAs-GaAs	IG; Y-C	65 (cw)	90 (cw)	SL (4.2°)
Mitsubishi, 1985 [64]	3	GaAlAs-GaAs	IG	100 (cw)	150 (cw)	SL (3.6°)
Xerox/Spectra Diode, 1986 [65]	10	GaAlAs-GaAs	IG (Y-C) stripe GG	200 (cw)	575 (P)	SL (3°)

* GG = gain-guided; IG = index-guided; Y-C = Y-coupled.
† P = pulsed.
‡ DL = double-lobe; SL = single-lobe.

– 353 –

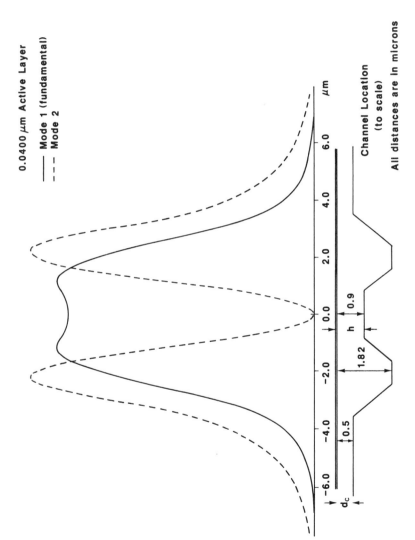

Figure 13.9 Schematic diagram of the near-field intensity for the twin-channel laser (TCL) with gain in the interelement region. Note the high field intensity in the region between the channels for the in-phase mode. (After [75].)

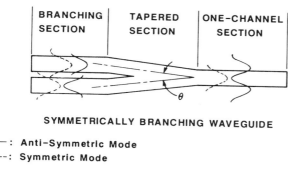

BRANCHING SECTION | TAPERED SECTION | ONE-CHANNEL SECTION

SYMMETRICALLY BRANCHING WAVEGUIDE

——— : Anti-Symmetric Mode
----- : Symmetric Mode

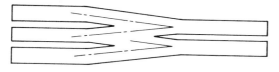

DOUBLE-Y SYMMETRICALLY BRANCHING WAVEGUIDE

Figure 13.10 Schematic diagram of interferometric structures that can be used for lateral mode selection in laser array structures.

[68]. Mode selectivity is accomplished because the in-phase mode adds coherently at each Y-junction, while the out-of-phase mode has destructive interference, since the single waveguides after the Y-junction can support only the fundamental mode. Similar interferometric and mode-selective techniques have been used in the development of optical modulators [74].

At the present time it is not clear which technique will be most useful for achieving stable, fundamental array mode operation. The gain-coupling concept works well for two or three elements. However, array mode selection described by the difference in gain, Δg, between the first and second array modes rapidly decreases as the number of array elements increases beyond two or three [75]. The Y-junction approach does not appear to have the same limitations. However, preliminary experimental results indicate that the main output beam is significantly broadened (2–3 times the diffraction limit), and in some cases there is a significant side-lobe level. Furthermore, this approach increases the number of interfaces within the laser cavity, which can lead to problems with reliability.

It is clear that the final questions and answers for semiconductor laser array research are far from being complete, and there remain many exciting challenges that could lead to devices with powers and output characteristics approaching other solid-state lasers. In particular, methods for improving array mode discrimination, for example, for large arrays, will need to be developed, and device reliability with respect to mean time to failure (MTTF) and changes in the output characteristics needs to be explored.

13.5 TWO-DIMENSIONAL HIGH-POWER LASER ARRAYS

There has been a significant amount of research activity in the past few years in the area of very high power diode lasers [76–81]. The activity has been driven by the significant reductions in threshold current density of GaAlAs/GaAs lasers that can be achieved with metal-organic chemical vapor deposition (MOCVD), utilizing a quantum well design. Threshold current densities as low as 200–300 A/cm^2 with external efficiencies exceeding 80% have been achieved using GRIN-SCH quantum well lasers [76, 77]. CW powers of ~6–9 W have been achieved from single laser bars. Such power levels correspond to a maximum of 11 W/cm from a single laser bar. Table 13.4 lists some of the more recent results on very high power diode laser arrays.

In order to increase the output power from laser array structures, researchers have investigated the use of a two-dimensional laser array. One particular configuration, referred to as the "rack-stack approach," is schematically shown in Figure 13.11. In essence, the approach involves stacking a linear array of edge emitters into a two-dimensional array. The two-dimensional arrays are fabricated [79] by (1) cleaving linear arrays of laser

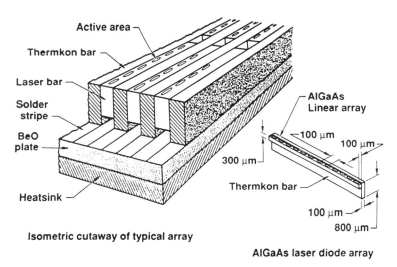

Figure 13.11 A schematic diagram of the two-dimensional rack-stack laser array architecture. (Courtesy of R. Solarz, Lawrence Livermore.)

TABLE 13.4. Summary of high-power laser array results*

Laboratory	Array type	Maximum output power	Power eff.	Slope eff.	Power density
General Electric	ID array [77]	80 W (200-μs pulse; 10–100 Hz)	20%	0.9 W/A	80 W/cm
McDonnell-Douglas	Broad stripe [78–80] ($W = 300~\mu m$; $L = 1200~\mu m$)	6 W (cw)	38%	0.91 W/A	200 W/cm
	2D: 4 bars, 8 mm	15 W (cw)	15%	—	50 W/cm^2
	2D: 5 bars, 8 mm	320 W (0.3% DF)	—	—	2560 W/cm^2
Spectra Diode	ID array [48]	8 W (cw)	—	—	—
	1D array	134 W (150-μs pulse)	49%	1.26 W/A	134 W/cm

* All high-power laser structures are fabricated using MOCVD and a quantum well design.
  DF: Duty Factor

diodes from a processed wafer, (2) mounting the bars on heat sinks, and (3) stacking the heat sinks into a two-dimensional array.

As shown in Table 13.4, the main players in this business are General Electric, McDonnell-Douglas, and Spectra Diode Labs. The largest stacked [80] two-dimensional array has been manufactured by McDonnell-Douglas and has an active area with five laser bars 8 mm in length. The array was operated with 150-μs pulses to the limit of the driver [80] at pulse repetition rates of 20–666 Hz. Approximately 2.5 kW/cm^2 was obtained at 20 Hz ($P_{av} \approx 300$ W) and 0.9 kW/cm^2 ($P_{av} \approx 92$ W) at 666 Hz. Higher output powers will be obtained as a result of achieving the ultimate limits in threshold current density and optical losses in the individual cavities with advances in nonabsorbing mirror techniques and active cooling techniques. Some recent progress has been seen in the latter with the use of etched silicon grooves for fluid flow, which function as radiator elements to remove the heat.

13.6 A GLIMPSE AT THE FUTURE

So far we have discussed the present state of the art. Now let us take a short glimpse at what the future may bring. A good summary of the state of the art and future trends is provided in recent SPIE proceedings on high power laser diodes [89, 90].

One important trend is the increasing use of MOCVD technology for the fabrication of high power single element lasers. Researchers from STC and Spectra Diodes [90] have used Ridge Waveguide designs in combination with a GRIN-SCH to fabricate laser diodes with single mode powers in excess of 150 mW cw.

As laser technology matures, it might be possible to fabricate laser diodes with very long lengths (~ 1 mm) that would emit significantly more power than the present structures. Furthermore, variation of the geometry along the length of the laser could lead to vastly improved lateral mode discrimination at high power levels. In the past several years, several observations of improved high-power performance using a semiconductor laser with a cavity having a nonuniform geometry along the length of the device have been reported [63, 65, 68, 82, 90]. In the future, such structures may be optimized to provide a controlled spot size at high power levels. Lastly, longer laser cavities will also lead to higher operating power levels and higher operating temperatures.

The use of external cavities for both controlling and combining the output of several individual lasers could significantly enhance the state of the art. In particular, optical fiber cavities [83] appear to be very attractive. Furthermore, external optoelectronic waveguides could be used for controlling the phase of laser array structures and possibly as phase conjugate mirrors. External cavities have been used to control the output array mode with phase plates [84] and fiber ring resonators [85]. Such structures offer the potential for high output power in a stable spatial and spectral mode.

Lastly, the possibility of fabricating a two-dimensional surface-emitting laser array has been explored in the last several years. Most recently, a 160-element array of surface-emitting lasers operating at 1.3 μm have been fabricated and have achieved powers in excess of 750 mW cw [86]. With improving technology, these types of arrays are expected to emit powers in excess of several watts and power densities approaching 1 kW/cm^2. A recent development in monolithic two-dimensional arrays is the use of grating structures for outcoupling. Structures of this type were initially proposed by Zory and Comerford [87] and have recently been demonstrated by researchers from David Sarnoff Laboratories [88]. Such structures offer the potential for very narrow output beam widths ($\sim 0.06°$) and spectral control.

Acknowledgments

I wish to express my gratitude for the many valuable technical contributions by former TRW colleagues. In particular, I wish to thank A. Burghard, K. Burghard, L. Davis, G. Evans (David Sarnoff), T. Holcomb, C. S. Hong (Boeing), C. Morrison (IBM), J. Niesen, E. Rezek, J. Yang, and L. Zinkiewicz for many valuable technical inputs.

In addition, I thank D. Botez (TRW) for a critical reading of the manuscript and many important suggestions.

REFERENCES

1. N. Chinone, *J. Appl. Phys.*, **48**: 3237 (1978).
2. P. A. Kirby, A. R. Goodwin, G. H. B. Thompson, D. F. Lovelace, and S. E. Turley, *IEEE J. Quantum Electron.*, **13**: 720 (1977).
3. R. Lang, *IEEE J. Quantum Electron.*, **15**: 718 (1979).
4. S. Wang, C. Y. Chen, A. S. Liao, and L. Figueroa, *IEEE J. Quantum Electron.*, **17**: 453 (1981).
5. K. Aiki, N. Nakamura, T. Kurada, and J. Umeda, *Appl. Phys. Lett.*, **30**: 649 (1977).
6. M. Nakamura, *IEEE Trans. Circuits Syst.*, **26**: 1055 (1979).
7. D. Botez, *IEEE Spectrum*, **22**: 43 (1985).
8. C. H. Henry, P. M. Petroff, R. A. Logan, and F. R. Merritt, *J. Appl. Phys.*, **50**: 3721 (1979).
9. R. W. H. Englemann and D. Kerps, *Proc. IEEE Int. Conf. Semicond. Lasers*, Ottawa-Hull, Canada, 1982, p. 26.
10. H. Yonezu, I. Sakuma, T. Kamejima, M. Ueno, K. Iwanoto, I. Hino, and I. Hayashi, *Appl. Phys. Lett.*, **34**: 637 (1979).
11. I. Ladany, M. Ettenberg, F. Lockwood, and H. Kressel, *Appl. Phys. Lett.*, **30**: 87 (1976).
12. H. Namazaki, S. Takamija, M. Ishii, and W. Susaki, *J. Appl. Phys.*, **50**: 3743 (1978).
13. K. Mitsuishi, N. Chinone, H. Sata, and K. Aiki, *IEEE J. Quantum Electron.*, **16**: 728 (1980).
14. M. Ettenberg, H. S. Sommers, H. Kressel, and H. F. Lockwood, *Appl. Phys. Lett.*, **18**: 571 (1971).
15. H. L. Casey and M. B. Panish, *Heterostructure Lasers*, Academic, New York, 1978, p. 282.
16. D. Botez, *RCA Rev.*, **39**: 577 (1978).

17. H. C. Casey, M. B. Panish, W. O. Schlosser, and T. L. Paoli, *J. Appl. Phys.*, **45**: 322 (1974).
18. G. H. B. Thompson, *Physics of Semiconductor Laser Devices*, Wiley, New York, 1980, Chap. 5.
19. M. Wada, K. Hamada, H. Himuza, T. Sugino, F. Tujiri, K. Itoh, G. Kano, and I. Teramoto, *Appl. Phys. Lett.*, **42**: 853 (1983).
20. K. Hamada, M. Wada, H. Shimuzu, M. Kume, A Yoshikawa, F. Tajiri, K. Itoh, and G. Kano, *Proc. IEEE Int. Semicond. Lasers Conf.*, Rio de Janeiro, Brazil, 1984, p. 34.
20a. R. J. Fu, C. J. Hwang, C. S. Wang, and B. Lolevic, *Appl. Phys. Lett.*, **45**: 716 (1984).
21. K. Endo, H. Kawamo, M. Ueno, N. Nido, Y. Kuwamura, T. Furese, and I. Sukuma, *Proc. IEEE Int. Semicond. Laser Conf.*, Rio de Janeiro, Brazil, 1984, p. 38.
22. D. Botez, J. C. Connolly, M. Ettenberg, and D. B. Gilbert, *Electron. Lett.*, **19**: 882 (1983).
23. B. Golstein, J. K. Butler, and M. Ettenberg,* *Proc. CLEO Conf.*, Baltimore, MD, 1985, p. 180.
24. Y. Yamamoto, N. Miyauchi, S. Maci, T. Morimoto, O. Yammamoto, S. Yomo, and T. Hijikata, *Appl. Phys. Lett.*, **46**: 319 (1985).
25. S. Yamamoto, H. Hayashi, T. Hayashi, T. Hayakawa, N. Miyauchi, S. Yomo, and T. Hijikata, *Appl. Phys. Lett.*, **42**: 406 (1983).
26. D. Ackley, *Electron. Lett.*, **20**: 509 (1984).
27. J. Yang, C. S. Hong, L. Zinkiewicz, and L. Figueroa, *Electron. Lett.*, **21**: 751 (1985).
28. J. Ungar, N. Bar-Chaim, and I. Ury, *Electron. Lett.*, **22**: 280 (1986).
29. D. Botez, *IEEE J. Quantum Electron.*, **17**: 2290 (1981).
30. T. Kuroda, M. Nakamura, K. Aiki, and J. Umeda, *Appl. Opt.*, **17**: 3264 (1978).
31. T. Hayakawa, N. Miyauchi, S. Yamamoto, H. Hayashi, S. Yano, and T. Hijika, *J. Appl. Phys.*, **53**: 7224 (1982).
32. R. S. Speer and B. M. Hawkins, *IEEE Trans. Components, Hybrid Manuf. Tech.*, **3**: 480 (1980).
33. J. S. Esher, H. H. Berg, G. L. Lewis, T. V. Robertson, and H. A. Wey, *Trans. Electron. Develop.*, **19**: 1463 (1982).
34. K. Uomi, T. Nakatsuka, T. Ohtoshi, Y. Ono, N. Chinone, and T. Kajimora, *Proc. IEEE Int. Semicond. Laser Conf.*, Rio de Janeiro, Brazil, 1984, p. 128.
35. H. Yonezu, M. Ueno, T. Kamejima, and I. Hayashi, *IEEE J. Quantum Electron.*, **15**: 775 (1979).
36. H. Kumabe, T. Tumuka, S. Nita, Y. Seiwa, T. Sugo, and S. Takamija, *Jap. J. Appl. Phys.*, **21**: 347 (1982).
37. H. Blauvelt, S. Margalit, and A. Yariv, *Appl. Phys. Lett.*, **40**: 1029 (1982).
38. D. Botez and J. C. Connally, *Proc. IEEE Int. Semicond. Laser Conf.*, Rio de Janeiro, Brazil, 1984, p. 36.
39. H. Matsubara, K. Isshiki, H. Kumabe, H. Namazaki, and W. Susaki, *Proc. CLEO.* Baltimore, MD, 1985, p. 180.
40. Y. Suzuki, Y. Horikoshi, M. Kobayashi, and H. Okamoto, *Electron. Lett.*, **20**: 384 (1984).
41. M. Yamaguchi, M. Nishimoto, H. Kitumara, S. Yamazaki, I. Moto, and K. Kobayashi, *Proc. CLEO.*, Baltimore, 1988, p. 180.
42. C. B. Morrison, D. Botez, L. M. Zinkiewicz, D. Tran, E. A. Rezek, and E. R. Anderson, *Proc. SPIE Conf. High Power Laser Diodes and Applications*, **893** (1988).
43. Y. Sakakibara, E. Oomura, H. Higuchi, H. Namazaki, K. Ikeda, and W. Susaki, *Electron. Lett.*, **20**: 762 (1984).
44. K. Imanaka, H. Horikawa, A. Matoba, Y. Kawai, and M. Sakuta, *Appl. Phys. Lett.*, **45**: 282 (1984).
45. M. Kawahara, S. Oshiba, A. Matoba, Y. Kawai, and Y. Tamara, *Proc. Opt. Fiber Commun.* (OFC' 87), January 1987, paper ME1.
46. S. Oshiba, A. Matoba, H. Hori Kawa, Y. Kawai, and M. Sakuta, *Electron. Lett.*, **22**: 429 (1986).
47. M. Arvind, H. Hsing, and L. Figueroa, *J. Appl. Phys.*, **63**: 1009 (1988).

48. D. Scifres, D. F. Welch, G. Harnagel, P. S. Cross, H. Kung, W. Streifer, and J. Berger, *Proc. SPIE Conf. High Power Laser Diodes and Applications*, **893**, January 1988.
49. D. R. Scifres, R. D. Burnham, and W. Steifer, *Appl. Phys. Lett.*, **33**: 1015 (1978).
50. D. Botez and J. C. Connally, *Appl. Phys. Lett.*, **43**: 1097 (1983).
51. D. E. Ackley and R. G. Engelmann, *Appl. Phys. Lett.*, **39**: 27 (1981).
52. D. R. Scifres, R. D. Burnham, W. Streifer, and M. Bernstein, *Appl. Phys. Lett.*, **41**: 614 (1982).
53. D. R. Scifres, C. Lindstrom, R. D. Burnman, W. Streifer, and T. L. Paoli, *Appl. Phys. Lett.*, **19**: 160 (1983).
54. F. Kappeler, H. Westmeier, R. Gessner, M. Druminski, and K. H. Zschauer, *Proc. IEEE Int. Semicond. Laser Conf.*, Rio de Janeiro, Brazil, 1984, p. 90.
55. J. P. Van der Ziel, H. Temkin, and R. D. Dupuis, *Proc. IEEE Int. Semicond. Laser Conf.*, Rio de Janeiro, Brazil, 1984, p. 92.
56. L. Figueroa, C. Morrison, H. D. Law, and F. Goodwin, *Proc. Int. Electron Devices Meeting*, 1983, p. 760.
57. L. Figueroa, C. Morrison, H. D. Law, and F. Goodwin, *J. Appl. Phys.*, **56**: 3357 (1984).
58. C. Morrison, L. Zinkiewicz, A. Burghard, and L. Figueroa, *Electron. Lett.*, **21**: 337 (1985).
59. Y. Twu, A. Dienes, S. Wang, and J. R. Winnery, *Appl. Phys. Lett.*, **45**: 709 (1984).
60. S. Mukai, C. Lindsey, J. Katz, E. Kapon, Z. Rav-Noy, S. Margalit, and A. Yariv, *Appl. Phys. Lett.*, **45**: 834 (1984).
61. D. F. Welch, D. Scifres, P. Cross, H. Kung, W. Streifer, R. D. Burnham, and J. Yaeli, *Electron. Lett.*, **21**: 603 (1985).
62. N. Dutta, L. A. Kozzi, S. G. Napholtz, and B. P. Seger, *Proc. Conf. Lasers Electro-Optics* (*CLEO*), Baltimore, MD, 1985, p. 44.
63. M. Taneya, M. Matsumoto, S. Matsui, Y. Yano, and T. Hijikata, *Appl. Phys. Lett.*, **47**: 341 (1985).
64. J. Ohsawa, S. Himota, T. Aoyagi, T. Kadowaki, N. Kaneno, K. Ikeda, and W. Susaki, *Electron. Lett.*, **21**: 779 (1985).
65. D. F. Welch, P. S. Cross, D. R. Scifres, W. Streifer, and R. D. Burnham, *Proc. CLEO*. San Francisco, CA, 1986, p. 66.
66. J. K. Butler, D. E. Ackley, and D. Botez, *Appl. Phys. Lett.*, **44**: 293 (1984).
67. E. Kapon, J. Katz, and A. Yariv, *Opt. Lett.*, **10**: 125 (1984).
68. K. L. Chen and S. Wang, *Electron. Lett.*, **21**: 347 (1985).
69. W. Streifer, A. Hardy, R. D. Burnham, and D. R. Scifres, *Electron. Lett.*, **21**: 118 (1985).
70. S. Chinn and R. J. Spier, *IEEE J. Quantum Electron.*, **20**: 358 (1985).
71. J. Katz, E. Kapon, C. Lindsey, S. Margalit, U. Shreter, and A. Yariv, *Appl. Phys. Lett.*, **42**: 521 (1983).
72. E. Kapon, C. P. Lindsey, J. S. Smith, S. Margalit, and A. Yariv, *Appl. Phys. Lett.*, **45**, 1257 (1984).
73. D. Ackley, *Electron. Lett.*, **20**: 695 (1984).
74. T. R. Ranganath and S. Wang, *IEEE J. Quantum Electron.*, **13**: 290 (1977).
75. L. Figueroa, T. Holcomb, K. Burghard, D. Bullock, C. Harrison, L. Zinkiewicz, and G. Evans, *IEEE J. Quantum Electron.*, **22**: 241 (1986).
76. L. J. Mawst, M. E. Givens, C. A. Zmudzinski, M. A. Emanuel, and J. J. Coleman, *IEEE J. Quantum Electron.*, **QE-23**: 696 (1987).
77. P. S. Zory, A. R. Reisinger, R. G. Walters, L. J. Mawst, C. A. Zmudzinski, M. A. Emanuel, M. E. Givens, and J. J. Coleman, *Appl. Phys. Lett.*, **49**: 16 (1986).
78. R. G. Walters, P. L. Tihanyi, D. S. Hill, and B. A. Soltz, *Proc. SPIE Conf. High Power Laser Diodes and Appl.*, **893**, January 1988.
79. M. S. Zediker, D. J. Krebs, J. L. Levy, R. R. Rice, G. M. Bender, and D. L. Begley, *Proc. SPIE Conf. High Power Laser Diodes and Appl.*, **893**, January 1988.
80. C. Krebs and B. Vivian, *Proc. SPIE Conf. High Power Laser Diodes and Appl.*, **893**, January 1988.

81. A. R. Reisinger, P. A. McDonald, J. R. Shealy, E. P. Jochym, F. Worth, J. M. Gilman, J. W. Sprague, and P. S. Zory, *Proc. CLEO*, 1987, p. 118.

82. M. Takayoshi, T. Okada, K. Kaise, and O. Yoneyama, *Proc. CLEO Conf.*, 1986, p. 140.

83. R. Rediker, R. P. Schloss, and L. J. Van Ruyen, *Appl. Phys. Lett.*, **46**: 133 (1985).

84. S. Thaniyavarn and W. Dougherty, *Proc. Topical Meeting Semicond. Lasers*, February 1987, p. 158.

85. L. Goldberg and J. F. Weller, *Proc. Opt. Fiber Conf.*, January 1988, paper TUH4.

86. J. N. Walpole and Z. L. Liau, *Proc. CLEO*. 1986, p. 64.

87. P. S. Zory and L. D. Comerford, *IEEE J. Quantum Electron.*, QE-11: 451 (1975).

88. N. W. Carlson et al., *Appl. Phys. Lett.*, **52**: 1037 (1988).

89. L. Figueroa (ed.), *Proc. SPIE Conf. High Power Laser Diodes and Appl.*, **893** (1988).

90. L. Figueroa (ed.), *Proc. SPIE Conf. Laser Diode Technol. Appl.*, **1043** (1989).

14

Photodetectors for Long-Wavelength Lightwave Systems

J. C. Campbell

14.1 INTRODUCTION

In a lightwave system the function of the optical receiver is to transform the input optical signal, which consists of a series of light pulses, back into the original electrical format, usually a binary stream of voltage pulses. In order to determine whether, in a specific time slot, the signal from the photodetector corresponds to a "1" or a "0", the receiver must perform a series of functions including detection, amplification, equalization, filtering, and retiming [1–4]. Figure 14.1 shows a block diagram of the components in a typical lightwave receiver. First, the receiver must convert the light signal into current; this function is performed by the photodetector. The photocurrent must then be amplified to a usable level. The first stage of amplification is achieved with a low-noise transistor preamplifier, and the remainder of the amplification is provided by the postamplifier. Often the output of the amplifier stage is distorted and it is the function of the equalizer to remove the signal distortions, thus providing a reasonable pulse shape to the

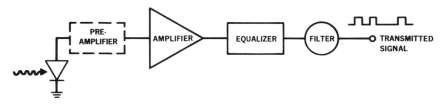

Figure 14.1 Block diagram of a typical lightwave receiver.

filter section. Finally, the filter maximizes the signal-to-noise ratio while introducing minimal distortion to the signal itself. The performance of the receiver is usually measured in terms of its sensitivity, that is, the minimum received optical power required to achieve a given bit error rate (BER). Typically, the sensitivity is quoted at 10^{-9} BER in dBm (0 dBm = 10^{-3} W) of optical power. Better receiver sensitivities permit wider repeater spacings and greater loss margins, which ultimately result in lower systems costs.

The combination of the photodetector and preamplifier is the heart of a receiver and is referred to as the *front end*. The front end is the source of the dominant noise components added to the signal. As a result, its design is the critical factor in determining the receiver sensitivity. To date, the front ends of lightwave receivers have consisted of either a PIN (*p-i-n*) photodiode or an avalanche photodiode (APD) in combination with a low-noise transistor such as a silicon bipolar transistor or a GaAs field-effect transistor (FET) [5–8]. The relative merits of PIN photodiodes and avalanche photodiodes for lightwave receivers have been the subject of numerous papers [7–10]. In both cases, optimum performance requires the photodetector to have high quantum efficiency, low dark current, and sufficient bandwidth to achieve the desired bit rate. Differences arise as a result of their dissimilar noise characteristics and ease of implementation. The sensitivity of receivers that utilize PIN photodiodes is usually limited by the noise of the following transistor preamplifier and not by the noise of the PIN itself. As a result, the characteristics of different types of transistor preamplifiers for this application have been studied in some detail [5, 6, 11–14]. In many cases, particularly at higher bit rates, the internal gain of the APD yields higher receiver sensitivity [9]. However, the random nature of the multiplication process is an additional source of noise [15]. In general, the receiver sensitivity can be improved by increasing the gain of the APD, but only to the point where the multiplication noise is comparable to the noise of the transistor preamplifier. In addition to higher noise, the APD has the disadvantages that it requires higher bias voltage than the PIN and the bias voltage and temperature must be precisely maintained. This results in more complex and expensive receiver circuitry.

Selection of the "best" device or combination of devices (e.g., PIN photodiode versus APD) for a particular application ultimately leads to an evaluation of the trade-off between cost and performance, which in turn requires a knowledge of the parameters and operating characteristics of the devices. The purpose of this chapter is to describe the physical processes that determine the characteristics of photodetectors and compatible transistor preamplifiers for long-wavelength (1.0 μm < λ < 1.65 μm) lightwave receivers. The performance of state-of-the-art devices and receivers will also be documented. Section 14.2 describes PIN photodiodes. Section 14.3 contains a brief description and comparison of some of the transistors that have been used in the preamplifier sections of lightwave receivers. Section 14.4 is devoted to structures that today would be referred to as novel; however, in the future these devices may become some of the basic building blocks of lightwave systems. Included are integrated front-end circuits, phototransistors, and multiple-wavelength structures for wavelength-division multiplexing and wavelength sensing. The evolution of avalanche photodiodes for long-wavelength lightwave systems is presented in Section 14.5. This section also includes descriptions of some of the most promising APD structures, a summary of "champion" APD receiver sensitivities, and a comparison of PIN and APD receivers.

14.2 PIN PHOTODIODES

The photodetector that has been used most widely in lightwave receivers is the PIN (or *p-i-n*) photodiode. The most important characteristics for evaluating a PIN photodiode are quantum efficiency, dark current, capacitance, and bandwidth. The quantum efficiency is a measure of how well the photodiode converts the incident optical signal into current, and the dark current is important primarily because it is a source of noise. The capacitance can be a factor in the frequency response through the RC time constant, but more significantly the photodiode capacitance contributes to the total front-end capacitance, which is a critical parameter in determining the receiver sensitivity [1]. Finally, the bandwidth must be sufficient to accommodate the transmission bit rate.

14.2.1 Quantum Efficiency

One of the design goals for a PIN photodiode is to achieve quantum efficiencies as close to 100% as possible. Normally, owing to their relatively high index of refraction, semiconductors reflect approximately 30% of the incident light. This can be significantly reduced by depositing an antireflection (AR) coating at the point where light enters the photodiode. These AR coatings may consist of single or multiple layers of dielectric materials such as SiN, Al_2O_3, or Pb_2O_3. Typically, these films reduce the reflection to less than 5%, and reflections as low as 0.03% have been achieved [16].

Once the photons have entered the photodiode, they must be absorbed and the photogenerated carriers must be collected. The primary absorption processes in a semiconductor are free carrier absorption, band-to-band absorption, and band-to-impurity absorption. Since only the band-to-band absorption contributes to the photocurrent, the other two absorption processes can be considered loss mechanisms. Fortunately, in most cases band-to-band absorption is much larger than either of the other two absorption mechanisms. In a properly designed photodiode most of the absorption occurs in the depletion region of the *pn* junction. The electric field in this region quickly separates the photogenerated electron–hole pairs and sweeps them away to the neutral n and p regions. However, carriers that are created within a diffusion length of the edge of the depletion region will also contribute to the photocurrent. For absorption to occur at all, the energy of the photon must exceed the semiconductor bandgap. The absorption process is characterized by the absorption coefficient α, which is defined as follows: $1 - \exp(-\alpha x)$ is the probability that a photon will be absorbed in a distance x from the edge of the absorbing layer. The quantum efficiency is defined as

$$\eta = \frac{\text{number of photogenerated carrier pairs collected}}{\text{number of incident photons}}$$

$$= \frac{I_{ph}/q}{P_0/hv} = (1 - R)(1 - e^{-\alpha d}) \tag{1}$$

where q is the electronic charge, hv is the photon energy, R is the Fresnel reflection coefficient at the semiconductor–air interface, P_0 is the intensity of the incident optical signal, and d is the width of the absorption region. The conversion efficiency is often expressed in terms of the responsivity R_0, which is defined as

$$R_0 = \frac{I_{ph}}{P_0} = \frac{\eta q}{hv} \tag{2}$$

Figure 14.2 shows the absorption coefficient of several semiconductors as a function of wavelength. For wavelengths that correspond to photon energies greater than the bandgap energy, the absorption coefficient is usually greater than 10^4 cm^{-1}. This means that most of the light is absorbed within a few micrometers from the point where it enters the absorbing layer. It can be seen from these curves that owing to their relatively high bandgap energies, GaAs and Si are not suitable as photodetectors for wavelengths beyond $\lambda = 1.0$ μm. In fact, much of the recent research on photodetectors has been devoted to developing long-wavelength ($1.0\ \mu m < \lambda < 1.65\ \mu m$) photodetectors for second- and third-generation lightwave systems that would work as well as the excellent silicon devices that were used in the first-generation systems ($0.8\ \mu m < \lambda < 0.9\ \mu m$). It can be seen in Figure 14.2 that germanium and the quaternary compound $In_xGa_{1-x}As_yP_{1-y}$ are potentially good materials for long-wavelength photodetectors. $In_xGa_{1-x}As_yP_{1-y}$ has the property that by changing the crystal composition its bandgap can be adjusted continuously to provide photoresponse in the long-wavelength range while maintaining a lattice match to InP. The particular composition $In_{0.53}Ga_{0.47}As$ (referred to below as InGaAs) has the advantage that it exhibits high photoresponse across the entire long-wavelength range [17, 18]. This is one reason that InGaAs has become the material of choice for long-wavelength PIN photodiodes. In addition, the high mobility of InGaAs [19] makes it an excellent candidate for FETs and monolithic integrated structures.

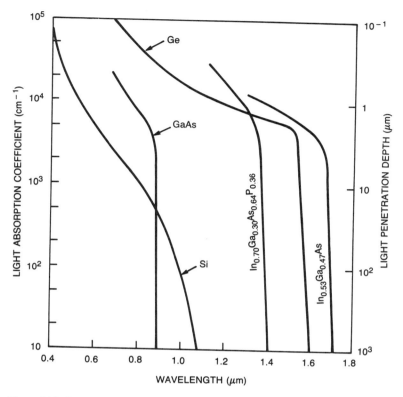

Figure 14.2 Optical absorption coefficient and penetration depth versus wavelength for Si, Ge, GaAs, $In_{0.70}Ga_{0.30}As_{0.64}P_{0.36}$, and $In_{0.53}Ga_{0.47}As$.

Figure 14.3a shows a back-illuminated mesa-type InGaAs PIN photodiode. Back-illuminated structures [20] have the advantage that none of the photosensitive area is masked by the contact and hence the entire device area is photosensitive. This permits smaller device areas, which can result in lower capacitance and lower dark current. Most of the early InGaAs photodiodes utilized the mesa structure [21–26] because it is quick and easy to fabricate and it is usually easier to achieve low dark currents with the mesa structure than with a planar structure [27, 28] such as the one shown in Figure 14.3b. On the other hand, the mesa structure may be more difficult to manufacture. Furthermore, the planar structure will probably be more reliable because the edges of the *pn* junction in the mesa-type structure are exposed and are thus vulnerable to contamination and atmospheric degradation. Passivation of mesa structures with organic materials has proved effective for short periods, but the long-term stability of this type of passivation is suspect [29, 30].

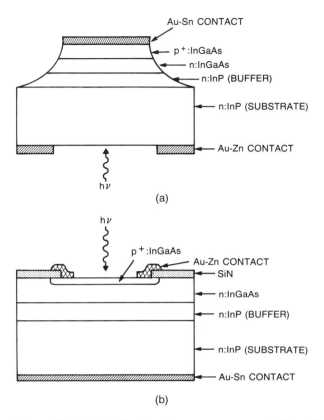

Figure 14.3 (a) Back-illuminated mesa-structure and (b) top-illuminated planar-structure $In_{0.53}Ga_{0.47}As$ PIN photodiodes.

Figure 14.4 shows the responsivity and quantum efficiency of a back-illuminated mesa-type PIN photodiode with and without an AR coating [24]. Across the wavelength range of current interest for lightwave systems the quantum efficiency is close to the theoretical maximum of 70%. The short-wavelength cutoff is due to absorption in the InP substrate, and the long-wavelength cutoff reflects the bandgap energy of InGaAs. Commercial devices usually have an AR coating that increases the quantum efficiency to greater than 90%.

14.2.2 Dark Current

The dark current is important because it is a source of noise. In fact, the primary noise component of the photodiode itself is the shot noise of the dark current. The sources of dark current in a PIN photodiode are diffusion current, generation–recombination in the depletion region, and tunneling (Fig. 14.5). The relative magnitudes of these components in InGaAs photo-

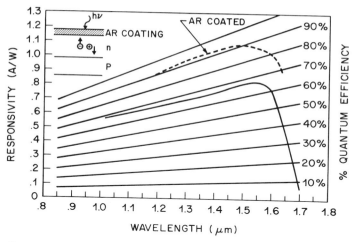

Figure 14.4 Responsivity and quantum efficiency of a back-illuminated mesa-structure $In_{0.53}Ga_{0.47}As$ PIN photodiode with (dashed curve) and without (solid curve) an antireflection coating.

diodes and their temperature and voltage dependencies have been analyzed and measured by Forrest [31]. The diffusion component of the dark current is due to minority carriers that are thermally generated in the p and n neutral regions and subsequently diffuse into the depletion region. It is given by the Shockley equation [32].

$$I_{diff} = I_s \left[\exp\left(\frac{qV}{kT}\right) - I \right] \tag{3}$$

where k is Boltzmann's constant, T is the junction temperature, V is the applied voltage, and I_s is the saturation current. Assuming complete ionization of impurities, the saturation current can be written as [33]

$$I_s = qAn_i^2 \left(\frac{D_p}{L_p N_d} + \frac{D_n}{L_n N_a} \right) \tag{4}$$

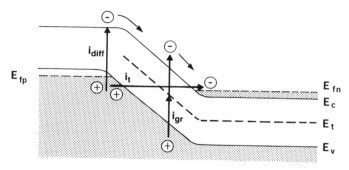

Figure 14.5 Primary components of the dark current of a PIN photodiode.

where A is the junction area, n_i is the intrinsic carrier concentration, D_n and D_p are the minority carrier diffusion constants, L_n and L_p are the minority carrier diffusion lengths, and N_d and N_a are the densities of ionized donors and acceptors, respectively.

The generation–recombination current in the depletion region is due to the net generation of carriers by emission from traps near the center of the bandgap [34]. This process occurs as follows: First the trap captures an electron from (emits a hole to) the valence band. Later this electron is thermally excited to the conduction band. At thermal equilibrium these capture and emission processes would be balanced by equal and opposite emission and capture processes. However, in the depletion region the generated carriers are quickly swept away and net generation results. In a depletion region of width W this current is approximately

$$I_{\mathrm{gr}} = \frac{qn_i A W}{\tau_e} \left[\exp\left(\frac{qV}{2kT}\right) - 1 \right] \tag{5}$$

where τ_e is the effective carrier lifetime [35].

The third effect that contributes to the dark current is tunneling [22, 34, 36–38]. As shown in Figure 14.5, at high reverse bias a large number of empty conduction band states on the n side of the junction line up directly across from filled valence band states on the p side of the junction. If the barrier is narrow enough, quantum tunneling of electrons can occur. Two factors that affect the width of the barrier are the bandgap energy and the reverse bias. The smaller the bandgap, the narrower the tunneling barrier will be. The barrier width also becomes narrower as the reverse bias is increased. For direct bandgap semiconductors the tunneling current is given by [34]

$$I_t = \gamma A \exp\left[-\frac{2\pi\theta m_0^{1/2} E_g^{3/2}}{qhE_m} \right] \tag{6}$$

where m_0 is the free electron mass, h is Planck's constant, and E_m is the maximum electric field in the depletion layer. The parameter θ is a dimensionless quantity given by $\theta = \alpha(m_c^*/m_0)^{1/2}$, where m_c^* is the electron effective mass. For band-to-band tunneling the constant α is approximately unity and the prefactor $\gamma = (2m_c^*/E_g)^{1/2}(q^3 E_m V/h^2)$. It has been shown by deep-level transient spectroscopy that the dominant tunneling mechanism in InGaAs photodiodes is not band-to-band but involves thermal activation of deep centers and subsequent tunneling of carriers from the traps into band states [39]. The parameters α and γ will differ somewhat for trap-assisted tunneling of this type, but the band-to-band values provide a good approximation [31].

In InGaAs PIN photodiodes at low bias voltages (<20 V) generation–recombination in the depletion region is responsible for most of the dark current, and at higher voltage tunneling dominates. For both mesa and planar homojunction InGaAs photodiodes the dark current is typically in the range $(2-5) \times 10^{-5}$ A/cm^2 at 10-V reverse bias. While dark currents of this magnitude are suitable for most lightwave applications, it has been shown that the dark current of planar structures can be lowered even further by using a wide-bandgap cap layer to reduce the surface leakage current. Kim et al. [40] have achieved a dark current of 0.1 nA (at -10-V bias) with an InGaAsP cap layer. A somewhat lower value of 0.05 nA (5×10^{-7} A/cm^2) was obtained with a semi-insulating InP cap layer [41]. The lowest dark current reported to date was achieved with an InP/InGaAs/InP double heterostructure grown by chloride vapor-phase epitaxy [42]. These photodiodes exhibited a dark current of only 4 pA (4×10^{-8} A/cm^2) and 50 pA at 5-V and 10-V reverse bias, respectively.

14.2.3 Capacitance

The depletion-layer capacitance of a PIN photodiode is given by the equation for the capacitance of an abrupt, one-sided pn junction [33],

$$C = \frac{\varepsilon_s A}{W} = \left(\frac{q\varepsilon_s N_a N_d}{2(N_a + N_d)(V_{bi} + V)} \right)^{1/2} \tag{7}$$

where ε_s is the semiconductor permitivity, W is the width of the depletion region, N_a and N_d are the acceptor and donor impurity densities, respectively, V_{bi} is the built-in potential, and V is the applied voltage. It is evident from Eq. (7) that the capacitance is the same as that of a parallel-plate capacitor with plate separation W and stored charge equal to that in the space charge region of the pn junction. Although the capacitance can affect the bandwidth through the RC time constant, in most cases the bandwidth is limited by the transit time of the photogenerated electrons or holes through the depletion region. More important, however, the diode capacitance adds to the total capacitance of the front end, which has a significant impact on the receiver sensitivity. Typically, devices with diameters <100 μm have capacitances <0.2 pF.

14.2.4 Bandwidth

There are three effects that determine the bandwidth of a PIN photodiode. If some of the photons are absorbed outside the depletion region, the photogenerated carriers must diffuse to the edge of the depletion region in order to be collected. This is usually a very slow process; a typical diffusion time is 4 ns/μm [43]. Consequently, it is important to design the photodiode structure to minimize the amount of absorption outisde the depletion region. If the resistance or capacitance is excessive, the RC time constant may limit the speed of response. However, for most practical structures the RC time constant is less than 10 ps.

The ultimate limitation is the transit time—the time it takes the photo-generated carriers to drift through the depletion region. To determine the transit time limit we consider the back-illuminated PIN photodiode structure shown in Figure 14.3a. Light enters through a transparent wide-band-gap window, which is often the substrate. The pn junction is located a distance W from the point where absorption commences. It is assumed that the electric field strength is sufficient throughout the depletion region for the drift velocities of both electrons and holes to be saturated. Under these conditions the time-dependent photocurrent is given by

$$i_{ph}(t) = \frac{qv_n}{W} N(t) + \frac{qv_p}{W} P(t) \tag{8}$$

where v_n and v_p are the electron- and hole-saturated drift velocities and $N(t)$ and $P(t)$ are the photogenerated electron and hole concentrations, respectively, in the depletion region. For a monochromatic light impulse consisting of N_0 photons, $N(t)$ and $P(t)$ are given by

$$N(t) = N_0\left\{\exp\left(\frac{-\alpha Wt}{\tau_n}\right) - \exp\left(-\alpha W\right)\right\}[u(t) - u(t - \tau_n)] \tag{9a}$$

and

$$P(t) = N_0\left\{1 - \exp\left[-\alpha W\left(1 - \frac{t}{\tau_p}\right)\right]\right\}[u(t) - u(t - \tau_p)] \tag{9b}$$

where

$$\tau_n = \frac{W}{v_n} \equiv \text{transit time for electrons}$$

$$\tau_p = \frac{W}{v_p} \equiv \text{transit time for holes}$$

The impulse function $u(t)$ has the following characteristics:

$$u(t) = \begin{cases} 0 & t < 0 \\ 1 & t \geq 0 \end{cases}$$

The transit-time-limited frequency response of the photocurrent can be obtained by taking the Fourier transform of $N(t)$ and $P(t)$.

$$N(\omega) = \frac{\alpha i\omega\tau_n[1 - \exp\left(-\alpha W\right)] - \alpha W \exp\left(-\alpha W\right)[1 - \exp\left(-i\omega\tau_n\right)]}{i\omega\tau_n(\alpha W + i\omega\tau_n)} \tag{10a}$$

$$P(\omega) = \frac{\alpha i\{\alpha W \sin \omega\tau_p - \omega\tau_p[1 - \exp\left(-\alpha W\right)]\} + \alpha W(1 - \cos \omega\tau_p)}{i\omega\tau_p(\alpha W - i\omega\tau_p)} \tag{10b}$$

In the limit of very shallow penetration of the light signal, these equations reduce to the standard normalized frequency factor for transit-time-limited response, namely [44],

$$|F(\omega)| = \frac{\sin \left(\omega\tau_p/2\right)}{\omega\tau_p} \tag{11}$$

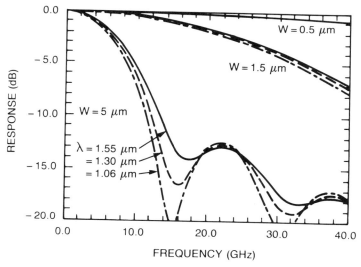

Figure 14.6 Transit-time-limited frequency response of an InGaAs PIN photodiode. (After Bowers et al. [45].)

Using Eqs. (8)–(10), the transit-time-limited frequency response of an InGaAs photodiode has been calculated for incident wavelengths of $\lambda = 1.06 \, \mu m$ ($\alpha = 1.8 \, \mu m^{-1}$), $1.30 \, \mu m$ ($\alpha = 1.15 \, \mu m^{-1}$), and $1.55 \, \mu m$ ($\alpha = 0.66 \, \mu m^{-1}$) and for depletion layer widths of 0.5, 1.5, and 5.0 μm [45]. For this calculation both the electron and hole velocities are taken as 7×10^6 cm/s [46, 47]. The curves in Figure 14.6 show that for this particular geometry the increased penetration of longer wavelengths results in better frequency response. Furthermore, for a depletion layer width of 1.5 μm a 3-dB bandwidth in excess of 20 GHz is projected. This is consistent with a report by Burrus et al. [48] of 22 GHz for a back-illuminated mesa-structure InGaAs PIN photodiode.

14.3 TRANSISTOR PREAMPLIFIERS

The choice for a low-noise transistor companion to the InGaAs PIN photodiode has not been as straightforward as that for the photodiode itself. At present, the three leading candidates are silicon bipolar transistors [11, 12], GaAs MESFETs [13, 14], and silicon MOSFETs [5, 6]. For most receiver front ends the photodiode contributes negligible noise and, as a result, the receiver sensitivity is determined primarily by the characteristics of the transistor preamplifier. Smith and Personick [1] have derived analytical expressions for the sensitivities of receivers that utilize bipolar or FET preamplifiers.

In both cases, assuming Gaussian statistics and an operating wavelength of 1.3 μm, **P**, the average received optical power required to achieve a bit error rate (BER) of 10^{-9}, is related to the equivalent input noise current $(\langle i^2 \rangle_c)^{1/2}$ by the expression

$$\eta \mathbf{P} \approx 6(\langle i^2 \rangle_c)^{1/2} \tag{12}$$

where η is the quantum efficiency of the photodiode. For the bipolar front end the base current can be adjusted to minimize the noise current at a particular bit rate. At the optimum bias point the total noise current referred to the input is given by

$$\langle i^2 \rangle_c = \frac{8kTC_T}{\beta^{1/2}(I_2 I_3)^{1/2} B^2} \left(1 + \frac{I_2/I_3}{(2\pi BC_T R_L)^2}\right)^{1/2}$$
$$+ 4kTR_b \left(\frac{BI_2}{R_L^2} + (2\pi)^2 (C_d + C_s)^2 I_3 B^3\right) \tag{13}$$

where C_T is the total input capacitance (neglecting the Miller effect), B is the bit rate, β is the common emitter current gain, I_2 and I_3 are Personick integrals derived from expressions for the relative shapes of the input and output pulses, R_L is the load resistance, R_b is the base resistance, and C_d and C_s are the capacitance of the photodetector and the stray capacitance, respectively. In most instances the contribution from the base resistance is negligible and the minimum noise varies quadratically with the bit rate. In this case an appropriate figure of merit for the bipolar front end is $\beta^{1/2}/C_T$. Clearly, optimum receiver performance is achieved with high gain and low capacitance.

For an integrating or high-impedance front-end design utilizing a FET preamplifier, the noise current due to the FET and the bias resistor, referred to the input, is given by

$$\langle i^2 \rangle_c = \left[\frac{4kT}{R_L}\left(1 + \frac{\Gamma}{g_m R_l}\right) + 2qI_{\text{gate}}\right] I_2 B + 4kT\Gamma \frac{(2\pi C_T)^2}{g_m} I_3 B^3 \tag{14}$$

where g_m is the transconductance, I_{gate} is the gate-leakage current, C_T is the total front-end capacitance (including the photodiode capacitance, the gate capacitance of the FET, and the interconnect capacitance), and Γ is the noise figure of the FET. In most cases the second term in Eq. (14) is dominant and the noise current increases as B^3. Also, it can be seen that the figure of merit is g_m/C_T^2. High receiver sensitivity thus requires minimization of the total capacitance while keeping the transconductance g_m high.

Theoretically, lightwave receivers that utilize a FET preamplifier should exhibit somewhat better sensitivities at the lower bit rates (< 500 Mb/s); however, since the sensitivity of FET receivers increases as B^3 while that of bipolar receivers increases as B^2, bipolars should have an advantage at the highest bit rates. So far, the highest sensitivities at all bit rates have been achieved with GaAs FETs. An advantage for silicon bipolar transistors, on

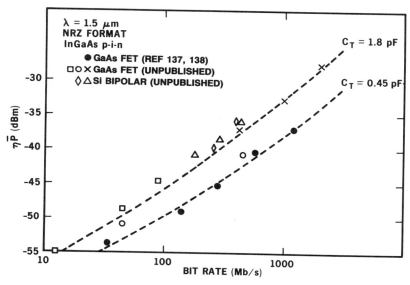

Figure 14.7 Experimental receiver sensitivities that have been achieved with GaAs FET and silicon bipolar transistors in front-end circuits. The dashed curves are calculated sensitivities for C_T equal to 1.8 pF and 0.45 pF.

the other hand, is that they are available in integrated circuits. Figure 14.7 shows some of the "champion" receiver sensitivities that have been achieved with FET and bipolar front ends. The solid lines are calculated receiver sensitivities for total front-end capacitances of 1.8 pF and 0.45 pF.

Recently, silicon MOSFETs having a gate length of 0.5 μm have shown promise of superior performance [5, 6]. The figure of merit for GaAs FETs is approximately twice that of the silicon MOSFETs, but when the total noise characteristics including channel noise, gate-induced noise, and noise due to the hot electron effect are considered, these two transistors should be roughly equivalent. Of course, the silicon MOSFETs also have the advantage of permitting integration of the front-end circuitry and very low gate leakage current.

14.4 NOVEL PIN STRUCTURES

Other PIN structures that show promise for long-wavelength applications are the PIN/FET integrated circuit and the heterojunction phototransistor (HPT). Functionally, these two structures are quite similar. The PIN/FET circuit combines a PIN photodiode with a FET, while the HPT is a PIN photodiode integrated in the common-collector configuration with a wide-bandgap-emitter bipolar transistor preamplifier [49, 50]. The primary motivation for integrating these front-end devices onto a single chip is the virtual elimination of the interconnect capacitance. Recall that the figure of merit of both bipolar and FET front ends is inversely proportional to the total front end capacitance.

14.4.1 Integrated PIN/FET Receivers

The work on integrated PIN/FET circuits can be subdivided into materials and structure. To date, most of the work has concentrated on the AlGaAs/GaAs crystal system because of its more advanced laser and FET technology. The InP/InGaAs crystal system, on the other hand, shows promise for the second- and third-generation lightwave systems at 1.3 and 1.55 μm. The structures can also be classified according to whether they are vertically or horizontally integrated. For horizontal integration the photodiode material is embedded in a semi-insulating substrate and the transistor circuit is fabricated by standard techniques on an adjacent portion of the substrate. In the vertically integrated structures the epitaxial layers for the photodiode and the transistors are grown sequentially, separated by a high-resistivity isolation layer. None of the integrated structures have achieved sensitivities comparable to the best hybrid circuits, but improvements in material technology and processing technology should result in steady improvements.

Kolbas et al., [51] reported one of the first GaAs structures. Their horizontally integrated structure consisted of a GaAs PIN photodiode with a transimpedence preamplifier of depletion mode MESFETs. A response time of 40 ns was achieved. By reducing the parasitic capacitances, Wada et al. [52] achieved a response time of 1 ns. The transimpedance amplifier gain of their circuit was 1060 V/A and its equivalent noise current was 13 pA/Hz$^{1/2}$.

The earliest work on integration of PINs and FETs was an InGaAs PIN and an InGaAs JFET reported by Leheny et al. [53, 54]. A schematic cross section of this structure is shown in Figure 14.8. The transconductance of the JFET was 1 mS (50 mS/mm) and the combined gain was 30. More recently Hata et al. [55] developed a planar InGaAs integrated PINFET that was fabricated by beryllium implantation. They report a transconductance of 10 mS and a total photocurrent gain of 5.8. The use of InGaAs for both the PIN photodiode and the FET is advantageous, since InGaAs has the potential for superior FET performance in addition to being the material of choice for long-wavelength PIN photodiodes. Results on InGaAs FETs have demonstrated that transconductance values as high as 300 mS/mm can be achieved [56, 57]. Another approach has been integration of an InGaAs PIN photodiode with an InP MISFET [58, 59], the advantage being that shorter gates can be formed in the MIS structures. Using a monolithically integrated front end consisting of an InGaAs PIN photodiode and an InP MISFET, Kasahara et al. [58] achieved a receiver sensitivity of −34.5 dBm at 100 Mb/s. While this initial result is encouraging, it is still approximately 5 dB worse than the best hybrid circuits.

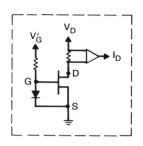

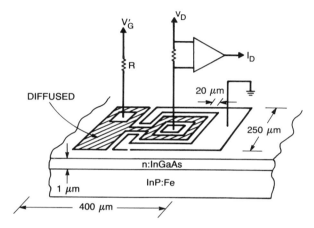

Figure 14.8 An integrated $In_{0.53}Ga_{0.47}As$ PIN/JFET front-end circuit. (After Leheny et al. [54].)

14.4.2 Phototransistors

Early work on heterojunction phototransistors did not show much promise for lightwave applications because appreciable gains could be achieved only at much higher signal levels ($> 1 \mu W$) than those encountered in lightwave systems [60–67]. However, Campbell et al. [68, 69] demonstrated high current gains (> 100) at low light levels (~ 1 nW) with an InP/InGaAs HPT (Fig. 14.9). This was due in large part to the benefits offered by the wide-bandgap InP emitter [70] and to improvements in the characteristics of the InGaAs/InP base-emitter junction. In most cases the response time of an HPT at low light levels is limited by the charging time of the emitter capacitance [71, 72]. The realization of high cutoff frequencies, therefore, requires that the junction capacitances, particularly the base-emitter capacitance, be minimized. Pulse response measurements on a small-area low-capacitance (emitter capacitance ~ 1.5 pF and collector capacitance ~ 0.05 pF) InGaAs/InP HPT indicated that cutoff frequencies greater than 2 GHz can be obtained [69]. Further improvements in bandwidth will probably require an

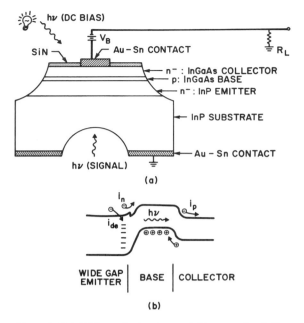

Figure 14.9 (a) Schematic drawing and (b) energy-level diagram of an $InP/In_{0.53}Ga_{0.47}As$ heterojunction phototransistor. (From Campbell and Ogawa [49].)

asymmetric structure that would permit even further reduction of the base-emitter capacitance.

Models for the noise characteristics of HPTs have been developed by Campbell and Ogawa [49] and by Brain and Smith [50]. Using experimental device parameters it was shown in both papers that the sensitivity of an optical receiver with an InGaAs/InP HPT front end can be as good as that of the best hybrid PIN/FET combination reported to date. These calculations have also shown that optimization of the performance of an HPT in a lightwave receiver for a particular bit rate, like that of a hybrid PIN–bipolar amplifier combination, can be achieved by supplying a dc bias current in addition to the photogenerated signal current. Campbell and Dentai [73] demonstrated that instead of the bias current being provided electrically through a base contact it can be generated optically by a light-emitting diode on the same chip as the HPT. This has the advantage of eliminating the capacitances associated with the base contact.

14.4.3 Wavelength-Selective Photodetectors

Another novel PIN photodetector that may prove useful in lightwave systems is the dual-wavelength photodetector [74–77]. Wavelength-division multiplexing is a technique that may be used to improve the efficiency and increase the utility of lightwave systems by expanding the information transmission rate without increasing the bit rate. This approach could lead to lower systems costs through simplified circuitry and a reduction in the number of required repeaters, as well as provision of additional system capabilities such as two-way transmission or simultaneous transmission of anolog and digital signals. The dual-wavelength photodetector is capable of simultaneously detecting and demultiplexing optical signals from two different wavelength bands without the need for additional optical components such as gratings, dielectric filters, or prisms. Wavelength discrimination is accomplished in this multilayer structure by an $In_xGa_{1-x}As_yP_{1-y}$ quaternary layer (labeled Q in Fig. 14.10) and an InGaAs ternary layer (T). Photons in the shorter-wavelength band are absorbed in the larger-bandgap quaternary layer Q; longer-wavelength photons pass through Q to the lower-bandgap ternary layer T, where they are absorbed. Crosstalk levels of -19 and -30 dB have been measured in the short- and long-wavelength regions, respectively. A wavelength-division-multiplexed system consisting of a dual-wavelength InGaAsP light-emitting diode and the dual-wavelength demultiplexing photodiode has also been built [78]. At 33 Mb/s the measured sensitivities of the two channels at 10^{-9} BER were -38.5 and -39.7 dBm.

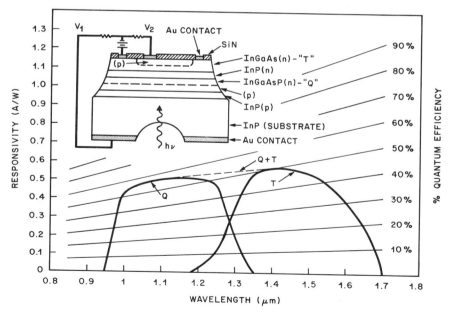

Figure 14.10 Schematic cross section and responsivity of a dual-wavelength photodetector. (After Campbell et al. [79].)

A new application for the dual-wavelength photodetector, wavelength sensing, was reported by Campbell et al. [79]. It has been demonstrated that this photodetector can be used to determine simultaneously both the wavelength and the signal level of an optical input. The signal wavelength is determined by monitoring the ratio of the two outputs from this photodetector at the crossover of its responsivity curves (Fig. 14.10); the signal level is determined from the sum of the two outputs. One of the key parameters for a photodetector of this type is the rate of change of the ratio signal with wavelength. Initial measurements show that the ratio signal changes by up to 20% for a 20-Å shift in wavelength. Possible applications for this type of photodetector include mode control of single-frequency injection lasers, direct demodulation for lightwave systems using frequency-shift-keyed transmission, wavelength-division multiplexing, and optical logic. Wood et al. [80] demonstrated a multiple quantum well photodetector whose voltage maximum exhibits a strong wavelength dependence. Using this photodetector the detected wavelength can be determined to a precision of 0.03 Å.

14.5 AVALANCHE PHOTODIODES

The internal gain of avalanche photodiodes (APDs) can provide improved receiver sensitivity, particularly at higher bit rates (≥ 500 Mb/s). First-generation lightwave systems, which operated in the wavelength range from 0.8 to 0.9 μm, utilized the "near-perfect" silicon avalanche photodiodes [81]. Unfortunately, the operating wavelengths of more recent lightwave systems are beyond the long-wavelength cutoff of silicon photodetectors. As a result, much of the work on detectors in recent years has concentrated on developing long-wavelength APDs that could perform as well as their silicon counterparts.

The earliest long-wavelength APDs were homojunction devices similar to those shown in Figure 14.3 fabricated from $In_xGa_{1-x}As_yP_{1-y}$ [35–38] or $Al_xGa_{1-x}As_ySb_{1-y}$ [82–84]. Figure 14.11 shows a typical dark current characteristic for one of these photodiodes. For low to intermediate voltages (< 30 V) the dark current is due primarily to generation–recombination in the depletion region. However, just below breakdown, the normal operating point for APDs, these devices exhibit large dark currents and soft breakdowns. The source of this excess dark current has been identified as tunneling [22, 36–38, 84]. Unfortunately, due to the small bandgap and small effective electron mass of the III–V compounds, which are photosensitive in the wavelength range of interest, this is an intrinsic problem.

The structure shown in Figure 14.12 was first proposed by Nishida et al. [85] to effectively eliminate the tunneling component of the dark current. The *pn* junction and thus the high field region is located in a wide-bandgap

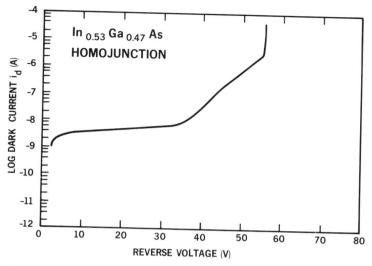

Figure 14.11 Dark current of an $In_{0.53}Ga_{0.47}As$ PIN photodiode.

material such as InP, where tunneling is insignificant, and absorption occurs in an adjacent narrow-bandgap material such as InGaAs. This structure is often referred to as an SAM APD because it utilizes *s*eparate *a*bsorption and *m*ultiplication regions. For this structure to operate properly, three boundary conditions must be satisfied:

1. The electric field in the multiplication region must be high enough to produce useful gain ($E \gtrsim 4.5 \times 10^5$ V/cm).
2. The electric field at the absorbing layer interface must be low enough that tunneling in the absorbing layer is negligible ($E \lesssim 2.5 \times 10^5$ V/cm).

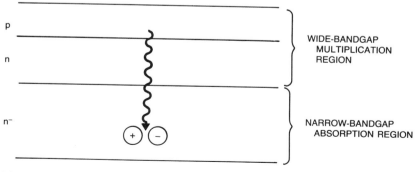

Figure 14.12 Avalanche photodiode structure with separate absorption and multiplication regions (SAM APD).

3. The depletion region must extend far enough into the absorbing region that diffusion effects are eliminated ($\gtrsim 2 \ \mu$m).

Forrest et al. [86] have shown that the first two conditions can be expressed in terms of the sheet charge densities σ_m and σ_a in the multiplication and absorption regions, respectively. That analysis shows that the boundary conditions listed above impose relatively narrow constraints on the carrier concentrations and layer thicknesses of these SAM APD structures. Nevertheless, workers at several laboratories around the world have successfully demonstrated SAM APDs having very low dark currents and good avalanche gain [87–90]. Figure 14.13 shows the dark current and gain of an InP/InGaAs SAM APD. Near the breakdown voltage the dark current has been reduced by nearly a factor of 100 compared to the InGaAs homojunction APD. At $M_0 = 10$ the total dark current is approximately 25 nA and the primary multiplied dark current is ≤ 1 nA. Gains as high as 60 have been reported for these APDs.

Unfortunately, the frequency response of these SAM ADPs is often poor. Figure 14.14 shows a typical frequency response curve for a planar InP/InGaAs SAM APD [91]. The bandwidth is only a few hundred megahertz, which poses a serious problem since APDs should prove must useful for bit rates ≥ 500 Mb/s. Forrest et al. [92] have shown that the pulse response of these SAM APDs contains both fast and slow components.

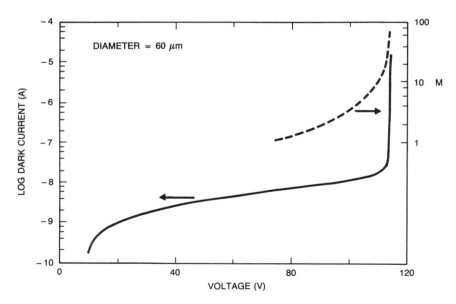

Figure 14.13 Dark current (solid line) and gain (dashed line) of an SAM avalanche photodiode.

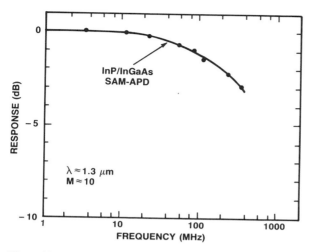

Figure 14.14 Frequency response of an InP/In$_{0.53}$Ga$_{0.47}$As SAM avalanche photodiode. (After Shirai et al. [91].)

The fast component has a time constant of < 200 ps. The slow component decreases with increasing bias voltage and can be as low as a few nano-seconds or as long as a microsecond. It has been determined that the origin of the slow component is charge accumulation at the interface between the absorption and multiplication layers. That is, as the photogenerated holes are swept toward the multiplication region, many of them get trapped in the potential well formed by the valence band discontinuity at the heterojunction interface. These trapped holes may eventually escape by thermionic emission, but that is a slow process.

Many solutions have been proposed to improve the bandwidth of these avalanche photodiodes. Capasso et al. [93] reported that for the AlInAs/InGaAs heterojunction the valence band discontinuity is smaller than that of the InP/InGaAs heterojunction. They successfully fabricated SAM APDs from AlInAs/InGaAs, which achieved low dark current (~ 40 nA at $M_0 = 1$) and high-speed response (10 GHz gain–bandwidth product). In addition, a receiver sensitivity of -36 dBm was achieved at 420 Mb/s ($\lambda \approx 1.3$ μm). However, in this material the electron ionization coefficient is greater than that of the hole [94]. This means that to achieve low noise, depletion and absorption should occur on the p side of the pn junction. This is difficult to implement in III–V compounds.

Another approach is to introduce a transition region between the multiplication and absorption regions. The most sophisticated structure of this type, proposed by Capasso et al. [95], uses a variable-period superlattice of InP and InGaAs between the InP multiplication region and the InGaAs

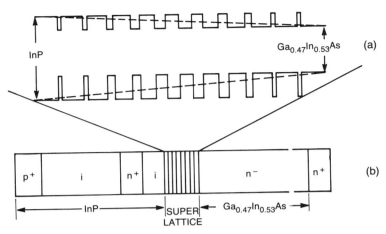

Figure 14.15 (a) Band diagram of a chirped-period superlattice. The dashed lines represent the average bandgap seen by the carriers. (b) Schematic drawing of an SAGM avalanche photodiode utilizing the chirped period superlattice as the transition region. (After Capasso et al. [95].)

absorption region (Fig 14.15). The effect of varying the period of the superlattice is to grade the effective bandgap energy. A maximum bandwidth of 900 MHz has been achieved.

Forrest et al. [92] proposed a somewhat simpler approach. The transition region can have a uniformly graded bandgap or, as has been demonstrated at several laboratories, it can consist of one or more intermediate-bandgap InGaAsP layers [96–99]. This type of structure will be referred to below as an SAGM APD. The G has been inserted to denote the presence of a "grading," or transition, layer. The motivation, as shown in Figure 14.16, is to replace the one large valence band discontinuity by two or more smaller steps. The effect on the response time can be illustrated as follows: The transient hole current due to the emission of holes that have been trapped at the heterojunction interfaces is given by

$$J(t) = q \frac{dN}{dt} \tag{15}$$

If recombination of trapped holes via deep interface traps is neglected, N, the number of trapped holes, is given by

$$N(t) = N_0 \exp(-e_h t) \tag{16}$$

where N_0 is the number of trapped holes at $t = 0$ and e_h is the hole emission rate, which, for thermionic emission, can be written as

$$e_h = B \exp \frac{-\varepsilon_b}{kT} \tag{17}$$

where B is a constant and ε_b is the effective valence band barrier height. It is

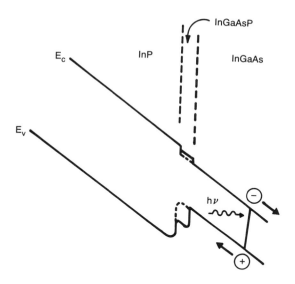

Figure 14.16 Band diagram of an In/InGaAsP/InGaAs avalanche photodiode with separate absorption, grading, and multiplication regions (SAGM APD).

clear from Eqs. (15)–(17) that the hole current due to interface trapping is an exponential of an exponential in the barrier height. The result is that small changes in the barrier height can result in large improvements in the frequency response. This is a somewhat oversimplified picture. At each interface the crystal growth process results in a slight grading of the interface, which also tends to reduce the effective barrier height. So the addition of a transition region also results in more grading and less interface trapping as well as enhanced emission.

Figure 14.17 shows some of the device structures of this type that have been investigated. Figure 14.17a shows a mesa structure [87–90, 97, 100]. One of the advantages of this structure is ease of fabrication. Since it does not require a guard ring, there are no masking steps for diffusion or ion implementation. Also it is usually easier to achieve low dark current and low capacitance with the mesa than with the planar structure. On the other hand, for mass production and reliability, planar structures are preferred. A "standard" planar SAGM APD is shown in Figure 14.17b. The guard rings are fabricated by either low-temperature diffusion [101, 102] or ion implantation [91, 99, 103]. A recessed guard ring [104], which reportedly enhances its effectiveness, is shown in Figure 14.17c. The APD in Figure 14.17d operates on the same principle, but its fabrication was achieved by liquid-phase epitaxial overgrowth [103, 105].

Excellent device characteristics have been reported for each of these structures. Quantum efficiencies for devices without AR coatings are typically 60–70%, and with AR coatings >90% is relatively routine. The primary multiplied dark current is less than 1 nA, and dc multiplication values as high as 60 have been reported.

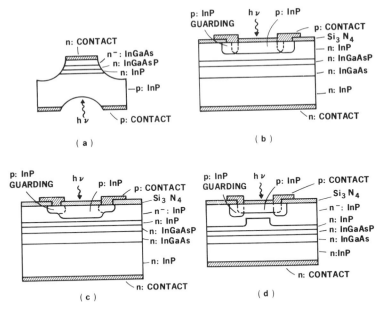

Figure 14.17 Mesa-structure and planar-structure SAGM avalanche photodiodes.

The improvement in speed afforded by the SAGM APD structure is illustrated in Figure 14.18. Figure 14.18a shows the pulse response ($M_0 = 4$) of a mesa-type SAGM APD with two intermediate-bandgap transition layers. The pulse width (FWHM) is 80 ps, which represents an improvement of two to three orders of magnitude over the standard SAM APD. This pulse response is a convolution of the width of the laser excitation pulse (50 ps), the RC time constant (35 ps), and the effective transit time including the avalanche buildup time (~ 45 ps). The sum-of-squares approximation gives an estimated pulse width of 75 ps, which is close to the measured value. This indicates that the effect of hole trapping at the heterojunction interfaces has effectively been eliminated. The frequency response at $M_0 = 12$ is shown in Figure 14.18b. The 3-dB bandwidth is >5 GHz, which yields a gain–bandwidth product of 60 GHz.

The frequency response of this type of APD is given by the following expressions [106]:

$$
\frac{i_s(\omega)}{i_s(0)} = \frac{1}{M_0[1 - \exp(-\alpha W_a)]} \left(\frac{1}{1 - \omega^2 LC + j\omega RC} \right)
$$
$$
\times \left[T_n(\omega) + T_p(\omega) + \frac{T_{ns}(\omega)}{[1 + j\omega(M_0 - 1)\tau_m](1 + j\omega/e_h)} \right] \quad (18)
$$

where W_a is the width of the absorbing region, M_0 is the dc gain, L is the inductance (primarily due to the bond wire), C is the junction capacitance,

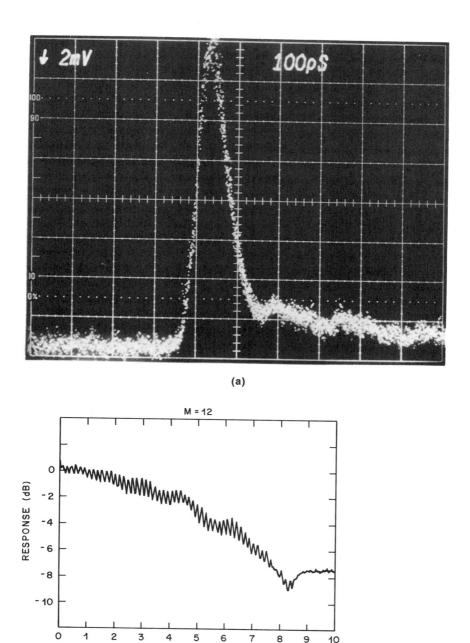

Figure 14.18 (a) Pulse response (at $M_0 = 4$) and (b) frequency response (at $M_0 = 12$) of an InP/InGaAsP/InGaAs SAGM avalanche photodiode.

R is the sum of the diode series resistance and the load resistance, and $(M_0 - 1)\tau_m$ is the avalanche buildup time [107]. The terms $T_n(\omega)$, $T_p(\omega)$, and $T_{ns}(\omega)$ are the transit time factors for the primary electrons and holes and the secondary electrons, respectively. They can be written as

$$T_n(\omega) = \frac{v_n}{W}\left[\frac{1 - \exp\left(-j\omega W_a/v_n\right)}{j\omega}\right.$$
$$\left. - \exp\left(-\alpha W_a\right)\frac{1 - \exp\left(-j\omega W_a/v_n + \alpha W_a\right)}{j\omega - \alpha v_n}\right] \tag{19}$$

$$T_p(\omega) = \frac{v_p}{W}\left\{\frac{1 - \exp\left(-\alpha W_a\right)}{j\omega} + \frac{j\omega/e_h + \exp\left(-j\omega(W - W_a)/v_p\right)}{1 + j\omega/e_h}\right.$$
$$\left. \times\left[1 - \exp\left(-\frac{j\omega W_a}{v_p - \alpha W_a}\right)\right]\left(\frac{1}{j\omega + \alpha v_p} - \frac{1}{j\omega}\right)\right\} \tag{20}$$

and

$$T_{ns}(\omega) = \frac{v_n}{W}(M_0 - 1)\left\{\exp\left(-\frac{j\omega(W - W_a)}{v_p}\right)\left[1 - \exp\left(-\frac{j\omega W}{v_n}\right)\right]\right.$$
$$\left. \times\left[1 - \exp\left(-\frac{j\omega W_a}{v_p - \alpha W_a}\right)\right]\left(\frac{1}{j\omega} - \frac{1}{(\alpha v_p + j\omega)}\right)\right] \tag{21}$$

where W is the width of the depletion region. It is assumed in these expressions that the absorbing layer is completely depleted.

The bandwidth versus dc avalanche gain of a mesa-type InP/InGaAsP/InGaAs SAGM APD with three transition layers is shown in Figure 14.19 [108]. In the high-gain regime ($M_0 \geq 15$), a gain–bandwidth-limited response is observed; the gain–bandwidth product is 70 GHz. This high gain–bandwidth product was achieved by minimizing the width of the multiplication region. Since the gain–bandwidth product arises from the regenerative nature of the multiplication process, for a given gain the bandwidth-limiting effects of carrier feedback can be reduced by decreasing the width of the multiplication region. For the SAGM APD structure this change in the width of the multiplication region must be accompanied by an increase in the carrier concentration to avoid excess dark current [86]. At lower gains, a bandwidth ceiling of 8 GHz is observed. The flattening out of the bandwidth at lower gains can be caused by either the RC time constant, hole trapping, or the transit times. In this case the low-gain bandwidth is limited almost entirely by the secondary electron transit time term.

Figure 14.20 shows the spatial uniformity of the gain of a mesa-type SAGM APD grown by liquid-phase epitaxy. At unity gain the response is very uniform across the whole photosensitive area. At a higher gain ($M_0 = 10$), some structure, which corresponds to variation in the gain of 5–10%, is evident. It has been shown that this nonuniformity in the gain is due to

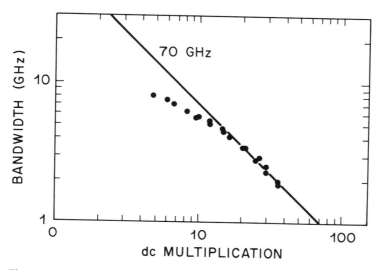

Figure 14.19 Measured bandwidth of a mesa-structure InP/InGaAsP/InGaAs SAGM avalanche photodiode with two intermediate-bandgap transition layers as a function of gain.

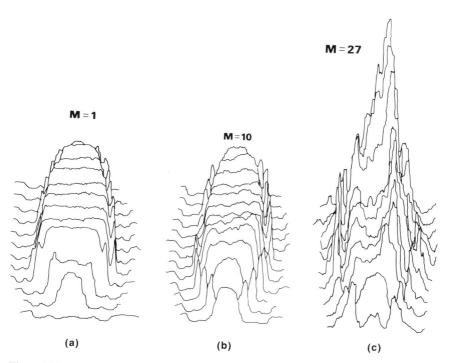

Figure 14.20 Gain profiles of an InP/InGaAsP/InGaAs SAGM-APD at (a) $M_0 = 1$, (b) $M_0 = 10$, and (c) $M_0 = 27$.

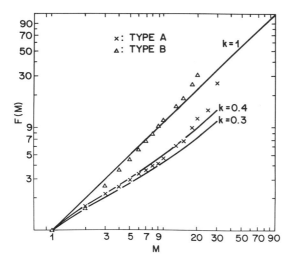

Figure 14.21 Measured excess noise factor versus gain for type A and type B SAGM avalanche photodiodes.

either variations in the thickness of the multiplication region or doping [109–112]. At even higher gains (Fig. 14.20c), these nonuniformities often become even more pronounced, creating "hot spots." This type of nonuniformity can have a significant effect on the noise characteristics of an APD. The noise of an APD can usually be characterized by shot noise multiplied by an excess noise factor $F(M_0)$ that is due to the random nature of the multiplication process [113, 114]. $F(M_0)$ is a function of the dc gain M_0 and k, the ratio of the hole and electron ionization coefficients. There have been several reports that for multiplication in InP, k is in the range 0.3–0.4 [87, 115–119].

Figure 14.21 shows the measured excess noise $F(M)$ of two SAGM APDs. The device labeled "type A" exhibited uniform gain for $M > 15$. At higher gains a "hot spot" appeared and $F(M)$ exceeded the values expected on the basis of the accepted values of k. Even higher excess noise was observed on a device (type B) that exhibited significant nonuniformity even at low multiplication values. It is evident from these results that the excess noise increases as the gain becomes nonuniform. In the extreme, this can limit the maximum usable gain and degrade the receiver sensitivity [120].

14.5.1 Germanium Avalanche Photodiodes

In parallel with the often dramatic improvements in APDs fabricated from group III–V compounds there has been steady incremental improvement in the characteristics of germanium APDs. The first high-speed germanium APDs were reported by Melchior and Lynch in 1966 [121]. Owing to its ease of fabrication, almost all of the early germanium APDS utilized an n^+p structure [121, 122]. These planar APDs exhibited high gains, relatively soft gain–voltage characteristics, spatially uniform response, and high quantum efficiencies. However, the high dark currents and high excess noise of the n^+p germanium APDs limited their use in lightwave receivers.

The high dark current was due in large part to the absence of an adequate surface passivation such as the thermally grown oxide used in silicon technology. This problem has not yet been eliminated, but passivation techniques have been developed that greatly reduce the surface leakage current, thus providing more stable device characteristics and overall greater reliability [122, 123]. Torikai et al. [123] reported that dark currents as low as 250 nA at 90% of breakdown can be accomplished reproducibly using encapsulated thermal oxidation, which includes SiO_2 deposition and subsequent annealing in an O_2 ambient.

In germanium the ionization rate of holes β is approximately twice that of electrons α. As a result, minimum multiplication noise is achieved when injection into the multiplication region consists entirely of holes [113, 114]. In the conventional n^+p structure, however, at longer wavelengths ($\lambda >$ 1.3 μm) the incident light signal is absorbed primarily in the p region owing to the reduced absorption coefficient of germanium in this wavelength range. In this case, electron injection predominates and the performance is degraded. If the thickness of the n^+ is increased in order to increase hole injection, the quantum efficiency decreases due to surface and bulk recombination of carriers as they diffuse toward the multiplication regions. An n^+np structure [124–126] and a p^+n structure [127, 128] have been developed for use in 1.3-μm lightwave systems that give a 30–40% improvement in the effective k value. This reduction in noise has resulted in improved receiver sensitivities and longer lightwave system transmission spans [129]. At 1.55 μm the absorption length of germanium is greater than 10 μm. In order to achieve depletion widths comparable to this relatively long absorption length and still maintain reasonable breakdown voltages, a p^+nn^- structure was developed [130, 131]. Small-area devices of this type show promise for future single-mode lightwave systems.

14.5.2 Novel Avalanche Photodiodes

A number of novel APD structures have been proposed that could, at least in theory, provide improved performance by "artificially" increasing the ratio of the ionization coefficients. One structure of this type, shown in Figure 14.22, incorporates a superlattice into the multiplication region [132, 133]. The difference between the conduction and valence band discontinuities is used to enhance the electron ionization rate relative to that of the holes. Using a GaAs/AlGaAs APD of this type, Capasso et al. achieved some enhancement in the ratio of the ionization coefficients [133]. A similar structure proposed by Williams et al. utilizes a graded bandgap superlattice to produce a "staircase"-like band structure [134, 135]. This structure could

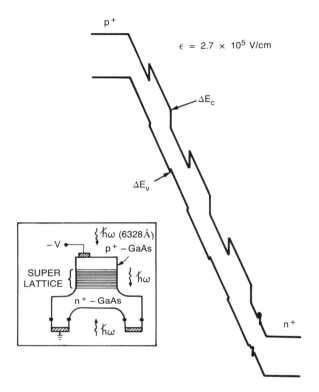

Figure 14.22 Superlattice avalanche photodiode. (After Capasso et al. [134].)

provide even greater enhancement of the ionization rates ratio, which would result in noise performance comparable to that of silicon APDs.

Of course, the important criterion for judging a photodetector is how well it performs in a lightwave receiver. Figure 14.23 is a summary of the best receiver sensitivities that have been achieved with InP/InGaAsP/InGaAs SAGM APDs, germanium APDs, and $In_{0.53}Ga_{0.47}As$ PIN photodiodes in the front-end circuits of long-wavelength lightwave receivers. To date, SAGM APDs have achieved 3–5 dB better sensitivities than germanium APDs.

There are two ways to compare the sensitivities of receivers utilizing PIN photodiodes with those having APDs. Figure 14.23 shows the best results that have been achieved with each type of receiver. It is clear that the APD has an edge, but a small one. However, these results are somewhat deceptive because the detectors were in different receivers. The PIN results were obtained with a very low capacitance (<0.6 pF) hybrid front end

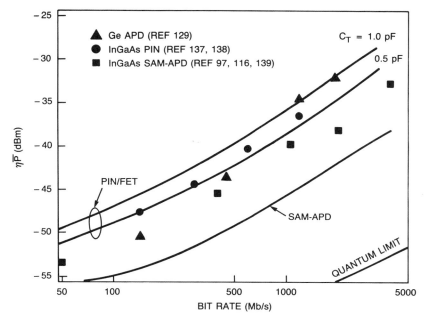

Figure 14.23 Experimental receiver sensitivities.

[136, 137], whereas the capacitances of the APD front ends were two to three times higher. Figure 14.24 shows the net improvement when a PIN was replaced directly with an APD in the same receiver [138]. It can be seen that the largest improvements are obtained at the higher bit rates, and the improvements are substantial. However, as the total front-end capacitances of these receivers are reduced, this differential will shrink a little.

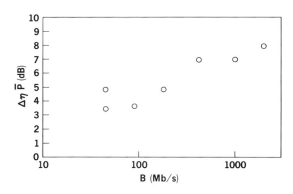

Figure 14.24 Comparison of receiver sensitivities when PIN photodiodes were replaced with an InP/InGaAsP/InGaAs SAGM avalanche photodiode in the same receiver.

14.6 CONCLUSION

Photodetectors and compatible low-noise transistor preamplifiers for long-wavelength optical receivers are the subject of numerous research and development efforts. This work is directed at creating new materials, processing techniques, and device structures as well as improving the parameters of existing conventional devices. There is no single best choice for the front-end components. The "optimum" devices will be determined by the system requirements and constraints such as bit rate, fiber length, cost, and the degree of circuit complexity that can be tolerated. Fortunately, a large number of satisfactory devices are available, so initial implementation of long-wavelength lightwave systems can proceed.

REFERENCES

1. R. G. Smith and S. D. Personick, "Receiver design for optical communications," in *Semiconductor Devices for Optical Communications*, H. Kressel (Ed.), Springer-Verlag, New York, 1980.
2. T. V. Muoi, "Receiver design for high-speed optical-fiber systems," *J. Lightwave Technol.*, **LT-2**: 243–267 (1984).
3. D. R. Smith and I. Garrett, "A simplified approach to digital optical receiver design," *Opt. Quantum Electron.*, **10**: 211–221 (1978).
4. K. Ogawa, "Considerations for optical receiver design," *IEEE J. Selected Areas. Commun.*, **SAC-1**: 524–532 (1983).
5. K. Ogawa, "Noise caused by GaAs MESFETs in optical receivers," *Bell Syst. Tech. J.*, **60**: 923–928 (1981).
6. K. Ogawa, B. Owen, and H. J. Boll, "Long-wavelength optical receiver with a short-channel MOSFET," Conf. Lasers and Electro-Optics, Washington, DC., 56–58 (1981).
7. K. Ogawa and J. C. Campbell, "Comparison of photodetector/low-noise amplifier combinations for long-wavelength receivers," 3rd Int. Conf. Integrated Opt. Opt. Fiber Commun., San Francisco, CA, 1981.
8. D. R. Smith, R. C. Hooper, and I. Garrett, "Receivers for optical communications," *Opt. Quantum Electron.* **10**: 293–300 (1978).
9. R. G. Smith and S. R. Forrest, "Sensitivity of avalanche photodetector receivers for long-wavelength optical communications," *Bell Syst. Tech. J.*, **61**: 2929–2946 (1982).
10. J. C. Campbell, "Photodetectors and compatible low-noise amplifiers for long-wavelength lightwave systems," *Fiber Integrated Opt.*, **5**: 1–21 (1984).
11. J. L. Hullett and T. V. Muoi, "A feedback receiver amplifier for optical transmission systems," *IEEE Trans. Commun.*, **COM-24**: 1180–1185 (1976).
12. J. E. Goell, "Input amplifiers for optical PCM receivers," *Bell Syst. Tech. J.*, **53**: 1771–1793 (1974).
13. M. J. N. Sibley, R. T. Unwin, and D. R. Smith, "The design of *p-i-n* bipolar transimpedance pre-amplifiers for optical receivers," *J. Inst. Electron Radio Eng.*, **55**: 104–110 (1985). February 1985.
14. D. R. Smith, R. C. Hooper, K. Ahmad, D. Jenkins, A. W. Mabbitt, and R. Nicklin, "*p-i-n/* FET hybrid optical receiver for longer-wavelength optical communication systems," *Electron. Lett.*, **16**: 69–71 (1980).
15. A. S. Tager, "Current fluctuations in semiconductors under the conditions of impact ion-

ization and avalanche breakdown," *Fiz. Tver. Tela.*, **6**: 2418–2427 (1964). [*Sov. Phys. Solid State*, **8**: 1919–1925 (1965)].

16. G. Eisenstein and L. W. Stulz, "High quality antireflection coatings on laser facets by sputtered silicon nitride," *Appl. Opt.*, **23**: 161–164 (1984).

17. T. P. Pearsall and R. W. Hopson, "Growth and characterization of lattice-matched epitaxial films of $Ga_xIn_{1-x}As/InP$ by liquid-phase epitaxy," *J. Electron. Mat.*, **7**: 133–146 (1978).

18. K. J. Bachman and J. L. Shay, "An InGaAs detector for the 1.0–1.7 μm wavelength range," *Appl. Phys. Lett.*, **32**: 446–448 (1978).

19. H. Morkoc, T. J. Andrews, Y. M. Houng, R. Sankaran, S. G. Bandy, and G. A. Antypas, "Microwave $In_xGa_{1-x}As_yP_{1-y}/InP$ F.E.T.," *Electron. Lett.*, **15**: 448–449 (1978).

20. C. A. Burrus, A. G. Dentai, and T. P. Lee, "InGaAsP p-i-n photodiodes with low dark current and small capacitance." *Electron. Lett.*, **15**: 655–657 (1979).

21. R. F. Leheny, R. E. Nahory, and M. A. Pollack, "$In_{0.53}Ga_{0.47}As$ p-i-n photodiodes for long-wavelength fiber-optic systems," *Electron. Lett.*, **15**: 713–715 (1979).

22. S. R. Forrest, R. F. Leheny, R. E. Nahory, and M. A. Pollack, "$In_{0.53}Ga_{0.47}As$ photodiodes with dark current limited by generation-recombination and tunneling," *Appl. Phys. Lett.*, **37**: 322–324 (1980).

23. F. Capasso, R. A. Logan, A. Hutchinson, and D. M. Manchon, "InGaAsP/InGaAs heterojunction p-i-n detectors with low dark current and small capacitance for 1.3 μm–1.6 μm fiber systems," *Electron. Lett.*, **16**: 893–895. (1980).

24. T. P. Lee, C. A. Burrus, A. G. Dentai, and K. Ogawa, "InGaAsP p-i-n photodiodes with low dark current and small capacitance," *Electron. Lett.*, **16**: 155–156 (1980).

25. G. H. Olsen, "Low-leakage, high efficiency, reliable VPE InGaAs 1.0–1.7 μm photodiode," *IEEE Electron Dev. Lett.*, **EDL-2**: 217–219 (1981).

26. D. G. Jenkins and A. W. Mabbitt, "The reliability of GaInAs photodiodes and GaAs FETs for use in PIN-FET fibre optic receivers," *Proc. IEEE Specialists Conf. Light Emitting Diodes and Photodetectors*, Ottawa-Hull, Canada, 152–153 (1982).

27. N. Susa, Y. Yamauchi, H. Ando, and H. Kanbe, "Planar type vapor-phase epitaxial $In_{0.53}Ga_{0.47}As$ photodiode," *IEEE Electron. Dev. Lett.*, **EDL-1**: 55–57 (1980).

28. S. R. Forrest, I. Camlibel, O. K. Kim, S. J. Stocker, and J. R. Zuber, "Low dark-current, high efficiency planar $In_{0.53}Ga_{0.47}As.InP$ pin photodiodes," *IEEE Electron. Dev. Lett.*, **20**: 283–285 (1981).

29. V. Diadiuk, C. A. Armiento, S. H. Groves, and C. E. Hurwitz, "Surface passivation techniques for InP and InGaAsP p-n junction structures," *IEEE Electron. Dev. Lett.*, **EDL-1**: 177–178 (1980).

30. H. Nickel and E. Kuphal, "Surface-passivated low dark current InGaAs pin photodiodes," *J. Opt. Commun.*, **4** 63–67 (1983).

31. S. R. Forrest, "Performance of $In_xGa_{1-x}As_yP_{1-y}$ photodiodes with dark current limited by diffusion, generation-recombination, and tunneling," *IEEE J. Quantum Electron.*, **QE-17**: 217–226 (1981).

32. W. Shockley, "The theory of p-n junctions in semiconductors and p-n junction transistors," *Bell Syst. Tech. J.*, **28**: 435–489 (1949).

33. S. M. Sze, *Physics of Semiconductor Devices*, Wiley, New York, 1969.

34. J. L. Moll, *Physics of Semiconductors*, McGraw-Hill, New York, 1964.

35. A. S. Grove, *Physics and Technology of Semiconductor Devices*, Wiley, New York, 1967.

36. S. R. Forrest, M. DiDomenico, R. G. Smith, and H. J. Stocker, "Evidence for tunneling in reverse-bias III–V photodetector diodes," *Appl. Phys. Lett.*, **36**: 580–582 (1980).

37. H. Ando, H. Kanbe, M. Ito, and T. Kaneda, "Tunneling current in InGaAs and optimum design for InGaAs/InP avalanche photodiodes," *Jap. J. Appl. Phys.*, **19**: L277–L280 (1980).

38. Y. Takanashi, M. Kawashima, and Y. Horikoshi, "Required donor concentration of epitaxial layers for efficient InGaAsP avalanche photodiodes," *Jap. J. Appl. Phys.*, **19**: 693–700 (1980).

39. R. Trommer and H. Albrecht, "Confirmation of tunneling current via traps by DLTS measurements in InGaAs photodiodes," *Jap. J. Appl. Phys.*, **22**: L364–L366 (1983).

40. O. K. Kim, B. V. Dutt, R. J. McCoy, and J. R. Zuber, "A low dark-current, planar InGaAs *p-i-n* photodiode with a quaternary InGaAsP Ca layer," *IEEE J. Quantum Electron.*, **QE-21**: 138–143 (1985).

41. J. C. Campbell, A. G. Dentai, G. J. Qua, J. Long, and V. G. Riggs, "Planar InGaAs PIN photodiode with a semi-insulating InP cap layer," *Electron. Lett.*, **21**: 447–448 (1985).

42. S. Kagawa, J. Komeno, M. Ozeki, and T. Kaneda, "Planar $Ga_{0.47}In_{0.53}As$ PIN photodiodes with extremely low dark current," *Proc. Conf. Optical Fiber Commun.*, San Diego, CA, 1985, p. 92.

43. T. P. Lee and Tingye Li, "Photodetectors," in *Optical Fiber Communications*, S. E. Miller and A. G. Chynoweth (Eds.), Academic, Orlando, FL, 1979.

44. H. W. Ruegg, "An optimized avalanche photodiode," *IEEE Trans. Electron. Dev.*, **ED-14**: 239–251 (1967).

45. J. E. Bowers, C. A. Burrus, and R. S. Tucker, "22-GHz bandwidth InGaAs/InP PIN photodiodes," Picosecond Electron. Opt. Conf., Incline Village, NV, 1985.

46. T. H. Windhorn, L. W. Cook, and G. E. Stillman, "Temperature dependent electron velocity-field characteristics for $In_{0.53}Ga_{0.47}As$ at high electric fields," *J. Electron. Mat.*, **11**: 1065–1082 (1982).

47. K. Brennan and K. Hess, "Theory of high field transport of holes in GaAs and InP," *Phys. Rev. B*, **29**: 5581—5590 (1984).

48. C. A. Burrus, J. E. Bowers, and R. S. Tucker, "Improved very-high-speed InGaAs PIN punch-through photodiode," *Electron. Lett.*, **21**: 262–263 (1985).

49. J. C. Campbell and K. Ogawa, "Heterojunction phototransistors for long-wavelength optical receivers," *J. Appl. Phys.*, **53**: 1203–1208 (1982).

50. M. C. Brain and D. R. Smith "Phototransistors in digital optical communication systems," *IEEE J. Quantum Electron.*, **QE-19**: 1139–1148 (1983).

51. R. M. Kolbas, J. Abrokwah, J. K. Carney, D. H. Bradshaw, B. R. Elmer, and J. R. Baird, "Planar monolithic integration of a photodiode and a GaAs preamplifier," *Appl. Phys. Lett.*, **43**: 821–823 (1983).

52. O. Wada, H. Hamaguchi, S. Miura, M. Makiuchi, K. Nakai, H. Horimatsu, and T. Sakurai, "AlGaAs/GaAs *p-i-n* photodiode/preamplifier monolithic photoreceiver integrated on semi-insulating GaAs substrate," *Appl. Phys. Lett.*, **46**: 981–983 (1985).

53. R. F. Leheny, R. E. Nahory, M. A. Pollack, A. A. Ballman, E. D. Beebe, J. C. DeWinter, and R. J. Martin, "Integrated InGaAs PIN-FET photoreceiver," *Electron. Lett.*, **16**: 353–355 (1980).

54. R. L. Leheny, R. E. Nahory, J. C. Dewinter, R. J. Martin, and E. D. Beebe, "An integrated PIN/JFET photoreceiver for long wavelength optical systems," *Tech. Digest, Int. Electron. Dev. Mtg.*, Washington, D.C., 1981, pp. 276–279.

55. S. Hata, M. Ikeda, T. Amano, G. Motosugi, and K. Kurumada, "Planar InGaAs/InP PINFET fabricated by Be ion implantation," *Electron Lett.*, **20**: 947–948 (1984).

56. C. L. Cheng, A. S. H. Liao, T. Y. Chang, R. F. Leheny, L. A. Coldren, and B. Lalevic, "Submicrometer self-aligned recessed gate InGaAs MISFET exhibiting very high transconductance," *IEEE Electron Dev. Lett.*, **EDL-5**: 169–171 (1984).

57. C. L. Cheng, A. S. H. Liao, T. Y. Chang, E. A. Caridi, L. A. Coldren, and B. Lalevic, "Silicon oxide enhanced Schottky gate $In_{0.53}Ga_{0.47}As$ FET's with a self-aligned recessed gate structure," *IEEE Electron. Dev. Lett.*, **EDL-5**: 511–514 (1984).

58. K. Kasahara, J. Hayashi, K. Makita, K. Taguchi, A. Susuki, H. Homura, and S. Matushita, "Monolithically integrated $In_{0.53}Ga_{0.47}As$-PIN/InP-MISFET photoreceiver," *Electron Lett.*, **20**: 314–315 (1984).

59. B. Tell, A. S. H. Liao, K. Brown-Goebeler, T. J. Bridges, E. G. Burkhardt, T. Y. Chang, and N. S. Bergano, "Monolithic integration of a planar embedded InGaAs PIN detector with InP depletion mode FETs," *IEEE Trans. Electron. Dev.*, **ED-32**: 2319–2321 (1985).

60. Zh. I. Alferov, F. A. Akhmedov, V. I. Korolkov, and V. G. Nitkitin, "Phototransistor utilizing a GaAs-AlAs heterojunction," *Sov. Phys. Semicond.*, 7: 780–782 (1973).
61. H. Beneking, P. Mischel, and G. Schul, "High-gain wide-gap-emitter $Ga_{1-x}Al_xAs$-GaAs phototransistor," *Electron. Lett.*, 12: 395–396 (1976).
62. M. Konagai, K. Katsukawa, and K. Takahashi, "(GaAl)As/GaAs heterojunction phototransistors with high current gain," *J. Appl. Phys.*, 48: 4389–4394 (1977).
63. R. A. Milano, T. H. Windhorn, E. R. Anderson, G. E. Stillman, R. D. Dupuis, and P. D. Dapkus, "$Al_{0.5}Ga_{0.5}As$-GaAs heterojunction phototransistors grown by metalorganic chemical vapor deposition," *Appl. Phys. Lett.*, 34: 562–564 (1979).
64. K. Tabatabai-Alavi, R. J. Markunus, and C. G. Fonstad, "LPE-grown InGaAsP/InP heterojunction bipolar phototransistor," *Tech. Digest, Int. Electron. Dev. Mtg.*, Washington, D.C., 1979, pp. 643–645.
65. M. Tobe, Y. Amemiya, S. Sakai, and M. Umeno, "High-sensitivity InGaAsP/InP phototransistors," *Appl. Phys. Lett.*, 37: 73–75 (1980).
66. P. D. Wright, R. J. Nelson, and T. Cella, "High-gain InGaAsP-InP heterojunction phototransistors," *Appl. Phys. Lett.*, 37: 192–194 (1980).
67. D. Fritzsche, E. Kuphal, and R. Aulbach, "Fast response InP/InGaAsP heterojunction phototransistors," *Electron. Lett.*, 17: 178–180 (1981).
68. J. C. Campbell, A. G. Dentai, C. A. Burrus, and J. F. Ferguson, "InP/InGaAs heterojunction phototransistors," *IEEE J. Quantum Electron.*, **QE-17**: 264–269 (1981).
69. J. C. Campbell, C. A. Burrus, A. G. Dentai, and K. Ogawa, "Small-area high-speed InP/InGaAs phototransistor," *Appl. Phys. Lett.*, 39: 820–821 (1981).
70. H. Kroemer, "Theory of a wide-gap emitter for transistors," *Proc. IRE*, 45: 1535–1537 (1957).
71. L. E. Tsyrlin, "Response of a phototransistor," *Sov. Phys. Semicond.*, 11: 1127–1129 (1977).
72. R. A. Milano, P. D. Dapkus, and G. E. Stillman, "An analysis of the performance of heterojunction phototransistors for fiber optic communications," *IEEE Trans. Electron. Dev.*, **ED-29**: 266–274 (1981).
73. J. C. Campbell and A. G. Dentai, "InP/InGaAs heterojunction phototransistor with integrated light emitting diode," *Appl. Phys. Lett.*, 41: 192–193 (1983).
74. J. C. Campbell, T. P. Lee, A. G. Dentai, and C. A. Burrus, "Dual-wavelength demultiplexing InGaAsP photodiode," *Appl. Phys. Lett.*, 34: 401–402 (1979).
75. J. C. Campbell, A. G. Dentai, T. P. Lee, and C. A. Burrus, "Improved two-wavelength demultiplexing InGaAsP photodetector," *IEEE J. Quantum Electron.*, **QE-16**: 601–602 (1980).
76. S. Sakai, M. Umeno, and Y. Amemiya, "Optimum designing of InGaAsP/InP wavelength demultiplexing photodiodes," *Trans. IECE Jap.*, E 63: 192–197 (1980).
77. S. Sakai and M. Umeno, "Wavelength demultiplexing photodiode with very high isolation ratio," *Jap. J. Appl. Phys.*, 22: L338–L339 (1983).
78. K. Ogawa, T. P. Lee, C. A. Burrus, J. C. Campbell, and A. G. Dentai, "Wavelength division multiplexing experiment employing dual-wavelength LEDs and photodetectors," *Electron. Lett.*, 17: 857–859 (1981).
79. J. C. Campbell, C. A. Burrus, J. A. Copeland, and A. G. Dentai, "Wavelength-discriminating photodetector for lightwave systems," *Electron. Lett.*, 19: 672–674 (1983).
80. T. H. Wood, C. A. Burrus, A. H. Gnauck, J. M. Wiesenfeld, D. A. B. Miller, and D. S. Chemla, "Wavelength-selective voltage-tunable photodetector made from multiple quantum wells," *Appl. Phys. Lett.*, 47: 190–192 (1985).
81. H. Melchior, A. R. Hartman, D. P. Schinke, and T. E. Seidel, "Planar epitaxial silicon avalanche photodiode," *Bell Syst. Tech. J.*, 57: 1791–1807 (1978).
82. F. Capasso, M. V. Panish, and S. Sumski, "The liquid-phase epitaxial growth of low net donor concentration (5×10^{14}–5×10^{15}/cm^3) GaSb for detector applications in the 1.3–1.6 µm region," *IEEE J. Quantum Electron.*, **QE-17**: 273–274 (1981).
83. H. D. Law, R. Chin, K. Nakano, and R. A. Milano, "The GaAlAsSb quaternary and

GaAlSb ternary alloys and their application to infrared detectors," *IEEE J. Quantum Electron.*, **QE-17**: 275–823 (1981).

84. N. Tabatabaie, G. E. Stillman, R. Chin, and P. D. Dapkus, "Tunneling in the reverse dark current of GaAlAsSb avalanche photodiodes," *Appl. Phys. Lett.*, **40**: 415–416 (1982).

85. K. Nishida, K. Taguchi, and Y. Matsumoto, "InGaAsP heterojunction avalanche photodiodes with high avalanche gain," *Appl. Phys. Lett.*, **35**: 251–252 (1979).

86. S. R. Forrest, R. G. Smith, and O. K. Kim, "Performance of $In_{0.53}Ga_{0.47}As/InP$ avalanche photodiodes," *IEEE J. Quantum Electron.*, **QE-18**: 2040–2048 (1982).

87. N. Susa, H. Nakagome, O. Mikami, H. Ando, and H. Kanbe, "New InGaAs/InP avalanche photodiode structure for the 1–1.6 μm wavelength region," *IEEE J. Quantum Electron.*, **QE-16**: 8064–869 (1980).

88. N. Susa, H. Nakagome, H. Ando, and H. Kanbe, "Characteristics in InGaAs/InP avalanche photodiodes with separated absorption and multiplication regions," *IEEE J. Quantum Electron*, **QE-17**: 243–250 (1981).

89. V. Diadiuk, S. H. Groves, C. E. Hurwitz, and G. W. Iseler, "Low dark-current, high gain GaInAsP/InP avalanche photodetectors," *IEEE J. Quantum Electron.*, **QE-17**: 260–263 (1981).

90. O. K. Kim, S. R. Forrest, W. A. Bonner, and R. G. Smith, "A high gain $In_{0.53}Ga_{0.47}As/InP$ avalanche photodiode with no tunneling leakage current," *Appl. Phys. Lett.*, **39**: 402–404 (1981).

91. T. Shirai, T. Mikawa, T. Kaneda, and A. Miyauchi, "InGaAs avalanche photodiodes for 1 μm wavelength region," *Electron. Lett.*, **19**: 535–536 (1983).

92. S. R. Forrest, O. K. Kim, and R. G. Smith, "Optical response time of $In_{0.53}Ga_{0.47}As/InP$ avalanche photodiodes," *Appl. Phys. Lett.*, **41**: 95–98 (1982).

93. F. Capasso, B. Kasper, K. Alavi, A. Y. Cho, and J. M. Parsey, "New low dark current, high speed $Al_{0.48}In_{0.52}As/In_{0.53}Ga_{0.47}As$ avalanche photodiode by molecular beam epitaxy for long wavelength fiber optic communication systems," *Appl. Phys. Lett.*, **44**: 1027–1029 (1984).

94. F. Capasso, K. Mohammed, K. Alavi, A. Y. Cho, and P. W. Foy, "Impact ionization rates for electrons and holes in $Al_{0.48}In_{0.52}As$," *Appl. Phys. Lett.*, **45**: 968–970 (1984).

95. F. Capasso, H. M. Cox, A. L. Hutchinson, N. A. Olsson, and S. G. Hummel, "Pseudo-quaternary GaInAsP semiconductors: a new $In_{0.53}Ga_{0.47}As/InP$ graded gap super-lattice and its applications to avalanche photodiodes," *Appl. Phys. Lett.*, **45**: 1193–1195 (1984).

96. Y. Matsushima, A. Akiba, K. Sakai, Y. Kushiro, Y. Noda, and K. Utaka, "High-speed-response $In_{0.53}Ga_{0.47}As/InP$ heterostructure avalanche photodiode with InGaAsP buffer layers," *Electron. Lett.*, **18**: 945–946 (1982).

97. J. C. Campbell, A. G. Dentai, W. S. Holden, and B. L. Kasper, "High-performance avalanche photodiode with separate absorption, grading, and multiplication regions," *Electron. Lett.*, **19**: 818–820 (1983).

98. K. Yasuda, T. Mikawa, Y. Kishi, and T. Kaneda, "Multiplication-dependent frequency responses of InP/InGaAs avalanche photodiodes," *Electron. Lett.*, **20**: 373–374 (1984).

99. Y. Sugimoto, T. Torikai, K. Makita, H. Ishihara, K. Minemura, and K. Taguchi, "High-speed planar-structure InP/InGaAsP/InGaAs avalanche photodiode grown by VPE," *Electron. Lett.*, **20**: 653–654 (1984).

100. R. Trommer, *Proc. 9th Eurp. Conf. Opt. Commun.*, 1983, pp. 159–162.

101. H. Ando, Y. Yamauchi, and N. Susa, "High-speed planar InP/InGaAs avalanche photodiode fabricated by vapor phase epitaxy," *Electron. Lett.*, **19**: 543–544 (1983).

102. Y. Matsushima, Y. Noda, Y. Kushiro, N. Seki, and S. Akiba, "High sensitivity of VPE-grown InGaAs/InP-heterostructure APD with buffer layer and guard-ring structure," *Electron. Lett.*, **20**: 235–236 (1984).

103. Y. Yasuda, Y. Kishi, T. Shirai, T. Mikawa, S. Yamazaki, and T. Kaneda, "InP/InGaAs buried-structure avalanche photodiodes," *Electron. Lett.*, **20**: 158–159 (1984).

104. H. Ando, Y. Yamauchi, and N. Susa, "Reach-through type planar InGaAs/InP avalanche photodiode fabricated by continuous vapor phase epitaxy," *IEEE J. Quantum Electron.*, **QE-20**: 256–264 (1984).

105. M. Kobayashi, S. Yamazaki, and T. Kaneda, "Planar InP/GaInAsP/GaInAs buried-structure avalanche photodiode," *Appl. Phys. Lett.*, **45**: 759–761 (1984).

106. J. C. Campbell, W. T. Tsang, G. J. Qua, B. C. Johnson, and J. E. Bowers, "Wide-bandwidth InP/InGaAsP/InGaAs avalanche photodiodes grown by chemical beam epitaxy," *Tech. Digest Int. Electron Dev. Mtg.*, Washington, D.C., 1987, pp. 233–236.

107. E. B. Emmons, "Avalanche-photodiode frequency response," *J. Appl. Phys.*, **38**: 3705–3714 (1967).

108. J. C. Campbell, W. T. Tsang, G. J. Qua, and J. E. Bowers, "InP/InGaAsP/InGaAs avalanche photodiodes with 70 GHz gain–bandwidth product," *Appl. Phys. Lett.*, **51**: 1454–1454 (1987).

109. F. Capasso, P. M. Petroff, W. B. Bonner, and S. Sumski, "Investigation of microplasmas in InP avalanche photodiodes," *IEEE Electron Dev. Lett.*, **EDL-1**: 27–29 (1980).

110. N. Magnea, P. M. Petroff, F. Capasso, R. A. Logan, and P. W. Foy, "Microplasma characteristics on LPE grown InP-$In_{0.53}Ga_{0.47}As$ long wavelength avalanche photodiodes with separated multiplication and absorption regions," *Appl. Phys.*, **23**: 66–68 (1984).

111. T. Takanohashi, T. Shirai, S. Yamazaki, and S. Komiya, "Inhomogeneous distribution of avalanche multiplication in InP APDs," *Jap. J. Appl. Phys.*, **23**: 207–271 (1984).

112. W. S. Holden, J. C. Campbell, and A. G. Dentai, "Gain uniformity of InP/InGaAsP/InGaAs avalanche photodiodes with separate absorption, grading, and multiplication regions," *IEEE J. Quantum Electron.*, **QE-21**: 1310–1313 (1985).

113. R. J. McIntyre, "Multiplication noise in uniform avalanche diodes," *IEEE Trans. Electron Dev.*, **ED-13**: 164–168 (1966).

114. R. J. McIntyre, "The distribution of gains in uniformly multiplying avalanche photodiodes: theory," *IEEE Trans. Electron. Dev.*, **ED-19**: 703–713 (1972).

115. M. Ito, T. Kaneda, K. Nakajima, Y. Toyoma, and T. Kotani, "Impact ionization ratio in $In_{0.73}Ga_{0.27}As_{0.57}P_{0.43}$," *Electron. Lett.*, **14**: 418–419 (1978).

116. S. R. Forrest, G. F. Williams, O. K. Kim, and R. G. Smith, "Excess-noise and receiver sensitivity measurement of $In_{0.53}Ga_{0.47}As$/InP avalanche photodiodes," *Electron. Lett.*, **17**: 917–919 (1981).

117. R. Yeats and K. Von Dessonneck, "Detailed performance characteristics of hybrid InP-InGaAsP APDs," 3rd Int. Conf. Integrated Opt. Opt. Fiber Commun., San Francisco, CA, p. 104.

118. T. Shirai, S. Yamazaki, F. Osaka, K. Nakajima, and T. Kaneda, "Multiplication noise in planar InP/InGaAsP heterostructure avalanche photodiodes," *Appl. Phys. Lett.*, **40**: 532–533 (1982).

119. B. L. Kasper, J. C. Campbell, and A. G. Dentai, "Measurements of the statistics of excess noise in separate absorption, grading and multiplication (SAGM) avalanche photodiodes," *Electron. Lett.*, **20**: 796–797 (1984).

120. Y. K. Jhee, J. C. Campbell, W. S. Holden, A. G. Dentai, and J. K. Plourde, "The effect of nonuniform gain on the multiplication noise of InP/InGaAsP/InGaAs avalanche photodiodes," *IEEE J. Quantum Electron.*, **QE-21**: 1858–1861 (1985).

121. H. Melchior and W. T. Lynch, "Signal and noise response of high speed germanium avalanche photodiodes," *IEEE Trans. Electron. Dev.*, **ED-13**: 829–838 (1966).

122. H. Ando, H. Kanbe, T. Kimura, Y. Yamaoka, and T. Kaneda, "Characteristics of germanium avalanche photodiodes in the wavelength region of 1–1.6 μm." *IEEE J. Quantum Electron.*, **QE-14**: 804–809 (1978).

123. T. Torikai, I. Hino, H. IwaSaki, and K. Nishida, "Encapsulated thermal oxidation for Ge-APDs passivation," *Jap. J. Appl. Phys.*, **21**: 1776–1778 (1982).

124. S. Kagawa, T. Mikawa, and T. Kaneda, "Germanium avalanche photodiodes in the 1.3 μm wavelength region," *Fujitsu Sci. Tech. J.*, **18**: 397–418 (1982).

125. T. Kaneda, S. Kagawa, T. Mikawa, T. Toyama, and H. Ando, "An n^+-n-p germanium avalanche photodiode," *Appl. Phys. Lett.*, **36**: 572–274 (1980).

126. T. Mikawa, S. Kagawa, T. Kaneda, T. Sakurai, H. Ando, and O. Mikami, "A low-noise n^+-n-p germanium avalanche photodiode," *IEEE J. Quantum Electron.*, **QE-17**: 210–216 (1981).

127. T. Kaneda, H. Fukuda, T. Mikawa, T. Banba, Y. Toyama, and H. Ando, "Shallow-junction p^+-n germanium avalanche photodiodes (APDs)," *Appl. Phys. Lett.*, **34**: 866–868 (1979).

128. S. Kagawa, T. Kaneda, T. Mikawa, Y. Banba, Y. Toyama, and O. Mikami, "Fully ion-implanted p^+-n germanium avalanche photodiodes," *Appl. Phys. Lett.*, **38**: 429–431 (1981).

129. J. Yamada, A. Kawana, T. Miya, H. Nagai, and T. Kimura, "Giga-bit/s optical receiver sensitivity and zero-dispersion single-mode fiber transmission at 1.55 μm," *IEEE J. Quantum Electron.*, **QE-18**: 1537–1546 (1982).

130. T. Mikawa, S. Kagawa, and T. Kaneda, "Germanium reachthrough avalanche photodiodes for optical communication systems in the 1.55 μm wavelength region," *IEEE Trans. Electron Dev.*, **ED-31**: 971–977 (1984).

131. M. Niwa, Y. Tashiro, K. Minemura, and H. Iwasaki, "High-sensitivity hi-lo germanium avalanche photodiode for 1.5 μm-wavelength optical communication," *Electron Lett.*, **20**: 552–553 (1984).

132. R. Chin, N. Holonyak, G. E. Stillman, J. Y. Tang, and K. Hess, "Impact ionization in multi-layered heterojunction structures," *Electron. Lett.*, **16**: 467–469 (1980).

133. F. Capasso, W. T. Tsang, A. L. Hutchinson, and G. F. Williams, "Enhancement of electron impact ionization in a superlattice: a new avalanche photodiode with a large ionization rates ratio," *Appl. Phys. Lett.*, **40**: 38–40 (1982).

134. G. F. Williams, F. Capasso, and W. T. Tsang, "The graded bandgap multilayer avalanche photodiode: a new low noise detector," *IEEE Electron Dev. Lett.*, **EDL-3**: 71–73 (1982).

135. F. Capasso, W. T. Tsang, and G. F. Williams, "Staircase solid state photomultipliers and avalanche photodiodes with enhanced ionization rate ratio," *IEEE Trans. Electron Dev.*, **ED-30**: 381–390 (1983).

136. D. R. Smith, R. C. Hooper, P. P. Smyth, and D. Wake, "Experimental comparison of a germanium avalanche photodiode and InGaAs PINFE receiver for longer wavelength optical communication systems," *Electron. Lett.*, **18**: 453–454 (1982).

137. M. C. Brain, P. P. Smyth, D. R. Smith, B. R. White, and P. J. Chidgey, "PINFET hybrid optical receivers for 1.2-Gbit/s transmission systems at 1.3- and 1.55-μm wavelength," *Electron. Lett.*, **20**: 894–895 (1984).

138. B. L. Kasper, J. C. Campbell, A. H. Gnauck, A. G. Dentai, and J. R. Talman, "SAGM avalanche photodiode receiver for 2 and 4 Gbit/s," *Electron. Lett.*, **21**: 982–984 (1985).

PART 4

Optical Transmitters and Receivers

CHAPTER 15

Semiconductor Laser Transmitters

P. W. Shumate

15.1 INTRODUCTION

The conversion of a low-level electrical signal (e.g., ECL or TTL for a digital application, or ~1 V for an analog application) to a corresponding light-intensity envelope in the time domain is accomplished at the transmitter. Such *direct modulation* is intended to affect only the average optical power, and any phase or frequency information imparted to the optical carrier frequency itself is not used at the receiver. Important increases in receiver sensitivity and selectivity can be derived using intentional optical phase or frequency modulation. Such "coherent" transmission is dealt with in Chapter 25.

For directly modulated lightwave transmitters, two semiconductor devices—light-emitting diodes (LEDs) and injection laser diodes (ILDs or LDs) — are suitable in terms of size, speed, efficiency, electrical characteristics, and so on. Because of low coupling efficiency between an LED and typical transmission fibers, particularly single-mode fiber, and their broad spectral width (50–150 nm FWHM), LEDs are generally most useful for shorter-distance, lower-bit-rate applications (e.g., maximum system bit rate length product of a few gigabit-kilometers per second).

Conversely, lasers couple light efficiently into all types of fibers because of their output beam characteristics. They also offer intrinsically higher modulation speeds and narrow spectral widths (< 10 nm). Combined with 10–30 dB additional launched power, lasers are useful over a significantly wider bit rate × length region (e.g., maximum bit rate × length product of > 100 Gb-km/s).

This chapter deals with laser transmitters, mostly for digital applications although analog is also considered. Issues of packaging, coupling, and controlling temperature will be discussed, because for most applications they play critical roles in the overall performance of the transmitter. First, however, laser properties necessary for transmitter design will be discussed.

15.2 LASER CHARACTERISTICS AND MODULATION

A typical laser light output versus current (L–I) transfer characteristic for several temperatures is shown in Figure 15.1. When the diode current is below threshold (I_{th}), cavity and mirror losses exceed the gain derived from stimulated emission and lasing oscillation does not occur. The corresponding light output is the incoherent, spontaneous emission characteristic of an LED. The frequency response of this subthreshold output is also LED-like, being inversely proportional to the spontaneous recombination lifetime of the injected minority carriers.

Beyond threshold, lasing results in efficient conversion of input current to output light, seen as the high differential quantum efficiency (%) or "slope" efficiency (W/A) of the L–I curve. In this region, the emission is spectrally narrow and has an extended high-frequency response.

Since threshold is determined by the gain/loss balance in the laser, I_{th} responds sensitively to temperature- or aging-induced changes in either gain

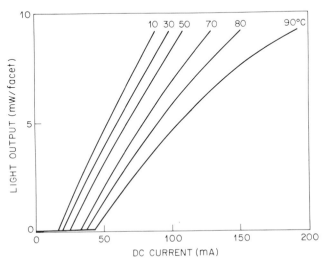

Figure 15.1 Injection laser output as a function of current and temperature. Data shown are for a 1.3-μm InGaAsP device.

or loss. Temperature can affect gain or loss through numerous mechanisms (e.g., current spreading and heterobarrier heights). Empirically, threshold current is found to rise with increasing temperature according to

$$I_{th} \propto \exp \frac{T}{T_0} \tag{1}$$

where the characteristic temperature T_0 ranges from 40 to 70 K for InGaAsP lasers (although research structures with higher T_0 values have been reported) and as high as 180 K for AlGaAs lasers.

Long-term aging of lasers usually results in irreversible increases in threshold due to a variety of causes. For example, if the series resistance of the contact to the laser increases (degrades) with time, the added Joule heating raises the threshold. On the other hand, increased losses in the cavity due to facet oxidation, facet damage, growth of crystalline defects ("dark lines"), and so on, increase the threshold. These mechanisms and others may result in lower slope efficiency as well. Therefore, changes in laser characteristics must be allowed for in the design of a transmitter.

Above threshold, the frequency response of the output is determined by the solution of coupled rate equations that describe the charge-carrier and photon populations. For our purposes, it is sufficient to recognize that the response is substantially higher than below threshold, as shown in Figure 15.2. The high-frequency resonance peak, which has the effect of flattening the

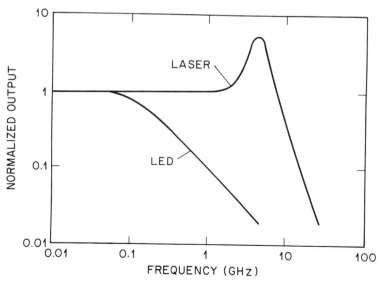

Figure 15.2 Small-signal frequency responses of light-emitting and laser diodes with negligible parasitic effects.

overall response of the laser, increases in frequency with diode current according to

$$f_r \propto \left(\frac{I}{I_{th}} - 1\right)^{1/2} \tag{2}$$

where I is the quiescent bias point (above threshold) for the small-signal modulation. The proportionality factor in Eq. (2) is $(1/2\pi\tau_p\tau_c)^{1/2}$, where τ_p is the photon lifetime in the cavity (order of 10 ps) and τ_c is the minority carrier lifetime in the cavity (order of 1 ns). The width of the resonance peak depends on the laser structure, and the overall frequency response depends on parasitics such as capacitance due to blocking junctions or contacts [1]. Chapter 12 treats high-speed modulation in detail.

When operated continuously (cw) above threshold, multi-longitudinal-mode lasers show a spectrum characterized by several wavelengths (modes) spaced according to the dimensions of the Fabry-Perot cavity as in Figure 15.3a. Single-longitudinal-mode (SLM) lasers show only a single frequency determined by grating dimensions [e.g., distributed-feedback (DFB) or dis-

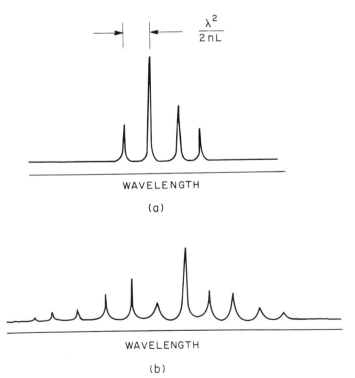

Figure 15.3 Output spectrum of laser showing (a) few longitudinal modes when operated continuously and (b) additional modes when modulated at high speed.

tributed-Bragg-reflector (DBR) lasers]. When the diode forward current is modulated at high frequencies, many multilongitudinal-mode lasers develop additional modes as in Figure 15.3b. In addition to resulting in a wider spectrum, the modes have a probability of exchanging power with each other. Whether or not additional modes appear, modulation induces "noisy" statistical power fluctuations (modal partition noise, or MPN) among the modes already present [2, 2a]. Because these fluctuations in amplitude and wavelength are chromatically time-dispersed during transmission, jitter and eye closure are observed at the receiver, leading to a sensitivity penalty [3].

Fast changes in laser current also induce unwanted spectral effects by modulating the refractive index of the cavity. The resulting "chirp" of the emitted wavelength is chromatically dispersed, leading again to receiver power penalties at high bit rates. Chirp can be reduced by injection locking (slaving the modulated laser to a second, continuously operated laser), either from light injected into the cavity through one of the mirrors [4] or from light injected laterally from adjacent laser cavities [5]. Chirp-suppressing circuit techniques have been reported and will be discussed in Section 15.3.

For digital applications, it is important to maximize the energy in the 1 or mark level and minimize the energy in the 0 or space level. The ratio of these two intensities is the extinction or contrast ratio γ. High values of γ cannot be attained arbitrarily, however, since the 1 level is usually subject to peak-power limitations of the laser set by reliability considerations, and the 0 level is determined by laser turn-on dynamics or spontaneous emission near threshold. Suppose a laser initially has zero bias (current or voltage) applied to it. If the current is now rapidly increased to a level corresponding to the desired power for the 1 state, a delay between the leading edge of the current pulse and the light output is observed. The delay is accompanied by a transient overshoot (ringing) in the light output, again arising from the dynamics of the coupled carrier and photon populations in the cavity. The delay t_d is given by

$$t_d = \tau_{\text{th}} \ln \frac{I}{I - I_{\text{th}}} \tag{3}$$

where I is the magnitude of the current pulse and τ_{th} is the carrier recombination lifetime at threshold (usually 2–5 ns). The delay can be reduced by making the drive current large with respect to threshold (usually impractical) or by biasing the laser close to threshold, in which case Eq. (3) becomes

$$t_d = \tau_{\text{th}} \ln \frac{I}{I - (I_{\text{th}} - I_{\text{bias}})} \tag{4}$$

It is seen that the delay vanishes when $I_{\text{bias}} = I_{\text{th}}$. The amplitude of the overshoot or ringing (whose frequency is nearly equal to the laser resonance frequency and whose damping time constant is twice the recombination lifetime [6]) becomes small as the laser is biased near threshold. Therefore, to minimize delay and overshoot, both of which cause serious pattern-dependent distortion, a near-threshold dc bias is added to the pulse current, and operation is as shown in Figure 15.4.

This bias, however, or an above-threshold bias to extend the frequency response as in Eq. (2), results in a nonzero 0 light level, reducing the extinction ratio and incurring a sensitivity penalty at the receiver [7]. Therefore a bias current must be critically chosen that is optimum for the application. Also to be considered is the transmitter circuit's capability for maintaining this bias at the optimum point under all operating conditions.

So far we have considered only modulation of the laser current. For high-bit-rate transmitters, problems with MPN, chirp, delay, overshoot, or high levels of 0 light can be circumvented by using an external modulator between a continuously operated laser and the fiber. These are waveguide devices usually fabricated in materials having large electrooptic coefficients such as $LiNbO_3$, or in III–V materials. Group III–V modulators are very important because they can be monolithically integrated with the laser. Such an integrated device was reported that significantly reduced chirp at 450 Mb/s [8].

Currently, however, modulators are more often fabricated by diffusing titanium into $LiNbO_3$ to create waveguides through the resulting refractive

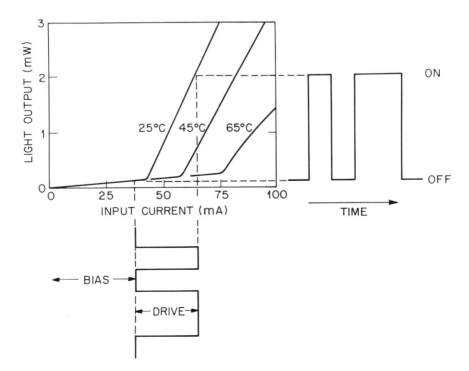

Figure 15.4 Transfer characteristic for typical high-speed pulsed operation of a laser.

index difference. Mach-Zender interferometers and directional couplers are suitable for achieving high extinction ratios and low-voltage operation with single-mode lasers [9, 10, 10a]. Difficulties that accompany their use have to do with losses in coupling to the laser and fiber and with packaging the combination. Modulators designed for low-voltage operation at 1.3–1.5 μm tend to be long (~ 2 cm).

For analog applications, operation of a laser as a linear or "class A" device is straightforward: The laser is quiescently biased near the midpoint of its lasing region, and modulation takes place about this point. This is analogous to the operation depicted in Figure 15.4 except for the placement of bias level. Modulation current is limited by the need to remain above threshold for negative excursions or by peak-power or linearity considerations for positive excursions. Linearity, as measured by harmonic distortion or two-tone intermodulation products, is usually the key consideration.

Effects that induce nonlinearities in the $L-I$ characteristics of lasers arise mostly from transverse modal instabilities in the active region, reflections, junction heating, or intensity-dependent gain-saturation effects. The most severe nonlinearities, seen as obvious "kinks" in the $L-I$ characteristic, are usually due to modal instabilities. Their occurrence renders the laser unsuitable for most analog applications. Nonlinearities are often related to the laser design, except those induced by reflections, which depend on coupling, packaging, and connectors. Therefore the choices of laser structure and packaging are foremost considerations.

Laser noise is also of major importance. Analog applications are generally broadband and require high signal-to-noise ratios (SNRs). This combination leads to shot-noise-limited detection at the receiver. Therefore any excess noise generated by the laser directly determines the maximum received SNR (or carrier-to-noise ratio, CNR, where applicable).

Excess noise (i.e., excess above the intrinsic shot noise associated with a given optical power level) arises from the amplification of carrier and photon fluctuations in the cavity. It can reach tens of decibels near threshold [6, 11] and has been analyzed extensively [12]. In addition to this well-defined mechanism, other effects such as modal instabilities, reflections, and self-pulsations also cause excess noise levels to be as high as 30 dB. Strong temperature dependence is often observed. For example, temperature-induced wavelength change results in longitudinal mode hopping as Fabry-Perot boundary conditions are periodically satisfied [13, 13a].

The magnitude of these effects can be quantified as excess noise level or, more conventionally, as relative intensity noise (RIN). RIN is the total mean-square noise power per unit bandwidth divided by the mean optical output (squared) of the laser:

$$\text{RIN} = \frac{\langle p_n^2 \rangle}{p^2 B} \tag{5}$$

with units, commonly, of decibels per hertz. By straightforward calculation of RIN at a noiseless photodetector, the shot-noise limit (excess noise = 0 dB) is − 156 dB/Hz at 1 mW and scales inversely with power. Analogous to other modulation parameters, RIN depends on operating current level [14]:

$$\text{RIN} \propto \left(\frac{I}{I_{\text{th}}} - 1\right)^{-3} \tag{6}$$

As with distortion, noise depends on the laser structure and reflections. Noise associated with modal partitioning is minimized or eliminated by using a spectrally stable laser structure such as a DFB or DBR design. Recent results have shown that both DFB and stable single-longitudinal-mode Fabry-Perot lasers have desirable noise and distortion properties for analog applications [15, 16].

If multimode fiber is used, however, it is desirable to have many longitudinal modes to minimize speckle effects in the fiber. This form of modal noise and the distortion that accompanies it (since it depends indirectly on laser current) are quite serious for analog applications. They arise from interference effects along the length of the fiber and subsequent selective attenuation of the resulting patterns, such as might occur at splices or connectors. To reduce modal noise, it has been found advantageous to "dither" the laser current at a frequency much higher than the modulation to reduce the coherence of the laser and thus reduce speckle [17, 17a]. Such dithering also reduces noise that arises from reflections back into the laser cavity itself [18] and RIN resulting from mode hopping [19].

Since reflections pose a problem, insertion of an optical isolator between the laser and fiber can be considered, although cost and complexity are increased. Laser-to-fiber coupling schemes can also be selected for minimizing reflections.

Analog techniques are sometimes applied to save the cost of pulse-code modulation or time-division multiplexing. Direct baseband intensity modulation, however, may be accompanied by the noise and distortion problems mentioned above. Where PCM is undesirable, and where noise and/or nonlinearities limit performance, subcarrier modulation using a distortion-resistant modulation format can be used effectively.

The analog signal is first impressed upon a higher-frequency RF subcarrier, using ordinary modulation techniques [20]. The laser is linearly modulated by the resulting subcarrier. At the reciever, the electrical subcarrier is recovered at the photodetector and processed ("detected") by whatever means are appropriate for the modulation technique used at the transmitter. In order to obtain immunity from laser noise and distortion, modulation schemes such as FM and PFM are commonly used. With these, the postdetection SNR is greater than the predetection CNR (which characterizes the output of the photodetector). For FM, this "FM improvement" is proportional to the square of the subcarrier frequency deviation.

By modulating the laser with a composite of subcarriers, the photodetected signal can then be processed using bandpass amplifiers to achieve simple demultiplexing. The optical power is divided among the subcarriers according to the modulation index of each. Conventional care must be taken to examine the intermodulation consequences of the various subcarrier frequency assignments.

15.3 TRANSMITTER CIRCUITS

15.3.1 Digital

Based on the discussions above, digital transmitter circuits should:

1. Supply high-speed current pulses
2. Supply dc bias
3. Maintain high extinction ratio
4. Maintain constant output power

A block diagram of such a transmitter is shown in Figure 15.5. The high-speed drive is shown here simply summed with the dc bias at a tee. In anticipation of later parts of this section, the laser output is monitored by the light emitted from the back mirror (the front mirror is coupled to the fiber). The resulting photocurrent is used with a regulator circuit from which the bias is derived to perform functions 3 and 4, above. Other functions of importance to overall transmitter design include

5. Providing logic-level compatibility
6. Providing laser protection
7. Providing alarms for abnormal operation

High-speed current pulses are usually supplied by a differential current switch as shown in Figure 15.6. It can easily be designed to supply either constant or controllable peak current. It can be driven at gigahertz rates using controlled rise time single-ended or differential input signals, and it is

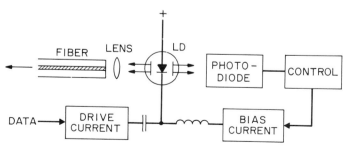

Figure 15.5 Block diagram of directly modulated laser transmitter.

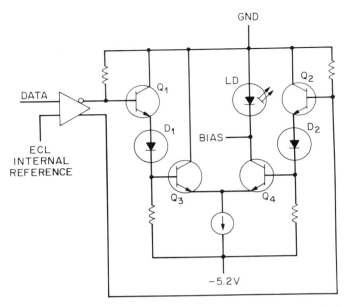

Figure 15.6 Differential current switch typical of modulator circuit for lasers.

readily made compatible with standard high-speed logic families. Transistors Q_1 and Q_2 along with diodes D_1 and D_2 shift the logic-level input more negative to prevent Q_4 from saturating, even if the voltage requirement of the laser rises (e.g., due to high end-of-life current, increased series resistance, etc.).

Current-source drivers have also been found advantageous at high bit rates because they minimize rise time degradation due to lead inductance [21]. For the typical circuit shown in Figure 15.6, operation between the ground rail and a negative voltage is convenient, as many lasers have their p contact connected to the package (i.e., grounded). Current-source drivers using a common base/common emitter configuration may also be found. In Figure 15.7a, the laser bias is set by the base current of Q_1, while the high-speed drive is introduced via Q_2. A variation on this means of bias control is shown in Figure 15.7b. A *pnp* common-emitter stage determines the bias while the *npn* common-emitter switch shunts a portion of current supplied by Q_1 around the laser.

A logic gate is used at the input in Figure 15.6. Logic gates are often internally compensated to ensure logic-level tracking and highest noise immunity over the full operating temperature range and thus provide a simple means for obtaining compatibility with associated circuitry. By using a D flip-flop instead, input data can be retimed to correct for pulse-shape degradation ahead of the transmitter.

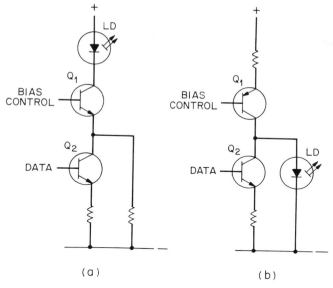

(a)　　　　　　　　　(b)

Figure 15.7 Alternate current-driving circuits that provide for control of laser bias.

For operation at gigabit-per-second rates, GaAs MESFETs are often substituted for silicon bipolar transistors. Because of the high pinch-off voltages, additional gain may be inserted between the input and the differential switch as shown in Figure 15.8. Single-ended shunt or series drivers using

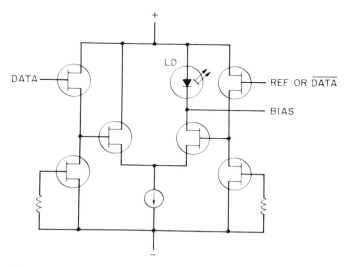

Figure 15.8 High-speed modulator using GaAs FETs.

GaAs MESFETs for maximum speed are also used, but control of the bias or driver current is less straightforward than with the differential switch. The single-ended circuits lend themselves to pole-zero compensation or "speed-up" networks to cancel the effect of packaging capacitance.

Transient effects such as chirp can be minimized at the circuit level by shaping the drive pulse. A dual-pulsing scheme was analyzed wherein a low-amplitude pulse of width equal to the period of the relaxation oscillation preceded the main drive pulse [22]. Calculations indicated that with proper timing and amplitude, the transient overshoot can be minimized or eliminated, thus reducing chirp. In other work, a network with a resonant characteristic that compensated that of the laser was placed between the driver and laser [23]. Chirp in a 1.5-μm DFB laser was reduced by a factor of 3.

Another method combined device and circuit techniques. One section of a two-cavity 1.3-μm DFB laser was operated normally, modulated at 100 MHz, while the second cavity, biased below threshold, was driven at 100 MHz out of phase with the first section. In this way, a complementary change in the refractive index nearly canceled the chirp induced in the lasing section. This approach achieved a 0.002-nm shift under modulation, compared with several hundredths of a nanometer for ordinary DFB lasers [24].

15.3.2 Bias Control

We consider now the control of laser bias to compensate for changes in threshold. Without compensation, a constant bias results in unacceptable changes in output power, extinction ratio, and turn-on delay. For example, Table 15.1 shows the effects of ambient temperature changes on a laser with a T_0 of 60 K operated at constant 20-mA bias and 30-mA drive currents. Equations (1) and (4) are used, with τ_{th} assumed to be 3 ns and slope efficiency assumed to be constant.

Although the changes in power might be tolerable in some systems, the turn-on variation that arises from operating below threshold at 55°C would permit a maximum bit rate of only about 30 Mb/s if 5% jitter were allowed.

TABLE 15.1. Effect of ambient temperature on a laser operated at constant bias (20 mA) and drive (30 mA)

Temperature	10°C	25°C	55°C
Threshold current	16 mA	20 mA	33 mA
Power/facet	5.7 mW	5 mW	2.8 mW
Turn-on delay	0 ns	0 ns	1.7 ns
Extinction ratio	<10:1	>20:1	>20:1

Also it is seen that the extinction ratio falls to approximately 8:1 at 10°C due to operation above threshold, although no turn-on delay occurs. Clearly it is necessary to provide control of the bias in response to operating conditions. For some applications, open-loop control based simply on a measurement of the ambient temperature and (1) may be adequate. Presently, however, it is more desirable to use closed-loop control, which monitors the output power, so that aging-induced changes in threshold are also compensated.

Simple regulation of the average power is adequate when digital data are scrambled or encoded to ensure an equal density of marks and spaces (50% duty cycle). A negative feedback (NFB) regulator of this type is shown in Figure 15.9. Adjustment of the mean-power reference level (using V_{DC}) and peak drive current (using V_{DRIVE}) are adjusted initially at 25°C to set the power level while optimizing extinction ratio and turn-on delay. The photodiode that monitors the light output is usually placed in the laser package to intercept the back-mirror output, which is otherwise unused. Alternatively, the back-mirror light can be brought outside the package through simple optics to a separately packaged photodiode. Other monitoring alternatives include placing a tap in the output fiber [25] or monitoring the front-mirror light that is not coupled into the fiber.

In some cases, data interrupts or long strings of 1s or 0s cannot be eliminated, and it is undesirable to encode the data. In the circuit of Figure 15.10, the time average of the input data pattern is compared with the time average of the light output so that only power-level variations are sensed

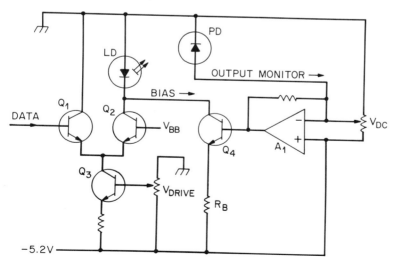

Figure 15.9 Simple average-power bias regulator for control of laser bias.

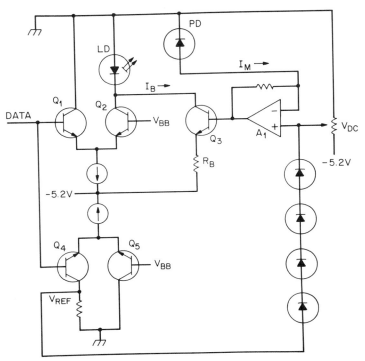

Figure 15.10 Average-power bias regulator that is independent of duty cycle of the input data stream.

[26]. Transistors Q_4 and Q_5 are used to regenerate the logic-level data to obtain well-shaped pulses with a 0-V offset to minimize drift. This duty-cycle reference, which is independent of power variations, is compared with the power level at A_1. Thus only mean-power changes are acted upon for bias control, and the absence of data will not result in cw operation of the laser at the half-power setpoint as would be the case with Figure 15.9.

For absences of data longer than the time constant of the control circuit, however, the bias returns to its original setup value, which may not reflect the prevailing temperature or age of the laser. (In this sense, the circuit remains duty-cycle-dependent.) When the data reappear, there will be a period equal to the loop time constant in which there may be pattern-dependent effects. The effect of interrupts can be minimized for the circuits of either Figure 15.9 or Figure 15.10 by using long-term memory to hold the last value of bias current when the duty cycle is 50%. More complex control schemes can also be used. For example, the bias can be "dithered" at a low frequency when the input level is 0 so that synchronous detection of the resulting small signal can be used to determine the actual threshold location [27].

It is instructive to examine the operation of simple feedback loops such as these [28]. We can determine how effectively changes in threshold are compensated and how extinction ratio changes, particularly if the slope efficiency of the $L-I$ characteristic is not constant. For simplicity, we will take the combination of A_1 and Q_3 as a current amplifier of gain A in Figure 15.10, and treat all inputs as currents. The bias current I_B is given by

$$I_B = A(-I_M + \beta I_{REF} + I_{DC}) \tag{7}$$

where I_M is the monitor photocurrent, βI_{REF} is the duty cycle β times the peak data-pattern reference current, and I_{DC} is the dc offset that establishes the setup value of bias current. The monitor photocurrent is given in terms of the peak 1 and 0 power levels of the laser as

$$I_M = r\bar{P} = r[\beta P_1 + (1 - \beta)P_0] \tag{8}$$

where r is the net responsivity of the detector, which is the product of the detector's actual responsivity (typically 0.5–1 A/W) times the fraction of the laser power intercepted by the detector.

To relate P_0 and P_1 to the bias and drive currents of the laser, we model the $L-I$ characteristic of the laser as in Figure 15.11. For a bias current $I_B \leq I_{TH}$, P_0 and P_1 are given in terms of the slope efficiencies below and above threshold, s_0 and s_1, respectively, as

$$P_0 = s_0 I_B \tag{9a}$$

and
$$P_1 = s_0 I_{TH} + s_1(I_B + I_D - I_{TH}) \tag{9b}$$

The bias current is found from the solution of Eqs. (7)–(9):

$$I_B = \frac{\beta I_{REF} + I_{DC} + r\beta[I_{TH}(s_1 - s_0) - s_1 I_D]}{1/A + r\beta(s_1 - s_0) + rs_0} \tag{10}$$

which is in the form of the usual solution for an NFB loop, with the final simplification obtained by taking the forward or open-loop gain A very large (actually, the loop gain $A[r\beta(s_1 - s_0) + rs_0] \gg 1$). The resulting expression for I_B can then be examined for its variation with the $L-I$ parameters. For example, the partial derivative with respect to threshold current is

$$\frac{\partial I_B}{\partial I_{TH}} = \frac{\beta(s_1 - s_0)}{\beta(s_1 - s_0) + s_0} = \frac{1}{1 + s_0/\beta(s_1 - s_0)} \tag{11}$$

which we see is unity (perfect compensation) if s_0 is zero. If s_0 is nonzero, however, Eq. (11) is less than unity (compensation lags changes in threshold), with the error being on the order of s_0/s_1. From Eq. (11) we also see that if

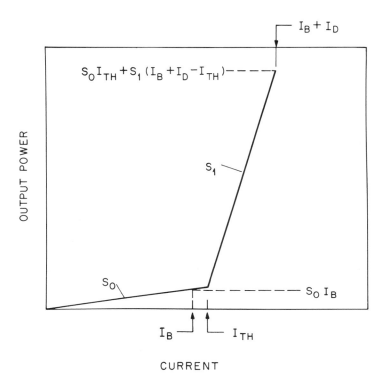

Figure 15.11 $L-I$ characteristic used to model negative feedback bias control.

the duty cycle β goes to zero, then Eq. (11) also goes to 0 (no correction), as noted in the discussion of the circuit of Figure 15.10.

Finally, we examine the extinction ratio γ, the ratio of peak powers in the 1 and 0 states. With Eqs. (10) and (11), this can be solved exactly, leading to tedious expressions. Instead, let us use Eqs. (9a) and (9b) with the result simplified by assuming $I_B \approx I_{TH}$ to get

$$\gamma = \frac{P_1}{P_0} = 1 + \frac{s_1}{s_0}\frac{I_D}{I_{TH}} \qquad (12)$$

from which it is seen that aging- or temperature-induced increases in threshold current or subthreshold spontaneous emission s_0, and/or decreases in lasing slope efficiency all tend to reduce γ. These effects are seen schematically in Figure 15.12, portraying constant average output power and constant drive current. Figure 15.12a depicts the ideal laser ($s_0 = 0$) for which bias under single-loop control precisely tracks changes in threshold and γ remains constant. In Figure 15.12b, τ is degraded and the possibility of a turn-on delay is incurred. In Figure 15.12c, only γ is degraded. Impairments such as

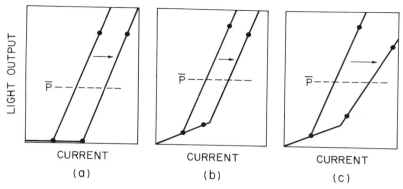

Figure 15.12 Illustration of the causes for poor extinction ratio when (b) threshold rises or (c) slope efficiency decreases in nonideal laser, using single-loop regulation of average power. Ideal case (no spontaneous emission) is shown in (a). Bullets indicate 0 and 1 operating points.

these have been observed in gain-guided lasers, which tend to develop higher levels of subthreshold emission P_0 than index-guided structures [29].

From this discussion, it is clear that both bias and drive currents must be varied to compensate for laser changes if significant spontaneous emission occurs near threshold.

More complex "dual-loop" feedback circuits monitor the 1 and 0 light levels or the 1 and average-power levels to adjust bias and drive under NFB control. Since the 0 level is small in most cases, it is usually desirable to deduce the 0 level from the latter approach. For example, from Eq. (8) we find P_0 to be, given $\bar{P}$ and P_1,

$$P_0 = \frac{\bar{P} - \beta P_1}{1 - \beta} \tag{13}$$

Again, since P_0 is usually small, the numerator of Eq. (13) is seen to be the difference between two large quantities. Equation (13) is also duty-cycle-dependent. Therefore determination of the 0 level has some of the problems associated with measuring it directly.

The approach shown in Figure 15.13 maintains the average and peak power constant without explicitly determining the 0 level or placing conditions on the duty cycle [30]. The duty-cycle reference voltage V_{REF} supplied to the control amplifiers A_1 and A_2 is derived as in Figure 15.10. Amplifier A_2 acts with a peak detector (A_3) to compensate for changes in slope efficiency only, regulating the 1 level as follows: Since the peak detector is ac coupled to the photodiode, its dc (zero) reference is the average light level and its peak-to-peak input is P_1. Duty-cycle variations are compensated at A_2 using V_{REF}, so amplifier A_2 acts only on P_1, regulating it through the pulse drive current determined by the base bias A_2 supplies to Q_3. It is clear that once P_1 is fixed, regulation of the average power by A_1 as described

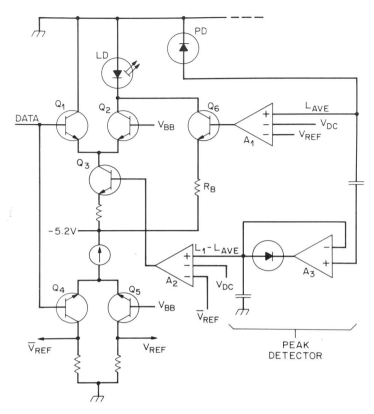

Figure 15.13 Dual-loop bias regulator that is independent of duty cycle. (Reprinted with permission of AT & T Bell Laboratories.)

earlier results in bias corrections that compensate only for changes in P_0, and hence threshold.

We have discussed the control aspect of laser bias, but often associated with it are two other requirements—transient protection and alarms. Lasers are destroyed by currents that result in optical power-density levels from the order of a few megawatts per square centimeter (AlGaAs) to a few tens of megawatts per square centimeter (InGaAsP) but can be degraded in more subtle ways by smaller transients. Therefore, transient protection prevents the laser bias (or drive) from responding to the transmitter power being turned on or off or, more important, to the removal or insertion of a transmitter assembly into a powered-up equipment frame. Protection is often accomplished by preventing bias and/or drive current from being applied until the transmitter power rails have become stable, as determined by R–C time constants. When more than one power rail is used, it is prudent to consider carefully all turn-on/turn-off combinations (including discharge paths of capacitors) that might induce an unintentional current through the laser.

Alarm circuitry alerts the user that the optical power is outside preset limits, the input data pattern has disappeared, or the laser bias current has reached some value that corresponds to end of life. Power monitoring can readily be derived from the photodiode, including a real-time output pattern monitor if desired. Data pattern and laser bias are easily monitored by examining the duty-cycle monitor and by measuring the voltage across R_B in Figures 15.9 and 15.10.

15.3.3 Analog

For linear operation, control circuitry such as that shown in Figure 15.9 can be used to stabilize the bias in the lasing region of the $L–I$ curve, usually at a point around which the laser exhibits maximum linearity. The time constant should be selected to be less than the lowest-frequency component in the modulating signal.

There are two means for minimizing the effects of laser and/or circuit nonlinearities: negative feedback and predistortion. Negative feedback as illustrated in Figure 15.14 is used in the conventional sense, where the input electrical signal is compared with the photocurrent derived from sampling the laser's output using a photodetector. Since photodetectors have excellent linearity (distortion products 60 dB or more below the fundamental [31]), nonlinearities above this level in both the laser and drive circuit can be corrected.

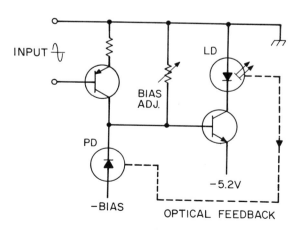

Figure 15.14 Negative-feedback linearization of a laser.

Negative feedback linearization has been applied effectively with lasers at video frequencies (e.g., <20 MHz) [17, 32]. At higher frequencies, however, NFB is limited by the large gain–bandwidth (GBW) product required and complex stability considerations. These, then, are the fundamental problems with NFB for most bandwidths currently of interest.

The need for high GBWs can be circumvented by predictively predistorting the modulating current [33]. For a given device, if its $L–I$ characteristic is stable, predistortion or preemphasis can be added using a nonlinear circuit between the input signal and the laser. Predistortion that results in a linearized output is adjusted during transmitter assembly while monitoring of distortion products. For nonlinearities that change with time (e.g., are induced by aging or temperature), predistortion has limited effectiveness because it ordinarily would not track these changes. (Active control of the laser temperature can minimize this problem. An adaptive predistortion technique has also been reported [34].) In the case of preemphasis, however, a nonlinearity such as reduced modulation index at low frequencies (to compensate for junction-heating nonlinearities) is introduced at the transmitter [35]. The preemphasis is not trimmed for each chip but corrects for a generic nonlinearity and thus is not sensitive to aging effects. The inverse transfer function is applied at the receiver.

In view of the difficulty of linearization schemes for lasers, particularly for high-frequency applications, it is best to start with either a laser designed for adequate linearity (e.g., DFB designs) and carefully suppress reflections or to use a subcarrier approach as described earlier.

15.4 LASER/SUBSYSTEM PACKAGING

The basic requirements of a laser package are to provide

1. A low-thermal-impedance attachment for the laser
2. A stable laser-to-fiber optical interface
3. Stable back-mirror power monitoring
4. A controlled (possibly hermetic) environment

Depending on the application, the package may also be required to provide

5. A thermoelectric device for active laser cooling
6. High-speed (GHz) electrical connections
7. Electrical matching components and/or active circuitry
8. Other optical devices such as taps, isolators, or modulators
9. Means for laser linewidth control such as external cavities

The laser–fiber interface is very critical, particularly for single-mode fibers. Degradation of this interface (e.g., via fiber motion relative to the laser) can result in transmitter failure through loss of output power. Hence, much effort has been directed to designing stable interfaces while at the same time attaining maximum coupling efficiency.

The radiation patterns of injection lasers are directional compared with other sources such as LEDs. Half-power (−3 dB) angles of 30° or less are typical of index-guided devices, compared with 120° for Lambertian sources. In addition, the near-field dimensions of lasers (mode-field diameters or MFDs) and the corresponding areas are usually much smaller than the area of the fiber core. For example, the MFDs of 1.3-μm single-transverse-mode lasers and single-mode fibers are the order of 2 μm and 10 μm, respectively. Therefore, a lens can further reduce the divergence angle by magnifying the laser's MFD to match the fiber, as illustrated in Figure 15.15. This conserves brightness [power/(area × solid angle) = constant] [36, 37] while increasing the coupling efficiency. (As also indicated in the figure, when the area of the source *exceeds* the area of the core, an increase in coupling efficiency is not possible without violating conservation of brightness. The additional area focused into the fiber MFD is accomplished by increasing its solid angle, which is not guided by the fiber.) The details of coupling sources to fibers have been treated in detail elsewhere [38, 39].

Numerous lensing schemes have been described for coupling to single- and multimode fibers. The lenses can either be formed on the fiber end itself [40–45] or placed between the laser and fiber [46–49]. The latter approach

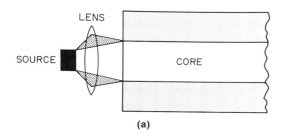

(a)

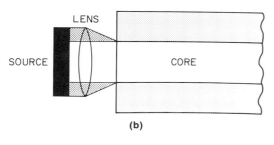

(b)

Figure 15.15 (a) Lens used to increase the coupling efficiency between a laser and fiber. Shaded area is energy that would lie outside the fiber numerical aperture without the use of a lens. (b) When the source area is larger than the fiber core, the energy brought into the core dimensions with a lens is outside the fiber NA and not coupled.

is often used when a hermetic interface is placed between the laser and fiber as part of the packaging strategy.

Even without a lens, the alignment between fiber and laser is critical. Displacements in the transverse directions (parallel to the plane of the laser mirror) of ± 5–10 μm and ± 1–1.5 μm are typical for 1-dB loss in coupling for multimode and single-mode fibers, respectively. Axial tolerances are substantially larger. With lensing integral to the fiber, the 1-dB transverse displacements are typically reduced by a factor of 2 or more. For separate lenses, there always remains one critical interface, such as between the laser and the first lens. However, it is possible to design the assembly sequence so that the last-adjusted interface has reduced alignment sensitivity (loose tolerances) because it takes place where the beam is expanded. Figure 15.16 shows schematically butt, integral lens, and confocal lens coupling schemes. It has been shown that if two or three lenses are arranged confocally between the laser and fiber, alignment tolerances can be increased significantly [47, 49]. The trade-offs between coupling efficiency and tolerances have been addressed either theoretically or empirically in several of the references cited.

For cases where the tolerances are several micrometers, such as multimode fiber or single-mode multiple-lens schemes, alignment and locking (securing the aligned fiber relative to the laser) can be accomplished with precise but straightforward operations, using fixtures with ordinary micrometers and filled epoxies or solders [50]. A photomicrograph of a typical epoxy

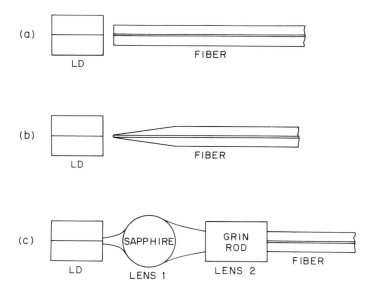

Figure 15.16 Schematic illustrations of (a) butt coupling, (b) integral lens, and (c) confocal lens coupling of laser to fiber.

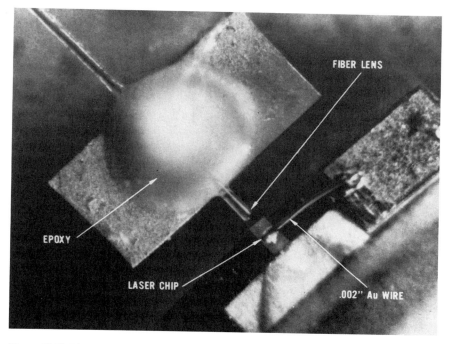

Figure 15.17 Photograph of fiber locked in place near laser with epoxy.

lock is shown in Figure 15.17. The fiber is positioned in three dimensions for maximum coupling efficiency, and epoxy is deposited over the fiber away from the laser and cured [26].

For single-mode packages using lenses integral to the fiber, alignment fixtures ensuring smooth, backlash-free positioning at submicrometer displacements are much more complex, and vibration isolation is usually required. Even more critical is locking the aligned fiber. Where submicrometer dimensional stability is required, thermal expansions, shrinking of either epoxy or solder, and reflection-induced changes in output power can result in significant assembly problems.

In view of the sensitivity of noise and spectral characteristics to reflections as discussed earlier, lenses or windows are sometimes antireflection coated, and fiber ends or monitor photodiodes are beveled or tilted. Different fiber lensing schemes have been observed to result in different working distances, with reflection problems minimized by use of the longer distances derived by using a tapered fiber end [51] or a short piece of lensed multimode fiber fused to a single-mode fiber [52].

Several single- and multimode packages currently in production use rectangular microwave-HIC packages with either pins or butterfly leads. Figures

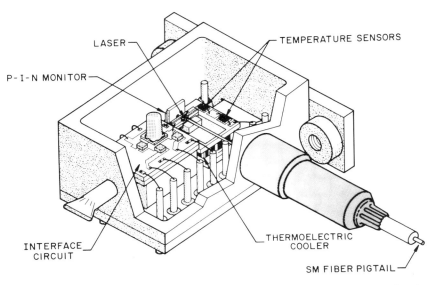

Figure 15.18 Single-mode laser package with an internal thermoelectric cooler. (Reprinted with permission of AT & T Bell Laboratories.)

15.18 and 15.19 show two such packages [53, 54]. Both contain the laser, the monitor, an interface circuit (Fig. 15.18) or the transmitter circuit (Fig. 15.19), an integral thermoelectric cooling device, and a temperature sensor for regulation of the cooler. The integrally lensed fiber is aligned and attached within 30 μm of the laser facet, the distance depending on the specific lensing scheme. The fiber exits the package via a hermetic feedthrough.

Axially symmetric package designs are often used when discrete lenses separate the laser–fiber interface; the laser diode is sealed hermetically in a windowed subassembly along with the first lens [55–58]. The fiber and second lens are assembled, aligned, and locked to the subassembly outside the hermetic seal. The axial geometry reduces sensitivity to differential thermal expansions. Figure 15.20 illustrates one of these concentric package designs. Due to the presence of several optical interfaces, attention must be given to antireflection coating. Thermoelectric cooling, if used with these concentric packages, is usually applied externally.

At bit rates above approximately 500 Mb/s, the reactive components associated with the laser package become important. Packaged lasers can be modeled to understand their operation at high speeds and the effects of parasitics [1, 59]. Location of the driver circuit adjacent to the laser chip, if possible, minimizes parasitics. If this cannot be accomplished inside the laser package, extreme care must be used to minimize lead inductance when bringing the laser connection through the package and interconnecting it to

Figure 15.19 Single-mode laser package with an internal thermoelectric cooler and high-speed drive circuitry. (Reprinted with permission of Rockwell International Corp., Collins Transmission Systems.)

the driver. Integration of the driver circuit, or the driver circuit and laser (optoelectronic integration), leads potentially to highest-speed operation, reduced power dissipation, and improved reliability [60, 61]. Device models of the circuit and laser that are accurate at high speeds are important so

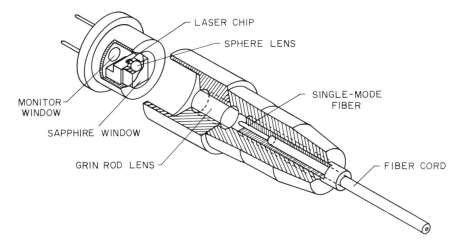

Figure 15.20 Concentric single-mode laser package in which the hermetically sealed diode is coupled to the fiber using confocal lenses. (Reprinted with permission of Nippon Telephone & Telegraph Corp.)

that compensation can be designed, if necessary, to attain speeds limited only by the fundamental parameters of the devices themselves.

15.5 THERMOELECTRIC COOLING

Table 15.2 shows approximate sustained and worst-case high-temperature ambients for several applications in telephony. Also shown are corresponding temperatures expected at device package sites (allowing for the thermal path from the ambient to the package, often the order of $10°/W$), and representative failure-rate objectives in FITs.* Sustained temperatures are prin-

---

* Lifetimes are usually characterized in terms of mean time to failure (MTTF). Other measures of reliability are failure rates expressed in fraction/unit time (usually %/khr) or FITs (failures per billion hours). The advantage of these units is that they can be summed with the rates for other components to approximate the total failure rate of a system. An approximate relationship between these units is FITs $\approx 1 \times 10^4 \times (\%/\text{khr}) \approx 1 \times 10^9/\text{MTTF}$, so that 1-million-hour devices can also be characterized as having a failure rate of 0.1%/khr or 1000 FITs. It is this approximation that is used here, for simplicity. The relationship is approximate since semiconductor devices display a log-normal failure distribution and the failure rate depends on the standard deviation σ of the distribution. A small value of σ can modify the failure rate significantly, reducing it from the value just given. In the extreme, σ near zero means that the probability of failure is negligible until near the median life, at which time all failures would take place. This is contrary to popular failure models such as the random or exponential, where the probability of failure is the same at all times. Therefore it is important to understand the log-normal model, its applications and implications [66].

**TABLE 15.2. System and optoelectronic device
temperatures in telephony applications***

	Long-term	Short-term	LD failure rate
Central office	38°C	49°	—
O-E devices	50°C	60°C	< 10,000 FITs
Loop plant	50°C	65°C	—
O-E devices	70°C	85°C	1000 FITs
Submarine	20°C	20°C	—
O-E devices	30°C	30°C	300–700 FITs

* The difference arises from the thermal impedance between
the "ambient" and the site of the device. The junction temperature, when relevant, is higher by yet an additional 10–20°. Order-
of-magnitude failure-rate objectives for the laser are also given
in Refs. 62–65.

cipally of importance for determining device failure rates whereas short-term
temperatures must be considered in determining whether the system will
function properly during these extremes. System reliability depends on both
aspects.

Extrapolations of accelerated-aging experiments indicate that for many
lasers failure rates at actual operating temperatures will be higher than required for some of these applications. It is commonly accepted that the degradation of semiconductor devices varies with temperature according to an
Arrhenius relationship:

$$\text{Acceleration factor} = \exp\left[\frac{E_a}{k}\left(\frac{1}{T'} - \frac{1}{T''}\right)\right] \quad (14)$$

where E_a is an activation energy determined from accelerated-aging data
(usually in the range 0.4–0.9 eV) and k is Boltzmann's constant (8.625 ×
10^{-5} eV/K). In working with lasers, the laser heat sink or stud temperature
is commonly used with Eq. (14), whereas with silicon devices the junction
temperature is more often used.

An adequate lifetime at elevated temperatures can be obtained, however,
by using thermoelectric devices to reduce the laser temperature. Peltier, or
thermoelectric, coolers (hereafter TECs) are arrays of bulk n- and p-type
semiconductor material connected electrically in series and thermally in
parallel, as in Figure 15.21. Current I_{TEC} passed through the device results
in a temperature differential across the TEC, leading to a pumping of heat
Q_C from the cold side to the hot side. The heat rejected at the hot side, Q_H,
is the sum of Q_C plus the input power to the TEC ($I_{TEC}V$ or I_{TEC}^2R, where
V is the voltage developed across the series-connected elements, which have
a total resistance R). Thus the device acts like a Carnot refrigerator with a

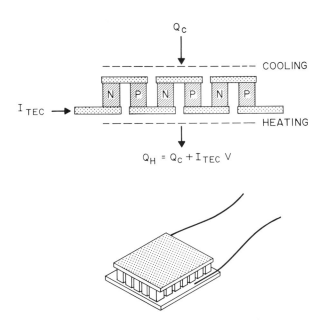

$$Q_C$$

COOLING

$$I_{TEC} \longrightarrow$$

HEATING

$$Q_H = Q_C + I_{TEC} V$$

Figure 15.21 Arrangement of semiconductor elements in a thermoelectric cooler.

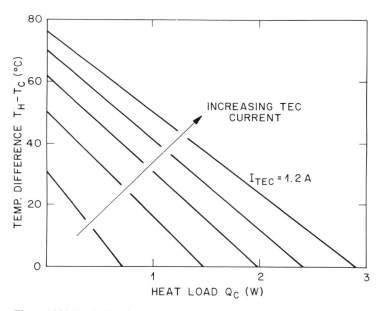

INCREASING TEC
CURRENT

$$I_{TEC} = 1.2\ A$$

Figure 15.22 Typical performance data describing the ΔT attainable for a given heat load and cooler input current.

coefficient of performance given by

$$E = -\frac{\text{heat removed}}{\text{work input}} = -\frac{Q_c}{I_{\text{TEC}}V} \qquad (15)$$

For design purposes, cooling parameters can be represented in many ways, and this information is available from TEC manufacturers. Figure 15.22 shows a typical representation of this type relating parameters of interest for laser cooling: heat load and temperature differential versus cooler current.

The principle of operation of a TEC is illustrated in Figure 15.23. For the n-type material, as thermalized electrons in the metal move into or out of the conduction band, heat on the order of kT is first absorbed and then rejected. An analogous process takes place for the metal–p-type junctions. Semiconductors rather than metals are used for the thermoelectric elements because they have higher Peltier coefficients π (related to thermoelectric power α through absolute temperature T: $\pi = \alpha T$, the so-called first Kelvin relation), thus developing larger temperature differentials in response to current flow. They also can be designed to have low thermal conductivity K if semiconductors are compounded of heavy elements such as bismuth, tellurium, or antimony. Low thermal conductivity is required to prevent the individual elements from thermally "shorting" themselves. On the other hand, high electrical conductivity σ is also desirable to minimize Joule self-heating. These different, usually competing requirements on the material properties are collected in a "figure of merit" $Z = \alpha^2\sigma/K$, which is to be maximized for best TEC operation [67].

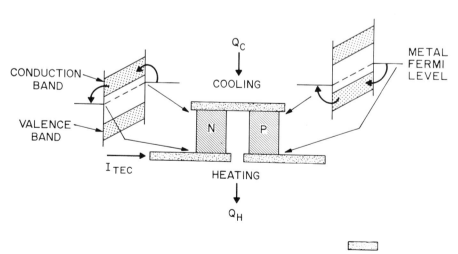

Figure 15.23 Simplified band diagram showing heating and cooling effects at the hot and cold sides of TEC.

To return to the issue of laser reliability, suppose a laser is believed to have a useful life of 250,000 hr at 25°C, with degradation characterized by an activation energy of 0.7 eV. If a 100,000-hr lifetime is to be assured, the highest sustained temperature is calculated from Eq. (14) to be about 35.4°C. Therefore, to use this device in a 50°C central office ambient, a TEC providing at least 14.6° cooling, or ΔT, should be used. Since 15° is relatively easy to obtain, the designer would probably choose to cool the laser further, perhaps to 25°C, to ensure margin and/or to ensure proper operation during short-term temperature excursions. To obtain 25°C at the laser, the ΔT across the TEC would exceed $>25°$ since the power dissipated by the TEC would raise the 50°C package-site ambient. The overall thermal problem must be modeled, so that the temperatures, currents, and size of the TEC (number of elements, cross sections, etc.) can be optimized for the application [67, 68].

Many lasers currently require thermoelectric cooling to meet central office objectives. To attain 1000 FITs (1×10^6 hr) for loop applications, these same lasers would require cooling to below 25°C. For the example in the last paragraph, the laser should be cooled to about 10°C. Taking the thermal resistance between the ambient and package site as 10°/W, and the IV input to the TEC as 1 W, we quickly discover that TEC might be required to provide a ΔT of 65° [$T_{hot} - T_{cold} = 65°C + 1$ W $(10°/W) - 10°C$]. This equals or exceeds the capability of single-stage TECs operated near 300 K, thus raising doubts that such a laser could be applied in a loop application with the requirements given in Table 15.2. Obviously, lasers with a longer lifetime would be needed.

Reliability requirements for submarine applications are met (1) through having low sustained ambient temperatures and (2) by highly selective screening of lasers having the most stable characteristics determined from results of severe burn-in or "purging" procedures [69, 70]. Thermoelectric cooling is not generally considered.

Thermoelectric cooling can be used as part of the feedback control scheme. Laser degradation, characterized by an increase in threshold, can be compensated by decreasing the laser temperature according to Eq. (1). Closed-loop control of the TEC can be performed by monitoring light output as discussed earlier for control of bias.

This scheme has limited use with AlGaAs lasers, all of which have high T_0 and many of which have high thresholds arising from gain-guided designs. Large amounts of cooling would be required for long-term compensation: ΔT's of as much as 70°C to compensate for a 50% rise in threshold, a typical end-of-life criterion. For InGaAsP lasers, however, more modest ΔT's of about 25°C will accomplish the same compensation. (It should be kept in mind that these ΔT's must be in addition to any cooling needed to achieve adequate reliability, although reliability may be enhanced by the very operation of closed-loop temperature control.)

Additional benefits that arise from active cooling are stabilization of differential quantum efficiency (DQE) and wavelength. For example, the variation of DQE with temperature (Fig. 15.1) can be circumvented by keeping the laser temperature constant. Then single-loop feedback maintains constant extinction ratio and turn-on characteristics, as discussed previously. The emission wavelengths of $1.3-1.55$-μm Fabry-Perot lasers change approximately $4-5$ nm/$°$C. If wavelength is stablized by active cooling, dispersion penalties can be reduced for multi-longitudinal-mode 1.3-μm lasers used in long-distance applications or for 1.5-μm lasers used with dispersion-shifted fiber, which equates to larger repeater spacings.

Finally, temperature control is mandatory in coherent applications due to the large change in frequency with temperature (tens of gigahertz per degree, depending on wavelength and laser design).

There are caveats to active cooling, however. Since a TEC is a heat pump, it rejects its own aforementioned IV dissipation. For typical designs with the TEC inside the laser package, thus minimizing parasitic heat loads, approximately 1 W is rejected at the hot side for ΔT's of $40-50°$. For designs using a TEC external to the laser package, higher dissipation is usually required. This places additional burdens on package and system thermal design and on powering requirements.

Another caveat is the reliability of the combined TEC, power supply, power control circuitry, and laser. Properly screened TECs operated free of moisture condensation appear to have failure rates on the order of 500 FITs [71] to 2000 FITs [72] in typical applications. Power supply and control circuitry can contribute on the order of 1000 FITs, depending on specific details. Therefore, the TEC and power converter place a lower limit of perhaps 1000–3000 FITs before the laser itself is considered. Because of the low thermal conductivity of TEC material, the thermal resistance of a nonfunctioning TEC may be very high, on the order of 100$°$/W. As a result, a cooling failure may result in system failure rather than merely high-temperature operation of the laser. Table 15.3 shows representative failure rates for cooled and uncooled lasers of currently prevailing technology.

TABLE 15.3. Representative failure rates for lasers with and without thermoelectric cooling in a 50°C ambient temperature*

	Laser	Laser + TEC	LED
Typical	20,000 FITs	2500 FITs	300 FITs
Best	1000 FITs	1000 FITs	1 FIT

* Power converter failure rate is not included, as redundancy can be used if necessary. LED failure rates under similar ambient conditions are included for reference.

15.6 TRANSMITTER IMPAIRMENTS, SUBSYSTEM RELIABILITY, AND TESTING

Ideally, the output of a digital laser transmitter will be a pattern of rectangular light pulses of constant amplitude and high extinction ratio exactly filling the time slot T. These ideal pulses have rise and fall times much less than T, no over- or undershoot at the edges, and negligible turn-on delay. These characteristics should not depend on the data pattern, including long strings of 1s or 0s, or on temperature ($+2$ to $+60°C$ or -40 to $+85°C$, depending on the application), power-supply voltage (± 5 to $\pm 10\%$, also depending on the application), or age of the components. In fact, departures from any of these ideal characteristics result in penalties in system performance, usually specified in terms of bit error rate, and can thus be expressed as a power penalty (additional power), in decibels, required at the receiver to maintain constant BER performance.

Analog transmitters have many similar requirements, but with specifications on distortion, noise, and frequency response rather than BER.

Output-power degradation can result from either laser or packaging deficiencies. The impact of an increase in threshold or a decrease in slope efficiency was discussed earlier. Another source of laser-related power degradation arises from shifts in the emission pattern that alter the coupling to either the fiber or the feedback monitor. These "beam-steering" effects can also arise from moisture on the laser facet, usually the result of active cooling below the local dew point.

Finally, particularly for single-mode laser packages, creep or differential expansion in the materials serving to align the laser and fiber can lead to large changes in coupled power with either temperature or time. Even for fiber alignment secured with solder, long-term creep effects related to packaging stresses are possible particularly at temperatures much above 25°C. Power stability is first assured by screening procedures applied to the lasers themselves. These may include

1. Examination of the laser's $L-I$ and IV characteristics and their derivatives, often as a function of temperature
2. Examination of the laser's far-field pattern as a function of power level and temperature
3. Burn-in at elevated temperature and current, selection being based on a shift in threshold
4. Burn-in at elevated temperature and constant lasing output power, selection being based on a shift in threshold

These and other screening procedures have been discussed in the literature [69, 70]. Completed packages may then be screened by procedures including

1. Reexamination of the $L-I$ characteristics, especially as a function of tem-

perature since beam-steering effects are enhanced by coupling to the fiber

2. Elevated-temperature burn-in and/or thermal cycling

Given a stable package and an adequate feedback control circuit for power control, output power is in fact constant, and end of life is determined by criteria applied to changes in the laser itself, usually an increase in threshold (e.g., 50% or 100% increase in bias current, or 20% or 50% increase in bias + drive current). Specific end-of-life criteria vary with the manufacturer and application and are usually determined as part of a process to ensure that the *overall system* meets service-continuity objectives. Long-term aging of completed packages and transmitters at several temperatures, supplemented by field data, provide information for assigning failure rates to each possible mechanism.

Power stability of transmitters is generally analyzed using either pseudo-random or 1010 input data patterns and is tested over the anticipated temperature and power-supply ranges, including allowable shifts in input logic levels. Most other tests require examination of the light output "eye" (Fig. 15.24) as sensed by a fast photodetector and oscilloscope using a pseudo-

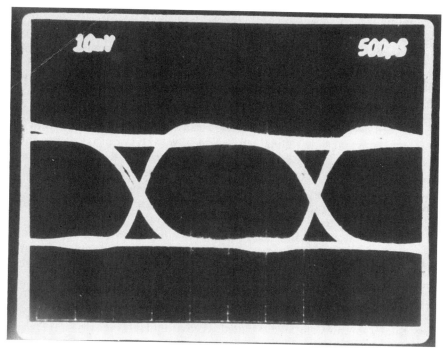

Figure 15.24 A 400-Mb/s transmitter eye pattern. Rise time ≈ 500 ps. fall time ≈ 800 ps, jitter ≈ 100 ps, overshoot ≈ 10%.

random bit input pattern. This eye, which ideally should be rectangular, is obtained by synchronizing the oscilloscope to the basic clock rate and is a superposition of all possible sequences of 1s and 0s (allowed by the generator). Such a sequence, possibly with added insertions of long blocks of 1s or 0s, induces and displays any pattern-dependent effects that may arise from the circuit or device, in addition to pattern-independent effects such as rise time and overshoot. In combination with a receiver and a bit-error-rate test set (BERTS), detailed examination of output degradations can be investigated, as even subtle changes in the eye pattern may affect the error rate.

An important laser parameter is the spectral width of its emission, usually given as the full width at half-maximum amplitude, or FWHM. By observing the output on an optical spectrum analyzer, FWHM may be estimated from the −3-dB points, calculated from −15-dB or −20-dB points (with the assumption that the distribution is, for example, Gaussian), or obtained in a statistical manner from the squares of the amplitude-weighted wavelengths of the longitudinal modes about the mean value. A typical objective for FWHM is 4–6 nm, determined from an analysis of dispersion penalties permitted for the intended bit rate and length of fiber. As was mentioned earlier, the spectrum often becomes wider under modulation. Therefore, FWHM may sometimes be specified under modulated conditions at the intended bit rate. For systems operating in the gigabit-per-second realm, however, single-longitudinal-mode lasers become necessary for long fiber lengths.

There are degradations that are observed only after transmission through fiber. As discussed earlier, chromatic dispersion produces significant eye degradation at the receiver when modal-partition noise, chirp, or reflection effects are present. These degradations can be quantized in terms of power penalty when the BER versus received power function of the receiver in response to a "perfect" input signal is known. In some cases, modal noise can be observed at the transmitter by using a spectrometer to monitor the output of only one longitudinal mode. Excessively noisy output, as measured using a high-speed detector and spectrum analyzer of RF voltmeter, indicates partitioning effects. Reflection effects are often characterized by substantial shifts in the power distribution among the longitudinal modes as the coupling or temperature is changed, and are observed using a scanning optical spectrometer.

A related degradation occurs when speckle mode patterns form in multimode fiber, particularly one driven by a nearly monochromatic laser source. Changes in the laser temperature or the fiber position, for example, can induce changes in the fiber mode pattern. If a mode-selective loss, such as a connector or splice, lies in the fiber path, then "noisy" changes in received power are observed that result in burst errors. In service, such degradations are usually intermittent, often related to environmental changes.

Turn-on jitter is important if it becomes a substantial fraction of a time slot (e.g., >20% of time slot) because it reduces the eye opening at the receiver and because jitter in a regenerated system is cumulative. The eye is reduced in amplitude because displacement in time of an equalized 1 results in a mark that has less amplitude at the decision point of the bit interval.

A BERTS used as a sensor can be a powerful tool for examining subtle degradations. When operating at a low but measurable bit error rate, such as 1×10^{-6}, small changes in output characteristics may induce fractional-decibel changes in receiver sensitivity that quickly show up in the bit error rate. Combinations of two parameters relevant to transmitter performance, such as the positive and negative power-supply voltages, may be examined by plotting a locus of constant bit error rate as a function of the two parameters, one the ordinate and the other the abscissa. The resulting figure, sometimes called a "shmoo" after a cartoon character of somewhat similar shape, indicates the region within which proper operation is obtained. Sometimes irregularities in a section of a shmoo plot can reveal circuit or device problems; that is, a region of marginal operation that might otherwise escape detection. Under computer control, shmoo plotting is simplified significantly.

Analog transmitter measurements usually include frequency response and two-tone intermodulation measurements, the techniques for which are well known and found widely in manufacturers' application notes. These techniques will not be discussed here. The measurement of relative intensity noise is also straightforward using a high-speed detector and spectrum analyzer or RF voltmeter. Other performance measurements for analog transmitters are similar to those for digital transmitters.

REFERENCES

1. R. S. Tucker, "High-speed modulation of semiconductor lasers," *J. Lightwave Technol.*, LT-3: 1180–1192 (1985).
2. P. L. Liu and K. Ogawa, "Statistical measurements as a way to study mode partition in injection lasers," *J. Lightwave Technol.*, LT-2: 44–48 (1984).
2a. M. M. Choy, P. L. Liu, P. W. Shumate, T. P. Lee, and S. Tsuji, "Measurements of dynamic photon fluctuations in a directly modulated 1.5 μm InGaAsP distributed feedback laser," *Appl. Phys. Lett.*, 47: 448–450 (1985).
3. Y. Okano, K. Nakagawa, and T. Ito, "Laser partition noise evaluation for optical fiber transmission," *IEEE Trans. Commun.*, COM-28: 238–243 (1980).
4. C. Lin and F. Mengel, "Reduction of frequency chirping and dynamic linewidth in high-speed directly modulated semiconductor lasers by injection locking," *Electron. Lett.*, 20: 1073–1075 (1984).
5. F. Kappeler, "Dynamic single-frequency low-chirp operation of a laterally coupled waveguide (LCW) laser," *Electron. Lett.*, 20: 1040–1041 (1984).
6. G. H. B. Thompson, *Physics of Semiconductor Laser Devices*, Wiley, New York, 1980, Chapter 7.
7. S. D. Personick, "Receiver design for digital fiber optic communication systems, II," *Bell Syst. Tech. J.*, 52: 875–886 (1973).
8. M. Yamaguchi, K. Emura, M. Kitamura, I. Mito, and K. Kobayashi, "Frequency chirping suppression by a distributed feedback laser diode with a monolithically integrated loss modulator," *Conf. Opt. Fiber Commun.*, San Diego, February 1985, paper WI-3.

9. R. V. Schmidt, in *Integrated Optics: Physics and Applications*, S. Martellucci and A. N. Chester (Eds.), Plenum, New York, 1983, pp. 181–210.

10. R. C. Alferness, C. H. Joyner, L. L. Buhl, and S. K. Korotky, "High-speed traveling-wave directional coupler switch modulator for $\lambda = 1.32 \, \mu\text{m}$," *IEEE J. Quantum Electron.*, **QE-9**: 1339–1341 (1983).

10a. R. C. Alferness, "Guided-wave devices for optical communication," *IEEE J. Quantum Electron.*, **QE-17**: 946–959 (1981).

11. K. Sato, "Intensity noise of semiconductor laser diodes in fiber optic analog video transmission," *IEEE J. Quantum Electron.*, **QE-19**: 1380–1391 (1983).

12. Y. Yamamoto, "AM and FM quantum noise in semiconductor lasers—Part I: theoretical analysis," and "Part II: Comparison of theoretical and experimental results for AlGaAs lasers," *IEEE J. Quantum Electron.*, **QE-19**: 34–58 (1983).

13. N. Chinone, K. Takahashi, T. Kajimura, and M. Ojima, "Low-frequency mode-hopping noise of index-guided semiconductor lasers," 8th IEEE Int. Semicond. Laser Conf., Ottawa, September 1982, paper 25.

13a. N. Chinone, T. Kuroda, T. Ohtoshi, T. Takahashi, and T. Kajimura, "Mode-hopping noise in index-guided semiconductor lasers and its reduction by saturable absorbers," *IEEE J. Quantum Electron.*, **QE-21**: 1264–1269 (1985).

14. H. Melchoir, H. Jaeckel, and R. Welter, *Integrated Optics—Physics and Applications*, Plenum, New York, 1981, pp. 383–393.

15. K. Y. Lau and A. Yariv, "Ultra-high speed semiconductor lasers, " *IEEE J. Quantum Electron.*, **QE-21**: 121–138 (1985).

16. W. I. Way, M. M. Choy, P. L. Liu, and K. Kobayashi, "Gigahertz linearity under large-signal intensity modulation and noise characteristics of DFB-DC-PBH lasers," Conf. Lasers and Electro-Optics, Baltimore, May 1985, paper ThZ-4.

17. K. Sato, S. Tsuyuki, and S. Miyanaga, "Fiber optic analog transmission techniques for broadband signals," *Rev. Electron. Commun. Lab.*, **32**: 586–597 (1984).

17a. K. Sato and K. Asatani, "Fiber optic analog video transmission using semiconductor laser diodes," in *Optical Devices and Fibers—1982*, Y. Suematsu (Ed.), North-Holland, New York, 1982, pp. 351–368.

18. K. E. Stubkjaer and M. B. Small, "Noise properties of semiconductor lasers due to optical feedback," *IEEE J. Quantum Electron.*, **QE-20**: 472–478 (1984).

19. A. Ohishi, N. Chinone, M. Ojima, and A. Arimoto, "Noise characteristics of high-frequency superposed laser diodes for optical disk systems," *Electron. Lett.*, **20**: 821–822 (1984).

20. M. Schwartz, *Information Transmission, Modulation and Noise*, McGraw-Hill, New York, 1980, Chapters 4 and 5.

21. M. Uhle, "The influence of source impedance on the electrooptical switching behavior of LEDs," *IEEE Trans. Electron Dev.*, **ED-23**: 438–441 (1976).

22. R. Olshnasky and D. Fye, "Reduction of dynamic linewidth in single-frequency semiconductor lasers," *Electron. Lett.*, **20**: 928–929 (1984).

23. L. Bickers and L. D. Westbrook, "Reduction of laser chirp in 1.5 μm DFB lasers by modulation pulse shaping," *Electron. Lett.*, **21**: 103–104 (1985).

24. K. Kaede, I. Mito, M. Yamaguchi, M. Kitamura, R. Ishikawa, R. Lang, and K. Kobayashi, "Spectral chirping suppression by compensating current in modified DFB-DC-PBH LDs," Conf. Opt. Fiber Commun., San Diego, CA, February 1985, paper WI-4.

25. M. A. Karr, F. S. Chen, and P. W. Shumate, "Output power stability of GaAlAs laser transmitter using an optical tap for feedback control," *Appl. Opt.*, **18**: 1262–1265 (1979).

26. P. W. Shumate, Jr., F. S. Chen, and P. W. Dorman, "GaAlAs laser transmitter for light-wave transmission systems," *Bell Syst. Tech. J.*, **57**: 1823–1836 (1978).

27. D. W. Smith and M. R. Matthews, "Laser transmitter design for optical fiber systems," *IEEE J. Selected Areas Commun.*, **SAC-1**: 515–523 (1983).

28. R. G. Schwartz and B. A. Wooley, "Stabilized biasing of semiconductor lasers," *Bell Syst. Tech. J.*, **62**: 1923–1936 (1983).

29. M. Dean and B. A. Dixon, "Aging of the light-current characteristics of proton-bombarded AlGaAs lasers operated at 30°C in pulsed conditions," 3rd Int. Conf. Integrated Opt. Opt. Fiber Commum., San Francisco, April 1981, paper MJ7.

30. F. S. Chen, "Simultaneous feedback control of bias and modulation currents for injection lasers," *Electron. Lett.*, **16**: 7–8 (1980).

31. T. Ozeki and E. H. Hara, "Measurement of nonlinear distortion in photodiodes," *Electron. Lett.*, **12**: 80–81 (1976).

32. M. Nakamura, T. Tamura, and T. Ozeki, "Elimination of laser noise and distortion induced by reflected waves," 3rd Int. Conf. Integrated Opt. Opt. Fiber Commun., San Francisco, April 1981, paper MJ2.

33. K. Asatani and T. Kimura, "Linearization of LED nonlinearity by predistortions," *IEEE J. Solid State Circuits*, **SC-13**: 133–138 (1978).

34. M. Bertelsmeier and W. Zschunke, "Linearization of light-emitting and laser diodes for analog broadband applications by adaptive predistortion," 4th Conf. Integrated Opt. Opt. Fiber Commun., Tokyo, June 1983, paper 30C2-3.

35. K. Asatani, "Nonlinearity and its compensation of semiconductor laser diodes for analog intensity modulation systems," *IEEE Trans. Commun.*, **COM-28**: 297–300 (1980).

36. M. Born and E. Wolf, *Principles of Optics*, 6th ed., Pergamon, Oxford, 1980, Sec 4.8.3.

37. M. C. Hudson, "Calculation of the maximum optical coupling efficiency into multimode optical waveguides," *Appl. Opt.*, **13**: 1029–1033 (1974).

38. M. K. Barnowski, "Coupling components for optical fiber waveguides," in *Fundamentals of Optical Fiber Communications*, M. K. Barnowski (Ed.), Academic, New York, 1976, Chapter 2.

39. W. B. Joyce and B. C. DeLoach, "Alignment of Gaussian beams," *Appl. Opt.*, **23**: 4187–4196 (1984).

40. H. Kuwahara, M. Sasaki, and N. Tokoyo, "Efficient coupling from semiconductor lasers into single-mode fibers with tapered hemispherical ends," *Appl. Opt.*, **19**: 2578–2583 (1980).

41. J. Sakai and T. Kimura, "Design of a miniature lens for semiconductor laser to single-mode fiber coupling," *IEEE J. Quantum Electron.*, **QE-16**: 1059–1066 (1980).

42. J. Yamada, Y. Murakami, J. Sakai, and T. Kimura, "Characteristics of a hemispherical microlens for coupling between a semiconductor laser and single-mode fiber," *IEEE J. Quantum Electron.*, **QE-16**: 1067–1072 (1980).

43. G. D. Khoe, J. Poulissen, and H. M. deVrieze, "Efficient coupling of laser diodes to tapered monomode fibres with high-index end," *Electron. Lett.*, **19**: 205–207 (1983).

43a. G. D. Khoe, H. G. Kock, D. Kuppers, J. H. F. M. Poulissen, and H. M. deVrieze, "Progress in monomode optical-fiber interconnection devices," *J. Lightwave Technol.*, **LT-2**: 217–227 (1984).

44. H. R. D. Sunak and M. A. Zampronio, "Launching light from semiconductor lasers into multimode optical fibers having hemispherical ends and taper-with-hemisphere ends," *Appl. Opt.*, **22**: 2344–2348 (1983).

45. W. B. Joyce and B. C. DeLoach, "Alignment-tolerant optical-fiber tips for laser transmitters," *J. Lightwave Technol.*, **LT-3**: 755–757 (1985).

46. M. Maeda, I. Ikushima, K. Nagano, M. Tanaka, H. Nakashima, and R. Itoh, "Hybrid laser-to-fiber coupler with a cylindrical lens," *Appl. Opt.*, **16**: 1966–1970 (1977).

47. M. Saruwatari and T. Sugie, "Efficient laser diode to single-mode fiber coupling using a combination of two lenses in confocal condition," *IEEE J. Quantum Electron.*, **QE-17**: 1021–1027 (1981).

48. M. Sumida and K. Takemoto, "Lens coupling of laser diodes to single-mode fibers," *J. Lightwave Technol.*, **LT-2**: 305–311 (1984).

49. K. Kawano, M. Saruwatari, and O. Mitomi, "A new confocal combination lens method for a laser-diode module using a single-mode fiber," *J. Lightwave Technol.*, **LT-3**: 739–745 (1985).

50. W. H. Dufft and I. Camlibel, "A hermetically encapsulated AlGaAs laser diode," *Proc. 30th Electron. Components Conf.*, San Francisco, April 1980, pp. 261–269.

51. G. Wenke and Y. Zhu, "Comparison of efficiency and feedback characteristics of techniques for coupling semiconductor lasers to single-mode fiber," *Appl. Opt.*, **22**: 3837–3844 (1983).

52. K. Mathyssek, R. Keil, and E. Klement, "New coupling arrangement between laser diode and single-mode fiber with high coupling efficiency and particularly low feedback effect," 10th Eur. Conf. Opt. Commun., Stuttgart, September 1984, paper 10A5.

53. W. A. Asous, G. M. Palmer, C. B. Swan, R. E. Scotti, and P. W. Shumate, "The FT4E 432 Mb/s lightwave transmitter," *Pocr. IEEE Int. Commun. Conf.*, Amsterdam, May 1984, pp. 787–789.

54. K. Y. Maxham, C. R. Hogge, S. J. Clandening, C. T. Chen, J. M. Dugan, S. K. Sheem, and D. O. Offutt, "Rockwell 135 Mb/s lightwave system," *J. Lightwave Technol.*, LT-2: 394–402 (1984).

55. J. Minowa, M. Suruwatari, and N. Suzuki, "Optical componentry utilized in field trial of single-mode fiber long-haul transmission," *IEEE J. Quantum Electron.*, **QE-18**: 705–717 (1982).

56. Y. Tachikawa and M. Saruwatari, "Laser diode module for analog video transmission," *Rev. Electron Commun. Lab.*, **32**: 598–607 (1984).

57. H. G. Kock, "Coupler comprising a light source and lens," U. S. Patent 4,355,323, issued Oct. 19, 1982.

58. M. Saruwatari, "Laser diode module for single-mode optical fiber," *Optical Devices and Fibers—1984*, Y. Suematsu (Ed.), North-Holland, New York, 1984, pp. 129–140.

59. B. W. Hakki, F. Bosch, S. Lumish, and N. R. Dietrich, "1.3 μm BH laser performance at microwave frequencies," *J. Lightwave Technol.*, LT-3: 1193–1201 (1985).

60. H. Matsueda and M. Nakamura, "Monolithic integration of a laser diode, photo monitor, and electric circuits on a semi-insulating GaAs substrate," *Appl. Opt.*, **23**: 779–781 (1984).

61. T. Horimatsu, T. Iwama, Y. Oikawa, T. Touge, O. Wada, and T. Nakagami, "400 Mb/s transmission experiment using two monolithic optoelectronic chips," *Electron Lett.*, **21**: 319–321 (1985).

62. K. Nakagawa, "Second-generation trunk transmission technology," *IEEE J. Selected Areas Commun.*, SAC-1: 387–393 (1983).

63. J. W. Olson and A. J. Schepis, "Description and application of the fiber-SLC carrier system," *J. Lightwave Technol.*, LT-2: 317–322 (1984).

64. D. Paul, K. H. Greene, and G. A. Koepf, "Undersa fiber-optic cable communications system of the future: operational, reliability and systems considerations," *J. Lightwave Technol.*, LT-2: 414–425 (1984).

65. H. Fukinuki, T. Ito, M. Aiki, and Y. Hayashi, "The FS-400M submarine system," *J. Lightwave Technol.*, LT-2: 754–760 (1984).

66. A. S. Jordan, "A comprehensive review of the lognormal failure distribution with application to LED reliability," *Microelectron. Reliab.*, **18**: 267–279 (1978).

67. H. J. Goldsmid, *Thermoelectric Refrigeration*, Plenum, New York 1964.

68. W. H. Clingman, "New concepts in thermoelectric device design," *Proc. IRE*, **49**: 1155–1160 (1961).

69. Y. Nakano, G. Iwane, and T. Ikegami, "Screening method for laser diodes with high reliability," *Electron. Lett.*, **20**: 397–398 (1984).

70. *AT & T Technical Journal*, volume 64, March 1985.

71. P. W. Shumate, unpublished results.

72. "Reliability and Failure Modes of Thermoelectric Heat Pumps," Marlow Industries, Inc., Garland, TX.

16

Optical Receivers

Tran Van Muoi

16.1 INTRODUCTION

The optical receiver is a critical element of an optical communication system since it often determines the overall system performance. The function of the optical receiver is to detect the incoming optical power and extract from it the signal (either analog or digital) that is being transmitted. It must achieve this function while satisfying certain system requirements such as desired level of signal-to-noise ratio and bit error rate.

Block diagrams of typical optical receivers for analog and digital transmission systems are shown in Figures 16.1 and 16.2, respectively. It can be

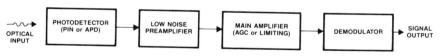

Figure 16.1 Block diagram of an analog optical receiver.

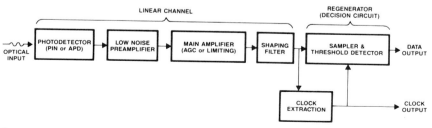

Figure 16.2 Block diagram of a digital optical receiver.

observed that in both cases the incident optical power is detected by a photodetector. The detected signal is then amplified by the amplifier chain, which includes a low-noise preamplifier and a postamplifier (normally with automatic gain control or limiting action). For analog systems, the signal is then passed through a demodulator to recover the transmitted information. For digital systems, the signal is passed through a shaping filter in order to minimize noise and interference (the filter in the analog case is included in the demodulator). A clock extraction circuit is then used to recover the timing clock, which is used to strobe the sampler and threshold detector. The transmitted data are thus regenerated.

The two elements that distinguish the optical receiver from receivers for other transmission media (for example, coaxial cable or microwave) are the photodetector and the low-noise preamplifier. Together they determine most of the receiver performance. The theory and practice of optical receiver design have been well developed and also well documented in the literature [1–15]. In this chapter we discuss the basics of receiver design with emphasis on the performance of the photodiode and preamplifier (hereinafter referred to as amplifier for simplicity). The impacts of and trade-offs between different receiver design requirements will also be highlighted.

The organization of this chapter is as follows. In Section 16.2 the characteristics and performance of photodiodes for optical fiber communication systems are summarized. The receiver design requirements are discussed in Section 16.3. Then in Section 16.4, various receiver–amplifier configurations are considered. The noise analysis and performance of optical receivers are then presented in Section 16.5. Receiver sensitivity and dynamic range considerations follow in Sections 16.6 and 16.7. Design of optical receivers for analog applications is discussed in Section 16.8. Then in Section 16.9, the technologies for fabricating optical receivers including hybridization and monolithic integration are reported. Finally future development trends are indicated in Section 16.10.

16.2 PHOTODETECTORS

The two most popular detectors for optical fiber communications are the PIN photodiode and avalanche photodiode (APD). Their basic properties are summarized in this section. More details on photodiodes as well as other types of photodetectors are presented in Chapter 14.

16.2.1 Structures and Materials

The simplest photodiode structure is a *pn* junction formed in a semiconductor with a bandgap energy less than the photon energy of the optical signal to be detected. Under normal operation the detector is reverse-biased.

Incident photons are absorbed in the depletion region and create electron–hole pairs that drift in opposite directions under the influence of the field in this region. The photocurrent generated in the external circuit is given by

$$I_p = \frac{\eta e P}{hv} = \mathcal{R} P \tag{1}$$

where P is the incident optical power, e is the electron charge (1.6×10^{-19} C), hv is the photon energy, η is the detector quantum efficiency, and $\mathcal{R}$ is the detector responsivity, which is related to the quantum efficiency by

$$\mathcal{R} = \frac{\eta e}{hv} \tag{2}$$

The photodiode response time is determined by the time taken for the generated photoelectrons and holes to reach the device terminal (through drift and diffusion) and the RC time constant. Since diffusion is a slow process, we want to minimize the number of carriers generated in the diffusion region by making the depletion region thick (or wide) enough to absorb most of the incident light. This can be achieved by adding a low-doped intrinsic layer between the pn junction, which results in the p-i-n (or PIN) structure. Two additional advantages of the PIN structure are higher quantum efficiency (carrier loss due to recombination through the slow diffusion process is minimized) and lower device capacitance (due to the wider depletion region).

In avalanche photodiodies (APDs) the reverse bias voltage is increased enough that a single photon-generated electron–hole pair can create additional carriers through an impact ionization process. The total (also referred to as secondary) signal current generated is thus multiplied by an avalanche gain G:

$$I_p = \frac{\eta e G P}{hv} = \mathcal{R} G P \tag{3}$$

Because the avalanche multiplication is a random process, avalanche excess noise is also produced. However, APDs still offer much better receiver sensitivity than PIN detectors.

For high quantum efficiency, the photodiode material must be chosen appropriately for the incident wavelength range. For short-wavelength (0.8–0.9 μm) detection, silicon is mostly (if not universally) used. Very high performance silicon PIN and APD detectors have been fabricated. For the long-wavelength (1.1–1.6 μm) region, the choice of materials at present includes germanium and several group III–V alloys such as InGaAsP and GaAlAsSb. Because of high dark current and avalanche excess noise, long-wavelength APDs are not as high performance as silicon APDs.

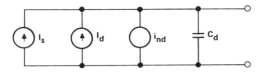

Figure 16.3 Photodiode equivalent circuit.

16.2.2 Photodiode Equivalent Circuit

The small-signal equivalent circuit of the photodiode is shown in Figure 16.3. The signal current source I_p is determined by the incident power from Eq. (3). Capacitance C_d is the device capacitance under the operating bias voltage. The dark current source I_d is given by

$$I_d = I_{du} + I_{dm}G \qquad (4)$$

where I_{du} is the unmultiplied dark current component and I_{dm} is the multiplied dark current component (which is multiplied by the avalanche gain process).

The photodiode noise due to dark current I_d is represented by a noise current source i_{nd} with one-sided spectral density given by

$$N_d(f) = 2eI_{du} + 2eI_{dm}G^2F(G) \qquad (5)$$

where $F(G)$ is called the avalanche excess noise factor and is given by

$$F(G) = kG + (1 - k)\left(2 - \frac{1}{G}\right) \qquad (6)$$

The parameter k in the above expression is the effective ratio of the ionization constants of holes and electrons of the APDs (k is defined such that $k \leq 1$). Sometimes for ease of mathematical manipulation, $F(G)$ is approximated by

$$F(G) = G^x \qquad (7)$$

where x is called the excess noise factor exponent.

It can be noted that both I_{du} and I_{dm} dark current components are required for complete APD characterization, since it is often the case that dark current I_d is dominated by I_{du} while dark current noise i_{nd} is dominated by I_{dm}.

16.3 RECEIVER DESIGN REQUIREMENTS

The primary aim in receiver design for optical communication systems is to minimize the optical signal power required at the optical detector input for a desired receiver performance goal. For digital communication systems, this performance goal is often stated in terms of a received bit error rate (BER), normally required to be in the 10^{-9}–10^{-12} range. For analog communication systems, the receiver performance goal is characterized by the signal-to-noise ratio (SNR) and signal-to-distortion ratio (SDR) of the received analog signal output. The optical power required at the receiver input in order to satisfy the above requirements is a measure of the receiver sensitivity. Receiver sensitivity performance depends mainly on the noise level of the optical receiver. Thus the goal of receiver design is to maximize the sensitivity (i.e., minimizing the optical power input requirement) by minimizing the receiver noise level. Consequently, we maximize the allowable loss (attenuation) that the optical signal can suffer in propagating from the optical transmitter to the optical receiver. As a result, the allowable distance between the transmitter and receiver can be as large as possible.

However, for practical system implementation, other receiver design requirements have to be taken into consideration. For example, the incident optical power at the input of the receivers in a telephone trunking network or a local area network will vary from receiver to receiver. This power variation is due to differences in transmitter optical output levels and in various loss contribution factors that the optical signal suffers in propagating from the transmitter to the receiver (such as fiber length and loss, number of connectors and splices and their total loss, and loss of optical coupling and multiplexing devices). Thus the receiver has to operate satisfactorily not only at the minimum incident optical power (for the longest or most lossy link) but also over a range of power level (up to the level determined by the shortest or least lossy link). The maximum allowable optical power level at which the receiver can operate satisfactorily is determined by its nonlinear distortion and saturation characteristics.

The difference (in dB) of the maximum and minimum allowable optical power levels (in dBm) is referred to as the receiver dynamic range. From the discussion above, it is obvious that in addition to the maximum sensitivity requirement, another receiver design goal is to make the receiver dynamic range as large as possible.

An example is shown in Figure 16.4 for a 400-Mbit/s optical receiver. The sensitivity level in terms of average input optical power required for an error rate of 10^{-9} is $\bar{P}_{min} = -35$ dBm. The maximum average input optical power for the same BER is $\bar{P}_{max} = -10$ dBm. Outside the input power range defined by these two limits, the BER is worse than 10^{-9}. The receiver dynamic range in this case is 25 dB optical.

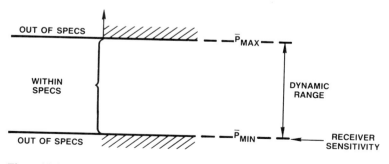

Figure 16.4 Receiver sensitivity and dynamic range definition.

Since receiver sensitivity and dynamic range are the two most important parameters in characterizing receiver performance, we will consider them in more detail in this chapter. Other receiver design requirements include low power consumption, high reliability, bit sequence independency, and fast acquisition time. Depending on the particular application, these requirements may become significant. For example, in submarine communications, high reliability and low power consumption are of considerable importance.

16.4 RECEIVER CONFIGURATION

Optical receivers can be classified into three basic configurations: simple resistor termination, high impedance, and transimpedance design.

16.4.1 Simple Resistor Termination

In this configuration, the photodiode is simply loaded into a 50-Ω input impedance amplifier (or test equipment). The coupling can either be dc (Fig. 16.5a) or ac (Fig. 16.5b) where resistor R_b provides the detector bias.

The receiver bandwidth in this case is limited by the photodiode intrinsic bandwidth, the input RC time constant, and the amplifier bandwidth. As far as photodiodes are concerned, they have been fabricated with bandwidths up to 25 GHz (PIN detector) and even to 100 GHz (Schottky barrier detectors). For a load impedance of 50 Ω and an input capacitance (detector and amplifier capacitance) of 1 pF, the RC time constant is 50 ps, which corresponds to a bandwidth of about 7 GHz. In addition, commercially available amplifiers with a baseband bandwidth of over 9 GHz can be obtained. Thus with this simple configuration a very wide bandwidth optical receiver can be readily implemented. In addtion it should be noted that if the RC time constant turns out to be the bandwidth-limiting factor, the load resistance R can be reduced. For example, in the ac-coupled case of Figure 16.5b, bias

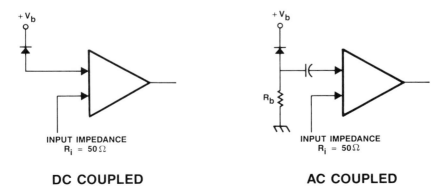

DC COUPLED　　　　　　　　　　**AC COUPLED**

Figure 16.5 Receiver amplifier configuration with simple 50-Ω termination.

resistance R_b can be reduced to 50 Ω, making the effective load resistance $R = 25$ Ω. The bandwidth limit due to the RC time constant is thus increased to 14 GHz. The trade-off here is that the signal gain is reduced due to the smaller load.

The disadvantage of the resistor termination configuration is that the receiver noise level is high compared to the other two approaches (high impedance and transimpedance design). Therefore, relatively large optical power is required at the detector input in order to obtain a reasonable SNR. However, because of its simplicity and extremely wide bandwidth, it is very useful in instrumentations where high sensitivity is not required. For example, this configuration is often used to characterize detector bandwidth or optical transmitter response time.

16.4.2 High-Impedance Design

As we will see later, the receiver noise level can be reduced by increasing the load resistance. This is the reason behind the high-impedance design [6]. In this configuration the bias resistance is increased enough that its noise contribution becomes negligible. In addition, the amplifier input impedance is increased to lower its noise level. For example, with a high-impedance FET front-end receiver amplifier, the effective load resistance can be as high as 1 MΩ to obtain low-noise performance. The problem here is that the front-end bandwidth is very limited. Even with an input capacitance of 1 pF, a load resistance of 1 MΩ means an input RC time constant of 1 μs, which corresponds to a bandwidth of about 160 kHz! Thus the front-end amplifier would tend to integrate the input signal waveform. Because of this fact, the high-impedance design is also known as the integrating front-end design. It is therefore obvious that the receiver amplifier should be followed by a differentiating equalizer in order to compensate for the integrating effect of the front end.

The structure of the high-impedance receiver is shown in Figure 16.6. Its equivalent circuit is shown in Figure 16.7, where the equalizer is a simple RC differentiator network. The frequency response plots of the front end, the equalizer, and the complete receiver are shown in Figure 16.8. It can be

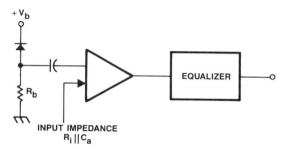

Figure 16.6 High-impedance receiver amplifier.

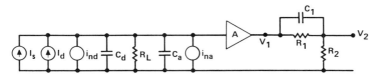

Figure 16.7 Equivalent circuit of high-impedance receiver amplifier.

seen that the equalizer "pick-up" frequency has to be matched with the front-end roll-off frequency. Thus if the front-end bandwidth changes due to device variations or aging, the equalizer "pick-up" frequency may have to be changed adaptively to maintain good equalization. This tracking may complicate the receiver design.

However, the major drawback of the high-impedance receiver is in its limited dynamic range. It can be observed from Figure 16.8 that in the front-end amplifier the low-frequency components (around f_1 and lower) are amplified much more than the midband and higher frequency (around f_2) components. Consequently the maximum input optical power that the receiver can tolerate is low because of saturation in the front-end amplifier due to the high gain of the low-frequency components.

16.4.3 Transimpedance Design

The transimpedance design avoids the equalization and limited dynamic range problem by using negative feedback [16]. Its configuration and equivalent circuit are shown in Figure 16.9, where R_f is the feedback resistance and C_f is its associated parasitic capacitance. The transimpedance frequency

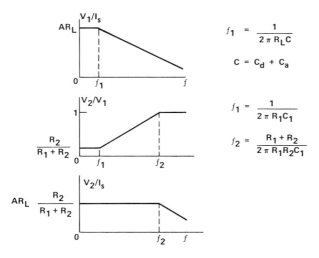

Figure 16.8 Frequency response of high-impedance receiver amplifier.

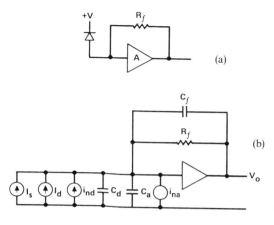

Figure 16.9 Transimpedance receiver amplifier. (a) Configuration; (b) equivalent circuit.

response is shown in Figure 16.10. Its bandwidth is given by

$$f = \frac{A}{2\pi R_f (C + AC_f)} \tag{8}$$

where A is the open-loop voltage gain and C is the total input capacitance (detector and amplifier input capacitance).

Thus the amplifier bandwidth can be made as large as required by increasing A or by reducing R_f. As we will see later, the amplifier noise level increases when R_f is reduced. Thus it is desirable to obtain the wide-bandwidth requirement by increasing A while still keeping R_f high. However, for the wide-bandwidth requirement the maximum usable value of A is limited for two reasons. First, for wideband operation, A is limited by the finite gain bandwidth product and propagation delay of transistor devices. Second, as can be seen from Eq. (8), even if A is infinite the amplifier bandwidth is still limited by the $R_f C_f$ time constant. Thus if C_f is significant, R_f has to be reduced to satisfy the bandwidth requirement. Consequently, the feedback resistance value is much lower than the corresponding load resistance value

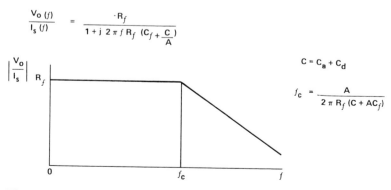

Figure 16.10 Frequency response of transimpedance receiver amplifier.

in the high-impedance design. As a result the transimpedance amplifier design has poorer receiver sensitivity (typically by 1–2 dB) than the high-impedance design.

The advantage of the transimpedance design is that a much wider dynamic range can be obtained. As can be seen from Figure 16.10, all frequency components within the signal bandwidth are amplified equally (recall that it is not the same situation in the high-impedance amplifiers). In addition, the feedback resistance R_f is smaller than the corresponding load resistance R_L in the high-impedance design. Therefore the transimpedance amplifier can tolerate a much higher optical power level at the input. In general the dynamic range improvement is equal to the product of the open-loop voltage gain A and the ratio R_L/R_f. Typically the transimpedance design can offer 20–30 dB more dynamic range than the high-impedance design.

The choice of whether to use the high-impedance or transimpedance design is dependent on whether sensitivity or dynamic range is more important. In general when APDs (particularly silicon APDs for the short-wavelength region) are used, the receiver amplifiers are of the transimpedance type. This is because the avalanche gain can partly compensate the increased amplifier noise. For receivers using PIN detectors, the high-impedance design is used when high receiver sensitivity is the major concern.

16.5 RECEIVER NOISE ANALYSIS AND PERFORMANCE

The receiver noise level is a crucial parameter in determining receiver sensitivity. In this section we discuss the noise analysis of optical receivers and present results on state-of-the-art receiver amplifier noise performance.

The receiver noise can be represented by an equivalent noise current source i_{na} at the input (where the photodiode is connected) [17]. Its one-sided spectral density as a function of frequency is denoted $N(f)$. Once $N(f)$ is determined, the receiver noise is completely characterized. The determination of $N(f)$ and the input equivalent noise power $\langle i_{na}^2 \rangle$ has been well documented in the literature [5, 13]. We will only summarize the results in this section. Interested readers are referred to the indicated references.

The receiver noise analysis is dependent on whether the front-end amplifying device is a field-effect transistor (FET) or a bipolar junction transistor (BJT).

16.5.1 FET Front-End Amplifiers

For FET front-end preamplifiers, the equivalent input noise current spectral density can be shown to be [5]

$$N(f) = \frac{4kT}{R} + 2eI_L + \frac{4kT\Gamma}{g_m}(2\pi C_T)^2 f^2 \tag{9}$$

where I_L is the total leakage current (including both FET gate leakage and unmultiplied dark current component I_{du} of photodiode), g_m is FET transconductance, C_T is total input capacitance, Γ is a numerical constant, k is the Boltzmann constant, and T is absolute temperature.

Equation (9) applies for both high-impedance and transimpedance receiver design. The term R_F is either the feedback resistance R_f (TZ design) or detector load resistance R_L (HZ design). The parameter Γ is a noise factor associated with thermal channel noise and gate-induced noise in the FET. Its value is typically 0.7 for silicon FETs, 1.03 for short-channel silicon MOSFETs, and 1.75 for GaAs MESFETs [18].

The total capacitance C_T is given by

$$C_T = C_{ds} + C_{gs} + C_{gd} + C_f \tag{10}$$

where C_{ds} is the detector and input stray capacitance; C_{gs} and C_{gd} are the FET gate-to-source and gate-to-drain capacitance, respectively; and C_f is the parasitic feedback capacitance of R_f.

The input equivalent noise power for a digital receiver operating at bit rate B can therefore be shown to be given by

$$\langle i_{na}^2 \rangle = \frac{4kT}{R_F} BI_2 + 2eI_L BI_2 + \frac{4kT}{g_m}(2\pi C_T)^2 B^3 I_3 \tag{11}$$

where I_2, I_3 are weighting integrals originally defined by Personick [6] that are dependent only on the incident optical pulse shape and the equalized output pulse shape. For incident optical pulse with non-return-to-zero (NRZ) format and an equalized pulse shape with a full raised cosine spectrum, we have $I_2 = 0.562$ and $I_3 = 0.0868$. For the return-to-zero (RZ) format, we have $I_2 = 0.403$ and $I_3 = 0.0984$.

The first term in Eq. (11) is the thermal noise due to R_F (R_L in HZ or R_f in TZ design). As mentioned previously, its noise is reduced when its value increases. In the HZ design, its value is chosen high enough that its noise contribution is negligible. A typical value of R_F for the HZ design is 1 MΩ. In the TZ design, its value is reduced to satisfy the bandwidth requirement as discussed before. Thus its value is reduced as the bit rate increases (typically its value changes from 100 kΩ for 45 Mb/s to 6 kΩ for 420 Mb/s).

The input equivalent noise power levels of typical FET front-end receiver amplifiers calculated from Eq. (11) are shown in Figure 16.11. The contributions of noise components due to R_F, I_L, and FET are plotted separately for various parameter values.

It can be observed that as the data rate B increases, the effect of leakage current I_L becomes less important and the FET noise becomes more dominant (particularly for the HZ design). Thus the amplifier noise is proportional to B^3 for high bit rates.

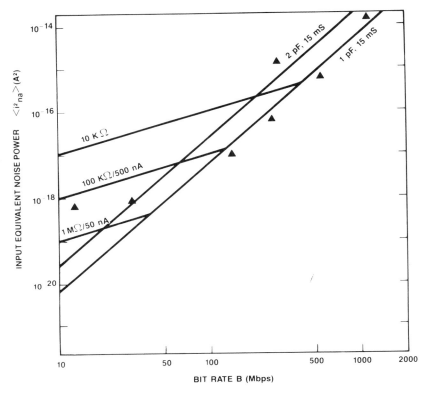

Figure 16.11 Noise level of FET front-end receiver amplifiers.

We can also note that at high bit rates the amplifier noise is proportional to C_T^2. Thus reduction of input capacitance is critical in obtaining low-noise performance. Photodiode and FET capacitance can be reduced by decreasing their dimensions, while parasitic capacitance can be reduced by using hybrid or monolithic technology. Reported noise levels of FET front-end amplifiers fabricated by hybrid technology are also shown in Figure 16.11 as experimental data points [18–29]. It can be observed that total input capacitance of under 1 pF can be obtained.

A popular FET front-end receiver amplifier circuit configuration [18–23] is shown in Figure 16.12. The input stage is a cascode circuit (common-source FET followed by common-base BJT) for wide bandwidth. For the HZ design, load resistor R_L also provides the detector bias. For the TZ design, R_L is not present and the detector is biased through the feedback resistor R_f. Most of the experimental data shown in Figure 16.11 are obtained from this circuit fabricated on a hybrid substrate.

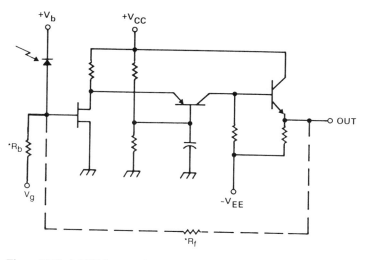

Figure 16.12 A FET front-end receiver amplifier circuit.

16.5.2 Bipolar Front-End Receiver Amplifiers

For receiver amplifiers with a bipolar transistor front end, the equivalent input noise spectral density can be shown to be [5]

$$N(f) = \frac{4kT}{R_F} + 2eI_b + \frac{2eI_c}{g_m}(2\pi C_T)^2 f^2 + 4kTr_{bb'}(2\pi C_{dsf})^2 f^2 \qquad (12)$$

where I_b and I_c are the base and collector bias current, respectively; $r_{bb'}$ is the base spreading resistance; and $g_m = I_c/V_T$ is the BJT transconductance, with $V_T = kT/e$ (= 25.6 mV at room temperature).

The capacitance C_{dsf} in Eq. (12) is defined as

$$C_{dsf} = C_{ds} + C_f \qquad (13)$$

The total capacitance C_T in this case is given by

$$C_T = C_{ds} + C_\pi + C_\mu + C_f \qquad (14)$$

where C_π and C_μ are the small-signal hybrid-π model capacitances of the transistor.

The input equivalent noise power is thus given by

$$\langle i_{na}^2 \rangle = \frac{4kT}{R_F}BI_2 + 2eI_bBI_2 + \frac{2e}{I_c}(2\pi V_T C_T)^2 B^3 I_3 + 4kTr_{bb'}(2\pi C_{dsf})^2 B^3 I_3$$

$$(15)$$

It can be observed that the amplifier noise level is a function of the bipolar bias current. This dependence is illustrated in Figure 16.13. As bias

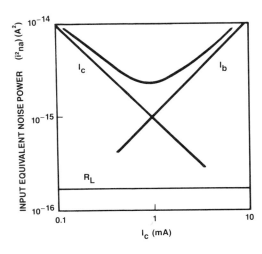

Figure 16.13 Variation of bipolar front-end receiver amplifier noise level with collector bias current.

current I_c increases, the base current shot noise contribution increases while the collector current shot noise decreases. Therefore an optimum bias current exists where the total amplifier noise is a minimum. In deriving this optimum bias current the variation of C_T with I_c needs to be taken into account. To a first order, C_T can be approximated by the expression

$$C_T = C_0 + \alpha I_c \tag{16}$$

The optimum collector bias current can be shown to be given by

$$I_c = 2\pi C_0 V_T B \sqrt{\frac{\beta I_3}{I_2}} \tag{17}$$

At this optimum bias current the amplifier noise power is given by

$$\langle i_{na}^2 \rangle = \frac{4kT}{R_F} BI_2 + 8\pi kTC_0 B^2 \left\{ \frac{I_2 I_3}{\beta} \left[1 + (2\pi\alpha V_T B)^2 \frac{\beta I_3}{I_2} \right] \right\}^{1/2}$$
$$+ 4e\alpha C_0 (2\pi V_T)^2 B^3 I_3 + 4kT r_{bb'} (2\pi C_{dsf})^2 B^3 I_3 \tag{18}$$

The last two terms in Equation (18) are important only at very high bit rates. The dominant term (at least for the HZ design) is normally the second term, which is proportional to B^2. Thus as bit rate increases, the bipolar transistor noise (at optimum bias) increases more slowly than FET noise.

A simple but popular circuit for a transimpedance BJT receiver amplifier [30] is shown in Figure 16.14. This circuit can be readily integrated using silicon bipolar technology.

The noise performance of state-of-the-art receiver amplifiers are summarized in Figure 16.15. Calculated noise power as well as experimental data points are shown for both GaAs MESFET and BJT front-end amplifiers.

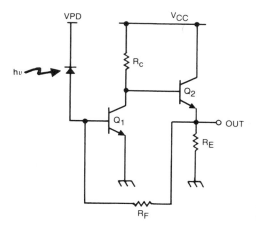

Figure 16.14 A popular bipolar front-end transimpedance receiver amplifier circuit.

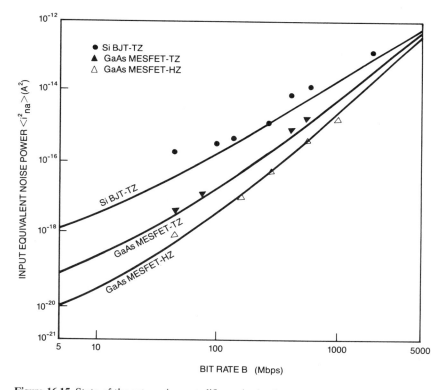

Figure 16.15 State-of-the-art receiver amplifier noise level.

In addition, both HZ and TZ designs are presented. It can be observed that the HZ GaAs MESFET design offers the lowest noise up to at least several gigabits per second. However, this noise advantage decreases with increasing bit rate. Thus the TZ design is preferred beyond 1 Gb/s because of its higher dynamic range.

16.6 RECEIVER SENSITIVITY ANALYSIS AND PERFORMANCE

Once the receiver amplifier input equivalent noise power is known, the receiver sensitivity can be readily evaluated. The analysis is well documented in the literature. The results are summarized in this section.

The receiver sensitivity is often quoted in terms of incident average optical power $\bar{P}$ for a BER of 10^{-9}. It is also often quoted in terms of the detected average optical power $\eta\bar{P}$, which is the optical power that is actually useful in the photodetection process. In other words, the detected optical power is equal to the incident optical power only when the detector quantum efficiency is 100%. The detected optical power is used because it can be measured simply by monitoring the detected photocurrent. Thus it is a very accurate measurement since no optical power measurement is involved. Generally speaking, the incident optical power measurement is useful for comparing the sensitivity of complete optical receivers while the detected optical power measurement is more useful in comparing the performance of different receiver amplifiers.

16.6.1 Quantum Limit of Receiver Sensitivity

In optical receivers, even for the ideal case of zero receiver noise (i.e., zero detector dark current and zero amplifier noise), the detection sensitivity is still limited by the quantum nature of the photodetection process (i.e., by the shot noise due to the incident optical power). The receiver sensitivity in this ideal case of negligible receiver noise is called the quantum limit. Using the Poisson statistics, we can readily show that in order to detect the presence of an incident optical pulse with an error probability of 10^{-9} the pulse must contain at least 21 photons. Thus for a digital optical transmission system with equal probability of 1 and 0 bits (where an optical pulse is sent only if a 1 bit is transmitted), we need an average of 10.5 photons per bit at the receiver input for a BER of 10^{-9}.

The quantum limit in terms of incident average optical power for a BER of 10^{-9} is therefore

$$\bar{P}_{\mathrm{QL}} = 10.5h\nu B \qquad (19)$$

The quantum limit is thus proportional to the bit rate. It represents the best possible receiver sensitivity that can be achieved and is often used as a reference baseline to which the actual experimental results are compared.

16.6.2 PIN Receiver Sensitivity Analysis

For PIN receivers the total noise at the receiver is normally dominated by the amplifier noise (it may not be the case for analog receivers with a high SNR requirement). Thus Gaussian noise statistics is applicable and the decision threshold level can be set halfway between the 1 and 0 signal level. An "eye pattern" diagram at the input of the threshold detector (cf. Fig. 16.2) is shown in Figure 16.16. The receiver sensitivity in terms of detected optical power is simply given by

$$\eta \bar{P} = Q \frac{h\nu}{e} \sqrt{\langle i_{na}^2 \rangle} \tag{20}$$

where Q is a parameter determined by the desired BER through the equation

$$\text{BER} = \frac{1}{\sqrt{2\pi}} \int_Q^\infty \exp\left(-\frac{x^2}{2}\right) dx \tag{21}$$

For 10^{-9} BER, we have $Q = 6$. Values of Q for other BERs can be determined from mathematical tables.

The physical meaning of Q is also shown in Figure 16.16. Q is simply the ratio of the average signal level and the rms noise at the threshold detector input.

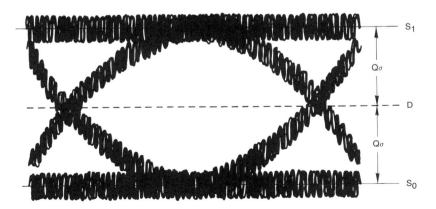

Figure 16.16 Eye pattern of PIN photodiode receivers.

It can be observed from Eq. (20) that the receiver sensitivity is directly proportional to the receiver equivalent rms noise current. Thus for PIN receivers, a low-noise amplifier is necessary. This is normally achieved by following the PIN detector with a hybrid GaAs MESFET front end amplifier. The resulting amplifier is normally referred to as a PINFET receiver.

16.6.3 APD Receiver Sensitivity Analysis

For APD receivers, the analysis is complicated by the fact that the avalanche gain statistical distribution is given by a mathematically rather complex expression. It is not symmetrical about the average avalanche gain value but is skewed toward the high gain values. Thus in order to derive closed-form expressions for receiver sensitivity, the avalanche gain distribution is approximated by a Guassian process. Extensive numerical analysis through computer simulation has shown that the Gaussian approximation gives a fairly accurate receiver sensitivity (well within 1 dB of the exact calculation) although it tends to overestimate the optimum avalanche gain and underestimate the optimum threshold level [8].

For APD receivers the shot noise due to the incident optical power is a significant portion of the total receiver noise. Thus the noise level of a 1 bit (incident optical pulse present) is higher than the noise level of a 0 bit (no optical pulse). Therefore the decision threshold level has to be set below the halfway mark of the 1 signal level so that equal error probability is obtained from the 1 and 0 bits. This fact is illustrated from the eye pattern diagram shown in Figure 16.17.

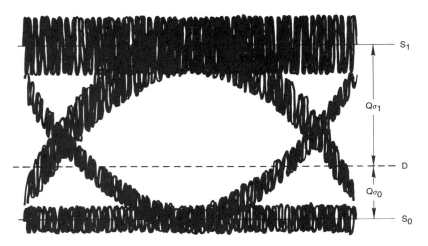

Figure 16.17 Eye pattern of avalanche photodiode receivers.

When the analysis is carried out, the receiver sensitivity as a function of the avalanche gain is given by

$$\eta \bar{P} = Q \frac{hv}{e} \left[QeBI_1 F(G) + \left(\frac{\langle i_{na}^2 \rangle}{G^2} + 2eI_{dm}F(G)BI_2 \right)^{1/2} \right] \qquad (22)$$

where I_1 is another Personick weighting integral ($I_1 \approx 0.5$ for both the NRZ and RZ formats).

As an example, the receiver sensitivity calculated from the above expression is plotted in Figure 16.18 for a transimpedance receiver operating at 400 Mb/s. The sensitivity of three APDs are shown: silicon APDs at 0.85 μm, germanium APDs and InGaAs APDs at 1.3 μm.

It can be observed that there exists an optimum avalanche gain where the required optical power is a minimum. If the avalanche gain increases beyond this value, the avalanche excess noise becomes dominant and degrades the receiver sensitivity.

In general the avalanche gain is optimum when the shot noise due to the incident optical power and dark current I_{dm} component is comparable

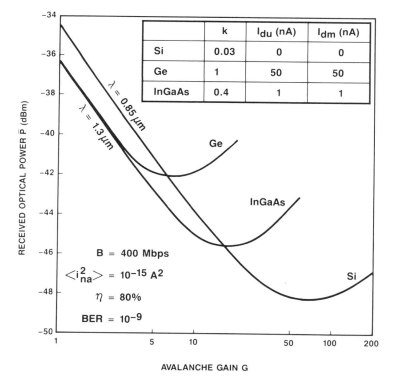

Figure 16.18 Receiver sensitivity of avalanche photodiodes as a function of avalanche gain.

to the rest of the receiver noise. Thus a receiver amplifier with a high noise level requires a higher optimum avalanche gain. By the same argument, the optimum avalanche is reduced when the APD dark current (I_{dm}) is high and/or the avalanche excess noise is high. It is also important to note that the higher the optimum avalanche gain, the better the receiver sensitivity improvement that the APD offers compared to a PIN detector.

For example, silicon APDs typically have $I_{dm} = 1$ pA and $k = 0.03$. Thus the optimum avalanche gain is quite high (approximately 80 in Fig. 16.18) and the sensitivity improvement over the PIN detector is about 14 dB. On the other hand, for germanium APDs, which have high dark current $(I_{dm} = 50-100$ nA) and excess noise $(k = 1)$, the optimum avalanche gain is quite small (about 8 in Fig. 16.18) and the sensitivity improvement is much less (about 6 dB in Fig. 16.18 over a PIN detector with the same dark current). Recently, InGaAs APDs have been developed that offer lower noise than germanium APDs. However, they still fall short of the silicon APD performance.

16.6.4 Experimental Results of Receiver Sensitivity

Some of the best results of receiver sensitivity at short wavelength reported in the literature [30–37] are summarized in Figure 16.19. In this figure (and

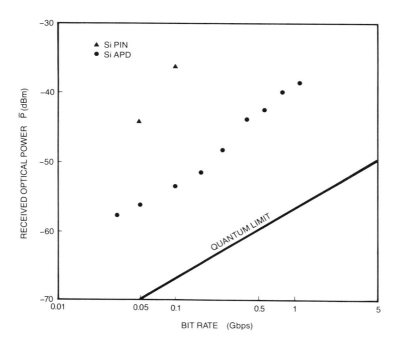

Figure 16.19 Receiver sensitivity of short-wavelength (0.85 μm) optical receivers.

also in Fig. 16.20) the sensitivity is shown in terms of the average received optical power $\bar{P}$. When silicon APDs are used, because of their low noise, the dependence of receiver sensitivity on the amplifier noise level is less critical than in the PIN detector case. Thus most of the APD sensitivity results shown in Figure 16.19 are obtained with the transimpedance design for wide dynamic range. It can be observed that sensitivity within 10–15 dB of the quantum limit can be obtained.

For the long-wavelength (1.1–1.6 μm) region, experimental sensitivity results are shown in Figure 16.20 for four different optical receivers: HZ PINFETs, TZ PINFETs [18–29], germanium APDs [38–45], and InGaAs APDs [46–51]. For simplicity we do not distinguish between HZ and TZ amplifier designs for the APD case. The sensitivity results that are obtained at 1.3 μm are shown as completely filled triangles (HZ PINFETs), PINFETs circles (germanium APDs), and squares (InGaAs APDs). Sensitivity results at wavelengths other than 1.3 μm are also included by scaling to the equivalent optical power at 1.3 μm. For example, a receiver sensitivity of −40 dBm at

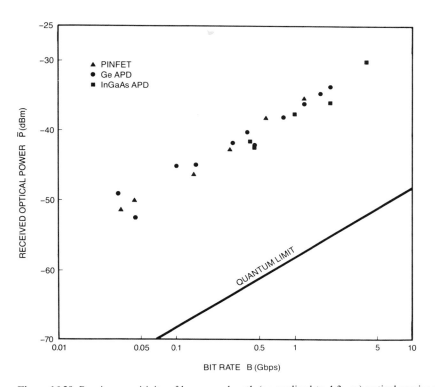

Figure 16.20 Receiver sensitivity of long-wavelength (normalized to 1.3 μm) optical receivers.

1.55 μm is scaled to an equivalent sensitivity of -39.2 dBm at 1.3 μm. Most of the sensitivity results plotted in Figure 16.20 are listed in the references.

It can be recalled from Section 16.2 that germanium APDs have much higher dark current than InGaAs PIN detectors. Thus, in general, since the effect of detector dark current noise is less critical at high bit rates, PINFET receivers offer better sensitivity than germanium APDs at low bit rates, and vice versa. Depending on particular detector performance together with amplifier configuration (HZ or TZ) and noise level, the crossover point can range from several hundred megabite per second up to 1 Gb/s. It can also be noted that since dark current is a very sensitive function of temperature, we must take temperature effects into account in comparing the sensitivity of PINFET and germanium APD receivers for practical applications.

With the recent development of InGaAs APDs, better receiver sensitivity can be obtained. Sensitivity improvements of 2–3 dB have been achieved over PINFET and germanium APD receivers, as shown in Figure 16.20, with potentially more. However, it can be observed that sensitivity of long-wavelength optical receivers is still about 20 dB away from the quantum limit.

It is also important to note that the gain–bandwidth product of germanium APDs and InGaAs APDs is typically in the range of 10–20 GHz (compared to over 200 GHz for silicon APDs). Thus for high-bit-rate applications (for example, over 1 Gb/s), the useful avalanche gain can be limited by this gain–bandwidth product. Thus the full sensitivity improvement offered by APDs over the PIN detector cannot be obtained.

16.7 DYNAMIC RANGE CONSIDERATION

We have already mentioned the trade-off between receiver sensitivity and dynamic range of the HZ and TZ designs. This fact is illustrated more clearly in Figure 16.21. As we reduce the load (or feedback) resistance, the amplifier noise increases and hence the receiver sensitivity is degraded (the minimum required power increases). However, the maximum allowable optical power increases at a much faster rate. As a result, the receiver dynamic range is improved.

The receiver dynamic range is determined by saturation and nonlinearity in the amplifier. Thus it is dependent on the amplifier bias condition and supply voltage. The receiver normally has an automatic gain control (AGC) or a limiting amplifier so that the signal at the threshold detector (or demodulator) input is kept at a constant level as the incident optical power varies over its dynamic range. When APDs are used, the avalanche gain can also be incorporated in another AGC loop so that as the incident optical signal increases from its minimum level, the avalanche gain is reduced from its optimum value to keep the total detected photocurrent constant. Because

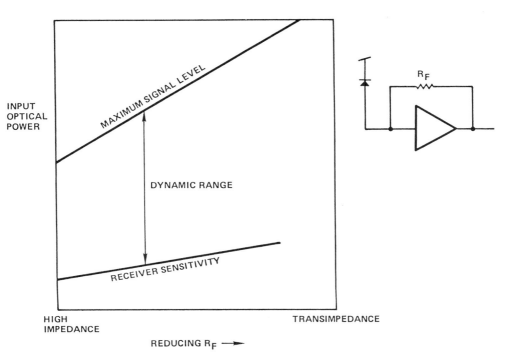

Figure 16.21 Trade-off between receiver sensitivity and dynamic range.

of this added flexibility, APD receivers offer higher dynamic ranges than PIN receivers.

The dynamic range of hybrid PINFET receivers is typically 15–20 dB for the HZ design and 25–30 dB for the TZ design. It can be recalled from the previous discussion that the HZ dynamic range is limited by the enhanced gain of the low-frequency signal components causing saturation in the front-end amplifier. Thus the HZ PINFET dynamic range can be improved by selecting a line coding with negligible or little low-frequency components. Suitable line coding formats include Manchester code, mBnB code, and others.

When required, the PINFET receiver dynamic range can be extended to well over 40 dB by additional control circuitry. For example, the detected photocurrent can be shunted into another path away from the front-end amplifier input as the incident optical power increases. This shunting path can be provided by a voltage-controlled resistor such as a Schottky diode or a microwave PIN diode. With this technique, a dynamic range improvement of 20 dB can be obtained, yielding a total dynamic range of 40–50 dB [24, 25, 27].

An alternative approach to obtaining a higher dynamic range is to use APD receivers, particularly with a transimpedance amplifier design. Silicon

APD receivers at short wavelengths typically provide a dynamic range of about 40–50 dB. For germanium APDs and InGaAs APDs at long wavelengths, the dynamic range is in the range of 25–40 dB.

The discussion in this chapter has been limited to receiver sensitivity and dynamic range. Other receiver characteristics must also be taken into account depending on the applications. They include bit rate transparency, bit sequence independency, fast acquisition time (for data bus and local area network applications), low supply voltage and power consumption, and high reliability (for submarine transmission). The interested reader can refer to the literature for discussions on these characteristics [e.g., 5].

16.8 ANALOG OPTICAL RECEIVERS

Most of the analysis and discussion so far have concentrated on digital optical receivers. This is because digital transmission technique is ideally suited for optical fiber communications due to the large available bandwidth. However, in certain applications where simplicity and low cost are the main concern, analog transmission techniques over fiber are also used. For example, video signals are often transmitted and distributed by simple vestigial sideband (VSB) or frequency modulation (FM).

For analog receivers we need to determine the required incident optical power for a desired signal-to-noise ratio (SNR) or carrier-to-noise ratio (CNR). The calculation can be carried out on the basis of the amplifier noise analysis discussed previously. The expressions for the input equivalent noise spectral density [Eqs. (9) and (12)] are directly applicable. Thus the input equivalent noise power of the amplifier, $\langle i_{na}^2 \rangle$, can be evaluated by integrating the noise density over the required receiver bandwidth. The total receiver noise can then be obtained by summing the amplifier noise with the shot noise due to the incident optical power and the detector dark current (remembering that the shot noise may not be negligible even for PIN detectors).

The signal-to-noise ratio (or carrier-to-noise ratio) detected by the photodiode is given by

$$\text{SNR} = \frac{1}{2} \frac{(m\mathscr{R}G\bar{P})^2}{\langle i_{nc}^2 \rangle + 2e\mathscr{R}\bar{P}G^2F(G)B} \tag{23}$$

where m is the modulation index, B is the receiver bandwidth, and $\langle i_{nc}^2 \rangle$ is the receiver circuit noise given by

$$\langle i_{nc}^2 \rangle = \langle i_{na}^2 \rangle + 2e[I_{du} + I_{dm}G^2F(G)]B \tag{24}$$

From this equation, the minimum required optical power and the optimum avalanche gain for a given SNR requirement can be evaluated.

As an example, a single-channel video transmission system using FM subcarrier modulation is considered. In this case Eq. (23) provides an expression for the CNR at the FM demodulator input. The video SNR at the demodulator output can then be obtained from the CNR via standard FM demodulation analysis. The FM improvement factor (the difference between CNR at demodulator input and weighted video SNR at demodulator output) is dependent on the frequency deviation and the intermediate-frequency filter bandwidth. It is taken to be 35 dB in this example.

The incident optical power required is plotted as a function of the desired SNR in Figure 16.22. The optical receiver is a TZ PINFET receiver operating at 1.3 μm. Parameter values assumed in the calculation are shown in this figure. It can be observed that for low SNR requirements, the receiver circuit noise dominates and the required optical power increases as the square root of SNR (e.g., a 10-dB increase in SNR requires 5 dB in optical power). However, for high SNR values, the shot noise of the incident optical power is dominant and the required power is directly proportional to the SNR (e.g., a 10-dB increase in SNR requires a 10-dB increase in optical power).

Apart from the receiver noise, other noise and degradation sources need to be taken into account in analog applications. For example, the detected SNR may be limited by the optical source excess noise or modal noise. In

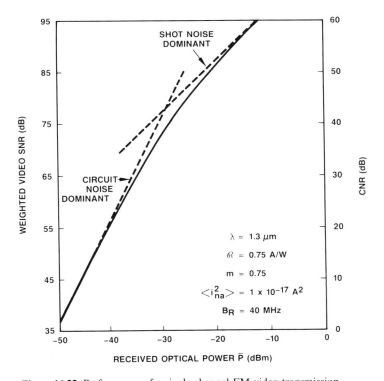

Figure 16.22 Performance of a single-channel FM video transmission.

addition, nonlinear distortion has to be taken into account whenever necessary. For well-designed systems, however, most of the signal distortion is caused by the optical transmitters and not by the optical receivers.

16.9 IMPLEMENTATION TECHNOLOGIES: HYBRIDIZATION AND MONOLITHIC INTEGRATION

It can be recalled from Figures 16.1 and 16.2 that the optical receiver consists of many functional blocks. In the early optical transmission systems, the optical receivers were often built on printed circuit boards using discrete electronic components and off-the-shelf commercial integrated circuits. For example, in the Atlanta fiber system experiment at 45 Mb/s carried out by the Bell Systems in 1976, the optical receiver was built on two separate circuit boards and the optical transmitter on one circuit board, each board measuring 6.5 in. × 4.5 in. [36]. As the transmission speed increases to take advantage of the single-mode fiber bandwidth, there is a need to hybridize and/or to integrate the optical receiver as much as possible. The obvious advantages that can be obtained are high performance, small size, high reliability, and potentially low cost.

As an example, the regenerator in the FT4E 432-Mb/s transmission system developed by AT&T Bell Laboratories contains a number of specially developed and dedicated hybrid and monolithic integrated circuits (a regenerator consists of an optical receiver, an optical transmitter, and associated supervisory and performance monitoring circuits). Even then the regenerator is still a big circuit board measuring 31 cm × 20 cm × 5 cm. This is acceptable for terrestrial transmission systems. However, for certain stringent applications such as submarine communications, we still need to increase the reliability and reduce the regenerator size even further. Thus total monolithic integration of the regenerator is desirable.

For optical receivers (and optical transmitters), we need to distinguish between two types of integration: electronic integration and optoelectronic integration, as shown in Figure 16.23.

Electronic integration refers to the case where the optical detector (or source) is still a discrete device but all the electronic circuitry in the receiver (or transmitter) is integrated in either one or several integrated circuits (ICs). Optoelectronic integration refers to the case where the detector (or source) is integrated together with the preamplifier (or source driver) in an optoelectronic integrated circuit (OEIC). Thus the receiver (or transmitter) consists of an OEIC followed (or preceded) by either one or several ICs. In the ultimate case, all the receiver (or transmitter) circuitry is integrated together with the detector (or source). The complete receiver (or transmitter) in this case consists of only one OEIC.

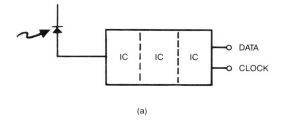

(a)

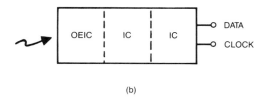

(b)

Figure 16.23 Monolithic integration of optical receivers. (a) Electronic integration; (b) optoelectronic integration (OEIC).

Optoelectronic integration is under intense investigation by many laboratories around the world. However, it is still a less mature technology than electronic integration. It is considered separately in this book. Thus we conclude this section by discussing several examples of electronic integration.

For data rates up to 100 Mb/s, the level of integration can be quite complex. An example is shown in Figure 16.24, where all the receiver circuitry including a phaselock loop (PLL) for clock extraction and a Manchester decoder is integrated in a single silicon bipolar IC [52]. The receiver sensitivity obtained with a low-cost silicon PIN detector is -33 dBm for a 50-Mb/s Manchester-coded signal. The application in this case is low-cost data links for interrack connections.

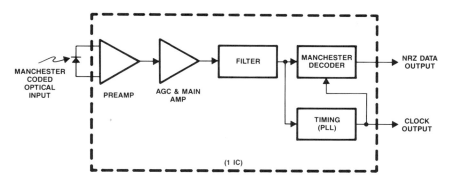

Figure 16.24 A 50-Mb/s electronic integrated optical receiver and Manchester decoder.

An example of an integrated optical receiver for submarine applications is shown in Figure 16.25. It was developed by AT&T Bell Laboratories for the TAT-8 systems at 295.6 Mb/s. The optical receiver consists of four ICs: a preamplifier IC, an AGC main amplifier IC, a timing IC, and a decision circuit IC (not including passive *RLC* filters and a SAW filter). The receiver sensitivity achieved with an InGaAs PIN detector at 1.3 μm is -36.3 dBm for 10^{-9} BER [53].

Another highly integrated optical receiver is used in the FS-400 submarine transmission system developed by NTT of Japan. The receiver consists of three ICs: an amplifier IC, a timing IC, and a decision circuit IC (again not counting passive filters). The receiver sensitivity obtained is -37.5 dBm at 446 Mb/s for a 10^{-11} BER with germanium APDs [54].

The highest-speed integrated optical receiver reported so far is a 1.8 Gb/s receiver developed by NTT of Japan [55]. The preamplifier and AGC post-amplifier chain consists of five ICs, and the decision circuit is in one IC. The integration of the clock extraction circuit has not been reported. A receiver sensitivity of -33 dBm is achieved at 1.8 Gb/s with germanium APDs.

All of the above examples have used silicon bipolar technology. However, other IC technologies are also under investigation. Fast CMOS technology has been used for bit rates under 50 Mb/s. A preamplifier IC and an AGC amplifier IC have also been fabricated with NMOS technology. A receiver sensitivity of -28 dBm is obtained at 800 Mb/s with a PIN detector at 1.3 μm with fine line NMOS [56]. In addition, the GaAs technology is maturing rapidly and is a very promising alternative. In fact, its low

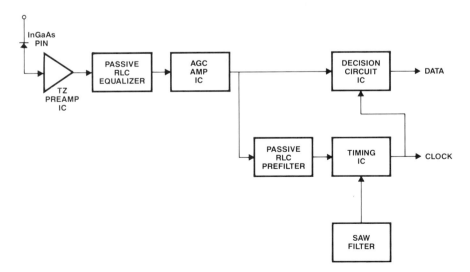

Figure 16.25 An electronic integrated receiver at 296 Mb/s for submarine applications.

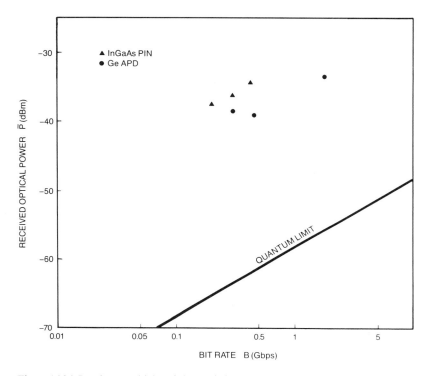

Figure 16.26 Receiver sensitivity of electronic integrated optical receivers at 1.3-μm wavelength.

power consumption may be advantageous over silicon bipolar technology. GaAs amplifiers and decision circuits have been reported with excellent performance up to 2 Gb/s.

The results of receiver sensitivity obtained with integrated receivers are summarized in Figure 16.26. It can be observed by comparison with Figure 16.20 that electronic integrated receivers offer sensitivity comparable to that obtained by hybrid technology. However, the sensitivity offered by OEIC receivers is still much lower, indicating the immaturity of this technology at this time.

16.10 FUTURE DEVELOPMENT TRENDS

We have already indicated that a continuing development trend is monolithic integration. As the transmission speed increases higher and higher into the multigigabit-per-second data rate region, electronic integration is necessary for performance improvement. In addition, optoelectronic integration is actively pursued by many research laboratories around the world.

It has also been mentioned that the noise level of long-wavelength APDs is still not as good as that of silicon APDs at short wavelengths. For high-data-rate applications, the performance of germanium and InGaAs APDs is also limited by their gain–bandwidth product. As a result, a great deal of effort is being spent in searching for a high-performance APD for the long-wavelength region. For example, APDs fabricated from material systems other than InGaAsP (for example, AlGaAsSb and CdHgTe) have been reported. In addition, detector structures other than APDs that may be more suitable for optoelectronic integration are also under intense investigation (for example, photoconductive detectors and optical FETs).

It can be concluded that the theoretical aspects of optical receiver design are now well understood. The active area of research and development is in the implementation technology to satisfy the demand of ever-increasing transmission capacity.

REFERENCES

1. S. D. Personick, "Receiver design for optical fiber systems," *Proc. IEEE*, **65**: 1670–1678 (1977).
2. I. Garrett, "Receivers for optical fibre communications," *Radio Electron. Eng.*, **51**: 349–361 (1981).
3. D. R. Smith et al., "Receivers for optical fibre communication systems," *Telecommunication J.*, **48**: 680–685 (1981).
4. K. Ogawa, "Considerations for optical receiver design," *IEEE J. Selected Areas Commun.*, **SAC-1**: 524–532 (1983).
5. T. V. Muoi, "Receiver design for high speed optical fiber systems," *IEEE/OSA J. Lightwave Technol.* **LT-2**: 243–267 (1984).
6. S. D. Personick, "Receiver design for digital fiber optic communication systems, parts I and II," *Bell Syst Tech. J.*, **52**: 843–886 (1973).
7. T. V. Muoi and J. L. Hullett, "Receiver design for multilevel digital optical fiber systems," *IEEE Trans. Commun.*, **COM-23**: 987–994 (1975).
8. S. D. Personick et al., "A detailed comparison of four approaches to the calculation of the sensitivity of optical fiber system receivers," *IEEE Trans. Commun.*, **COM-25**: 541–548 (1977).
9. T. V. Muoi and J. L. Hullett, "Receiver design for optical PPM systems," *IEEE Trans. Commun.*, **COM-26**: 295–300 (1978).
10. D. R. Smith and I. Garrett, "A simplified approach to digital optical receiver design," *Opt. Quantum. Electron.*, **10**: 211–221 (1978).
11. R. C. Hooper and B. R. White, "Digital optical receiver design for non-zero extinction ratio using a simplified approach," *Opt. Quantum. Electron.*, **10**: 279–282 (1978).
12. D. R. Smith et al., "Receivers for optical communications: a comparison of avalanche photodiodes with PINFET hybrids," *Opt. Quantum. Electron.*, **10**: 293–300 (1978).
13. R. G. Smith and S. D. Personick, "Receiver design for optical fiber communications systems," in H. Kressel (ed.), *Semiconductor Devices for Optical Communication*, Springer-Verlag, New York, 1980, chap. 4.
14. R. G. Smith and S. R. Forrest, "Sensitivity of avalanche photodetector receivers for long wavelength optical communication," *Bell Syst. Tech. J.*, **61**: 2929–2946 (1982).

15. T. V. Muoi, "Receiver design for digital fiber optic transmission systems using Manchester (biphase) coding," *IEEE Trans. Commun.*, **COM-31**: 608–619 (1983).

16. J. L. Hullett and T. V. Muoi, "A feedback receive amplifier for optical transmission systems," *IEEE Trans. Commun.*, **COM-24**: 1180–1185 (1976).

17. J. E. Goell, "Input amplifiers for optical PCM receivers," *Bell Syst. Tech. J.*, **53**: 1771–1793 (1974).

18. K. Ogawa, "Noise caused by GaAs MESFETs in optical receivers," *Bell Syst. Tech. J.*, **60**: 923–928 (1981).

18a. K. Ogawa and E. L. Chinnock, "GaAs FET transimpedance front-end design for a wideband optical receiver," *Electron. Lett.*, **15**: 650–652 (1979).

19. D. R. Smith et al., "PIN photodiode hybrid optical receivers," *Proc. 5th Eur. Conf. Opt. Commun.*, pp. 13.4.1–13.4.4, 1979.

20. D. R. Smith et al., "pin/FET hybrid optical receiver for longer wavelength optical communication systems," *Electron. Lett.*, **16**: 69–71 (1980).

21. D. R. Smith et al., "pin/FET hybrid optical receiver for 1.1–1.6 μm optical communication systems," *Electron. Lett.*, **16**: 750–751 (1980).

22. R. C. Hopper et al., "PINFET hybrid optical receivers for longer wavelength optical communication systems," *Proc. 6th Eur. Conf. Opt. Commun.*, pp. 222–225, 1980.

23. D. Gloge et al., "High speed digital lightwave communication using LEDs and PIN photodiodes at 1.3 μm," *Bell Syst. Tech. J.*, **59**: 1365–1382 (1980).

24. D. P. M. Chown, "Dynamic range extension of PINFET optical receivers," *Proc. 7th Eur. Conf. Opt. Commun.*, pp 14.5.1–14.5.3, 1981.

25. G. F. Williams, "Wide dynamic range fiber optic receivers," *Tech. Dig. ISSCC*, pp. 160–161, February 1982.

26. D. R. Smith et al., "Experimental comparison of a germanium avalanche photodiode and InGaAs PINFET receiver for longer wavelength optical communication systems," *Electron. Lett.*, **18**: 453–454 (1982).

27. B. Owen, "PIN-GaAs FET optical receiver with a wide dynamic range," *Electron. Lett.*, **18**: 626–627 (1982).

28. K. Ogawa et al., "A long wavelength optical receiver using a short channel Si MOSFET," *Bell Syst. Tech. J.*, **62**: 1181–1188 (1983).

29. M. C. Brain et al., "PINFET hybrid optical receivers for 1.2 Gbit/s transmission systems operating at 1.3 and 1.5 μm wavelength," **20**: 894–896 (1984).

30. R. G. Smith et al., "Optical detector package", *Bell Syst. Tech. J.*, **57**: 1809–1822 (1978).

31. J. E. Goell, "An optical repeater with high impedance input amplifier," *Bell Syst. Tech. J.*, **53**: 629–643 (1974).

32. T. Ueno et al., "A 40 Mbit/s and 400 Mbit/s repeater for fiber optic communication," *Proc. 1st Eur. Conf. Opt. Commun.*, pp. 147–150, September 1975.

33. P. K. Runge, "An experimental 50 Mb/s fiber optic PCM repeater," *IEEE Trans. Commun.*, **COM-24**: 413–418 (1976).

34. T. Ogawa et al., "Low noise 100 Mbit/s optical receiver," *Proc. 2nd Eur. Conf. Opt. Commun.*, pp. 357–363, September 1976.

35. R. W. Berry et al., "Optical fiber system trials at 8 Mbit/s and 140 Mbit/s," *IEEE Trans. Commun.*, **COM-26**: 1020–1027 (1978).

36. T. L. Maione, "Practical 45 Mbit/s regenerator for lightwave transmission systems," *Bell Syst. Tech. J.*, **57**: 1837–1856 (1978).

37. S. M. Abbott and W. M. Muska, "Low noise optical detection of a 1.1 Gb/s optical data stream," *Electron., Lett.*, **15**: 250–251 (1979).

38. J. Yamada et al., "High speed optical pulse transmission at 1.29 μm wavelength using low loss single mode fibers," *IEEE J. Quantum Electron.*, **QE-14**: 791–800 (1978).

39. J. Yamada et al., "Characteristics of Gbit/s optical receiver sensitivity and long span single mode fiber transmission at 1.3 μm," *IEEE J. Quantum Electron.*, **QE-18**: 718–727 (1982).

40. J. Yamada et al., "Gigabit/s optical receiver sensitivity and zero dispersion single mode fiber transmission at 1.55 μm," *IEEE J. Quantum Electron.*, **QE-18**: 1537–1546 (1982).

41. T. Mikawa et al., "Small area germanium avalanche photodiode for single mode fibre at 1.3 μm wavelength," *Electron. Lett.*, **19**: 452–453 (1983).

42. H. Nishimoto et al., "Injection-locked 1.5 μm InGaAsP/InP lasers capable of 450 Mbit/s transmission over 106 km," *Electron. Lett.*, **19**: 509–510 (1983).

43. D. A. Frisch et al., "High capacity 1.3 μm unrepeatered optical fibre transmission system, trials for submarine applications," **20**: 154–155 (1984).

44. H. Toba et al., "Injection locking technique applied to a 170 km transmission experiment at 445.8 Mbit/s," *Electron. Lett.*, **20**: 370–371 (1984).

45. S. D. Walker and L. C. Blank, "Ge APD/GaAs FET/OP-AMP transimpedance optical receiver design having minimum noise and intersymbol interference characteristics," *Electron. Lett.*, **20**: 808–809 (1984).

46. S. R. Forrest et al., "Excess noise and receiver sensitivity measurements of InGaAs/InP avalanche photodiodes," *Electron. Lett.*, **17**: 917–918 (1981).

47. J. C. Campbell et al., "High performance avalanche photodiode with separate absorption grading and multiplication regions," *Electron Lett.*, **19**: 818–820 (1983).

48. Y. Matsushima et al., "High sensitivity of VPE-grown InGaAs/InP heterostructure APD with buffer layer and guardring structure," *Electron. Lett.*, **20**: 235–236 (1984).

49. Y. Sugimoto et al., "High speed planar structure InP/InGaAsP/InGaAs avalanche photodiode grown by VPE," *Electron Lett.*, **20**: 653–654 (1984).

50. B. L. Kasper et al., "A 130 km transmission experiment at 2 Gb/s using silica-core fiber and a vapor phase transported DFB laser," *Proc. 10th Eur. Conf. Opt. Commun.*, postdeadline paper no. 6, September 1984.

51. A. H. Gnauck et al., "4 Gb/s transmission over 103 km of optical fiber using a novel electronic multiplexer/demultiplexer," *Tech. Dig.* (Postdeadline paper), Opt. Fiber Commun. (OFC), February 1985.

52. G. W. Sumerling and P. J. Morgan, "A low cost transparent fiber optic link for inter-rack communication," *Proc. 6th Int. Conf. Comput. Commun.*, London, September 1982.

53. D. G. Ross et al., "A highly integrated regenerator for 295.6 Mbit/s undersea optical transmission," *J. Light. Technol.* **LT-2**: 895–900 (1984).

54. M. Aiki, "446 Mbit/s integrated repeater," *J. Light. Technol.* **LT-3**: 392–399 (1985).

55. K. Nakagawa et al., "1.6 Gb/s optical transmission experiment with monolithic integrated circuits," *Proc. Int. Conf. Commun.*, pp. 771–774, June 1984.

56. A. A. Abidi, "Gigahertz transresistance amplifiers in fine line NMOS," *IEEE J. Solid State Circuits*, **SC-19**: 986–994 (1984).

PART 5

Applications of Optoelectronics in Lightwave Systems

CHAPTER 17

Optical Communications: Single-Mode Optical Fiber Transmission Systems

Chinlon Lin

17.1 INTRODUCTION

Research and development in the field of optical fiber transmission technology saw dramatic progress in the first 10 years (1970–1980). From the first low-loss optical fiber and first room-temperature cw semiconductor injection laser in the laboratory to the field trial of optical fiber transmission systems, the field is unique in the rapid deployment of laboratory advances in component and system technology to commercial systems. Transmission systems based on optical fiber communication technology have indeed revolutionalized the field of communications and are leading the change toward a totally digital network which promises a variety of new functions and features based on the concept of broadband integrated services digital networks (BISDN). The broadband ISDN promises to provide services including low-speed telemetry, high-speed data, high-quality audio, high-definition TV distribution, and two-way video in addition to the POTS (plain old telephone service) that we have today on paired copper wires to the home.

In this chapter we discuss the technology choices, the important features, and limitations of the backbone of this evolving fiber-based digital communication network: single-mode optical fiber transmission systems. For in-depth discussions on the individual technology such as the characteristics of optical fibers, optical sources and detectors, optical transmitters and receivers, and passive optical components, refer to other chapters in this book. Also, for analog optical fiber transmission technology, consult references in Chapter 15.

17.2 OPTICAL FIBER TRANSMISSION SYSTEMS—BASIC BUILDING BLOCKS

Figure 17.1 shows the schematic block diagram of an optical fiber transmission system. There are four basic parts in a typical optical fiber transmission system:

1. The transmission medium—the optical fiber/cable
2. Active optical device modules—the optical transmitters and optical receivers
3. Passive optical components and interconnection elements—connectors, splices, couplers, attenuators, WDM (wavelength-division multiplexing) components, etc.
4. Electronics—electronic multiplexers/demultiplexers, signal-processing circuits, supervisory/maintenance electronics, etc.

Note that the electronics part of the system is not unique to fiber transmission technology; in contrast, the optoelectronics parts of the system are unique to optical communications. In system design and development, since all these system parts will interact with each other, they all need to be considered in deciding system performance objectives. For example, the fiber loss, transmitter output power level and receiver sensitivity, splice and connector losses, and so on all affect the system's power budget. Likewise, an optical transmitter's spectral properties, noise characteristics, modulation

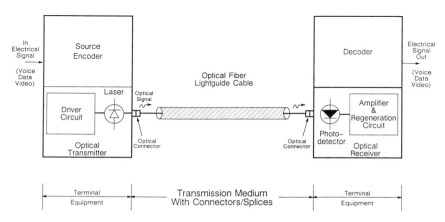

Figure 17.1 Schematic block diagram of an optical fiber transmission system.

response, the connector's reflection level, the fiber's dispersion, the crosstalk level of the WDM components, and so on can all contribute to system impairment.

In general, the active optoelectronic devices and the transmission fibers together determine the most essential characteristics of the particular optical fiber transmission system. On the other hand, system flexibility is often made possible or enhanced only with the use of various passive optical components.

17.3 MULTIMODE VS. SINGLE-MODE OPTICAL FIBER TRANSMISSION SYSTEMS

Most of the earlier optical fiber transmission systems installed were based on multimode optical fiber technology using first short-wavelength (0.8-μm) AlGaAs semiconductor laser diodes and then long-wavelength (1.3-μm) InGaAsP semiconductor laser diodes [1, 2]. However, compared with single-mode optical fiber systems, multimode optical fiber systems have two significant disadvantages:

1. Modal dispersion is large and difficult to control [3]. High-bandwidth (e.g., 4–10 GHz-km) multimode fibers are not easy to make because precise graded-index profile parameter control in manufacturing is needed. Moreover, the length dependence of the bandwidth in a concatenated multimode fiber link is not simple to predict. Furthermore, profile dispersion, the wavelength dependence of the index profile, causes the bandwidth to peak in a narrow wavelength region; therefore, multimode fibers optimized for one wavelength region may not be good for other wavelength regions. This is undesirable for wavelength-division multiplexing (WDM), thus limiting the upgradability of multimode systems.
2. Modal noise causes significant system penalty [4]. Modal noise, due to time-varying modal-distribution-related interference effects, which can cause serious system penalty, is difficult to predict or quantify because it depends on the location and characteristics of the interconnection components in the multimode fiber link and the optical source coherence, thus requiring extra care in system design, deployment, and maintenance.

17.3.1 Single-Mode Optical Fibers for Long-Distance Transmission

Mainly due to the factors just listed, multimode optical fiber systems are typically limited to short-distance, low-bit-rate optical transmission applications. In addition to more complex manufacturing control and system design, multimode fibers have higher losses than single-mode fibers because of the increased Rayleigh scattering due to higher dopant concentration in the core. Thus, for long-haul transmissions, multimode fibers have distinct disadvantages both in loss and in bandwidth, although the splicing and connector losses are lower. Early fiber transmission system installations (pre-1980) were mostly based on multimode fibers, although research and development of single-mode fiber transmission technology was well under way in many laboratories.

Shortly after 1980, though, there was a clear trend to using all-single-mode fiber systems in long-haul (interoffice and trunking) applications, as the technology moved toward higher bit rates and longer repeater spacing systems. The first half of the 1980s saw a rapid growth of the use of single-mode fiber transmission systems and the associated technology; this was accompanied by an accelerated expansion in the fiber industry to meet the growing demand for single-mode optical fiber transmission systems. As a consequence, the rate of single-mode optical fiber production and system deployment was rather fast, resulting in a situation that some would consider as reaching a saturation level. With the signs of saturation following the preceding years' exponential growth, new single-mode fiber systems are being added to the interoffice and trunking network at a slower rate, but characterized by the new era of high-speed gigabit-per-second systems using more advanced technology such as single-frequency distributed feedback (DFB) laser diodes [5]. It is obvious that almost all of the new long-haul optical communication systems today and in the future will be based on single-mode fibers.

There is still a substantial demand, however, for multimode-fiber-based, low-bit-rate, short-distance intrabuilding systems for local area optical networks. In such applications, modal dispersion and modal noise do not pose significant problems. Instead, the coupling and branching efficiency of large-core multimode fibers may become more important than the fiber's transmission loss/dispersion impairment. Together with inexpensive LEDs or low-cost lasers, multimode fiber-based systems are useful for data/voice/ video transmission in intelligent building-type local area network (LAN) applications. In fact, for some less demanding intrabuilding applications, even inexpensive large-core plastic multimode fibers (which have higher losses than silica glass fibers) could be used with inexpensive LEDs operating in the visible wavelength region to accomplish the transmission objective. However, for long-distance high-bandwidth optical fiber transmission systems, single-mode optical fibers are clearly the transmission medium of choice.

17.3.2 Single-Mode Optical Fiber Transmission for Subscriber Loops

While the long-haul trunking portion of the digital communication network continues to deploy new high-speed single-mode optical fiber transmission systems, the exponential demand growth period is probably over. It is generally agreed that the next large-scale deployment of optical fiber transmission systems will be in the distribution part of the network, bringing optoelectronic information transmission technology to subscribers' homes and business premises. Bringing fibers directly to the home to provide the new broadband services together with narrowband services is indeed the next wave, the next challenge of optical communications.

Although subscriber loop applications in the near term can be accomplished using multimode optical fiber systems due to the short distance (1–5 km typically), current thinking is that single-mode optical fibers should be used in the loop, for the following reasons:

1. The very-high-bandwidth potential of single-mode fibers ensures future upgradability by either high-speed TDM or WDM.
2. Single-mode fiber technology is compatible with OEIC (optoelectronic integrated circuits), which is promising for future low-cost high-reliability system applications.
3. Single-mode fibers have lower loss and are less expensive than multimode fibers; in future coherent transmission systems, single-mode fibers must be used.

All these considerations favor the use of single-mode fibers even for relatively short distribution distances as a once-for-all transmission medium of choice, making the entire digital network almost a totally single-mode-fiber network.

In this chapter we discuss the essential aspects of single-mode optical fiber transmission systems and technology and their applications in long-distance trunk and interoffice transmission as well as in the loop and distribution part of the network.

17.4 SINGLE-MODE FIBER TRANSMISSION: OPERATING WAVELENGTH CONSIDERATIONS

17.4.1 Fiber Transmission Loss and Dispersion— Choice of Operating Wavelength Region and Optical Sources

Figure 17.2a shows the loss spectra of a low-loss silica fiber; the OH^- peak near 1.4 μm usually separates the two low-loss valleys, although the best low-OH^- fibers may have practically no water peak so that the entire loss spectrum shows a wide-open loss window spanning from 1 to 1.7 μm. Figure

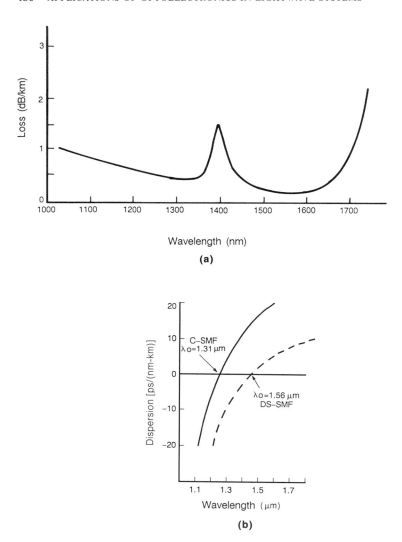

Figure 17.2 (a) Typical loss spectra of a low-loss single-mode fiber. (b) Chromatic dispersion characteristics of a conventional single-mode fiber (C-SMF) and a dispersion-shifted single-mode fiber (DS-SMF).

17.2b shows the chromatic dispersion characteristics of a conventional single-mode fiber (C-SMF) and a dispersion-shifted single-mode fiber (DS-SMF), with zero dispersion at 1.31 and 1.56 μm, respectively.

It is clear that the low-loss transmission windows are the 1.3-μm region shorter in wavelength than the OH$^-$ peak and the broad low-loss window in the 1.56-μm spectral range. Typically these two wavelength regions are used for long-distance single-mode fiber transmission. The minimum loss is about 0.37 dB/km near 1.3 μm and 0.18 dB/km near 1.56 μm for high-quality single-mode optical fibers. Assuming, for example, a loss budget of 30 dB (considering transmitter power, receiver sensitivity, and system margin) for the fiber link alone, the allowable transmission distance is about 80 km near 1.3 μm and 160 km near 1.56 μm. This illustrates the clear advantage of the lower-loss transmission of 1.56-μm wavelength sources for these single-mode fibers.

17.4.2 Loss-Limited vs. Dispersion-Limited Transmission

In reality, though, the above conclusion is valid only when the transmission is loss-limited and the dispersion penalty is of little concern. As can be seen from Figure 17.2b, the chromatic dispersion of a regular (conventional) single-mode fiber (C-SMF) in the lowest-loss 1.56-μm region is typically 10–20 times that in the 1.3-μm region. The intersymbol interference due to the fiber's chromatic dispersion can quickly limit the high-bit-rate optical signal transmission to a distance much shorter than the loss-limited distance. Since the single-mode fiber's zero-dispersion wavelength is typically around 1.3 μm, optical sources operating in the 1.3-μm loss window region experience much smaller pulse broadening or other dispersion-induced penalty compared with operation in the 1.56-μm region. As a consequence, the dispersion-limited distance in 1.56-μm spectral region transmission is much shorter than that for the 1.3-μm spectral region.

To take advantage of the lowest fiber loss near 1.56 μm without introducing a large dispersion penalty, one can either (1) use a dispersion-shifted single-mode fiber (DS-SMF) with its zero-dispersion wavelength shifted to the 1.56-μm spectral region or (2) use a narrow-spectral-width single-frequency DFB laser diode as the optical source. Either of these two approaches will reduce the chromatic dispersion-related transmission penalties. The DS-SMF curve in Figure 17.2b shows the chromatic dispersion characteristics of a dispersion-shifted single-mode fiber, which crosses zero in the 1.56-μm spectral region. In contrast, the C-SMF curve shows the dispersion of a conventional single-mode fiber with its zero dispersion wavelength in the 1.31-μm region.

Figures 17.3a and b illustrate the spectral characteristics of a single-frequency [or single-longitudinal-mode (SLM)] semiconductor laser diode in contrast with the regular multi-longitudinal-mode (MLM) Fabry-Perot cavity (F-P) laser diodes. The broad spectral width of the MLM laser diodes contributes to the dispersion penalty due to pulse broadening in the transmission fiber and due to mode partition noise (MPN), to be discussed later.

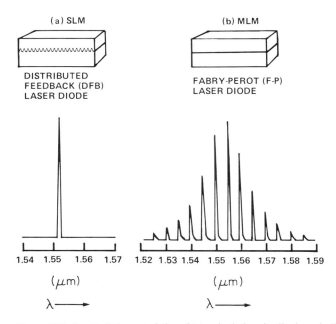

Figure 17.3 Spectral characteristics of (a) a single-longitudinal-mode (SLM) laser diode and (b) a multi-longitudinal-mode (MLM) laser diode. The laser structures are typically distributed-feedback (DFB) type cavity for SLM laser diodes and Fabry-Perot (F-P) cavity for MLM laser diodes.

17.4.3 Optical Source and Fiber Combinations for Single-Mode Fiber Systems

Table 17.1 shows some examples of possible source (MLM F-P laser diodes vs. SLM DFB laser diodes, 1.3-μm vs. 1.56-μm wavelength) and fiber (regular 1.3-μm zero-dispersion fiber, C-SMF, vs. 1.56-μm dispersion-shifted fiber, DS-SMF) combinations in long-distance single-mode optical fiber transmission systems. The majority of today's single-mode fiber transmission systems use C-SMF fiber (zero dispersion, 1.31 μm) and Fabry-Perot-type MLM laser diodes operating near 1.3 μm. However, 1.3-μm SLM DFB lasers are beginning to be used in Gb/s systems. These DFB laser-based single-mode fiber systems represent the first use of SLM (sometimes called single-frequency) laser diodes in commercial optical fiber communication applications. There are also WDM single-mode fiber systems using both 1.3-μm Fabry-Perot MLM laser diodes and 1.56-μm SLM DFB laser diodes in conventional single-mode fibers (C-SMF) with the zero-dispersion wavelength at 1.3 μm.

The main use of 1.56-μm single-mode fiber systems with MLM laser diode sources is typically for long-repeater-spacing but lower-bit-rate (less

TABLE 17.1. Examples of possible source–fiber combinations for long-haul system applications

	Source characteristics		Fiber characteristics	
Laser type	Wavelength λ_s (nm)	Fiber type	Zero dispersion region (nm)	Loss at λ_s (dB/km)
1. MLM F-P LD	1300	C-SMF	1310	0.40
2. SLM DFB LD	1300	C-SMF	1310	0.40
3. SLM DFB-LD	1560	C-SMF	1310	0.21
4. MLM F-P LD	1560	C-SMF	1310	0.21
5. MLM F-P LD	1560	DS-SMF	1550	0.24
6. SLM DFB-LD	1560	DS-SMF	1550	0.24

Note: 1 and 5 are multi-longitudinal-mode lasers operating at the fibers' zero-dispersion wavelength; 2 and 6 are single-longitudinal-mode lasers operating at the fibers' zero-dispersion wavelength; 3 and 4 are single- and multi-longitudinal-mode lasers, respectively operating at a high dispersion region of the fiber. Most of the single-mode fiber systems use 1 and 2, and some use 3 for upgrading existing systems by wavelength-division multiplexing.

than Gb/s) applications. The long repeater spacing is of particular interest for repeaterless island-hopping undersea optical fiber systems [6]. For the next generation of undersea optical fiber systems, 1.56-μm single-mode fiber transmission systems based on dispersion-shifted single-mode fibers (DS-SMF) or single-longitudinal-mode (SLM) DFB lasers are being developed, as such systems promise to require a greatly reduced number of undersea repeaters [7].

Obviously if one designs a system using both a 1.56-μm SLM DFB laser diode source and DS-SMF fibers with the minimum dispersion wavelength near 1.56 μm, the loss and dispersion limitations would be simultaneously minimized. The transmission distance/capacity will thus be maximized, if we assume that the same optical source power and receiver sensitivity are available at 1.56 μm as at 1.3 μm. Such DFB laser/DS-SMF systems operating at 1.56 μm are just beginning to be considered for some special transmission applications [7].

17.5 EXAMPLES OF SINGLE-MODE OPTICAL FIBER TRANSMISSION SYSTEMS

There are a large number of single-mode optical fiber transmission systems installed since 1980 that are carrying traffic in telephone networks around the world. Here we describe a few examples of actual single-mode optical fiber transmission systems that have been installed or are being planned, together with their essential system parameters. These examples serve to illustrate some of the main features [8–10] of existing single-mode optical fiber transmission systems.

17.5.1 SL Undersea Lightwave System

The SL Undersea Lightwave System [11] is a large-capacity digital optical fiber transmission system capable of spanning the world's largest oceans and seas. The system is being developed and manufactured for use in telecommunications systems planned for both the Atlantic and Pacific oceans.

The SL single-mode fiber system has been selected to be the main part of the 8th Trans-Atlantic Telephone (TAT-8) cable system spanning more than 5600 km and connecting the North American and European continents, placed in service in 1988. Near the European continent, the SL system, developed by AT&T, has a branching repeater connected by the British STC NL2 system to England and by the French Submarcon S280 system to France [11]. See Figure 17.4.

It is fair to say that although there was strong interest in realizing the potential of single-mode fiber transmission in the late 1970s, and terrestrial single-mode optical fiber systems were deployed long before 1988, the SL undersea lightwave system was the first practical system proposal that really stimulated the research and development of single-mode optical fiber

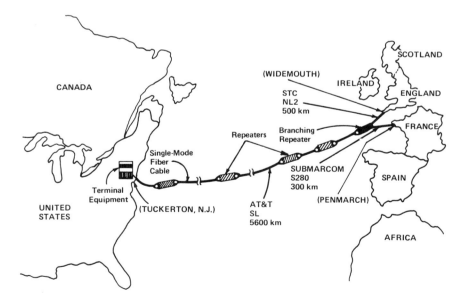

Figure 17.4 Route map of the Trans-Atlantic Telephone cable system (TAT-8) based on the SL single-mode fiber-based lightwave system connecting the North American and European continents.

transmission technology. The proposal for TAT-8 [11] started before 1978; at that time single-mode optical fiber transmission technology was in its infancy. However, the advantages of an optical transmission system based on low-loss and high-bandwidth single-mode optical fibers were clearly significant; there was simply no competition from copper cable systems or multimode optical fiber transmission technology. The proposal and the development work of the SL undersea single-mode fiber system helped stimulate the research and development of single-mode fiber transmission technology.

It is also important to realize that because of the high-reliability objective—25 years of system lifetime with fewer than three repairs—the SL undersea lightwave system effort was also principally responsible for the serious and challenging task of researching and developing high-performance, high-reliability optoelectronic devices [11].

The TAT-8 is designed to carry a traffic of over 20,000 voice circuits across the Atlantic connecting the United States and Europe. To achieve this goal, the SL system uses two single-mode optical fiber pairs with a transmission bit rate of 296 Mb/s, operating near the minimum dispersion wavelength of 1.3 μm. Table 17.2 lists some of the important system parameters for the SL undersea lightwave system. The key features of the system are discussed in the following paragraphs.

a. Transmission Medium. Conventional single-mode fibers with a depressed cladding design are used. The minimum dispersion wavelength is around

TABLE 17.2. Some of the SL undersea lightwave system parameters

Transmission medium	Depressed-cladding single-mode fiber
	$\lambda_0 \approx 1310$ nm
	Cutoff wavelength 1250 nm
	Loss at 1300 nm = 0.4 dB/km
Transmission rate	296 Mb/s (NRZ)
Optical transmitters	1310 nm $\pm$ 20 nm BH structure MLM laser diode
	Rms spectral width = 2 nm
	Output power > -2 dBm
	Four per regenerator connected to a 1 × 4 optical relay switch for redundancy
Optical receiver	InGaAs PIN photodiode with an integrated silicon transimpedance amplifier
	Sensitivity > -31 dBm at 10^{-9} BER
	Dynamic range > 18 dB
System life	25 years, <3 repairs in 25 years

1310 nm. Average fiber cable loss is about 0.4 dB/km; the cutoff wavelength is around 1250 nm. Chromatic dispersion of the fiber over the 1290–1330-nm region is less than 2.0 ps/(nm-km). In addition to these basic optical parameters, the fiber cable and the splicing have to meet the strength and reliability requirements to allow recovery and laying of the fiber-optic cable under the worst sea conditions at depths of 5.5 km and wave heights of 4–6 m [11].

b. Optical Transmitter/Receiver/Regenerators. The SL system's major redundancy feature is the transmitter array of up to four laser transmitters together with an optical relay switch and a transposition network that allow the signal to be switched to a standby line, bypassing a repeater section in case of failure. Each transmitter uses a high-reliability multi-longitudinal-mode InGaAsP BH laser [12] with central wavelength in the 1290–1330-nm range, having an rms spectral width of about 2 nm. The laser has a threshold current typically less than 25 mA and a wideband modulation response up to 2 GHz, adequate for Gb/s applications. The laser–fiber coupling loss is typically 4 dB; with the laser operating at 5 mW, the single-mode fiber pigtail output power is typically 2 mW cw and 1 mW (0 dBm) average under modulation with 296 Mb/s NRZ (non-return-to-zero) digital pseudorandom pulses. The transmitter was designed to operate up to 800 Mb/s, although the SL system was initially targeted at 296 Mb/s operation. To meet the system's stringent reliability objective, the transmitter array module has been designed to accommodate up to four transmitters per regenerator. Elaborate screening, characterization, and burn-in of each transmitter are required, even with the redundancy scheme.

The receiver uses a passivated and antireflection-coated InGaAs planar *p-i-n* (PIN) photodiode, together with a silicon bipolar integrated transimpedance amplifier for high-sensitivity and high-reliability operation. At 1300 nm and 296 Mb/s the receiver has a better than −34 dBm sensitivity at 10^{-9} bit error rate (BER), and a dynamic range of greater than 18 dB. The receiver package uses a multimode fiber pigtail for more efficient light coupling. An InGaAs PIN photodiode is used instead of a germanium avalanche photodiode (APD) in this case, because in germanium APDs, in addition to the need for noise and gain control, the high bias voltage required poses design constraints and increased degradation. Production germanium APDs have problems with dark current, passivation, and excess noise; the advent of InGaAs APDs promises significant improvements, and they are beginning to be used in high-performance systems.

One other key component, as part of the redundancy design for the high-reliability objective, is the 4×1 single-mode fiber optical switch used to switch the laser source in the SL regenerators. Preferentially etched silicon V-groove chips, originally developed for fiber array splicing, are used for fiber alignment in the design of the electromechanical moving-fiber optical switch. The assembled switch with an insertion loss less than 1.8 dB has been achieved; the reliability objective must also be met to ensure switchover ability during the 25-year system lifetime.

c. System Transmission Parameters. The overall system length is about 5600 km; a total of 130 repeaters are used. To meet the system BER requirement of less than 4.4×10^{-8}, the individual repeater spans are designed to have an end-of-life BER performance of better than 10^{-9}. This span requirement translates into a loss and dynamic range budget.

The optical output power at the optical relay switch (which is connected to four laser transmitters) is greater than -2 dBm (counting the transmitter output of 0 dBm and the switch loss of less than 1.8 dBm). The receiver sensitivity of the regenerators is better than -34 dBm. The regenerator is to be operated at a nominal receiver power greater than -31 dBm. Thus there is a 29-dB power budget.

Table 17.3 shows the power budget; the available 29 dB is allocated as follows:

23 dB for fiber link loss including loss of fiber and splices
1 dB for system penalties due to fiber dispersion, optical feedback, timing error, and mode partition noise
5 dB system margin for system aging, degradation, etc.

With the 23 dB of power budgeted for the fiber link, and with a 0.45-dB/km fiber link loss, this results in a deep-sea single-mode fiber span length of about 50 km.

For a 6000-km system length and a 50-km repeater span, more than 100 repeaters are needed for the TAT-8. A longer repeater span means that fewer undersea repeaters are needed for the entire system. This amounts to significant cost savings and reliability improvements. The next generation of undersea lightwave systems, such as the SL system being proposed for TAT-9, therefore, begins to consider the single-mode fiber system operating in the 1560-nm region to take advantage of the lower fiber loss (0.2 dB/km as compared with 0.4 dB/km at 1310 nm). These next-generation undersea single-mode fiber transmission systems are described in the following section.

TABLE 17.3. Power budget of SL undersea lightwave system repeater

Laser transmitter power > -2 dBm	
Receiver sensitivity > -31 dBm	
Power budget available per repeater span	29 dB
1. Fiber link loss* (including fiber and splice loss)	23 dB
2. Power penalty (due to dispersion, timing error, mode partition noise, feedback)	1 dB
3. System margin (for aging of system components)	5 dB

  * Repeater span > 50 km for 0.45-dB/km loss.

17.5.2 Next-Generation Undersea Fiber System Based on 1560-nm Operation

Due to the projected growth in the demand for telecommunications service, the next-generation undersea optical fiber transmission system may be needed by 1992–1993, after the 1988 installation of the TAT-8 system. The new generation of the transatlantic optical fiber system will undoubtedly consider the use of the newer technology to achieve the goal of higher capacity transmission at reduced cost [7]. Since the total distance and hence total fiber cable length are fixed, the cost reduction must come primarily from reducing the number of expensive undersea repeaters while maintaining the same high reliability.

To reduce the number of repeaters, the most effective approach is to operate the system in the 1560-nm wavelength region of minimum loss for single-mode silica glass optical fibers. As was mentioned above, since the loss at 1560 nm could be as low as one-half the loss at 1310 nm, for the same power budget (assuming the same transmitter power, receiver sensitivity, and splice loss), one can have a repeater span twice that at 1310 nm, namely, over 100 km. The number of repeaters for a 6000-km system length, for example, would thus be reduced from 120 to 60. This would represent a very significant cost reduction indeed.

These 1560-nm single-mode optical transmission systems will be based on one of the following options, which are basically what we have already outlined in Section 17.4 and Table 17.1.

Option A. Using A Single-Longitudinal-Mode (SLM) DFB Laser Source. In this option one uses conventional single-mode fibers (C-SMF), which have minimum dispersion near 1300 nm but lowest loss in the 1560-nm spectral region. To take advantage of the lowest loss without suffering the intersymbol interference or mode-partition-noise penalty due to the high dispersion at 1560 nm, a single-longitudinal-mode distributed feedback laser diode (SLM DFB-LD) is used. The SLM DFB-LD under modulation can have an effective spectral width of less than 0.2 nm (due to chirping, to be discussed later). For a fiber dispersion of 17 ps/(nm-km) at 1560 nm, and a 100-km span length achievable at this lowest-loss wavelength, the dispersion-induced pulse broadening will be less than 340 ps, posing no significant problem for systems operating above Gb/s bit rates. Of course, here we assume that the SLM DFB lasers are well behaved so that a large side-mode suppression ratio (30 dB or more) is maintained under modulated conditions and there is no mode-partitioning noise due to insufficiently suppressed side modes.

Actually single-mode fiber system transmission experiments at various bit rates based on such C-SMF and SLM DFB LDs have already been reported [12]; furthermore, a few such systems have been developed for terrestrial applications. One example of these systems will be discussed later.

Option B. Using Dispersion-Shifted Single-Mode Fibers (DS-SMF). In this approach one uses multi-longitudinal-mode (MLM) laser diodes as optical sources at 1560 nm to take advantage of the fiber's low loss. The penalty due to the wide MLM laser spectral width (2–4 nm) would be too high if a conventional single-mode fiber (C-SMF) with large chromatic dispersion at 1560 nm is used. Therefore, one has to use dispersion-shifted single-mode fibers (DS-SMF), which have both the dispersion minimum and the loss minimum in the 1560-nm region. Such SMF transmission systems based on DS-SMF and MLM-LDs have been demonstrated in the laboratory [13]; the potential for 100–150-km-long repeater spacing long-haul undersea transmission or for repeaterless one-hop undersea transmission between islands is clearly evident.

Option C. Using Both DS-SMF and SLM DFB Laser Diodes. This uses the combination of the key technologies from both Option A and Option B, for the ultimate, very long repeater spacing and very high speed (high-Gb/s) single-mode fiber transmission systems. The laboratory demonstration of such systems does not require any new development once the previous options are developed. The actual applications of such high-performance single-mode long-haul transmission systems based on both SLM DFB-LDs and DS-SMF will probably be mainly for undersea transmission and not so much for terrestrial transmission, at least in the near term. This is simply because so many conventional non-dispersion-shifted fibers have already been installed and represent a substantial long-term investment in the terrestrial digital network.

The next generation of undersea lightwave systems based on single-mode fiber transmission technology will choose either Option A or Option B, with Option C a good possibility. As to the technology state of the art, the development of the DS-SMF fiber technology is more mature; in comparison, the development of the 1560-nm region laser diodes is not as mature, especially the SLM DFB laser diodes for the 1560-nm spectral region. Nevertheless, research and development of these devices is continuing at a very impressive pace, and the important reliability studies are well under way. In a few years, all the technology and system options will very likely become readily available to system designers. The difference in the state-of-the-art technology is mainly due to the expected reliability problems one needs to solve, which, in general, are much more difficult to tackle in active devices than in passive transmission media such as glass optical fibers. It is expected that in the future, SLM DFB laser diodes in both the 1300- and 1560-nm spectral regions will be developed and manufactured with low cost and high reliability just as the 1300-nm MLM laser diodes used in present single-mode fiber transmission systems, allowing system designers to have greater flexibility in optical fiber transmission design.

17.5.3 A High-Speed Terrestrial Single-Mode Optical Fiber System, F1.6G

While most of the earlier fiber transmission systems operate in the 45–565-Mb/s bit rate range, the trend is toward Gb/s bit-rate systems for future terrestrial long-haul applications. This is because the relative cost per Gb/s per kilometer (or, equivalently, per voice circuit per mile) is much cheaper with higher-bit-rate systems when the volume need exists, such as in long-haul trunking and interoffice transmission.

Recently, a few terrestrial single-mode optical fiber transmission systems that operate above 1 Gb/s have been developed and have become available for commercial applications. Examples include AT&T's FT series G 1.7-Gb/s system, Fujitsu's 1.6-Gb/s optical fiber transmission system, NEC's 1.12- and 1.6-Gb/s systems, Rockwell International's LTS-21130, a 1.13-Gb/s system, and many other similar systems.

Most of these commercial Gb/s single-mode transmission systems use SLM-DFB lasers operating at 1300 nm, although 1300-nm MLM laser diodes are also used in some cases. In this section we look at the system parameters and the technology used in NTT's F1.6G system as an example of high-speed Gb/s terrestrial single-mode fiber systems.

Table 17.4 lists the important system parameters for NTT's F1.6G system [14]. The high-capacity (close to 24,000 voice circuits per fiber) transmission is achieved at an operating bit rate of 1.8 Gb/s (information bit rate of ~ 1.6 Gb/s). The single-mode fibers used are C-SMF with zero-dispersion wavelength in the 1310-nm region. A laser module containing an SLM DFB-LD (wavelengths of 1301–1317 nm) has been designed with a YIG plate optical isolator (isolation based on the nonreciprocal Faraday rotation of the optical field's polarization in a magnetic field) in the package. Low-reflection FC/PC (physical contact) connectors (reflection less than -30 dB) are used. The isolator provides 25-dB isolation. The combination of the isolator and the low-reflection connectors ensures minimal reflection-induced noise effects. The optical output power is about 0 dBm. An InGaAs APD is

---

TABLE 17.4. Some of NTT's F1.6G single-mode system parameters

Information rate	1.5888 Gb/s
Line rate	1.8209 Gb/s
Line code	Scrambled 10B1C code, RZ
Fiber	Single-mode fiber ($\lambda_0 = 1.31\ \mu$m)
Optical source	1.3-μm InGaAsP/InP DFB laser diode
Optical output power	0 dBm
Optical detector	InGaAs APD
Average received power	At BER of 1×10^{-11}, > -30 dBm
Repeater spacing	>40 km
Jitter	$2°$ rms/repeater; $15°$ rms/2500 km

used as the photodetector, which, as mentioned above, has better dark current and noise characteristics than germanium APDs. The receiver has a sensitivity of −30 dBm at 1.8 Gb/s. High-speed silicon integrated circuit technology is used. The F1.6G system has been developed and designed to have a repeater spacing of >40 km, allowing the upgrading from the existing F-400M, a 400-Mb/s system. The F1.6G system has been installed for field trial in a route of approximately 120 km from Tachikawa in Tokyo to Kofu in Yamanashi Prefecture.

In a related experiment, both SLM DFB and Fabry-Perot-type MLM laser diodes were used in the F1.6G transmission, and the results were compared. Figure 17.5 shows the bit-error-rate performance curves of the F1.6G repeater designed with these two different lasers. With the DFB laser diodes,

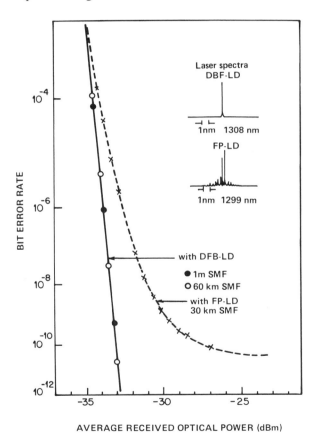

AVERAGE RECEIVED OPTICAL POWER (dBm)

Figure 17.5 Bit-error-rate performance of NTT's F1.6G repeater designed with an SLM DFB laser diode source and an MLM F-P laser diode source in the transmitter section. An InGaAs APD is used in the receiver. The bit-error-rate "floor" was seen with the MLM F-P laser diode, with only 30 km of C-SMF ($\lambda_0 \approx 1310$ nm) at the transmission line rate of 1.8 Gb/s; this floor effect is attributed to mode partition noise (MPN).

the side-mode suppression ratio was better than 25 dB, and the 10-dB down spectral width was 0.5 nm. With such a narrow output spectrum near 1308 nm, the bit-error-rate performance with the DFB laser after 60 km of single-mode fiber transmission showed no degradation due to fiber dispersion. In contrast, the bit-error-rate performance of a repeater using a regular multi-longitudinal-mode Fabry-Perot laser diode even after only 30 km of transmission readily showed the dispersion penalty due to mode-partition noise. The bit-error-rate floor, as seen in Figure 17.5, is characteristic of mode-partition noise, to be discussed later. The distinct advantage of using SLM DFB laser diodes instead of MLM laser diodes for Gb/s transmission systems is thus clearly demonstrated. It may be possible to use MLM laser diodes in Gb/s systems, but this would require a very tight selection of multi-mode laser spectral width and center wavelength, to match very closely the fiber's dispersion minimum (to minimize mode-partition noise). This could mean a less flexible system design, shorter fiber span between repeaters, or less margin for device degradation.

Beyond the 1300-nm distributed feedback laser diode based 1.6-Gb/s system, NTT is also developing, for future upgrading, a 1560-nm F1.6G system that will use both SLM DFB-LDs near 1560 nm and DS-SMF (as in case 6 of Table 17.1 or Option C of the previous section), to have a repeater spacing of > 80 km.

17.5.4 Advanced Single-Mode Fiber Transmission Experiments

While the previous sections describe actual single-mode optical fiber transmission systems that have been developed for commercial applications, many system transmission experiments have been performed in the laboratory environment to study the capability and the limitations of a particular system configuration and the associated optoelectronics/fiber technology. In this section we describe briefly several representative single-mode fiber transmission experiments, which serve to illustrate the various possible system design concepts, system capabilities, and the limits of a given system technology.

Most laboratory transmission experiments are carried out in a well-controlled environment, tested with bit-error-rate test sets, with uncabled fibers on reels, with selected optoelectronic devices in the transmitters and receivers, and with key system components fine-tuned for optimum performance over a narrow operating range of parameters. Usually there is little system margin, and the long-term reliability or degradation concerns are not considered. Therefore, deploying such a laboratory-demonstrated transmission system in the real world requires a substantial development effort, and realizable system performance would be less than what is described here. Nevertheless, such laboratory transmission experiments provide the basis for establishing the technological limits and challenge researchers to advance the technology

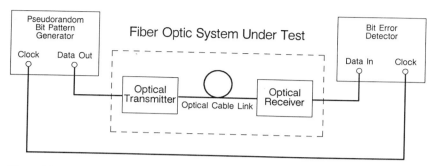

Figure 17.6 Schematic diagram of a typical setup for bit-error-rate (BER) measurement in an optical fiber transmission experiment.

frontier. Figure 17.6 shows a schematic of a typical bit-error-rate (BER) measurement setup for optical fiber transmission experiments; in actual measurements, data (curves) on bit error rate versus received optical power (as in, for example, Fig. 17.5) as a function of various system parameters (fiber length, transmission bit rate, source spectral width, extinction ratio, reflection level, etc.) of interest are studied.

a. Over 200-km Transmission In Dispersion-Shifted Single-Mode Fibers.
Long-span single-mode fiber transmission experiments at the loss minimum wavelength of 1.55 μm at relatively low bit rates of 140 and 34 Mb/s have been reported [13]. It was shown that in conventional single-mode fibers (C-SMF) which have their zero-dispersion wavelength at 1.3 μm, and with an MLM-LD source at 1.55 μm (3-nm rms spectral width), transmission at 34 Mb/s is already dispersion-limited. The fiber loss was 0.2 dB/km; the achieved dispersion-limited distance was 176 km. When a dispersion-shifted single-mode fiber (DS-SMF) was used (loss 0.23 dB/km including splices), loss-limited transmission was obtained over 220 km at 140 Mb/s and 233 km at 34 Mb/s. The germanium APD used had a sensitivity of -52.7 dBm at 1.55 μm and 34 Mb/s; at 140 Mb/s the sensitivity was -46.5 dBm. The launched laser power was $+5$ dBm, so there was a substantial power budget for the fiber span, allowing over 200 km of transmission.

b. 4-Gb/s, >100-km Transmission Experiments. While 2.4-Gb/s transmission systems are under development for commercial applications, laboratory transmission experiments in the 4–8-Gb/s range have already been reported [15, 16]. These were started after short-cavity high-speed InGaAsP laser diodes capable of being directly modulated at 4 Gb/s rates became available [17–20].

In one 4-Gb/s transmission experiment reported by AT&T Bell Laboratories, transmission over 103 km of C-SMF fiber at 1.54 μm was demonstrated [15]. The optical source used was a 1.54-μm vapor-phase-transported buried heterostructure (VPTBH) DFB laser directly modulated with 4-Gb/s NRZ pulses. The laser was not really high-speed; its small-signal 3-dB bandwidth was only 2.1 GHz. However, with a preequalizer LC circuit [15, 17], the combined bandwidth was improved to 4.3 GHz. The averaged optical power launched into the lensed single-mode input fiber was −1.6 dBm. The receiver employed a separate absorption, grading, and multiplication region (SAGM) InGaAs APD and a high-impedance GaAs FET preamplifier with low input capacitance. The overall receiver sensitivity was −30.6 dBm at 4 Gb/s for 10^{-9} bit error rate. A C-SMF with an average loss of 0.21 dB/km and dispersion of 17 ps/(nm-km) at 1550 nm was used; the total fiber link loss was 23 dB for the 103-km C-SMF fiber, splices, and connectors. BER curves for the experiment [15] show that in order to achieve optimum performance it was necessary to reduce the laser linewidth broadening (due to chirping, to be discussed later) by operating with a low extinction ratio of 3.5:1; the low extinction ratio caused a 3.9-dB power penalty. Still, there was a 1.9-dB dispersion penalty due to the pulse broadening associated with the resultant chirped linewidth.

In another 4-Gb/s transmission experiment reported by NEC [16], transmission over 74 km at 1.3 μm and over 120 km at 1.54 μm was demonstrated. NEC used double-channel planar buried heterostructure (DC-PBH) DFB laser diodes at both 1.3- and 1.54-μm wavelengths for the optical transmitters. The lasers had a double-mesa structure with low parasitic capacitance; the bandwidth is about 5 GHz. Return-to-zero pattern electrical pulses of 4 Gb/s were used for direct modulation of these DFB-LDs. The laser output was coupled by a GRIN rod lens into the transmission single-mode fiber. The power in the fiber was about 1.3 dBm. The receiver used a high-impedance type of front-end circuit; the photodiode used was a high gain–bandwidth product (50 GHz) planar InGaAs APD with a capacitance less than 0.3 pF. The receiver sensitivity was −31.5 dBm at 1.3 μm and −32.4 dBm at 1.54 μm. A conventional single-mode fiber (C-SMF) with zero dispersion at 1.31 μm and a dispersion of 17 ps/(nm-km) at 1.54 μm was used in the transmission experiments. Fiber losses at 1.3 and 1.54 μm were 0.37 and 0.21 dB/km, respectively. The achieved transmission distance was 74 km at 1.3 μm and 120 km at 1.54 μm. Table 17.5 lists the power budgets for these 4-Gb/s transmission experiments. Note that the side-mode suppression ratio (SSR) of these DFB lasers was typically better than 30 dB, and the chirped linewidth under 4 Gb/s RZ modulation was about 0.5 nm. Thus the dispersion penalty is very small at 1.3 μm but high at 1.54 μm [because of 17 ps/(nm-km) × 0.5 nm × 120 km = 1020 ps, or about 1 ns of pulse spreading]. It was also found necessary to bias the DFB-LD at 1.1 times threshold current at 1.3 μm and at twice the threshold at 1.54 μm for optimized overall operation and

TABLE 17.5. Power budgets for a 4-Gb/s single-mode fiber transmission experiment

Laser wavelength (nm)	1304	1540
Fiber length (km)	74	120
Optical output power (dBm)	+ 1.3	+ 1.6
Fiber link loss (dB)	27.6	25.2
	(0.37 dB/km)	(0.21 dB/km)
Receiver sensitivity* (dBm)	− 31.5	− 32.4
After transmission (dBm)	− 30[†]	− 24.5[‡]
Power penalty (dB)	1.5	7.9
Level margin (dB)	3.7	0.9

* $I_b = I_{th}$.
[†] $I_b = 1.1 I_{th}$.
[‡] $I_b = 2 I_{th}$.

take the corresponding extinction ratio penalties. Such optimization seems to be more important in high-speed systems using SLM DFB laser diodes. The trade-off between higher current biasing (for large side-mode suppression, low chirp broadening, and low mode-partition noise) of SLM DFB lasers and the resultant low extinction ratio is necessary. More discussion on this will be presented later.

c. 8-Gb/s and 10-Gb/s Transmission Experiments. The advent of truly high-speed semiconductor laser diodes with a small-signal 3-dB bandwidth in the 7–15-GHz range [19, 20] has made possible transmission experiments at very high Gb/s rates. Direct modulation of high-speed semiconductor laser diodes at 8 Gb/s [19] and 16 Gb/s [21, 22] has been successfully demonstrated; very wide bandwidth (> 20 GHz) PIN photodetectors have also been developed [23]. These paved the way to system transmission experiments at 8 Gb/s and higher bit rates.

In one experiment, a multi-longitudinal-mode high-speed constricted-mesa laser diode with its central wavelength temperature-tuned to match the fiber's dispersion minimum of 1307 nm was directly modulated with 8-Gb/s NRZ pulses [24]. Biased at 50 mA and modulated with a 22-mA peak-to-peak current, the average optical power of the MLM laser launched into a lensed single-mode fiber was 1.3 dBm. The on/off ratio was 6:1. A total of 30 km of C-SMF of the depressed-cladding design was used in the experiment. The total loss of the fiber, slices, and connectors was 13.6 dB. A SAGM InGaAs APD receiver with a sensitivity of − 15.5 dBm at 8 Gb/s for an error rate of 10^{-9} was used. The bit error rate versus received power results from the 30-km C-SMF transmission experiment showed a 1.5-dB penalty due to fiber dispersion and a 1.6-dB penalty due to the 6:1 extinction ratio.

In a recent experiment [25], transmission of 10-Gb/s RZ pulses from a a narrow-linewidth SLM DFB-LD at 1.55 μm, and an external modulator [24] to avoid the effect of chirp-induced line broadening in directly modulated DFB-LDs, transmission over 68 km at 8 Gb/s was demonstrated. It was reported that while there was no chirping in this external modulated scheme, the information bandwidth corresponding to the 8-Gb/s transmission limited the transmission distance due to dispersion.

In a recent experiment [25], transmission of 10-Gb/s RZ pulses from a 1.55-μm phase-shifted DFB laser diode in an 80-km-long dispersion-shifted single-mode fiber was achieved. At 10-Gb/s modulation, the laser input optical power into the fiber was -0.5 dBm, the side-mode-suppression ratio was 38 dB, and the chirp width was 0.8 nm at -20-dB point. A high-speed InGaAs APD with a 70-GHz gain-bandwidth product (gain M = 10, maximum bandwidth 7.5 GHz) was used with a low-impedance GaAs FET preamplifier to realize a wideband front end. The 10-Gb/s data streams were demultiplexed to 5 Gb/s, and the receiver sensitivity was -21.4 dBm at 5 Gb/s. The average fiber loss was 0.22 dB/km.

d. 16-Gb/s Laser Modulation Experiments. Laboratory research on the system technology of very high Gb/s single-mode fiber transmission continued beyond the 8-Gb/s transmission experiment described above. Based on the high-speed MLM constricted-mesa semiconductor laser diodes with 3-dB bandwidth in the 12–16-GHz range, direct modulation with 16-Gb/s NRZ pulses has been successfully demonstrated [21, 22]. The modulation experiments showed that just as in the 8-Gb/s transmission experiment, relatively high bias current has to be used and adjusted properly for the laser to respond properly at 16-Gb/s modulation. Figure 17.7 shows the laser modulation response to the 16-Gb/s NRZ modulation current pulses as a function of dc bias level. Note that when the bias current was too low, significant undesirable pulse patterning effects were observed. On the other hand, when the bias was too high, even though the optical pulses followed the modulation current waveform properly, a low extinction ratio (light on/off ratio) resulted. Thus there was an optimum bias current range that led to the optimum pulse response without too low an extinction ratio. It was also found that to obtain 100% modulation at the optimum bias current, high modulation current (e.g., 80 mA for a differential quantum efficiency of 0.2 mW/mA per facet) drive at 16 Gb/s would be needed. This would correspond to a laser power of 16 mW at the "on" level. This points to an important fact: for very high bit rate modulation, not only is a *high-speed and high-power laser* required, but also *high-speed and large-amplitude current pulses* are needed for deep modulation. An alternative to the high-speed large-amplitude current requirement is to operate the laser with a low extinction ratio and take the corresponding power penalty.

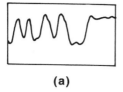

(a)

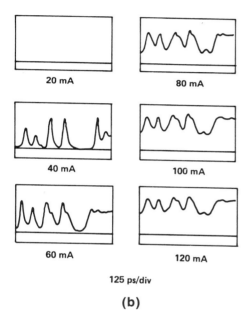

(b)

Figure 17.7 Direct modulation response of a high-speed semiconductor laser diode to NRZ modulation current at 16 Gb/s as a function of dc bias level. (a) 16-Gb/s NRZ modulation current pulses. (b) Laser output versus bias level. Note that the laser pulses follow the modulation within a range of dc bias currents where the extinction (on/off) ratio may have to be compromised.

17.6 WAVELENGTH-DIVISION MULTIPLEXING IN SINGLE-MODE OPTICAL FIBER SYSTEMS

While the development of high-capacity single-mode optical fiber transmission systems continues to move in the direction of higher and higher speed, high Gb/s bit rates, to upgrade an existing transmission link capacity one can either go to a higher transmission bit rate or use additional optical channels sharing the same optical fiber for increased bandwidth transmission. The optical wavelength-division multiplexing (WDM) approach [26–29] uses several simultaneous optical channels transmitting through the same optical fiber. With WDM schemes, for an established route where fiber cable installation was completed long ago, one can upgrade the system capacity by adding more optical channels without installing new fiber cable.

It appears that upgrading by going to higher speed or higher bit rates (time-division multiplexing, essentially) is likely to be considered first, because it is a more straightforward approach if the higher-speed system is already available. However, when the economic considerations are such that adding another optical channel at the same bit rate is more cost-justified, or if the higher-bit-rate system equipments are simply not yet available (for example, limited by the high-speed electronics required), then WDM will be deployed. This will depend on the relative maturity and cost of WDM technology versus high-speed optoelectronics technology.

Figures 17.8a and b show two typical WDM system configurations. Figure 17.8a shows a typical application—multichannel transmission in one direction in a single fiber [26–29]. Figure 17.8b shows the case for bidirectional transmission in the same fiber with two lasers at different wavelengths [26, 27].

The technology of WDM devices has been the subject of research and development for years [26–28]. Several multimode fiber transmission systems have used WDM technology to increase total capacity, usually at low transmission bit rates. Single-mode WDM optical fiber transmission systems operating at two to four wavelengths in the 1.3- and 1.55-μm regions have also been developed [28] for multichannel transmission at higher bit rates, although there was no significant deployment of WDM single-mode fiber systems until recently.

Wavelength-division multiplexing can be accomplished by using various dispersive optical components such as dielectric interference filters, gratings, fused biconical tapered fiber couplers with wavelength-dependent transmission, and microoptic components such as graded-refractive-index (GRIN) rod lenses with wavelength-dependent coatings [26, 27]. Important practical requirements of WDM components include low loss, low crosstalk between channels, low reflection, high stability under operating environmental conditions, small size, and low cost. Specific applications will determine the WDM technology of choice; for example, multiplexing two channels such as 1.3 μm and 1.55 μm could be accomplished more readily than multiplexing 4–10 channels in the same spectral region [28].

Actually WDM technology is of great interest not just as an alternative to a higher-speed upgrading approach but also because it offers system architectural flexibility. The future trend of WDM technology is to reduce optical channel spacing for densely multiplexed multichannel systems. This has been called dense WDM [30], or high-density (HD) WDM, which is between typical "coarse" WDM (two to five channels, channel spacing 20–100 nm) and OFDM (optical frequency-division multiplexing, 50–1000 channels with channel spacing in the GHz range). Dense WDM, with channel spacing between 1 and 10 nm, offers the possibility of 10–100 optical channels per fiber without the coherent detection technology that OFDM requires. Such WDM technology, with tunable filters multiwavelength DFB laser array

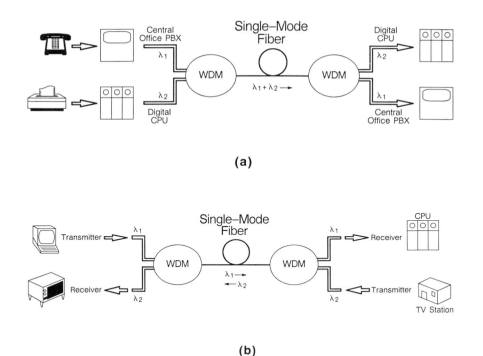

(a)

(b)

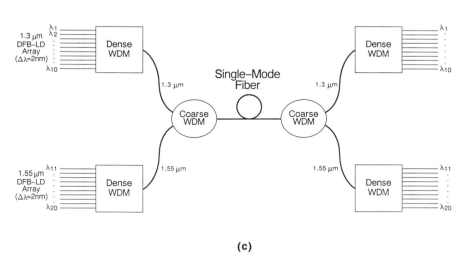

(c)

Figure 17.8 Examples of wavelength-division multiplexing (WDM) in single-mode fiber transmission applications. (a) Two-channel, unidirectional transmission; (b) two-channel, bidirectional tranmission; and (c) a high-capacity multichannel system based on SLM DFB laser arrays and high-density (HD) WDM.

technology [31] can provide very high capacity transmission for a given single-mode fiber link, or a multinode multichannel distribution network using direct-detection system components. Such an HD-WDM system could become an important part of future single-mode fiber transmission networks [31]. Figure 17.8c illustrates the concept of such a high-capacity HD-WDM DFB laser-based single-mode fiber system.

17.7 TRANSMISSION LIMITATIONS AND FACTORS CAUSING SYSTEM IMPAIRMENTS

As discussed above, the most obvious transmission limitations are due to fiber transmission loss and dispersion. However, in addition to this transmission loss and dispersion, in a single-mode fiber transmission system there are other important system degradation factors associated with the interaction of the optoelectronic devices and components with the transmission media that can cause system impairments. In this section we discuss these limitations and a few of these important factors (mode-partition noise in MLM and SLM lasers, chirping in SLM lasers, reflection-induced noise, etc.) and the corresponding system penalties. Figure 17.9 shows the general features of single-mode fiber transmission distance versus bit rate bounded by these limiting factors.

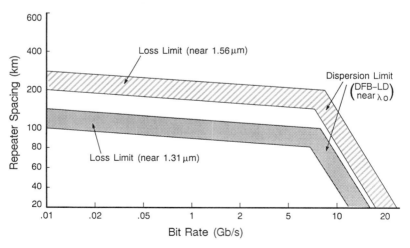

Figure 17.9 Repeater spacing or fiber distance versus bit rate in single-mode optical fiber transmission with boundaries/limitations due to fiber loss and dispersion near λ_0.

17.7.1 Transmission Loss and Dispersion

We discussed the different implications of the loss-limited and dispersion-limited transmission briefly in an earlier section. Here we elaborate on that discussion.

Loss-limited system operation means that the power-budgeting consideration limits the transmission distance. For example, if

$$P(\text{transmitter output power}) = P_t \, [\text{dBm}]$$

$$P(\text{receiver sensitivity}) = P_r \, [\text{dBm}]$$

and if there is no dispersion-related penalty or limit, then the maximum transmission distance is

$$L_m = \frac{(P_t - P_r) \, [\text{dB}]}{\text{average fiber link loss } [\text{dB/km}]}$$

The fiber link loss includes fiber loss, splice loss, and the loss due to connectors, WDM components, fiber couplers, and any other interconnecting components. For example, assuming we have a single-mode fiber system operating at 1300 nm with $P_t = 3$ dBm, $P_r = -34$ dBm, and an average fiber link loss of 0.41 dB/km (C-SMF at 1300 nm), the maximum loss-limited transmission fiber length is $L_m = (37/0.41)$ km $= 90$ km. Here, the system margin is not considered.

Dispersion-limited transmission occurs when the power budget allows for transmission over a length L_m, but the achievable transmission distance is shorter than the L_m from loss calculation alone, because the dispersion-induced signal pulse broadening $\Delta\tau$ is causing significant intersymbol interference between transmitted neighboring pulses. This is usually the case with higher-bit-rate systems with large overall dispersion. Significant intersymbol interference happens when the pulse broadening $\Delta\tau$ is larger than half the bit period T, that is, when

$$\Delta\tau > \tfrac{1}{2}T \tag{1}$$

The pulse broadening $\Delta\tau$, when λ_s is not at λ_0, is

$$\Delta\tau = \Delta\lambda_s D L \tag{2}$$

where D is the fiber dispersion, L is the transmission fiber length, and $\Delta\lambda_s$ is the laser source (rms) spectral width.

For example, suppose the loss-limited transmission distance is 90 km, as in the example above. But now suppose that the particular system is to be operated at 2 Gb/s (500-ps pulse period), using an MLM laser diode with an rms spectral width of 4 nm at a wavelength of 1280 nm (due to laser wavelength tolerance, it does not match the fiber's λ_0 exactly) where the fiber dispersion is 2 ps/(nm-km); we then have a case where the dispersion limits the transmission distance to approximately

$$L \cong (500 \text{ ps}/2)/(2 \text{ ps/nm} \times 4 \text{ nm}) \text{ km} = 250/8 \text{ km} = 31 \text{ km}$$

which, of course, is much less than the 90 km obtained from the loss consideration alone. As a rule of thumb, loss-limited transmission is usually possible only at relatively low bit rates (< 150 Mb/s) where the dispersion effects are negligible.

17.7.2 Mode-Partition Noise in MLM Lasers

In Fabry-Perot cavity MLM laser diodes, there is a certain degree of power distribution fluctuation among the various longitudinal modes. From theoretical analysis, in MLM lasers the statistical nature of the transient buildup of the various longitudinal modes leads to a fluctuating power distribution among the modes from one optical pulse to the next in a modulated pulse train. This pulse-to-pulse variation of the optical power distribution means that more power is in longitudinal mode A for one pulse while for the next pulse there is more power in longitudinal mode B. Figure 17.10 illustrates the mode-partition effect and the consequence of dispersive fiber transmission of such optical signal pulses. The pulse-to-pulse fluctuating optical power distribution when coupled with the chromatic dispersion of a long single-mode fiber will cause pulse-to-pulse signal waveform fluctuations, because

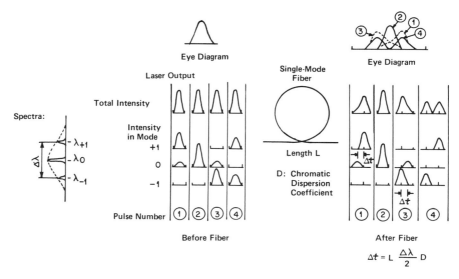

Figure 17.10 Simplified picture of mode-partition noise (laser's modal intensity distribution fluctuating from pulse to pulse) and the transmission of such optical signal pulses in a dispersive single-mode fiber (different arrival times for different longitudinal modes of the laser diode).

the wavelength-dependent delays for different longitudinal modes lead to pulse-to-pulse changes in the resultant received signal waveform. This in turn leads to eye closing and a bit-error-rate floor that is independent of the optical power received. This is called mode-partition noise, or MPN [32, 33]. It has been shown that the system penalty caused by MPN leads to the bit rate–distance product (BL) limit in the following expressions:

$$BL \leq \frac{130}{\Delta \lambda_s D \sqrt{k}} \quad \text{Gb/(s-km)} \tag{3}$$

where k is the "mode-partition noise factor" ($0 < k < 1$, typically between 0.4 and 0.7 for MLM laser diodes) [33].

If there were no mode partition, that is, if the longitudinal-mode power distribution profile were constant from pulse to pulse (no mode partition, $k = 0$), then the system penalty would be due to the pulse broadening alone, the broadening being associated with the fixed (not pulse-to-pulse varying) spectral width and the fiber dispersion. If this is the limiting factor, the transmission limitation is described by [33]

$$BL \leq \frac{341}{\Delta \lambda_s D} \quad \text{Gb/(s-km)} \tag{4}$$

Comparing the two expressions, Eqs. (3) and (4), it is clear that the MPN could be the dominant limiting factor when mode partitioning is severe (i.e., k factor large) in the laser source. Indeed, this has been observed in several system experiments using MLM laser diodes; the MPN cannot be improved by going to higher optical powers, so there is a distinct bit-error-rate "floor" effect due to MPN [33, 34], as is shown in Figure 17.5.

Mode-partition noise is easily understood qualitatively, as shown in Figure 17.10, but is not easily quantified even though there have been several studies [33–36] concerning MPN. The difficulty lies in the fact that a single parameter such as the mode-partition-noise factor k [33] is insufficient for describing MPN for all situations, as MPN depends not only on the laser diode design but also on the laser diode biasing and modulation conditions; it may also be affected by system component reflections. Thus, while it is possible to characterize mode-partition statistics in a particular laser diode under a specific operating condition, it is preferable to study the MPN with the laser transmitter in the actual system operating environment. Often MPN is inferred from an actual system transmission experiment; the bit-error-rate floor in a typical BER versus received optical power curve (see Fig. 17.5) is attributed to MPN, and some data fitting is used to extract the MPN information.

To minimize MPN, one needs to select the lasers, have a low reflection coupling to the transmission fiber link, and adjust for optimum laser biasing and modulation conditions. Typically, with a narrow time-averaged spectral width and low laser relative intensity noise, MPN is also minimized. In any case, it is clear that the use of MLM laser diodes in general causes significant MPN in single-mode fiber transmission unless the operating laser wavelength is very close to the dispersion minimum. In Gb/s systems, MLM laser diodes should probably be used only near the minimum dispersion region of the single-mode fiber, except for short-distance transmission. SLM laser diodes are undoubtedly the primary optical sources of choice for Gb/s long-haul single-mode fiber systems.

17.7.3 Mode-Partition Noise in SLM Laser Diodes

Compared with MLM laser diodes, which have a large spectral width and significant MPN, SLM lasers such as DFB laser diodes have a much smaller spectral width and mode partitioning. With SLM DFB laser diodes, if the side-mode suppression is above 30 dB, there usually seems to be little mode-partition problem due to the residual side modes. On the other hand, one cannot assume that the use of SLM DFB laser diodes eliminates the problem of MPN entirely. For SLM lasers with a weaker side-mode suppression (say less than 25 dB), MPN due to the strongest side modes may cause system penalty at high Gb/s rates, depending on the fiber length and dispersion. Thus, the issue of MPN in "imperfect" SLM DFB laser diodes for the high Gb/s systems needs further study, especially for the Gb/s systems having a source wavelength operating outside the low dispersion region. Note that the SLM DFB laser could be imperfect due to improper fabrication or become so due to time-dependent degradation as part of the aging process.

In SLM DFB laser diodes, the mode-partition effect is greatly reduced because the laser's built-in grating gives rise to distributed feedback of the selectively reflected optical signal, which leads to the desired SLM selection (see Chapter 11). Proper design and fabrication of a DFB laser diode is needed to have the side-mode intensity suppressed to -30 dB or below relative to the selected main mode. A side-mode suppression ratio (SSR) of 30 dB or better *under modulation* is thus desirable to minimize the dispersion or mode-partition penalty due to the side modes. This is especially true for Gb/s systems operating in the high-dispersion wavelength region, for example, with a 1560-nm DFB laser in a C-SMF with zero dispersion near 1300 nm. The degree of side-mode suppression required would in general also depend on the total dispersion and system bit rate.

Experimental measurement of mode partition in several DFB laser diodes has shown that the mode partition k factor is typically very small (k less than 0.02) in these DFB lasers [37]. In a 100-km, 280-Mb/s single-mode fiber transmission experiment using C-SMF with DFB laser at 1550 nm, a 20-dB side-mode suppression ratio was enough to have negligible MPN penalty. For higher-bit-rate transmission, the required SSR for low MPN would be significantly higher. For 2–8-Gb/s systems, an SSR of 30–35 dB or higher may be required for the SLM laser diode under desired modulation condition to have negligible MPN or dispersion penalty due to the added spectral width associated with the side modes. In the case when MPN and the pulse broadening due to side modes are negligible, the remaining limitation will be due mainly to the dispersion broadening of the chirped (single) longitudinal mode. The subject of chirping and its system implications will be discussed in the next section.

17.7.4 Chirping in SLM Lasers and System Implications

a. Chirping and Dynamic Line Broadening in SLM Lasers. It was recognized very early that the broadening of the individual longitudinal modes of a laser diode under fast modulation is due to the transient phenomenon of chirping [38, 39]. Chirping refers to a time-dependent frequency (or wavelength) shift during the transient turn-on (and off) in a laser diode under fast excitation. The wavelength shift is caused by the time-varying refractive index excursion associated with the carrier density change, which is a natural consequence of nonconstant excitation, such as in short-pulse pumping or high-speed digital modulation. The resultant refractive index change leads to a time-varying longitudinal-mode frequency. The chirp could be a red-shift chirp ($C > 0$, wavelength increasing with time) or a blue-shift chirp ($C < 0$, wavelength decreasing with time) depending on the nature of the pulse excitation. Typically both red-shift and blue-shift chirps are observed in different time-varying parts of a square-wave pulse. In the case of very high speed excitation when only a single short pulse of less than a few hundred picoseconds is obtained (such as a single 1 digit in an 8-Gb/s RZ or NRZ pulse stream), red-shift chirp predominates because of the nature of pulse generation. This can be observed in time-resolved spectral studies [39, 40].

Chirping results in a dynamic linewidth broadening of the individual longitudinal modes, which is broadened from the cw linewidth of, for example, less than 0.01 nm to about 0.2 nm, depending on the biasing and modulation conditions. Since in direct detection optical transmission systems, the laser diode transmitters are directly current-modulated, chirping and the resultant line broadening are commonly observed in both MLM and SLM laser diodes. However, in MLM laser diodes, the effect of overall spectral envelope broadening is more important than the individual longitudinal mode broadening, so the latter effect (due to chirping) is usually neglected.

In good SLM lasers, since there is only one dominant longitudinal mode, chirp broadening becomes the most important factor other than the side-mode supression ratio. Figure 17.11 shows typical *time-averaged* spectra of a SLM DFB laser diode under both cw and high-speed modulation conditions; note the dynamic line broadening and the asymmetrical shape of the chirp for the spectra under modulation. It has been experimentally shown that chirping, a frequency shift with time, is in fact a deep frequency

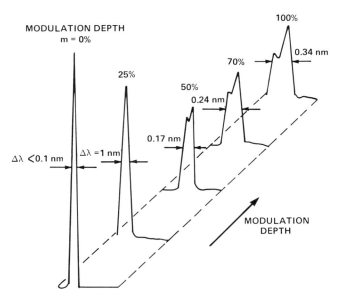

Figure 17.11 Typical time-averaged spectra of an SLM DFB laser diode under cw operation and high-speed modulation, showing the effect of direction modulation on the line broadening due to chirping.

modulation, induced by intensity modulation, in which many modulation frequency sidebands are generated and the intensity modulation causes the spectral asymmetry [41, 41a].

This chirped linewidth imposes a very significant transmission limitation in a single-mode fiber transmission system using an SLM DFB laser diode. If there are no side modes, the limitation is due only to the chirped width of the only longitudinal mode. For transmission near the fiber's zero-dispersion wavelength, there is negligible dispersion penalty. However, if the dispersion is, for example, 18 ps/(nm-km) (as for a 1560-nm DFB laser in a C-SMF), then 0.2 nm of chirped laser linewidth gives rise to 3.6 ps/km of pulse broadening, which for 200 km of C-SMF transmission amounts to a total pulse broadening of about 0.7 ns, limiting the transmission bit rate to less than 1 Gb/s. So, this is indeed a significant limitation. As described above, chirp limitation has been observed in several Gb/s system transmission experiments, where a high dc laser bias was often needed (and low extinction ratio tolerated) to minimize the chirp-related dispersion penalty.

b. Chirp Reduction and Controlled Chirped Optical Transmission in DS-SMF.
The transient chirping phenomenon cannot be eliminated in directly modulated laser diodes because it is intrinsic; in directly modulated laser diodes the current modulation forces the carrier density to vary with time in order

for the generated laser emission to have properly modulated optical pulses. Nevertheless, there are several ways to reduce the degree of chirping:

One can use an external modulator and run the laser diodes cw to eliminate the laser chirp; in this case the simplicity and other advantages of dc modulation of laser diodes are lost, unless it is an integrated modulator.

Special low-chirp laser diode designs (such as two-section laser diode with chirp control or external cavity SLM laser diode for lower chirp) could be considered; this is usually more complicated to implement.

Special chirp reduction schemes such as modulation pulse shaping [42] or injection locking [43, 43a] of a modulated laser with a cw master laser can be used. The former is of limited application; the latter is complicated to implement unless an integrated version can be made.

Thus, it is possible to reduce the chirp and the chirp-induced penalty in a dispersive transmission. If the limitation is really down to the chirp issue in a system, then perhaps the simplest solution is to operate near the zero-dispersion wavelength so that the chirped linewidth contributes very little to the pulse broadening. However, in a multiwavelength WDM system, there will be some SLM lasers operating in the dispersive regions of the spectrum, and it would be difficult to reduce the chirp penalty for all the laser wavelengths. For multiple-channel WDM with several SLM DFB laser diodes in the same transmission single-mode fiber (such as in Fig. 17.8), a scheme of chirped pulse transmission has been proposed [44]. The idea is that with a DS-SMF the fiber dispersion for laser wavelength λ_s *shorter* than the zero-dispersion wavelength ($\lambda_s < \lambda_0$) can be used for compression of *red-shift chirped* pulses to achieve nearly "dispersionless" (meaning no resultant pulse broadening) transmission over 100–200-km fiber lengths in the 2–8-Gb/s range. This is a situation where the *fiber dispersion and chirp* are used to *compensate each other* and the input optical pulse is compressed (instead of broadened) in the first half of the fiber link and then rebroadened to the input pulsewidth and beyond. The fiber length over which the output pulse width rebroadens to its input pulse width, called the dispersionless transmission length L_0, could be 100–300 km, depending on the bit rate, the original pulse width, the dispersion, and the chirp parameters [44].

Optical pulse compression experiments in long single-mode fiber transmission have been reported [45, 46]. There were also reports of single-mode fiber transmission experiments in which the pulse compression improved the transmission bit-error-rate characteristics [47]. For this proposed chirped pulse transmission scheme to be deployed in practice, the use of DS-SMF in single-mode fiber transmission networks is a prerequisite, and the need for multichannel WDM has to exist. Presently installed single-mode fibers are the conventional type (C-SMF); in the near-term future, use of dispersion-shifted single-mode fibers (DS-SMF) will probably be mainly in undersea fiber transmission and in some selected terrestrial routes. Thus, single-mode fiber transmission systems based on all three key technologies simultaneously, that is, (a) DS-SMF, (b) SLM DFB-LD, and (c) dense-WDM, at Gb/s bit rates, will probably find use only in long-term future applications.

In the meantime, chirping and mode-partition noise in SLM DFB laser diodes will continue to be the most important limitations in recently installed SLM DFB-LD Gb/s C-SMF single-mode fiber transmission systems. For this reason, adjusting the laser transmitter bias and modulation conditions to optimize overall system performance (which could correspond to reduced chirp, increased SSR, lower MPN, higher speed, lower pulse-patterning effect, but lower extinction ratio too) is definitely required as part of the system design and testing task.

17.7.5 Reflection-Induced Noise and Penalty

It is a well-known fact that reflection feedback into an oscillator can affect the oscillator characteristics. Likewise, depending on the phase and magnitude of optical reflection back into a laser cavity, the laser output characteristics will be disturbed, changed, and modified. Reflection-induced noise in semiconductor laser diodes has received much attention [48–50], although by nature reflection effects on the laser oscillation behavior are rather complicated and difficult to discern.

In direct detection optical fiber transmission systems, reflection from the laser transmitter fiber pigtail end, the near-end connectors, and so on have been observed to introduce noises and system penalties [48]. As an alternative to detailed experimental studies that attempt to comprehend the whole complex issue of reflection effects on the laser source under all possible reflection conditions, one can measure directly the induced system penalty for a given reflection feedback noise. Such measurement is most important in single-mode fiber systems using SLM laser diodes. The reflection effect on MLM laser diode characteristics may be just as significant, but the system penalty is less significant in MLM systems unless the reflection introduces unpredictable mode jumps.

Figure 17.12 shows the experimental results of a 1-Gb/s NRZ transmission experiment with a 1550-nm DFB laser diode in a C-SMF with zero dispersion at 1310 nm [50]. The observed power penalty increases with the increase in the reflection level. Under the particular experimental conditions, a reflection level of −19 dB was found to cause 1-dB penalty; a reflection level of −10 dB, as could be caused by air-gap reflection from two noncontacting connector ends, would cause a penalty of more than 4 dB. When the optical feedback was large, an increase in noise and jitter and degradation of response speed were observed in the received signal. The effect of the reflection was also dependent on the modulation condition; a larger modula-

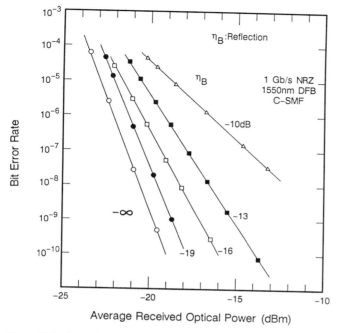

Figure 17.12 Experimental results of an 1-Gb/s NRZ transmission experiment with a 1550-nm SLM DFB laser in a C-SMF (zero dispersion near 1310 nm), showing the observed reflection-induced power penalty at different reflection levels.

tion signal improved the tolerance to reflections. For various types of SLM DFB lasers at different Gb/s bit rates and system transmission wavelengths, the acceptable reflection level will be different. In general, a reflection level of −30 dB (0.1% reflection) is probably satisfactory for most applications; in contrast, a reflection level of −12 dB (about 8% reflection typical of an air gap in noncontacting connectors or other components) will probably always cause significant system penalty.

To make sure there is little reflection-induced noise, some systems use both an optical isolator within the DFB laser package and low-reflection physical-contact connectors (such as the case of NTT's F1.6G system described above) to minimize and prevent reflection-induced penalties. Also, it has been found that lasers with low relative intensity noises are less susceptible to reflection-induced effects [49].

Clearly, as the system bit rate gets into high Gb/s ranges with the increase of SLM laser diodes, the issues of reflection tolerance in SLM lasers, isolator, isolation requirement, low reflection coupling components, and so on will have to be addressed. Such important system component and device considerations will certainly become more and more commonplace for the very high capacity multi-Gb/s systems where system impairments must be minimized.

17.8 FUTURE DIRECTIONS IN SYSTEM TECHNOLOGY AND APPLICATIONS

17.8.1 Future System Technology

Since single-mode fibers have practically unlimited bandwidth [51] and very low losses, system technology is really not limited by this excellent transmission medium. Rather, the mature single-mode fiber technology is relatively advanced and waiting to be fully utilized. In contrast, optoelectronics and electronic IC technology, impressive as they are today, still cannot fully utilize the remarkable transmission medium—single-mode fiber—to its full potential.

a. Very High Speed and High-Density WDM Direct-Detection Systems. One of the future directions of system technology is ultrahigh-capacity transmission by combining very high speed capability and high-density WDM. The rapid progress in both high-speed optoelectronics and IC technology has been very impressive. The speed limits of laser diodes and photodetectors seem to be beyond at least 20 Gb/s; the current limitations seem to exist mainly in the electronic IC technology. The VHSIC (very high speed IC) technology based on the group III–V compound semiconductors and novel structures is under intense research and development and seems to hold great promise for the future. The wide low-loss window of single-mode fibers promises tens or hundreds of simultaneous optical channels, each channel capable of operating at 10–20-Gb/s bit rates with direct-detection technology, transmitting through the same fiber by high-density WDM (see Fig. 17.8c and imagine each optical channel carrying a 20-Gb/s information bit rate), resulting in ultrahigh-capacity (100 Gb/s to 1 Tb/s) transmission.

b. OEIC and Coherent Transmission Technology. In addition to multichannel direct detection systems based on high-speed high-density WDM technology there are two main directions of future system technology:

1. OEIC (optoelectronic integrated circuit) technology combining various optical devices and electronic circuits on a single chip to perform complex functions on a monolithic chip or a hybrid chip
2. Coherent transmission technology, using coherent modulation and demodulation techniques to improve receiver sensitivity and achieve tunable high-density multichannel OFDM transmission

For details on these two subjects, see Chapters 23 and 24.

17.8.2 Future System Applications—Fiber to the Home and Intelligent Buildings

So far we have described the important essential aspects of single-mode fiber transmission system technology and have given examples of actual system applications. In the past 10 years or so, the emphasis was mainly on deploying single-mode optical fiber transmission systems for the digital long-haul transmission network including backbone and interoffice trunking applications. See Figure 17.13.

Now the new wave of optical fiber transmission system applications is in the loop part of the network. This includes both the "loop feeder" (typically 10 km or shorter) and the "distribution" portions of the subscriber network. The distribution portion of the subscriber loop is typically 1–5 km long, with the majority (80% or more) of the distribution distances being less than 2 km.

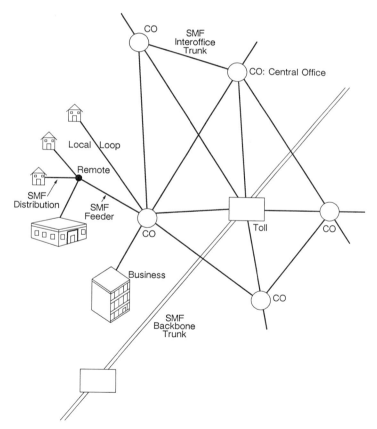

Figure 17.13 Schematic of a single-mode optical fiber transmission network including long-haul and interoffice transmission and loop distribution.

Thus, the fiber loss and dispersion considerations take on a different signif-icance; in fact, other than the consideration that single-mode fibers should be used for once-for-all installation and nearly unlimited future upgrading, fiber loss and dispersion are not as important as the network architectural considerations, coupling and distribution losses, maintenance and service concerns, and so on. There are a number of studies and field trials being conducted worldwide [52, 53] on "fiber to the home" systems and networks, to determine the service, architecture, economics, and maintenance/installa-tion issues as well as the evolution strategies, for deploying optical fiber transmission technology in the subscriber loops. Presently, two-way trans-mission to and from the home (downstream and upstream) with single-mode fiber transmission at about 600 Mb/s and 150 Mb/s is envisioned, as shown in the example of Figure 17.14 [53], although there are certainly many other versions and possibilities (including analog video transmission over fibers, broadcast video plus switched video, etc.) There are a number of technology, architecture, and economics issues that challenge the system engineers to make fiber to the home a reality.

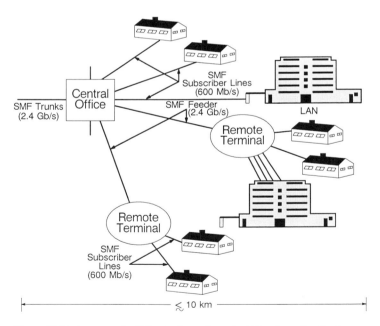

Figure 17.14 An example of the single-mode optical fiber loop distribution part of the network including loop feeders and subscriber lines. Many other network architectures (star, bus, etc.) and transmission technologies (e.g., analog video transmission over fiber/coax hybrids for broad-cast, subcarrier multiplexing of digital signals plus analog video over fiber, etc.) are also being considered for interim fiber to the home systems.

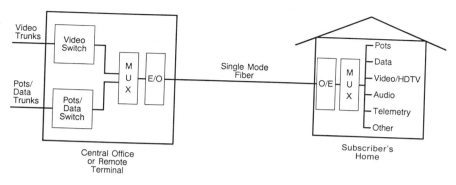

Figure 17.15 An example of the broadband integrated digital services network (B-ISDN) for fiber to the home, based on advanced high-speed single-fiber transmission technology. Interim strategies based on analog video transmission over fiber for broadcast, and switched video services are being considered for earlier deployment.

Undoubtedly, with the progress in these system trials and the simultaneous progress in optical fiber transmission technology, a broadband integrated services digital network (B-ISDN) may become a reality in the 1990s, providing a variety of services including POTS (plain old telephone service), telemetry (e.g., home security and energy management), high-speed data (for electronic mail, banking, shopping, library information retrieval, etc.), and high-quality audio and video (including HDTV) as illustrated in Figure 17.15. For more detailed discussions on the broadband services (e.g., video on demand) envisioned for homes and "intelligent building," see Chapter 19. In the next 10–20 years, fiber to the home systems and technologies providing a variety of advanced broadband services (e.g., two-way multimedia communication) will be the pioneers of the information age.

17.9 CONCLUSION

The field of optical fiber communications technology and system applications has been a very remarkable one in the history of science and technology. The past and future impacts on society are just beginning to take shape. Optical transmission systems and networks based on the optoelectronics technology of silica-glass fibers and semiconductor-based optical sources and detectors will form the backbone of the broadband integrated digital services network. Interestingly, the gradual evolution of single-mode optical fiber transmission systems from long-haul applications into subscribers' homes and intelligent local area networks will make all these look like a communications revolution of the information age, because while the system technology is essentially the same, the total impact is clearly far more significant when fiber to the home is realized. This is a challenging task for those of us working in the field; it is also an interesting path in the progress of human civilization which we will all witness.

REFERENCES

1. I. Jacobs and J. R. Stauffer, "FT3—a metropolitan trunk lightwave system," *Proc. IEEE*, **68**: 1286–1290 (1980).
2. T. Li, "Advances in optical fiber communications: an historical perspective," *IEEE J. Selected Areas Commun.*, **SAC-1**: 356–372 (1983).
3. R. E. Epworth, "The phenomenon of modal noise in analogue and digital optical fiber systems," Proc. 4th Eur. Conf. Opt. Commun., Genoa, Italy, September 1978, pp. 492–501.
4. T. Kanada, "Evaluation of modal noise in multimode fiber-optic systems," *J. Lightwave Technol.*, **LT-2**: 11–18 (1984).
5. Y. Suematsu, S. Arai, and K. Kishino, "Dynamic single-mode semiconductor lasers with a distributed reflector," *J. Lightwave Technol.*, **LT-1**: 161–176 (1983).
6. J. J. McNulty, "A 150-km repeaterless undersea lightwave system operating at 1.55 μm," *J. Lightware Technol.*, **LT-2**: 787–791 (1984).
7. R. E. Wagner, "Future 1.55-μm undersea lightwave systems," *J. Lightwave Technol.*, **LT-2**: 1007–1015 (1984).
8. T. Kimura, "Single-mode digital transmission technology," *Proc. IEEE*, **68**: 1263–1268 (1980).
9. I. Garrett and C. J. Todd, "Components and systems for long-wavelength monomode fiber transmission," *Opt. Quantum Electron.*, **14**: 95–143 (1982).
10. W. Albrecht, C. Baack, G. Elze, et al., "Optical digital high-speed transmission: general consideration and experimental results," *IEEE Trans. Microwave Theory Tech.*, **MTT-30**: 1535–1547 (1982).
11. P. K. Runge and P. R. Trischitta, "The SL undersea lightwave system," *J. Lightwave Technol.*, **LT-2**: 744–753 (1984).
12. M. Hirao, K. Mizuishi, and M. Nakamura, "High reliability semiconductor lasers for optical communications," *IEEE J. Selected Areas Commun.*, **SAC-4**: 1494–1501. (1986).
13. L. C. Blank, L. Bickers, and S. D. Walker, "Long-span optical transmission experiments at 34 and 140 Mb/s," *J. Lightwave Technol.*, **LT-3**: 1017–1026 (1985).
14. K. Nakagawa, K. Aoyama, J. Yamada, and N. Yoshikai, "Field experiments on the F-1.6G optical fiber trunk transmission system," *Conf. Rec., IEEE Globecom 86*, Houston, 1986, pp. 1205–1209.
15. A. H. Gnauck, B. L. Kasper, R. A. Linke, et al., "4 Gb/s transmission over 103 km of optical fiber using a novel electronic multiplexer/demultiplexer," *J. Lightwave Technol.*, **LT-3**: 1032–1035 (1985).
16. S. Fujita, I. Takano, N. Henmi, et al., "4 Gb/s long span transmission experiments employing high-speed DFB-LDs and InGaAs-APDs," *Proc. 12th Eur. Conf. Opt. Commun.*, Barcelona, 1986, pp. 507–510.
17. C. Lin, B. L. Kasper, A. H. Gnauck, and C. A. Burrus, "Direct 4 Gb/s modulation of 1.3 and 1.5 μm short-cavity DCPBH InGaAsP lasers," *Proc. 10th Eur. Conf. Opt. Commun.*, Stuttgart, 1984, p. 236.
18. U. Koren, G. Eisenstein, J. E. Bowers, et al., "Wide-bandwidth modulation of three-channel buried-crescent laser diodes," *Electron. Lett.*, **21**: 500–501 (1985).
19. J. E. Bowers, B. R. Hemengway, A. H. Gnauck, and D. P. Wilt, "High-speed InGaAsP constricted-mesa lasers," *IEEE J. Quantum Electron.*, **QE-22**: 833–844 (1986).
20. C. B. Su, V. Lanzisera, R. Olshansky, et al., "15-GHz direct modulation bandwidth of vapor-phase regrown 1.3 μm InGaAsP BH lasers under cw operation under room temperature," *Electron. Lett.*, **21**: 577–579 (1985).
21. C. Lin and J. E. Bowers, "High-speed large-signal digital modulation of a 1.3 μm InGaAsP constricted-mesa laser at a simulated bit rate of 16 Gb/s," *Electron. Lett.*, **21**: 906–908 (1985).
22. A. H. Gnauck and J. E. Bowers, "16 Gb/s direct modulation of an InGaAsP laser," *Electron. Lett.*, **23**: 801–803 (1987).

23. C. A. Burrus, J. E. Bowers, and R. S. Tucker, "Improved very-high-speed packaged InGaAs PIN punch-through photodiode," *Electron. Lett.*, **21**: 262–263 (1985).

24. A. H. Gnauck, J. E. Bowers, and J. C. Campbell, "8 Gb/s transmission over 30 km of optical fiber," *Electron. Lett.*, **22**: 600–602 (1986).

25. S. Fujita, N. Henmi, I. Takano, et al., "A 10 Gb/s-80 km optical fiber transmission experiment using a directly modulated DFB-LD and a high-speed InGaAs-APD," *Tech. Digest (Postdeadline Papers), Conf. Opt. Fiber Commun.*, New Orleans, 1988, PD-16.

26. G. Winzer, "Wavelength multiplexing components—a review of single-mode devices and their applications," *J. Lightwave Technol.*, **LT-2**: 369–378 (1984).

27. H. Ishio, J. Minowa, and K. Nosu, "Review and status of wavelength-division-multiplexing technology and its application," *J. Lightwave Technol.*, **LT-2**: 448–463 (1984).

28. M. Bauer, H. Dunwald, and W. Herrmann, "A four-channel WDM system for monomode fiber," *Proc. 10th Eur. Conf. Opt. Commun.*, Stuttgart, 1984, pp. 174–175.

29. N. A. Olsson, J. Hegarty, R. A. Logan, et al., "68.3 km transmission with 1.37 Tbit km/s capacity using WDM of ten single-frequency lasers at 1.5 μm, "*Electron. Lett.*, **21**: 105–106 (1985).

30. H. Toba, K. Inoue, and K. Nosu, "A conceptional design on optical FDM distribution systems with optical tunable filters," *IEEE J. Selected Areas Commun.*, **SAC-4**: 1458–1467 (1986).

31. C. Lin, H. H. Kobrinski, A. Frenkel, and C. A. Brackett, "A wavelength-tunable 16 optical channel transmission experiment at 2 Gb/s and 600 Mb/s for broadband subscriber distribution," *Proc. 14th Eur. Conf. Opt. Commun.*, Brighton, UK, 1988, pp. 251–254.

32. Y. Okano, K. Nakagawa, and T. Ito, "Laser mode partition noise evaluation for optical fiber transmission," *IEEE Trans. Commun.*, **COM-28**: 238–243 (1980).

33. K. Ogawa, "Analysis of mode partition noise in laser transmission systems," *IEEE J. Quantum Electron.*, **QE-18**: 849–855 (1982).

34. S. Yamamoto, H. Sakaguchi, and N. Seki, "Repeater spacing of 280 Mb/s single-mode fiber-optic transmission system using 1.55 μm laser source," *IEEE J. Quantum Electron.*, **QE-18**: 264–273 (1982).

35. K. Ogawa and R. Vodhanel, "Measurement of mode partition noise of laser diodes," *IEEE J. Quantum Electron.*, **QE-18**: 1090–1093 (1982).

36. P. L. Liu and K. Ogawa, "Statistical measurements as a way to study mode partition in injection lasers," *J. Lightwave Technol.*, **LT-2**: 44–48 (1984).

37. S. Yamamoto, H. Sakaguchi, and N. Seki, "Measurement of mode partition in DFB lasers," *J. Lightwave Technol.*, **LT-4**: 672–679 (1986).

38. K. Kishino, S. Aoki, and Y. Suematsu, "Wavelength variation of 1.6 μm wavelength buried heterostructure GaInAsP/InP lasers due to direct modulation," *IEEE J. Quantum Electron.*, **QE-18**: 343–351 (1982).

39. C. Lin, T. P. Lee, and C. A. Burrus, "Picosecond frequency chirping and dynamic line broadening in InGaAsP injection lasers under fast excitation," *Appl. Phys. Lett.*, **42**: 141–143 (1983).

40. M. Osinski and M. J. Adams, "Picosecond spectra of gain-switched quaternary lasers," *Opt. Commun.*, **47**: 190–192 (1983).

41. C. Lin, G. Eisenstein, C. A. Burrus, and R. S. Tucker, "Fine structure of frequency-chirping and FM sideband generation in single-longitudinal-mode semiconductor lasers under 10 GHz direct intensity modulation," *Appl. Phys. Lett.*, **46**: 12–14 (1985).

41a. G. P. Agarwal, "Power spectrum of directly modulated single-mode semiconductor lasers: chirp-induced fine structure," *IEEE J. Quantum. Electron.*, **QE-21**: 680–686 (1985).

42. L. Bickers and L. D. Westbrook, "Reduction of laser chirp in 1.5 μm DFB lasers by modulation pulse shaping," *Electron. Lett.*, **21**: 103–104 (1985).

43. C. Lin and F. Mengel, "Reduction of frequency chirping and dynamic linewidth in high-speed directly modulated semiconductor lasers by injection locking," *Electron. Lett.*, **20**: 1073–1075 (1984).

43a. C. Lin, F. Mengel, and J. K. Andersen, "Frequency chirp reduction in a 2.2 Gb/s directly modulated InGaAsP semiconductor laser by cw injection," *Electron. Lett.*, **21**: 80–81 (1985).

44. C. Lin, "Controlled chirped 'single-frequency' laser pulse transmission in dispersion-shifted single-mode fibers for WDW Gb/s optical communication," *Tech. Digest, 10th Eur. Conf. Opt. Commun.*, Stuttgart, 1984, paper 10B6, p. 244.

45. C. Lin and A. Tomita, "Chirped picosecond injection laser pulse transmission in single-mode fibers in the minimum chromatic dispersion region," *Electron. Lett.*, **19**: 837–838 (1983).

46. K. Iwashita and K. Nakagawa, "Dynamic spectrum broadening effect on a single-mode fiber pulse transmission," *Tech. Digest, 4th Int. Conf. Integrated Optics Opt. Fiber Commun.*, Tokyo, 1983, paper 28C4, pp. 318–319.

47. K. Iwashita, K. Nakagawa, K. Nakano, and Y. Suzuki, "Chirp pulse transmission through a single-mode fiber," *Electron. Lett.*, **18**: 873–874. (1983).

48. W. C. Young, L. Curtis, N. K. Cheung, et al., "Practical method of measuring reflection-induced power penalties in single-mode fiber transmission systems," *Electron. Lett.*, **23**: 56–57 (1987).

49. M. Shikada, S. Takano, S. Fujita, et al., "Evaluation of power penalties caused by feedback noise of DFB laser diodes," *Tech. Digest, OFC/IOOC'87*, Reno, 1987, Paper TuB4, p. 46.

50. S. Sasaki, H. Nakano, and M. Maeda, "Bit-error-rate characteristics with optical feedback in a 1.5 μm DFB semiconductor laser," *Proc. 12th Eur. Conf. Opt. Commun.*, Barcelona, 1986, pp. 483–486.

51. D. Marcuse and C. Lin, "Low dispersion single-mode fiber transmission—the question of practical versus theoretical maximum transmission bandwidth," *IEEE J. Quantum Electron.*, **QE-17**: 869–879 (1981).

52. K. Hashimoto and K. Nawata, "High-speed digital subscriber loop systems using single-mode fiber," *Conf. Rec., Globecom '86*, Houston, 1986, pp. 364–369.

53. P. Kaiser, "Single-mode fiber technology for the subscriber loop," *Proc. 11th Eur. Conf. Opt. Commun.*, Venice, 1985, vol. II, p. 125.

CHAPTER 18

Optical Fiber Communication Systems: Local Area Networks

Nobuyuki Tokura and Masaki Koyama

18.1 OVERVIEW OF LOCAL AREA NETWORKS

With the extensive introduction of information processing equipment into offices, factories, educational institutions, and laboratories, there has in recent years been a growing need for facilities that permit various resources within a building or site to communicate with one another. Local area networks (LANs) have appeared in response to this demand. Typical LANs are shown in Figure 18.1. A LAN is a group of information-processing machines, with communication capabilities, linked together by transmission lines. The typical LAN communication service territory is shown in Figure 18.2. LANs are normally positioned between computer buses and public communications networks.

Communication between information-processing machines requires a "protocol," a set of conventions that defines the format and contents of messages to be exchanged between communicating processes. A typical protocol is the OSI reference model (Reference Model for Open Systems Interconnection) [1] established by the International Standards Organization (ISO). This protocol is a seven-layer reference model, as shown in Figure 18.3. The hierarchical structure permits the protocol layers to be developed independently of each other. This section describes the IEEE Computer Society LAN reference model [2] as it applies to optical communication.

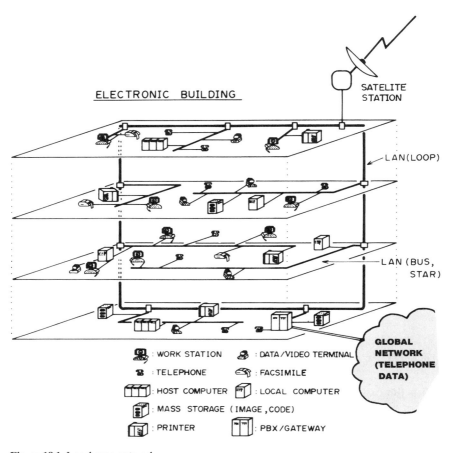

Figure 18.1 Local area network.

The physical layer of the LAN hierarchical model corresponds to the OSI reference model. However, the data link layer of the OSI LAN hierarchical model, which is affected by the physical layer, is divided into two sublayers: one under the influence of the physical layer and the other independent of it. The former is called the MAC (media access control) layer and the latter the LLC (logical link control) layer. When applying optical transmission technology to a LAN, prime consideration should be given to the physical and MAC layers, which are not involved in conventional electrical transmission lines.

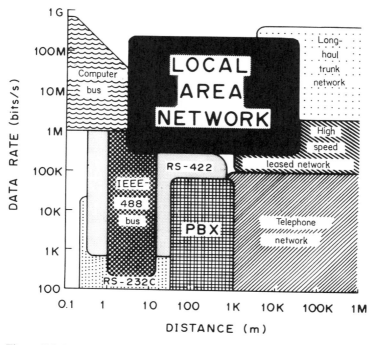

Figure 18.2 LAN communication service territory.

OSI REFERENCE MODEL
LAYERS

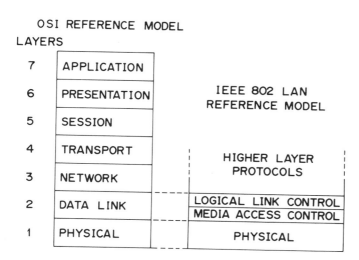

		IEEE 802 LAN REFERENCE MODEL
7	APPLICATION	
6	PRESENTATION	
5	SESSION	
4	TRANSPORT	HIGHER LAYER PROTOCOLS
3	NETWORK	
2	DATA LINK	LOGICAL LINK CONTROL
		MEDIA ACCESS CONTROL
1	PHYSICAL	PHYSICAL

Figure 18.3 Relationship between LAN reference model and OSI reference model.

18.2 PHYSICAL LAYER

18.2.1 Transmission Medium

The wavelength of light available for current optical fiber communications is $0.7-1.8$ μm (frequencies of several hundred terahertz). This wavelength, which is 1000 times shorter than that of millimeter waves, has the potential for providing ultralarge-capacity transmission systems. However, coherent transmission techniques and appropriate optical devices are required for its implementation, and these are now actively being investigated.

In current optical fiber communication systems employing LEDs or LDs as light sources, pure light is not used as a carrier. Instead, noisy, broad spectral width lightwaves are intensity-modulated for transmission. Thus, the present optical fiber communication systems do not make full use of the ultrawide band of the optical fiber. Still, optical fibers offer the advantages of lower loss and larger bandwidth than conventional pair cables or coaxial cables. A comparison of optical fibers with other transmission media is shown in Table 18.1, with the frequency response characteristics of various transmission lines presented in Figure 18.4.

Compared with conventional electrical transmission lines, optical fibers provide higher-speed and longer-distance digital transmission and are easier to install because of their small diameter and light weight. Furthermore, they are immune to electromagnetic interference.

TABLE 18.1. Comparison of communication media

COMMUNICATION MEDIA		DATA RATE	DISTANCE	MULTI -DROP	IMMUNITY TO EMI*	COST	APPLICATION
TWISTED PAIR CABLE		✕	✕	◯	✕	◯	PBX
COAXIAL CABLE	BASEBAND	◯	△	◯	△	△	ETHERNET
	BROADBAND	◯	◯	◉	◯	△	CATV
OPTICAL FIBER CABLE		◉	◉	✕	◉	△	LONG-HAUL NETWORK
AIR	RADIO	△	◯	◉	✕	◉	RADIO BRODCAST
	LIGHT	△	△	◉	✕	◉	————

good ← ◉ ◯ △ ✕ → poor

EMI⁺: electromagnetic interference

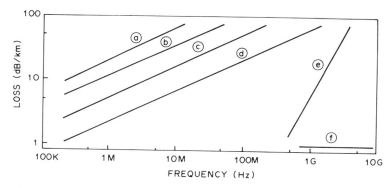

Figure 18.4 Transmission media loss. (a) Pair cable (intracity line type); (b) pair cable (intercity line type); (c) coaxial cable (Ethernet); (d) coaxial cable (long-haul line type); (e) multimode fiber (GRIN type); (f) single-mode fiber.

18.2.2 Optical Fiber Communication System Components

The configuration of a typical optical fiber communication system is shown in Figure 18.5. The functions and features of the system components are discussed in the following paragraphs.

a. Coder/Decoder. The coder/decoder converts information into a code suitable for optical transmission, and vice versa. The receiver circuit can be simplified due to the wide transmission band, and code conversion procedures involving multiple timing components (Manchester code, CMI code, etc.) are frequently used. In burst transmission, code conversion with little mark density variation is used to reduce the dc fluctuations of the received signal.

b. Optical Source. The optical source is an electrooptical converter. A laser diode is suitable for high-speed transmission, while a light-emitting diode is

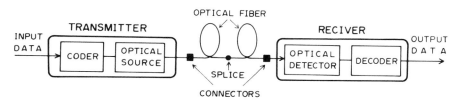

Figure 18.5 Optical fiber transmission.

best for low-speed transmission. This is because the LD optical source provides a high optical switching rate and high power, while the LED optical source provides only a low switching rate and low power due to the carrier storage effect. As far as light-emitting area is concerned, an LD is smaller than an LED. As a result, both LDs and LEDs can easily be connected to an optical waveguide with a wide incidence area, but the LED is difficult to connect to a narrow waveguide.

c. **Optical Fibers.** For optical fibers that use silica as the dielectric medium, the refractive index difference between the optical waveguide (core) and its surrounding (cladding) is from 10^{-3} to 10^{-2}.

The low-loss region for silica fibers is in the $0.7-1.8$-μm wavelength region in the infrared. Minimum loss of 0.2 dB/km at a wavelength of 1.55 μm has been reported. This loss is equivalent to that for transmission through air on a clear day. Thus, silica fibers are well suited for long-distance transmission.

Fibers using plastic as the dielectric medium feature low loss in the visible region, larger core diameter, and ease of bending. However, these fibers have a transmission loss more than 100 times greater than that of silica fibers, so they can be used only for short-distance transmission. Since plastic fibers permit visible lightwaves to be transmitted and the core diameter to be larger, a display LED can be used as the optical source to implement low-cost transmission channels.

The band characteristics that restrict the optical fiber transmission rate depend on the mode of transmission, material used, and structural dispersion.

(1) Multimode fibers. Multimode fibers for low-speed transmission are divided into two types: step-index (SI) and graded-index (GRIN). Each has a core diameter of several tens of micrometers. The transmission band of the SI fiber is narrow because of mode, material, and structural dispersion limitations. The core of the GRIN fiber has a parabolic refractive index profile that reduces mode dispersion limitations. This feature makes the bandwidth of GRIN fiber wider than that of SI fiber.

A multimode fiber is easily connected to an optical source because of its large core. For example, when an LED with a large light-emitting area is used as the optical source with multimode fiber, the loss at the connection is greatly reduced.

(2) Single-mode fibers. Silica single-mode fiber for high-speed transmission is of the SI type and has a core diameter of $5-10$ μm. The bandwidth of this fiber is restricted by material and structural dispersion. The band is widest at about 1.3 μm, the minimum dispersion wavelength, where material dispersion and structural dispersion cancel each other.

To implement low-loss connection with a fiber having a small core, the light-emitting area of the optical source must be small. Laser diodes have a small emission area and therefore are suitable for use with single-mode fibers. On the other hand, LEDs will have much lower coupled powers in single-mode fibers, since their emission surfaces have a diameter of several tens of micrometers or more.

d. Fiber Connection

(1) Optical connector. Because of the very small fiber core diameter, the connector alignment accuracy must be about 2 μm or better for single-mode fibers and about 6 μm or better for multimode fibers to obtain a connection loss of 0.7 dB or less. The optical connector normally has a loss of 0.5–1 dB, which is equivalent to the amount of transmission loss in a silica fiber 100 m or more in length. Accordingly, the number of connectors to be inserted in the transmission line must be minimized.

(2) Optical splice. In optical fiber installation, it is frequently necessary to permanently splice fibers together. An electric arc is commonly used for silica fiber splicing. The loss at a splice formed by this technique is about 0.5 dB or less. Efforts are now being made to implement automated splicing.

e. Optical Detector.
A PIN or avalanche photodiode is used for photo-detection. Silicon is used as the receiving element for wavelengths of 0.9 μm or less, with germanium or InGaAs used for wavelengths of 1.8 μm or less. The PIN diode has low sensitivity but helps simplify the optical receiver circuit. On the other hand, the avalanche photodiode is highly sensitive but requires a high-voltage generator for avalanche amplification, resulting in more complicated circuitry.

18.2.3 Transmission Level Design

In optical fiber communication systems, the difference between the optical source output and the minimum optical power level at the optical receiver is the allowable limit of loss in the transmission line. Transmission losses occur at fibers, connectors, splices, and optical couplers that branch or combine optical signals. The arrangement of these elements must be such that system topology is maintained. The system design must also be such that the total loss does not exceed the allowable limit described above. Particularly, in the design of a LAN where the transmission distance is short, sufficient consideration should be given to the use of connectors, splices, and optical couplers.

a. Optical Branching and Coupling.
When optical fibers are used as the transmission channel, node-to-transmission line connection is the first consideration. Unfortunately, optical signal access at nodes is rather difficult due to

the immunity of optical fibers to electromagnetic interference. Three access methods are available, as shown in Figure 18.6.

(1) Active access. The optical signal is converted into electrical form at nodes, and branching or insertion is effected according to the state of the electric signal, with the nodes linked together by separate point-to-point transmission lines (optical link). Use of this optical link facilitates implementation of multinode or high-speed long-distance transmission. However, signals that bypass the node pass through the active circuits (electric repeater circuits), resulting in decreased system reliability.

(2) Passive access. Branching or insertion of an optical signal by lightwaves at a node requires an optical tap or optical coupler. Use of optical couplers increases the reliability of the transmission line, since other optical signals bypass the node in the passive optical coupler. Figure 18.7 [3] shows

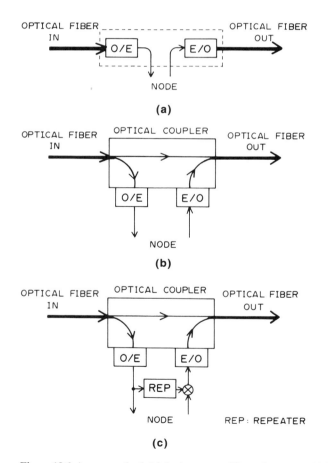

Figure 18.6 Access method. (a) Active access; (b) passive access; (c) hybrid access.

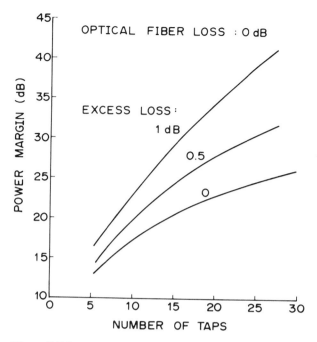

Figure 18.7 Required power margin versus number of optical taps. (From Limb [3].)

the number of optical taps versus required power margin when an optical tap that minimizes the level difference between optical source and optical receiver is used. The excess loss (loss excluding signal distribution loss) is as great as 1–5 dB; therefore serial connection of many nodes (about 20 or less) is difficult to implement. The number of taps can be increased by the use of an optical star coupler (with small excess loss per node) and a set of optical taps.

Typical optical couplers include mirrors [4], buried waveguides [5], directional couplers [6], and slab waveguides [7]. These are shown in Figure 18.8. Efforts are now under way to develop optical couplers that have low excess loss and are independent of mode or polarization.

(3) Hybrid access. The hybrid access combines active access and passive access. More specifically, under normal conditions, electrical repeat operation is performed at each node; but when a node is faulty, node bypassing is effected through a passive optical coupler with a high degree of reliability. Thus, hybrid access provides a highly reliable multinode network. However, it has the disadvantage of requiring complicated node circuits.

b. Topology. An optical fiber LAN composed of optical fiber transmission lines and nodes can have any of three topologies—star, bus, or loop—as

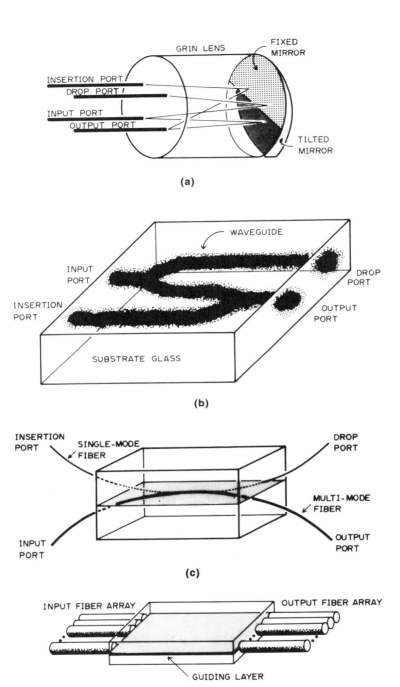

Figure 18.8 Optical coupler. (a) Mirror type; (b) buried-waveguide type; (c) directional coupler type; (d) slab-waveguide type.

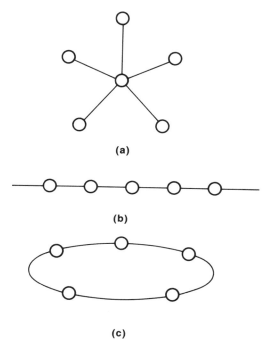

Figure 18.9 Topologies for LANs. (a) Star, (b) bus, (c) loop (ring). Lines are indicated by lines, and nodes by circles.

shown in Figure 18.9. Transmission line node connection (access) also comes in three types, as described in the previous section. Further, there exist certain congenialities between the topologies and media access control (MAC). A reference table for the topologies, optical access, and MAC in optical fiber LANs is given in Table 18.2. An example of MAC suitable for any topology is Anarchy [8], which employs a flood routing algorithm.

TABLE 18.2. Relation between topology and access method

TOPOLOGY	OPTICAL ACCESS			MEDIA ACCESS CONTROL			
	PASSIVE	ACTIVE	HYBRID	CSMA /CD	TOKEN PASSING	TDMA	FLOOD ROUTING
STAR	◯	◯	◯	◯	◯	◯	◯
BUS	✕	◯	◯	◯	◯	◯	◯
LOOP (RING)	✕	◯	◯	✕	◯	◯	◯
MESH	✕	◯	✕	✕	✕	✕	◯

◯ : AVAILABLE   ✕ : UNAVAILABLE

(1) Star. The star network consists of a node in the center of the network and other nodes extending radially from that node. Node-to-node communication is performed via the center node. For active access, the center node allows for compensation of transmission loss, and certain transmission control functions can be added. Thus, the center node provides concentrated network control and circuit-switching services.

When the center node utilizes passive access, node-to-node communication is of the broadcast form. In other words, light emitted from one node is distributed to all the other nodes by the star coupler at the center node. Loss in the star coupler is 10 log N (dB), where N is the total number of nodes in the network. Because this loss is added to the loss in the transmission line, the number of nodes in a network is limited. Thus far, an optical star coupler with 100 nodes has been developed [7].

(2) Bus. The bus network is constructed as a node-interconnected single bidirectional transmission line (single bus) or as two node-interconnected transmission lines (double bus, U bus) designed for transmission in opposite directions. Node-to-node communication is based on time-division multiplexing (TDM) or wavelength-division multiplexing (WDM).

The single bus is difficult to couple to optical fibers with low loss and low reflection at the optical connection. Therefore it is not often used. For double buses or U buses, repeaters must be inserted every several nodes, since the current optical taps (optical branch/coupler) do not have sufficient low-loss characteristics. Few optical buses have been implemented because of the unavailability of low-loss taps.

(3) Loop. In the loop topology, nodes are connected by unidirectional transmission lines in a circular configuration. As with the bus topology, the loop uses TDM for node-to-node communication. However, this topology requires an additional procedure so that a signal sent from a node is removed when it loops back. In a single loop, node failure or a transmission line break produces a complete communication system failure. To avoid this, node bypasses or double loops incorporating loop reconstruction capabilities are used.

18.3 MAC LAYER

In LANs, the nodes share the transmission medium for communication. This is accomplished by media access control (MAC), which allocates the transmission medium between nodes so that transmission signals do not interfere with each other. MAC is classified as either contention assignment or controlled assignment. A typical contention assignment is CSMA/CD, which is suitable for packet transmission. Among the controlled assignment strategies are token passing, suitable for packet transmission, and time-division multiple access (TDMA), suitable for continuous transmission.

18.3.1 *CSMA/CD*

The CSMA/CD protocol is a procedure in which nodes compete independently for use of the transmission channel. In this protocol, for efficient use of the transmission channel, control is based on channel traffic and signal collisions in the channel. Ethernet [9] using coaxial cable is a typical LAN to which this procedure is applied.

The CSMA/CD protocol is shown in Figure 18.10.

A. Send procedure
  1. A node having data to send checks the condition of the transmission channel. (Carrier sense: CS.)
  2. If the node determines that the channel is idle, it starts sending data with the destination address added. If the channel is busy, the node waits for a fixed period of time and performs step 1 again.
  3. The transmitting node checks to see if the transmitted signal has collided with signal from another node. (Collision detection: CD.)
  4. If all the data have been sent without collision, the node completes transmission. If a collision is detected during transmission, the node immediately stops transmission, waits for the period determined by the retransmission algorithm, and then performs step 1 again.
B. Receive procedure
  1. All nodes receive any signal on the transmission channel at all times.
  2. A node accepts received data only if they are normal (without any error and in the specified data format) and addressed to that node.

Thus, in CSMA/CD-based networks, nodes stop transmission upon detection of a signal collision, resulting in reduced ineffective signal generation and increased channel utilization efficiency.

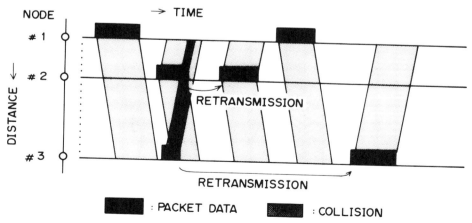

Figure 18.10 CSMA/CD.

A node detects a signal sent from another node after the lapse of the node-to-node propagation delay time; therefore, it can transmit within the propagation delay time. Accordingly, the collision probability is increased under high offered load conditions. Furthermore, the longer the node-to-node propagation delay time for data transmission, the more frequently collisions occur. Thus, under CSMA/CD, the channel utilization efficiency is significantly reduced in large-area transmission at high speed and under high load (Fig. 18.11).

One problem with applying the CSMA/CD protocol to optical fiber LANs is the need for collision detection of signals differing much more in level than in the case of electrical LANs. Many approaches to collision detection have been proposed, including collision detection after electrooptical conversion, collision detection in terms of time difference, and high-sensitivity optical signal collision detection.

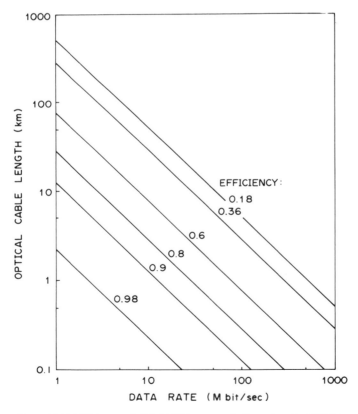

Figure 18.11 Efficiency of CSMA/CD protocol. Data length, 2000 bits; propagation speed, 5 μs/km.

18.3.2 Token Passing

The token-passing protocol is a procedure for assigning a token (right to transmit) to nodes in a certain priority sequence (Fig. 18.12). One example of a network with a fixed token-passing sequence is the loop (ring). Networks with an undetermined priority sequence include the bus and the star. For these topologies, the priority sequence is logically determined and a destination address is attached to the token.

A typical loop network employing the token-passing protocol is shown in Figure 18.13.

A. Send procedure
    1. Upon receipt of a token, a node having data to send holds the token.
    2. The node transmits the token after sending the data with a destination address added.
    3. The node clears the sent data after they have made a round trip. (In some cases, the data are cleared at the destination node.)
B. Receive procedure
    1. Each node receives all transmitted data and repeats them to the next node.
    2. A node accepts data only if they are normal (without any error and in the specified format) and addressed to that node.

Token transmission is usually accomplished in one of two ways:

1. The node transmits the token after receiving the header of the data it sent (single token type: IEEE-802.5 token ring protocol [10]).

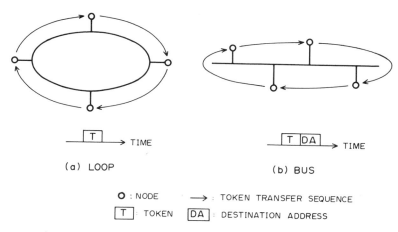

(a) LOOP   (b) BUS

O : NODE   → : TOKEN TRANSFER SEQUENCE
T : TOKEN   DA : DESTINATION ADDRESS

Figure 18.12 Token-passing scheme.

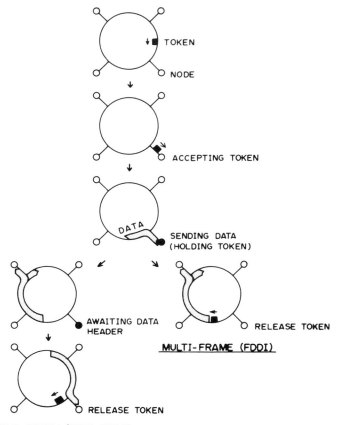

Figure 18.13 Token passing.

2. The node sends the token immediately after it completes data transmission (multiframe type: ANSI FDDI protocol [11]).

Method 1 facilitates detection of a transmission channel failure since the channel is filled with data sent from token-holding nodes. However, when the length of the sent data is short, idle data must be transmitted until the header is received. Method 2 eliminates the necessity of sending idle data irrespective of the sent data length. As a result, for high-speed or long-distance loops, the FDDI protocol provides higher channel utilization efficiency.

The token-passing protocol must include a procedure for restoring a destroyed token (due to noise, node failure, etc.). Furthermore, in a bus or star network employing this protocol, the token-passing sequence must be changed when a new node participates in the network or an old node leaves it, since the priority sequence has been logically determined. Thus, the token-passing protocol involves complicated procedures.

18.3.3 TDMA

In the TDMA protocol, one control node exercises concentrated control over use of the transmission channel and the other nodes operate according to the directions from that control node. The control node generates a frame synchronization signal, time-divides the frame, and allocates the time slots to each node. Thus, the control node exercises all required channel control to provide increased channel utilization efficiency and stable operation even at high load.

A demand assign-TDMA (DA-TDMA) protocol that allocates time slots upon request from a node is shown in Figure 18.14.

1. The sender node sends a request for time slot allocation to the control node.
2. The control node, upon receipt of an allocation request, selects unused time slots and changes them from "unused" to "in use" to allow them to be used. The control node then sends the allotted time slot numbers to both the sender and receiver nodes.

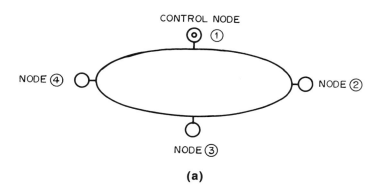

(a)

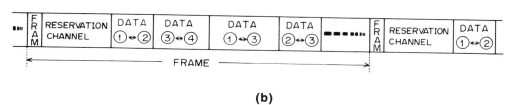

(b)

Figure 18.14 DA-TDMA. (a) Topology; (b) frame structure.

3. The sender and receiver nodes start transmission using the time slots allotted by the control node.

4. Upon completion of transmission, the node sends end of transmission to the control node.

5. The control node, upon receipt of end to transmission, changes the time slot from "in use" to "unused."

The DA-TDMA protocol permits continuous transmission of sound, video, and other signals (circuit-switching services) through fixed time slot allocation. It also enables some time slots allocated to all nodes to be controlled under a token-passing protocol (packet-switching service). Thus, this protocol has the potential to provide a variety of communication services. The control node must be highly reliable, because a control node failure would produce a complete communication system failure.

The DA-TDMA protocol has been applied to satellite communication because channel utilization efficiency is not decreased even when propagation delay is great. A high-speed optical fiber LAN can easily be implemented by connecting the nodes together with optical links and employing this protocol. However, the current optical interfaces do not have sufficient reliability.

18.4 APPLICATIONS

Various optical fiber LANs that take full advantage of optical fiber transmission have been developed (Tables 18.3 and 18.4). However, not very many optical fiber LANs make effective use of wavelength-division multiplexing techniques.

18.4.1 Optical Fiber LANs Using the CSMA/CD Protocol

When the CSMA/CD protocol is applied to an optical fiber transmission channel, the first consideration is how to detect collisions between sent signals. Due to the differences in line length between two transmission channels, number of connectors and splices, and coupler losses, the node-to-node loss variation amounts to about 10 dB. Therefore one optical signal could be up to 10 times greater in level than another. It is difficult to detect a collision between such signals. Two typical approaches to collision detection are discussed in this section.

a. Converting an Optical Signal Collision into Electrical Form. Fibernet-II is a system that applies the electric signal collision detection approach to

TABLE 18.3. Optical fiber local area network

Characteristic	Fibernet-II	Siecor FiberLAN	Hubnet	SIELOCnet	SOLARnet
Topology	Active star	Passive star	Star (rooted tree)	Passive	Passive star
Access control	CSMA/CD	CSMA/CD	Hub	CSMA/CD	CSMA/CD
Bit rate	10 Mb/s	10 Mb/s	50 Mb/s	16 Mb/s	32 Mb/s
Span	2.5 km	2.8 km	Open	600 m	2 km
Max. no. of nodes	1000	4000	65,536	—	9000
Fault tolerance	—	Passive components	—	Passive components	Passive components
Optical access	Active	Passive	Active	Passive	Passive
Fiber type	Step index	—	Graded index	Step index	Graded index
Fiber core/O.D.	100 μm/140 μm	100 μm/140 μm	100 or 50 μm/140 μm	100 μm/140 μm	50 μm/125 μm
Wavelength	850 nm	820 nm	—	830 nm	850 nm
Optical source	LED	LED	LED	LED	LED
Optical detector	PIN	PIN	PIN	PIN	APD
Coding	Manchester	Manchester	—	Scrambled NRZ	Manchester

TABLE 18.4. Optical fiber local area networks

Characteristic	Loop 6770	D-Net	SIGMA NETWORK	OPALnet-II
Topology	Loop	Loop bus	Loop	Loop
Access control	Token passing	Locomotive	TDMA	TDMA and token passing
Bit rate	32 Mb/s	1 Gb/s	32 Mb/s	100 Mb/s
Span	2 km	—	2 km	1 km
Max. no. of nodes	126	—	64	100
Fault tolerance	Dual fiber, bypass, loopback	Passive components	Dual fiber, bypass, loopback	Dual fiber, optical bypass, loopback
Optical access	Active	Passive	Active	Hybrid
Fiber type	Graded index	—	Graded index	Graded index
Fiber core/O.D.	50 μm/125 μm	—	50 μm/125 μm	50 μm/125 μm
Wavelength	850 nm	—	830 nm	850 nm
Optical source	LED	—	LED	LD
Optical detector	PIN	—	PIN	APD
Coding	RZ	—	DMI	8B1C

an active star coupler. Siecor FiberLAN applies the same approach to a passive star coupler. These systems can be linked to an Ethernet (10 Mb/s) that uses coaxial cables. Hubnet is a system that solves the problem of detecting collision occurring within an active star coupler.

(1) Fibernet-II [12]. Fibernet-II has a star configuration, as shown in Figure 18.15, and uses an active star coupler as the center node. In this system, optical signals from the nodes are coupled after they are individually converted into electric decision signals. Collisions can easily be detected by monitoring the amplitude of the electric signals. Upon detection of a collision, the center node stops relay operation and sends a collision acknowledgement signal to all nodes. The use of one active star relay can extend the transmission distance to 2.5 km, which is the maximum transmission distance for Ethernet. However, if the common active element at the center node fails, the whole communication system is disabled.

(2) Siecor FiberLAN [13]. Siecor FiberLAN has a star configuration, as shown in Figure 18.16, employs a passive star coupler as the center node, and incorporates additional collision-detection capabilities. The optical signal from each node passes through the passive star coupler, where collision detection is performed. The optical signal from each node is branched at the

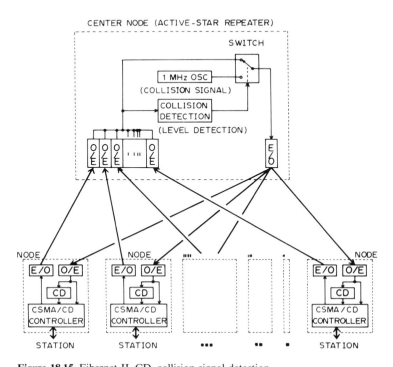

Figure 18.15 Fibernet-II. CD, collision signal detection.

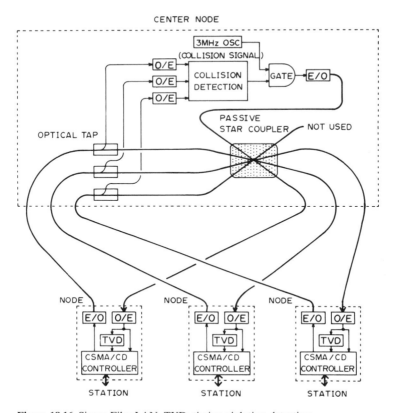

Figure 18.16 Siecor FiberLAN. TVD, timing violation detection.

star coupler input, and the branched signals are individually converted into electric signals for optical level detection. If the system detects two or more optical signals at one time, it concludes that a collision has occurred. To notify all nodes that a collision has occurred, a higher-level collision acknowledge signal is sent out to the passive star coupler. In this system, collision-detection malfunction does not cause complete communication disablement.

(3) Hubnet [14]. Hubnet comes in two topologies, star or rooted tree (hierarchical configuration), as shown in Figure 18.17. When signals collide within the active star repeater, only the signal that first arrives at the repeater is repeated to all nodes. This system can be regarded as incorporating Ethernet in the active star repeater [15]. This feature eliminates propagation delay leading to degraded carrier sense performance and prevents signal collision. Thus, Hubnet differs slightly from other CSMA/CD-based networks. More specifically, each node does not need carrier sense (which is

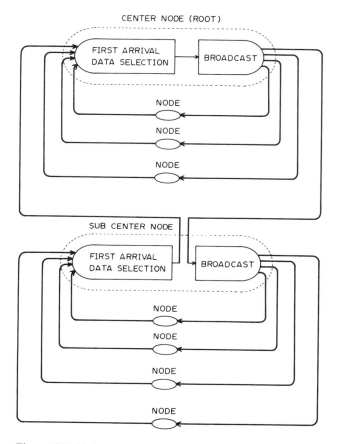

Figure 18.17 Hubnet.

performed by the active star repeated) and can transmit at any time. When a transmission node does not receive its echo from the center node, the node concludes that a collision has occurred.

Unlike CSMA/CD-based networks, Hubnet maintains high channel utilization efficiency for high-speed, high-load transmission in a large area. Consequently, Hubnet is well suited for high-speed, long-distance optical fiber transmission.

b. Detecting Optical Signal Collision. SIELOCnet is a system that performs time-domain optical signal collision detection, while SOLARnet performs amplitude-domain optical signal collision detection. These networks have a star configuration and use a passive star coupler as the center node (Fig. 18.9a). They provide excellent reliability.

(1) SIELOCnet [16]. SIELOCnet concludes that a collision has occurred when a signal is received before the signal sent from a node returns to that node through the optical star coupler, that is, before the echo comes back. In this system, at least one node may not detect a particular signal collision. Therefore, a node that detects a particular collision must notify the other nodes of that collision. When transmission is performed in a narrow area and at high load, signal collisions are difficult to detect, the probability of missing a collision being more than 1%.

(2) SOLARnet [17]. To increase collision detection sensitivity, SOLARnet generates a stable electrical level and compares it with the received signal. To implement this, the sender uses a Manchester coder (1 → [L][H], 0 → [H][L]; L = low level, H = high level) and the receiver incorporates a partial response circuit ([L] → [L][−L], [H] → [H][−H], [0] → [0][0]; 0 = zero level), as shown in Figure 18.18a.

The partial response circuit provides an eye pattern with two and three levels, as shown in Figure 18.18b. The received data can be identified only by the two-level portion. The zero level in the center of the three-level portion is stable, so a reference level can be set up around this level. This setup enables collision detection with a high degree of sensitivity, since interference with the collision signal causes no level variation or decision errors.

SOLARnet with a transmission rate of 32 Mb/s can detect collision of optical signals whose levels differ by 8 dB or more, and a 100-node passive star network is already in use.

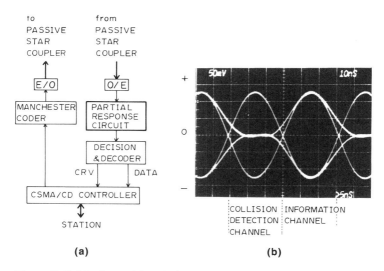

(a) **(b)**

Figure 18.18 SOLARnet. (a) Node block diagram. CRV, coding rule violation. (b) Eye diagram of partial response circuit output.

18.4.2 Optical Fiber LAN Using the Token-Passing Protocol

Most optical fiber LANs using the token-passing protocol are high-speed systems. The reason is that compared with CSMA/CD-based systems, such LANs have higher channel utilization efficiency in high-speed transmission and facilitate implementation of high-speed optical fiber transmission.

Loop 6770 and D-Net are examples of systems that apply the token-passing protocol to an active access loop and passive access loop bus, respectively.

a. Loop 6770 [18]. Loop 6770 is a typical active-access, double, optical fiber loop, as shown in Figure 18.9c, and has a transmission rate of 32 Mb/s. Under the token-passing protocol employed, a token is transmitted after data transmission is completed. This format is similar to that proposed in the FDDI protocol [11]. This network is provided with various failure recovery functions, such as node electric bypassing (optical repeater using battery backup), switching to a reserve loop, and loopback (forming the double loop into one loop when both loops fail).

Other examples include ILLINET (32 Mb/s) and H-8644 (32 Mb/s), with many of these systems being used as high-speed loops.

b. D-Net [19]. D-Net is a passive-access optical fiber loop bus, as shown in Figure 18.19. The sender line is of the bus configuration, while the receiver line is star-shaped to reduce excess losses on the receiver side.

The D-Net protocol is defined as being between the CSMA/CD and token-passing protocols. An R_G node connected to the loop bus generates a signal called a *locomotive* to relocate the timing. Each transmitting node concatenates sent data successively to the tail of the locomotive. When signals from two or more nodes collide (carrier sense), the downstream node suspends transmission until no signal comes from the sense tap. A node receives

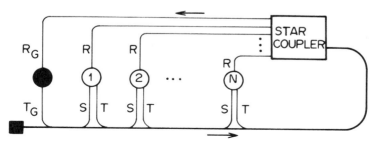

Figure 18.19 D-net. 1–N, nodes; R_G, locomotive generator; S, sense tap; T, transmit tap; (From Tseng and Chen [19].)

the signal distributed by the star coupler and accepts it if it is addressed to that node.

Under this protocol, sent signals are delivered in order along the loop, as in token passing. As a result, if a suitable locomotive timing frequency is selected, it is possible to achieve high channel-utilization efficiency in high-speed transmission. A 1-Gb/s D-Net has offered six-TV-channel packet transmission.

18.4.3 Optical Fiber LAN Using the TDMA Protocol

TDMA is used in many optical fiber LANs, since it maintains high channel-utilization efficiency in high-speed transmission. This protocol is applied to many optical loops. SIGMA NETWORK and OPALnet-II are examples of systems that apply active access and hybrid access, respectively.

a. SIGMA NETWORK [20]. SIGMA NETWORK is configured in the same manner as Loop 6770 but differs from it in communication protocol. This network uses active access, has a transmission rate of 32 Mb/s, and is of double-loop construction. Failure recovery functions include node electric bypassing, switching to a reserve loop, and loopback.

The protocol used in SIGMA NETWORK is DA-TDMA. The link information needed to time slot allocations to the nodes and control node is communicated by packetized data. The communication of link information is performed by the use of a fixed time domain in the frame.

b. OPALnet-II [21]. OPALnet-II is a hybrid-access, double optical fiber loop with a transmission rate of 100 Mb/s. This system performs node bypassing at all times and, if node failure occurs, provides node bypassing with high reliability. In this system, bypass signal interference is reduced by reversing the phase of the node bypass and insertion signals in normal operation.

OPALnet-II makes effective use of the transmission channel by utilizing the time slot allocation capabilities of the TDMA protocol. Any time slots not used for circuit-switching services are dynamically concatenated in 1.5-Mb/s steps as a packet-switching channel (token-passing protocol). The packet signal information include user, time slot allocation (link), and loop control and supervision. This LAN allows circuit-switching and packet-switching capacities to be changed with high flexibility.

REFERENCES

1. H., Zimmermann, "OSI reference model—the ISO model of architecture for open systems interconnection," *IEEE Trans. Commun.*, **COM-28**(4): 425–432 (1980).
2. IEEE Computer Society, *IEEE Project 802 Local Area Network Standards.* 1981.

3. J. O. Limb, "On fiber optic taps for local area networks," *Proc. ICC'84,* May 1984, pp. 1130–1136.

4. F. H. Levinson and S. W. Granlund, "Single GRIN-lens directional couplers," *Bell Syst. Tech. J.,* **63** (3): 431–439 (1984).

5. E. Okuda I. Tanaka, and T. Yamasaki, "Planar gradient-index glass waveguide and its applications to a 4-port branched circuit and star coupler," *Appl. Opt.,* **23**(11): 1745–1748 (1984).

6. T. H. Wood and M. S. Whalen, "Effective nonreciprocal evanescent-wave optical-fiber directional coupler," *Electron Lett.,* **21**(5), 175–176 (1985).

7. Y. Fujii, N. Suzuki, and J. Minowa, "A 100 input/output star coupler composed of low-loss slab-waveguide," *Tech. Digest, IOOC'83* Paper 29C2-4, pp. 342–343, June 1983.

8. R. Neff and D. Senzig, "A local network design using fiber optics," *IEEE Proc. COMPCON,* Spring 1981, pp. 64–69,

9. "Ethernet, A local area network data link layer and physical layer specifications—Version 2.0," DEC, Intel Corp., and Xerox Corp., November 1982.

10. IEEE Computer Society, "Draft IEEE standard 802.5 token ring access method and physical layer specifications," IEEE Project 802 LAN Standards, Working Draft, February 1984.

11. American National Standards Institute, "Fiber distributed data interface token ring media access control," proposed draft X3T9/83-X3T9.5/83-16 REV. 5, June 1984.

12. R. V. Schmidt, E. G. Rowson, R. E. Norton, S. B Jackson, Jr., and M. D. Bailey, "Fibernet II: a fiber optic Ethernet," *IEEE J. Selected Areas Commun.,* **SAC-1**(5): 702–710 (1983).

13. R. P. Kelley, J. R. Jones, B. J. Bhatt, and P. W. Pate, "Transceiver design and implementation experience in an Ethernet-compatible fiber optic local area network," *IEEE Proc. INFOCOM84,* April 1984, pp. 2–7.

14. P. I. P Boulton and E. S. Lee, "Hubnet: a 50 Mb/s glass fiber local area network," *Proc. LAN'82,* September 1982, pp. 15–17.

15. A. Albanese, "Star network with collision-avoidance circuits," *Bell Syst. Tech. J.,* **62**(3): 631–638 (1983).

16. S. Moustakas and H. H. Witte, "Passive optical star bus with collision detection for local area networks," *IOOC'83 Tech. Dig.,* 29C2-3, June 1983, pp. 340–341.

17. K. Oguchi and Y. Hakamada, "New collision detection technique and its performance," *Electron. Lett.,* **20**(25/26): 1062–1063 (1984).

18. M. Kiyono, M. Tada, K. Yasue, K. Takumi, and Y. Narita, "C & Cnet loop 6770—a reliable communication medium for distributed processing systems," *Proc. LAN' 82,* September 1982, pp. 47—50.

19. C. W. Tseng and B. U. Chen, "D-net, a new scheme for high data rate optical local area networks," *IEEE J. Selected Areas Commun.,* **SAC-1**: 493–499 (1983).

20. K. Kawakita, K. Hiyama, and O. Takada, "An integrated services local area fiber network—SIGMA NETWORK," *IOOC'83 Tech. Digest,* 29C2-2, June 1983, pp. 338–339.

21. N. Tokura, K. Oguchi, Y. Kimura, and Y. Oikawa, "100 Mb/s optical loop network for on-premises use," *Conf. Rec. GLOBECOM'84,* Paper 2.6, November 1984, pp. 64–68,

CHAPTER 19

Future Applications of Optical Fiber Networks

Elmer H. Hara

19.1 INTRODUCTION

Application of single-mode (SM) optical fiber transmission technology to terrestrial long-distance telecommunication trunk lines is almost universal now, and by 1990 transoceanic submarine cables across the Atlantic and Pacific oceans should be operational as well. Optical fiber cables are cost-effective because of the long repeater spacing, which can exceed 50 km, and the self-checking electronic circuitry built into the system itself. Downtime caused by lightning strikes on optical fiber terrestrial trunk lines is greatly reduced compared to copper-wire-based systems because the optical fiber waveguide is an insulator and there are fewer repeater amplifiers.

The use of SM fiber cables is not limited to intercity trunking but is also being extended to interoffice telephone trunks between telephone central offices, and to T1 (1.544-Mb/s) trunk line services from the central office to office buildings. Almost none of these trunks require a repeater, and the small cable diameters compared to copper cables of equivalent transmission capacity make the optical fiber cable the preferred choice in congested conduits of cities. Of course, the immunity of optical fiber cables to electromagnetic interference (EMI) and radio-frequency interference (RFI) also reinforces the choice of fiber.

Application of optical fiber cables to the telephone subscriber loop is a problem in economics and politics as well as in the selection of the appropriate technology. On the last point of technology, most experts agree on use of SM fiber based on a centrally switched star network where the central office (CO) is the central node [1]. Most network designers favor the use of subnodes that provide switching services to a group of neighboring subscribers [2] as in the case of current telephone practices, whereas a few favor a completely centralized scheme [3]. The latter preference arises from installation and maintenance considerations, or from the viewpoint of providing new services such as demand access video service, which is discussed later. However, the use of SM fibers is accepted as the most logical choice today [4].

Coherent optical transmission techniques have progressed rapidly during the past few years [5], but their practical implementation is probably 15–20 years away, and there is a window of opportunity for commercial implementation of a broadband integrated services digital network (ISDN) to the subscriber using SM fiber technology.

Establishment of the optical fiber broadband ISDN, which provides telephone, data, audio (radio), cable TV, pay TV, and videotelephone services over a single optical fiber network on a national scale is being pushed by many European countries, notably by the Federal Republic of Germany and France. In Japan, NTT (Nippon Telegraph and Telephone Corporation) also has intentions of fiberizing the telephone subscriber loop to offer broadband ISDN services. However, a regulation that requires cable TV service providers to own the transmission plant has been used by the Ministry of Posts and Telecommunications to prevent NTT's service trials in broadband integrated services.

Canada and the United States have similar regulatory problems, although in the United States, telephone companies can provide transmission facilities to the cable TV operator. This approach of separation of carrier and content should work to the advantage of cable TV operators, provided they are assured of an exclusive franchise in some new video services such as the demand access video service that is discussed below in detail. Actually, without a new service that can generate substantial revenue over and above that supplied by telephone, data, cable TV, and pay TV, the optical fiber broadband ISDN for subscribers may be difficult to realize because of the large capital investment that is needed. It is expected that the demand access video service will provide more than the extra revenue that is required.

Independent of developments in optical fiber technology, the intelligent building market has been growing rapidly, and $3 billion a year in equipment sales is projected for the United States by 1990 [6]. In addition to providing cost-effective telephone and data services through PBXs (private branch exchanges) and local area networks (LANs), an intelligent building also offers efficient energy management of heating, ventilating, and air conditioning (HVAC), lighting, and many other services including access control and video security monitoring [7]. Up to 35% of the cost of wiring a building may be saved by combining telephone, lighting and climate controls, fire alarms, and security into an integrated system [8].

With current technology, the overall construction cost of intelligent buildings is about 2% higher, and as a result rental charges are higher [6, p. 65]. Nevertheless, for reasons of prestige and higher staff efficiency, many corporations choose to occupy an intelligent building.

The area of telecommunication services for buildings, factories, and campuses is of particular interest to equipment and system vendors because the free competition in the interconnect market offers ample opportunities to firms that have the networks and systems that can provide telephone, data, video, and HVAC services at competitive prices. Although the technologies are available, no organization appears to have yet mastered them sufficiently to offer a universal telecommunication network and systems package for intelligent buildings on a turnkey basis.

This could be considered a failure of optical fiber technology, which promises much in performance but has not really delivered it at commercially attractive costs for such applications. In the following sections, a design approach to optical fiber networks for intelligent buildings is discussed with an eye to meeting their needs. A similar approach is taken for broadband ISDN services to the subscriber as well, with the thought of extending the technology base developed for the intelligent building into the subscriber loop, taking into consideration the need of extra revenue-producing services that may provide a good return on the large capital investment required for fiberization of the subscriber loop.

19.2 INTELLIGENT BUILDINGS

A measure of the intelligence of a building can be obtained by listing all the telecommunication services and building signal and control functions that are contained in it. Table 19.1 is a limited list of such services and functions and also includes equipment that is interconnected in some manner. Software functions such as energy management are also included. The actual number of entries in Table 19.1 can exceed 100, and an IQ of a building might be measured by simply comparing the number of services it offers against the list. Since some services, functions, or equipment interconnections are more important than others, a weighting function with a range of, for example, 1–5 can be used to make the measure a more precise representation of the IQ of a specific building such as a rental office building, hotel, apartment building, or hospital. Factory sites and campuses can also be included as special extended cases of the intelligent building.

19.2.1 Fourth Utility Services

The telecommunication system and services provided in an intelligent building might be referred to as the *fourth utility*, after the three utilities of electricity (wiring), water (plumbing), and air (HVAC). If a building is to be

TABLE 19.1. Intelligent building communication services—The fourth utility

Telephone, data	Heating control
Telex, Fax	Climate control
Computer terminal	Air conditioning
Personal computer	Lighting control
Computer mainframe	Elevator control
Word processor	Public address
Filing (optical disk)	Paging
Image filing (video disk)	Energy management
Electronic mail	Load management
Message service	CCTV, cable TV
Photocopying	Pay TV
Management data base	Videophone
Fire detection	High-speed data
Access control	Temperature sensor
Thermostat	Lock monitor

labeled intelligent, most if not all of the services, equipment interconnections, and functions listed in Table 19.1 should be provided. Of particular concern to a building owner would be energy management and fire alarm and security monitoring, whereas for a tenant in an office building, telephone, data services, and climate control will be major concerns. Keeping these listed requirements in mind, we can consider the features needed in a communication network for an intelligent building.

19.2.2 Fourth Utility Network Features

The desirable features of a fourth utility network for an intelligent building are listed in Table 19.2. More than 10 digital channels should be provided

TABLE 19.2. Fourth utility network features

Low cost per channel

Large number of channels
    10 or more digital channels per desk
    Broadband channel

Transparency
    ISDN/OSI compatibility
    TTL level connection option
    Freedom of choice of office and switching equipment
    Freedom of choice of bit rates

Reliability
    Simple circuitry, low component count
    Standby power for telephone and fire alarms

to each office because office automation equipment, climate control, telephone, and data line requirements will readily fill 10 channels. Transparency of the network is important because there is a wide selection of equipment available. The telephone set is assumed to be a four-wire digital electronic type.

The network should be cost-effective, and when expansion is required, extra capacity should be readily available. In the event of an electric power failure, standby power should be provided to maintain telephone service and fire alarms as a minimum. Availability of extra capacity will also reduce the costs associated with moving of personnel.

With respect to the large number of services and interconnection requirements listed in Table 19.1, the network should deliver 10 or more data channels to an office. In anticipation of the realization of ISDN as a standard, the services must also be based on digital signals.

The basic bit rate used in office automation is increasing every year. Digital PBXs already offer 64 kb/s interconnections on a nonblocking basis, and today there are many occasions where 1.544 Mb/s is needed. Such is the case for university campuses where there are a number of computer mainframes. This demand is expected to increase significantly over the coming years because of the proliferation of PCs (personal computers) and other computer-based equipment such as word processors. The introduction of CD (compact disk) type storage media having gigabyte capacities for PCs will accelerate this process, particularly when erasable optical disks are introduced in place of today's hard disks and floppy disks based on magnetic recording. Actually, WORM (write-once, read-many) type CD memories will be sufficient for many users, because most do not care to erase stored information, no matter how old the records are.

The maturing video disk technology will also increase the use of video imagery if a suitable broadband capacity is provided in the network. The optical disk for document storage in the form of facsimile or word processor digital formats will then be used side by side with video disks that store images of documents and photographs. The capability to carry broadband signals such as 1.544-Mb/s and baseband analog video, including high-definition television (HDTV) is therefore desirable in a fourth utility network.

The issue of network "transparency" is most important to owners and tenants of rental office buildings because the network should not dictate to tenants what equipment they might use. It is in this area of "transparency" that many commercial networks fail to meet user needs today. In other words, the connection to the network should be on a low-cost basis and the signal protocol-independent.

To be transparent, compatibility with ISDN standards conforming with the concept of open system interconnection (OSI) is of course required. From the standpoint of low-cost interconnection, simple transistor–transistor logic (TTL) level connections using screw terminals or plastic telephone jacks should be available. In order to be transparent, the network should also accept any bit rate from direct current to a maximum design bit rate such as 64 kb/s. This is important for building signal and control systems because there are many different bit rates in use today. The transparent network will then provide the freedom to choose the types of building signal and control systems, terminals, and switching equipment.

A combination of copper-wire pairs and coaxial cables can satisfy the requirements listed above, but the bulkiness will mean extra conduit and installation costs, and problems that might arise from EMI and RFI could present significant postinstallation expenditures. These problems can be solved by optical fiber technology, and their component costs are decreasing rapidly. Today, in the long-haul telephone trunk lines, single-mode optical fiber cables are chosen for their low cost and high transmission capacity. It is expected that by 1990 the same might be said of the graded index (GI) multimode components, which are most suited for use in fourth utility networks for intelligent buildings.

19.2.3 Fourth Utility Network Design

Table 19.3 lists the basic design of an optical fiber network for an intelligent building that can provide the features described in Table 19.2. It is based on a centrally switched star configuration with one network for digital signals and a second network for broadband signals, which may include high-bit-rate (e.g., 1.544-Mb/s) digital signals. Routing of the signals is achieved in the switching room with a PBX, centralized LAN, and broadband matrix switch. Network transparency is provided for digital signals by the use of an asynchronous high-speed TDM system. Separation of the digital and broadband networks avoids the extra cost of modulators, demodulators, and filters required in the frequency division multiplex (FDM) approach for a single network.

The advantages of choosing a centrally switched star network are listed in Table 19.4. The maximum transmission bit rate required of the network

---

TABLE 19.3. Fourth utility network design

Optical fiber star networks
  Centrally switched
  Asynchronous TDM digital network
  Separate broadband network
Switching room
  Digital PBX for telephone
  Centralized LAN for data
  Broadband matrix switch

TABLE 19.4. Advantages of the centrally switched star network

The maximum bit rate required is that of the highest-speed terminal equipment.

The comparatively low bit-rate requirement means that lower-cost optoelectronic modules and electronic circuitry can be used.

Multimode GI fibers can be used. This means that lower-cost connectors can be used. In the near future when 650-nm video disk laser diodes become available in large quantities, plastic fibers (e.g., 484/500) can be used for distances of up to 200–300 m.

---

is set by the terminal equipment with the highest bit-rate requirement. Normally 64 kb/s is more than adequate, but, as mentioned earlier, 1.544 Mb/s will be needed more and more in the future. The higher bit rate can be handled by the broadband network in the present design. In contrast, ring and bus networks require operational bit rates 100–1000 times larger than the basic bit-rate requirement of the terminal equipment. Even then, at present no commercial LAN can guarantee a throughput of 1.544 Mb/s per terminal, and many will have trouble providing a throughput of 64 kb/s when traffic is heavy. The need for high bit rates by ring and bus networks means that electronic and optoelectronic modules and interface circuits will be comparatively expensive.

Because of the lower bit-rate requirements, the centrally switched star network can use GI multimode fibers such as 50/125 (i.e., 50-μm core diameter/125-μm cladding diameter) or 100/140 sizes because the maximum transmission distances do not often exceed 1 km, which readily ensures the availability of transmission bandwidths in excess of 50 MHz even when a 100/140 GI fiber is used. The dimensional precision of connectors can also be relaxed because of the larger core diameters compared to that of single-mode fibers (e.g., 8 μm). Losses associated with the connectors and splices can also be in the neighborhood of 2 dB because a transmission distance of 1 km introduces less than 6 dB of loss at a wavelength of 850 nm. For the two preceding reasons, low-cost connectors and splicing techniques can be employed when installing a star network. The low bit-rate requirement also allows the use of low-cost light-emitting diodes in the optoelectronic modules. As a result, these modules need not be expensive.

In the near future when low-cost video disk laser diodes lasing at 650 nm become available, the use of plastic fibers (e.g., 484/500) can be considered because the transmission loss is about 150 dB/km at that wavelength, and the transmission bandwidth is about 10 MHz/km, which means that distances of 200–300 m can be reached with a transmission bandwidth larger than 30 MHz. When mass-produced, the cost of plastic fiber cable will be lower than copper telephone wire, and because of the large core size and ease of assembly, connectors can be as low as 25¢ each. Indeed, with such possibilities, the centrally switched star network design will soon be difficult to ignore.

The basic configuration of the fourth utility network is shown in Figure 19.1. The configuration is a centrally switched star network. Four fibers connect each office or workstation to a centrally located switching room. Two of the fibers carry all the digital data signals that have comparatively low bit rates (e.g., 64 kb/s) by using an asynchronous time-division multiplex (TDM) scheme. For broadband signals such as video or high-speed digital data (e.g., 1.544 Mb/s), a second pair of fibers is used. A pair each is used to provide full bidirectional transmission. The reason for using a second network for broadband signals is to avoid the use of modulators, demodulators, and filters required when a frequency-division multiplex (FDM) method is used to place the broadband signal on the same fiber as that for digital data signals, which occupy the baseband portion of the radio-frequency spectrum. For short distances, the cost of using extra fibers is lower than the cost of using the modulator, demodulator, and filters. At present, most wavelength division multiplex (WDM) devices are too expensive to consider their use. From the standpoint of installation cost of the four-fiber cable, the conduit and wiring (fibering) labor costs should be equal to or less than that required for telephone systems using copper wires.

For transmission of the video signals, standard optoelectronic baseband video transmission modules can be used. There is a large selection available today, with some transmitter/receiver sets priced below $300. Two complete sets will provide full bidirectional connection to the switching room.

For TDM transmission of digital signals, high-speed asynchronous sampling can be used. Such a system can readily multiplex 20 or more 64-kb/s digital signals onto a single optical fiber waveguide by sampling each digital bit a minimum of 10 times. The resulting TDM bit stream includes frame codes to properly identify each of the signal channels at the demultiplexer. Such an asynchronous TDM system is equivalent to the old mechanical commutator–decommutator arrangement for multiplexing many telegraphic lines [9].

Figure 19.2 shows a block diagram of an asynchronous TDM transmission system. For 20 channels of signals with maximum bit rates of 64 kb/s, the aggregate bit rate is approximately 13 Mb/s when each signal

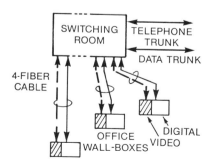

Figure 19.1 Fourth utility network.

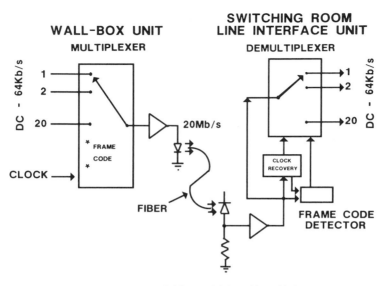

Figure 19.2 Asynchronous time-division multiplexer/demultiplexer.

bit is sampled at least 10 times. Synchronization of the multiplexer and de-multiplexer is achieved by a simple frame code that takes advantage of the large transmission bandwidth available in a fiber. Two such systems make a full duplex link. The result in effect is the provision of 20 four-wire copper lines from an office to the switching room.

In devising a frame code to identify the signal channels at the demulti-plexer, a very simple comma code can be formulated once the assumption is made that there is little or no restriction to the bandwidth available in the transmission line. The result is a frame code that has an aggregate bit rate of approximately 7 Mb/s, which brings the total multiplexed bit rate for 20 dc-to-64-kb/s channels to about 20 Mb/s.

Table 19.5 shows the generation of a comma code for the case of 10 data channels. The multiplexer channels are sampled sequentially at high speed from channels 1 to 18. The dummy channel 6, which is set at 0, ensures that the frame code, which is 1111110, is not generated by the data channels.

By inserting a dummy channel that is set at 0 in the midst of the data channels, the uniqueness of the frame code is assured. This kind of approach is extravagant in its use of transmission capacity because more than one-third of the signal is taken up by the frame code. However, optical fiber transmission systems have ample capacity, and no problem is encountered even when there are 20 data channels. The demultiplexer can detect the frame code sequence with a shift-register/comparator and reset the demultiplexing operation to start from channel 1 and thereby distribute the asynchronously sampled signals to the correct channels.

TABLE 19.5. Multiplexer frame code generation

Data channels	Logic levels
1	0 or 1
2	0 or 1
3	0 or 1
4	0 or 1
5	0 or 1
6 (dummy)	0
7	0 or 1
8	0 or 1
9	0 or 1
10	0 or 1
11	0 or 1
Frame code channels	
12	1
13	1
14	1
15	1
16	1
17	1
18	0

The advantage of the asynchronous system is the freedom from the additional buffering and synchronizing electronic circuitry that is normally required in fully synchronized TDM systems, which are standard in the telephone hierarchy, because no synchronization is needed between the sampling frequency and the digital signals. Circuit simplicity is the result, and more than ten 64-kb/s data lines can be offered cost-effectively to each office or workstation. In fact, because of the high-speed asynchronous sampling, any digital signal ranging from dc to 64 kb/s can be multiplexed onto the same optical fiber line by this method. The digital signals need not all be derived from the same reference clock either. This means that although the signals may share the same fiber network, each system can operate independently. Analog signals such as those from building control systems can also be transmitted by first converting them to pulsed analog forms such as pulsed freqency modulation (PFM) signals, by using voltage-controlled oscillators (VCOs). Such flexibility in digital bit rates and the ability to carry analog signals are important for intelligent buildings where the communication needs are diverse.

The option also exists to submultiplex one or more extra channels in the same manner to provide 10 or more control channels (e.g., dc to 300 b/s) to be used in conjunction with the dc-to-64-kb/s channels. The design in fact has been commercialized to provide twenty dc-to-80-kb/s and forty dc-to-300-b/s channels. By the use of suitable input–output circuitry, connections to two-wire analog telephones, four-wire digital telephones, RS-232C and RS-422 ports, and TTL levels can be provided. At present, 20 RS-232C connections can be offered at a cost of $300 per channel. In the future the cost might be reduced to much less than $100 per channel by using VLSI circuits. The optical fiber cable is not included in these costs.

In order to respond to the expected need for 1.544-Mb/s channels to each office in the future, the inclusion of a single 1.544 Mb/s channel is being considered for the design shown in Figure 19.2. By introducing suitable buffers and timing circuits, 1.544 Mb/s can be transmitted synchronously by using two or three of the available channels. The asynchronous feature can be retained if an analog frame code relying on pulse height is used. In that case, a single channel for dc-to-2.5-Mb/s signals can be added without exceeding an aggregate bit rate of 25 Mb/s.

19.2.4 Fourth Utility Network Switching Room

Figure 19.3 shows the block diagram of the switching room, which is located centrally. For clarity, the broadband facility is shown separately in Figure 19.4. The centrally located switching room houses the various systems that provide intelligence to a building. Communication with the outside world is achieved through the trunk lines connected to the PBX. Each system such as the PBX, LAN, and HVAC system operates independently, but they share the optical fiber lines. Dedicated hardwire connections can also be made at the interface section.

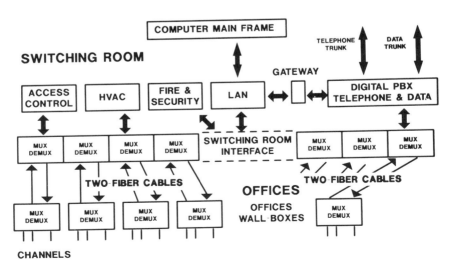

Figure 19.3 Switching room: digital signals.

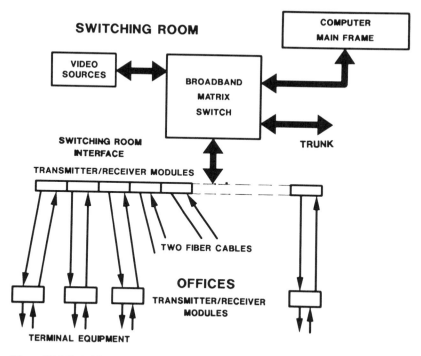

Figure 19.4 Switching room: broadband signals.

The switching room houses a digital PBX for telephone and a centralized LAN for data. By assigning the telephone traffic to the PBX and data traffic to the LAN, cost-effective use of equipment is realized because the PBX and LAN are designed specifically for telephone and data, respectively, and many LAN designs offer lower connection costs than the PBX. Of course, when the data communication requirements are comparatively small, both telephone and data can be switched by the PBX. All digital PBXs today have excellent features dealing with data signals. The PBX and LAN can be chosen to suit the needs of the user, keeping in mind that four-wire digital telephones are preferred and that the maximum data rate is set at 80 kb/s. The RS-232C ports might be used to connect the data signals to the LAN. The $2B + D = 144$ kb/s ($B = 64$ kb/s, $D = 16$ kb/s) ISDN signal can be handled by decomposing it into one 16-kb/s and two 64-kb/s signals at the input–output section of the asynchronous TDM system. Experiments have shown that the dc-to-80-kb/s channel has no problem in carrying 144 kb/s, and therefore probably only one channel need be used for the $2B + D$ signal.

The asynchronous TDM system, in effect, delivers 40 four-wire cables from each office to the switching room. The only restrictions on their use is to limit the digital bit rate to less than 80 kb/s for half of the cables and to less than 300 b/s for the other half and to use TTL signal levels. The system designer is therefore looking at 40 independent copper-wire communication channels that are available at the switching center for interconnecting offices. The optical fiber network thus becomes transparent to the fourth utility system designer. Dedicated hardwire connections can be made at the interface section, or independent building control systems such as access control and smoke and fire detection can be installed. It is envisaged that in an intelligent building, a control center for the entire building will be located adjacent to the switching room. This control center can house a centralized HVAC system, which can reduce energy costs through the intelligent use of sensors and control units placed throughout the building. This becomes possible because of the availability of many wire connections to each office. Because 20 dc-to-80-kb/s and 20 dc-to-300-b/s channels are available at each asynchronous TDM unit, a single unit might be shared by three or four offices, depending on the communication requirements.

The network of the broadband facility in the switching room (Fig. 19.4) in effect delivers two coaxial cables from each office. Direct connections between two video terminals in separate offices can be made at the video interface section. For monitoring of sensitive areas in a building, closed-circuit television (CCTV) cameras and monitors can be connected directly as well. High-bit-rate signals such as 1.544 Mb/s can be carried readily by the broadband network and routed suitably to their destination through the broadband matrix switch or connected directly to other offices at the video interface section. Control signals for the broadband matrix switch can be routed through the digital network.

A number of commercial firms (for example, Central Dynamics, Ltd., Montreal, Canada; Dynair Electronics, Inc., San Diego, California; The Grass Valley Group, Inc., Grass Valley, California) offer analog broadband matrix switches. Because of the large bandwidth requirements, whether the signal is analog or digital, these matrix switches necessarily use spatial (circuit) switches. The cost per crosspoint is about $40. A 100 × 100 matrix switch will therefore cost $400,000, which is expensive. Fortunately, the requirement for broadband signals is not expected to be too large in the immediate future, and direct connections or small (e.g., 5 × 5) matrix switches can be used as the need arises. However, a development effort is needed to reduce the cost per crosspoint to less than $1 to meet future needs. The most promising technology in this direction is probably the optoelectronic switch [10].

19.2.5 Advantages of the Fourth Utility Network and Intelligent Buildings

It should be noted that the LAN is located in the switching room where all the interface units are also located. What has been done is to extend the connecting "wires" from each terminal device in an office to the switching room. This star network approach raises the security of the data lines connected to the LAN to that of a telephone line connected to the PBX, because the LAN proper is now located in the switching room. Maintenance of the LAN is also carried out readily for the same reason. If a corporation in the intelligent building wishes to have its own LAN and/or PBX independently, then the present design allows the placing of a number of separate LANs and PBXs in the switching room because of the star network configuration and asynchronous TDM system. When the requirement lapses because of a corporate move, the extra PBX and LAN can be removed without disturbing the other occupants. These are features not available in current commercial products.

Using the digital data network for all signaling and controlling requirements such as fire alarms, access monitoring, and climate control has the advantage of eliminating extra networks, and this should result in lower conduit and wiring costs. The question of reliability of the network can be answered by built-in checking functions for circuit integrity and suitable fault indicators. Redundancy of communication channels for fire alarms required by fire codes can also be provided by connecting each fire sensor to two different multiplexers. An alternative approach would be to provide more fire sensors using the large channel capacity that is available to each office. Since the fire alarm system is independently established in a central location, malfunction of other building control systems will not affect it. Breakdowns in one of the asynchronous TDM transmission systems will affect only a few offices, which are connected through one arm of the star network. Fire sensors located in adjacent offices can provide ample protection while repairs are carried out.

It must be emphasized that although integration of the communication network is achieved through the optical fiber star network, the functions of building sensing and control systems will remain independent. This is in contrast to some intelligent building system designs that have building sensing and control systems integrated into a single entity using a computer as the master controller. Such an approach contains the uneasiness of having all systems in a single unit, because in the event of a computer breakdown the entire building's "IQ" can suddenly drop to an unacceptably low level.

For the occupant of an intelligent building, higher productivity can be expected through personalized climate and lighting control made possible through the large telecommunication capacity. Office automation should also improve productivity, particularly through the use of PCs and optical disk document filing systems, which will significantly reduce or eliminate the need for photocopying and also shorten transmittal time of documents. Office productivity studies have indicated that much office work is occupied with searching for documents or information. Computer-controlled electronic document handling without the medium of paper should make them readily available on a computer terminal.

Personal security for the occupant can be much better through good access control and CCTV monitoring. In case of emergencies such as fires, the location can be pinpointed accurately with the centralized fire-monitoring system and evacuation routes can be indicated either through the public address system or by route indicator lights controlled by the fire and security system located in the switching room.

Office moves can also be made more readily because of the abundance of network connections that are available. Most connection changes will be a matter of unplugging and plugging in plastic jacks for telephones and terminals and making appropriate alterations to the switching software. In the worst case, only input–output cards in the multiplexer/demultiplexer need be added or changed.

For the building owner, probably the energy efficiency derived from better sensing and control, and the lower insurance rates obtained from the superior fire and security, access control, and CCTV monitoring, will be attractive. By good use of a computer, maintenance, inventory control, and work scheduling should realize lower operating costs as well. The higher rental income made possible because of the prestige and real benefits such as higher productivity for tenants is of course very attractive for rental office building owners from the viewpoint of marketability.

The fourth utility optical fiber network discussed here can satisfy most, if not all, foreseeable communications requirements of an intelligent building because of its transparency and flexibility. The same approach can be applied to the case of the subscriber loop using single-mode fibers and WDM devices. The latter becomes necessary because of the longer transmission lines that make the use of these devices cost-effective. By developing a competence in the fourth utility interconnect market today, we should be able to meet the challenge that will arise if and when the move to fiberize the subscriber loop on a national scale materializes. The following sections describe the technologies that are providing the push for this fiberization and discuss the impact of broadband ISDN in the subscriber loop.

19.3 BROADBAND ISDN

The basic services to be provided by the broadband ISDN are telephone, data, cable TV, and pay TV. Data can include services such as fire and security monitoring, meter reading, and PC interconnection. The network is based on single-mode optical fiber technology in the form of a centrally switched star network similar to that of the current telephone network configuration. Figure 19.5 shows the basic configuration. Connection to a subscriber is through a single single-mode fiber. Wavelength-division multiplex devices are used to provide bidirectional transmission and multichannel services. Table 19.6 shows one possible arrangement of the signal formats. For

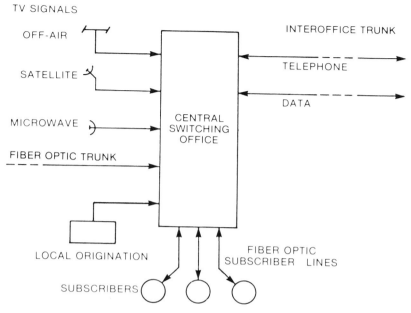

Figure 19.5 Broadband ISDN subscriber loop.

the near future (1990–1995) the ISDN standard $2B + D = 144$ kb/s can provide one telephone and one data channel with the possibility of submultiplexing the data channel. Television service in the digital format is still too expensive because of high component costs. Therefore, for the near future, four channels of TV signals in analog frequency-division multiplex (FDM) format will most likely be used.

TABLE 19.6. Broadband ISDN signal formats

Service	Downstream*	Upstream*
Telephone and data	WDM λ_1	WDM λ_2
Near future (synchronous TDM)	$2B + D = 144$ kb/s	$2B + D = 144$ kb/s
Future (asynchronous TDM)	$4B, 4D$	$4B, 4D$
	$T_1 = 1.544$ Mb/s	$T_1 = 1.544$ Mb/s
	Aggregate: 25 Mb/s	Aggregate: 25 Mb/s
Television	WDM λ_3	
Near future (analog FDM)	4×5 MHz $= 20$ MHz	
Future (digital TDM)	4×100 Mb/s $= 400$ Mb/s	
HDTV	WDM λ_4	
Future (analog)	1×30 MHz $= 30$ MHz	
Far future (Digital)	1×400 Mb/s $= 400$ Mb/s	

* $B = 64$ kb/s, $D = 16$ kb/s

As digital TV sets begin to proliferate and digital codecs become low in price through mass production, the signal format will become digital. One unresolved problem is the picture-in-picture presentation capability of the digital TV set. The number of TV channels delivered to a subscriber may need to be as high as 12 channels ($=4$ sets $\times$ 3 channels/set), which will result in an aggregate bit rate of

$$1.2 \text{ Gb/s} = (100 \text{ Mb/s})/\text{channel} \times 12 \text{ channels}$$

The need for multiple-signal channels is met by the four WDM wavelengths.

Telephone and data services can be expanded in the future (1995–2000) by using the asynchronous TDM system described earlier. This will remove some of the headaches of synchronizing signals arising from different services. Also TV sets will be capable of receiving digital signals, and, through mass production, the overall transmission system cost will undoubtedly become comparable to or lower than that of analog systems. At about the same time we might expect HDTV services to be offered in an analog format and eventually, in the far future (2000–2005), a digital format should become cost-competitive.

A number of technologies are providing the push toward establishment of broadband ISDN. The following section discusses some features of the key technologies.

19.3.1 Digital TV Sets

Digital TV sets are already on the market today, and in the near future they promise to be lower in cost than conventional analog TV sets because of the component count, which can be as much as 30% lower. Through digital processing and the use of single-frame memories, digital TV sets can provide improved pictures. For example, ghosts can be eliminated, and, for large projection screens, scan lines can be doubled to produce pseudo-HDTV pictures. In the process of doubling the lines, aside from interpolating visual information between the original lines, frame-to-frame temporal interpolation can also be performed to further improve the picture. In addition, the frame presentation rate can be converted to 60 frames/s on a noninterlace basis to remove many of the picture quality problems such as flicker and poor color definition that are associated with the current method of interlace scanning. Display of multiple-channel pictures on the same screen is also a feature offered in the commercial digital TV set today [11].

The frame-memory capacity of the digital TV set can be used to provide still-picture service by transmitting a single frame to the subscriber from a suitable source such as the video disk. This approach has the potential to supersede the present Telidon and videotext services.

19.3.2 High-Definition Television (HDTV)

The HDTV technology is just emerging from the laboratory [12]. Video pictures matching the quality of 35-mm slide films can be projected onto large screens, and recordings have been made on video disks (i.e., 45 min on a CD disk). Small HDTV cameras using charge-coupled devices (CCDs) have been assembled. By the early 1990s, practical applications of HDTV technology will probably be seen in a number of locations, and the merging of cinema film production and TV program production might become possible. Some concern over the impact of HDTV and optical disk technology on the photographic film industry is already being expressed by photographic film manufacturers because optical disk recording of high-definition images may eventually take over a large portion of the service provided by photographic film today.

Large-screen HDTV creates an impression of three-dimensionality that is impressive when viewing sporting events. Because of the higher resolution, close-up shots of players are no longer necessary and the entire team of players can be shown. A new dimension of sports entertainment can therefore be offered, allowing every viewer to be a coach or linesman calling for the plays or judging the off-sides. Of course, close-up shots of key players can still be offered on a split-screen basis to further enhance the entertainment value.

The same can be said for theater stage events such as ballet, theater, and symphony performances. Combined with digital stereophonic sound, performances can be enjoyed as if the viewer were seated in the best seat of the house, with minimum or no movement of the camera. Split-screen close-ups of principal performers will further enhance the enjoyment.

Still pictures in HDTV might also be used to transmit the images of fine art objects such as famous paintings or sculptures to the home. Because of the high definition, text material as it appears on standard $8\frac{1}{2}$ in. × 11 in. paper or books can also be viewed on HDTV sets. This means that all written material can be offered to viewers, and interactive educational programs on an individual basis become possible with those HDTV sets that have frame-storage capability.

19.3.3 Video Disks

Video disk sales are showing a healthy annual growth in Japan today, and some have compact disk digital stereophonic audio channels as well. The erase/write/read video disk with multiple digital stereophonic audio channels will most likely be commercially available in the early 1990s and the current video tape recorder (VTR) market may be threatened by the superior picture and sound qualities offered by the video disk.

The large storage capacity of 27,000 video frames per disk can also be used as a source for still pictures or text material. This opens the possibility of providing interactive educational instruction through video disks.

Not only is the optical disk technology used in video disks well established now in CD music recordings, but rapid growth is also being seen in use of the medium as storage for computer data. In fact, compact disk units for replacement of floppy disk units in personal computers are already on the market and application to large-scale document storage systems has started.

It is this video disk technology that will most likely provide the basis for extra revenue, over and above that obtainable through standard telephone, data, cable TV, and pay TV services, and thereby support the extra capital requirement for establishing broadband ISDN. The following section discusses the impact of this technology.

19.4 DEMAND ACCESS VIDEO SERVICES

A large capital investment is required for replacing the copper-wire telephone plant with an optical fiber network, and the telephone, data, radio, cable TV, and pay TV services offered through the single optical fiber network to each subscriber may not generate sufficient revenue.

Southern Bell has indicated that for new housing developments, fiber will be used if the plant cost is about $2500 per subscriber provided cable TV service carrier charges are included in the revenue. In Japan, NTT had a target of $2000 per subscriber in developing components for a graded-index multimode subscriber system. They have already completed the work and are also about to complete development of single-mode fiber components. Examination of the component requirements for broadband ISDN generates an impression that the cost per subscriber is more in the range of $4000–5000. This additional capital cost of $2500–3000 must be backed by extra service revenue.

It is my contention that demand access video services offered through the optical fiber network will generate more than enough additional revenue to make the fiberization of the subscriber loop appear as an attractive investment.

By demand access video services [3] I mean the provision of TV programs and videotext information services according to subscribers' demands on a 24-hr basis. This is realized by taking advantage of the centralized switching feature of the optical fiber network and placing video disk jukeboxes that have multiple disks and multiple playback heads at the central switching office. Figure 19.6 shows a schematic diagram of the video disk jukebox system. The subscriber's request is processed by the central processing unit, and the appropriate cross-point of the matrix switch is closed

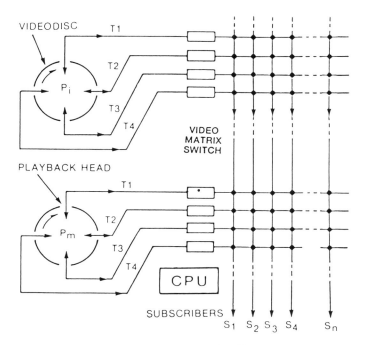

Figure 19.6 Demand access video system. The video disk jukebox has multiple playback heads on multiple disks. The subscriber uses a keypad to enter requests to the CPU for the desired disk and starting time, which is available within 15 min for a 1-hr program.

to provide the program from the selected disk at the selected time. With eight playback heads on a disk that has a 1-hr program, starting times can be spaced 7.5 min apart, 24 hr a day. The waiting-time-to-start is less than 7.5 min, and this time can be used to show advertisements or public relations material. A countdown clock may indicate the time remaining to the start of the program with synthesized voice announcements included as well. Most, if not all, subscribers can be served from a single playback head for a given program and time by this arrangement. The starting times for various disks can be distributed evenly throughout the hour to avoid overloading the CPU with a large number of coincident requests for programs.

Table 19.7 shows the types of demand access video services that might be provided. The term "channel" refers to a class of video sources. The computer program can readily specify whether a program is to be supplied on a "per-channel" monthly charge basis or on a per-program charge basis and can maintain an account for each subscriber. Of course, the per-channel service may not necessarily make available some special programs that are specifically offered on a per-program charge basis. Demand access video service is offered in addition to standard cable TV service. Telidon-II refers

TABLE 19.7. Demand access video service

	TV service			
	Per "channel"		*Per program*	
Subscription options	*Advertisement*	*No advertisement*	*Advertisement*	*No advertisement*
Subscription cost	Low	Medium	Medium	High
Programs	Delayed "Broadcast," cinema, sports, performing arts, special events			

	Still-picture service			
	Per "channel"		*Per page*	
Subscriptions options	*Advertisement*	*No advertisement*	*Advertisement*	*No advertisement*
Subscription cost	Low	Medium	Medium	High
Material	Computer-aided instruction, Telidon-II, library, fine arts, archives			

to second-generation Telidon service based on the transmission of single TV frames.

Each program can be offered with or without advertisements, and rates can be set on a monthly basis for a class of programs or on a per-program basis. Variable rates depending on the number and types of programs viewed in a given month can be considered. Setting of the rates is a computer program problem that can be modified readily to optimize the return on investment. For example, per-program charges may be set on a logarithmic scale, each program becoming progressively cheaper until finally, when an aggregrate sum of, for example, $30 is reached for a month, the remaining programs are provided at no additional cost.

It should be noted that news programs take on a sense of immediacy by being available 24 hr a day with clearly indicated update times. Videotext service is also possible with digital TV sets, and since the time required to transfer a single picture to a subscriber from the video disk is less than a second, no traffic problems will be encountered at the switching center even if this service becomes very popular.

The types of programs and services that might be offered are limited only by the imagination of the entrepreneur, switching system capacity, and video disk jukebox capacity. Real-time broadcast services in the form of conventional cable TV, pay TV, and HDTV services may also be offered in competition with the demand access video service.

19.5 IMPACT OF DEMAND ACCESS VIDEO SERVICE

The technologies of optical fibers, digital TV, and video disks combined to offer demand access video services will have a significant impact not only on the form of entertainment enjoyed by subscribers but also on the infrastructure of the broadcasting, cable TV, telecommunications, and cinema production industries. The following sections comment on the possible forms this impact might take.

19.5.1 Narrowcasting

Since programs can be provided on an individual basis, a wide spectrum of preferences in terms of content and viewing time can be accommodated. What we know today as broadcasting will turn into what has been termed narrowcasting. All the cinema films and TV programs new and old can be offered. Special film or program festivals can be held to boost sales; for example, provision of ethnic (foreign) movies may be a profitable service.

19.5.2 Precision Advertising

Because of the nature of narrowcasting where the viewer can pick and choose the content and time, the audience for advertising will become fragmented. This need not have a negative impact on the advertising industry, because advertising can now become a more precisely gauged business. For example, because program selection, including that for cable TV service, is controlled at the central switching office by a computer, the advertiser can specify the type of viewer by selecting the program (e.g., symphony performance), time slot (e.g., Friday 9:00 to 11:00 P.M.), residential district, and total number of viewers.

Within a short time, the advertiser will have sufficient statistics on the effectiveness of his advertising and should be able to optimize his return on investment for subsequent advertisements. The system will also allow small businesses to place advertisements for local residents. The overall impact for advertising will probably be an increase in small-scale regional advertising, with its cost effectiveness measurable with more confidence than is possible today.

19.5.3 TV Broadcaster Transformation

The role of the TV broadcaster will have to change because transmission and distribution will be in the hands of the optical fiber network owner and many programs will be provided as video disks. The role of the broadcaster might be concentrated in the areas of real-time program production and cinema film production using HDTV technology. In particular, real-time HDTV broadcasting will be an important function because uncompressed video disk recording of HDTV programs of long duration (e.g., 1 hr) may not become practical for some time. Of course, the films will be offered to the entity that provides the demand access video service and also supplied to movie theaters and cinema chains, which might be established specifically for this purpose.

Since almost 100% of today's urban households have telephones, fiberization of the subscriber loop can bring cable TV, pay TV, and demand access video services to every home. This means that TV broadcasting stations will no longer be needed in the cities and the TV spectrum might be reused for other communication services such as cellular radio. Rural TV services can be provided through a direct broadcast satellite if fiberization of the subscriber loop is not practical in areas of low population density.

19.5.4 Cable TV Operator Transformation

The cable TV industry can still provide conventional cable TV and pay TV services by receiving signal feeds from the broadcaster or program supplier. However, in competition with demand access video service, the cable TV and pay TV business may not be as attractive. Therefore, demand access video service might be considered a natural extension of cable TV and pay TV services for the cable TV operator.

The issue of transmission plant ownership will arise for the cable TV operator because telephone companies are the most likely entities to install the broadband ISDN subscriber loop. Judging from the current popularity of videotape shops, we can expect the demand access video service to be a very profitable operation. Therefore, in exchange for the right to provide demand access video services, cable TV operators might forgo the ownership of cable TV distribution plants. For Canada and the United States, establishment of the broadband ISDN subscriber loop will be an ideal opportunity to reestablish the principle of separation of the carrier and content provider.

19.5.5 Collection of Copyright Fees

Since the programs are offered on a demand basis, copyright fees and actors' fees can be collected from each individual viewing. This is not the case for those movie rights that are sold to videotape shop suppliers, where only a one-time lump sum is received by the movie producer. In the case of demand access video services, the movie producer and actor can retain the right to collect fees and gain a higher profit from their investment.

19.6 DISCUSSION

The ownership of the broadband ISDN subscriber plant must be resolved at an early date if orderly development of telecommunication services, as well as universal access to such services, is to be maintained. One solution might be to have a consortium of cable TV operators, telephone companies, other common carriers, broadcasters, and the government own the plant jointly by sharing the large capitalization required to establish it. Such an approach makes sense when the broadband ISDN is looked upon as a public communications highway to every home. Separation of carrier and content can be made clearly. The content provider as well as the user who is the subscriber can pay a tariff for using the roadway.

Content suppliers such as cable TV operators might be licensed to offer demand access video services, and the telephone companies may continue their monopoly position of telephone service to the home to ensure universal access. Long-haul transmission can be left as a free market where competition will depend on the alternative trunking service provided between cities. Data transmission within an urban area poses a problem because most needs will be equivalent to that of a telephone, in which case the telephone company would be the logical choice to deal with such traffic.

The TV broadcaster could concentrate on TV program and cinema film production and the narrowcast advertising business that rides on the demand access video service. In order to preserve the broadcaster's role as an advertising service provider, this function might be legislated to be their domain.

Establishment of the broadband ISDN subscriber loop will necessarily mean a change in the infrastructure of telecommunication institutions as we understand them today. Only by anticipating the impact of new technologies correctly and selecting appropriate policies will we be able to maximize the benefits for the public, who deserve a cost-effective communications highway to the home together with the enhanced services that such a highway promises to deliver.

REFERENCES

1. G. Mogensen "An overview of broadband systems—applications and tradeoffs", *IEEE J. Selected Areas Commun.*, **SAC-1**(3): 420–427 (1983).
2. T. Miki and J. Tamagata, "Fiber optic transport system for the present and future local networks", *IEEE J. Selected Areas Commun.*, **SAC-4**(9): 1531–1537 (1986).
3. E. H. Hara and R. I. MacDonald, "A demand access video system," *Offic. Tech. Record, Can. Cable Television Ann. Convention*, Toronto, June 1982, pp. 69–74.
4. P. Cochrane, R. D. Hall, J. P. Moss, R. A. Betts, and L. Bickers, "Local-line single-mode optics—viable options for today and tomorrow," *IEEE J. Selected Areas Commun.*, **SAC-4**(9): 1438–1445 (1986).
5. Special Issue on Coherent Communications, *J. Lightwave Technol.*, **LT-5**(4): (1987).

6. M. Walsh, "Towers with minds of their own," *Time*, June 24, 1985, p. 64.

7. S. Gannes, "The bucks in brainy buildings," *Fortune*, Dec. 24, 1984, pp. 132–144.

8. P. Houston, "A big move into 'smart buildings'," *Business Week*, June 24, 1985, p. 104J.

9. I. S. C., "Telegraph," *The New Encyclopedia Brittanica*, 15th ed., Vol. 18, 1975, pp. 71–72.

10. E. H. Hara, "Optoelectronic switching," First Int. Conf. Integrated Opt. Circuit Eng., Cambridge, Mass., Oct. 23–25, 1984, *Proc. SPIE*, 517: 234–241 (1984).

11. Nikkei Electronics Report, "Production of digital TV sets: will it take off?," *Nikkei Electronics*, No. 355, Nov. 5, 1984, pp. 87–90 (in Japanese).

12. T. Fujio, "The present state of HDTV; what it takes and what should be done," 2nd Int. Colloquim on New Television Systems: HDTV'85, Ottawa, Canada, May 13–16, 1985, sponsored by Department of Communications, Canadian Broadcasting Corporation and National Film Board, Paper 1.1.

CHAPTER 20

Free-Space Optical Communication Systems

Joseph Katz

20.1 INTRODUCTION

The main advantage of using optical frequencies in free-space communication systems is the high directivity of the propagating optical beam. (In the context of this chapter, the term "free space" refers to unguided optical communications; exoatmospheric environments are referred to as "space.") When transmitted from an antenna, electromagnetic waves spread (diffract) as they propagate. The diffraction angle is directly proportional to the radiation wavelength and inversely proportional to the antenna size. Due to the much shorter wavelength of optical signals compared to microwave signals, optical beam diffraction is much smaller, making it possible to communicate over much larger distances. In addition, under some conditions it is possible to take advantage of the quantum nature of the optical radiation and achieve highly energy efficient communication schemes.

The wavelength region of the electromagnetic spectrum that is commonly referred to as "optical" extends approximately from 0.1 μm to 100 μm. In order to realize an efficient optical communication system, a wavelength region where efficient and reliable light sources and detectors are available should be chosen. In addition, the wavelength should not be too long, since that would result in larger diffraction of the propagating beam (or, alternatively, the need for large-aperture transmitting telescopes). The above considerations narrow down the region of operation to the visible—near-infrared region (0.5–1.5 μm) of the spectrum. At the present time the major candidates for light sources are semiconductor laser devices (lasers and laser arrays) and (frequency-doubled) Nd: YAG lasers. Available detectors include photomultiplier tubes, photodiodes, and avalanche photodiodes.

The subject of free-space optical communications is too broad to be adequately treated in its entirety in one chapter. Several books and conference proceedings are available; see, for example, Refs. 1 and 2. Thus the main emphasis here will be on those aspects that are both unique to unguided optical communications and more directly related to optoelectronic technology. Other topics will generally be mentioned briefly, and readers will be referred to a more complete discussion elsewhere (in some cases to other chapters of this book).

The outline of this chapter is as follows. Section 20.2 serves as a basic introduction to some of the general aspects of free-space optical communications. Section 20.3 discusses the subject of optical sources for free-space optical communication systems. The first topic is the power combining of semiconductor lasers. Because of their many attractive features, semiconductor lasers are the preferred choice as light sources for free-space optical communications. However, the power available reliably from a single device in stable single-mode operation is intrinsically limited to about 100 mW cw (see Chapter 13 of this volume), which is insufficient for many applications. Since power levels in the range of 100 mW to 1 W are required in many applications (the actual amount depends, of course, on other system parameters), power combination of several lasers may offer a solution to this problem. Because of some disadvantages of semiconductor lasers, the Nd:YAG laser (pumped by semiconductor laser sources) is also a viable candidate, and it will also be discussed. Section 20.4 reviews several topics related to the detection of optical signals. These include the issues of background noise and receiver sensitivity, novel detector (device) implementations that may enhance this sensitivity, and an improved heterodyne receiver configuration that is important in applications where the laser sources have noise in excess of the unavoidable shot noise (as is the case with semiconductor lasers). Finally, Section 20.5 describes some systems applications where optical communication links are either in use or can be employed in the future. These range from interbuilding links to communicating with spacecraft outside the solar system.

20.2 FREE-SPACE OPTICAL COMMUNICATIONS

A basic free-space optical communication system is shown schematically in Figure 20.1; it consists of an optical transmitter emitting P_t watts of power into a solid angle of Ω_t steradians and an optical receiver with a collection aperture area of A_r m^2. The distance R between the transmitter and the receiver can range from a few tens of meters (for interbuilding links), through

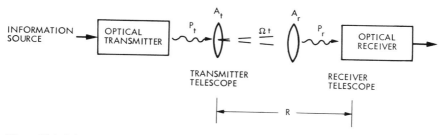

Figure 20.1 Schematic configuration depicting the basic parameters of a free-space optical communication system. (See text for explanation of the various parameters.)

several tens of thousands of kilometers (for intersatellite links), to several astronomical units—and possibly even farther—in deep-space applications. In addition to the transmitter and the receiver, optical communication systems include subsystems to perform functions such as beam manipulation (acquisition, tracking) and interfacing with the host environment (e.g., spacecraft carrying the optical communication system). The complexity of these additional subsystems varies. In some applications it can become the dominant factor in determining the cost of the overall communication system.

20.2.1 Free-Space Propagation Formula

In free-space communications, and under idealized conditions (i.e., perfect alignment of transmitter and receiver, lossless optical components and detectors), the received power P_r is determined by the formula

$$P_r = P_t \frac{A_r}{\Omega_t R^2} \tag{1}$$

where P_t is the transmitted power, A_r is the area of the receiving telescope, Ω_t is the solid angle into which the transmitted beam is emitted, and R is the distance between the transmitter and the receiver (i.e., the range of the communication link). For diffraction-limited telescopes, Ω_t is given by

$$\Omega_t = \frac{\lambda^2}{A_t} \tag{2}$$

where λ is the wavelength of the electromagnetic wave employed in the communication channel and A_t is the area of the transmitting telescope. Combining Eqs. (1) and (2) we find the free-space propagation formula for a diffraction-limited transmitter telescope:

$$P_r = P_t \frac{A_t A_r}{\lambda^2 R^2} \tag{3}$$

In order for Eq. (3) to represent the correct functional dependence on the various parameters, the receiving and transmitting apertures have to be in the farfield of each other (i.e., the conditions $R \gg A_t/\lambda$ and $R \gg A_r/\lambda$ should be satisfied).

When real-life (nonideal) conditions are taken into account, the right-hand side of Eq. (3) has to be multiplied by several inefficiency factors, some of which are listed below. First, when the transmitter telescope is not dif-fraction-limited, $\Omega_t > \lambda^2/A_t$, and Eq. (1) must be used instead of Eq. (3). This translates into an inefficiency factor of $\lambda^2/\Omega_t A_t$. Second, the optical elements in the transmitter and receiver terminals have some loss factors associated with them (e.g., due to obscuration caused by secondary mirrors or field stops of the telescopes, due to finite reflection at each optical surface, and subunity transmission factors through each element). Since the overall optical chain of the system usually contains many optical elements, the combined effects of all these ingredients can result in a significant loss factor. Additional losses are caused by the finite (and unavoidable) pointing errors (i.e., the inaccuracy in the alignment of the optical transmitter and receiver). When these effects are taken into account it can be shown that there is an optimum value for the transmitter telescope diameter, which depends on the root-mean-square pointing error θ_e (for diffraction-limited telescopes, the optimum diameter equals approximately $0.6 \, \lambda/\theta_e$). The fact that the photodetector at the receiver has a quantum efficiency smaller than unity must also be taken into account. Finally, when the path of the optical beam is not entirely in space, the propagation medium (e.g., atmosphere, water) can have adverse effects on the optical signal in terms of increased attenuation, beam spread, and other effects [3]. This subject will be discussed in more detail in Section 20.5.

20.2.2 System Considerations

When considering an optical communications system in its entirety, many functions that are not directly related to light generation and detection have to be taken into account. Some of these factors, schematically illustrated in Figure 20.2. will be discussed in the following paragraphs.

Unlike optical fiber communications, in order to maintain a free-space optical link it is not sufficient to just have the transmitter and the receiver operational; they must also be spatially aligned. Thus an important step in establishing the free-space link involves the process of beam acquisition. The transmitted beam has to be pointed in the direction of the receiver. A beacon source (usually a laser operating at a different wavelength from that of the transmitter) at or near the receiver is often used to designate the correct direction. Alternatively, in duplex systems each transmitter can serve as a beacon to the other terminal. The optical receiver has to be pointed in the

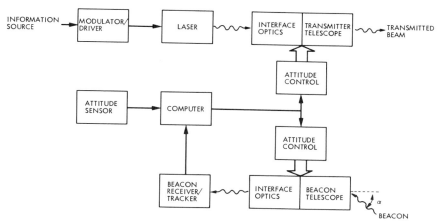

Figure 20.2 Simplified block diagram of a free-space optical communication terminal. [*Notes*: (1) In some systems there may be only one telescope for transmission and reception; (2) α is the point-ahead angle (see text)].

approximate direction of the transmitter, and then some scanning has to be performed until the transmitted beam is within the field of view of the tracking detector. (Most receivers employ separate detectors to assist in the beam-manipulation functions. These detectors usually have a larger field of view and a lower sensitivity than the optical detector that detects the information signal.) Beam acquisition can be simplified by using detector arrays with a large number of elements (e.g., CCD with or without light-amplification stages) that can "stare" into a large field of view. Once the beam is acquired, the alignment is maintained using the spatial information derived from the tracking detector, which is usually a quadrant detector. Recent developments of randomly accessible detector arrays may make it possible to employ the same detectors for the acquisition and tracking functions.

The next question that needs to be addressed is the actual implementation of the beam-pointing operations. Since the high directivity of the optical beams (typically in the microradian regime) usually exceed the attitude-control capabilities of the host environment (e.g., satellite, spacecraft), separate subsystems are needed to perform the various functions associated with beam manipulation. (In addition, overall system design requirements may actually demand the capability of independent manipulation of the optical beam.) The hardware that is used for these purposes includes, for example, gimbaled mounts and steerable mirrors. There are also some long-range research efforts that may result in electronic implementation of beam pointing; this subject is described in more detail in the next section.

In many applications the relative angular positions of the transmitter and the receiver can change significantly during the time that it takes the light to propagate between them. This "point-ahead" angle is another factor that must be taken into account; the signal or beacon beam has to be launched in the direction where the corresponding receiver will be by the time the light gets there. Relative velocities between the transmitter and the receiver can also result in Doppler shifts, a phenomenon that can cause a degradation in the performance of coherent receivers (see Section 20.4).

Since the system performance is adversely affected by background radiation, narrowband optical filters are often required, especially in noisy environments. Several different physical phenomena and engineering methods can be employed to fabricate filters with bandpasses down to the sub-angstrom regime and with sufficiently large fields of view [4; see also *Optical Engineering*, Vol. 20, November–December 1981].

Another critical technology involves optical telescopes, especially in long-range applications where large telescopes are required. Research in new materials and fabrication techniques as well as novel methods for telescope construction and wave front correction can significantly reduce both the cost and the weight of optical telescopes [5]. This is especially true in the area of large-aperture, low-quality telescopes (commonly known as "photon buckets"), which can be used in noncoherent optical systems (see Section 20.4), affording a significant reduction in the overall system cost.

Before concluding this section it is important to note that environmental conditions, mainly increased radiation levels, in space-based systems can adversely affect the optoelectronic (and electronic) components of the system. Examples of observed effects include an increase in the laser threshold and a reduction in its efficiency, and an increase in the noise levels in the optical detectors. More information on this subject can be found in Ref. 6.

20.3 SOURCES

Semiconductor injection lasers are excellent candidates as light sources in free-space optical communications because of their high reliability and efficiency and their small size and weight. Their major drawback is that a single device typically cannot radiate all the needed power (typically of the order of a few hundred milliwatts to a watt) in a stable single-mode operation. The main topic of this section is power combining of semiconductor injection lasers. We also discuss Nd:YAG lasers as alternative light sources that avoid several problems associated with semiconductor laser devices. Finally, we briefly review the possibilities of employing integrated optoelectronic technology for performing beam-manipulation functions electronically. The CO_2 laser, though a highly efficient source, will not be considered here because of its long wavelength—10.6 μm (which results in large diffraction angles). Furthermore, it is a gas laser and hence is intrinsically less reliable [7].

20.3.1 Incoherent Power Combining of Semiconductor Injection Lasers

The power of several semiconductor lasers can be combined either coherently or incoherently. Before we proceed it should be mentioned that the relevant figure of merit of the light source in optical communication systems is not its power but its brightness, defined as the power divided by the product of the source area and the solid angle into which the light is emitted. The reason is that higher source brightness, and not just more source power, translates into higher intensity (P_r/A_r) at the receiver.

The main advantage of incoherent power combining is that it is conceptually a simpler approach, avoiding the more complicated issues of coherent interaction in multioscillator systems. The simplest method of incoherent power combining is aperture sharing, shown schematically in Figure 20.3a. As can be seen from this figure, the source area is increased in direct proportion to the power level, and since the solid angle into which the light is emitted is unchanged, the source brightness remains unchanged.

There are two basic methods for increasing the source brightness while still employing noncoherent power combining; these involve lasers that emit light at different polarizations or different wavelengths (a combination of the two is also possible). Since the light emitted by a semiconductor laser is linearly polarized (in the direction of the junction plane), the power of two lasers can be combined by aligning their junction planes perpendicular to each other and combining the orthogonal polarizations (for example, with a polarizing beam splitter). With added system complexity, this method can be extended to a somewhat larger number of lasers. Power from lasers emitting at different wavelengths can be combined by using either wavelength-selective elements (e.g., filters) or wavelength-dispersive elements (e.g., gratings). As an example, the second method (also known as the inverse prism method) is shown schematically in Figure 20.3b. In this case, however, the increased spectral extent of the source may dictate the use of wider-bandpass filters at the receiver, thus increasing the amount of background noise admitted to the detector. In addition, special efforts and power have to be expended in order to stabilize the wavelengths of all the emitters.

On a more basic level, by incoherently combining the power of several discrete devices we lose one of the basic advantages of the semiconductor laser (compared to other laser types): that it is a semiconductor device. Instead, the entire transmitter head becomes a sophisticated optical system that includes, in addition to the lasers themselves, many other optical components.

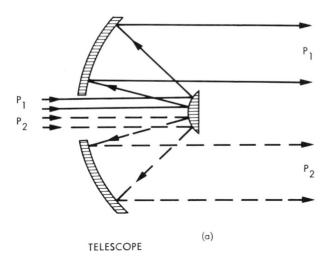

TELESCOPE

(a)

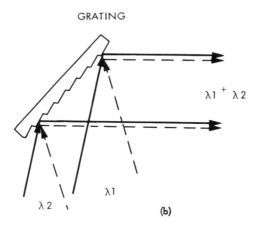

(b)

Figure 20.3 Schematic configuration of methods for noncoherent power combining. (a) Aperture sharing; (b) inverse prism.

20.3.2 Coherent Power Combining of Semiconductor Injection Lasers

Coherent power combination is usually more complicated to implement. The potential payoff is mainly an increased source brightness due to the phaselocking of the lasers (similar to microwave-phased antenna arrays).

This effect is demonstrated in Figure 20.4, where radiation patterns of coherent and noncoherent arrays are compared. It can be seen that for an N-element array, the on-axis ($0°$) intensity increases by a factor of N in the first case, but an N^2-fold improvement is achieved in the second case (provided that the array elements operate in phase). In addition, since all phaselocked lasers have by definition the same spectrum, the overall spectral width of the source is usually reduced. This allows the use of a narrower-bandpass filter at the receiver and thus lower levels of background radiation noise.

In this section we will look into several aspects of coherent power combining of semiconductor lasers. First the three main generic methods will be described and compared, and then we will look into some implementation issues, including device configuration and coupling mechanisms.

Schematic configurations of the three basic methods of coherent power combining are shown in Figure 20.5. All these methods have been analyzed and demonstrated in various degrees of sophistication [8]. Their common feature is the establishment of coherent interaction between all the elements of the array.

In the case of mutual coupling (Fig. 20.5a), no laser in the array has a privileged status. There is a certain amount of coupling among the lasers, which, under certain conditions, results in their synchronization. The necessary amount of coupling depends on the separation $\Delta\lambda_i/\lambda$ between the intrinsic wavelengths of oscillation of the interacting lasers and the center wavelength of the laser transition linewidth. For (nominally) identical lasers, operating with approximately equal amplitudes, the following approximate

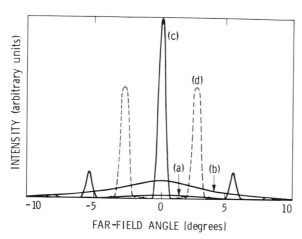

Figure 20.4 Far-field radiation patterns of (a) single laser and (b) 10-element incoherent linear arrays, (c) 10-element coherent array (in-phase operation) and (d) 10-element coherent array (out-of-phase operation).

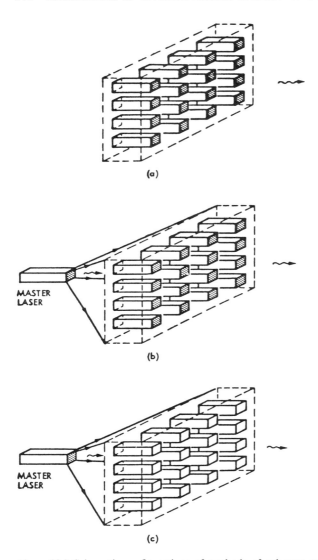

Figure 20.5 Schematic configurations of methods of coherent power combining. (a) Mutual locking; (b) injection locking; (c) coherent amplification.

condition must be satisfied [9]:

$$\frac{1}{\lambda}\left|\sum_i \Delta\lambda_i\right| < \frac{\kappa}{k} \tag{4}$$

where κ is the coupling coefficient between adjacent lasers and k is the wavevector of the mode in the laser material. Since the left-hand side of

Eq. (4) represents a random walk process, it becomes more difficult to phase-lock the array as the number of elements increases.

Injection locking (Fig. 20.5b) is obtained by the same physical mechanism as mutual phaselocking. In this case, however, there is a single master laser oscillator. Portions of its emitted radiation are coupled simultaneously into all the other lasers in the array, forcing them to oscillate at its frequency. There is no coupling among the lasers in the array and no feedback from the array into the master laser. The condition for injection locking can be expressed as

$$2 \left| \frac{\Delta\lambda}{\lambda} \right| < \frac{\delta E_m}{QE_s} \tag{5}$$

where $\Delta\lambda$ is the difference between the wavelengths of the master laser and the array element laser: E_m and E_s are the electric field strengths of the master and slave lasers, respectively: δ is the fraction of the master laser field coupled into the slave laser; and Q is the figure of merit of the slave laser cavity.

In the case of coherent amplification (Fig. 20.5c) the elements of the array are only gain elements (i.e., amplifiers) without feedback. (This can be accomplished, for example, by applying antireflection coating to the array mirrors.) Light generated in the master laser is split and fed simultaneously into all the gain elements, where a traveling-wave amplification is employed. The amplified outputs of the amplifiers are automatically phaselocked (provided, of course, that the output of the master laser is coherent over its near-field pattern).

An intermediate case between injection locking and coherent amplification occurs when the gain elements in the array have some feedback but not enough for lasing, that is, when they operate as regenerative amplifiers. This method is more complicated to implement, mainly because of the high degree of stability required of the regenerative amplifiers.

These three methods are quite similar in terms of power efficiency, but coherent amplification has the advantage of avoiding the generally stringent conditions required for establishing phaselocking. Mutual phaselocking is more amenable than the other methods to monolithic implementation (as will be shown). The issue of reliability is more complex. Both in injection locking and coherent amplification a failure of the master oscillator implies a failure of the entire source. In mutual phaselocking the situation may not be so critical. However, failure of a single device may alter the modal shape of the entire array to such an extent as to render it useless.

The various methods just described can be implemented in different device configurations. One possibility is to have the interacting lasers be discrete devices that are placed in a common cavity where spatial filters help discriminate against unwanted modes [10]. Although experimental results (Fig. 20.6a; [11]) have been obtained so far in arrays with a relatively small

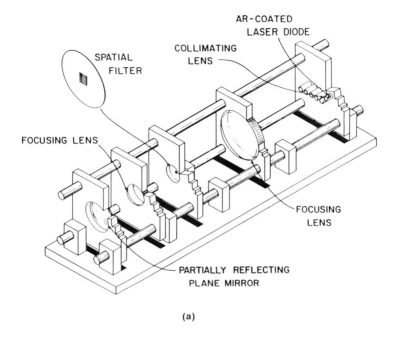

(a)

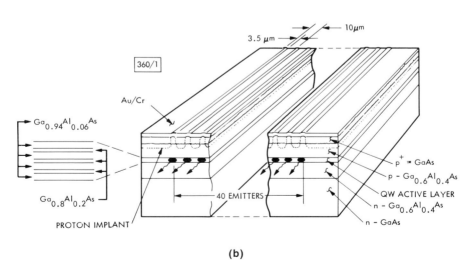

(b)

Figure 20.6 Several experimental implementations of coherent laser arrays. (a) External cavity [11]; (b) single-contact monolithic array [14]; (c) separate contact monolithic array [12].

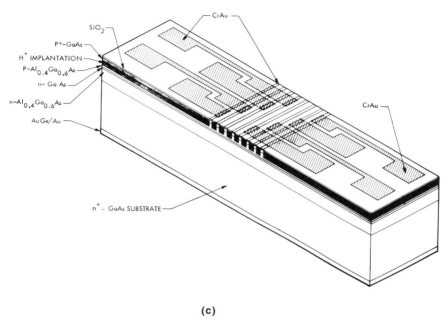

(c)

Figure 20.6 (*continued*)

number of elements, their extension to a larger number is straightforward. Among the advantages of this method are reduced thermal loads (since individual devices need not be placed in close proximity to each other) and less anisotropic beam patterns (two-dimensional arrays can be easily arranged). The main disadvantage is increased system complexity, for example, the need to maintain alignment of many optical elements. However, the results of Rediker et al. [11] seem to indicate that this problem may be less severe than originally anticipated.

Much interest has been devoted recently to monolithic implementations of one-dimensional (linear) laser arrays [12], some of which are shown in Figures 20.6b and 20.6c. The main motivation is to increase the amount of power emitted into a stable fundamental radiation mode while still preserving the advantages associated with a semiconductor light source. The lasers are fabricated in close proximity to one another (typically within 10 μm), so that a directional waveguide type of evanescent coupling is established between their fields. (Diffraction coupling, which is used mostly in external cavity configurations, can also be applied in monolithic devices [13].) Significant power levels (over 8 W) have been achieved [14], but the problem of modal control (i.e., forcing the array to oscillate in the particular array mode that produces the desired single-lobed far-field pattern) has not yet been completely solved and is still under investigation [15]. An important issue is the ability to extend the monolithic implementation into two-dimensional configurations that may be required for increased output power and beam symmetry. Recent developments in surface-emitting lasers [16] may point the way to solving this problem. However, even linear monolithic arrays can be useful devices on their own merits. Their beam anamorphicity can be corrected with relatively simple optical systems, and they can also be used as building blocks in discrete element array configurations, in both coherent or noncoherent power-combining schemes.

20.3.3 Nd:YAG Lasers Pumped by Semiconductor Laser Sources

As seen in the previous section, the task of combining the radiation from several lasers in order to increase the overall source power poses several nontrivial technological problems that must be overcome, especially if the power combining is to be performed coherently. In addition, semiconductor laser sources are limited in the peak power that they can emit due to considerations of catastrophic mirror degradation. (This problem can be mitigated, but not completely solved, by using techniques such as nonabsorbing mirrors [2].) Most efficient modulation schemes in free-space optical links involve low duty-cycle pulsed operation of the source, because concentrating its energy into high peak power pulses helps overcome the adverse effects of background radiation noise. Due to their peak power limitations, semiconductor lasers in low duty-cycle operation can reliably emit, on the average, much less power than their cw (or high duty-cycle) rating.

These limitations can be overcome by employing a different type of laser as the radiation source. Among all the possible alternatives, the Nd:YAG [17] laser seems to be the most attractive choice. First, it is a solid-state laser and its technology is mature. Lasers emitting cw power levels and pulse energies well in excess of what is required in most communication applications are commercially available. Second, their wavelength (1.064 μm) and, more specifically, its second harmonic produced with frequency doubling (0.532 μm) fall in a region where high-quantum-efficiency detectors are available.

Traditionally, the main problems encountered in the use of Nd:YAG lasers for optical communications applications were the low reliabilities (< 1000 hr) and low efficiencies of their lamp pumps. Recently, however, there has been considerable interest in using semiconductor laser devices (individual lasers or arrays) as pump sources [18]. (Solar pumping is another alternative that will not be discussed here [18].) The most efficient pumping band for Nd:YAG lasers is at 0.81 μm. It is relatively easy to fabricate AlGaAs laser sources that emit their radiation at this wavelength. (Some temperature stabilization may be required, though, to prevent excessive wavelength shifts

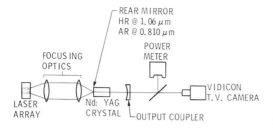

Figure 20.7 Schematic diagram of semiconductor laser-pumped Nd: YAG laser with tightly focused end-pump geometry [19].

that reduce the pump absorption efficiency.) Even though the Nd:YAG radiation can be emitted in a pulsed form (employing Q-switching, cavity dumping, or modelocking), the pump is operated continuously. Thus an alternative way to consider the Nd:YAG laser is as a highly efficient dc-to-pulsed converter to the light that is (efficiently) generated by the semiconductor laser sources. In addition, the beam quality requirements from the semiconductor laser arrays are less stringent when they are used indirectly as pump sources and not directly as the communications light sources.

Since the power levels required for optical communication are relatively small (compared to the available power from the Nd:YAG technology), it is possible to employ end-pumping rather than side-pumping geometry, thus further improving the source efficiency. In this configuration it is possible to optimize the various factors determining the system efficiency and achieve more than 50% conversion efficiency from the pump output power to the output Nd:YAG power. (This includes the $0.8 \mu m/1.064 \mu m = 0.76$ pump photon conversion efficiency.) Using a commercially available semiconductor laser array, 80 mW cw at $1.064 \mu m$ has been recently achieved from a Nd:YAG source at an overall (wall-plug) efficiency of 8% [19]. The end-pump experiment configuration is shown schematically in Figure 20.7. Issues that still need to be addressed are the scaling up of the output power levels by approximately an order of magnitude and the extent of efficiency degradation in the frequency-doubling and modulation processes. An alternative approach that may be technically feasible is to employ a low-power $1.064-\mu m$ semiconductor laser as the modulation source and to use one or more Nd:YAG stages for subsequent amplification [20].

20.3.4 Electronic Beam Steering

The main advantage of free-space optical communications, namely, the increased directivity of the transmitted beam, also carries the penalty of the need for highly accurate tracking, pointing, and acquisition systems. Whereas typical beam widths in the microwave regime are of the order of a fraction of a degree, optical communication systems have to handle beam widths down to the sub-arcsecond region.

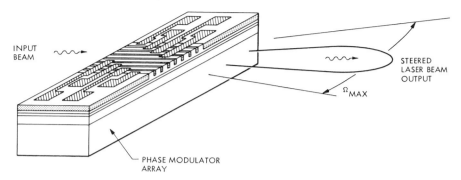

Figure 20.8 Schematic implementation of electronic beam steering with a monolithic phase modulator array.

Today virtually all beam-manipulation functions are performed electro-mechanically via combinations of gimbals, mirrors, and lenses. Although the role of these will probably not be eliminated in the future, there are advantages to the electronic implementation of at least some of these functions.

One method involves control of individual phases in a fashion similar to the control of microwave phased arrays [21]. A possible device configuration is shown in Figure 20.8. Light from the source (either a single device or an array) is fed into an array of reverse-biased electrooptic waveguide phase modulators. Applying a different voltage (and hence a different phase shift) to each element of the modulator array results in beam steering.

Another method involves incorporating the steering mechanism within the semiconductor laser array itself. When a different current flows through each laser stripe, an asymmetry of the gain profile is established across the device, which can translate into a shift of the far-field pattern of the source. Although this method is more amenable to monolithic implementation, it suffers from the inevitable conflicts encountered when trying to achieve optimum conditions for both light-generating and light-steering functions within the same device. Still open are the feasibility issues of monolithic implementation of both light generation and steering and extension of electronic beam-steering concepts to two-dimensional configurations.

20.4 RECEIVERS AND DETECTION

Most aspects of optical detection are discussed elsewhere in this book. In this section we concentrate on several issues that are more unique or more important to free-space optical communication systems. First we discuss the problem of background noise radiation, which is usually not a relevant factor in optical fiber communication. Next we review the issue of receiver sensitivity in the two generic receiver configurations (i.e., direct-detection and coherent receivers). Finally we present two methods for increasing the system sensitivity so that it can operate over a larger range or operate in the same range with more relaxed system constraints (or higher data rates). The first method involves solid-state detectors for single-photon counting, and the second method, which pertains to coherent detection, involves a novel receiver configuration that makes it possible to eliminate the local oscillator excess noise and approach the quantum-limited sensitivity.

20.4.1 Background Noise

In addition to the various noise mechanisms and sources that exist in optical-fiber-based communication systems, the free-space channel has the added ingredient of background radiation noise; in many cases it is the dominant noise component [22]. Except for some special cases where the exact noise statistics have to be fully taken into account, its effects can be translated into those caused by increased detector dark current.

The effect of background (and dark current) noise on the receiver performance depends on its mode of operation. Most optical receivers measure the current generated by the detected photons. In these receivers the received background power is translated into a shot noise current with white spectral density and a mean square value of

$$\overline{i_b^2} = 2q\frac{2\eta P_b}{hf} B \quad [A^2] \tag{6}$$

where P_b is the received background noise, B is the electronic bandwidth of the system, η is the detector quantum efficiency, q is the electronic charge, h is Planck's constant, and f is the light frequency.

The second type of receiver bases its decisions on actually counting the number of detected photons (both signal- and noise-generated). In these receivers the noise is expressed as the number of detected photons n_b:

$$n_b = \frac{\eta P_b t_s}{hf} \tag{7}$$

where t_s is the relevant observation time (e.g., bit time in binary modulation formats).

The magnitude of the total background power P_b at the receiver aperture depends on the particular background observed by the receiver. It can generally be expressed as

$$P_b = \left[N_b(\lambda)\Omega_r + \sum_i E_i(\lambda) \right] A_r \, \Delta\lambda \quad [W] \tag{8}$$

The first term in Eq. (8) is due to extended background sources, that is, background sources whose angular extent is larger than the receiver field of view. These include, for example, the sky or a planet in a situation where Ω_r is smaller than the angle that is subtended by that planet at the receiver. The second term is due to discrete background sources, that is, background sources whose angular extent is smaller than the receiver field of view (e.g., stars). $N_b(\lambda)$ is the spectral radiance of the background in watts per square centimeter per steradian per micrometer, Ω_r is the receiver field of view in steradians, $E_i(\lambda)$ is the spectral irradiance from discrete sources in watts per

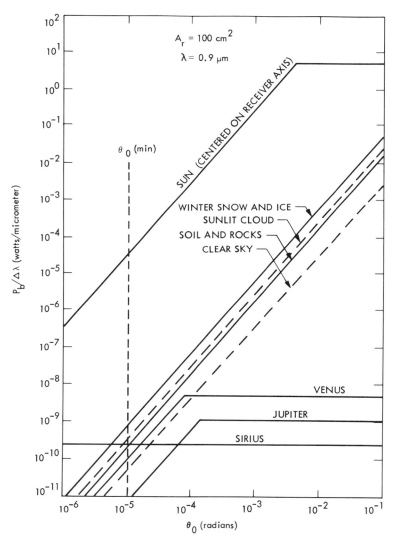

Figure 20.9 Received background noise power due to different noise sources as a function of the receiver field of view. [*Notes*: (1) Noise sources indicated by dashed lines do not exist in spaced-based systems; (2) the vertical line at $\theta_0 = 10$ μrad indicates the approximate lower limit in atmospheric reception (see Section 20.5).]

square centimeter per micrometer, and $\Delta\lambda$ is the bandpass of the receiver optical filter.

Figure 20.9 shows the background noise power as a function of the receiver field of view for several noise sources. It should be noted that there can be many orders of magnitude difference between the background noise in different scenarios. It should also be noted that the curves in Figure 20.9 are for a particular wavelength. At other wavelengths the spectral distribution of each source has to be taken into account. Spectral distributions of several sources are shown in Figure 20.10.

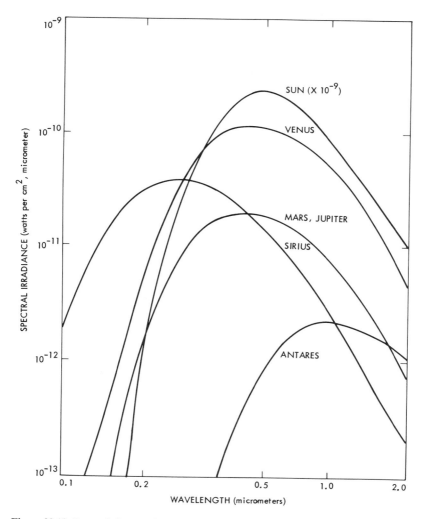

Figure 20.10 Spectral characteristics of irradiance from several background noise sources.

20.4.2 *Receiver Sensitivity*

One of the key parameters in the communications link analysis is the minimum amount of received power, $P_r(\min)$, that is required at the receiver. Among the factors that determine $P_r(\min)$ are the data rate, the noise in the system, the required error performance, and the coding and modulation formats used in the system.

There are two major generic configurations of optical receivers. In direct-detection receivers, the received signal (and noise) impinge directly on the photosensitive area of the photodetector. In coherent receivers, on the other hand, the incoming signal is combined with a (strong) field of a local oscillator before reaching the detector, thus making it possible to eliminate the effects of all the noise sources in the system except the intrinsic noise of the local oscillator. The two variations of this last scheme are called heterodyne detection (when there is a difference between the frequencies of the local oscillator and the received signal) and homodyne detection (when the two frequencies are identical). Since in order to realize the full benefits of coherent detection we need to employ a stable local oscillator that tracks the received signal and maintain the coherence of the signal field over the entire area of the receiving aperture (by using diffraction-limited telescopes), it is preferable to use direct detection whenever possible. Furthermore, coherent detection is usually not possible in atmospheric links due to atmospheric distortion of the propagating beam wave front. However, in high-background-noise environments, coherent detection must be used when the background-limited direct detection sensitivity is not sufficient.

Several criteria can be employed to evaluate the sensitivity of the optical receiver. The most common practice is to define it as the amount of received power required to obtain a signal-to-noise ratio of unity at the output of the detector. This definition is appropriate to non-photon-counting receivers, because in this case the signal-to-noise ratio is a parameter sufficient to characterize all aspects of the system performance.

Using this definition, the sensitivity of a direct-detection receiver operating in a background-noise-limited situation is given by [23]

$$P_r(\min) = 2\left(\frac{P_b hf B}{\eta}\right)^{1/2} \tag{9}$$

The ideal heterodyne receiver, on the other hand, is limited only by the shot noise of the local oscillator, and its sensitivity is given by [23]

$$P_r(\min) = 2\,\frac{hf B}{\eta} \tag{10}$$

which is much smaller than the value calculated from Eq. (9).

An alternative criterion for the receiver sensitivity—and one that is more meaningful from the communications theoretical point of view—is the information capacity of the channel. This parameter sets the absolute upper bound on data transfer rates that can be reliably accommodated by the system. Although it is usually defined in units of bits per second, in several cases it is more illuminating to employ the units of bits per detected photon, in which case the results convey the energy efficiency of the system. With coherent receivers, the channel capacity is limited to 1.44 and 2.89 bits per detected photon for the heterodyne and homodyne schemes, respectively. However, the channel capacity of direct-detection photon-counting receivers using single-photon detectors under low-background conditions can be much higher; it is usually limited by considerations of modulation and coding complexities to approximately 5–10 bits per detected photon [24].

20.4.3 Solid-State Single-Photon Detectors

In direct-detection optical receivers operating in low-background conditions, the system sensitivity can be maximized by employing single-photon counters, that is, detectors that owing to a photodetection event of an individual photon produce an output that can be used without any further amplification. It is clear that in order to meet this requirement the detector must have internal gain that is high enough to overcome the thermal noise of the system. This condition can be expressed as

$$\frac{qG}{2t_r} \gg \left(\frac{4K_B T_e B}{R_i}\right)^{1/2} \tag{11}$$

The term on the left-hand side of Eq. (11) is the approximate peak of the single photoelectron current pulse (q is the electron charge, G is the average detector internal gain, and $2t_r$ is the pulse width), and the term on the right-hand side of Eq. (11) is the thermal (Johnson) noise of the system to which the detector is connected (K is the Boltzmann constant, B is the bandwidth, R_i is the input resistance, and T_e is the equivalent noise temperature of the amplifier). Typically, $t_r \approx 1/B$, $R_i = 50 \, \Omega$, and $T_e \approx 900$ K (corresponding to an amplifier with a noise figure of 4.8 dB). Equation (11) is then reduced to

$$G \gg \frac{12{,}500}{\sqrt{B \, [\text{GHz}]}} \tag{12}$$

So far the only commercially available photodetectors that can satisfy this requirement are photomultiplier tubes (PMTs) [25]. However, because of the relatively low quantum efficiency ($\sim 20\%$ at 0.8 μm; $\sim 0\%$ above 1.1 μm), high operating voltages (approximately kilovolt), and low reliability of PMTs, there is an incentive to develop a solid-state device equivalent.

Basic theoretical calculations have indicated [26] that under certain conditions avalanche photodiodes (APDs) can be operated in a "Geiger tube" mode where they can be used as single-photon counters. This phenomenon has also been experimentally verified [27]. In order to operate in this mode, which is different from its common mode of operation, the APD has to be cryogenically cooled to reduce its dark current. It is then biased at a voltage level above its avalanche breakdown voltage [but below its dielectric (Zener) breakdown point]. Under these conditions, a single carrier (which is either photogenerated or thermally generated) has a high probability of producing a sustained avalanche. When this occurs, the diode voltage drops to the avalanche breakdown voltage, thus producing a voltage pulse that can easily be detected without any further amplification. (Equivalent gains of 10^7–10^8 can thus be realized.) Then the avalanche has to be quenched and the device bias reset so that the APD will be ready for a new detection process. Since quantum efficiencies of solid-state detectors are higher than those of PMTs, and since the probability that a single carrier will ignite a sustained avalanche can exceed 50%, APDs operating in this mode may be an attractive alternative to PMTs, especially with the added bonus of their improved size, ruggedness, reliability, and operating voltages.

The major problem with APDs operating in this mode is carrier trapping during the phase in which the device is in sustained avalanche. After the avalanche is quenched and the bias is reset, detrapping of these carriers initiates new sustained avalanche processes that are translated to false detection events in the communication system. The exact consequences of this phenomenon have not yet been fully evaluated, and this topic is still under research.

An alternative approach to "imitating" PMTs in semiconductor devices is the use of novel epitaxial growth techniques (e.g., molecular beam epitaxy) to tailor the bandgap of the semiconductor material in APDs so that only one type of carrier (e.g., electrons) undergoes multiplication [28]. This method shows promise in avoiding the excess noise that plagues common APDs, but it remains to be seen if the method can be extended to the high gains that are needed for single-photon counting.

20.4.4 High-Sensitivity Coherent Receivers

A prerequisite for achieving the ultimate sensitivity of the ideal coherent receiver [Eq. (10)] is that the only noise associated with the local oscillator is the common quantum shot noise inherent in ideal coherent lasers. However, semiconductor lasers also have excess noise that can be an order of magnitude larger than the quantum shot noise. Thus when they are used as local oscillators, the sensitivity of the "traditional" heterodyne receiver configuration (Fig. 20.11a) is limited by the excess noise, resulting in a sensitivity degradation of approximately 5 dB. An alternative scheme was developed in 1983 [29], in which the receiver structure, shown in Figure 20.11b, employs a balanced mixer configuration. When all the parameters of the system are properly adjusted, the local oscillator excess noise is canceled out, and an almost quantum-limited performance level can be obtained, as shown in Figure 20.12.

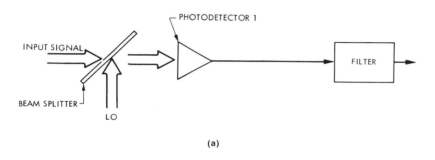

(a)

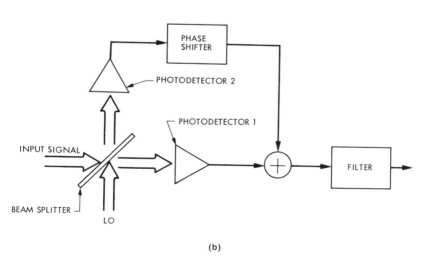

(b)

Figure 20.11 Block diagrams of coherent receivers. (a) Conventional receiver; (b) balanced receiver [29].

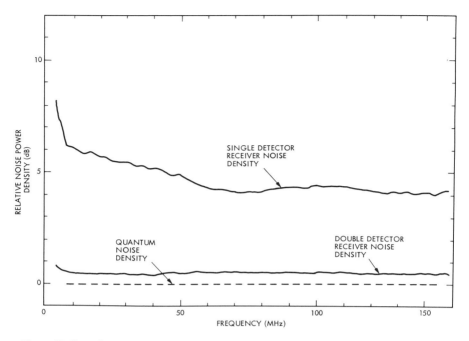

Figure 20.12 Performance comparison between Figure 20.11a and b in terms of noise levels [29].

20.5 SYSTEMS AND APPLICATIONS

In the first few years following the invention of the laser there was great interest shown in optical communication systems. However, only in recent years has the relevant technology become sufficiently mature that practical systems for space applications can be seriously contemplated. The purpose of this section is to highlight several of the possible applications of free-space optical communications and to briefly describe the systems that can be employed in these applications. This section is not intended to serve as an exhaustive review of all the applications and systems that have been proposed, designed, or demonstrated. Instead we try to concentrate on several generic applications, each with its own (almost) unique characteristics.

20.5.1 Atmospheric and Marine Communications

This subsection describes several optical communication links where part or all of the beam path is in an environment other than space—atmosphere or water. These include terrestrial links, for example, between buildings, communication links between airplanes, communication links to submarines, ground reception of optical signals from space, and more.

At optical frequencies the atmosphere is a weather-dependent channel. For example, clouds will completely block any optical communication, except in cases of extremely short ranges (or, of course, in systems employing very high power lasers). Even without the presence of clouds, an optical beam that propagates through the atmosphere suffers attenuation (with an extinction coefficient that is approximately proportional to the inverse of the visibility range) and several other degradation phenomena that are caused by atmospheric turbulence (e.g., scintillation, beam spread, and beam wander). These turbulence effects set a practical lower limit to the receiver field of view. For example, even under relatively clear atmospheric conditions, reducing the receiver field of view to below approximately 10 μrad will result in unacceptable performance degradation due to fading losses.

There are nevertheless applications where terrestrial point-to-point optical communication is more attractive or less costly than other communication options. Examples for such applications are temporary links between two sites in urban environments that are already saturated by microwave links and communications that require a higher degree of security than what can be obtained in other frequencies (e.g., between airplanes). Several such systems are commercially available [30]. They are typically capable of transmitting information (video, voice, or data) at multimegahertz rates for distances of up to 10 miles in clear weather. Other systems are being developed for related military applications [31].

The main motivation for developing optical links for communication with submarines is the fact that the blue-green region of the electromagnetic spectrum is the only spectral range above 100 Hz where there is a significant "window" in the high attenuation characteristics of seawater (attenuation of up to 50 dB is still expected in systems under consideration even at these wavelengths). The main problem in this application is that due to the high degree of turbidity of the propagation medium, radiation reaching the receiver appears to be arriving over a solid angle of approximately 1 sr centered around the zenith, regardless of the actual direction from which the light is coming. This requires the use of receivers with a large field of view to collect all the signal, and hence there is a huge increase in background radiation noise detected by the system. (In particular, there is a high probability that the sun will be in the receiver field of view during a substantial fraction of the operating time.) Since the seawater turbidity also destroys the signal coherence, heterodyne detection is not practical, and very narrowband optical filters with a large field of view must be used to enable operation under these high-background conditions. More details on this application can be found in Ref. 32.

20.5.2 Intersatellite (Near-Earth) Communications

Intersatellite communications is one of the more promising applications for optical communication systems, and most of the research and development activity in space-based optical communication systems is conducted in this area [33]. The two main generic systems involve low earth orbit to geostationary earth orbit (LEO–GEO), and GEO–GEO links. The LEO–GEO systems will probably be used for high data rate transfer (~1 Gb/s) from LEO satellites collecting data on earth resources, weather, and so on to GEO satellites serving as relay stations for transmitting the data back to earth. (Transmission from the satellites to earth is usually envisioned to be accomplished via a microwave link.) This configuration increases the amount of possible real-time communication as compared with a direct LEO–earth link.

The main applications envisioned for optical GEO–GEO links are in the area of commercial telephone and television transmission, tracking and data relay satellites for various space missions, and military communication networks. Because of the relatively short distances involved in intersatellite communications (10^3–10^5 km), high data rate systems can be implemented with a modest amount of transmitted power (10–100 mW) and relatively small (10–50 cm) transmitter and receiver telescopes. Thus optical systems may offer smaller size and weight and require less power than similar links in the microwave region.

As an example, consider a communications link between two GEO satellites that have an angular separation of 120°. This corresponds to a link range of approximately 73,000 km, which is a "typical" maximum value for near-earth applications. Assuming 20-cm-diameter transmitter and receiver telescopes, and a 10-mW semiconductor laser source operating at a wavelength of 0.85 μm, the received power under the idealized conditions of Eq. (3) will be approximately 2.5 nW. If an overall system efficiency of 10% is taken into account, the actual detected power will be ten times smaller, corresponding to a photon arrival rate of approximately 10^9 sec^{-1}. Simple receivers usually require approximately 1000 detected photons/bit to maintain an acceptable error rate performance (typically 10^{-9}). The actual number may even be higher under adverse background conditions. This implies that the maximum data rate that can be sustained by such a system is of the order of 1 Mb/s. Since most intersatellite applications involve much higher data rates, we will consider the following factors that can result in increased data rate transmission well into the gigahertz regime:

1. Increasing the telescope diameters by a factor of 2 will result in a 16-fold improvement. However, the larger diameters (which result in higher transmitted beam directivity) may present a problem from the standpoint of beam-pointing requirements and overall system dimensions.
2. Increasing the source brightness by a factor of 10 increases the data rate by the same factor.
3. Employing more signal modulation and coding formats in direct-detection systems or using coherent receivers (in high-background scenarios)

can yield up to a 100- to 1000-fold improvement in the receiver sensitivity and hence also in the maximum possible data rate. A more detailed example on this topic is discussed in Section 20.5.3.

Of course, in some systems (e.g., closely spaced GEO satellites) the range may be shorter, thus alleviating even further the difficulty in achieving high data rates.

One of the main problems facing optical communication systems in near-earth applications is the possibility of having the sun in the receiver field of view. (Since the solid angle subtended by the sun near earth is relatively small, 6.7×10^{-5} sr, this will happen only for short periods of time.) From Figure 20.9 we can calculate that for a 20-cm-diameter telescope with a 10-μrad field of view and a 10-Å optical filter, the received optical background is approximately 0.1 μW. This noise level translates into an equivalent shot-noise process with an rms value of 10 nW for a 1-GHz system. Thus if we require the link to operate even under these conditions, and assuming that a signal-to-noise ratio of approximately 10 is required to maintain reliable communications, received power levels of the order of 0.1 μW will be required. (As discussed above, smaller SNR values may be sufficient with more elaborate modulation and coding schemes.) A smaller problem may be caused by having the sunlit earth in the receiver field of view. From Figure 20.9 we see that this contributes a background noise radiation level that is 4 orders of magnitude smaller. In general, we can see from Eq. (8) that received background noise levels can be reduced by reducing the receiver field of view and by employing narrower bandpass optical filters.

20.5.3 Interplanetary (Deep-Space) Communications

Typical requirements from interplanetary links are usually different from those of near-space links. The link distance is much longer, ranging from 1 astronomical unit (AU) to above 10 AU (1 AU = 1.5×10^{11} m is the average sun–earth distance), and the data rates are correspondingly lower (typically 10^4–10^6 bits/s). One of the problems with deep-space links is that earth-based receivers limit the potential system performance mainly due to atmospheric effects and increased sky background radiation. However, recent developments in spaceborne structures, such as the Large Space Telescope [34], and in particular the proposed permanent space station [35], make it possible to realistically design receiving terminals with large telescopes (5–20 m in diameter) that orbit the earth, thus solving this problem. Furthermore, space-based reception also reduces the amount of received background radiation by avoiding the sky background (Figure. 20.9).

While no deep-space optical links exist as yet, a 1984 design study demonstrates the potential benefits of interplanetary optical communication [36]. The study involved adding a hypothetical optical communications module to the Venus Radar Mapper (VRM) spacecraft (see Fig. 20.13). It was found that even in this high background noise environment, a data rate of 4 Mb/s could be maintained from Venus to earth (~ 1 AU) using a 1-W laser (at 0.5 μm) and a 20-cm-diameter transmitter telescope. For that particular study the Mt. Palomar telescope ($A_r \approx 20$ m^2) was assumed as the receiving station (current microwave deep-space links are received by 64-m-diameter antennas.) From Eq. (3), and assuming an overall 10% system efficiency, we find that the received signal power is approximately 10 pW, corresponding to a rate of 2.5×10^7 detected photons/s (or ~ 6 photons/bit). The detected background noise count from Venus (assuming a 10-μrad field of view, 10-Å filter, and 8-dB attenuation at the atmosphere and the receiver optics) can be as high as 6×10^8 sec^{-1}.

Since in this case the average noise exceeds the average signal, special emphasis must be placed on the signal modulation. Heterodyne detection is not practical in this case because atmospheric effects limit the area over which the signal beam remains coherent to a size that is much smaller than the receiver telescope aperture. Employing a direct detection receiver and concentrating the signal energy into high-peak pulses, one can overcome the background noise. A popular choice of modulation format is the M-ary pulse position modulation (PPM) format. In a PPM format a preset time interval is divided into M time slots, and $\log_2 M$ bits of information are conveyed by transmitting a signal pulse in one of these time slots. For $M = 128$, seven bits are conveyed by each pulse, and thus the basic time slot in a 4-Mb/s link is $7/(128 \times 4 \times 10^6) = 13.7$ ns. The average number of background detected photons in each time slot is $13.7 \times 10^{-9} \times 6 \times 10^8 = 8.2$. The average number of signal detected photons in the signal time slot is $(7 \times 2.5 \times 10^7)/(4 \times 10^6) = 43.8$.

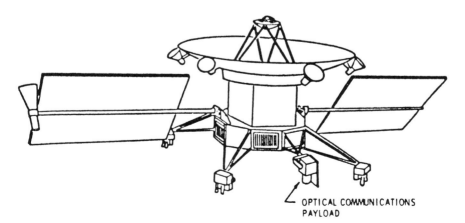

Figure 20.13 VRM spacecraft with optical communications experimental payload.

Communication theoretical calculations show that a 128-ary PPM system operating under these background noise conditions requires approximately three detected photons/bit to achieve an uncoded bit error rate of 0.01 (this value is typical for deep-space links, and it can be substantially improved with coding). This means that in the designed system there is a link margin of approximately 3 dB with respect to the actual number of six detected photons/bit.

20.5.4 Interstellar Optical Communications

The last example in this chapter describes a rather exotic communications link—from a spacecraft in the vicinity of Barnard's Star (one of the prime candidates for having a solar system) back to earth [37]. Even though such a mission is not under active consideration at the present time, this example serves as a useful illustration for the potential capabilities of optical communications technology. Such a link would be virtually impossible to implement via microwave communications.

The basic parameters of the system (modified from Ref. 37) include 20-W optical power at a wavelength of 0.5 μm transmitted through a 1-m telescope. The link range is approximately 6 light-years ($\sim 5.7 \times 10^{16}$ m), and the optical signal is received by a 5-m telescope orbiting the earth. The technology for both the laser source and the telescopes exists today.

From Eq. (3) we find that the average received optical power is 3.8×10^{-19} W, which corresponds to a photon arrival rate of approximately 1 photon/s. With a modest amount of modulation and coding complexity, we can achieve data transfer efficiency of 5 bits/photon; thus the communications link will be operating at the respectable rate of 5 bits/s. At this rate it will take only 6.5 h to transmit one Voyager class video image (117,000 bits). While many factors have not been included in this analysis, it should also be stressed that even higher values than the assumed transmitted power and optical telescope diameters are possible with the present technology.

REFERENCES

1. R. M. Gagliardi and S. Karp, *Optical Communications*, Wiley, New York, 1976.
2. *Proc. SPIE*, **295** (*Control and Communication Technology in Laser Systems*, K. Yong, Ed.), 1981; Proc. ESA Workshop on Laser Applications and Technology, Les Diablerets, March 26–30, 1984 (ESA SP-202, May 1984).
3. J. H. Shapiro and R. C. Harney, "Simple algorithms for calculating optical communications through turbulence," *Proc. SPIE*, **295**: 41–54 (1981), and references therein.
4. J. B. Marling, J. Nilsen, L. C. West, and L. L. Wood, "An ultrahigh-Q isotropically sensitive resonance transition," *J. Appl. Phys.*, **50**: 610–614 (1979).
5. P. N. Swanson, S. Gulkis, T. B. H. Kuiper, and M. Kiya, "Large deployable reflector (LDR): a concept for an orbiting submillimeter infrared telescope for the 1990s," *Opt. Eng.*, **22**: 725–731 (1983), and references therein.

6. J. A. Holzer and B. C. Passenheim, "Performance of laser systems in radiation environments," *Opt. Eng.*, **18**: 562–567 (1979).

7. J. H. McElroy, N. McAvoy, E. H. Johnson, J. J. Degan, F. E. Goodwin, D. M. Henderson, T. A. Nussmeier, L. S. Stokes, B. J. Peyton, and T. Flattan, "CO_2 laser communication system for near-earth space applications," *Proc. IEEE*, **65**: 221–251 (1977); E. Bonek and H. Lutz, "CO_2 laser communication technology for intersatellite data links," *ESA J.*, **5**: 83–98 (1981).

8. J. Katz, "Power combining of semiconductor lasers: a review," TDA Progress Report 42–70, pp. 87–94, Jet Propulsion Laboratory, Pasadena, CA, 1982, and references therein.

9. J. Katz, "Phase locking of semiconductor injection lasers," TDA Progress Report 42–66, pp. 101–114, Jet Propulsion Laboratory, Pasadena, CA, 1981.

10. E. M. Philipp-Rutz, "Single laser beam of spatial coherence from an array of GaAs lasers: free running mode," *J. Appl. Phys.*, **46**: 4552–4556 (1975).

11. R. H. Rediker, R. P. Schloss, and L. J. van Ruyven, "Operation of individual diode lasers as a coherent ensemble controlled by a spatial filter within an external cavity," *Appl. Phys. Lett.*, **46**: 133–135 (1985).

12. J. Katz, "Phase locked arrays of semiconductor injection lasers: a review" (Invited), Intl. Conf. on Lasers and Appl. (LASERS '85), Las Vegas, Dec. 2–6, 1985, and references therein.

13. J. Katz, S. Margalit, and A. Yariv, "Diffraction coupled phase locked semiconductor laser array," *Appl. Phys. Lett.*, **42**: 554–556 (1983).

14. D. F. Welch, B. Chan, W. Streifer, and D. R. Scifres, "High-power, 8 W cw, single-quantum-well laser diode array," *Electron. Lett.*, **24**: 113–115 (1988)

15. J. Katz and W. K. Marshall, "Grain saturation in supermodes of phase-locked semiconductor laser arrays," *Electron. Lett.*, **21**: 974–976 (1985).

16. K. Iga, F. Koyama, and S. Kinoshita, "Surface emitting semiconductor lasers," *IEEE J. Quantum Electron.*, **QE-24**: 1845–1855 (1988).

17. R. B. Chesler and J. E. Geusic, "Solid state ionic lasers," in *Laser Handbook*, **1**, F. T. Arecchi and E. O. Schultz-Dubois (Eds.), North-Holland, Amsterdam, 1972, pp. 325–368.

18. M. Ross, P. Freedman, J. Abernathy, G. Matassov, J. Wolf, and J. D. Barry, "Space optical communications with the Nd:YAG laser," *Proc. IEEE*, **66**: 319–344 (1978).

19. D. L. Sipes, Jr., "A highly efficient Nd:YAG laser end pumped by a semiconductor laser array," *Appl. Phys. Lett.*, **47**: pp. 74–76 (1985); see also T. Y. Fan and R. L. Byer, "Diode laser-pumped solid-state lasers," *IEEE J. Quantum Electron.*, **QE-24**: 895–912 (1988).

20. "Answer for space communications: divide laser oscillator, amplifier," *Photonics Spectra*, **16**: 36, 38 (1982).

21. J. Katz, J. R. Lesh, W. K. Marshall, and D. L. Robinson, "Electro optic phase modulator array," Final Report, JPL D-1885, Jet Propulsion Laboratory, Pasadena, CA, 1984.

22. V. A. Vilnrotter, "Background sources in optical communications," JPL Publication 83-72, Jet Propulsion Laboratory, Pasadena, CA, 1983.

23. A. Yariv, *Optical Electronics*, 3rd ed., Holt, New York, 1985, Chap. 11.

24. R. J. McEliece, "Practical codes for photon communication," *IEEE Trans. Inf. Theory*, IT-27: 393–398 (1981), and references therein.

25. R. W. Engstrom, *Photomultiplier Handbook*, RCA Corp., Lancaster, PA, 1980.

26. R. J. McIntyre, "On the avalanche initiation probability of avalanche photodiodes above the breakdown voltage," *IEEE Trans. Electron. Devices*, **ED-20**: 637–641 (1973).

27. D. L. Robinson and D. A. Hays, "Photon detection with cooled avalanche photodiodes: theory and preliminary experimental results," TDA Progress Report 42-81, Jet Propulsion Laboratory, Pasadena, CA, 1985, and references therein.

28. F. Capasso, W. T. Tsang, and G. F. Williams, "Staircase solid-state photomultipliers and avalanche photodiodes with enhanced ionization rates ratio," *IEEE Trans. Electron Devices*, **ED-30**: 381–390 (1983).

29. G. L. Abbas, V. W. S. Chan, and T. K. Yee, "Local oscillator excess noise suppression for homodyne and heterodyne detection," *Opt. Lett.*, **8**: 419–421 (1983).

30. General Optronics, New Jersey, and American Laser Systems, California, data sheets on various systems.

31. J. D. Barry and G. S. Mecherle, "LPI optical communication system," *Proc. 1984 IEEE Military Commun. Conf.*, Los Angeles, CA, Oct. 21–24, 1984, pp. 17.5.1–4.

32. T. F. Wiener and S. Karp, "The role of blue/green laser systems in strategic submarine communications," *IEEE Trans. Commun.*, **COM-28**: 1602–1607 (1980).

33. W. Englisch, "Study on intersatellite laser communication links," Final Report, ESA contract No. 3555/NL/PP (SC), 1979.

34. D. N. B. Hall (Ed.), *The Space Telescope Observatory*, NASA CP-2244, 1982.

35. R. R. Lovell and C. L. Cuccia, "Geostationary platforms in the space station era," *Telecommunications*, pp. 90–104, October 1984; "Definition studies shape space station," *Res. Develop.* December 1984, p. 45.

36. J. R. Lesh, "Optical communications data transfer from Venus," *Proc. 1984 IEEE Military Commun. Conf.*, Los Angeles, Oct: 21–24, 1984, pp. 17.6.1–3.

37. E. C. Posner, "Prospects for very deep space optical communication using photon counting links," *Proc. IEEE Natl. Telecommun. Conf.*, Houston, Nov. 30–Dec. 4, 1980, pp. 57.1.1–2.

CHAPTER 21

Optical Fiber Sensor Technology

A. Dandridge, J. H. Cole, W. K. Burns,
T. G. Giallorenzi, and J. A. Bucaro

21.1 INTRODUCTION

The ability of optical fibers and optical systems employing fibers to sense various environmental and device parameters has been exploited to create a new class of optical fiber sensors. Sensitivity, size, weight, EMI immunity, cost, and geometric versatility are among the many features offered by fiber sensors, and these features have made fiber sensors very competitive with older, well-established sensor technologies. Over 70 optical fiber sensors have been demonstrated [1–3], and recent progress in packaging has permitted field use and to some extent commercialization of many of them. This chapter is intended to provide an overview of the technology and to illustrate representative sensor structures.

21.2 SENSOR CLASSES

Optical fiber sensors have been fabricated using a wide variety of configurations, which may be grouped into one of the four main sensor classes depicted in Figure 21.1. This figure also serves to illustrate that many different mechanisms can be employed to convert a particular perturbation into a modulation of light. Currently, the fiber sensor types of broadest use are the intensity modulation classes shown at the top of the figure. In an intensity modulation sensor using external sensing elements, light transmitted by the optical fiber is modulated by an external sensing element. A second fiber may be used to

INTENSITY MODULATION (EXTERNAL)   FIBER INTENSITY MODULATION

PHASE MODULATION IN FIBER

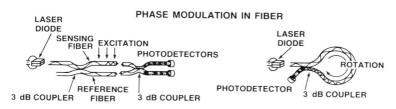

Figure 21.1 Fiber sensor classes. The four principal classes of optical fiber sensors are illustrated in this figure. In the intensity modulation (external) sensors, light exits the fiber and is modulated by some transduction mechanism external to the fiber. For the other three classes, transduction directly affects some property (intensity or phase) of the light propagating inside the fiber.

recapture the modulated light, or the light may be reflected from the sensing element back into the original fiber. The usual form of the external sensing device is either a material or mechanical structure that is sensitive to the external perturbation to be detected. For example, external temperature-sensing elements have been fabricated using semiconductor liquid crystals [4] and fluorescent crystals [5]. In the case of the fluorescent crystal temperature probe, a crystal attached to the fiber tip is excited via an input fiber. Fluorescence from two emission bands possessing different sensitivities to temperature is recaptured by the fiber and the two intensities are ratioed. This ratio provides a unique determination of temperature. Accuracies of $0.1°C$ and sensing ranges of several hundred degrees are typical for these types of devices. Other examples of external intensity modulation sensors are fiber linear and rotary position sensors [6]. Light from the fiber is directed onto a template that is encoded with absorbing and reflecting areas. The template is affixed to the part whose position is to be measured. Light reflected from the template is captured initially by an array of fibers and is detected, providing a direct digital readout of position. Pressure [7], pH [8], and pCO_2 [9] are other examples of phenomena that have been accurately determined using this class of device.

Internal intensity modulation sensors are configured so that light propagating within the fiber never passes through a path external to the fibers. In this class, the light intensity is modulated by varying some characteristic of the fiber itself. Radiation sensors [10] fall into this sensor class and work on the principle that ionizing radiation creates absorption centers in the fiber that attenuate the light propagating through the fiber. The attenuation, which is proportional to the amount of radiation absorbed, is used to give both real-time and quasi-archival radiation measurements. In addition to radiation, liquid-level [11], pollution [12, 13], and refractive index [13] sensors have been fabricated by stripping a portion of the cladding from the fiber so that light is coupled out of the fiber by the presence of the liquid. The amount of outcoupled light is controlled by the presence of the liquid (liquid-level sensor) and by the properties of the liquid. In this latter case, sensing changes in the liquid characteristics can be employed to measure temperature, refractive index, or liquid composition.

An important variant of the internal intensity sensor is the microbend sensor [14, 15]. Microbends are small lateral deformations of the optical fibers that cause light to switch from propagating to radiation modes. For sensor applications, mechanical deformers with "bump" periodicities optimized for maximum coupling loss are attached to the fibers and transform particular perturbations into a modulation of light in the fiber. Depending on the deformer design and the material of which it is constructed, acoustic, magnetic, temperature, acceleration, displacement, and many other types of microbend sensors have been fabricated. Dynamic displacements of these "bumps" on the order of 10^6 μm can be detected with a properly designed microbend sensor.

Two types of interferometric sensors account for the majority of phase modulation fiber sensors. In the Mach-Zehnder interferometer configuration (lower left of Figure 21.1), one of the interferometer fiber arms (signal arm) is exposed to the perturbation while the other arm (reference arm) is either desensitized to the perturbation via proper coatings or shielded from it. The perturbation causes a differential phase shift between light propagating in the signal and reference arms that is converted into an optical intensity variation by means of an optical 3-dB fiber coupler. In these devices, sensitivity may be enhanced by increasing the length of sensing fiber or by optimizing the fiber coatings. Tens of meters of fiber usually yield sensitivities superior to existing competing technologies. An attractive attribute of this type of sensor is its applicability to sensing a broad range of environmental parameters. This derives from the fact that coating materials applied as jackets to the optical fibers usually play the major role in determining the fiber's sensitivity to a particular field or parameter. Experimental studies on the acoustic response of coated fibers [16] have shown dramatic increases in sensitivities. The discovery that strains induced in the coating material by an external perturbation communicate directly and in a very controllable way to the glass fiber have led to the demonstrations [1] of very sensitive acoustic devices and a variety of sensors including magnetic, ultrasonic, thermal, chemical, current, acceleration, and electric field sensors.

The Sagnac fiber sensor class (lower right, Figure 21.1) involves a single-fiber arrangement for the fiber interferometer. In this configuration two counterrotating optical beams propagate within the single fiber. Nonreciprocal effects influence the counterrotating beams differently and can be sensed with this interferometric configuration. Angular rotation is the most important nonreciprocal effect that can be detected with Sagnac fiber sensors [17–20], and many laboratories have fabricated sensitive gyroscopes using this configuration. The details of operation of the various fiber sensor configurations are presented in the following sections.

21.3 INTENSITY MODULATION SENSORS

When a single-valued and reproducible relationship exists between an applied perturbation and the modulation of light levels propagating in a fiber, the intensity modulation may, in principle, be used to measure the magnitude of the perturbation. This has become the basis for developing optical fiber sensors. The characterization and optimization of this modulation are critical factors in the construction of sensitive fiber sensors. Consider a fiber into which W_0 of optical power has been launched. After transversing the fiber and any elements placed in its path, the light received by a detector at the output end of the fiber sensor is $W_0 T$, where T is the power transmission coefficient of the fiber sensor. When a perturbation ΔP is applied to the sensor, the optical power W_0 transmitted through the fiber will change by $W_0 \Delta T$, and the output detector signal current will be given by the equation

$$i_s = \eta \, \frac{eW_0}{hv} \left(\frac{\Delta T}{\Delta P} \right) \Delta P \tag{1}$$

where η is the detector quantum efficiency, e is the charge of an electron, h is Planck's constant, v is the frequency of light, and $\Delta T/\Delta P$ is the transduction coefficient for the sensor. In Eq. (1), $\Delta T/\Delta P$ describes the relationship between the perturbation P and its effectiveness in modulating the light signal. With it, one can predict the minimum detectable threshold for the sensor.

The mean square noise of a sensor detection system is given by

$$i_N^2 = \left(2e^2\eta \, \frac{W_0 T}{hv} \right) \Delta f + 4k_B T_\theta \, \frac{\Delta f}{R} + \langle i_{amp}^2 \rangle + \langle i_{source}^2 \rangle \tag{2}$$

where Δf is the detection bandwidth, k_B is Boltzmann's constant, T_θ is the temperature, and R is the circuit resistance. The first term in this equation is the shot noise of the detector, the second term is the system Johnson noise, the third term arises from additive amplifier noise, and the last term arises from source amplitude noise. Optical sensors are usually configured so that there is sufficient optical power for the shot noise term to dominate the intrinsic sensitivity of the sensor. Shot noise performance may not always be achievable in practice, however. When a noisy optical source is used, increasing the optical power does not necessarily ensure that shot noise performance will be realized. Solid-state semiconductor sources, LEDs and lasers, have amplitude noise components in their outputs that could dominate performance. In Figure 21.2 the relative amplitude noise of common diode lasers is presented. For most sensor applications, sufficient optical power is usually available so that one can usually operate at about the 120-dB relative noise level. For operating frequencies below about 100 Hz, which are common for many sensors, laser amplitude noise would dominate over shot noise, and care must be exercised in the design of the optical circuits for low-frequency operations. The relative amplitude noise for LEDs is generally 10 dB or more below that of semiconductor lasers, and therefore shot noise level performance is usually achievable (except at very low frequency where the source/detector $1/f$ noise eventually dominates over shot noise). With these limitations in mind, we will determine the intrinsic threshold detectability of a single-fiber sensor. The signal-to-noise ratio (SNR) can be deter-

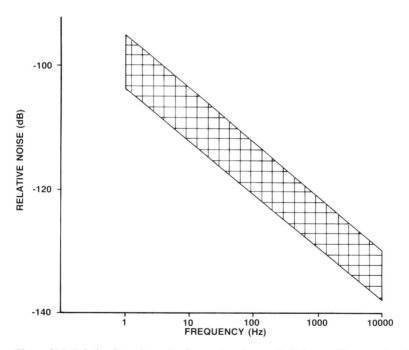

Figure 21.2 Relative intensity noise for semiconductor diode lasers. The cross-hatched area depicts the typical range of intensity noise for most diode lasers as a function of frequency. Amplitude noise decreases as a function of frequency in diode lasers.

mined using Eqs. (1) and (2) and is given as

$$\text{SNR} = \frac{i_s^2}{i_n^2} = \frac{\eta W_0}{2h\nu \, \Delta f \; T} \left(\frac{\Delta T}{\Delta P} \Delta P\right)^2 \tag{3}$$

The shot-noise-limited detection threshold for SNR = 1 is given from Eq. (3) by

$$\Delta P_{\min} = \left(\frac{2h\nu \, \Delta f \; T}{\eta W_0}\right)^{1/2} \left(\frac{\Delta T}{\Delta P}\right)^{-1} \tag{4}$$

From Eqs. (1) and (4) we see that the minimum detectable threshold and the sensor's sensitivity are completely determined by the transmission coefficient $\Delta T/\Delta P$ and the optical power. Examination of intensity modulation sensors in this section will therefore concentrate on the determination of $\Delta T/\Delta P$ for various sensor configurations. The representative sensors to be examined will include fiber-to-fiber sensors, attenuation sensors, and photoelastic sensors.

21.3.1 Fiber-to-Fiber Sensors

A number of conceptually simple sensors may be realized by either moving two fibers relative to each other or placing a movable object between the fibers. Modulation of the light is directly related to the motion of the fibers or movable sensor elements. In Figure 21.3, one fiber is fixed in position whereas the second fiber is arranged in a cantilever beam configuration [21] and is free to move in response to a perturbation force. The fibers are typically separated by a few tens of micrometers if multimode fibers are used and a few micrometers if single-mode fibers are employed. The end of the

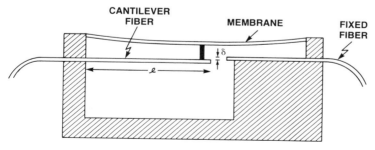

Figure 21.3 Cantilever beam optical fiber sensor configuration. In this configuration, the position of one fiber is fixed, whereas the second fiber is free to move in response to some perturbation. The movable fiber may be affixed to a membrane as shown to amplify the effects of a given perturbation.

free fiber may be attached to a diaphragm or coated with a sensing material. A force F will deflect the fiber a distance $\Delta\delta$ given by

$$\Delta\delta = \frac{Fl^3}{3EI} \tag{5}$$

where E is the Young's modulus of glass, $I = \pi d^2/16$ is the moment of inertia of the beam taken about its neutral point, and l is the length of the fiber beam. The transmission coefficient can be derived for this sensor configuration using Eq. (5):

$$\frac{\Delta T}{\Delta P} = \frac{\Delta T}{\Delta\delta}\left(\frac{\Delta\delta}{\Delta P}\right) = \frac{l^3}{3EI}\left(\frac{\Delta T}{\Delta\delta}\right) = \frac{16l^3}{3E\pi d^2}\left(\frac{\Delta T}{\Delta\delta}\right) \tag{6}$$

$\Delta T/\Delta\delta$ is a function of the core radius and the type of fibers employed. For small displacements, $\Delta T/\Delta\delta$ for fibers with Gaussian output beams is given by

$$\frac{\Delta T}{\Delta\delta} \simeq 4p_0^2 \, \delta T \tag{7}$$

where $p_0^2 = 1/\Omega_b^2 + \Omega_b^2 k^2/R^2$, k is its optical wavenumber ($k = 2\pi/\lambda$), δ is the total fiber-to-fiber offset, Ω_b^2 is the Gaussian beam width, and R is the radius of curvature of the beam's phase front.

Choosing a small-core fiber increases sensitivity; however, the choice of small-core fibers requires higher levels of fiber alignment, which usually leads to compromising by choosing a modest core size fiber ($\sim 50-100 \ \mu$m). While these cantilever-type sensors generally have limited sensitivity, they do appear to be well suited for measuring modest pressure, temperature, displacement, and acceleration levels in cases where a miniature sensor probe is important. It should also be noted that the sensitivity may be increased roughly an order of magnitude above that given by Eq. (6) if a graded-index (GRIN) lens is attached to the optical fiber ends and used to magnify the fiber motion.

The second principal type of fiber-to-fiber moving sensor was demonstrated by Spellman and McMahon [7] when they developed a Schlieren or grating sensor as shown in Figure 21.4. This sensor modulates light coupled from one multimode fiber to another by means of a pair of opposed gratings that move relative to one another. Light from the input fiber is usually collimated with a 1/4 pitch Selfoc lens transmitted through the gratings, and refocused via a second Selfoc lens. In order to create a linear response to the applied perturbation, the rulings of one grating are usually displaced by half the width of a stripe (or one-fourth the grating periodicity). Since the Schlieren sensor produces a maximum modulation when the position of the second grating has been displaced by one-fourth the width of the stripe, the sensitivity no longer depends on the fiber diameter and is increased by the ratio of the fiber core diameter to the stripe width. Equation (6) can then be

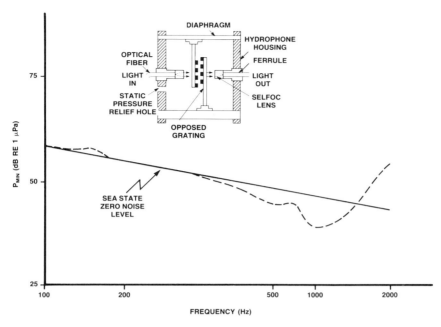

Figure 21.4 Typical acoustic response of a Schlerien acoustic sensor (dashed line). These sensors have been demonstrated with sea-state zero response.

used, but now d is not the fiber diameter but the stripe width. The sensitivity of the Schlieren sensor with 5-μm stripes can be 20 times greater than that of a 100-μm-diameter cantilever fiber sensor.

In Figure 21.4, the minimum detectable pressure levels measured with a Schlieren hydrophone are presented. At frequencies below 1 kHz, sea-state zero pressures (50 dB relative to 1 μPa at 1 kHz) can be detected, whereas above 1 kHz the response decreases sharply essentially due to the smaller diaphragm displacements at higher frequencies.

21.3.2 Attenuation Sensors

Many different types of sensors can be realized by monitoring the attenuation induced in a fiber via the perturbation to be sensed. These include microbend- and macrobend-induced losses and intrinsic material losses as well. In a microbend loss sensor, an optical fiber is periodically deformed by mechanical means. This deformer is usually a pair of ridged plates, wire grids, etc. The fiber deformation has two critical parameters, the spatial period of the bending, Λ_f, and the amplitude of the deformation, x. The transduction coefficient of the sensor, $\Delta T/\Delta P$, used in Eq. (4) can be written in terms of the maximum displacement of the fiber deformation, x, as follows:

$$\frac{\Delta T}{\Delta P} = \frac{\Delta T}{\Delta x}\left(\frac{\Delta x}{\Delta P}\right) \tag{8}$$

In this equation, $\Delta T/\Delta x$ depends on the sensitivity of the optical fiber to microbending losses, and $\Delta x/\Delta P$ depends on the mechanical design of the sensor.

A periodic deformation along the fiber axis causes strong mode coupling, which redistributes the light among core modes and couples some light from the core to the radiation and clad modes. Thus, by monitoring the power in certain modes, an applied perturbation can be detected. For communication applications, microbending losses have deliberately been minimized by making strongly guiding fibers (high Δn) with thick cladding and soft concentric primary coating. Optical fibers having the opposite characteristics, particularly very weakly guiding fibers (small Δn), exhibit high microbending loss. Increased microbending sensitivity is also expected when high-order guided modes are excited. These higher-order modes are easily coupled to radiation modes (loss) and are very susceptible to any fiber deformation.

From the theory of mode coupling [22, 22a] it is well known that when a microbend of periodicity Λ_f is imposed along the fiber axis, light power is coupled between modes with propagation constants k_i and k_j such that

$$\Delta k = k_i - k_j = \pm\frac{2\pi}{\Lambda_f} \tag{9}$$

It has been shown [23] that the difference in adjacent mode propagation constants is

$$\Delta k = k_{m+1} - k_m = \left(\frac{b}{b+2}\right)^{1/2}\left(\frac{2\sqrt{\Delta}}{a}\right)\left(\frac{m}{M}\right)^{(b-2)/(b+2)} \tag{10}$$

where m is the mode label, M is the total number of modes, and Δ is a parameter related to the refractive index profile, $\Delta = (n_1^2 - n_2^2)/n_1^2$. Equation (9) becomes

$$\Delta k = k_{m+1} - k_m = \frac{2\Delta^{1/2}}{a}\left(\frac{m}{N}\right) \tag{11}$$

for step-index fibers ($b = \infty$) and

$$\Delta k = k_{m+1} - k_m = \left(\frac{2\Delta}{a}\right)^{1/2} \tag{12}$$

for parabolic index fibers ($b = 2$).

For step-index fibers, the separation of modes in k-space depends on the order of the mode m. From Eqs. (9) and (10) we see that high-order modes (large M) can be coupled with small periodicity Λ_f. In a parabolic index fiber all modes are equally spaced in k-space (to within the WKB approximation) and Δk is independent of m. The same bending periodicity couples all adjacent modes efficiently in parabolic index fibers. This critical mechanical wavelength Λ_c is given by

$$\Lambda_c = \frac{2\pi a}{(2\Delta)^{1/2}} \tag{13}$$

For typical sensing fiber parameters ($\delta = 0.15$, $a = 50\ \mu m$), Λ_c for the parabolic fiber is ~ 1 mm [Eq. (11)]. As can be seen from Eq. (11) for a fiber sensor that operates by switching low-order modes, much longer deformer periodicities are required for the step-index fibers. For sensors that operate by switching high-order modes to radiation modes, however, somewhat shorter deformer periods are required for the step-index case. In any case, the parabolic fiber with its single value of Λ_c has ideal mode properties for microbend sensors. Indeed, large enhancement of the bending loss can be achieved in a parabolic fiber when the bending periodicity is Λ_c as given by Eq. (13). In this case efficient coupling takes place from one propagating mode to the next and from higher-order modes to radiation modes. As can be seen in Figure 21.5, Fields et al. [24] observed a strong resonant peak indicating the existence of a critical periodicity at 1.6 mm, in agreement with the prediction of Eq. (13) ($\Delta = 0.008$, $2a = 65\ \mu m$) for this fiber.

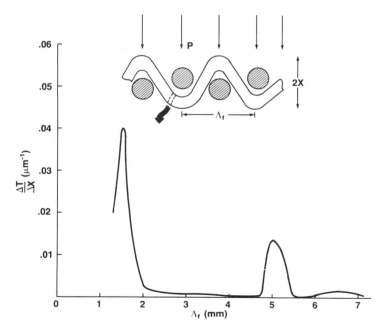

Figure 21.5 Transduction response of a microbend sensor as a function of microbend periodicity for a graded-index fiber.

Having discussed the optical loss term $(\Delta T/\Delta x)$ in Eq. (8), we now consider the mechanical factor $\Delta x/\Delta P$. In most microbend sensors, force is applied to the fiber indirectly through couplers (pistons, diaphragms, etc.) that are used as force multipliers. If A is the area of the coupler against which the perturbation pressure P is applied, Eq. (8) can be written as

$$\frac{\Delta T}{\Delta P} = \frac{\Delta T}{\Delta x} AC_m \tag{14}$$

where C_m is the mechanical compliance of the sensor. The mechanical term AC_m in Eq. (14) depends on the sensor design. For example, in the case of an acoustic sensor, C_m can be found from the equivalent acoustic circuit of the sensor [25]. When the compliance of the surrounding fluid is substantially larger than the mechanical compliance of the sensing fiber element C_f, the compliance of the sensor C_m is proportional to C_f:

$$C_m \propto C_f = G\left(\frac{\Lambda_f^3}{Ed^4}\right)\left(\frac{1}{N_d}\right) \tag{15}$$

where N_d is the number of deformer intervals and G is a constant that depends on how the fiber is loaded and suspended. For example, for a fiber clamped at its ends and deformed by a load at the center, $G = 1/3\pi$.

From Eqs. (14) and (15) we see that high sensitivity is obtained with highly compliant fibers. This can be achieved with a large deformer periodicity Λ_f. Also, for large C_m, N_d, and d, the Young's modulus E should be small. However, the Young's modulus of typical glasses used for optical fibers does not vary very much, certainly not by an order of magnitude. Also, very small values of d, the fiber diameter, are prevented due to light coupling and fiber mechanical strength considerations. And finally, even though utilizing a small number of deformer intervals N_d increases C_1 proportionally, it decreases $\Delta T/x$ by the same amount. Thus, the sensor sensitivity is approximately independent of N_d. In practice, the optimum value of N_d is determined from sensor mechanical design considerations such as the position of the mechanical resonance and thus the required sensor bandwidth.

An important advantage of the microbend sensor is that the optical power is contained within the fiber. The transduction mechanism is compatible with multimode fibers, and low detection thresholds are possible. Bucaro et al. [26] reported detection thresholds for acoustic sensors of 60 dB relative to 1 μPa, demonstrating that these sensors are generally more sensitive than fiber-to-fiber types.

Macrobend loss sensors are realized when a weakly guiding fiber is tightly wound around a mandrel that expands and contracts in response to the excitation. In this case, a uniformly wound fiber experiences added radiation losses that are a strong function of mandrel radius.

If α is the loss due to bending, the fiber transimission coefficient T is simply

$$T = 10^{-\alpha l} \tag{16}$$

The transduction coefficient is then

$$\frac{\Delta T}{\Delta P} = -2.3l(10^{-\alpha l})\frac{\Delta \alpha}{\Delta P} = -2.3l(10^{-\alpha l})\frac{R_m}{3K_m}\left(\frac{\Delta \alpha}{\Delta R_m}\right) \tag{17}$$

where R_m and K_m are the mandrel radius and bulk modulus, respectively.

This strong dependence on R_m is generally predicted theoretically [27] and is of the form

$$\alpha = A'e^{-B'R_m}R_m^{1/2} \tag{18}$$

where A' and B' are functions that depend on the core and clad refractive indices, the core radius, and the light wavelength. Figure 21.6 shows α computed from Eq. (18) for a single-mode fiber for various values of numerical aperture (NA). As can be seen, α is higher for lower NA, although (for a given bend radius) higher values for $\Delta\alpha/\Delta R_m$ are obtained for fibers of larger NA.

We can calculate the minimum detectable pressure by inserting Eq. (17) into Eq. (4):

$$P_{\min} = \left(\frac{2h\nu T\,\Delta f}{\eta W_0}\right)^{1/2}\left(\frac{3K_m}{2.3R_m}\right)\left(\frac{10^{\alpha/2}}{l}\right)\left(\frac{\Delta\alpha}{\Delta R_m}\right)^{-1} \tag{19}$$

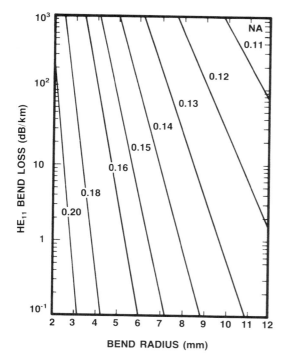

Figure 21.6 Theoretically calculated power loss for single-mode fibers as a function of bend radius. These curves are plotted for various fiber numerical apertures with $V = 2.4$ and $\lambda = 0.84\ \mu$m.

Inspection of Eq. (19) shows that there exists an optimum length l_{op} that minimizes P_{min}. Maximizing Eq. (19) with respect to l gives

$$l_{op} = \frac{0.87}{\alpha} \tag{20}$$

Thus, a trade-off exists between total fiber length l, bend loss α, and bend radius R_m (which determines α). For small mandrel radii, the bend loss is high and l_{op} is small. Alternatively, if long fiber lengths are desired for extended elements or for shaded windings, α must be low, which requires a large R_m.

The most noteworthy example of an intrinsic loss sensor is a radiation sensor [10]. In this case,

$$\Delta P_{min} = \left(\frac{2h\nu \, \Delta f \, T}{\eta W_0}\right)^{1/2} \frac{\Delta P/\Delta\alpha}{2.3 l \times 10^{-\alpha l}} \tag{21}$$

where α is the total loss in the fiber and ΔP is the change in radiation dosage needed to change the signal by an amount equal to the shot noise. For flint glass fibers, $(\Delta P/\Delta\alpha)^{-1}$ can range between 1 dB/(km-rad) to 10 dB/(km-rad) and typical dosages that are detectable are 1–10 mR.

21.3.3 Photoelastic Sensors

The photoelastic effect is based on the change of index of refraction of a material when a stress σ is applied. This change in index of refraction is manifested as a change in the birefringence of the material. When the photoelastic material is placed between crossed polarizers, each oriented at 45° to the axis of the stress, and light is transmitted through the polarizers and photoelastic material along a direction perpendicular to the axis of the stress in the material, then the optical transmission is given by

$$T = A \sin^2 \left(\frac{\pi t \sigma}{f_\sigma} + C\right) \tag{22}$$

where t is the thickness of the photoelastic material traversed by the light, A and C are constants, and f_σ is the material stress optical coefficient. If σ_0 is the amount of stress required to go from maximum to minimum transmission, then it can be written as $\sigma_0 = f_\sigma/2t$. The transduction coefficient for this sensor is then given by

$$\frac{\Delta T}{\Delta\sigma} = 2A\sigma_0 \sin \left(\frac{2\sigma}{\sigma_0} + C\right) \tag{23}$$

For small stress levels, optimum sensitivity is achieved when $C = \pi/4$. The value of σ for an applied force depends on the sensor design. In most acoustic sensors employing this type of sensor, mechanical concentration of the applied force is achieved through the use of piston arrangements. In the case of acoustic sensors, sea-state zero-level sensitivities have been achieved with these devices whereas accelerometers using this type of sensor have achieved sensitivity levels of a few micro-g.

21.4 MACH-ZEHNDER INTERFEROMETRIC SENSORS

Single-mode optical fiber interferometers can be configured in a number of ways—similar to their bulk analogs. Common configurations are Mach-Zehnder, Michelson, Sagnac, and polarimetric (birefringent). The Sagnac and Mach-Zehnder are by far the most common forms currently in use and will be discussed in some detail here. A separate section (Section 21.8) is devoted to the Sagnac interferometer both because its sensitivity and noise issues are fundamentally different from those of the others and because of the widespread interest in its use as a gyroscope.

A schematic of a Mach-Zehnder interferometer is shown in Figure 21.7. A laser beam is split, part being sent through a "reference" fiber arm and part through a "sensing" fiber arm that is immersed in, and sensitive to, its environment. After passing through the fibers, the two beams are recombined and allowed to interfere on the surface of a photodetector. There results in the photodetector current a signal that is directly related to the environmentally induced phase shift $\Delta\phi$.

Besides the laser and fiber, key elements of the Mach-Zehnder interferometer are the two beamsplitters. The fiber equivalent of a bulk optical dielectric beamsplitter is a fiber–fiber coupler, which in turn is the optical equivalent of a microwave directional coupler. In the fiber case, the fiber cores, where the optical energy is confined, are brought close enough together that light can transfer from one fiber to the other, yielding coupling. Accessing the fiber cores can be accomplished by polishing away the surrounding fiber cladding or by a fusion process in which the fibers are fused together in a single piece. In the case of polishing, the fibers are mounted separately on the surfaces of two glass blocks and polished. The blocks are then brought together to form the coupler. Both approaches provide the desired 50–50 (3-dB) coupling, although to date only the fusion process has demonstrated stability over a wide temperature range.

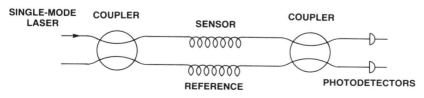

Figure 21.7 Schematic of a Mach-Zehnder interferometer.

Interferometers allow the measurement of extremely small phase shifts generated by the field to be detected (e.g., magnetic or acoustic) in the optical fiber. The optical phase delay (in radians) of light passing through a fiber is given by

$$\phi = nkl \tag{24}$$

where n is the refractive index of the fiber core, k is the optical wavenumber in vacuum $(2\pi/\lambda)$, and l is the length of the fiber. Small variations in the phase delay are found by differentiation:

$$\frac{\Delta\phi}{\phi} = \frac{\Delta l}{l} + \frac{\Delta n}{n} + \frac{\Delta k}{k} \tag{25}$$

The first two terms are related to physical changes in the fiber caused by the perturbation to be measured. Accordingly, they describe the transduction mechanism whereby the fibers can act as sensors. Generally, changes in pressure, temperature, electric field, and so on produce different contributions to $\Delta\phi$ via the Δl and Δn terms. The design of fibers optimized for these effects and the specifics of transduction will be discussed in Sections 21.5 to 21.7. The last term in Eq. (25) takes into account any wavelength (or frequency) variation associated with the laser source. This term has been specifically included to facilitate our discussion of certain demodulation issues such as phase noise.

If we consider the interference of two lightwaves as in an interferometer, the amplitudes of the two fields are

$$E_1 = E_{10} \sin(\omega t - \phi_1) \quad \text{and} \quad E_2 = E_{20} \sin(\omega t - \phi_2) \tag{26}$$

where ϕ_1 and ϕ_2 are the total path length in terms of optical phase in each arm of the interferometer and ω is the optical frequency. The resultant is given by

$$\begin{aligned} E_1 + E_2 = {}&\sin \omega t \, (E_{10} \cos \phi_1 + E_{20} \cos \phi_2) \\ &- \cos \omega t \, (E_{10} \sin \phi_1 + E_{20} \sin \phi_2) \end{aligned} \tag{27}$$

Rewriting this expression, the electric field is

$$E = E_0 \sin(\omega t - \phi) \tag{28}$$

where

$$E_0^2 = E_{10}^2 + E_{10}^2 + E_{20}^2 E_{20} \cos(\phi_2 - \phi_1) \tag{29a}$$

$$\tan \phi = \frac{E_{10} \sin \phi_1 + E_{20} \sin \phi_2}{E_{10} \cos \phi_1 + E_{20} \cos \phi_2} \tag{29b}$$

If $E_{10} = E_{20}$, then

$$E_0^2 = 2E_{10}^2[1 + \cos(\phi_2 - \phi_1)] \tag{30}$$

and as a consequence the interferometer is sensitive only to differential changes in ϕ_2 and ϕ_1. Defining $\Delta\phi = \phi_2 - \phi_1 = nkl$, we have the classic interference equation, the resultant intensity being given by

$$E_0^2 = 2E_{10}^2(1 + \cos \Delta\phi) \qquad (31)$$

To make a sensitive interferometric sensor, the phase shift $\Delta\phi$ for the particular field to be measured must be maximized, the nonlinear output of the interferometer demodulated, and the interferometer operated such that noise contributions that would mask the signal are minimized. The phase shift may be maximized by making the signal arm sensitive to the field while desensitizing the reference fiber. In Sections 21.6 and 21.7, two such sensors are dealt with in detail—acoustic and magnetic. The following section will deal with demodulation and noise in fiber interferometers.

21.4.1 Demodulation

The purpose of the demodulator is to transform the optical output of the sensor into an electrical signal proportional to the amplitude of the optical phase and thus the signal field. Three basic approaches have been employed: (1) active homodyne [28], in which the interferometer is locked at maximum sensitivity (quadrature) with feedback circuitry and a phase shifter; (2) passive homodyne [29], in which interferometer signal outputs are derived that have a $\pi/2$ phase shift allowing direct demodulation, and (3) heterodyne techniques [30]. Many techniques using these approaches or variants such as synthetic heterodyne [31] have been demonstrated.

Limited space precludes the description of all of the techniques, so as an example the active homodyne will be briefly discussed. If we consider a small phase shift $\Delta\phi$, the output of the interferometer will be maximum when the static phase shift in the interferometer is equal to $\pi/2$ [see Eq. (33)]. This is the quadrature point. This static phase shift can be produced by changing Δl, which in turn can be achieved by wrapping a length of the fiber (typically in the reference arm) around a piezoelectric cylinder to which the appropriate voltage is applied. Similarly, by changing the source wavelength [32, 32a] (i.e., Δk) [seee Eq. (25)], for example, by tuning the injection current of a diode laser, the quadrature condition can be met. This configuration has the advantage that the laser and electronics can be located remote from the actual interferometric sensor, that is, with no electrical leads, and only fiber intensity leads going to the sensor. Differential thermal effects between the arms of the interferometer will cause the static phase shift to drift; consequently, it is necessary to form a feedback loop and to apply a correction voltage to either the piezoelectric cylinder or the laser drive to keep the interferometer locked at quadrature. The circuitry required for this is extremely simple, the main component being an electronic integrator. This scheme can be operated in two modes: low gain and high gain. In the low-gain mode, the gain in the feedback loop is reduced such that only slow phase drifts (e.g., thermal drifts below, say, 20 Hz) are compensated for, and the signal of interest (>100 Hz) can be observed at the photodiode output. The advantage of this scheme is that the high-frequency response of the demodulator is limited only by the photodetector response, which is typically >1 MHz. However, for signals of amplitude in excess of 0.1 radian, one has to consider the nonlinear response of Eq. (31), which when expanded in terms of Bessel functions indicates considerable distortion and nonlinearity in the demodulated output. In the high-gain mode, the gain is increased so that both the low-frequency drifts and the signal frequencies are compensated for. The signal can then be observed by monitoring the voltage fed back to the feedback element. The high-gain system has the advantage of greater dynamic range, less harmonic distortion, and relative insensitivity to changes of laser output and of coupler splitting ratio.

Active homodyne techniques have generally been superseded by passive homodyne techniques (using either 3×3 couplers [33] or laser modulation [34]), which are more appropriate for multisensor configurations and multiplexing. Nevertheless, the above serves to show some of the basic principles of demodulator operation.

21.4.2 Noise

The typical single-mode fiber interferometer constructed with high-quality components (e.g., couplers) is an intrinsically "quiet" device. Most noise problems are associated with (1) perturbations causing unwanted Δl and Δn contributions to $\Delta \phi$ and (2) noise associated with the optical source that is used to measure the $\Delta \phi$ induced by the signal field. The first noise source is basically related to fiber coatings and packaging issues. As indicated in Section 21.5, properly designed coatings can be used in the signal arm to maximize $\Delta \phi$ due to the signal field while reducing the effects from other fields. Noise from the optical source can be separated into two terms, intensity noise contributions and phase noise contributions.

a. Intensity Noise. If we consider the output of the interferometer after the demodulator, the output voltage V_o can be represented by

$$V_o = V_1(1 + \cos \phi) + V_2 \tag{32}$$

Here V_1 is the amplitude of the output dependent on the optical phase of the interferometer and V_2 is a dc term due to either unbalanced splitting ratios of the couplers (or uneven losses in the fiber arms) or lack of coherence of the optical source resulting in a loss of fringe visibility V [$=(V_{max} - V_{min})/(V_{max} + V_{min})$]. Changes in $\Delta \phi$ due to perturbation of the signal field results in an output ΔV_o

$$\Delta V_o = -V_1 \sin \phi \, \Delta \phi \tag{33}$$

which at the quadrature (maximum sensitivity) point ($\phi = \pi/2$) is equal to

$$\Delta \phi_{\min} = \frac{\Delta V_0}{V_1} \quad \text{and} \quad \frac{\Delta I}{I} = \frac{\Delta V_o}{V_1 + V_2} \tag{34}$$

Thus

$$\frac{\Delta V_0}{V_1} = \frac{\Delta I}{I}\left(1 + \frac{V_2}{V_1}\right) \quad \text{and} \quad \Delta \phi_{\min} = \frac{\Delta I}{I}\left(\frac{1}{V}\right) \tag{35}$$

In terms of laser properties, the coherence length (and path difference D), as well as the intensity noise itself, determines the sensor noise floor. To generalize, as in Eq. (2), ΔV_n can be considered to be composed of not only the laser intensity noise ΔV_1 but also a shot-noise term ΔV_{n2}, a Johnson noise term ΔV_{n3}, and an additive amplifier noise ΔV_{n4} such that

$$\Delta V_n = (\Delta V_{n1}^2 + \Delta V_{n2}^2 + \Delta V_{n3}^2 + \Delta V_{n4}^2)^{1/2} \tag{36}$$

In typical fiber interferometers powered by single-mode GaAlAs lasers ($\lambda \approx 830$ nm), typically the ΔV_{n1} noise term dominates [35]. A plot of the intensity noise as a function of frequency is given in Figure 21.2. The relative intensity noise is given in terms of decibels [$20 \log (\Delta I/I)$]. Thus, typical values of $\Delta I/I$ and V indicate values of $\Delta \phi_{\min} \approx 10^{-5}$ rad. However, this noise term can be reduced either by using electronic feedback schemes (using the integral detector in the laser diode) or by using the concept of a balanced optical detector. For example, by subtracting the complementary outputs of the Mach-Zehnder interferometer, as much as 40 dB intensity noise rejection can be achieved [36]. The two output signals of the interferometer V_A and V_B are 180° out of phase and can be written as

$$V_A = I(1 + \Delta I) + VI(1 + \Delta I) \, \Delta \phi$$

and

$$V_B = I(V + \Delta I) - VI(1 + \Delta I) \, \Delta \phi \tag{37}$$

for the case where one or two couplers have a 50:50 splitting ratio. Here V indicates the fringe visibility with the interferometer at quadrature so that

$$V_A \simeq I(1 + \Delta I + V \, \Delta \phi) \quad \text{and} \quad V_B = I(1 + \Delta I - V \, \Delta \phi) \tag{38}$$

Consequently,

$$V_A - V_B = 2VI \, \Delta \phi + 2VI \, \Delta I \, \Delta \phi \approx 2VI \, \Delta \phi \tag{39}$$

and the first-order laser intensity noise term is eliminated. An example of the performance of this technique is shown in Figure 21.8, where intensity noise in an interferometer has been reduced by an order of magnitude.

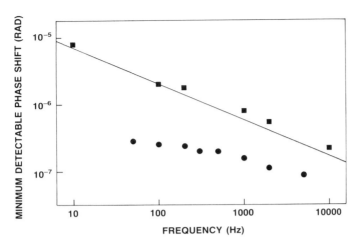

Figure 21.8 Minimum detectable phase shift as a function of frequency for the interferometer (1 Hz bandwidth). Solid line, calculated from laser intensity noise; (■) experimental values from V_A, (●) experimental values from $V_A - V_B$.

b. Phase Noise. As indicated earlier, the phase delay in the optical fiber is sensitive to the source wavelength [see Eq. (25)]. This can result in a noise source if the frequency v is unstable (Δv). However, light travels in both arms of the interferometer, and only the path length difference Δl is important. From Eq. (25), the phase shift $\Delta \phi$ resulting from frequency instability Δv is

$$\Delta \phi = \frac{\phi}{k} \Delta k = nl \, \Delta k \tag{40}$$

If the two arms of the interferometer have path lengths l_1 and l_2 such that $l_1 - l_2 = \Delta l_{12}$, then from Eq. (40) the minimum detectable phase shift $\Delta \phi_{min}$ is

$$\Delta \phi_{min} = \frac{2\pi n}{C} \Delta l_{12} \, \Delta v \tag{41}$$

where C is the speed of light in vacuum.

Shown in Figure 21.9 is the variation of $\Delta \phi_m$ with path difference for a number of different GaAlAs lasers [37]. Similar behavior has been observed with lasers operating at 1.3 μm and 1.55 μm. To ensure high-performance operation, it is necessary to operate the interferometer with less than 1 mm of path difference. Although construction of the interferometer with this path length mismatch is relatively easy, external perturbation (especially thermal) will tend to unbalance the paths. Again, coatings and packaging play an important role in reducing changes in Δl. Another method to reduce the noise due to the source frequency instability is to reduce the value of the instability Δv. A further method is to use a reference interferometer either

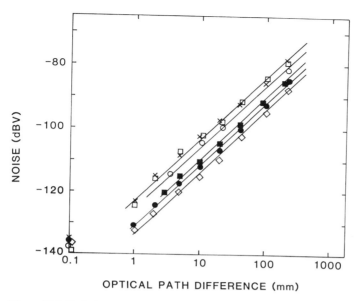

Figure 21.9 Variation of the noise output of the interferometer (at maximum sensitivity) in decibels as a function of path length difference at 2 kHz with a 1-Hz bandwidth. (×) General Optronics TB47, (●) Laser Diode Labs, (□) Hitachi HLP 1400, (○) Hitachi HLP 3400, (■) Hitachi HLP 2400U, (◇) Mitsubishi.

to frequency-stabilize the source or to subtract the output of the normalized reference interferometer from that of the signal interferometer [38]. Still another method to increase the frequency stability is to use optical feedback so as to narrow the line width of the laser source [39]. Experiment and theory indicate that the stability is increased by $(\Delta v / \Delta v_n)^{1/2}$, where Δv is the free running linewidth and Δv_n is the narrowed linewidth. However, as the amount of optical feedback is increased ($\geq 10^{-4}$), feedback into the laser cavity can degrade the device performance. Typical coherence lengths l_c of diode lasers (without feedback) range from 2 to several tens of meters. However, optical feedback has been shown to produce line broadening ($l_c \approx 30$ cm), satellite mode generation ($l_c \approx 10$ cm), and multimode operation (l_c of a few millimeters) [40]. Optical feedback has also been shown to cause longitudinal mode hopping, which results in large intensity and phase noise contributions [41]. Consequently, care to reduce optical feedback by using fiber termination and high quality fiber and components, for example, is extremely important.

21.4.3 Other Configurations

The foregoing analysis is applicable to all two-beam interferometers—Mach-Zehnder, Michelson, and polarimetric; each of these configurations has its own advantages and disadvantages in its use as a fiber sensor. Referenced to the Mach-Zehnder, for the same length of signal arm the Michelson has twice the sensitivity (owing to the double path); however, the phase noise limitation also doubles for an equivalent fiber length mismatch. The main disadvantage of the Michelson is that the complementary output is launched back into the laser, which for a low-loss system means (when laser–fiber coupling is included) that the optical feedback term is $\approx 10^{-1}$, which is far above the $\sim 10^{-6}$ ratio required for stable operation of semiconductor sources.

The polarimetric type of sensor is sensitive to changes in birefringence, caused by the signal field, of a single fiber. This approach can be considered to be a single-fiber Mach-Zehnder in the sense that the signal and reference arms are the two orthogonal polarization modes of the fiber. This configuration has a number of advantages in terms of common-mode noise rejection; however, the sensitivity is somewhat less than for the two-fiber interferometer.

Multiple beam interferometry has also been demonstrated with optical fibers. In principle, this approach is the most sensitive of all of the fiber interferometer configurations and gives excellent amplitude noise rejection. The effective path difference is extremely large, so the sensor is orders of magnitude more sensitive to laser phase noise contributions. Similar to the fiber Michelson, the complementary output is coupled directly back into the source, resulting in degradation of the source coherence and noise [40, 41].

21.5 COATED OPTICAL FIBER SENSORS

Interferometric optical fiber sensors are based on the principle that environmental perturbations shift the phase of light propagating in a single-mode fiber. As just discussed, this change of phase (compared to a phase reference) is ultimately converted by a demodulator to an electric signal proportional to the environmental perturbation. The key to the widespread applicability of this type of sensor is the fact that the shift in optical phase caused by a particular environmental parameter—pressure, temperature, magnetic or electric field, for example—can be effectively controlled by application of appropriate jacketing materials on the optical fiber [42]. The application of these coatings usually involves straightforward procedures and, more important, does not affect the optical performance of the optical fiber waveguide. This "dissociation" of the waveguide itself from the designer's attempt to control its sensitivity is perhaps the most important factor responsible for the rapid pace at which this generic sensor technology has been developed.

Consider a single-mode optical fiber responding to an environmental effect such as a change in temperature or pressure. As discussed in connection with Eq. (25), two effects contribute to the change of phase for light propagating through the fiber. The first is a change in the physical length of a given section of fiber, Δl. The second is a change in the optical index of refraction of the fiberguide glass, Δn, in part due to the strain optical effect, and (in the case of temperature) in part due to temperature-induced electronic polarizability changes.

For the usual case in which the perturbing field generates fiber strains whose wavelengths are large compared to the fiber diameter, the normalized phase shift is given by

$$\frac{\Delta\phi}{\phi} = \varepsilon_z - \frac{n^2}{2}\left[(P_{11} + P_{12})\varepsilon_r - P_{12}\varepsilon_z\right] + \left(\frac{\partial n}{\partial T_\theta}\right)_\varepsilon \Delta T_\theta \qquad (42)$$

where ε_z, ε_r are the axial and radial strains induced in the fiber core, the P's are the photoelastic constants, and $(\partial n/\partial T_\theta)_\varepsilon$ is the index variation of the core with temperature at constant strain. Thus, except for the case of temperature, the calculation of the phase response of an optical fiber sensor is reduced to the mechanical problem of calculating the strains communicated by the perturbation to the fiber core, and because these strains depend strongly on the fiber jacket parameters the sensitivity can be optimized by careful design of the coating.

For example, the application of an appropriate coating to the optical fiber produces a composite structure that can provide substantially increased strain levels in the fiber core. This is accomplished for the acoustic case by elastomeric coatings with high compressibility [16], for magnetic sensing with magnetostrictive coatings [43], for the thermal case with materials having a large linear expansion coefficient [44], and for electric field sensing with piezoelectric materials [45]. Figure 21.10 shows the sensor response of acoustic, magnetic, thermal, and electric field sensors as a function of coating thickness. These curves were developed utilizing the analysis of Lagakos et al. [46], Jarzynski et al. [47], Schuetz et al. [48], and DeSouza and Mermelstein [49], respectively, in which the mechanically induced strain is calculated for each respective case. The reader should refer to these references to better understand the differences among the calculations. For example, the orientation of the fiber with respect to the field direction is important for both the magnetic and electric field sensors.

Mach-Zehnder interferometric sensors offer great potential for a variety of applications since simply altering the fiber coating can change the sensitivity to a given perturbing field. The sensitivity of the electric field sensor illustrated in Figure 21.10 is moderate; however, the potential for wideband operation is dramatic. The sensitivity of long lengths of fibers coated with magnetostrictive coatings may operate at room temperatures and yet match the sensitivities of low-temperature SQUID devices. In the case shown in Figure 21.10, the moduli of the two magnetic coating examples are approximately equal, and the large difference in sensitivity is due mainly to the difference in their piezomagnetic strain constant. In Section 21.7 the discussion will center around metallic glasses, which have not yet been demonstrated as fiber coatings for reasonable lengths of fiber due to the nature of their

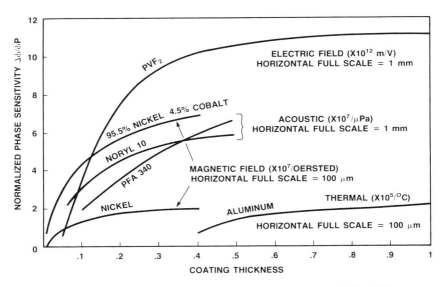

Figure 21.10 The normalized optical phase shifts as a function of outer jacket thickness on an 80-μm-outer-diameter glass fiber. The fiber response jacketed with a polyvinylidene fluoride (PVF$_2$) coating for electric field sensing, nickel and nickel cobalts for magnetic field sensing, Noryl 10 and Teflon PFA 340 for acoustic sensing, and aluminum for thermal sensing. In the case of the acoustic sensor, a 200-μm-outer-diameter silicone buffer layer was included in the calculation.

manufacture and thus may not be practical as fiber coatings. For the acoustic case (Section 21.6) the elastomeric coatings shown, when utilized with long lengths of fiber, can produce sensors with detection thresholds significantly below conventional technology. Finally, the case illustrated for a fiber coated for thermal sensitivity will provide detection thresholds of 10^{-6}°C with 1 cm of fiber. By noting the full-scale coating thicknesses of each material illustrated in Figure 21.10, it can be seen that sensors implemented with metal coating such as the magnetic or thermal sensors require much thinner coatings than the polymers utilized in the acoustic and electric field sensors. This is because of the relatively high Young's moduli of the metals compared to that of the polymers. The role of the elastic moduli on the acoustic response will be discussed further in Section 21.6.

As an example of the importance of individual sensor considerations, Figure 21.11 illustrates two cases where the $\Delta n / \Delta T_\theta$ term contributes substantially to the sensor response. The first case, Figure 21.11a, shows the thermal sensitivity of a metal-coated fiber. Below 200 Hz, the aluminum-coated fiber response is completely attributed to the $\Delta n / \Delta T_\theta$ term. Above 200 Hz, there is partial cancellation between the rapidly coupled thermally induced strain and the $\Delta n / \Delta T_\theta$ term, which is delayed by the conduction time required for the fiber surface temperature change to reach the fiber core.

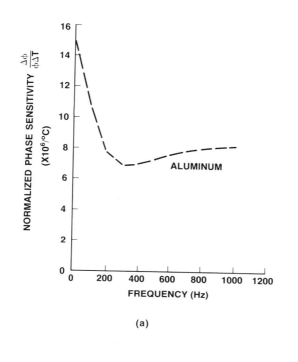

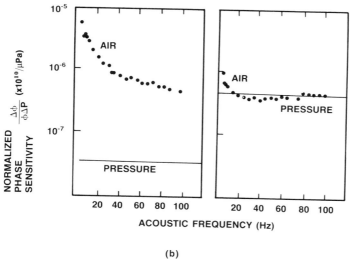

Figure 21.11 (a) The normalized thermal response of an aluminum-coated fiber as a function of frequency. (b) (Left) The normalized air acoustic sensitivity for an unjacketed fiber as a function of frequency with the calculated contribution of the pressure response. (Right) The normalized air acoustic response versus frequency for a polypropylene-jacketed fiber with the calculated contribution of the pressure response.

As the frequency is increased, the $\Delta n/\Delta T_\theta$ contribution is diminished and the fiber response approaches the level associated solely with the thermally induced strains. The directly induced index variation is also important in the acoustic response of optical fibers in air. Maurer et al. [50] measured the response of unjacketed fiber, as shown at the left in Figure 21.11b. These results clearly show the importance of the thermal contribution compared to the response due to pressure. The thermal contribution should vanish at sufficiently high frequencies [26], as is demonstrated at the right in Figure 21.11b, where the response of a polypropylene-coated fiber is plotted. Due to the low thermal conductivity of plastic, the thermal contribution does vanish at relatively low frequencies for this fiber. The reader is referred to Maurer et al. [50] and Bucaro et al. [26] for more comprehensive discussions of the air acoustic response.

21.6 ACOUSTIC SENSORS

One of the earliest interferometric sensors to be demonstrated was the acoustic sensor [51, 52], and since that time a significant amount of work has gone into optimizing the response of fibers for acoustic applications. This has involved the choice of the right elastomeric materials with respect to the dynamic behavior of their elastic moduli [46]. For fibers with typical coating thicknesses (< 1 mm), the acoustically induced strains, and therefore the acoustic sensitivity, are a complicated function of the elastic moduli. For such a composite fiber geometry, the axial stress carried by a particular layer is governed by the product of the cross-sectional area and the Young's modulus of that layer. Thus, for materials with a high Young's modulus, relatively small coating thicknesses are required to get a significant enhancement of the axial stress given that the bulk compressibility is sufficiently high to develop large compressive strains. For materials with a low Young's modulus, on the other hand, the degree to which the coating contributes to the axial strain is diminished and the glass plays the major role in determining the sensitivity. Accordingly, for typical coating thicknesses, high acoustic sensitivity requires a coating material with high Young's modulus and low bulk modulus. In Figure 21.10, the Noryl 10 and Teflon PFA 340 indicate the roles the Young's and bulk moduli play in determining the fiber sensitivity for typical fiber thicknesses. For thick coatings the acoustic response is driven by the material bulk modulus, which is lower for PFA 340 than for Noryl 10. However, for thinner coatings, the fiber acts as a composite structure, and higher Young's modulus polymer coatings such as Noryl will more effectively couple to the first term in Eq. (42), the axial strain term. Such thin coatings have the general advantage that more optical fiber and therefore higher acoustic sensitivity can be achieved in a restricted volume than with a thicker coating.

Appropriate coating of the optical fibers can also desensitize the fiber response to a given field. The sensitivity described by Eq. (42) can be minimized by designing a fiber coating that produces strains balancing each term. Such desensitization has been proposed and demonstrated, as shown in Figures 21.12a and b for the case of pressure [53]. Minimization of static temperature sensitivity has also been proposed [54]; however, unlike the pressure case, the glass compositions of the optical fiber must be modified to minimize the direct index modulation, $\Delta n/\Delta T_\theta$.

The foregoing discussion demonstrates that the phase sensitivity of the optical fiber can be determined by solving the appropriate mechanical problem in order to calculate the required strains. As the mechanical configuration varies, the sensor response is altered and must be calculated separately. Here, as an example of alternative mechanical configurations, the use of a cylindrical polymer mandrel for acoustic sensing is described. In the mandrel configuration, the optical fiber is wrapped around a cylinder of appropriate material, which will respond to the desired perturbing field. The fiber acts as a strain gauge measuring the radial strain of the mandrel, which is coupled to a phase shift in the fiber. The radial strain in the mandrel is directly related to the strains in the fiber, and the phase shift can be approximated as

$$\left(\frac{\Delta\phi}{\phi}\right)_{\text{mandrel}} = \left(1 + \frac{n^2}{2}\gamma_f(P_{11} + P_{12})\right)\frac{\Delta r}{r} \tag{43}$$

where $\Delta r/r$ is the mandrel radial strain due to the perturbing field and γ_f is the effective Poisson's ratio of the fiber. The mandrel radial strain is given by

$$e_r = e_1 = \frac{1}{E}\left[\sigma_1 - \gamma_m(\sigma_2 + \sigma_3)\right] \tag{44}$$

where E is the Young's modulus, γ_m is the Poisson's ratio of the mandrel material, and the σ_i are the stresses corresponding to the three directions $i = 1, 2, 3$.

Hydrostatic pressure conditions imply $\sigma_1 = \sigma_2 = \sigma_3 = P$, and therefore

$$e_r = e_z = -\frac{P(1 - 2\eta)}{E} = -\frac{P}{3B} \tag{45}$$

where P is the pressure, and the usual relationships between the bulk modulus B, the Young's modulus E, and Poisson's ratio have been utilized. Table 21.1 compares the calculated hydrostatic acoustic response for mandrel and typical fiber sensors for several polymer materials. The ratio of the coated fiber sensitivity to the mandrel sensitivity, $(\Delta\phi/\phi)_{\text{fiber}}/(\Delta\phi/\phi)_{\text{mandrel}}$, can be explicitly determined where the fiber coating thickness is large enough to dominate the acoustic response, allowing substitution of Eq. (45) into Eq. (42) as well as Eq. (43) and resulting in the ratio of 0.97. This ratio is higher than is indicated by Table 21.1 owing to the assumption of very thick fiber

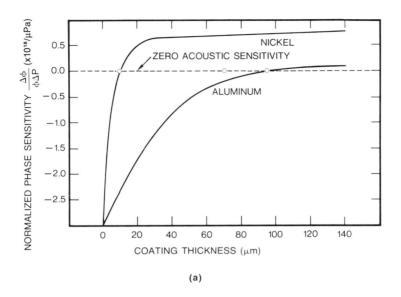

(a)

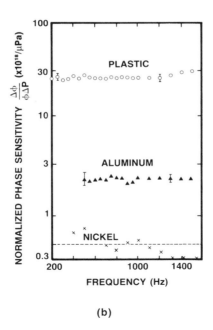

(b)

Figure 21.12 (a) The normalized acoustic response of two desensitized optical fibers as a function of coating thickness. (b) Normalized pressure sensitivity as measured for Hytrel-, aluminum-, and nickel-coated fibers. The dashed line indicates the calculated response for nickel.

TABLE 21.1. Calculated hydrostatic acoustic response for mandrel and coated fiber sensors

Material	$(\Delta\phi/\phi \; \Delta P)_{\text{fiber}}$	$(\Delta\phi/\phi \; \Delta P)_{\text{mandrel}}$	Ratio
Zytel 101 (nylon)	3.2×10^{-12}	5.0×10^{-12}	0.64
Noryl (10)	4.6×10^{-12}	7.8×10^{-12}	0.59
Polypropylene (7823)	2.9×10^{-12}	7.9×10^{-12}	0.37
Polyvinyl chloride	3.7×10^{-12}	5.6×10^{-12}	0.66
Hytrel 7296	3.1×10^{-12}	6.7×10^{-12}	0.46

jackets and as such represents a practical limit. The differential in response is due to the difference in sign of the second term, which is relatively small, in Eqs. (42) and (43).

This discussion addresses the low-frequency response of optical fiber sensors consisting of fibers jacketed with sensitive coatings and utilizes mandrel sensors to provide an example of alternative sensor designs. In the case of any optical fiber sensor, the dynamic response of the sensor must be understood, and even if only the low-frequency response is of interest, as addressed here, the position of the first mechanical resonance must be determined to ensure a uniform frequency response. If higher-frequency response is desired, then a more complete mechanical approach must be pursued. Jarzynski et al. [55], for example, have analyzed an elastomer-coated fiber in a loop configuration for acoustic sensor applications. As an example here, the phase response of a straight fiber is reviewed over a broad frequency regime.

As the spatial wavelength of the strains developed by a perturbing field approach the fiber length, a length resonance is observed beyond which the fiber can no longer dynamically respond in axial strain. Thus, the fiber response is constrained axially ($e_z = 0$) and if the spatial wavelengths of the developed radial strains is still large compared to the fiber radius, which guarantees isotropic radial strains, Eq. (42) is reduced to the fiber response of the axially constrained radially isotropic condition:

$$\left(\frac{\Delta\phi_c}{\phi}\right)_{\text{fiber}} = -\frac{n^2}{2}(P_{11} + P_{12})e_r \qquad (46)$$

Utilizing Eq. (44) with $\sigma_3 = 0$ (axially constrained), and assuming a fiber coating thickness large enough to dominate the acoustic response, the isotropic radial strains are given by

$$e_r = -\frac{P}{3B}\left(\frac{3B - E}{2E}\right) \qquad (47)$$

where the relationships between the Poisson's ratio of the fiber and the Young's and bulk moduli have again been utilized. Substituting Eq. (47) into

Eq. (46) and Eq. (45) into Eq. (42), the ratio $(\Delta\phi_c/\phi)(\Delta\phi_u/\phi) = 0.18$ for Hytrel-coated fiber is determined. This ratio is in good agreement with the Hytrel-coated fiber response, which has been more accurately determined and is plotted in Figure 21.13. The above analysis is not intended to be rigorous but attempts to illustrate simplifications that can be employed in estimating sensor performance. Readers interested in more rigorous analysis are referred to appropriate references.

Finally, as the frequency of the perturbing field is increased further and the spatial wavelength of the corresponding strains approach the fiber radius, the induced radial strain field becomes anisotropic [56] and the fiber becomes birefringent, supporting two phase velocities associated with the two principal indices of refraction n_1 and n_2.

$$\left(\frac{\Delta\phi_1}{\phi}\right)_{\text{fiber}} = -\left(\frac{n^2}{2}\right)(P_{11}e_1 + P_{12}e_1)$$

and
$$\left(\frac{\Delta\phi_2}{\phi}\right)_{\text{fiber}} = -\left(\frac{n^2}{2}\right)(P_{12}e_1 + P_{11}e_2) \tag{48}$$

The phase response of an optical fiber (Fig. 21.13) to an incident acoustic field and to a polyvinylidene fluoride (PVF$_2$) copolymer jacket in a modulator configuration [57] clearly shows the three regions of fiber response just discussed. The acoustic response of an uncoated fiber and a fiber coated with a silicone inner buffer and an outer Hytrel 7246 jacket is plotted in Figure 21.13a. In the low-frequency (unconstrained isotropic) regime characterized by Eq. (42), the e_z strain level developed in the core is substantially enhanced by the Hytrel elastomer compared to the uncoated fiber. Theory and experimental data on an uncoated fiber below 1 MHz demonstrate the axially constrained radially isotropic regime corresponding to Eq. (46). Finally, the two phase responses $\Delta\phi_1$ and $\Delta\phi_2$ associated with the principal indices of refraction are illustrated above 1 MHz [58]. Here the strains e_1 and e_2 are complicated functions of frequency and elastic and mechanical parameters. For example, the resonance peak corresponds to the $n = 2$ mode or the quadrupole resonance where two wavelengths are equal to the circumference of the fiber. The PVF$_2$ phase modulator response, shown in Figure 21.13b, again shows the unconstrained isotropic and axially constrained radially isotropic regimes, but in addition it shows experimental data through the transition between these regions. The calculation of the phase response of an optical fiber or modulator has been shown to reduce to the mechanical problem of calculating the induced strains at least at frequencies away from the transition regions. The strain calculations required to predict the phase response of a straight fiber, as illustrated in Figure 21.13, and many other configurations are readily available in the literature.

The discussion to this point has been in terms of normalized field sensitivity. An example is provided here to estimate actual sensor performance

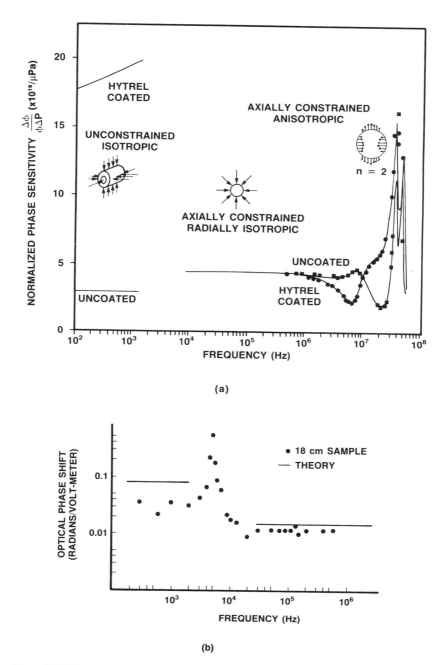

Figure 21.13 (a) The normalized acoustic response of Hytrel-coated and unjacketed fibers is shown over six decades of frequency. The fibers are 80-μm glass outer diameter with a 200-μm-outer-diameter silicone buffer layer and a Hytrel coating of 0.5-mm outer diameter. In the unconstrained isotropic regime, the large increase in response of the coating over an unjacketed fiber is clearly evident. In the axially constrained isotropic region, the response of the two fibers is similar. In the axially constrained anisotropic regime the unjacketed fiber shows a complicated response related to radial response and resonances excited in the fiber. (b) Optical phase shift induced in a PVF$_2$ fiber modulator illustrates the transition region between the unconstrained isotropic and axially constrained radially isotropic regions.

for the normalized sensitivity. From Figure 21.11, the acoustic field sensitivity $\Delta\phi/(\phi\,\Delta P) = 6.0 \times 10^{-12}$ cm^2/dyn for a 410-μm coating thickness of Teflon PFA 340. The actual phase shift of a sensor is $\Delta\phi/\Delta P = 2.3 \times 10^{-3}$ rad-cm^2/dyn, where $\phi = 2\pi n l/\lambda$ and the sensor parameters $l = 50$ m, $n = 1.458$, and $\lambda = 0.83$ μm have been assumed. Given a minimum detectable phase shift of the demodulator of $\phi_{mm} = 1 \times 10^{-6}$ rad, the threshold of pressure detection is $P_{min} = 4.3 \times 10^{-4}$ dyn/cm^2. Such a detection level compares favorably with conventional acoustic sensors.

21.7 MAGNETIC SENSORS

Two basic approaches exist for the measurement of magnetic fields with optical fibers: the Faraday rotation approach and the magnetostrictive approach. In the former approach, an external magnetic field H is applied longitudinally to a fiber to produce a rotation in the direction of linear polarization. Smith [59] demonstrated that this technique can be utilized to measure currents in the 0–1000-A range; however, the Verdet constant of most doped silica fibers is sufficiently small ($\sim 1.5 \times 10^{-2}$ min/A) that only large currents and magnetic fields can be detected by the Faraday approach. The Verdet constant can be increased by doping paramagnetic ions into optical fibers. Rare earth ions can be incorporated into silica to greatly enhance the Faraday effect, but constraints on solubility of the ions in the glass and the optical absorption introduced by the ions limit the enhancement that can be achieved in practice. Theoretical sensitivities of $\sim 10^{-4}$ gauss (G) per meter of fiber path length appear feasible. However, the Faraday approach requires both special materials and sophisticated fiber-drawing techniques to provide sensitive detection of small magnetic fields.

The alternative approach, employing magnetostrictive materials and fiber interferometry, is capable of much higher sensitivities approaching gauss per meter of fiber path length. The fundamental principle of sensor operation is associated with the measurement of the longitudinal strain produced in an optical fiber bonded to or jacketed by a magnetostrictive material. Magnetostriction is described as a change in dimension of a ferromagnetic material when it is placed along the axis of an applied magnetic field. For purposes of this treatment, only the longitudinal change is important, but volume and transverse changes also occur. The approach is totally analogous to the acoustic sensor except that a metallic jacket rather than a polymer jacket is used to stretch the waveguide. The measurement of small changes in optical path length is accomplished with a conventional Mach-Zehnder all-fiber interferometer such as that described earlier. Dandridge et al. [60] reported the first experimental results on fiber magnetic field sensors and found sensitivities as high as 10^{-6}–10^{-7} G per meter of fiber in a bulk nickel-clad fiber device operating at 1–10 kHz. Somewhat lower sensitivities were found experimentally in fibers coated with a thin nickel film or metallic glass. This is due to mechanical loading of the thin layer of magnetostrictive material by the rigid glass fiber.

To maximize the sensitivity of the sensor, it is necessary to maximize the magnetostriction, efficiently couple the induced strain to the fiber, and detect the phase shift in a low-noise interferometer. Magnetostrictive materials of interest fall into the categories of selected crystalline metals and metallic glasses. Magnetostrictive metals include iron, cobalt, and nickel and various alloys and compounds of the three elements. The magnetostrictive coefficient λ_s is given as a fractional dimensional change Δl divided by the total length l. For a polycrystalline jacket fabricated from these materials, λ_s is given by the expression

$$\bar{\lambda}_s = \frac{2\lambda_{100} + 3\lambda_{111}}{5} \tag{49}$$

Using Eq. (49), the λ_s for nickel is calculated to be 32.8×10^{-6}. For iron-nickel alloys, longitudinal magnetostriction can either be positive (expanding dimension) or negative (contracting dimensions) with 30% nickel yielding approximately zero dimensional change, 45% nickel exhibiting the highest positive magnetostriction, and pure nickel showing the largest contraction. For reasons of fabrication simplicity, corrosion resistance, and availability, pure nickel is a convenient material to use in fiber magnetostriction devices. However, its properties are greatly influenced by the presence of impurities, processing, and final heat treatment. Iron-cobalt alloys also offer large values of magnetostriction with 60–70% Co providing the maximum response for most applied fields. The highest sensitivities observed to date have used metallic glass strips of high magnetostriction to which fibers are bonded [61]. Commercially available metallic glasses produced by Allied Corporation such as alloys 2605 SC and 2605 S2 based on the FeBSi system appear to be extremely sensitive, and their sensitivity can be increased by appropriate field annealing. The dependence of the magnetostriction with the applied fields at low magnetic fields also makes metallic glasses very attractive materials in fiber magnetic sensors. The remainder of this section concerns optical fiber magnetic sensors using metallic glass as the transduction element.

A typical plot of strain versus magnetic field for a metallic glass is shown in Figure 21.14. In first-generation magnetostrictive fiber sensors, the term of interest is the first derivative of the function that relates the change in the magnetostriction to the applied time-varying magnetic field H_1 and the applied bias magnetic field H_0. The maximum sensitivity is obtained when the first derivative is a maximum. Although this approach works well for the detection of ac magnetic fields, where the interferometer noise in the frequency band of interest may be on the order of microradians, the noise at low frequencies ($\sim$rad) causes a severe limitation to the measurement of dc fields. Fortunately, the nonlinear magnetostrictive response allows a method [62] to avoid direct dc measurement of the magnetostriction. For magnetic fields

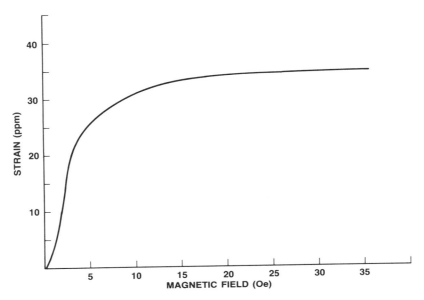

Figure 21.14 Direct current magnetostrictive strain as a function of magnetic field.

well below half the field for saturated magnetization (H_{sat}), the magnetostrictive response follows approximately a quadratic relation with respect to the magnetic field (see Fig. 21.14). Thus, the strain $\varepsilon = \Delta l/l$ is represented by

$$\frac{\Delta l}{l} = \varepsilon = CH^2 \tag{50}$$

where $C = \frac{3}{2}\lambda_s/H_A^2$, λ_s being the spontaneous magnetostrictive strain in each domain and H_A the anisotropy field. Equation (50) is a scalar equation if one assumes an isotropic case for the magnetostrictive response. If we let $H = H_{ac} + H_0$, where $H_{ac} = H_1 \cos \omega_m t$ is the ac magnetic field at the frequency ω_m, then

$$\frac{\Delta l}{l} = C[H_0 + H_{ac}]^2 = CH_0^2 + 2CH_0 H_{ac} \tag{51}$$

Thus,

$$\left(\frac{\Delta l}{l}\right)_{\omega_m} = 2CH_0 H_1 \tag{52}$$

where $(\Delta l/l)_{\omega_m}$ is the strain component oscillating at ω_m. Consequently, measurement of $(\Delta l/l)_{\omega_m}$ not only indicates the magnitude of H_1 if H_0 is known but also indicates the magnitude of H_0 if H_1 is known. Unlike the direct dc measurement of the magnetostriction, which is sensitive to dc thermal phase shifts, $(\Delta l/l)_{\omega_m}$ can be measured at frequencies where the interferometer noise floor is on the order of microradians. It should, however, be emphasized that Eqs. (49)–(52) are valid only in the range $H \ll H_{sat}/2$.

A typical optical fiber magnetometer is shown in Figure 21.15. The active length of the sensor (l) was 10 cm; the sensor constructed was of Metglas 2605 S2 and had a minimum detectable field of $\sim \gamma/\mathrm{Hz}^{1/2}$ (i.e., 10^{-5} Oe/Hz$^{1/2}$). To maximize the sensitivity of the device, λ_s was increased and H_A decreased by field annealing (5 min at 420°C in a 1200-Oe field [63]), and coupling of the Metglas strain to the fiber was maximized by using minimal amounts of adhesive to bond the fiber to the Metglas sample [64]. The sensor consisted of a pigtailed laser diode (LD) (stabilized with respect to the rear-mounted integral detector), fused couplers (FC), and two detectors (D1 and D2) to measure the interference signal. The reference arm of the interferometer was bonded to a piezoelectric (PZT) cylinder. The PZT is used in an active stabilization scheme to maintain the interferometer at the quadrature point (high-gain active homodyne demodulation). Also shown in Figure 21.15 are the signal-processing electronics (SPE) that form part of the feedback loop and also provide outputs to phase-sensitive detectors to measure both the magnetic signal at ω_m and a calibration signal applied to the PZT at ω'.

A signal corresponding to the dc magnetic field H_{dc} can be extracted by running the interferometer in either of two magnetic loop configurations, shown in Figure 21.16. In the open magnetic loop configuration (Fig. 21.16a), the output from the PSD tuned to the magnetic dither frequency ω_m was used as a direct measure of the dc magnetic field. That is, the output follows the magnetostriction curve shown in Figure 21.14, and in the quadratic region the output of the PSD is given by

$$(V)_{\omega_m} = \beta H_0 \qquad (53)$$

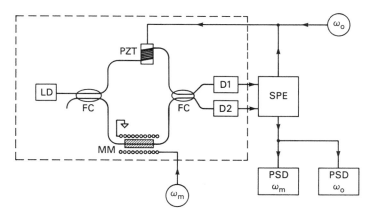

Figure 21.15 Block diagram of an all-fiber Mach-Zehnder interferometer including laser diode (LD), fused couplers (FC), optical detectors (D1, D2), magnetostrictive material (MM), piezo-electric cylinder phase shifter (PZT), signal-processing electronics (SPE), and phase-sensitive detectors (PSD).

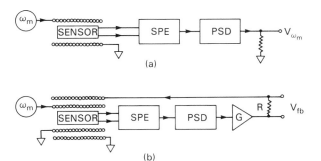

Figure 21.16 Two methods of extracting a voltage proportional to the DC magnetic field. (a) Open-loop configuration; (b) closed-loop configuration.

where

$$\beta = (1.1 \times 10^7)(gl_f\eta_e)(2CH_1) \quad [\text{V/Oe}] \tag{54}$$

g is the voltage-to-phase shift scale factor of the interferometer, l_f is the active fiber length, η_e is the efficiency with which strain in the ribbon is transferred to the fiber core, and the numerical factor comes from the $(\Delta n)kl_f$ term of the change in optical path length $\Delta(nkl_f)$ ($n = $ fiber core refractive index, $k = $ wavenumber in vacuum). In the closed magnetic loop configuration of Figure 21.16 [65, 65a], a nulling or feedback magnetic field H_{fb} is produced by the feedback solenoid to cancel the external dc field. Using standard feedback analysis, it can be shown that

$$H_{fb} = \frac{-G\beta}{1 + G\beta} H_{dc}H_0 \tag{55}$$

where G is the gain of the feedback loop. In the high gain limit $G\beta \gg 1$, Eq. (55) reduces to

$$H_{fb} = -H_0 \tag{56}$$

Hence, the current needed to produce H_{fb} in the feedback solenoid can be used as a direct measure of external field H_{dc}. The voltage across the feedback resistor R_f in Figure 21.17b is

$$V_{fb} = \frac{10^3}{4\pi} \left(\frac{R}{N}\right) H_0 \tag{57}$$

where N is the number of turns per meter of length of the feedback solenoid, R_f is in ohms, and H_0 is in oersteds. Equations (53) and (57) define the output voltages for the open-loop configuration and high-gain, closed-loop configuration, respectively.

It is clear that time-dependent changes in any of the factors contained in β will lead directly to long-term drift in the open-loop voltage V_{ω_m}. Some possible sources of drift include instability of the dither amplitude H_{ac}, aging of the magnetostrictive material C and the adhesive bond η_e, thermal strains l_f, and residual phase fluctuations of the optical interferometer g. In addition, hysteresis in the magnetostrictive response of the Metglas will cause C to change depending on the magnetic history of the metallic glass.

Drifts of V_{fb} can result, in lowest order, only from changes in either the feedback resistance R_f or the solenoid turns per length N. Of course, this assumes that any changes in $G\beta$ are small enough that the conditions $G\beta \gg 1$ remain satisfied, and it is therefore important to ensure that changes in the open loop are within the closed-loop locking capability.

Typical stability for the system described above is shown in Figure 21.17a, and the noise is shown in Figure 21.17b [66]. Clearly, optical fiber magnetometers with $1-0.1$ γ sensitivities and with frequency ranges from dc to

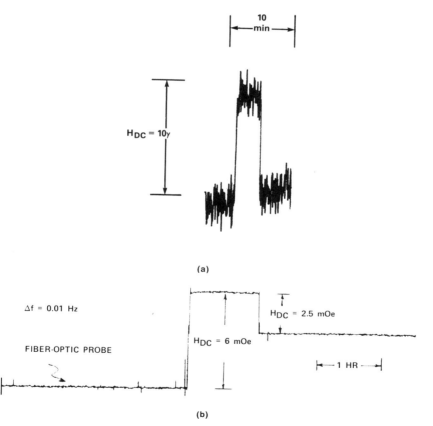

Figure 21.17 (a) Stability of the closed-loop optical fiber magnetometer. (b) Response of the open-loop magnetometer to a 10-γ (10^{-1}-Oe) step change in dc magnetic field (1-Hz bandwidth).

kilohertz are viable. However, sensitivities of $\sim 10^{-5}$ γ appear to be theoretically possible (assuming $l_f \approx 1$ km). Clearly, issues such as interferometer noise floor, demodulator linearity, sample loading, and Barkhausen noise will have to be addressed.

21.8 OPTICAL FIBER GYROSCOPE

The principle of rotation sensing with an optical fiber interferometer is based on a very small, nonreciprocal effect known as the Sagnac effect [67]. The basic Sagnac optical fiber interferometer is illustrated in Figure 21.18. Light from a laser is incident on a beamsplitter, which divides the beam into two equal parts. The beams are then coupled into each end of an optical fiber coil of length l and radius R. The optical beams traverse the coil in clockwise and counterclockwise directions. Upon exiting the fiber ends, they are recombined by the beamsplitter to form an interference pattern at the output. If the coil is rotated with an angular velocity Ω, then the phase difference between the output beams due to rotation is

$$2\phi_s = \frac{4\pi R l \Omega}{\lambda_0 c} \tag{58}$$

where λ_0 is the vacuum wavelength of the laser, c is the velocity of light, and $2\phi_s$ is the Sagnac phase shift. Equation (58) has been derived in several reviews [68, 69].

The size of the Sagnac effect can be estimated. For a fiber 500 m long wound on a 10-cm-radius coil with a source wavelength of 1 μm, a rotation rate of 0.1°/hr corresponds to a Sagnac phase shift of ~ 1 μrad. This very

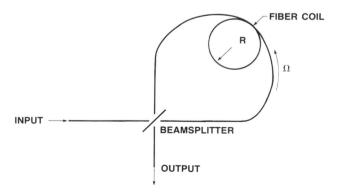

Figure 21.18 The basic setup of the optical fiber Sagnac interferometer.

small effect is similar to those encountered in other optical fiber sensing applications, such as acoustic and magnetic field measurements.

A Sagnac interferometer measures nonreciprocal effects. In other words, the sign of the effect, in this case a phase shift, depends on the direction of propagation of the light. After the beams traverse the fiber loop in opposite directions, the reciprocal phase shifts resulting from linear propagation cancel out upon interference, and the nonreciprocal Sagnac phase shifts add together. Needless to say, other known nonreciprocal effects, such as the Faraday effect and the nonlinear Kerr effect, may also provide nonreciprocal phase shifts that a Sagnac interferometer will measure.

To derive the Sagnac interferometer output equation, a nonreciprocal phase shift ϕ_{NR} is artificially introduced in the loop, as shown in Figure 21.19a. As indicated by the directional arrow in the figure, this phase shift is assumed to be experienced by the counterclockwise beam but not by the clockwise beam. The clockwise and counterclockwise beams then experience phase shifts of $\phi_s + kl$ and $\phi_{NR} - \phi_s + kl$, respectively, where kl is the component due to propagation through the fiber length l and $k = 2\pi n/\lambda_0$ is the optical propagation constant in the fiber of index n. For input intensity I_0, the input field is $E_1 = I_0^{1/2}$. The fiber inputs are calculated from the transfer matrix of Figure 21.20, which shows the amplitude transfer matrix for a -3-dB coupler. A consequence of this matrix is that the two equal outputs are $90°$ out of phase (see Fig. 21.20). This fact causes a sensitivity or quadrature problem.

After traversing the loop, the fiber outputs are

$$E_2 = \frac{-i}{\sqrt{2}} I_0^{1/2} \exp\left[i(\phi_{NR} - \phi_s + kl)\right] \qquad (59)$$

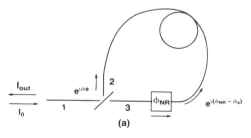

(a)

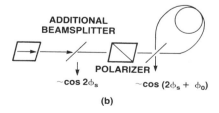

(b)

Figure 21.19 (a) The artificially introduced nonreciprocal phase shift (ϕ_{NR}) is used to derive the output equation for the basic Sagnac interferometer. (b) A setup using a polarizer to ensure that the counterpropagating beams possess the same polarization state. An additional beamsplitter is used before the polarizer. Output from the original Sagnac interferometer setup (after the polarizer) exhibits an additional phase bias, ϕ_0.

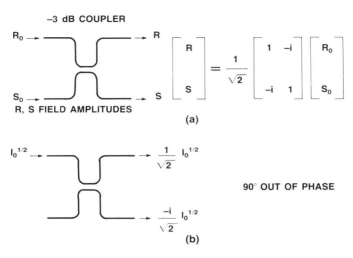

Figure 21.20 The amplitude transfer matrix for a -3-dB fiber coupler, which serves as the beamsplitter in the Sagnac interferometer.

and

$$E_3 = \frac{1}{\sqrt{2}} I_0^{1/2} \exp\left[i(\phi_s + kl)\right] \tag{60}$$

After again applying the coupler transfer matrix (see Fig. 21.20), the interferometer output at port 1 is

$$I_{\text{out}} = \tfrac{1}{2}I_0[1 + \cos(2\phi_s - \phi_{\text{NR}})] \tag{61}$$

For $\phi_{\text{NR}} = 0$ we see that for a small Sagnac phase shift $2\phi_s$ the derivative of Eq. (61), and thus the sensitivity of a small-signal measurement, is zero. This is the quadrature problem mentioned above. However, for $\phi_{\text{NR}} = \pi/2$ we obtain the output

$$I_{\text{out}} = \tfrac{1}{2}I_0(1 + \sin 2\phi_s) \tag{62}$$

Now the output curve is shifted $90°$, and the small-signal sensitivity is maximized. The problem becomes one of implementation of a $90°$, nonreciprocal phase shift. This can be done directly with a Faraday cell or by asymmetrically inducing a time-varying, but reciprocal, phase shift in the fiber loop or by changing the geometric configuration of the input coupler.

21.8.1 Nonreciprocal Noise Sources

The Faraday effect is a magnetically induced rotation of the optical polarization and is nonreciprocal. For a magnetic field H applied parallel to a fiber section of length l, the Faraday rotation of the linearly polarized optical polarization in the fiber is given by

$$\theta_F = V(\mathbf{H} \cdot \mathbf{l}) \tag{63}$$

where θ_F is the polarization rotation angle and V is the Verdet constant for the glass fiber. For a fiber loop in a constant magnetic field, the line integral $\int \mathbf{H} \cdot \mathbf{dl}$ around the loop is zero and the net optical rotation should also be zero. However, if the fiber loop is in a field gradient, or if the optical polarization changes its orientation around the loop, the Faraday effect can result in a nonreciprocal phase shift that the Sagnac interferometer will measure [70]. Since this effect is indistinguishable from a Sagnac signal, it results in noise and zero shifts in the gyroscope output. Both magnetic shielding and polarization-holding fibers have been used to reduce Faraday effects in fiber gyroscopes.

An additional nonreciprocal effect is the Kerr effect, which is an optical-intensity-induced nonreciprocity. At high optical intensities, the propagation constants for the counterpropagating beams become intensity-dependent. This is a nonlinear optical effect related to four-wave mixing. In particular, when the counterpropagating beams have unequal intensities, the propagation constants become unequal. We can write [71]

$$k_+ - k_- = \text{constant} (W_+ - W_-) \tag{64}$$

where $+$ and $-$ refer to clockwise and counterclockwise beams, k is the propagation constant, and W is the optical power. The nonlinear dependence of k on the electric field for the $+$ beam is given by

$$k_+^2 = \frac{\omega^2}{c^2} \left(n^2 + 4\pi \frac{P_+^{\text{NL}}}{E_+} \right) \tag{65}$$

where

$$P_+^{\text{NL}} = 3\chi(2|E_-|^2 + |E_+|^2)E_+ \tag{66}$$

Here ω is the optical frequency, P^{NL} is the nonlinear polarization, E is the electric field associated with the optical beam, and χ is the nonlinear susceptibility. As with the Faraday effect, the Kerr effect can cause a nonreciprocal phase shift indistinguishable from the Sagnac effect. It can be reduced in several ways. One way is to make $P_+ = P_-$, which is impractical in practice; another way is to use a laser or superluminescent diode source with a broad frequency spectrum. When summed over the wavelength components of such a source, the Kerr-induced phase shift averages to zero.

Successful fiber gyroscope operation assumes that the Sagnac effect is the only nonreciprocal effect present. In addition to the Faraday and Kerr effects, there are difficulties with fiber Sagnac interferometers that can result in nonreciprocal operation. These difficulties arise from the input coupler and the nature of single-mode fiber.

The basic tenet of reciprocal gyroscope operation is that the counter-propagating beams should possess identical optical polarization states and should travel a common path. With reference to the Sagnac beamsplitter, this means that the beams should enter and exit the beamsplitter through a common port [72]. The polarization state can be specified by a polarizer placed before the beamsplitter. This approach will require the setup shown in Figure 21.19b, which has an additional beamsplitter placed before the polarizer.

The setup with a polarizer provides the output Sagnac signal $2\phi_s$. Mathematical analysis shows that the signal obtained from the unused port of the original Sagnac beamsplitter (or coupler) has an additional phase bias or zero shift ϕ_0 added to the Sagnac signal $2\phi_s$. This additional phase bias depends on the detailed properties of the beamsplitter and can be environmentally sensitive. All high-sensitivity fiber gyroscopes today use this configuration (see Figure 21.19b).

Another source of nonreciprocal operation of a fiber gyroscope is polarization mode coupling in the fiber. A single-mode fiber actually supports two polarization modes that have very similar transverse mode structures but are polarized at 90° to each other. In ordinary single-mode fiber, slight ellipticities in the core cause these polarization modes to have slightly different propagation constants. Coupling between the polarization modes is caused by fiber bends or jacket pressure and is very sensitive to thermal and acoustic fluctuations.

Nonreciprocal gyroscope operation results when one beam travels one path between the polarization modes and the counterrotating beam travels a different one. Because of the rapid temporal fluctuations of the mode-coupling centers, the nonreciprocal output of the Sagnac interferometer due to this effect appears as noise and is referred to as polarization noise. The most straightforward solution to this problem is achieved by using high-birefringence polarization-maintaining fiber, which maintains a linear polarization state.

The discovery of Rayleigh backscattering as a noise source [73] was an important step in the development of optical fiber gyroscopes because its elimination lowered gyroscope sensitivities from the degree-per-second range to the degree-per-hour range. Rayleigh backscattering is a phenomenon that occurs when a light beam traverses any optical material. Light is reflected from microscopic density fluctuations in the material, and the net back reflected signal is the sum of the components from each segment of the fiber. Since this is a coherent summation and since the component phases are again influenced by thermal and acoustic perturbations along the fiber, the backscattered signal is normally a rapidly time-fluctuating quantity.

The problem for fiber gyroscopes is that the Rayleigh backscattered component mixes with the counterpropagating beam, and after interference it is indistinguishable from a Sagnac signal. With the He-Ne sources that were

TABLE 21.2. Contributions to gyroscope noise and nonreciprocal behavior

Source	Solution
Faraday effect	Polarization-holding fiber, magnetic shielding
Kerr effect	Equal-power counterrotating beams, broadband source
Nonreciprocal operation	Proper interferometer configuration, polarizer
Polarization noise	Polarization holding fiber, broadband source
Rayleigh backscattering	Broadband source

once customarily used in fiber gyroscopes, Rayleigh backscattering was a very large noise source.

Since Rayleigh backscattering is a coherent effect, the best way to reduce it is to reduce the coherence of the laser. This was done successfully by using either multimode GaAlAs laser diodes or superluminescent diodes [74]. The latter device provides gain like a laser but has antireflection-coated facets to avoid Fabry-Perot feedback. It ideally provides broadband emission over its entire spontaneous emission spectrum. The point of using a limited coherence source is that such a source has adequate coherence to measure the very small Sagnac phase shifts due to rotation, but its limited coherence reduces the magnitude of many noise sources, such as Rayleigh scattering, by moving the noise power to the dc part of the spectrum.

Table 21.2 summarizes the solutions to the noise sources and nonreciprocal effects discussed here.

21.8.2 Experimental Performance

Following the procedure leading to Eqs. (1)–(3), we can compute the quantum detection limit (shot noise) for the optical fiber gyroscope. If W is the average optical power on the detector, the mean square shot noise current is

$$i_N = \left(2e^2 W\eta \, \frac{\Delta f}{hv} \right)^{1/2} \qquad (67)$$

where the symbols are defined as in Eq. (2)—e is the electronic charge, η is the detector quantum efficiency, hv is the photon energy, and Δf is the system bandwidth. The signal current in the detector is

$$i_s = \frac{2\phi_s W\eta_e}{hv} \qquad (68)$$

where $2\phi_s$ is the Sagnac phase shift of Eq. (58). The minimum detectable rotation rate Ω_N in the presence of shot noise is obtained by equating signal and noise powers, which after using Eq. (58) yields

$$\Omega_N = \frac{\lambda_0 c}{2\pi R l} \left(\frac{hv \, \Delta f}{2\eta W} \right)^{1/2} \qquad (69)$$

Since shot noise has a white noise spectrum, we obtain a random drift co-efficient by dividing each side of Eq. (69) by the square root of bandwidth.

$$\frac{\Omega_N}{(\Delta f)^{1/2}} = \frac{\lambda_0 c}{2\pi R l} \left(\frac{h\nu}{2\eta W}\right)^{1/2} \tag{70}$$

Random drift as given by Eq. (70) is usually quoted in units of degrees/$\sqrt{\text{hr}}$, which expresses the fact that in a random process, system errors (in this case position) grow with the square root of time. When the random drift coefficient, as given by Eq. (70), is multiplied by the square root of the actual observation time, the position error in degrees is obtained.

Equation (70) is plotted versus fiber length for several values of coil radius, source wavelength, and input power to the fiber loop W_0 in Figure 21.21. The fiber attenuation coefficient α assumed for each wavelength is also shown in the figure. As fiber attenuation decreases with increasing wavelength, lower values of the random drift coefficient are achievable with longer fiber lengths.

The current status of high-sensitivity rotation detection in practice is illustrated by a fiber gyroscope demonstrated at the Naval Research Laboratory [17]. This device, diagramed in Figure 21.22, used polarization-holding fiber in the coil and Sagnac coupler. The fiber length was 430 m and it was wound on a 16-cm-radius drum.

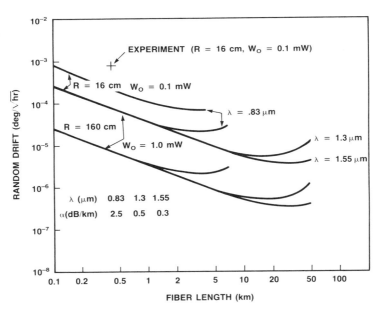

Figure 21.21 Calculated values of random drift coefficient versus fiber length for the wavelengths and fiber loses shown. R is the coil radius and W_0 is the optical power input to the fiber coil. The plus ($+$) sign is the experimental value.

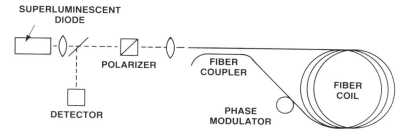

Figure 21.22 Gyroscope demonstrated at Naval Research Laboratory using polarization-holding fiber and a superluminescent diode.

The gyroscope noise level was measured under zero rotation conditions for signal-averaging times (time constant TC) of 1.25 and 40 sec. In this case, bandwidth and time constant are related by $\Delta f = 1/8TC$. The rms noise in the resulting traces is plotted versus time constant in Figure 21.23, showing the inverse square root dependence on TC that is characteristic of white noise. Also shown in Figure 21.23 is an experimentally determined detector noise limit, which is within a factor of 2 of the gyroscope noise, and a calculated minimum detectable rotation rate from Eq. (69), which is within a factor of 4 of the gyroscope noise. The corresponding experimental random drift coefficient is shown in Figure 21.21. This experiment demonstrates that the fiber gyroscope baseline stability is very cose to the fundamental limits imposed by thermal and shot noise in the detector and is also very close to the calculated theoretical limit imposed by shot noise. This and other experiments have shown, at least in a laboratory environment, that the optical fiber

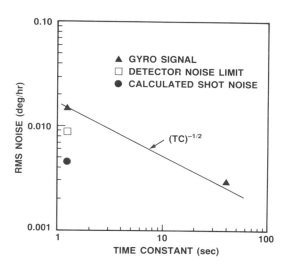

Figure 21.23 A plot of rms noise against time constant for the NRL gyroscope at zero rotation conditions. Also shown is the experimentally determined detector noise and the calculated limit due to shot noise.

gyroscope can operate at the level required for navigational quality strap-down inertial systems.

21.8.3 Quadrature and Dynamic Range

The most successful approach to forcing a fiber gyroscope to operate at the quadrature or maximum sensitivity point has been to introduce a time-varying reciprocal phase shift at an asymmetric location of the coil [72] (i.e., at one end). In the time domain, such a device has a nonreciprocal effect. This procedure was followed in the experiment described above as indicated by the phase modulator shown in Figure 21.22. The phase modulator is made by simply wrapping the fiber around a piezoelectric cylinder, which changes the fiber length as its radius is electrically varied. We assume phase modulation $\phi(t) = \phi_m \sin \omega_0 t$, where ω_0 is the modulation frequency. The fiber loop transit time is $\tau = nl/c$, where n is the fiber index. Then the fiber output electric fields are

$$E_2 \approx e^{-i\phi_s} \exp\left[i\phi_m \sin \omega_0(t - \tau)\right] \tag{71}$$

$$E_3 \approx e^{i\phi_s} \exp\left(i\phi_m \sin \omega_0 t\right) \tag{72}$$

and the output intensity is

$$I_{\text{out}} = \frac{I_0}{2}\left\{1 + \cos\left[2\phi_s + 2\phi_m \sin \frac{\omega_0 \tau}{2} \cos \omega_0\left(t - \frac{\tau}{2}\right)\right]\right\} \tag{73}$$

If we expand to obtain the component of output intensity at ω_0 we obtain

$$I(\omega_0) \approx I_0 \sin 2\phi_s \, J_1\left(2\phi_m \sin \frac{\omega_0 \tau}{2}\right) \sin \omega_0 t \tag{74}$$

where J_1 is a first-order Bessel function. This expression is proportional to $\sin 2\phi_s$ as we require.

Several attempts have been made to deal with the quadrature problem by changing the geometry of the input at the Sagnac beamsplitter. As noted before, this problem arises because for a single input, the outputs of a fiber coupler are 90° out of phase. One approach has been to divide the input beam with an additional beamsplitter and inject the inputs into each of the input ports of the Sagnac beamsplitter [75]. With an appropriate phase shift between the two inputs, the Sagnac beamsplitter outputs can be forced to have the same phase, and the gyroscope operates at quadrature. A similar result is achieved by using a coupler with three input and three output ports, which can be made by achieving coupling between three fibers instead of two [76]. If one fiber is used as an input and the other two are used as out-puts, the output phases are, by symmetry, identical, and the gyroscope operates at its maximum sensitivity point. Although these geometric concepts are appealing, they appear to violate the reciprocity condition discussed above, which requires a common input and output port. Presumably, as a consequence, these types of gyroscopes have not demonstrated performance equal to those which use the configuration of Figure 21.22.

The most serious limitation to the use of a passive fiber gyroscope to measure rotation is the limitation on linear dynamic range caused by an output proportional to $\sin 2\phi_s$. At large rotation rates the sine function is no longer linear, and, for $2\phi_s$ values of about 90°, sensitivity to changes in rotation vanishes. The problem is illustrated in Figure 21.24, where the calculated output of the gyroscope described above is shown. Between a noise floor of 0.1°/hr (TC = 1.25 s) and a linearity limit of 100 ppm, only about four decades of linear dynamic range are obtained. For rotations above 10°/s, the output is very nonlinear. Since most applications require five to six decades of linear dynamic range, up to at least 100°/s, this difficulty must be overcome. Two strategies for solving this problem are signal nulling and signal processing.

In the signal-nulling approach, an output phase shift is artificially generated of equal magnitude but opposite sign to the Sagnac phase shift so that the gyroscope is always operating at its zero point, where it is linear. The rotation rate is usually read from the nulling signal required. The dynamic range is limited only by the ability to generate the phase nulling signal. The most popular implementation of this approach is the use of an acoustooptic Bragg cell in the fiber loop, which frequency-shifts the light passing through it in one direction relative to the counterpropagating beam [77, 78]. This

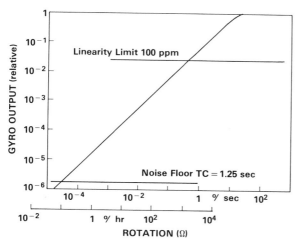

Figure 21.24 The calculated NRL gyroscope output for high rotation rates. Total linear dynamic range performance is limited to about four decades.

frequency shift is seen as a phase shift in the output, which is used to null the total output signal. A feedback loop from the detector is used to drive the Bragg cell and also provide a frequency output signal in which frequency is proportional to rotation.

The phase-nulling gyroscope effectively simulates a ring laser gyroscope, in that a frequency difference between counterpropagating beams Δf_b is proportional to rotation. This is expressed by the output equation

$$\Delta f_b = \frac{2R}{n\lambda_0} \Omega \tag{75}$$

An unfortunate difference between Eqs. (75) and (58) is that in Eq. (75) the fiber index of refraction n appears explicitly, whereas previously it did not. This fact causes thermal problems due to the temperature dependence of n and may make it more difficult for this type of gyroscope to operate stably over a broad temperature range.

Another approach to the linear dynamic range problem is to apply some type of signal processing to linearize the sine output. This can be done directly by inverting the sine output, or in some cases both $\sin 2\phi_s$ and $\cos 2\phi_s$ signals can be obtained and divided to obtain $\tan 2\phi_s$. An arctan operation can then be applied. Other approaches to obtain a direct Sagnac phase output are also being explored. The accuracy and speed with which these processes can be carried out is of great interest because the addition of extra optical components to the gyroscope is avoided.

21.8.4 Packaging

Nearly all gyroscope applications have relatively stringent size requirements ranging from volumes of 1 ft³ in airplanes to diameters of 2–3 in. in missiles. This makes packaging considerations important in any engineered device. There are three basic approaches under development: all-fiber, bulk optics, and integrated optics. The all-fiber approach, pioneered at Stanford University [79], maintains the optical beam in fiber all the way from the laser to the detector. It requires fiber components such as couplers, a polarizer, and a phase shifter. Its great advantage is that it has no optical alignments that can be degraded by vibration or thermal cycling.

In the bulk optic approach, bulk optical beamsplitters, polarizers, and, in some cases, Bragg cells have been used. The beam must then be focused into the fiber ends by means of lenses. This approach is technically difficult, because alignment must be achieved and maintained. However, it avoids the requirements for the sophisticated fiber components required in the all-fiber approach.

The final concept attempts to employ the technique of integrated optics, which guides light in channel waveguides on a small planar chip to replace the entire optical circuit except for the fiber itself. This is a demanding concept because several different active and passive devices are required. The material system generally chosen for this application has been titanium (Ti) diffused into lithium niobate ($LiNbO_3$) to form the higher-index optical waveguide. This system also provides electrical control of the guided light through electrically induced optical index changes (the electrooptic effect). At the GaAlAs wavelengths (0.83 μm) generally employed to date, this material has been plagued with optically induced index changes (the photorefractive effect), which leads to unstable behavior. However, at longer wavelengths of 1.3 μm and beyond, this effect becomes negligible. Overall throughput losses have also generally been higher than what can be achieved with other approaches, although experience with telecommunications devices indicates that excess losses can also be overcome.

Approaches to optical fiber gyroscope packaging will undoubtedly continue to evolve as development proceeds and components are improved in performance and reduced in size.

21.9 THE FUTURE FOR OPTICAL FIBER SENSORS

The development of optical fiber sensors has been pursued in earnest for less than a decade, and in that time laboratory models of many sensor types have demonstrated extremely competitive performance. At this point, these sensors appear to have found unique applications in instrumentation where extreme sensitivities, hostile environment operation, and/or probe compactness are principal prerequisites to their use.

For system controls, fiber sensors appear ideally suited for machinery and process monitoring. Although some machine manufacturers already offer optical fiber sensors as options, wide utilization awaits definitive demonstrations of their advantages. This is expected to be realized when fiber sensors are integrated with optical fiber telemetry to provide a "dielectric" control system that can operate in particularly noisy or hostile environments. Power plants and heavy machinery manufacturing facilities appear to be prime candidates for early applications. Additional opportunities exist in aircraft and ship flight and machinery control systems. Optical telemetry also provides unique capabilities to remotely sense hazardous environments with the required sensitivities. Detection of explosive gas accumulation in inaccessible and/or remote locations is easily accomplished with optical fibers while at the same time eliminating the dangers of sparking from electrical sensor or telemetry shorts.

Medical instrumentation based on optical fiber sensors has been developed and has begun to find clinical and laboratory use. While no explosion in market development is yet predicted in this area, it is generally realized that fiber sensors are ideally suited for invasive probing because of their small size and biochemical inertness. These sensors presently offer competitive capabilities and when further developed should be as commonplace as endoscopes.

The military is examining the use of fiber sensors and has actively pursued the development of several sensor types. Control sensors and monitor sensors (damage control, fuel status, etc.) are being evaluated and are expected to find application. The more sensitive surveillance and navigation sensors are still undergoing development and field testing.

In the latter case, there are many reasons why optical fiber gyroscopes are attracting substantial current interest. Fiber gyroscopes are all "solid-state" and have no moving parts, which implies reduced maintenance schedules compared to present-day spinning mass gyroscopes. Optical fiber gyroscopes appear to offer better sensitivity performance than the corresponding theoretical limits for ring laser gyroscopes, while avoiding some of the problems that have plagued ring laser development, such as optical lock-in and requirements for precision block and high-quality mirror fabrication. The last point leads to what should be the ultimate selling point for fiber gyroscopes and for other military sensor systems as well. They are generally constructed from inexpensive components and have the potential to be relatively inexpensive devices.

Considering the technical progress of the past several years, along with the appearance of so many important applications for fiber sensors, one can be certain that the use of fiber sensors will grow, first in specialty applications where some unique fiber sensor attribute makes them the device of choice and then eventually in common, widespread functions when the marriage of optical fiber sensors and telemetry is fully realized.

Acknowledgments

A significant portion of this chapter is based on work carried out at the Naval Research Laboratory. We wish to thank our colleagues who were participants in this research and who graciously made their materials available for this chapter. Particular thanks are extended to F. Bucholtz, N. Lagakos, G. H. Sigel, Jr., J. Jarzynski, C. A. Villarruel, K. P. Koo, A. B. Tveten, and R. P. Moeller.

REFERENCES

1. T. G. Giallorenzi, T. A. Bucaro, A. Dandridge, G. H. Sigel, J. A. Cole, S. C. Rashleigh and R. G. Priest, "Optical fiber sensor technology," *IEEE J. Quantum Electron.*, **QE-18**: 626–665 (1982).
2. A. L. Harmer, "Fiber optic sensors for industrial applications," *Proc. 2nd Opt. Fiber Sensor Conf.*, 1984, **17**: 17–22.

3. D. H. McMahon, A. R. Nelson, and W. B. Spellman, "Fiber optic transducers," *IEEE Spectrum*, 24–29 (1981).

4. T. C. Rozzel and C. C. Johnson, "Liquid crystal fiber optic RF probes," *J. Micro. Power*, **10**: 55–57 (1975).

5. T. Samulski and P. N. Shrivastova, "Photoluminescent thermometer probes: temperature measurements in microwave fields," *Science*, **208**: 193–196 (1980).

6. R. O. Stanton, "Digital optical transducers for helicopter flight control systems," *SPIE*, **412**: 122–129 (1983).

7. W. B. Spellman, Jr., and D. H. McMahon, "Schlierin multimode fiber-optic hydrophone," *Appl. Phys. Lett.*, **37**: 145–147 (1980).

8. J. L. Peterson, S. R. Goldstein and R. V. Fitzgerald, "Fiber optic pH probe for physiological use," *Anal. Chem.*, **52**: 864 (1980).

9. G. G. Vurek, J. L. Peterson, S. W. Goldstein and J. W. Severinghaus, "Fiber optic PCO_2 probe," *Proc. Fed. Am. Soc. Exp. Biol.*, **41**: 1484 (1982).

10. B. D. Evans, G. H. Sigel, J. B. Longworthy and B. J. Faraday. "The fiber optic dosimeter in the navigational technology satellite 2," *IEEE Trans. Nucl. Sci.*, **NS-25**: 1619–1624 (1978).

11. K. Spencer, M. D. Singh, H. Schulte, and H. J. Boehnel, "Experimental investigations on fiber optic liquid level sensors and refractometers," *Proc. 1st Opt. Fiber Sensor Conf.*, 1983, pp. 96–99.

12. M. Nagi, M. Shimizer, and N. Ohgi, "Sensitive liquid sensor for long distance leak detection," *Proc. 2nd Opt. Fiber Sensor Conf.*, 1984, pp. 207–210.

13. P. Extance, G. D. Pitt, B. J. Scott, and M. Verulls, "Turbidity monitoring using fiber optics," *Proc. 1st Opt. Fiber Sensor Conf.*, 1983, pp. 109–111.

14. J. N. Fields and J. H. Cole, "Fiber microbend acoustic sensor," *Appl. Opt.*, **19**: 3265–3267 (1980).

15. N. Lagakos, P. Macedo, T. Litovitz, R. Mohr, and R. Merster, "Fiber optic displacement sensor," in *Physics of Fibers* [*Advances in Ceramics*, Vol. 2, The American Ceramic Society, B. Bendow and S. S. Mitra (Eds.)], Columbus, Ohio, 1979, pp. 322–324.

16. J. A. Bucaro and T. R. Hickman, "Measurement of sensitivity of optical fibers for acoustic detection," *Appl. Opt.*, **18**, 938–940 (1979).

17. W. K. Burns, R. P. Moeller, C. A. Villarrue, and M. Abebe, "Fiber-optic gyroscope with polarization holding fiber," *Opt. Lett.*, **8**: 540–542 (1983).

18. H. C. Lefevre, R. A. Borgh, and H. J. Shaw, "All-fiber gyroscope with inertial-navigation short term sensitivity," *Opt. Lett.*, **7**: 454–457 (1982).

19. J. L. Davis and S. Ezekiel, "Closed loop low noise fiber-optic rotation sensor," *Opt. Lett.*, **6**: 505–507 (1981).

20. E. Udd and R. Cahill, "Phase-nulling fiber optic gyro development review," *SPIE*, **321**: 114–121 (1982).

21. W. G. Spellman, Jr., and R. L. Gravel, "Moving fiber optic hydrophone," *Opt. Lett.*, **5**: 30–31 (1980).

22. D. Gloge, "Optical power flow in multimode fibers," *Bell Syst. Tech. J.*, **51**: 1767–1783 (1972).

22a. D. Marcuse, *Theory of Dielectric Waveguides*, Academic, New York, 1974.

23. D. Gloge and E. A. Marcatili, "Multimode theory of graded core fibers," *Bell Syst. Tech. J.*, **52**: 1563–1578 (1973).

24. J. N. Fields, C. K. Asawa, C. P. Smith, and R. J. Morrison, "Fiber optic hydrophone," in *Physics of Fibers* [*Advances in Ceramics*, Vol. 2, The American Ceramic Society, B. Bendow and S. S. Mitra (Eds.)], Columbus, Ohio, 1981, pp. 529–533.

25. N. Lagakos, W. J. Trott, and J. A. Bucaro, "Microbending fiber-optic sensor design optimization," *Tech. Digest CLEO*, 1981, pp. 100–101.

26. J. A. Bucaro, N. Lagakos, J. H. Cole, and T. G. Giallorenzi, *Fiber Optic Transduction in Physical Acoustic*, Vol. 16, Academic, New York, 1982, pp. 385–457.

27. D. Marcuse, "Curvature loss formula for optical fiber," *J. Opt. Soc. Am.*, **66**: 216–220 (1976).

28. D. A. Jackson, R. G. Priest, A. Dandridge, and A. B. Tveten, "Elimination of drift in a single-mode optical fiber interferometer using a piezoelectrically stretched coiled fiber," *Appl. Opt.*, **19**: 2926–2929 (1980).

29. S. K. Sheem, T. G. Giallorenzi, and K. P. Koo, "Optical techniques to solve the signal fading problem in fiber interferometers," *Appl. Opt.*, **21**: 689–693 (1982).

30. J. A. Bucaro and J. H. Cole, "Acousto-optic sensor development," *Proc. EASCON 1979*, IEEE Pub. 79CH1476-1 AES, 1979, pp. 572–580.

31. J. H. Cole, B. A. Danner, and J. A. Bucaro, "Synthetic-heterodyne interferometric demodulation," *IEEE J. Quantum Electron.*, **QE-18**: 694–697 (1982).

32. A Olsson, C. L. Tang, and E. L. Green, "Active stabilization of a Michelson interferometer by an electro-optically tuned laser," *Appl. Opt.*, **19**: 1887–1889 (1980).

32a. A. Dandrige and A. B. Tveten, "A new method of phase compensation in interferometric fiber optic sensors," *Opt. Lett.*, **7**: 279–281 (1982).

33. K. P. Koo, A. B. Tveten, and A. Dandridge, "Passive stabilization scheme for fiber interferometers using (3 × 3) fiber directional couplers," *Appl. Phys. Lett.*, **41**: 616–618 (1982).

34. A. Dandridge, A. B. Tveten, and T. G. Giallorenzi, "Homodyne demodulation for fiber optic sensors using phase generated carrier," *IEEE J. Quantum Electron.*, **18**: 1647–1653 (1982).

35. A. Dandridge, A. B. Tveten, R. O. Miles, and T. G. Giallorenzi, "Laser noise in fiber-optic interferometer systems," *Appl. Phys. Lett.*, **37**: 526–528 (1980).

36. A Dandridge and A. B. Tveten, "Noise reduction in fiber optic interferometer systems," *Appl. Opt.*, **20**: 2337–2339 (1981).

37. A. Dandridge and A. B. Tveten, "Phase noise of single mode idode lasers in interferometric systems," *Appl. Phys. Lett.*, **38**: 530–532 (1981).

38. A. Dandridge, "Noise properties of stabilized single-mode lasers in fiber interferometers," *J. Lightwave Technol.*, **LT-1**: 517–518 (1983).

39. L. Goldberg, H. F. Taylor, A. Dandridge, J. F. Weller, and R. O. Miles, "Spectral characteristics of semiconductor lasers with optical feedback," *IEEE J. Quantum Electron.*, **18**: 555–564 (1982).

40. R. O. Miles, A. Dandridge, A. B. Tveten, H. F. Taylor, and T. G. Giallorenzi, "Feedback induced line broadening in CW channel substrate planar laser diodes," *Appl. Phys. Lett.*, **37**: 990–992 (1980).

41. R. O. Miles, A. Dandridge, A. B. Tveten, T. G. Giallorenzi, and H. F. Taylor,, "Low frequency noise characteristics of channel substrate planar laser diodes," *Appl. Phys. Lett.*, **38**: 848–850 (1981).

42. J. A. Bucaro, "Fiber coatings summary: an overview for sensors," SPIE Tech. Symp. East, Fiber Optics and Laser Sensors, Arlington, VA., Paper 412–02, April 4–8, 1983.

43. A Yariv and H. Winsor, "Proposal for detection of magnetostrictive perturbation of optical fibers," *Opt. Lett.*, **5**: 87–89 (1980).

44. G. B. Hocker, "Fiber optic sensing of pressure and temperature," *Appl. Opt.*, **18**: 1445–1448 (1979).

45. K. P. Koo and G. H. Sigel, Jr., "An electric field sensor utilizing a piezoelectric polyvinlidene fluoride (PVF_2) film in a single-mode fiber interferometer," *IEEE J. Quantum Electron.*, **QE-18**: 670–675 (1982).

46. N. Lagakos, E. U. Schmans, J. H. Cole, J. Jarzynski, and J. A. Bucaro, "Optimizing fiber coatings for interferometric acoustic sensors," *IEEE J. Quantum Electron.*, **QE-18**: 683–689 (1982).

47. J. Jarzynski, J. H. Cole, J. A. Bucaro, and C. M. Davis, "Magnetic field sensitivity of an optical fiber with magnetostrictive jacket," *Appl. Opt.*, **19**: 3746–3747 (1980).

48. L. S. Schuetz, J. H. Cole, J. Jarzynski, N. Lagakos, and J. A. Bucaro, "Dynamic response of single-mode optical fiber for interferometric sensors," *Appl. Opt.*, **22**: 478–483 (1983).

49. P. D. DeSouza, and M. D. Mermelstein, "Electric field detection with a piezoelectric polymer-jacketed single-mode optical fiber," *Appl. Opt.*, **21**: 4214–4218 (1982).

50. G. S. Maurer, W. S. Schuetz, J. H. Cole, and J. A. Bucaro, "Application of fiber-optic sensors to the detection of air-acoustic signals," *Opt. Lett.*, **7**: 503–505 (1982).
51. J. A. Bucaro, H. D. Dardy, and E. Carome, "Fiber optic hydrophone," *J. Acoust. Soc. Am.*, **62**: 1302–1304 (1977).
52. J. H. Cole, R. L. Johnson, and P. B. Bhuta, "Fiber optic detection of sound," *J. Acoust. Soc. Am.*, **62**: 1136–1138 (1977).
53. N. Lagakos and J. A. Bucaro, "Pressure desensitization of optical fibers," *Appl. Opt.*, **20**: 2716–2780 (1981).
54. N. Lagakos and J. A. Bucaro, "Minimizing temperature sensitivity of optical fibers," *Appl. Opt.*, **20**: 3276–3278 (1981).
55. J. Jarzynski, R. Hughes, T. R. Hickman, and J. A. Bucaro, "Frequency response of interferometric fiber optic coil hydrophones," *J. Acoust. Soc. Am.*, **69**: 1799–1808 (1981).
56. L. Flax, J. H. Cole, R. P. DePaula, and J. A. Bucaro, "Acoustically induced birefringence in optical fibers," *J. Opt. Soc. Am.*, **72**: 1159–1162 (1982).
57. J. Jarzynski, "Frequency response of a single-mode optical fiber phase modulator utilizing a piezoelectric plastic jacket," *J. Appl. Phys.*, **55**: 3243–3250 (1984).
58. R. D. DePaula, J. H. Cole, and J. A. Bucaro, "Broadband ultrasonic sensor based on induced optical phase shifts in single-mode fiber," *J. Lightwave Technol.*, **LT-1**: 390–393 (1983).
59. A. M. Smith, "Polarization and magneto-optic properties of single-mode fibers," *Appl. Opt*, **17**: 52–56 (1978).
60. A. Dandridge, A. B. Tveten, G. H. Sigel, Jr., E. J. West, and T. G. Giallorenzi, "Optical fiber magnetic field sensors," *Electron. Lett*, **16**: 408–409 (1980).
61. K. P. Koo and G. H. Sigel, "Characteristics of fiber optic magnetic field sensors employing metallic glasses," *Opt. Lett.*, **7**: 334–336 (1982).
62. K. P. Koo, A. Dandridge, A. B. Tveten, and G. H. Sigel, "A fiber optic DC magnetometer," *J. Lightwave Technol.*, **LT-1**: 524–525 (1983).
63. F. Bucholtz, K. P. Koo, and G. H. Sigel, "Field annealing of metallic glass ribbons for fiber optic sensors," 3rd Int. Conf. Optical Fiber Sensors, OFS 85 Paper THAA4 p124, San Diego, CA., February 1985.
64. F. Bucholtz, K. P. Koo, G. H. Sigel, and A. Dandridge, "Optimization of the fiber/metallic glass bond in fiber optic magnetic sensors," *J. Lightwave Technol.*, **LT-3**: 814–817, 1985.
65. K. P. Koo, G. H. Sigel, A. B. Tveten, and A. Dandridge, NRL Invention Disclosure no. 68077, 1983.
65a. A. D. Kersey, M. Corke, and D. A. Jackson, "Phase shift nulling DC field fiber optic magnetometer," *Electron. Lett.*, **20**: 573–574 (1984).
66. K. P. Koo, F. Bucholtz, A. Dandridge, and A. B. Tveten, "Stability of a fiber optic magnetometer," *IEEE Trans. Magnetics*, **MAG-22**: 141–144, 1986.
67. V. Vali and R. W. Shorthill, "Fiber ring interferometer," *Appl. Opt.*, **15**: 1099–1100 (1976).
68. S. Ezekiel and H. J. Arditty (Eds.), *Fiber-Optic Rotation Sensors and Related Technologies* (Springer Ser. Optical Sciences, Vol. 32), Springer-Verlag, New York, 1982.
69. R. A. Bergh, H. C. Lefevre, and H. J. Shaw, "Overview of fiber-optic gyroscopes," *J. Lightwave Technol.*, **LT-2**: 91–107 (1984).
70. K. Bohm, K. Petermann, and E. Weidel, "Sensitivity of a fiber-optic gyroscope to environment magnetic fields," *Opt. Lett.*, **7**: 180–182 (1982).
71. S. Ezekiel, J. L. Davis, and R. W. Hellwarth, "Observation of intensity induced nonreciprocity in a fiber-optic gyroscope," *Opt. Lett.*, **7**: 457–459 (1982).
72. R. Ulrich, "Fiber-optic rotation sensing with low drift," *Opt. Lett.*, **5**: 173–175 (1980).
73. C. C. Cutler, S. A. Newton, and H. J. Shaw, "Limitation of rotation sensing by scattering," *Opt. Lett.*, **5**: 488–500 (1980).
74. K. Bohm, P. Marten, K. Petermann, E. Weidel, and R. Ulrich, "Low-drift fiber gyro using a superluminescent diode," *Electron. Lett.*, **17**: 352–353 (1981).
75. S. C. Rashleigh and W. K. Burns, "Dual-input fiber-optic gyroscope," *Opt. Lett.*, **5**: 482–484 (1980).

76. S. K. Sheem, "Fiber optic gyroscope with 3×3 directional coupler," *Appl. Phys. Lett.*, **37**: 869–871 (1980).

77. J. L. Davis and S. Ezekiel, "Techniques for shot-noise-limited inertial rotation measurement using a multi-turn fiber Sagnac interferometer," *Proc. SPIE*, **157**: 131–136 (1978).

78. R. F. Cahill and E. Udd. "Phase nulling fiber optic laser gyro," *Opt. Lett.*, **4**: 93–95 (1979).

79. R. A. Bergh, H. C. Lefevre, and H. J. Shaw, "All-single-mode fiber-optic gyroscope," *Opt. Lett.*, **6**: 198–200 (1981).

22

Optoelectronic Information Processing: Laser Bar Code and Laser Printer Systems

Yuzo Ono and Nobuo Nishida

I LASER BAR CODE SYSTEMS

22.1 INTRODUCTION TO LASER BAR CODE SYSTEMS

Bar code reader equipment scans a bar code symbol printed on or attached to an article by using a laser beam and reads the code by detecting changes in the intensity of the light reflected from the bar code symbol as it is scanned by the beam. The equipment is used as an input device for a computer system that translates the code into information data for the user. Such computer systems are spreading rapidly to rationalize the handling of inventories and control sales activities. These systems, using laser scanners, are effectively used for retail store (especially for supermarket) POS (point-of-sale) systems, which control sales, stock and buying at the point of sale, and for automated control of an assortment of goods moving on belt conveyer systems, for example, to keep data on receipt and shipping of goods, as well as manufacturing processes.

22.2 FUNDAMENTALS OF LASER BAR CODE READER

22.2.1 Bar Code Reader System

a. Bar Code Symbol. A bar code symbol is composed of light and dark bars that have different light reflectivities, as shown in Figure 22.1. For common

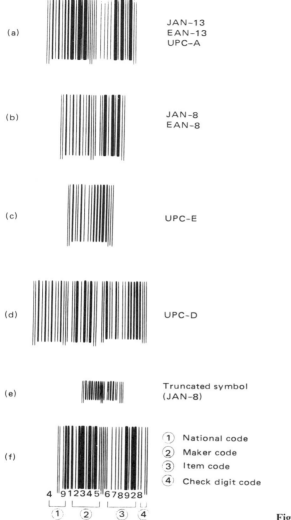

Figure 22.1 Bar code symbols.

articles, symbols are standardized throughout the world. The standardized codes are UPC (Universal Product Code, United States and Canada), EAN (European Article Number, Western Europe), and JAN (Japan Article Number, Japan, conforming to EAN). These symbols have fixed digits, except for version D in UPC. A standard version has 13 digits, and a shortened version has 8 or 6 digits. The minimum bar width is 0.26 mm. Symbols cut down in height, called truncated symbols, are used for small articles and for in-store marking, that is, for labels printed on-site.

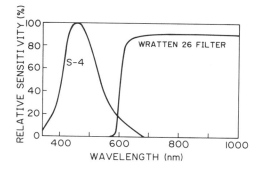

Figure 22.2 Spectral sensitivity for a photodetector and spectral transmissivity for Wratten 26 filter.

Optical characteristics for bar code symbols printed on articles are provided by a permissible maximum reflectivity for the dark bar with respect to the light bar reflectivity [1]. This is defined in terms of a PCS (print contrast signal) value as

$$\text{PCS value} = \frac{R_L - R_D}{R_L} \tag{1}$$

where R_L and R_D are reflectivities for light and dark bars, respectively. To measure the reflectivities and meet the standard definition, it is necessary to use the CIE (Commission Internationale de l'Eclairage) standard light source A for color measurements with a Wratten 26 filter and a JEDEC (Joint Electron Device Engineering Council) S-4 response photodetector. As a result, reflectivity is measured in the 600–650-nm wavelength region, as shown in Figure 22.2. A light source near the 600–650-nm wavelength region is preferred for the bar code reader.

b. System Configuration of Bar Code Reader. Figure 22.3 shows the bar code reader system configuration [2]. The laser beam emitted from a laser tube is deflected by the laser beam scanner over a bar code symbol attached to a moving article of merchandise. The light reflected from the bar code symbol

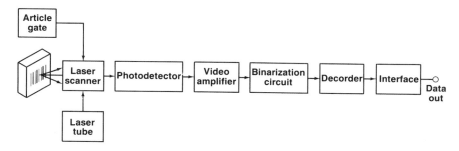

Figure 22.3 Point-of-sale (POS) scanner block diagram.

is detected and converted to an electric signal by a photodetector. This signal is amplified in the video amplifier and converted to a digital video signal, which is fed to a decoder to be interpreted and eventually output. An article gate detects the entering and leaving of articles and controls the scanning beam projection to prevent laser beam irradiation to human eyes.

22.2.2 Laser Scanning System

a. Requirements for the Optical System. In designing the bar code reader, the optical system performance is most important for the laser scanner and the collection and detection of reflected light. For the optical system, the following requirements must be met.

1. Exact reading must be possible regardless of the bar code symbol position and direction.
2. The scanning beam should have sufficient resolution and focal depth for bar code symbol reading.
3. Scanning reading should be done more than twice in one reading operation, to avoid reading mistakes. This should also be satisfied at a 250 cm/s maximum article moving speed for POS applications.
4. Since the equipment is used in a lighted shop, the reflected light signal must be detected effectively in the presence of strong backlighting.

In addition to these requirements, ease of handling, safety, compactness, and low cost are also factors to be considered.

b. Optical Configuration. Laser scanning optical systems are classified into three categories: (1) postobjective, in which the scanner is positioned after the objective lens (Fig. 22.4a); (2) preobjective, in which the scanner is posi-

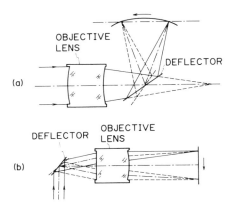

Figure 22.4 Two scanning systems. (a) Postobjective; (b) preobjective.

tioned preceding the objective lens (Fig. 22.4b); and (3) inner-objective, in which the scanner is the objective lens itself. For bar code reader systems, a postobjective optical system using a polygonal mirror scanner and an inner-objective optical system using a holographic laser scanner are used. The preobjective system is used for a laser beam printer with an $f\theta$ lens, as will be described later in the chapter (Section 22.5.2b).

For a POS bar code reader, equipment using a holographic laser scanner has been developed and put to practical use [3–6]. Its basic optical configuration is shown in Figure 22.5. Figure 22.6 is a photograph of the equipment.

A holographic laser scanner is a laser beam scanning device using a rotating disk on whose circumference many holographic lenses are attached. The principle is the same as that which applies when a laser beam is deflected as a conventional lens is moved across a laser beam except that the moving lens is realized by the holographic lens, which is a diffractive element. The divergent laser beam, which is focused by the lens, illuminates a hologram and is scanned by disk rotation. The scanned beam is focused on a predetermined scanning plane within the radius required for reading a bar code symbol by the hologram-focusing function. Backscattered light on the bar code symbol, modulated in accordance with the light and dark values of the symbol, is focused on an aperture in front of the photodetector via the retrodirective path by the hologram itself and is detected as an electric signal. As a result, backscattered light is always focused on the aperture, in spite of the hologram disk rotation, providing stable readout without signal shading. Furthermore, disturbance light, such as room lighting, incident on a hologram with various wavelengths and angles of incidence, is mostly diffracted out of the aperture. Readout at high signal-to-noise ratio is achieved. In the illustrated example, seven holograms whose diffraction angles in the disk

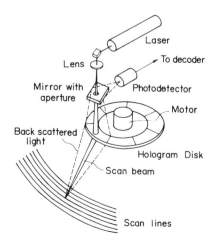

Figure 22.5 Optical configuration for POS scanner using a hologram disk.

Figure 22.6 POS scanner (NEC OBR-80-8).

radius direction differ from each other, are arranged on the hologram disk circle. Seven parallel scan lines are generated by the disk rotation only. In the actual equipment, auxiliary mirrors are arranged between the hologram disk and the scanning plane. The scanning lines are reflected in many directions and constitute a multidirectional scanning pattern to enable all-directional symbol reading.

Equipment using a polygonal mirror scanner requires a focusing lens to converge a scanning beam, a collecting lens to collect the backscattered light, and a color filter to exclude any light that disturbs the photodetector [7]. The holographic laser scanner contains all these functions.

c. Light Source. The wavelength of the light source must be in a range that gives the prescribed PCS value for various symbols. A He-Ne laser ($\lambda = 632.8$ nm), is usually used. Figure 22.7 shows spectral reflectivities for various printed symbols actually used. Purple and black printed on thermosensitive paper, which are used for the dark bars, show a high reflectivity at infrared wavelengths ($\lambda > 700$ nm), as shown in Figure 22.7, and their PCS values

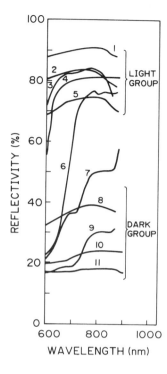

Figure 22.7 Spectral reflectivities for printed bar code symbols. 1, 2, white; 3, pink; 4, red; 5, orange; 6, black (thermosensitive paper); 7, purple; 8, brown; 9, blue; 10, green; 11, black.

are less than the prescribed permissible value. Therefore, for a symbol printed on thermosensitive paper, a light source must have a wavelength less than 680 nm to maintain a prescribed PCS value and less than 660 nm with practical equipment's operation margin concern. Except for this black, and for the purple symbol, the wavelength must be less than 750 nm to maintain a prescribed PCS value and must be less than 700 nm with practical equipment's operation margin concern. On the other hand, in the shorter wavelength range, the reflectivities of the red and orange group colors used for the light bar decrease and a wavelength longer than 610 nm is needed. For these reasons, He-Ne lasers are used for POS scanners. In the bar code readers used for inventory control, the label color is frequently restricted to black. In such a case, systems can be designed using a laser diode in the 780–830-nm wavelength region.

The light source wavelength is also restricted from the viewpoint of safety standards for laser use. Radiation power from the present POS scanner, using a He-Ne laser, is a little lower than 0.5 mW. On the basis of the American National Standard for the safe use of lasers [8], the present POS scanners just satisfy class 1. When the power exceeds class 1, the visible light is classified into class 2, while infrared light with its wavelength of > 700 nm is classified into class 3A. Since class 3A equipment is required to display not only a Caution label but also a radiation warning, it is not suitable for use in POS scanners. Therefore, a light source with a wavelength shorter than 700 nm is indispensable.

The use of a laser diode is very advantageous for making a POS scanner small and compact. Its transverse mode, of course, should be a single mode. In addition to the above-mentioned visible light conditions, when a holographic laser scanner is used, a single longitudinal mode is needed because the holographic scanner is a diffractive element and has a very large chromatic aberration. The diffraction angle θ_d for a hologram is

$$\theta_d = \sin^{-1} \frac{\lambda}{d} \qquad (2)$$

where d is the hologram grating pitch. When two modes, separated by 0.3 nm for example, are oscillating, the diffracted beam splits to 89 μm for a $d = 1.56$ μm ($\theta_d = 30°$ for $\lambda = 780$ nm) grating at a scan plane 300 mm from the hologram. This chromatic aberration is unacceptable for bar code readers because the focused spot size used by the reader is about 200 μm. A gradual wavelength change is acceptable, but frequent mode hopping is not, because it causes substantial changes in position on the scanning symbol.

d. Holographic Laser Scanner

(1) Principle of holographic laser scanner. The concept of using a hologram as a light deflector was proposed in the 1960s [9]. The first practical use of the holographic scanner is for a bar code reader.

As mentioned before, a holographic laser scanner is a laser beam scanning device using a rotating disk on whose circumference many holographic lenses are attached. It functions on the same principle as that of a laser beam being deflected toward the focal point when a conventional optical lens is moved across the beam.

The interferometric zone plate (IZP) hologram, generated by the interference between a divergent spherical wave and a reference plane wave as shown in Figure 22.8, has a function similar to that of a conventional plano-convex lens. When the IZP hologram is illuminated with a collimated beam, the diffracted scan beam is aberration-free and converges to the conjugate point for the recording point source. In this optical arrangement, the focused scan length is only the distance through which the hologram lens moves. In applications such as POS scanners, to obtain the required scan length using a reasonable size hologram, the scanning plane is set farther away than the hologram focal plane, as shown in Figure 22.9. The scan length increases in magnification by b/f, where b is the distance between the hologram and the scanning plane and f is the hologram focal length. The hologram, however, is illuminated with a spherical divergent wave to converge the diffracted laser beam onto the scanning plane. Such an illumination wave, however, is not

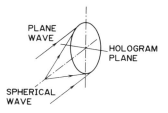

Figure 22.8 Optical configuration for generating IZP hologram.

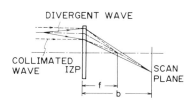

Figure 22.9 Optical configuration to magnify scan length.

the same as the original recording wave. Therefore, IZP holograms cause aberration, and the diffracted beam is distorted on the scanning plane.

(2) Aberration correction for the holographic scanner. To correct the aberration, three methods are used in practical equipment:

(i) Oblique incident reference wave method [4]. This method uses an oblique angle incident reference plane wave to shift a low-aberration deflection angle to a high diffraction angle. The deflection angle with low aberration does not increase.

(ii) Spherical reference wave method [3]. This method uses a spherical reference wave instead of a plane reference wave. The nth interference fringe radius $r_I(n)$ for the IZP is given as

$$r_I(n) = [2n\lambda f + (n\lambda)^2]^{1/2} \tag{3}$$

The first term in the square root in Eq. (3) gives the intrinsic focal length, and the second term gives the focal power distribution along the hologram radius. To correct the aberration, it has been found that the second term must be greater than zero, which corresponds to the geometric zone plate, and less than $(n\lambda)^2$, which corresponds to the IZP. When a hologram is generated by using N spherical waves in order, the resultant phase variation is

$$\phi_N(r) = \frac{2\pi}{\lambda} \sum_{k=1}^{N} [(r^2 + f_k^2)^{1/2} - f_k] \tag{4}$$

When $f_k = NF$, the nth interference fringe radius for the hologram is

$$r_N(n) = \left[2n\lambda F + \left(\frac{n\lambda}{N}\right)^2 \right]^{1/2} \tag{5}$$

The second term in the square root in Eq. (3) decreases to $1/N^2$. As a result, the aberration can be corrected by appropriate selection of f_k values. Over $\pm 20°$, aberration is corrected and a scan length of more than 50 cm is achieved.

(iii) Method using high diffraction angle IZP [5]. This method uses an uncorrected IZP. To obtain a wide scan angle, the off-axis part with a very high diffraction angle is used. When a high diffraction angle arrangement is used, the beam scan radius due to disk rotation is increased and a wide scan angle is achieved.

(3) Hologram disk. To use the holographic scanner for a bar code reader, hologram segments with different focal lengths and different scan positions can be arranged on the disk circumference. Three kinds of holograms are used—a thin-phase hologram using surface relief and volume-phase holograms using bleached silver halide or dichromated gelatin.

The surface relief hologram is first recorded on a photoresist film. A nickel mold is electroformed from the developed photoresist hologram. An embossed hologram disk is duplicated on a transparent thermoplastic sheet by a hot-pressing method using the mold. The diffraction efficiency of the embossed hologram is about 35%. Figure 22.10 shows the embossed hologram disk [3].

The volume-phase hologram is duplicated by contacting the master hologram and a silver halide photographic plate using coherent illumination [5]. The first-order and zeroth-order diffraction waves interfere with each other, and the interference fringe is recorded. The volume-phase hologram has a high diffraction efficiency, more than 50%, but lacks durability against humidity.

e. Scanning Patterns. To achieve exact reading regardless of bar code symbol position and its bar direction, many scanning patterns are used. Previously,

Figure 22.10 Embossed hologram disk.

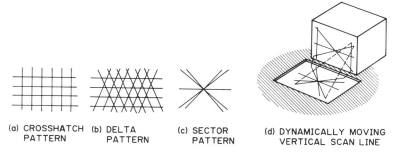

(a) CROSSHATCH (b) DELTA (c) SECTOR (d) DYNAMICALLY MOVING
   PATTERN   PATTERN   PATTERN   VERTICAL SCAN LINE

Figure 22.11 Scanning patterns. (a) Crosshatch pattern; (b) delta pattern: (c) sector pattern. (d) Dynamically moving vertical scan lines.

crosshatch [4] and delta [3] scan patterns, shown in Figures 22.11a and 22.11b, were used. Sector-scan patterns, shown in Figure 22.11c, are used to reduce the number of vertical scan lines by using two dynamically moving vertical scan lines as shown in Figure 22.11d, where the position of the intersection between vertical lines and the article moves as the article moves [7]. To read the truncated symbols over the entire window plane, a six-directional high-density scanning pattern [2] and a seven-directional sector scan (Fig. 22.11d) are used. To generate these scanning patterns, 6 to 10 mirrors are used in the scanner.

22.3 PRESENT AND FUTURE OF LASER BAR CODE SYSTEMS

The present laser bar code systems use a He-Ne laser as the light source because of the standards for symbol printing colors mentioned previously. Only a few special systems, not suitable for most POS systems, use a laser diode where an infrared wavelength can be used for selective photodetection in the background room illumination.

Point-of-sale systems using laser bar code readers are proliferating in supermarkets and other retail stores throughout the United States, Canada, Europe, and Japan. Bar code systems have also been introduced into automated factories, warehouses, and postal and delivery services.

The use of a laser diode as the light source is very advantageous in making a POS scanner small and compact. For this use, however, a visible wavelength in at least the 610–660 nm range is needed because of the spectral reflectivities of standard symbols. For a holographic scanner, a stable longitudinal mode laser is required because of the scanner's large chromatic aberration. In spite of these strict requirements, laser bar code readers based on a holographic scanner with a laser diode source will probably be introduced.

II LASER PRINTER SYSTEMS

22.4 INTRODUCTION TO LASER PRINTER SYSTEMS

A laser beam printer is a nonimpact output device for electronic computers that draws output results on a photoconductor using a laser beam and transfers the latent data to a printed image by means of electrophotography.

The laser beam printer features (1) high speed; (2) high resolution; (3) the availability of various fonts such as those needed for Chinese ideographs, graphics, and images; (4) the ability to print on plain (untreated) paper; and (5) low noise. It was first put into use as an output means for meeting EDP needs for high speed and large quantity. With the progress in individual processing, workstations, word processors, computer-aided design, and computer-aided manufacturing, relatively low speed laser printers, printing 10–40 pages/min, but small and easy to maintain, have been developed using laser diodes. Printing speeds range from 550 lpm (lines per minute) to 15,200 lpm.

22.5 PRINCIPLE OF LASER BEAM PRINTERS

22.5.1 Basic Construction of Laser Beam Printers

Figure 22.12 shows the basic construction of a laser beam printer. The equipment is made up of a laser beam scanning portion, an electrophotographic printing process portion, a paper or sheet feeder portion, and a printer controller portion. The electrophotographic printer and paper feeder are arranged around a photoconductor drum. Individual processes involving charging, laser beam exposure, development, image transfer, fixing discharging, and cleaning are successively repeated as the drum rotates. The optics portion is composed of a laser, a light modulator, a polygonal mirror scanner,

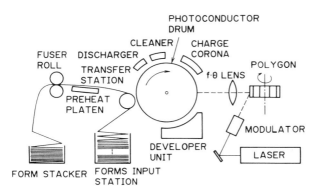

Figure 22.12 Basic construction of a laser beam printer.

an $f\theta$ lens, and so on. The laser beam is modulated to an intensity of 2 by the modulator in accordance with a dot command from a computer and scans the drum surface with a focused spot in the direction along the drum axis. Since the photoconductor drum rotates at a fixed speed, a two-dimensional electrostatic image can be drawn on the drum by a one-directional laser beam scan. The image is composed of small dots, 9.5–16 dots/mm; therefore, arbitrary printing fonts such as Chinese ideographs and graphic images can be printed.

22.5.2 Laser Scanning System

a. Light Source. The most important factor in selecting a laser source is the wavelength matching to the spectral sensitivity of the photoconductor drum. In earlier stages, high-speed equipment used gas lasers, such as He–Ne, He–Cd, and argon lasers [10, 11]. Laser diodes are used for relatively low speed equipment due to the low sensitivity of a photoconductor at infrared wavelengths [12]. High-speed equipment using a monolithic laser diode array composed of eight lasers has been developed [13]. Acoustooptic light modulators using TeO_2 or $PbMoO_4$ crystals are used for gas lasers. Laser diodes are directly modulated with drive current. Laser beams, for both gas lasers and diode lasers, are transformed by using a pair of cylindrical lenses or an anamorphic prism to obtain appropriate beam spot radii on the drum in both the major scanning direction by a laser scan and in the secondary scanning direction by a drum rotation.

In using a laser diode, since the output light power fluctuates with a change in lasing threshold due to ambient temperature variations, automatic power control must be accomplished by detecting the output power using a photodetector in a laser package. The output power also fluctuates with a change in pulse duty factor, because it causes a temperature change in the laser cavity. To avoid this light power fluctuation, the electric power dissipation is fixed using a bias current less than the lasing threshold. The wavelength also varies with a change in temperature; sensitivity of the photoconductor varies. Thus temperature control for a laser diode is needed when the sensitivity is not a constant value around the operating wavelength.

b. Laser Scanner. A large number of laser beam deflectors used in practical laser beam printers are polygonal mirror scanners. Holographic laser scanners are also used. In the earlier stages in the development of the laser beam printer, manufacturing costs for polygonal mirror scanners were very high. However, two major technologies were developed to overcome this problem. One is metal polygon manufacturing technology that employs a diamond cutting tool. The other is an optical tilt angle correction technology that has made it possible to increase the tilt angle accuracy limit for polygonal facets. A few laser beam printers use a holographic laser scanner, but the scan position for such a scanner shifts with a change in light source wavelength, as explained earlier. Since this problem has not yet been overcome, wavelength stabilization control is indispensable for a holographic scanner.

(1) fθ Lens. For polygonal mirror scanners in laser beam printers, two optical configurations, postobjective and preobjective, are used for the scanner. For a preobjective scanner, an $f\theta$ lens is used to scan a focused laser beam on a flat scanning plane with a fixed scan velocity. The image height for an $f\theta$ lens is

$$h = f\theta \qquad (6)$$

for a deflection angle θ, as shown in Figure 22.13. Image heights for an $f\theta$ lens and for a conventional lens are also shown in Figure 22.13. When the angular velocity of polygonal mirror scanner rotation is ω, the reflected deflection angle is $\theta = 2\omega t$. When an $f\theta$ lens is used, the linear velocity on an image plane is derived as

$$v = \frac{dh}{dt} = 2f\omega \qquad (7)$$

The fixed scan velocity makes it easy to control print position. The scan velocity linearity error is defined as

$$\varepsilon = \frac{h' - f\theta}{f\theta}\,100 \quad [\%] \qquad (8)$$

where h' is image height. Ordinarily, ε is designed for less than 0.5%. Such an $f\theta$ lens is easily constructed with a spherical doublet [14]. Print position control for a scanner with a postobjective lens will be described later.

(2) Polygonal mirror. For a polygonal mirror with n facets, the reflected deflection angle is normally $4\pi/n$. To focus a scan beam to a predetermined diameter a using an $f\theta$ lens with focal length f, the beam diameter

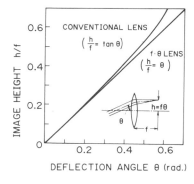

Figure 22.13 Image height for $f\theta$ lens.

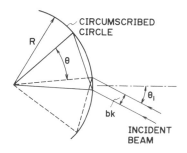

Figure 22.14 Effective rotation angle for polygonal mirror scanner.

b on the facet must be

$$b = \frac{4\lambda f}{\pi a} \qquad (9)$$

The effective deflection angle decreases due to beam truncation at both edges of the facet. When the safety factor for beam truncation is k, the effective scan length y for the polygonal mirror shown in Figure 22.14 is

$$y = 2f\theta = 2f \left\{ \frac{\pi}{n} - \frac{\beta}{2} - \sin^{-1} \left[\frac{4k\lambda f}{\pi Ra} - \sin \left(\frac{\pi}{n} + \frac{\beta}{2} \right) \right] \right\} \qquad (10)$$

where R is the circumscribed circle radius for the polygonal mirror and β is the beam separation angle between an incident beam and the nearest scanning beam, because they are on the same plane. Based on Eq. (10), the optimum polygonal mirror shape (that is, R and n) is designed using the parameters mentioned above. Drive power and mechanical stress are also considered in the design.

The parameters requiring high precision for a polygonal mirror scanner are (1) the facet-dividing angle, (2) the mirror facet tilt angle with respect to rotation axis, (3) mirror facet flatness, and (4) rotation speed fluctuation. The facet-dividing angle error is compensated for by detecting a scan line start time using a photodetector. One arc-second of error in facet tilt causes scan line displacement of $9.7 \times 10^{-6} f$ on the scanning plane, where f is the focal length of the $f\theta$ lens. For example, for $f = 400$ mm, the scan line displacement is 3.9 μm. The mirror facet tilt can be compensated for by using the optical relation that the mirror facet and the scan plane are conjugate to each other; that is, they are object point and image point, respectively [10]. Figure 22.15 shows the principle of a tilt angle correction method. The laser beam incident on a facet is focused in a direction perpendicular to the scanner rotation axis with cylindrical lens CL_1. The second cylindrical lens, CL_2, is arranged so that the facet and the scan line are in object point–image point relation with respect to the second cylindrical lens–scanning lens combination. In a practical system, CL_2 is a toroidal lens, as shown in Figure 22.16, to compensate for the changes in optical path length due to facet rotation.

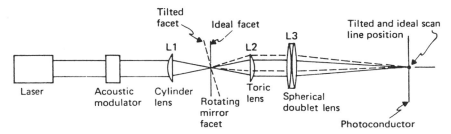

Figure 22.15 Principle of tilted facet correction. (Copyright 1977 by IBM Corp; reprinted with permission.)

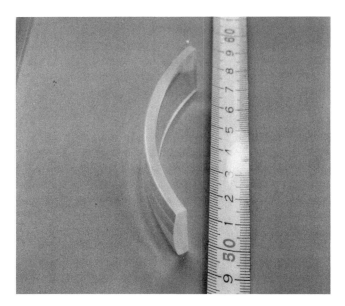

Figure 22.16 Toroidal lens.

c. Optical Configuration. Several examples of optical configurations for laser beam printers follow.

(1) IBM 3800 [10]. Figure 22.17 shows the optical configuration. This equipment was presented as the first laser printer in 1975. Print speed is 13,600 lines/min, and dot densities in the horizontal and vertical directions are 180 and 144 dots/in., respectively. The laser scanning system is composed of a He-Ne laser, an acoustooptic light modulator, a polygonal mirror, and an $f\theta$ lens. A toroidal lens is used to compensate for the scan line displacement. The modulator bandwidth is increased by reducing the width of the interaction region between the laser beam and the acoustic wave, using beam compression optics.

(2) Siemens ND-2 [11]. The speed and dot density are the same as those for the IBM 3800. The acoustooptic light modulator is driven with six different frequencies of acoustic waves and diffracts six parallel beams that are modulated independently. The scanner scans the six beams simultaneously. Therefore, the polygon rotation speed is only 3000 rpm. The scanning system has a postobjective configuration. One of the scan beams scans a grating and generates a clock pulse to control print position. The fixed print format recorded on a negative film is imaged onto the drum above the laser-beam-written data at the optical format overlay portion.

(3) Canon LBP-CX. The Canon LBP-10 developed in 1979 was the first laser beam printer to use a laser diode [12]. The LBP-CX is one of its very small successors. To realize the small size, scanner optics with a 100° deflec-

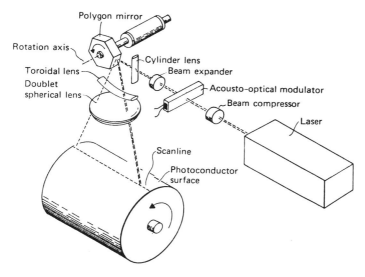

Figure 22.17 Laser subsystem for IBM 3800 laser beam printer. (Copyright 1977 by IBM Corp; reprinted with permission.)

tion angle have been developed using an $f\theta$ lens composed of a concave-convex spherical lens and a toroidal lens with a cylindrical inner surface, as shown in Figure 22.18 [15]. In the saggital plane, these two lenses and the cylindrical lens constitute facet tilt correction optics. The print speed is 8 pages/min for an A4 sheet, and resolution is 15.75 dots/mm.

(4) General Optronics Holoscan I [16]. In this device, a holographic laser scanner is used instead of a polygonal mirror scanner. The light source is a laser diode. The print speed is 28 pages/min for an 8.5 in. × 11 in. sheet, and the resolution is 11.8 dots/mm. Details of the optics have not been published.

(5) Optical system using a laser diode mounted on an actuator [17]. In postobjective scanning optics, although the optics are very simple, the focus

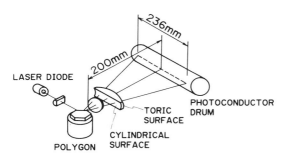

Figure 22.18 Wide-angle scanner for Canon LBP-CX.

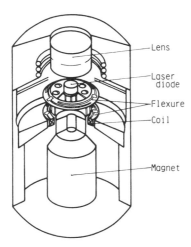

Figure 22.19 Actuator for laser diode.

plane forms a cylindrical surface and the scan velocity is not constant. To overcome these problems, a scanner was developed in which the focus error is compensated for by moving a laser diode mounted on an actuator in the direction of an optical axis synchronous with a polygon rotation. Figure 22.19 shows such an actuator for the laser diode. To compensate for the nonuniform scan velocity, a variable clock cycle has been introduced. As a result, a scan linearity error of less than 0.15% is achieved.

22.5.3 Electrophotographic Systems

a. Printing process. Figure 22.20 shows the printing process for a laser beam printer.

1. The photoconductor surface is charged to a polarity determined by the photoconductor characteristics (the positive polarity case is shown in Fig. 22.20) using a corona charger. The charged voltage is about 0.6–1 kV.
2. The photoconductor is then exposed by a scanning laser beam in accordance with a print dot pattern. Electrostatic voltage is decreased to ∼ 10 V, due to a discharge through an exposed photoconductor at an exposed position, and an electrostatic latent image is formed.
3. The electrostatic latent image is developed by selectively attracting toners charged at the same polarity as the first charging. The toner, which is a pigment to visualize the latent image, is attracted to the discharged position.
4. The toner image on the drum is contacted onto a plain paper and is transferred onto the paper surface by electrically attracting toners from the back of the paper.

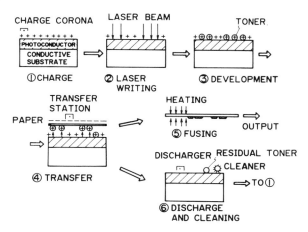

Figure 22.20 Printing process of electrophotography.

5. The toner is fused and then fixed by heating or pressing between rollers at the fuser unit.
6. The surface charge on the drum is discharged by applying an ac voltage or exposing the whole surface at the discharging unit.
7. Residual toner is removed by using a blade, brush, or sponge roller at the cleaning unit.

In the laser beam printer, reverse development is usually used for film overlay to print a fixed format using a negative film. The process outlined above is reverse development, where the toner is attracted to a position exposed with a laser beam and then discharged. Normal development is also used, but the polarity at the developing and transferring portions is opposite to that for the reverse development.

b. Photoconductor Drum. Photoconductor material requirements are (1) high sensitivity at the laser beam wavelength, (2) sufficiently large accumulated charge, (3) small dark decay charge, (4) sufficiently low residual voltage, and (5) excellent repeatability and print durability.

The spectral sensitivities for typical photoconductors in laser beam printers are shown in Figure 22.21 [18, 19]. Sensitivity is defined as the inverse of the energy density required to decrease a charged voltage by one-half. Both organic and inorganic materials are used.

(1) Organic photoconductor materials. Organic materials feature wide freedom in material design and have been extensively investigated. Organics have a shorter life than inorganics, but the materials are easily applied as a film coating. They are cost-effective and make good consumer articles. PVK-TNF (polyvinylcarbazor, 2, 4, 7-Trinitrofluorenone) is used for He-Ne

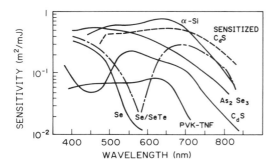

Figure 22.21 Spectral sensitivities for photoconductor materials.

lasers. A photoconductor using phthalocyanine has high sensitivity and has been widely investigated. To use it with an infrared laser diode, sensitivity must be improved in the infrared wavelength range. This is equivalent to making the bandgap smaller; as a result, it becomes sensitive to temperature changes.

(2) Inorganic photoconductor materials. Photoconductors using selenium, Cds, and ZnO are extensively used. In the selenium system, tellurium or arsenic is added to increase the infrared sensitivity. The Se–As system is stable and is used in monolayer film. The Se–Te system is susceptible to rubbing and is used with a protective layer. In the CdS and ZnO systems, copper or pigments, respectively, are added to improve sensitivity.

Amorphous silicon features excellent hardness and stability. Germanium is added to improve its infrared laser diode sensitivity. For widespread use of amorphous silicon, cost-effective manufacturing technology must be developed.

22.6 PRESENT AND FUTURE OF LASER PRINTER SYSTEMS

Laser beam printers will be widely used in a variety of applications, covering a wide range from high-speed, large equipment for line printers in EDP centers to compact intermediate-speed office equipment using cut sheets. A laser beam scanning system using a laser diode, of a small size compatible with the diode size, should be developed. Color printers and high-quality graphic printers are also required. In a compact-size intermediate-speed printer, it was said once that competition with printers using spatial light modulators such as liquid crystal shutter arrays or parallel light sources such as LED arrays was unavoidable because of its cost. The cost of laser scanner optics, however, has rapidly gone down with the development of manufacturing technology, while the cost of spatial light modulators is still high. A laser printer could be competitive with these systems.

REFERENCES

1. Japanese Industrial Standards Committee (Ed.), "Japanese industrial standard, Bar Code Symbol for Uniform Commodity Code, JIS B 9550-1078," Japanese Standards Association, Tokyo, 1978.
2. H. Miyazaki, et al., "Built-in type POS scanner using holographic deflector," *NEC Res. Dev.*, **75**: 56–63 (1984).
3. Y. Ono and N. Nishida "Holographic laser scanners using generalized zone plates," *Appl. Opt.* **21**: 4542–4548 (1982).
4. H. Ikeda, M. Ando, and T. Inagaki, "Aberration corrections for a POS hologram scanner," *Appl. Opt.*, **18**: 2166–2170 (1978).
5. L. D. Dickson, G. T. Sincerbox and A. D. Wolfheimer "Holography in the IBM 3687 supermarket scanner," *IBM J. Res. Div.*, **26**: 228–234 (1982).
6. K. Nishi, K. Kurahashi, and T. Kubo, "Multidirectional holographic scanner for point-of-sale bar-code symbol reader," *Opt. Eng.*, **23**: 784–787 (1984).
7. J. Godwin, "A simple low-cost scanner," *Laser Focus*, **17**: 91–93 (October 1981).
8. G. M. Wilkening (Ed.), American National Standard for the Safe Use of Lasers, ANSI Z136.1-1973, American National Standards Institute Inc., New York, 1973.
9. I. Cindrich, "Image scanning by rotation of a hologram," *Appl. Opt.*, **6**: 1531–1534 (1967).
10. J. M. Flescher, M. R. Latta, and M. E. Rabedeau, "Laser-optical system of the IBM 3800 printer," *IBM J. Res. Dev.*, **21**: 479–483 (1977).
11. W. Meye, "Optical character generation for a high-speed non-impact printer," *J. Photog. Sci.*, **25**: 183–186 (1977).
12. T. Kitamura and K. Masagi, "Laser beam printer LBP-10," *J. Inst. Image Electron. Eng. Jap.*, **8**: 158–166 (1979).
13. M. Otuki, et al., "High-speed electrophotographic printing with laser diode arrays," 2nd Int. Congr. Adv. Non-Impact Printing Technol., Advance Printing of Paper Summaries, (November 1984), pp. 37–38.
14. K. Minoura, M. Takeoka, and S. Minami, "Optical design for laser scanning lens," *Kogaku*, **10**: 348–355 (1981).
15. K. Matsuoka, K. Minoura, and S. Minami, "Optical design for laser scanning system using toric lens," *Proc. Autumn Meeting Jap. Soc. Appl. Phys.* 1981, p. 82.
16. Catalogue, General Optronics Corp., Edison, N. J.,
17. M. Araragi, et al., "Optical scanning system without f-θ lens for laser printer," *Trans. Inst. Electron. Commun. Eng. Jap.* **J68-C**: 171–177 (1985).
18. V. J. W. Weigh, *Angew. Chem.*, **89**: 386–406 (1977).
19. Y. Nakamura, et al., "A new α-Si:H photoreceptor drum: preparation, electrophotographic properties and application," *Photog. Sci. Eng.*, **26**: 188–193 (1982).

PART 6

Future Optoelectronic Technology and Transmission Systems

CHAPTER 23

Optoelectronic Integrated Circuits

U. Koren

23.1 INTRODUCTION

Semiconductor optical components have been subject to intense research efforts in recent years. These efforts have become apparent with the introduction of GaAs laser diodes for consumer products such as laser printers and audio laser compact disk players. The semiconductor materials used for optical devices are mostly group III–V compounds such as GaAs and InP and ternary or quaternary compounds that are lattice-matched to GaAs or InP and thus can be epitaxially grown on these wafers. Silicon and germanium are commonly used for optical detectors but are not suitable for optical sources such as lasers because they do not have direct energy bandgaps. The high optical gains that are required for lasers can be achieved only in direct bandgap materials such as the III–V compounds. Also, in principle, faster detectors can be obtained with direct bandgap semiconductors because the optical absorption lengths are inherently smaller.

The GaAs–GaAlAs system is useful in the wavelength range of 0.7–0.9 μm, while the InP–InGaAsP system is suitable for the 0.9–1,65 μm wavelength range. The relative maturity of GaAs technology is seen in low-cost mass production of lasers and transistors. It is thus very useful for component-rich applications such as data transfer local area networks (LANs), which require transmission over relatively short distances (typically, less than 3 km).

The development of very low loss optical fiber (less than 0.2 dB/km) in the 1.3–1.6 μm wavelength range has also led to intense research on the InP–InGaAsP material system. With these materials we have optical devices that are useful for fiber communication over long distances.

Monolithic integration of several or many optical and electronic components on a single GaAs or InP chip has been an attractive technical goal for some time [1–5], as it may offer the same rewards of low-cost mass production and reliability of complex assemblies that have been achieved with silicon microelectronics circuits. Furthermore, there are the inherent advantages of small physical dimensions and low parasitic impedances that are very important for high-performance systems operating at very high frequencies. This is the motivation for the development of a monolithic laser transmitter and the high-bit-rate receiver to be discussed in the following sections. Integration of active and passive waveguides (Section 23.4) may also lead, in the future, to complex switching, modulation, and signal-processing capabilities on semiconductor chips resembling the devices that are now being produced with $LiNbO_3$ integrated optics.

There are, however, several important difficulties that hinder the progress of optoelectronics integration on III–V materials.

1. *Insufficient basic silicon technology process.* The planar silicon technology such as oxidation, diffusion, and ion implanation are not sufficient because lasers and waveguides generally need layers of different compounds with different energy bandgaps (heterostructures). These can be obtained only by epitaxy.
2. *Material defects and epitaxial defects.* Basic starting material such as GaAs or InP wafers do not yet have the low defect density that has been achieved with silicon wafers. The need for epitaxial growth usually makes matters much worse. On the other hand, the goal of high-quality material with a low density of defects is essential for reliability and reasonable yields with integrated circuits.
3. *Conflicting material requirements between optical and electronic devices.* Lasers, detectors, and waveguides generally require heterostructures with layers that are epitaxially grown on the semiconductor substrates. These layers are of typically 2–5 μm in total thickness. The electronic circuits contain transistors that are made on the binary substrate or on epitaxial layers of different thicknesses and doping levels. Thus it is necessary to segregate the optical and electronic devices on different parts of the wafer and to interface them over the boundaries. The epitaxy is sometimes performed over grooved regions in the substrate so that total planarity is closely maintained. This is necessary to allow the high-resolution photolithography that is required for the fabrication of both lasers and field-effect transistors (FETs).
4. *Semiconductor laser requirements.* Very high yields and good reliability of the discrete components are obviously imperative requirements for practical large-scale integration. Semiconductor lasers operate at large current densities and large optical fields and are thus subject to failure. Yet good laser structures and high-quality semiconductor materials have

been developed to overcome this problem. Moreover, InP lasers are sensitive to temperature and generally need careful heat sinking and temperature control.

It can be seen that advanced material research plays a crucial role in the success of any optoelectronic integration. Thus, epitaxial growth techniques are receiving intense attention and are being used in the fabrication of optoelectronic devices. Liquid-phase epitaxy (LPE) was the first technique to be used for fabrication of double-heterostructure laser diodes and is still the most common. Crystal quality is usually good, but uniformity over large wafers and layer thickness control are poor. In the following sections it will be seen that many workers have also used metal-organic chemical vapor deposition (MOCVD), vapor-phase epitaxy (VPE), and molecular beam epitaxy (MBE) for the fabrication of lasers and other components for optoelectronic integration.

23.2 INTEGRATED LASER TRANSMITTER

An example of an advanced design for a GaAs monolithic laser transmitter as reported by Hirao et al. [6] is shown in Figure 23.1. The circuit is composed of a laser diode and a monitoring photodetector, with a driving circuit for the laser and a monitoring output circuit for the detector. The laser driving circuit (Fig. 23.1b) is designed with a differential constant-current driver (transistors Q_1, Q_2, Q_3) where the total current is determined by the saturation current of Q_3. The distribution of currents between Q_1 and Q_2 is controlled by the input buffer amplifier (Q_4–Q_{10}). This amplifier increases the responsivity of the circuit, allowing laser current modulation of 15 mA with a 0.8-V peak-to-peak swing of the input signal voltage. The driving circuit has been operated at high frequencies, and error-free 2-Gb/s operation has been demonstrated.

The driving and monitoring circuits are composed of metal semiconductor field-effect transistors (MESFETs) that are fabricated by double silicon implantation on the exposed semi-insulating GaAs substrate. The gate lengths are 1.5 μm, and transconductances of 60 mmho/mm are obtained. This is a planar technique that requires high-resolution precision photolithography. It is therefore necessary to minimize the height differences on the wafer, and this is done by growing the laser and detector layers in a groove that is preetched in the substrate. The laser layers are grown by a two-step MOCVD process, and layers outside the groove are removed by chemical etching to expose the original substrate for the FET processing. The schematic structure of the laser layers and laser–FET interface is shown in Figure 23.2. The bottom n layers of the laser are exposed and are connected by gold metalization to the FET that is located outside the groove.

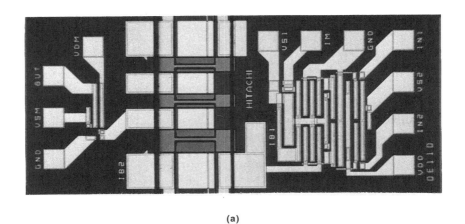

(a)

(b)

Figure 23.1 A monolithic GaAlAs–GaAs laser transmitter. (a) Top view of the integrated circuit; (b) schematic description of the laser driver circuit.

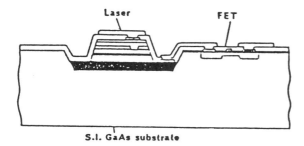

Figure 23.2 Schematic cross section of the laser diode and the FET structure.

One important problem that is associated with integration of lasers is the microfabrication of the laser mirrors (or other feedback mechanisms). For discrete laser devices the mirrors are conveniently formed by cleaving the laser wafer at two parallel cleavage planes to obtain bars, with many lasers located side by side on each bar. Since the laser optimal cavity length is 200–300 μm, this imposes a restrictive size that cannot be accepted for most integrated circuits.

In the case of the circuit of Hirao et al., there are two lasers and two detectors located collinearly, as shown in Figure 23.1a. The lasers (on the outside of the chip) have one mirror that is cleaved. The other mirror, located between the laser and the detector, is made by reactive ion beam etching (RIBE). This technique was used previously to obtain perpendicular etched facets [7]. An SEM picture of the RIBE mirror used by Hirao et al. is shown in Figure 23.3. It should be noted that the fabrication of a laser mirror with good control over orientation and smoothness of the surface is not an easy task. Wet chemical etching by itself is usually unsatisfactory, because preferential etching (at different orientation) occurs, and smooth surfaces are difficult to achieve. This usually degrades the laser performance and yields. Other

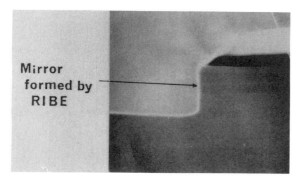

Figure 23.3 SEM photomicrograph of a cross section of a laser mirror formed by reactive ion beam etching.

approaches to this problem have been suggested. One such approach is to microcleave [8–10] the laser mirrors. Shown in Figure 23.4 is a SEM photomicrograph of a microcleaved triangular laser mirror [10] with a crescent-shaped active region. This mirror facet is obtained by wet etching to undercut the active region of the laser and form a suspended cantilever or bridge-shaped microstructure. This structure is subsequently cleaved by ultrasonic vibrations. This method produces high-quality mirrors, but it requires many processing steps and is difficult to reproduce with good yields. Another optional solution to the mirror problem is to get rid of the mirror entirely by using a grating for the laser feed-back mechanism as is done for a distributed feedback (DFB) laser [11]. An integrated optoelectronic circuit using a DFB laser has been reported by Kasahara et al. [12].

For the InP–InGaAsP material system there is also a strong motivation for the development of an integrated laser transmitter for long-distance optical fiber communication. The discrete laser performance and reliability of the InP–InGaAsP lasers are at present comparable to those of GaAs

← → 1 μm

Figure 23.4 SEM photomicrograph of a laser mirror formed by microcleaving.

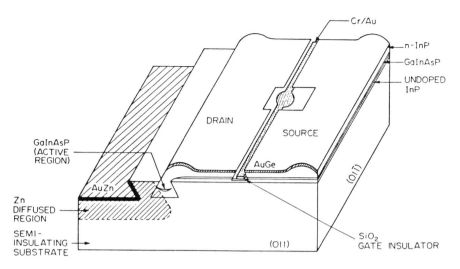

Figure 23.5 Schematic cross section of an integrated laser FET on semi-insulating InP substrate.

lasers. However, the FET technology on InP is not so well advanced. MESFETs are not suitable for the InP system because of difficulties in obtaining sufficiently high Schottky barriers. Therefore, metal-insulator-semiconductor field-effect transistors (MISFETs) and junction field-effect transistors (JFETs) have been used. An early demonstration [13] of an integrated circuit containing a 1.3-μm wavelength laser and a MISFET is shown in Figure 23.5. A laser with a buried crescent active region is located in a groove on a semi-insulating InP substrate. The MISFET channel is obtained with one of the LPE grown layers located outside the groove. A more recent structure reported by Kasahara et al. [12] is shown in Figure 23.6. This structure contains two MISFETs integrated with a DFB 1.3-μm laser

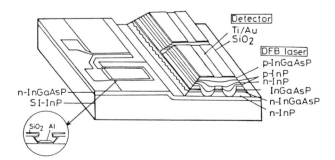

Figure 23.6 Schematic diagram of an integrated circuit consisting of a DFB laser, a detector, and two MISFETs on semi-insulating InP substrate.

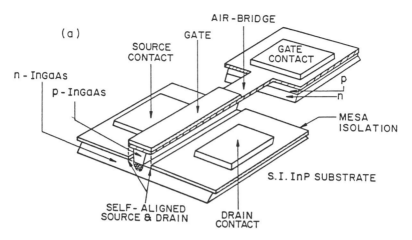

Figure 23.7 Schematic diagram of an InGaAs JFET on semi-insulating InP substrate.

and a monitoring detector. This device has been operated at up to 2 Gb/s.

The InP MISFET suffers from the commonly observed hysteresis of the source–drain current characteristics. This is a result of trap states in the insulator and high density of surface states at the semiconductor–insulator interface. JFETs have been proposed [14] to reduce this unwanted hysteresis effect. A schematic structure of a JFET reported by Cheng et al. [15] is shown in Figure 23.7. This JFET is fabricated on an InGaAs layer grown by molecular beam epitaxy on an InP substrate. InGaAs has very high electron mobilities and high peak drift velocities and is therefore a very promising material for fabrication of FETs. The JFET is made with a self-aligned mesa etching and contact deposition process. It includes a separate mesa for the gate pad, which is connected to the gate by a suspended oxide air bridge to reduce the parasitic capacitance of the gate pad. With this structure, gate capacitances of 0.8 pF/mm and transconductances of 110 mmho/mm have been reported. The cutoff frequency for this device is about 20 GHz. These results are superior to what is commonly obtained with GaAs MESFETs (for example, the results of Ref. 6) but at the expense of a more complex design, which is not planar and requires epitaxial growth for the channel layer.

23.3 RECEIVERS, REPEATERS, AND ARRAYS

Several reports [16–19] describe the integration of GaAs PIN (positive-intrinsic-negative) photodetectors with amplifier circuits composed of GaAs MESFETs. The motivation behind this work was to develop reliable compact front-end modules operating at moderate frequencies and sensitive enough for signal detection through a few kilometers of fiber transmission. It was found that the performance characteristics were sufficient for ordinary local area networks applications. A receiver front-end module used for a 400 Mb/s transmission experiment by Horimatsu et al. [16] is shown in Figure 23.8. This receiver contains a PIN photodiode and a transimpedance amplifier with six MESFETs and a feedback resistor. These components are fabricated on a GaAs substrate with epitaxial layers grown by MOCVD. A 4-km fiber transmission experiment was performed [16] with an integrated laser transmitter and the above-mentioned PIN amplifier circuit, and bit error rates (BER) of 10^{-9} sec^{-1} were obtained for received powers of -18 dBm at a transmission rate of 400 Mb/s.

For optical fiber transmission over longer distances, 1.33–1.55-μm wavelength detectors are required. The PINFET integrated receiver first reported by Leheny [19] et al. was made on an InP substrate employing a p diffusion for the detector and the JFET. Barnard et al. [20] have demonstrated an InGaAs FET integrated with a photoconductive detector. A more recent example of a PINFET integration with an InGaAs detector as reported by Tell et al. [21] is shown in Figure 23.9. This circuit contains two ion-implanted depletion mode MISFETs with self-aligned structures. The detector was made on a InGaAs layer grown by vapor-phase epitaxy in a channel that was preetched on a semi-insulating substrate. The PIN detectors were obtained by beryllium ion implantation, while silicon double implantation was used for the MISFETs. The load resistor was also made with the same silicon ion implantation. This integrated circuit has been tested at 90 Mb/s and had a sensitivity of -34 dBm with 10^{-9} sec^{-1} bit error rate.

It should be noted that for the transmission of very high frequency data over long distances, the performance of discrete devices such as InGaAs avalanche photodiodes, (APDs) [22, 23] is far superior to what has been reported until now with integrated circuits. The sensitivity of APDs is projected to be better than the PINFET circuits at gigabit-per-second frequencies [5]. However, with improved FET technology the advantages of low parasitic capacitance and low cost may enable integrated PINFET receivers to compete successfully with APDs for high-performance optical receivers.

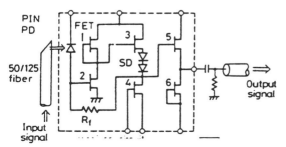

Figure 23.8 An integrated GaAs receiver circuit diagram.

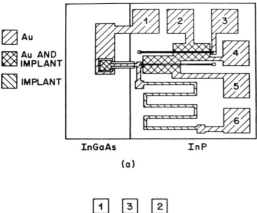

(a)

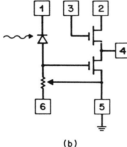

(b)

Figure 23.9 An integrated InGaAs–InP receiver circuit. (a) Chip layout; (b) circuit diagram.

Let us now briefly describe some other circuits that have been mono-lithically integrated. An optical repeater circuit has been demonstrated by Bar-chaim et al. [24]. This GaAs device contains an optical detector, a MESFET amplifier, and an output laser so that optical signals are amplified by the electronic circuit and converted back to high-power optical signal by the output laser.

Another important goal of optical device integration is the development of individually addressable arrays of light sources and detectors [25–28]. One motivation for this work is to replace many discrete devices with a single array at locations where large numbers of fibers are terminated, for example, at central office terminals of LANs where there is a high density of optical fibers converging from remote locations. A schematic description of a 1 × 12 array of PIN photodetectors on InP as reported by Brown et al. [27] is shown in Figure 23.10. The packaging of this device is of special interest. The spacing of the individual diodes corresponds to the spacing used for the fiber ribbon splicing using a V-grooved silicon chip for alignment. The same silicon block was also used for alignment of the fiber ribbon to the array. The packaged device with the array mounted vertically on the silicon V-groove block is shown in Figure 23.10c.

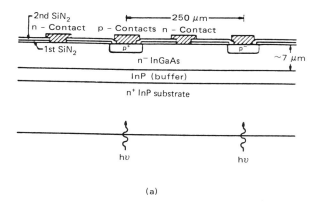

(a)

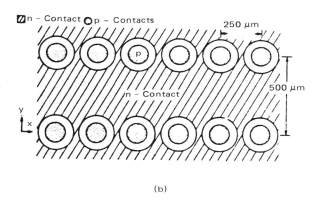

(b)

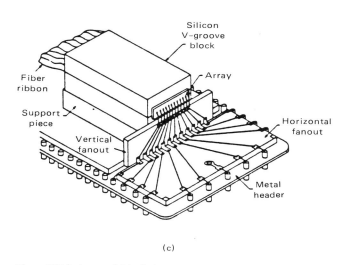

(c)

Figure 23.10 A monolithically integrated 1 × 12 PIN detector array. (a) Schematic cross section of the array; (b) top view of the array; (c) schematic diagram of the array after packaging.

23.4 *INTEGRATION OF ACTIVE AND PASSIVE WAVEGUIDES*

The terms *active waveguide* and *passive waveguide* are commonly used to indicate whether injection current is necessary in order to make the waveguide transparent to the propagating light. The "active" layer of a laser is highly absorbing at the laser wavelength (without the drive current), but a layer with a higher bandgap can be an effective "passive" waveguide that is reasonably transparent to the laser wavelength without any need for injection current. The use of intracavity passive waveguides in laser diodes is of interest [11, 29, 30] because it can improve laser spectral and noise characteristics. There is a wider potential for future development, however, for the integration of devices with additional functions such as passive filters, modulators, directional couplers, and switches, together with light sources and detectors on semiconductor substrates.

Two important problems that are related to crystal growth technology need to be addressed:

1. It is necessary to develop low-loss waveguide structures on III–V substrates.
2. Low-loss coupling techniques for different waveguides on a single substrate have to be found.

Several approaches have been suggested for integrated waveguides, including composite materials such as glasses and plastics [31] on semiconductors. The more conventional approach, however, is to use ridge or heterostructure waveguides of group III–V materials that are epitaxially grown. Figure 23.11a shows a ridge waveguide demonstrated by Hiruma et al. [32]. This waveguide was made by MOCVD epitaxial growth of very low doped GaAs–GaAlAs layers. The propagation loss is dominated by free-carrier absorption and by scattering from irregularities on the waveguide sidewall boundaries. Figure 23.11b shows the free-carrier absorption of the bulk and the loss measured for the waveguide without the GaAlAs clad layer (Fig. 23.11a). The cladding layer is added to prevent the optical field from penetrating the lossy n^+-substrate, and with this layer losses as low as 0.2 dB/cm have been obtained.

Such a low loss has not yet been realized in an integrated laser structure. Typically, losses of intracavity waveguides have been as high as 10–20 dB/cm, which practically limits the length of these passive waveguides to 1–2 mm. (As a rule of thumb, the waveguide and coupling loss should not be higher than the mirror loss of the laser.) An example of an InP integrated intracavity passive waveguide by Fujita et al. [29] is shown in Figure 23.12. The laser is composed of two parts; the short part (265 μm) is the active buried heterostructure laser, and the long part (1.5 mm) is a transparent passive waveguide.

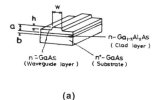

(a)

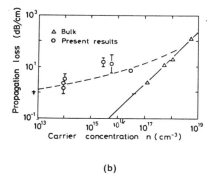

(b)

Figure 23.11 (a) Schematic diagram of a low-loss ridge waveguide on a GaAs substrate with a GaAlAs cladding layer. (b) Propagation loss at 1.3-μm wavelength as a function of carrier concentration for the bulk material and for the ridge waveguide without the cladding layer.

This structure is used to obtain single-longitudinal-mode (SLM) operation along with relatively narrow linewidths, as shown in Figure 23.12b. The dependence of linewidth on drive current is due to the dependence of the effective refractive indexes of the waveguides on the drive current, which determines a set of currents for which good phase-matching conditions occur. At these currents the laser linewidth achieves minimal values, as shown in Figure 23.12b

Murata et al. [30] reported an integrated optical waveguide with a distributed feedback (DFB) laser that is shown in Figure 23.13. This laser usually operates with a single longitudinal mode; moreover, the linewidth is reduced from 6 MHz to less than 1 MHz by the addition of the 1–1.5-mm-long passive waveguide.

The basic reason for improvement in linewidth with the passive waveguide is that the refractive index and thus the optical length of the active part are dependent on carrier concentration, which is subject to fluctuations. The passive part, however, has a constant optical length so the total (active–passive) optical length is subject to less fluctuation as the passive length is increased. Indeed, when external mirrors with large cavity lengths are used, linewidths as narrow as 30 kHz have been reported [30].

We can now describe some coupling schemes that are used for coupling active and passive waveguides. The first is called the integrated twin waveguide (ITG) design as demonstrated by Utaka et al. [33] for a distributed

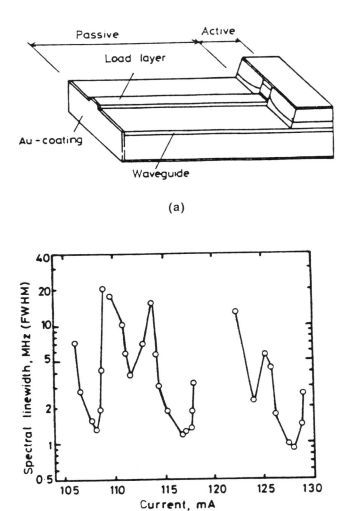

(a)

(b)

Figure 23.12 (a) Schematic diagram of an InGaAsP–InP laser with an integrated passive wave-guide. (b) Measured spectral linewidth as a function of the laser drive current.

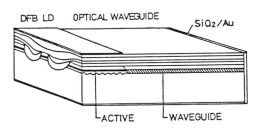

Figure 23.13 Schematic structure of a DFB laser integrated with a passive waveguide.

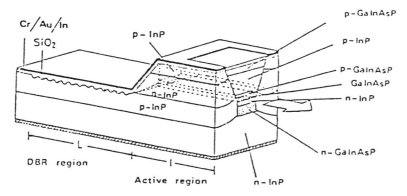

Figure 23.14 Schematic diagram of a 1.5-μm wavelength DBR laser with an integrated twin-guide configuration.

Bragg reflector laser (Fig. 23.14). In this case two layers with different wavelengths were grown by LPE. The top layer, which is the active laser layer, emits at 1.55 μm. The bottom layer has a bandgap corresponding to 1.3 μm and is thus transparent. The grating is etched on the bottom layer, as shown in Figure 23.14, to provide the DBR mirror. The coupling between waveguides is obtained by the evanescent fields of the modes, and phase matching is necessary for a good coupling. Phase matching is possible at certain laser currents, as discussed earlier. Another method of coupling passive and active waveguides for a DBR laser is shown in Figure 23.15. This is a butt-coupled DBR laser described by Abe et al. [34]. In this case there is no phase-matching requirement; however, good coupling can be obtained only if there is a good spatial overlap of the modes propagating in the waveguides as

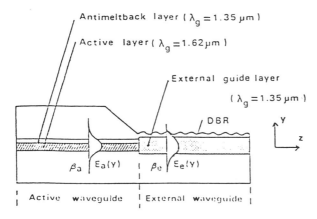

Figure 23.15 Schematic diagrams of a 1.5-μm wavelength butt-jointed built-in DBR laser.

shown schematically in Figure 23.15. This again calls for careful design of the thicknesses and bandgaps of the two waveguides. Generally a low-loss coupling is feasible but requires good control over layer thicknesses and refractive indices.

The progress in vapor-phase crystal growth as discussed in Section 23.1 is very important for integration of waveguides. As we have seen, there are critical requirements for precise control of layer thicknesses and bandgaps over large areas to obtain good coupling between waveguides. Also, very pure undoped materials are required for waveguides for low propagation losses. The vapor-phase crystal growth techniques such as MOCVD, VPE, and MBE offer new and promising technical solutions for these difficult requirements. New III–V devices based on optical waveguides for modulation, switching, and signal processing will probably emerge as a result of the advances now being made with these powerful crystal growth techniques.

REFERENCES

1. A. Yariv, "Components for integrated optics," *Laser Focus Magazine*, pp. 40–42 (December 1972).
2. N. Bar-chaim, S. Margalit, A. Yariv, and I. Uri "GaAs integrated optoelectronics," *IEEE Electron. Devices*, **ED-29**: 1372–1392 (1982).
3. T. Fukuzawa, M. Nakamura., M. Hirao, T. Kuroda, and J. Umeda, "Monolithic integration of a GaAlAs injection laser with a Schottky gate FET," *Appl. Phys. Lett.*, **36**: 181–183 (1980).
4. U. Koren, S. Margalit, T. R. Chen, K. L. Yu, A. Yariv, N. Bar-chaim, K. Y. Lau, and I. Uri, "Recent developments in monolithic integration of InGaAsP/InP optoelectronic devices," *IEEE J. Quantum Electron.* **QE-18**: 1653–1662 (1982).
5. S. R. Forrest, "Monolithic optoelectronic integration: a new component technology for lightwave communications," *J. Lightwave Technol.* **LT-30**: 1248–1264 (1985).
6. M. Hirao, S. Yamashita, T. P. Tanaka, and H. Nakano, "Monolithic integration of laser and driving circuits in GaAs system for high speed optical transmission," ECIO Conf., Berlin, Germany, May 1985; also T. P. Tanaka, M. Hirao, and M. Nakamusa, Opt. Fiber Commun. Conf. (OFC), San Diego, February 1985, paper TUC3.
7. L. A. Coldren, B. I. Miller, K. Iga, and J. A. Rentschler, "Monolithic two-section GaInAsP/InP active optical resonator devices formed by reactive ion etching," *Appl. Phys. Lett.*, **38**: 315 (1981).
8. H. Blauvelt, N. Bar-chaim, D. Fekete, S. Margalit, and A. Yariv, "AlGaAs lasers with microcleaved mirrors suitable for monolithic integration," *Appl. Phys. Lett.*, **40**: 289–290 (1982).
9. O. Wada, S. Yamakoshi, T. Fuji, S. Hiamatsu, and T. Sakurai, "AlGaAs/GaAs microcleaved facet laser monolithically integrated with photodiode," *Electron. Lett.*, **18**: 189–190 (1982).
10. U. Koren, Z. Rav-Noy, A. Hasson, T. R. Chen, K. L. Yu, L. C. Chin, S. Margalit, and A. Yariv, "Short cavity InGaAsP/InP lasers with dielectric mirrors," *Appl. Phys. Lett.*, **42**: 848–850 (1983).
11. S. Matsushita, "Recent progress of integrated optoelectronic devices and their applications," *Tech. Digest*, **2**: 21–28 (1985).
12. K. Kasahara, T. Terakado, A. Suzuki, and S. Murata, "Monolithically integrated high speed light source using 1.3 μm wavelength DFB-DC-PBH laser," *Tech. Digest*, **1**: 295–298 (1985).

13. U. Koren, K. L. Yu, T. R. Chen, N. Bar-chaim, S. Margalit, and A. Yariv, "Monolithic integration of very low threshold GaInAsP laser and MISFET on semi-insulating InP," *Appl. Phys. Lett.*, **40**: 643–645 (1982).
14. T. Y. Chang, R. F. Leheny, R. E. Nahory, E. Silberg, A. A. Ballman, E. A. Caridi, C. J. Harold, "JFETs using InGaAs material grown by MBE," *IEEE Electron Device Lett.*, **EPL 3**: 56–58 (1982).
15. J. Cheng, S. R. Forrest, C. L. Cheng, R. Stall, G. Guth, and P. H. Schmidt, "InGaAs/InP JFET for long-wavelength optoelectronic integration," Opt. Fiber Commun. Conf. (OFC) San Diego, 1985, paper WCG.
16. T. Horimatsu, T. Iwama, Y. Oikawa, T. Touge, O. Wada, and T. Nakagami, "400 Mbits/sec transmission experiment using two monolithic optoelectronic chips," *Electron. Lett.*, **21**: 319–321 (1985).
17. T. Horimatsu, M. Sasaki, H. Yamashita, T. Okiyama, T. Ohtsuka, K. Iguchi, H. Hamaguchi, and T. Nakagami, "High speed photoreceiver front end module with a monolithic PIN/FET and a GaAs amplifier, 10th Eur. Conf. Opt. Commun. (ECOC), Stuttgart, Germany, 1984, paper 9B2.
18. S. Miura, O. Wada, H. Hamaguchi, M. Ito, M. Makiuchi, K. Nakai, and T. Sakurai, "A monolithic integrated AlGaAs/GaAs PIN-FET photoreceiver by MOCVD," *IEEE Electron. Device Lett.*, **EDL-4**: 375–376 (1983).
19. R. F. Leheny, R. E. Nahory, M. A. Pollack, A. A. Ballman, E. D. Beebe, and J. C. DeWinter, *Electron. Lett.*, **16**: 353–354 (1980).
20. J. Barnard, H. Ohno, C. F. Wood, and L. F. Eastman, "Integrated double heterostructure GaInAs photoreceiver with automatic gain control," *IEEE Electron. Device Lett.*, **EDL-2**: 7–9 (1981).
21. B. Tell, A. S.H. Liao, K. Brown-Goebeler, T. Bridges, G. Burkhardt, T. Y. Chang, and N. Bergano, 'Monolithic integration of a planar embedded InGaAs PIN detector with InP depletion mode FETs," *IEEE Electron. Devices*, **ED-32**: 2319–2321 (1985).
22. J. C. Campbell, A. G. Dentai, W. S. Holden and B. L. Kasper, "High performance APD with separate absorption, grading and multiplication regions," *Electron. Lett.*, **19**: 818–820 (1983).
23. J. C. Campbell, W. S. Holden, J. F. Ferguson, A. G. Dentai, and Y. K. Jhee, "A high-speed InP-InGaAsP-InGaAs APD exhibiting a gain-bandwidth product of 60 GHz," *Tech. Digest*, **3**: 65–69 (1985).
24. N. Bar-chaim, K. Y. Lau, I. Uri and A. Yariv, "GaAlAs/GaAs integrated optical repeater," 7th Topical Meeting on Integrated and Guided Wave Optics (IGWO), Orlando, Florida, 1984, paper TWDI.
25. J. P. Van der Ziel, R. A. Logan, and R. M. Mikalyak, "A closely spaced array of 16 individually addressable BH GaAs lasers," *Appl. Phys. Lett.*, **41**: 9–11 (1982).
26. Y. Suzuki, Y. Nuguchi, K. Takahei, H. Nagai, and G. Iwane, "1.5 μm region BH laser array," *Jap. J. Appl. Phys.* **20**: L-229 (1981).
27. M. G. Brown, S. R. Forrest, P. H. S. Hu, D. R. Kaplan, M. Koza, Y. Ota, C. W. Potopowicz, and M. A. Washington, "Fully optically and electronically interfaced monolithic 1 × 12 array of InGaAs PIN photodiodes," IEDM Conf., San Francisco, 1984.
28. P. P. Deimel, J. Cheng, S. R. Forrest, P. H. S. Hu, R. B. Huntington, R. C. Miller, J. R. Potopowicz, D. D. Roccasecco, and C. W. Seaburg, "Individually addressable monolithic 1 × 12 light emitting diode array," *J. Lightwave Technol.*, **LT-3**: 988–991 (1985).
29. T. Fujita, J. Ohya, K. Matsuda, M. Ishino, H. Sato, and H. Serizawa, "Narrow spectral linewidth characteristics of monolithic integrated passive-cavity InGaAsP-InP semiconductor lasers," *Electron. Lett.*, **21**: 374–376 (1985).
30. S. Murata, I. Mitu, M. Shikada, and K. Kobayashi, "Narrow spectral linewidth DFB laser with a monolithically integrated optical waveguide," *Tech. Digest*, **1**: 299–302 (1985).

31. K. Furaya, B. I. Miller, L. A. Coldren, and R. E. Howard, "A novel deposit-spin waveguide interconnection for semiconductor integrated optics design and material," IGWO Conf. Pacific Grove, CA, January 1982.

32. K. Hiruma, H. Inoue, K. Ishida, and H. Matsumura, "Low loss GaAs optical waveguides grown by MOCVP method," *Appl. Phys. Lett.*, **47**: 186–188 (1985).

33. K. Utaka, K. Kobayashi, F. Koyama, Y. Abe, and Y. Suematsu, "Single-wavelength operation of 1.53 μn GaInAsP BH integrated twin waveguide laser with distributed Bragg reflector under direct modulation up to 1 GHz," *Electron. Lett.*, **17**: 368–369 (1981).

34. Y. Abe, K. Kishino, Y. Suematsu, and S. Arai, "GaInAsP/InP integrated laser with butt-jointed built-in distributed Bragg reflector waveguide," *Electron. Lett.*, **17**: 945–947 (1981).

CHAPTER 24

Coherent Optical Fiber Communication Systems—The Promise for the Future

I. W. Stanley and D. W. Smith

24.1 INTRODUCTION

The higher receiver sensitivity and enhanced tuning ability theoretically provided by coherent techniques offer the prospect of significantly improving upon the performance of present direct intensity detection single-mode optical fiber systems [1]. In current systems, the optical detector responds only to the presence of optical power from a laser in which the spectrum generated is similar to that of an intensity-modulated noise source. The optical linewidth of such a laser is many orders of magnitude greater than the bandwidth of the applied information, so the potential for exploiting its phase or frequency characteristics does not exist. The first requirement for a coherent system is therefore a laser with a very narrow spectral line.

In a coherent system the laser is used in two ways: first as the optical source in a transmitter producing a carrier wave that can be amplitude-, phase-, or frequency-modulated, and second as a continuous-wave (cw) oscillator in the receiver. When the power from the latter is combined with the incoming signal, the conversion gain provided by the square law photodetection process results in a receiver sensitivity that is limited by the shot noise of the local oscillator or incoming signal and not by the thermal noise of the receiver. Coherent detection thus provides a route for realizing a sensitivity almost equal to the quantum noise limit.

In optical systems the term *coherent* has come to be synonymous with heterodyne and homodyne detection regardless of whether the signal demodulation process itself is coherent. In homodyne detection, where the mean carrier and local oscillator frequencies are identical, the detection process is intrinsically synchronous. For heterodyne detection, where the frequencies differ, the demodulation scheme is electronic and can be either coherent or nonsynchronous. For both heterodyne and homodyne operation, however, the single-mode fiber transmission medium maintains a smooth wave front and provides the essential spatial coherence of the two waves at the detector that is necessary for efficient photomixing.

Theoretical studies [2–4] have shown that the potential improvement in receiver sensitivity ranges from almost 15 dB at less than 100 Mb/s to at least 20 dB at several gigabits per second. Such an improvement would allow greater regenerator spacing in inland or undersea systems or operation over the same regenerator spans at higher bit rates. The existence of a local oscillator in the receiver also suggests that it may be even more attractive in the context of a wideband network where the optical frequency-tuning ability of the local oscillator could be used to select individual optical carriers. The bandwidth provided by the monomode fiber in the wavelength range 1250–1650 nm is about 50,000 GHz, so the potential exists for transmitting many thousands of channels, each on a different carrier frequency. In this context, the additional sensitivity would allow greater optical distribution losses to be tolerated and a variety of optical circuit components to be exploited in a way that is not possible to the same extent using direct intensity-detection techniques.

In this chapter, the principles of heterodyne and homodyne detection will be presented, and an account will be given of the techniques used for practical realization at optical frequencies. Potential applications to both point-to-point links and ultrawideband multichannel networks will be outlined.

24.2 PRINCIPLES OF COHERENT OPTICAL TRANSMISSION

The principal differences between direct detection and coherent optical transmission links are shown in Figure 24.1; the components enclosed by the broken lines are absent for direct detection. The broken line at the transmitter encloses an optical modulator that can provide amplitude, phase, or frequency modulation of the cw output from the narrow spectral line laser; it may be possible in some instances to modulate the drive current to the laser instead. At the receiver, the optical power from the local oscillator is combined with the incoming signal and fed to the photodiode, which acts as an optical mixer and produces a conversion gain that results in improved receiver sensitivity.

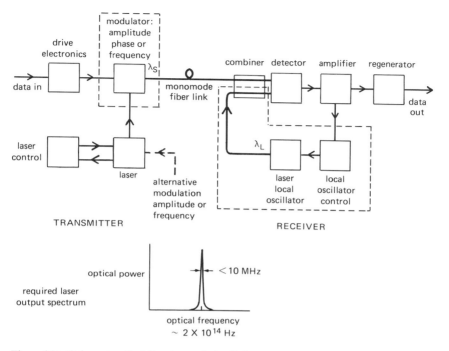

Figure 24.1 Coherent optical fiber transmission link.

24.2.1 *Photocurrent Generation in a Coherent Receiver*

The photocurrent generated by an incident photon stream of average power P is given by

$$i = \eta \frac{qP}{hf} \tag{1}$$

where P/hf is the average number of photons incident on the photodiode per unit time, I/q is the average number of electron–hole pairs collected across the junction region of the photodiode; and η is the quantum efficiency. If the constant of proportionality relating optical power and electric field is $1/z$, then $P = E^2/Z$ and

$$i = \eta \frac{qE^2}{Zhf} \tag{2}$$

If the signal carrier field is $E'_s = E_{0s} \cos w_s t$ and the local oscillator field is $E'_L = E_0 \cos w_L t$ then, neglecting dc terms, the sum frequencies which are too high to be passed by the detector, and writing $E = E'_s + E'_L$, the total peak current $\hat{i}$ in the photodetector is

$$\hat{i} = \eta q E_{0s} E_{0L} \frac{\cos (w_s t - w_L t)}{hfZ} \tag{3}$$

The root mean square current is then;

$$i_T = 2\eta q E_s E_L \frac{\cos(w_s t - w_L t)}{hfZ} \tag{4}$$

The rms carrier current at the intermediate-frequency stage in the heterodyne case is then

$$i_{T1} = \sqrt{2}\eta \frac{q E_s E_L}{hfZ} \tag{5}$$

The homodyne detection process recovers a dc component from the carrier at baseband, which is the rms current, so that

$$i_{T2} = 2\eta \frac{q E_s E_L}{hfZ} \tag{6}$$

In both cases the carrier photocurrent depends linearly on the optical signal field and is in effect amplified by a factor proportional to the electric field E_L produced by the local oscillator.

24.2.2 Carrier-to-Noise Ratio

For a photodetector in which the current generated is directly proportional to the incident power P, the mean square value of the shot noise $\overline{I_{sh}^2}$ is given [5] by

$$\overline{I_{sh}^2} = 2qBP \frac{\eta q}{hf} \tag{7}$$

Where B is the noise bandwidth defined by a low-pass filter having a rectangular passband. In terms of the electric field,

$$\overline{I_{sh}^2} = 2qBE^2 \frac{\eta q}{Zhf} \tag{8}$$

If the local oscillator field is very much greater than the incoming carrier, then $E = E_L$, and

$$\overline{I_{sh}^2} = 2qBE_L^2 \frac{\eta q}{Zhf} \tag{9}$$

The carrier-to-noise ratio (CNR) for heterodyne detection in the absence of any additional noise sources, including the thermal noise of the amplifier, is then

$$\text{CNR} = i_{T1}^2 R_L / \overline{I_{sh}^2} R_L \tag{10}$$

Thus,

$$\text{CNR(heterodyne)} = \eta \frac{E_s^2}{Zhf B} \tag{11}$$

and

$$\text{CNR(homodyne)} = 2 \frac{\eta E_s^2}{Zhf B} \tag{12}$$

Alternatively, the CNR can be written in terms of optical power as follows:

$$\text{CNR(heterodyne)} = \eta \, \frac{P}{hf B} \tag{13}$$

and

$$\text{CNR(homodyne)} = 2\eta \, \frac{P}{hf B} \tag{14}$$

Thus the CNR for homodyne detection is twice that of the heterodyne case. Consequently the theoretical sensitivity of a homodyne receiver is 3 dB greater than that for a heterodyne design.

If the thermal noise of the amplifier is significant, then the equivalent mean square amplifier noise current must be added to the denominator of Eq. (9).

The signal-to-noise ratio (SNR) of the system measured at the signal decision point in the regenerator depends on whether amplitude, phase, or frequency modulation is used. In addition, for heterodyne detection, it is affected by the type of intermediate-frequency demodulator and baseband filter selected for the receiver. In a digital system, all these factors influence the receiver sensitivity that can be obtained for a given bit error rate (BER).

24.2.3 Error Rate Considerations

When the dominant noise source is shot noise produced by a strong local oscillator field, Gaussian statistics for the photodetected signals can be applied. With synchronous demodulation, similar statistics apply at the input to the decision gate, and the Gaussian approximation BER expression can therefore be used [6]:

$$\text{Pe} = \frac{1}{2} \, \text{erfc} \, \eta \, \frac{kmPs}{2h_f B} \tag{15}$$

where Pe is the error probability, $k = 1$ for heterodyne detection and 2 for the homodyne case, and m depends on the choice of modulation format.

For nonsynchronous demodulation, such as envelope detection, the signal statistics at the decision gate are non-Gaussian, and in this case the BER approximation based on the Marcum Q function can be used [7], so that

$$\text{Pe} \approx \frac{1}{2} \, \exp\left(-\eta \, \frac{kmPs}{2h_f B} \right) \tag{16}$$

For bit error rates of about 1 in 10^9, Eq. (16) implies that the receiver sensitivity is approximately 1 dB less than that for the Gaussian case.

24.2.4 The Performance of Different Modulation Techniques

The use of heterodyne or homodyne in place of direct detection widens the choice of modulation that can be considered for optical fiber transmission. For radio systems it is well known that schemes that do not waste transmitter power in a large carrier component give better performance than simple on–off keying. It is therefore instructive to compare the performance of amplitude, phase, and frequency shift key modulation techniques.

a. Phase and Amplitude Shift Key Modulation. The peak-to-peak signal photocurrent generated in a homodyne receiver by an ASK signal is given by

$$i_{s1} - i_{s0} = 2\sqrt{P_s P_L}\,\frac{\eta q}{hf} \tag{17}$$

where i_{s1} is the current produced in the binary 1 state and i_{s0} is that resulting from the binary 0 condition. With the same receiver used for antipodal PSK ($\pi/2$) transmission,

$$i_{s1} - i_{s0} = 4\sqrt{P_s P_L}\,\frac{\eta q}{hf} \tag{18}$$

The factor of 2 in the peak-to-peak signal photocurrent represents a 6-dB improvement in receiver sensitivity in terms of peak received power or 3 dB if mean power is considered. The same relative advantage of phase over amplitude modulation occurs for heterodyne detection.

b. Frequency Shift Keying. In FSK, the two binary signal states are defined by different signaling frequencies with a separation that is normally very much greater than the bit rate. With two frequencies, the receiver sensitivity is approximately equivalent to ASK.

Although 2FSK will usually result in inferior performance compared to 2PSK, there is always the possibility of improving receiver sensitivity with FSK by increasing the number of signaling frequencies, thus trading the available optical bandwidth for sensitivity [8]. With an unlimited number of signaling frequencies, FSK would be about 7 dB better than PSK; with a more realistic number, say 8FSK, a sensitivity equivalent to 2PSK could be achieved but at the expense of greater receiver bandwidth.

24.2.5 Demodulation

The process of recovering the information impressed on a carrier wave is called detection or demodulation. If the demodulation process relies on detecting the transmitted phase information, then it is said to be coherent or synchronous. If knowledge of the instantaneous phase of the incoming carrier is not needed, then the demodulation process is nonsynchronous. The most efficient homodyne detection is intrinsically synchronous, but heterodyne signals can be demodulated using either coherent or nonsynchronous methods.

a. Coherent Demodulation. Coherent detection places the greatest demands on the phase stability of the optical source. With homodyne operation the optical local oscillator must be phaselocked to the transmitted signal, and with heterodyne detection it is necessary to lock an electrical oscillator to the intermediate frequency. For PSK transmission the carrier is normally suppressed, and there is therefore a requirement to transmit either a low-level pilot carrier or to generate a substitute carrier in the receiver from the modulation sidebands. With heterodyne detection the intermediate-frequency signal can be detected by either an electrical synchronous detector, which requires the generation of a local carrier at the intermediate frequency in the receiver, or by nonsynchronous means using envelope or square law detection for ASK signals. The nonsynchronous detection of PSK signals can be carried out using autocorrelation techniques.

b. Nonsynchronous Electrical Demodulation. Of the three modulation formats, ASK with nonsynchronous demodulation is least affected by laser phase noise. Here, such noise is of significance only when the IF bandwidth of the heterodyne receiver has to be broadened to allow for any frequency jitter in the intermediate frequency. Since the output signal-to-noise ratio of a square law or envelope detector is a nonlinear function of the input signal-to-noise ratio, a penalty will arise from increased shot noise power because of an excessive intermediate-frequency bandwidth; in practice, linewidths, in hertz, of the same order as the modulation rate, in bits per second, are required. FSK signals can be demodulated nonsynchronously by either an electrical limiter-discriminator or, in the case of 2FSK, by using separate filters for the mark and space frequencies followed by two envelope detectors; in either case, laser phase noise will result in a degradation of the BER. When a frequency deviation much larger than the bit rate is used, the requirements on phase noise for FSK with envelope detection are the same as for nonsynchronous ASK. With discriminator detection the laser phase noise is converted into amplitude noise [9]; in practice, the frequency deviation between the mark and space frequencies must be several times the source linewidth. Nonsynchronous demodulation of PSK signals using differential techniques is also affected by phase noise, and the full-width half-maximum (FWHM) laser linewidth must in this case be at least two orders of magnitude less than the modulation frequency [10].

24.2.6 Performance Comparison of Direct Detection and Carrier Wave Techniques

The performance of systems based on optical carrier techniques depends on the modulation format, the detection process, and the decision about whether a homodyne or heterodyne mode of operation is chosen. In addition, the choice of direct or external laser modulation and the local oscillator power available both significantly affect the performance. In this section the effect of these parameters will be discussed and receiver performance will be compared with experimental direct-detection systems.

a. Receiver Sensitivity. The potential performance of ideal coherent detection receivers, compared with the best reported PINFET and avalanche photodiode (APD) direct-detection receivers, is shown in Figure 24.2 over a range of bit rates. In this comparison it should be noted that the low-noise, high-impedance direct-detection receiver has a rising frequency noise term, which results in its performance degrading at 4.5 dB for each doubling in bit rate. For ideal coherent detection and for direct detection with an APD, a constant energy per bit is required, irrespective of data rate, which leads to the receiver sensitivity degrading at the rate of 3 dB as the bit rate is doubled. A summary of the theoretical performance of the various modulation and detection schemes is given in Table 24.1.

In addition to a direct improvement in receiver sensitivity, the use of angle modulation (PSK and FSK) could result in an extra 3 dB of available transmitter power if the laser output is peak-power-limited.

In practice, the improvements in sensitivity from using coherent detection may not, ultimately, be so large, because improvements in direct-detection receivers are still possible and perfect coherent detection may not always be achievable.

b. Effect of Local Oscillator Power. In practice, the amount of local oscillator power incident on the photodetector may be insufficient to achieve shot-noise-limited detection, and the receiver sensitivity will then be lower

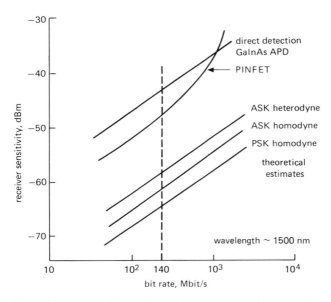

Figure 24.2 The sensitivity of direct detection and coherent receivers.

TABLE 24.1. Effect of bit rate and detection scheme on the sensitivity of a coherent receiver

Modulation type	Detection scheme	Theoretical receiver sensitivity (dB for 10^{-9} BER)			Transmitter power advantage (dB)
		140 Mb/s	1.2 Gb/s	2.4 Gb/s	
ASK heterodyne	Envelope	−59.5	−50.5	−47.5	0
	Synchronous	−60.5	−51.5	−48.5	0
ASK homodyne	Synchronous	−63.5	−54.5	−51.5	0
PSK heterodyne	Differential	−62.5	−53.5	−50.5	+3
	Synchronous	−63.5	−54.5	−51.5	+3
PSK homodyne	Synchronous	−66.5	−57.5	−54.5	+3
FSK heterodyne, large deviation, 2 FSK	Envelope	−59.5	−50.5	−47.5	+3
	Synchronous	−60.5	−51.5	−48.5	+3
FSK heterodyne unlimited range of signaling frequencies	Synchronous	−73.5	−64.5	−61.5	+3

than the theoretical estimates. This can result directly from restricted laser output power but can also arise if the loss in the optical path between the local oscillator and the receiver is high. Often, to ensure that there is low optical loss in the signal path, it is necessary to use a combiner with a low coupling coefficient at the receiver, which results in a higher loss in the local oscillator path. In addition, if excess intensity noise is generated by the local oscillator laser [11], better performance can be achieved by attenuating its output using a suitable coupling ratio. Consequently, a low-noise preamplifier is a prerequisite for performance that approaches the shot noise limit. Theoretically, by using a high-impedance PINFET [12] module in a homodyne receiver, shot-noise-limited detection can closely approach the optimum with only 1 μW of local oscillator power.

Alternatively, the effect of excessive local oscillator noise can be minimized by using a balanced receiver [13]. However, in practice it could be difficult to build such a receiver because it is necessary to cancel excess noise up to frequencies near the laser relaxation oscillation frequency of several gigahertz.

24.3 FIBER REQUIREMENTS

The most common single-mode fiber design has an overall diameter of 125 μm with a core diameter of about 8 μm. The attenuation characteristic of a typical fiber of this type [14] is shown in Figure 24.3; two low-loss

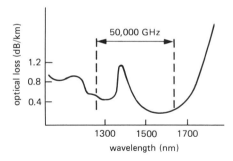

Figure 24.3 Optical loss and electrical bandwidth of single-mode fiber.

windows exist, one covering a range of about 1250–1350 nm and the other wavelengths spanning 1450–1700 nm. In coherent and direct systems based on narrow-line optical transmitters, the limit of transmission range is normally set by the fiber attenuation [15], and so the highest performance systems result from operating at a wavelength of about 1550 nm. The bandwidth theoretically available from the single-mode fiber taking both long-wavelength windows together is about 50,000 GHz, which is also indicated in Figure 24.3. An individual transmission based on a single optical frequency carrier would occupy only a small part of this very large bandwidth. For example, a 565-Mb/s digital transmission would require less than 1 GHz of optical bandwidth.

The transmission requirements are more demanding than those for a direct detection system. In particular, the polarization state of the signal arriving at the receiver must be matched to that produced by the local oscillator laser to obtain efficient mixing. Unfortunately, the polarization state received after a long optical fiber transmission path cannot always be predicted because mechanical movement and temperature fluctuations of the fiber cable will alter any residual birefringence within the fiber and influence mode coupling.

Therefore some attention must be given to the penalties incurred from polarization misalignment and related automatic control techniques. If the polarization states of the signal and local oscillator fields are not matched, there will be a penalty that depends on the relative polarization state of the received signal and local oscillator. For angular discrepancies of up to about ± 30°, the penalty is less than 0.5 dB.

24.3.1 Conventional Fiber

The conventional non-polarization-holding fiber can be considered for coherent transmission either by including a polarization control element within the receiver or by employing polarization diversity.

In practice it has been found that fluctuations in polarization state are fairly slow for cabled fiber. The polarization stability of a conventional single-

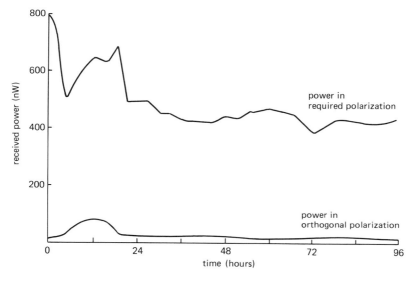

Figure 24.4 Polarization stability for single-mode fiber cable installed in duct.

mode fiber over a period of 96 hr is shown in Figure 24.4, both for fiber wound on a drum and for a cabled fiber in an underground duct [16]. For these measurements the source consisted of a single-frequency linearly polarized 1523-nm He-Ne gas laser, and the receiver contained a polarization-dependent beamsplitter to isolate the components. It is clear that large polarization changes do occur but only over periods of minutes or hours.

This suggests that the control element in the feedback system at the receiver can be similarly slow-acting. This allows a wide choice of possible compensation techniques; the fiber could, for example, be squeezed electromechanically or wound on a piezoelectric cylinder, although compact integrated waveguide electrooptic devices [17, 18] might prove to be the most attractive. The necessary compensator can be placed in either the signal or the local oscillator path; the latter is preferable if the device introduces significant optical attenuation. A control system would then optimize the performance of the system by making continuous adjustments.

The dual-polarization receiver could take the form shown in Figure 24.5. In this scheme the input signal is combined with a circular or obliquely polarized local oscillator beam and then passed through a polarizing beamsplitter; the two orthogonally polarized outputs are detected on separate photodiodes. In a heterodyne receiver the two outputs from the detectors are demodulated down to baseband by envelope detection before recombination. With this basic configuration there is the possibility of a performance degradation of 3 dB for certain input polarization states. If a method of polling the two receiver outputs is adapted that allows the largest signal

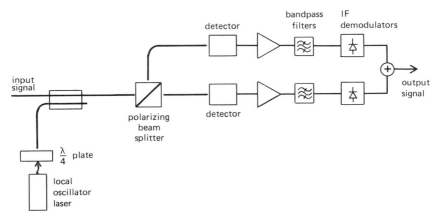

Figure 24.5 Polarization diversity receiver.

or an optimum combination to be obtained, then the penalty could be as small as 1 dB.

24.3.2 Polarization-Holding Fiber

If a fiber that maintained a defined state of polarization could be used, the need for a polarization controller would be removed. Linear-polarization-holding fibers depend on strong built-in birefringence to swamp external pressure- and bending-induced birefringence. There are a number of fiber designs that can be used; one successful design is the bow-tie fiber from Southampton University. In this fiber, stress-applying regions of borosilicate glass are located on two opposite sides of the silica core to give the required polarization selection by radially anisotropic stress [19, 20].

The introduction of a high degree of birefringence also results in a polarization mode dispersion of up to 5 nsec/km between the two orthogonal polarization modes. Therefore, if any power does couple between modes, for example, from axial misalignment of joints, the received signal polarization state will be a function of the source optical frequency [21] and hence its modulation. Although this problem can be avoided for certain fiber designs, such as the leaky mode fiber that allows only a single polarization mode to propagate [20], the present high attenuation of linear-polarization-holding fibers [22, 23] must be reduced from about 0.6 dB/km to less than 0.25 dB/km at 1500 nm in order to compete with conventional fiber for trunk transmission. There may, however, be a limited use for such fibers in interconnecting optical components over short distances in wideband networks where the frequent inclusion of polarization controllers would be impractical or expensive.

24.4 OPTICAL SOURCES

The optical source in the transmitter of a single-mode direct-detection optical system usually consists of a semiconductor laser diode with a nominal wavelength of 1300 or 1500 nm that is electrically switched on and off by a binary-coded data stream. The resulting optical spectrum often consists of a number of spectral lines spaced a few tenths to several nanometers in wavelength apart and covering an optical spectrum of several nanometers. Inasmuch as 1 nm at 1500-nm wavelength is equivalent to a frequency of 133 GHz, it is clear that for the normally used modulation rates up to 2000 Mb/s the optical linewidth of the laser is many orders of magnitude greater than the data bandwidth applied to it. For coherent systems a source is needed with a linewidth narrow compared with the spacing of the modulation sidebands; a cw source with similar characteristics is also required to provide the local oscillator power to the receiver.

24.4.1 The Gas Laser

A 1523-nm [24] He-Ne gas laser was used in many early tests because no suitable semiconductor laser was available. The quantum-limited linewidth of the gas laser is less than 1 Hz, although in practice frequency fluctuations caused by acoustic disturbances occur that result in a practical linewidth of about 10 kHz. The frequency of the laser line can be tuned over a range that is not much in excess of 200 MHz by adjusting the position of the laser mirrors using a piezoelectric transducer. This range is limited by the intrinsic gain–bandwidth characteristic of the laser and restricts the maximum intermediate frequency (IF) that can be used in a heterodyne system if He-Ne lasers are used for both the transmitter and the local oscillator in the receiver. To obtain a larger intermediate frequency, a different type of laser source with a wider tuning range is needed to replace either the transmitter or local oscillator source or both. Several alternative sources have been developed that complement the gas laser and, except for specialist and some measurement applications, may largely supersede it.

24.4.2 Narrow Spectral Line Semiconductor Laser Sources

The spectral output of a semiconductor laser often consists of a number of different frequencies or modes determined by the length of the laser cavity. The mode spacing would be about 1 nm (133 GHz), as shown in Figure 24.6a, for a typical optical cavity 400 μm long. In some instances the laser can be made to oscillate in a single longitudinal mode under continuous wave conditions but may revert to multimode when modulated. It is therefore necessary to induce the laser to oscillate in a single longitudinal mode

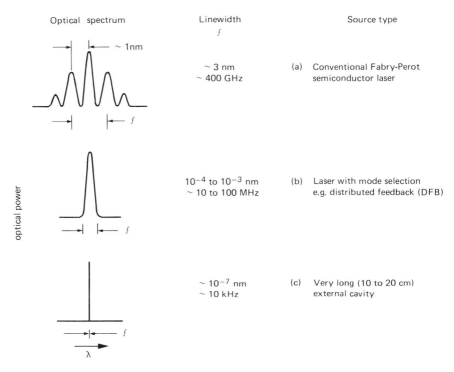

Figure 24.6 Laser spectra.

when modulated and then to narrow the optical width of the remaining mode so that it is small compared with the modulation bandwidth. Such a reduction implies a change by a factor of at least 10^4, from about 100 GHz to less than about 10 MHz.

A design that provides the spectrum shown in Figure 24.6b includes a frequency-selective region in the laser structure and is shown in Figure 24.7. A transverse grating structure running the length of the laser has been fabricated near the active region of the laser to provide distributed optical feedback [25, 26]. Wavelengths related to the grating period are preferentially reflected, and a single longitudinal mode of the required wavelength is obtained. Theoretical studies indicate that spectral linewidths of about 1 MHz should be attainable, but to date experimental devices have exhibited linewidths of rather less than 20 MHz [28].

The narrowest linewidths have been obtained for Fabry-Perot devices with wavelengths in the region of 850 nm compared to those with longer wavelengths; the linewidths for all, however, are broader than predicted by the Schawlow-Townes relationship [29]. This difference with theory has been accounted for by the inclusion of a multiplier [30] $(1 + \alpha)^2$ to give a FWHM

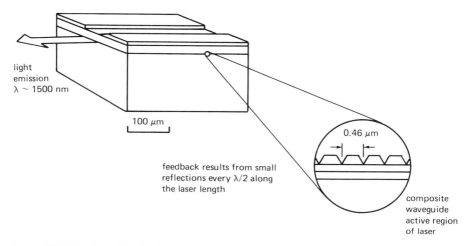

Figure 24.7 Distributed feedback (DFB) laser.

Lorentzian shaped linewidth of

$$\Delta f_L = \frac{A_0 h_f (\Delta f_c)^2}{P_0} (1 + \alpha)^2 \tag{19}$$

where P is the laser output power, Δf_c is the bandwidth of the Fabry-Perot cavity estimated from its physical parameters including the length and A_0 is a constant. The α factor is due to coupling between the intensity fluctuations and phase noise.

It is also possible to reduce the linewidth to a few tens of kilohertz by coupling the output from a conventional Fabry-Perot laser into a frequency-selective external reflecting cavity [31]. A 2–20-cm long optical cavity is necessary to produce an optical line a few kilohertz wide. By reflecting the collimated output beam from one facet back into the laser from a diffraction grating, the required line narrowing is obtained as shown in Figure 24.8.

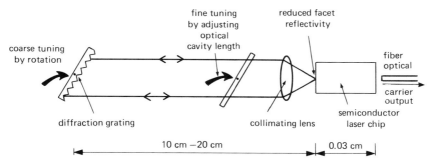

Figure 24.8 External cavity semiconductor laser.

TABLE 24.2. Spectral width of optical sources

a. Laser linewidth requirements

Modulation	Linewidth/bit rate %	Laser type
ASK, FSK	<20	DFB
DPSK	<0.3	At present external cavity
PSK	<0.1	External cavity

b. Coherent system requirements

Laser type	Modulation	Demodulation		Linewidth per bit rate %
		Het	Hom	
External cavity gas	ASK, FSK, PSK	Sync	Yes	<0.1
DFB	ASK, FSK	Env det*	No	<20

* Env det = envelope detection.

The spectral linewidth from the laser is inversely proportional to the square of the length of the external cavity, and so a distance of about 10 cm reduces the original linewidth to less than 10 kHz (Fig. 24.6c). The acoustic disturbances that act to broaden the spectral line of the He-Ne laser also affect the cavity structure and produce low-frequency jitter with megahertz deviations. Frequency deviations in this range can be reduced considerably by including a suitable AFC loop in the receiver. The requirements for various modulation formats in terms of a linewidth-to-bit rate ratio are summarized in Table 24.2.

In addition to linewidth reduction, it is necessary to stabilize the center frequency against longer-term drift caused by temperature variations. This could be achieved with an optoelectronic feedback loop that locks the laser output wavelength to, for example, the 1500-nm atomic or molecular absorption spectra of an ammonia absorption cell. Alternatively, both linewidth reduction and long-term stabilization can be obtained by injection locking a diode laser to a narrow-line reference source such as a He-Ne laser [32]. Either a gas or semiconductor laser can be used as the optical transmitting device or as the local oscillator in the receiver. In the transmitter application, the modulation method for the laser is of major importance.

24.4.3 Transmitter Design

The design of a transmitter for an optical carrier wave system is dictated primarily by the modulation format required. With amplitude modulation there is a choice of providing external modulation or using the laser drive current. For phase-shift keying (PSK) operation, external modulation is the natural choice, while frequency modulation is most easily obtained by changing the drive current to the laser. At high modulation rates, above about 1 Gb/s, the laser design itself also influences the choice. If the laser is speed-limited or exhibits an unacceptable wavelength shift (chirp) during the transient turn-on period [33], then the only choice may be to use a separate modulator.

The most commonly used techniques for fabricating external modulators are based on either III–V compound semiconductors or lithium niobate technology [18]. A device of the latter type is shown diagramatically in Figure 24.9. The refractive index of the optical path and therefore the optical phase can be changed by altering the applied electric field. To produce a 180° phase shift requires a potential of about 7 V and a device about 2 cm long. Amplitude modulators can also be fabricated using lithium niobate waveguide technology; these can take the form of either Mach-Zehnder interferometric devices or Y-switches.

Such modulators have been successfully operated using both digital [34, 35] and pulse-frequency modulation [36] techniques, but several decibels of additional optical loss is introduced into the signal path that could, in principle, be avoided by direct modulation of the laser diode. Direct optical frequency modulation of the laser is possible because the refractive index of the light-generating region within the laser is dependent on the magnitude of the injection current. This results from two effects. The first effect is thermal in origin and is significant for modulation frequencies below about 10 Mhz. The second arises from incomplete clamping of the carrier density above oscillation threshold. Both effects produce a current-dependent refractive index change [37], which in turn alters the laser frequency. For example,

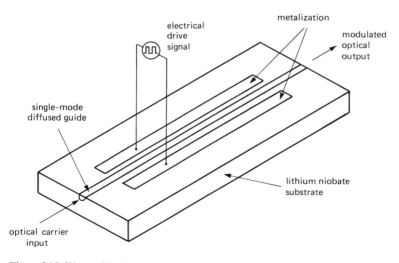

Figure 24.9 Waveguide phase modulator.

a 1-mA change in the current can produce a deviation of 1 GHz in the optical carrier frequency.

In practice the frequency deviation between the 0 and 1 frequencies results in a sideband spectrum that can be treated as two superimposed ASK signals. The implication is that at 140 Mb/s a linewidth on the order of 20 MHz would be adequate; higher bit rates imply a wider linewidth. The DFB lasers now under development are beginning to provide this performance, but a small penalty is incurred, because a small amplitude-modulation component is introduced by changing the laser drive current during the frequency modulation process. The laser can, of course, be amplitude-modulated specifically by switching the bias current between the lasing threshold and full output.

The choice of modulation type is governed by a combination of performance requirements and component limitations. Once defined, however, the modulation format, together with the choice of heterodyne or homodyne operation, greatly influences the design of the optical receiver.

24.5 RECEIVER DESIGN

One of the primary design objectives for a homodyne or heterodyne receiver is to obtain the maximum optical sensitivity at the required information rate with the lowest possible local oscillator power. This requirement implies that the effect of noise sources in the receiver must be minimized for the conditions under which the receiver operates. For example, the bandwidth of the receiver should be no greater than the minimum necessary to obtain error-free operation at the chosen information rate. In addition, the design must be tailored to minimize the effects of phase and amplitude noise from the local oscillator laser. Since the latter is of more significance in a homodyne receiver than in a heterodyne receiver, it is the homodyne case that will be described first.

24.5.1 Homodyne Receiver

The combined signal and local oscillator fields that are incident on the photodiode of a homodyne receiver produce an amplitude-modulated signal that is recovered directly at baseband [38]. The implication is that any direct-detection receiver design can be considered as the basis for a homodyne receiver. A receiver design based on the avalanche photodiode (APD) is unattractive because of the additional source noise associated with the avalanche multiplication process [39]. In the case of the PINFET receiver, the choice lies between a design in which the output from the PIN diode is fed to a high-impedance amplifier and one in which a transimpedance design is used. Most of the published coherent systems work, however, has been carried out using the high-impedance receiver, and so the former will not be considered further in the present context.

a. Design for a Homodyne Receiver. The basic design of a homodyne receiver is shown in Figure 24.10. The combined signal and local oscillator fields are recovered directly at baseband by a PINFET receiver with a bandwidth in hertz and approximately half the information rate in bits per second. Such an output will be obtained only when the local oscillator frequency is accurately phase-matched to the mean frequency of the incoming signal. In practice this requirement dictates the use of a phase-sensing feedback loop that is used to control the output from the local oscillator laser. In the example illustrated in Figure 24.10, the photodiode produces an output current that depends on the phase difference between the incoming signal and the local oscillator output. At point *A* the resulting signal is fed to a differential amplifier that allows only the low-frequency components resulting from drifts in the local oscillator or signal frequencies to pass. The resulting signal is then applied to one of the cavity mirrors of the He-Ne laser to maintain phaselocking between the two optical fields. The data signal is passed to the decision gate for regeneration and onward transmission.

In practice the major drawback of this circuit is that unwanted amplitude fluctuations in the incoming signal or from the local oscillator are treated in the same way as phase changes. With present lasers these fluctuations can be much larger than those resulting from the phase error signal used to generate the control voltage in the feedback loop. The phaselocking is therefore lost very easily, and a more sophisticated receiver design is necessary to overcome the limitations.

b. Balanced Homodyne Receiver. The purpose of the balanced receiver design is to provide a means of canceling amplitude fluctuations from the

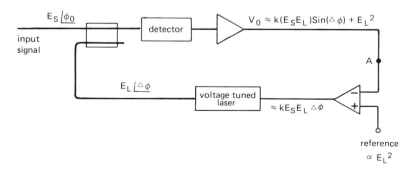

Figure 24.10 Basic homodyne receiver.

signal and local oscillator fields while simultaneously improving their sensitivity to the phase-modulated signal. The arrangement is shown in Figure 24.11.

The key addition to the circuit is the second phase detector connected to the unused output port of the coupler. This arrangement recovers the signal formerly wasted, and the sensitivity of the receiver is then doubled. Between ports A and B the phase-modulated signal undergoes a 90° phase retardation, and between ports A and D a 90° phase advance. The two optical detectors add the antiphase signals to produce the increased sensitivity. At the same time, amplitude fluctuations that appear at the input ports A or C are split equally between ports B and D and appear in phase. The balanced detectors then act to cancel the fluctuations.

In the ideal case the output from the balanced amplifier would be free of the effects of amplitude fluctuation. In practice, the frequency range over which amplitude fluctuations occur extends to gigahertz frequencies. It is impracticable to achieve good amplitude and phase balance over such a wide frequency range, and some residual amplitude fluctuations remain. At low frequencies the amplitude fluctuations arise mainly from acoustic disturbances to the system and variable reflections from air–glass interfaces. At high (gigahertz) frequencies, the noise arises primarily from the internal resonance effects of the laser design or are associated with the long external cavity. Provided the bandwidth of the phaselocked loop controlling the local oscillator exceeds about 10 MHz, however, the residual noise is that associated with laser-based resonances, and, by selection, its effects can be reduced to an acceptable level.

Of the other modulation techniques, narrowband FSK and, in particular minimum shift keying (MSK), where a degree of phase coherence can be

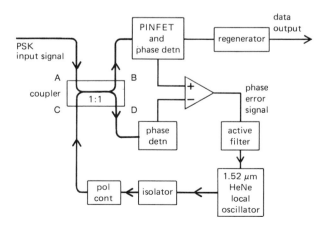

Figure 24.11 PSK homodyne receiver using balanced detector.

assumed, can be demodulated using techniques similar to those used with PSK. MSK is the particular variant of FSK in which the ratio of frequency deviation to bit rate is 1/2; it can be demodulated synchronously [39] and is therefore very efficient in its demands on the bandwidth of the heterodyne receiver.

For larger-deviation FSK, it is necessary to use a phaselocked loop with sufficient bandwidth to track the frequency modulation directly; the demodulated signal would be obtained from the error signal within the phaselocked loop. Such a scheme appears difficult at present for high data rates, although faster phaselocked loops are being developed for other applications.

In the case of ASK, the optoelectronic phaselocked loop described earlier is inappropriate, because the optical carrier is not in the required phase. Alternative schemes based on an injection-locked local oscillator or carrier amplification are possible but have yet to be investigated for fiber systems.

24.5.2 Heterodyne Receivers

The main design features of a heterodyne receiver are shown in Figure 24.12. The main difference from the homodyne designs described earlier lies in the inclusion of a bandpass filter following the PINFET receiver. The passband of the filter and its center frequency are dictated by the information rate and the choice of intermediate frequency; the latter also depends on the bandwidth of the receiver. As indicated earlier, for the local oscillator laser, He-Ne, long external cavity, or (DFB) lasers are possibilities. If a He-Ne laser is used for both the transmitter and the local oscillator, the intermediate frequency in the receiver is restricted to about 200 MHz by the tuning range of the lasers. The tuning range of the other types of lasers is much larger—approximately 7000 GHz for the long external cavity laser and 1000 GHz for the DFB laser.

To maintain the intermediate frequency constant, a control signal is derived from the frequency discriminator and is fed to the local oscillator laser [40]. The signal itself is recovered by the demodulator and low-pass filter and is subsequently regenerated.

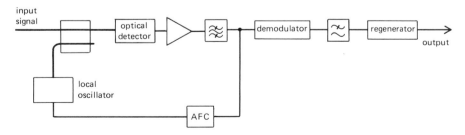

Figure 24.12 Basic heterodyne receiver.

The detailed implementation of the receiver depends on the modulation format and type of detection scheme adopted. The simplest arrangement is the one for an ASK signal with envelope detection, as shown in Figure 24.13a. A related technique can be used for FSK demodulation because the 0 and 1 states can be considered as two interleaved ASK signals with different carrier frequencies. The demodulation technique is shown in Figure 24.13b. The signal is split equally at the output from the photodetector amplifier and then applied to bandpass filters tuned to the 0 and 1 frequencies. After envelope detection the signals are added and then processed in the same way as ASK. If a reduction of 3 dB in sensitivity is acceptable, the receiver can be simplified by omitting either filter path 1 or 2 and detecting the presence of either the 0 or 1 state.

Synchronous detection is more efficient than envelope detection and increases the receiver sensitivity by approximately 1 dB, as shown in Table 24.1. In the case of ASK, the carrier can be extracted after the bandpass filter and mixed using the appropriate phase with the incoming carrier and signal sidebands as shown in Figure 24.13c. A similar scheme can be used for the demodulation of PSK signals when a vestigial carrier component is present, although the relative phase of the extracted carrier at the mixer is different from ASK. When the full PSK modulation of $\pi/2$ is used, no carrier is present at the receiver, and it is necessary to reinsert it, as shown in Figure 24.13d. The squaring circuit produces a component at twice the carrier frequency that is then divided down and phased to apply the correct information to the mixer; the subsequent signal processing is similar to the receivers described above.

In the case of differential phase-shift keying (DPSK) modulation, the squaring and division by two circuits can be replaced by a delay line demodulator, as shown in Figure 24.13e. The delay is 1 bit period; it is also possible to demodulate PSK using a similar technique if the delay is suitably chosen.

As an alternative to either homodyne or heterodyne operation, multiport detection is an attractive compromise that offers heterodyne performance but requires only the lower receiver bandwidth of a homodyne receiver and does not require the local oscillator to be phaselocked to the signal. In its most basic implementation, for ASK signals, the input is split into two paths and combined with separate in-phase and quadrature-phase local oscillator signals, which are then squared and added, as shown in Figure 24.14.

The two photocurrents are

$$I_{P1} \propto E_s E_L m(t) \cos \phi \tag{20}$$

and

$$I_{P2} \propto E_s E_L m(t) \sin \phi \tag{21}$$

where ϕ is the difference in phase between the signal and local oscillator.

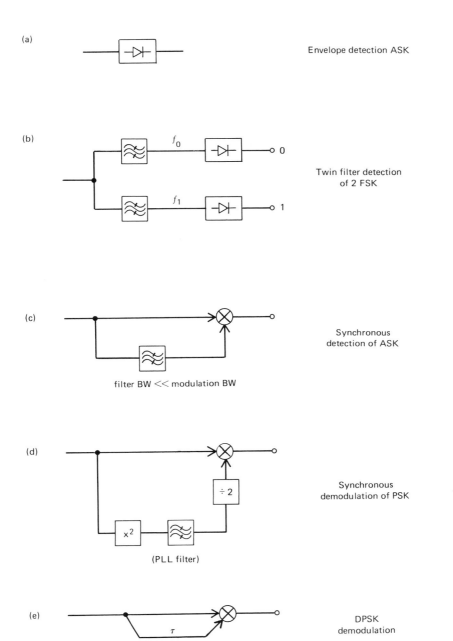

(a)   Envelope detection ASK

(b) f_0 0

Twin filter detection
of 2 FSK

f_1 1

(c)

Synchronous
detection of ASK

filter BW $\ll$ modulation BW

(d)

$\div 2$

Synchronous
demodulation of PSK

x^2

(PLL filter)

(e)

τ

DPSK
demodulation

Figure 24.13 Demodulation techniques. (a) Envelope detection ASK; (b) twin-filter detection of 2FSK; (c) synchronous detection of ASK; (d) synchronous demodulation of PSK; (e) DPSK demodulation.

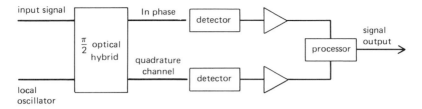

Figure 24.14 Receiver using multipoint detection principles.

Since $\cos^2 \phi + \sin^2 \phi = 1$, the combined output circuit of the two receivers is proportional to $E_s E_L$ and is therefore phase-insensitive.

This general approach can be extended to a three-phase receiver configuration, which has advantages in practical implementation; a receiver based on six ports has been demonstrated at 320 Mb/s [41, 42].

24.6 SYSTEM CONFIGURATION AND PERFORMANCE

A whole range of optical carrier transmission systems are possible depending on the selection of modulation format, detection scheme, and the selection of homodyne or heterodyne operation. The options are summarised in Figure 24.15.

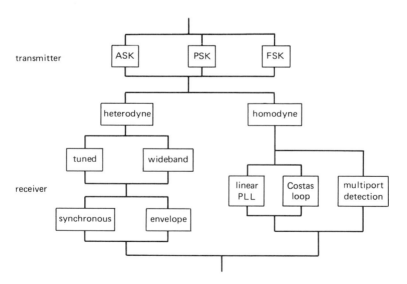

Figure 24.15 System options.

In addition to the options already covered, the possibility of using a tuned receiver [43] in place of a wideband design exists in the case of heterodyne operation. With the tuned receiver, the photodiode and preamplifier input capacitance are together shunted by an inductor, with the combination arranged to resonate at the intermediate frequency. Provided suitable equalization can be achieved, this scheme can offer the advantage of lower receiver noise, which in turn reduces the local oscillator power needed in the receivers with a high intermediate frequency.

In the homodyne case, an alternative to the linear phaselocked loop already described is provided by the Costas loop [44]. In the Costas loop receiver the optical input signal is split into two paths. In the first path the input signal is combined with an in-phase component, and in the second path it is combined with a quadrature phase component of the local oscillator wave. The electric outputs from the two detectors are multiplied together, and the product is used to control the frequency of the local oscillator. With this arrangement the phase information necessary for locking is obtained from the modulation sidebands of the signal. This has the major advantage of not requiring a continuous wave pilot quadrature carrier. However, splitting the optical input signal in present designs of Costas loop receivers incurs a 3-dB penalty.

A further subdivision of the system types can be made according to those capable of providing satisfactory performance with broad linewidth sources and those that specifically require a narrow line source. This subdivision is also shown in Table 24.2; narrow linewidth is defined as a linewidth-to-bit rate ratio of less than 0.1%, and a broad linewidth as one with that ratio 0.1–20%. For DPSK [45] it has been shown that a ratio of ~0.3% is needed for satisfactory performance. Sources with a spectral linewidth of 0.1% can be used with all modulation formats, but broad linewidth sources are only compatible with ASK or FSK modulation.

24.6.1 Homodyne Systems

Homodyne systems are synonymous with the use of narrow-line sources because of the stringent phaselocking requirements. To minimize the problems associated with the latter, early homodyne tests were carried out using the same He-Ne laser for both the transmitter and receiver, as shown in Figure 24.16. In the arrangement shown [34], the receiver sensitivity could be measured first using homodyne detection, and then, by breaking the local oscillator path link, in the direct-detection mode. With ASK modulation the homodyne receiver sensitivity of −56 dB was 11 dB better than that for direct detection. The use of PSK modulation in a similar test resulted in an improvement of 19 dB, which approaches the theoretical prediction of 22.5 dB.

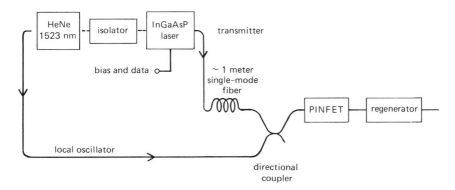

Figure 24.16 Homodyne experiment.

The development of homodyne techniques for use with separate transmitter and receiver lasers is still in its early stages. An example was shown earlier (Fig. 24.10) of a balanced photodiode receiver and optical phase-locked loop that has been used to recover a 140-Mb/s signal from a transmitter operating into a 30-km fiber link. A comparison is shown in Figure 24.17 with the bit error rate characteristics for the 30-km link and a 1-m test using the same laser for the transmitter and receiver and a theoretical curve. The error rate "bottoming" is attributable to limitations in the phase stability resulting from the limited loop gain available from the receiver used

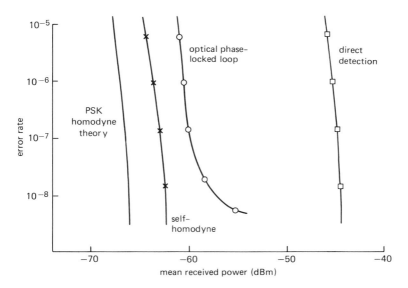

Figure 24.17 140-Mb/s PSK homodyne error rate.

TABLE 24.3 International activity on optical fiber Homodyne detection

				Transmitter			Receiver Sensitivity	
nm	*Modulation*	*Optical path*	*Data rate, Mb/s*	*Source*	*Modulator*	*L.O.*	*Measured*	*Theory*
				British telecom research labs				
1523	ASK	1 m	140	He-Ne	Diode	Shared	−59	−63
1523	PSK	1 m	140	He-Ne	LiNbO$_3$	He-Ne	−62	−66
1523	PSK	30 km	140	He-Ne	LiNbO$_3$	He-Ne	−57	−66
				Technical University of Denmark				
830	PSK	—	—	Diode	Diode	Shared	—	—

for the tests [38]. A summary of internationally published work on optical fiber homodyne systems is given in Table 24.3.

24.6.2 Heterodyne Systems

Optical fiber heterodyne systems have operated over fiber paths up to 109 km long with no detectable difference in performance between long and short paths [35]; no performance penalties were therefore introduced by the transmission fiber. With a transmitter based on a 1523-nm He-Ne laser and a lithium niobate waveguide phase modulator, the BER performance shown in Figure 24.18 was obtained. For PSK operation the receiver was based on the squaring and divide-by-2 carrier reinsertion technique described earlier; for DPSK the delay demodulator was used. The measured performance of the PSK system is within 4 dB of the theoretical shot noise limit; at least half of this difference is attributable to restricted local oscillator power. Imperfections in transmitter modulation and the dependence of laser output on the bit sequence accounts for the remaining difference.

Distributed feedback lasers can form the basis for a simple and practical heterodyne system [46, 47]. In the first example, the DFB laser was directly modulated at 100 Mb/s using a frequency deviation of 1.5 GHz. Since the frequency modulation characteristics of the laser are frequency-dependent, a coding scheme is adopted to minimize the low-frequency components in the transmitted signal. Manchester bipolar coding results in a doubling of the bandwidth requirement, however, which leads to a halving in SNR and receiver sensitivity. In this instance the detector was an InGaAs avalanche photodiode operating at a gain factor of less than 2 to obtain the optimum current-to-noise ratio [47]. The intermediate frequency of 600 MHz was

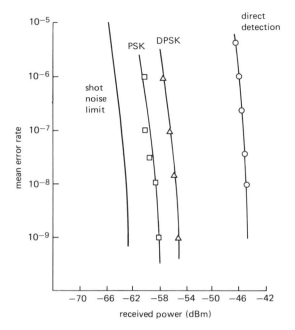

Figure 24.18 Heterodyne detection of PSK and DPSK at 140 Mb/s.

tuned to detect the 1 pulses in the bit stream and were demodulated using an envelope detector.

The system layout of the second example is shown in Figure 24.19; the DFB laser was again directly modulated but in this instance at 140 Mb/s. A modulation current of 1 mA produced a frequency deviation of 1.4 GHz, which was used as the basis for the transmitter. The output from a long external cavity diode local oscillator laser was combined with the incoming signal in a fiber coupler before being fed to the PINFET receiver. The intermediate frequency chosen was 420 MHz, and its bandwidth 300 MHz; an

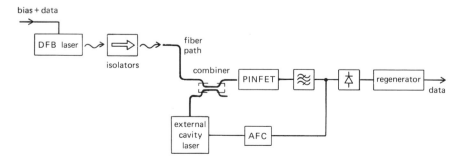

Figure 24.19 Experimental heterodyne system using DFB laser transmitter.

envelope detector followed by a 100-MHz low-pass filter completes the receiver.

If the receiver bandwidth is limited to the extent that it is necessary to use single-filter detection, the receiver cannot distinguish between FSK or ASK modulation, and amplitude demodulation techniques can be used at the intermediate frequency. For the output from the envelope detector to be insensitive to both phase noise and changes in the intermediate frequency, the ratio of the bit rate and IF linewidth to the intermediate frequency must be sufficiently small.

In addition, if the bandwidth of the intermediate frequency is too large, the performance is degraded by the threshold at which the demodulation becomes nonlinear. In contrast, if the IF bandwidth is made too small, the performance will be degraded by the IF linewidth. The nonlinear modulation characteristic of DFB lasers, referred to earlier, causes the width of the modulation spectrum to be pattern-dependent; such an effect can be treated as an effective broadening of the intermediate frequency linewidth.

It is interesting to note that in both experiments the DFB lasers were particularly sensitive to reflections, which dictated the use of some 40–60 dB of optical isolation. The results obtained for the two FSK direct feedback systems show that in both cases the receiver sensitivities obtained were approximately equivalent to that of a good direct detection receiver. The heterodyne system results are summarized in Table 24.4.

24.6.3 Multichannel Coherent Systems

The existence of a local oscillator in the receiver of a coherent system provides the opportunity for developing a tunable multichannel system analogous to the superheterodyne technique in radio transmission systems. The additional receiver sensitivity would permit a variety of optical components to be used in such a way that many terminals could be provided with a large number of channels on carriers closely spaced in optical frequency. An essential first step toward such systems is the experimental demonstration that tuning is possible in the context of an optical fiber system.

Transmission experiments using two channels at both 850 nm [48] and 1520 nm [49] have been reported, which show that optical tuning between channels is now practicable. The layout of the longer-wavelength system is shown in Figure 24.20. An external cavity semiconductor laser transmitter and a He-Ne laser transmitter, both operating at a wavelength close to 1520 nm, are used as transmitters. Both carriers are phase-modulated by external lithium niobate modulators; the correct state of polarization required at the input to these devices is set by fiber polarization controllers. Subsequent to modulation, the two optical channels are combined in the fused fiber coupler, which is connected to the heterodyne receiver by a short length of conventional fiber. At the receiver the incoming signals are combined with

TABLE 24.4. International activity on optical fiber heterodyne detection

λ, nm	Modulation	Optical path length	Data rate, Mb/s	Transmitter			Receiver		
								sensitivity	
				Source*	Modulator	L.O.*	Photodiode	Measured	Theory
British Telecom Research Labs									
1523	PSK	109 km	140	He-Ne	Lithium niobate	EC diode	PIN	−59	−63
1523	ASK	60 km	140	EC diode	Direct	EC diode	PIN	−54	−60
1540	FSK	199 km	140	EC diode	Direct	EC diode	PIN	−56	−60
CNET, France									
850	DPSK	3.9 km		EC diode	Lithium niobate	Shared	APD	—	—
Hinrich Hertz Labs, W. Berlin									
850	FSK	5 m		Diode	Direct	Diode	APD	−43	—

Wavelength	Modulation	Distance	Bit rate				Detector		
NEC, Japan									
1280	FSK	49.7 km	100	DFB	Direct	DFB	APD	−47	−56
1570	FSK	104.7 km	100	DFB	Direct	DFB	APD	−49	−57
Bellcore, U.S.									
1500	FSK	100 km	1000	DFB	Direct	EC diode	PIN	−37	—
AT & T, U.S.									
1500	DPSK	150 km	1000	EC diode	Lithium niobate	EC diode	Balanced PIN	−44.5	—
1500	DPSK	170 km	2000	EC diode	Lithium niobate	EC diode	Balanced PIN	−39	−53
Bellcore and Hitachi, U.S.									
1520	CPFSK	101 km	2000	DFB	Direct	EC diode	PIN	−39.2	−49
NEC Corporation, U.S.									
1550	CPFSK	204 km	1200	DFB	Direct	Monolithic DFB	EC DFB	−41.5	−54

* EC = external cavity

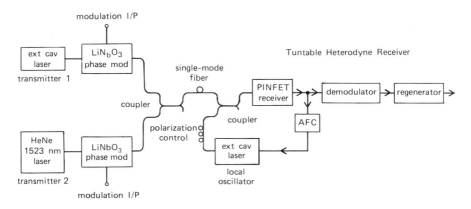

Figure 24.20 Experimental two-channel transmission system.

a local oscillator signal derived from a tunable external cavity laser; polarization matching is achieved by a fiber-based compensator in the local oscillator path. The combined signals are detected by a PINFET receiver to produce an electrical intermediate frequency equivalent to the optical difference frequency.

For simplicity, a residual carrier PSK modulation technique is used together with synchronous demodulator. By tuning the local oscillator in the receiver to be 566 MHz away from transmitter 1, one can select channel 1; the AFC loop maintains the intermediate frequency at 566 MHz. When the local oscillator tuning is adjusted to be 566 MHz away from transmitter 2, the second channel is obtained, with the AFC loop again maintaining the intermediate frequency at 566 MHz. When the local oscillator laser in the receiver is operated at a fixed wavelength, a beat signal will be generated for each of the multiplexed signals. The channel spacing and optical receiver bandwidth determine how many channels can be resolved at the output of the detector. Any one of those that are present within the bandwidth can then be obtained by electrical tuning; Figure 24.21 shows the output spectrum from the PINFET detector for two IF channels. The transmitters were

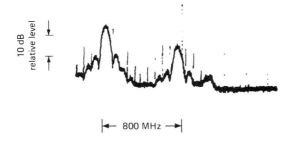

Figure 24.21 Simultaneous detection of two-frequency multiplexed 70-Mb/s phase-modulated channels.

independently phase-modulated by a 70-Mb/s signal, and the channel separation was adjusted to be approximately 650 MHz. This channel separation was chosen for illustrative purposes and is much wider than would normally be needed for a two-channel 70-Mb/s transmission.

The experiments showed that stable demodulation of either one of two channels spaced 1 GHz apart is feasible using optical heterodyne tuning. This spacing is compatible with that needed to eliminate intermodulation products and image band signals in the basic single-detector optical heterodyne receiver where the channel spacing is approximately 5 times the bandwith of the modulated carrier [50]. This limit could be reduced by optical prefiltering, "unlimited" local oscillator power, the use of a balanced (two-photodiode) receiver, or any combination of these techniques. With optical heterodyne detection and image-band filtering, the spacing could ultimately be equal to the modulation bandwidth. Homodyne detection should also enable a similar channel spacing to be obtained if intermodulation products are suppressed and ideal electrical baseband filtering is employed. Transmission of optical frequency multiplexed channels over long fiber lengths is still to be demonstrated, and, in particular, variations in the received states of polarization for the different channels require further study. For practical systems, especially those with narrow channel separations, the development of frequency-stabilized transmitter banks in an important consideration [51].

24.6.4 Discussion

The systems that at present provide the closest approach to theoretical expectations are those based on narrow-line optical sources. For most systems, the gas laser is unattractive and the long external cavity semiconductor source is large and acoustically sensitive. A heterodyne system based on "packaged" long external cavity lasers has, however, been operated continuously for several hours, and there is clearly much that can be done to make smaller and more rugged devices. In addition, the development of optical cavity devices based on fiber, lithium niobate, or III–V compound technologies should not be discounted. Possibly the ideal solution for some applications would be a narrow-line DFB laser diode. At present its wide linewidth, phase noise, and reflection sensitivity indicate that further development is needed before it becomes unambiguously attractive. The inclusion of longer grating structures or variants based around a distributed Bragg reflector (DBR) may be promising avenues of development. For example, the development of hybrid or integrated devices such as a DFB laser with an external cavity or a multisection DFB laser may lead to a promising solution.

At the receiver, the best results in terms of sensitivity have been achieved using PSK modulation and homodyne detection over a short fiber path. Matching this performance over longer transmission paths will depend critically on the development of high-performance optical phaselocked loops or developments associated with multiport receivers. At the opposite extreme of complexity, experiments with DFB lasers and large-deviation FSK, which feature single-filter detection, although less demanding in terms of source characteristics, have been unable to produce significant improvements over direct detection. As far as the total system budget is concerned, the option using narrow-deviation FSK has much to commend it, since direct frequency modulation by injection current eliminates losses in external phase modulators. Performance approaching that of PSK could, in principle, be achieved by an optimized form of narrow deviation FSK such as minimum shift keying (MSK).

24.7 FUTURE POSSIBILITIES AND APPLICATIONS

An evaluation of the potential for coherent systems depends on the ways in which the higher receiver sensitivity and optical tuning ability can be exploited. It is possible to consider such exploitation in terms of both the development of present-day point-to-point transmission systems and the possibilities that arise for new optical networks.

24.7.1 Development of Existing Systems

The additional receiver sensitivity can be utilized by increasing the regenerator spacing or bit rate in point-to-point links or in wideband networks to overcome the additional losses introduced by the optical components used in multichannel systems.

The three main factors that determine maximum regenerator spacing are laser output power, fiber loss, and receiver sensitivity. Lasers producing several hundred milliwatts of optical power near the shorter wavelength of 800 nm have already been demonstrated [52]. See also Chapter 13. It is reasonable to assume that similar performance will become realizable in due course at 1550 nm; lasers having high powers at the longer wavelength already exist in experimental form [53]. In the case of silica fiber, losses of less than 0.2 dB/km are now typical of premium commercial products, and 0.16 dB/km would appear to be a realistic target. The Brillouin scattering threshold [54] is only a few milliwatts in such fiber, and to minimize its effect a modulation technique with no carrier component, such as PSK, will have to be used to take full advantage of the low-loss fiber [55]. At the receiver, the increased power available from an advanced laser source acting as the local oscillator should allow the shot noise limit to be approached, perhaps to within 1 dB. With several hundred milliwatts available, the normal high-impedance PINFET receiver would be unnecessary and could be replaced by a variant based on a 50-Ω load impedance. Although intrinsically high, the thermal noise could still be neglected in the presence of such a high level of oscillator power. Under these conditions, very wide bandwidth (5–10-GHz) PINFET receivers, become possible and would make high-bit-rate homodyne and heterodyne operation feasible at 10 Gb/s and above. It is possible, for example, that an experimental PSK homodyne link with a 250-mW transmitter could be operated over a maximum of 500 km of fiber at 2.4 Gb/s, as shown in Figure 24.22; an operational system would be more conservatively rated. Perhaps the ultimate development for long-span transmission links lies in the combination of coherent techniques with ultralow-loss nonsilica fibers [14]. In the limit, a loss approaching 10^{-4} dB/km is predicted at a wavelength in the region of 4.5 μm, which would bring repeater spacings of several thousands of kilometers into the realm of possibility. At the time of writing, however, the best experimental loss for any of these fibers is about 1 dB/km, and so it will be a number of years before the prospects for such systems can be realistically assessed.

The possibility of operating at high bit rates over much shorter distances also exists. The sensitivity of a coherent receiver halves for each doubling in bit rate; with 0.15-dB/km fiber, a decrease of 20 km in repeater separation is implied. Clearly, if the required distance is very short, perhaps a few tens of kilometers or less, then the attenuation of the fiber becomes almost negligible compared with the losses introduced by other optical components in the system. Under these conditions it is worth examining the possibilities

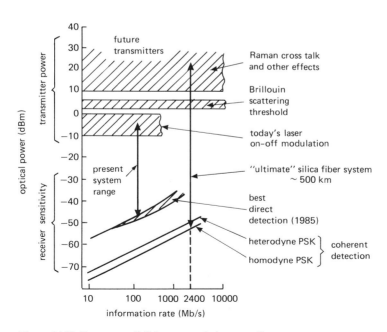

Figure 24.22 Future possibilities, transmission range?

for applying coherent techniques to wideband networks for metropolitan and local areas.

24.7.2 Optical Networks

In the context of local or metropolitan area networks, the single-mode fiber offers the prospect of almost unlimited bandwidth and very low loss over the short distances involved. In these networks, a large number of channels can be envisaged that could be fed singly or in combination to many users. Single-mode fiber, in conjunction with a variety of integrated optical devices, is potentially able to provide such channels over star or tree networks or any intermediate combination to suit a wide range of applications. Using direct-detection techniques with a channel spacing of 1 nm (133 GHz), a typical network might provide 32 channels to each of 128 terminals, as indicated in Figure 24.23a. A coherent variant in which the precision tuning ability provides a separation of perhaps 1 GHz and where the increasing receiver sensitivity is used to overcome the additional losses introduced by the optical components is shown in Figure 24.23b; in this instance up to 8192 terminals can be used to receive the 32 channels. An extension of these arguments could be applied to interactive networks in which it is possible to envisage the use of wavelength (frequency) switching where the frequency selected by the optical source or that generated by the transmitter is altered

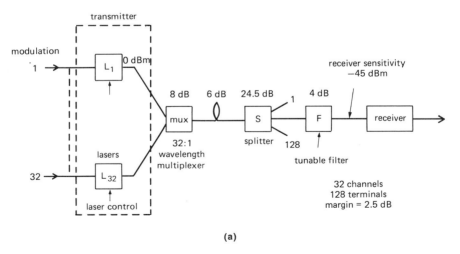

(a)

Figure 24.23a Wavelength multiplexed wideband local network using direct-detection (channel spacing > 1 nm). (i) Multiplexer losses are mostly independent of the number of channels being combined. (ii) Maximum number of channels distributed depends on resolution of the multiplexer. (iii) Return channels are wavelength-multiplexed over same fibers or on separate fibers.

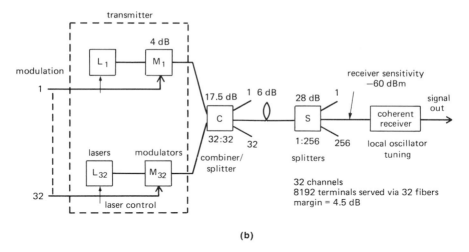

(b)

Figure 24.23b Wavelength-multiplexed waveband local network using coherent-detection channel spacing <0.01 nm. (i) Maximum number of terminals served depends on combiner losses and number of channels being combined. (ii) Return channels are wavelength-multiplexed over same fibers or on separate fibers.

according to the service or signal path required. Proposals have been made for direct-detection systems using optical filters and grating [55, 56], but topologies incorporating laser source and heterodyne receiver tuning are now appearing [57]. The possibility of incorporating combinations of frequency, time, and optical space switching exists, but the forms that they may take await further developments.

24.8 CONCLUSIONS

The higher receiver sensitivity and optical tuning ability expected from coherent systems has been realized in experimental systems. With the current progress in the development of optical components such as modulators and polarization controllers and the laboratory demonstration of 1–2-Gb/s links [58–60], it now appears to be only a matter of time before very high bit rate operation over long distances is demonstrated in the context of installed point-to-point transmission systems. A well-engineered, stable, and high-power narrow-line source is likely to be a key step in this development.

For wideband networks the high power requirement could be relaxed and the emphasis shifted to tunability and low cost, the latter being of particular importance for multichannel networks. The ultimate capacity of such wideband networks could be determined by the ability to produce a multiplex using many lasers operating at different wavelengths or a single, more powerful laser producing a multiple-carrier spectrum by, for example, wavelength shifting. It is not yet clear, however, what limits nonlinearities in the transmission medium and other components will impose on the number of available channels.

It is in wideband network applications that the combined advantages of high sensitivity and tunability in coherent systems could be exploited to the full [61–64]. Only in recent years have the possibilities offered by coherent techniques in new network topologies become apparent, which suggests that the present rapid pace of development is likely to continue for some time to come.

Acknowledgments

We thank the Director of Research of British Telecom for permission to make use of the information contained in this paper.

REFERENCES

1. R. C. Hooper, J. E. Midwinter, D. W. Smith, and I. W. Stanley, "Progress in monomode transmission technique in the United Kingdom," *J. Lightwave Technol.*, **LT-1**: 596–611 (1983).
2. Y. Yamamoto, "Receiver performance evaluation of various digital optical modulation and demodulation schemes in the 500 nm to 10,000 nm wavelength region," *IEEE J. Quantum Electron.*, QE-16: 1251–1259 (1980).
3. F. Favre, L. Jeunhomme, I. Joindot, M. Monerie, and J. C. Simon, "Progress towards heterodyne-type single mode fibre communications systems," *IEEE J. Quantum Electron.*, **QE-16**: 897-906 (1980).
4. I. Okoshi, K. Emura, K. Kikuchi, and R. T. Kersten, "Computation of bit-error rate of various heterodyne and coherent-type optical communication schemes," *J. Opt Commun.*, **2**: 89–96 (1981).
5. B. M. Oliver, "Signal to noise ratios in photoelectric mixing," *Proc. IRE*, **49**: 1960–1961 (1962).
6. S. D. Personick, "Receiver design for digital fiber optic communication systems," *Bell Syst. Tech. J.*, **52**: 843–886 (1973).
7. W. K. Pratt, *Laser Communication Systems*, Wiley, New York, 1969.
8. L. L. Jeromin and V. W. S. Chan, "Performance estimates for a coherent optical communication system," OFC'84 Conf. New Orleans, 1984, pp. 60–61.
9. S. Saito, Y. Yamamoto, and T. Kimura, "Signal to noise and error rate evaluation for an optical FSK-heterodyne detection system using semiconductor lasers," *IEEE J. Quantum Electron.*, **QE-19**(2): 180–193 (1983).
10. G. Jacobsen and I. Garrett, "A theoretical analysis of coherent optical communication receivers with non-negligible laser linewidths," Electromagnetics Institute Report No R297, Technical University of Denmark, 1984.
11. Y. Yamamoto, "AM and FM quantum noise in semiconductor lasers," *IEEE J. Quantum Electron.*, **QE-19**: 34–46 (1983).
12. D. R. Smith, A. K. Chatterjee, M. A. Z. Rejman, D. Wake, and B. R. White, "Pinfet hybrid optical receiver for 1100nm–1600nm optical communication systems," *Electron Lett.*, **16**: 750–751 (1980).

13. M. Ross, *Laser Receivers*, Wiley, New York, 1966.
14. G. H. Sigel and D. Tran, "Ultra low loss optical fibres—an overview," *SPIE Conf. Proc.*, 1984, p. 484.
15. I. W. Stanley, R. C. Hooper, D. W. Smith, and M. R. Matthews, "The performance of experimental monomode transmission links," *Proc. 8th Eur. Conf. Opt. Commun. (ECOC)*, Cannes, France, 1982, p. 447.
16. R. A. Harmon, "Polarisation stability in long lengths of monomode fiber," *Electron. Lett.*, **18**: 1058–1060 (1982).
17. R. C. Alferness, "Guided wave devices for optical communications," *IEEE J. Quantum Electron.*, **QE-17**: 946–958 (1981).
18. R. C. Booth, "Lithium niobate integrated optic devices for coherent optical fiber systems," *Thin Solid Films* (Switzerland), **126**: 167–176, Apr. 26, 1985.
19. D. N. Payne, J. A. Barlow, and J. J. R. Hansen, "Development of low and high birefringence optical fibers," *IEEE J. Quantum Electron.*, **QE-18**: 477–486 (1982).
20. M. P. Varnham, D. N. Payne, R. D. Birch, and E. J. Tarbox, "Single polarisation operation of highly birefringent bow-tie optical fibers," *Electron. Lett.*, **19**: 246–247 (1983).
21. C. J. Nielsen, "Bandwidth limiting effects in coherent optical communications systems based on polarisation preserving fibers," *Electron. Lett.*, **20**: 403–405 (1984).
22. I. Katsuyama, H. Matsumara, and T. Suganuma, "Low loss single polarisation fibers," *Appl. Opt.*, **22**: 1741–1747 (1983)
23. J. R. Simpson, R. H. Stolen, F. Pleibel, F. V. Di Marcello, and R. G. Huff, "Low-loss polarisation preserving fiber by preform deformation," Proc. OFC-OFS'85, San Diego, February 1985, paper PD 4-1.
24. W. J. Tomlinson and R. L. Fork, "Frequency stabilisation of gas lasers," *Appl. Opt.*, **8**: 121–129 (1969).
25. T. E. Bell, "Single frequency semiconductor lasers," *IEEE Spectrum*, **20**: 38 (1983).
26. L. D. Westbrook, A. W. Nelson, P. J. Fiddyment, and J. S. Evans, "CW operation of 1.5 μm distributed feedback ridge waveguide lasers," *Electron. Lett.*, **20**: 225–226 (1984)
27. T. Itaya, T. Matuoka, Y. Nakano, Y., Suzuki, K. Kuroiwa, and T. Megami, "New 1.5 μm-wavelength GaInAsP/InP distributed feedback laser," *Electron. Lett.*, **18**: 1006–1008 (1982).
28. L. D. Westbrook, A. W. Nelson, and C. Dix, "High quality InP surface corrugations for 1.55 μm InGaAsP DFB lasers fabricated using electron-beam lithography," *Electron Lett.*, **18**: 863–865 (1984).
29. M. W. Fleming and A. Mooradian, "Spectral characteristics of external cavity controlled semiconductor lasers," *IEEE J. Quantum Electron.*, **QE-17**: 44–59 (1981).
30. C. H. Henry, "Theory of the linewidth of semiconductor lasers," *IEEE J. Quantum Electron.*, **QE-18**: 259–264 (1982).
31. R. Wyatt and W. J. Devlin, "10 kHz linewidth 1.5 μm InGaAsP external cavity laser with 55 nm tuning range," *Electron. Lett.*, **19**: 110–112 (1983).
32. R. Wyatt, D. W. Smith, and K. Cameron, "Megahertz linewidth from a 1500 nm semiconductor laser with HeNe laser injection," *Electron. Lett.*, **18**: 292–293 (1982).
33. F. Koyama, A. Arai, Y. Suematsu, and K. Kishino, "Dynamic spectral width of rapidly modulated 1.58 μm GaInAsP/InP DFB twin guide laser," *Electron. Lett.*, **17**: 25–26 (1981).
34. D. J. Malyon, T. G. Hodgkinson, D. W. Smith, R. C. Booth, and B. E. Daymond-John, "PSK homodyne receiver sensitivity measurements at 1500 nm," *Electron. Lett.*, **19**: 144–146 (1983).
35. R. Wyatt, T. G. Hodgkinson, and D. W. Smith, "1520 nm PSK heterodyne experiment featuring an external cavity diode laser local oscillator," *Electron. Lett.*, **19**: 550–552 (1983)
36. D. J. Heatley and T. G. Hodgkinson, "Video transmission over cabled monomode fiber at 1.523 μm using PFM with 2-PSK heterodyne detection," *Electron. Lett.*, **20**: 110–112 (1984).
37. J. M. Osterwalder and B. J. Rickett, "Frequency modulation of GaAlAs injection lasers at microwave frequency rates," *IEEE J. Quantum Electron.*, **QE-16**: 250–252 (1980).

38. D. J. Malyon, "Digital fiber transmission using optical homodyne detection," *Electron. Lett.*, **20**: 281–283 (1984).

39. S. Haykin, *Communication Systems*, 2nd ed., Wiley, New York, 1983.

40. W. J. Liddell, "Optical phased lock loop (OPLL)," *IEE Colloq. Adv. Coherent Opt. Devices Technol.*, London, 1985, paper 9, *Digest* 1985/30.

41. J. A. Arnaud, "Enhancement of optical receiver sensitivities by amplification of the carrier," *IEEE J. Quantum Electron.*, **QE-4**: 893–899 (1968).

42. A. W. Davis and S. Wright, "A phase insensitive homodyne optical receiver," *IEE Colloq. Adv. Coherent Opt. Devices Technol.*, Paper 11, London, 1985, Digiest No. 1985/30.

43. K. Kikuchi, T. Okoshi, and K. Emura, "Realisation of nearly shot-noise-limited operation in a heterodyne-type PCM-ASK optical communication system," 8th Eur. Conf. Opt. Commun., Cannes, France, 1982.

44. H. K. Phillip, A. L. Scholtz, E. Bonek, and W. R. Leeb, "Costas loop experiment for a 10.6 μm communications receiver," *IEEE Trans.*, **COM 31**: 1000–1002 (1983).

45. G. Nicholson, "Probability of error for optical heterodyne DPSK system with quantum noise," *Electron. Lett.*, **20**: 1005–1007 (1984).

46. F. Mogenson, T. G. Hodgkinson, and D. W. Smith "FSK heterodyne system experiment at 1.5 μm using a DFB laser transmitter," *Electron. Lett.*, **21**: 518–519 (1985).

47. M. Shikada, K. Emura, S. Fujita, M. Kitamura, M. Arai, M. Kondo, and K. Minemura, "100 Mbit/s ASK heterodyne detection experiment using 1.3 μm DFB laser diodes," Paper TUK 6:62 Proc. OFC84. New Orleans, 1984.

48. E. J. Bachus, R. P. Braun, F. Bohnke, W. Eutin, H. Foisel, K. Heimes, and B. Strabel, "Two channel heterodyne type fiber optic transmission experiments," *Conf. Proc. ECOC 84*, Stuttgart, 1984, pp. 222–225.

49. D. W. Smith, T. K. Hodgkinson, D. J. Malyon, and P. Healey, "Demonstration of a tunable heterodyne receiver in a two channel optical FDM experiment," *IEE Colloq. Adv. Coherent Opt. Devices Technol.* London, 1985, Digest No. 1985/30.

50. P. Healey, "Effects of intermodulation in multichannel optical heterodyne systems," *Electron. Lett.*, **21**: 101–102 (1985).

51. F. Favre and D. Le Guen, "Autostabilisation technique for achieving highly stable resonant optical feedback in a fiber resonator loaded injection laser," *Electron. Lett.*, **19**: 1046–1048 (1983).

52. D. F. Scifres, R. D. Burnham, and W. Streifer, "Lateral grating array high power CW visible semiconductor laser," *Electron. Lett.*, **18**: 549–550 (1982).

53. T. Mito, M. Yamaguchi, S. Murata, M. Kitamura, and K. Kobayashi, "High power CW operation DFB-DC-PBH laser diodes with a first order grating," OFC' 85, San Diego, 1985, postdeadline paper 9.

54. D. Cotter, "Observation of stimulated Brillouin scattering in low loss single mode fibre at 1300 nm," *Electron. Lett.*, **18**: 495–496 (1982).

55. D. Cotter, "Transient stimulated Brillouin scattering in long single mode fibers," *Electron. Lett.*, **18**: 504–506 (1982).

56. C. Back, E. J. Bachus, and B. Strebel, "Future light carrier frequency technology in glass fiber networks," *Nachrichtentechni. Z.*, **35**: 686–689 (1982).

57. N. A. Olsson and W. T. Tsang, "An optical switching and routing systems using frequency tunable cleaved–coupled cavity semiconductor lasers," *IEEE J. Quantum Electron.* **QE-20** (4): 332–335 (1984).

58. R. S. Vodhanel, J. L. Gimlett, L. Curtis, and N. K. Cheung, "1 Gbit/s optical FSK heterodyne transmission experiments over 100 km of single mode fiber" *Electron. Lett.*, **22** (16), 829. (1986).

59. R. A. Linke, B. L. Kasper, N. A. Olsson, R. C. Alferness, L. L. Buhl, and A. R. McCormick, "Coherent lightwave transmission over 150 km fiber lengths at 400 Mbit/s and 1 Gbit/s data rates using DPSK modulation," *IOOC/ECOC*, Venice, Italy, 1985, *Postdeadline Proc.*, pp. 35–38.

60. A. H. Gnauck, R. A. Linke, B. L. Kasper, K. J. Pollock K. C. Reichmann, R. Valenzuela, and R. C. Alferness, "Coherent lightwave transmission at 2 Gb/s over 170 km of optical fiber using phase modulation," *Proc. OFC'87*, Reno, 1987, pp. 40–43.

61. J. L. Gimlett, R. S. Vodhanel, M. M. Choy, A. T. Elrefarie, N. K. Cheung, R. E. Wagner, and S. Tsuji, "2 Gb/s 101 km optical FSK heterodyne transmission experiment," *Proc. OFC'87*, Reno, 1987, pp. 44–47.

62. K. Manome, K. Emura, S. Yamazaki, S. Fujita, S. Takano, M. Shikada, and K. Minemura, "A 1.2 Gb/s CPFSK heterodyne detection transmission experiment with optimum system configuration for solitary laser diodes," *Proc. ECOC'87*, Helsinki, 1987, pp. 333–336

63. I. W. Stanley, G. R. Hill, and D. W. Smith, "The application of coherent optical techniques to wideband networks," *J. Lightwave Technol.* **LT-5**: 439–451 (1987).

64. C. Caspar, E. Grossmann, and B. Strebel, "Automatic' switching system in optical heterodyne technique, " *Proc. ECOC'87*, Helsinki, 1987, pp. 317–320.

PART 7

Impacts on the Information Society

CHAPTER 25

The Impact of Optoelectronics Technology on the Information Society

C. K. Kao

25.1 INTRODUCTION

A society in which the very existence of its members is intimately dependent on its ability to harness information can justifiably be called an "information society." How optoelectronics technology will influence this society is a rather broad subject that cannot be discussed intelligently without first defining a restricted framework. We will start by putting some specific interpretations on the semantics of the terms to be used and then prescribe some boundaries both for optoelectronics technology and for the fields and activities whose impacts are to be measured. A semiquantitative measure of these impacts can then be presented.

25.2 DATA, INFORMATION, AND KNOWLEDGE

The words data, information, and knowledge are used to describe different things on different occasions. An artificially constrained interpretation of the meanings of these three words will be used for this discussion. Here data, information and knowledge are defined as follows:

Data: Any recorded event
Information: Acquired data
Knowledge: Useful data

These interpretations can be justified along these lines: Any activity or event taking place anywhere in the world in our society can be a piece of data. If it is recorded on paper, on film, in computer memory, or in someone's head, or if it can be deduced from other recorded events, this activity can be recalled and is a piece of data. If an activity is not recorded, it is as if it had never taken place; therefore, it is not a piece of data. A pertinent example is a chess game between two persons. The fact that the game took place is incidental; it becomes a piece of data only if the moves or some aspects such as what, why, when or where of it were recorded.

Data become information if they are of interest to an observer who makes an effort to acquire them. The chess game could have been played by Bobby Fisher and Anatol Karpov in the world championship series, the ultimate of chess competitions, and, as such, been recorded for chess aficionados. The data on the match exist, but they become information only when someone wants the data for some purpose; hence, information is acquired data. To clarify this further, consider a news broadcast. A viewer might only be looking for the World Cup results of the day. The rest of the news is totally insignificant to the viewer; only the World Cup result—Italy 2, Argentina 1—is information.

Along this same line of argument, knowledge is defined as useful data. Data are increasing exponentially in our society each day. Most data have very limited usefulness; many are redundant. Our data, however, are usually filtered by interested persons and assembled into useful categories. Other interpretations can be made, and deductive or inductive rules can be established. From their original unorganized state, the data are in useful form, ready for would-be users. Of course, our whole civilization is essentially an educational process of sharing experience and developing knowledge. Hence, knowledge and useful data are synonymous. When someone uses the data, then knowledge is useful information.

Lest we lose the thread of our discussion on the impact of optoelectronics technology on our information society, it is important to connect data, information, and knowledge with optoelectronics. To make this connection, we need to summarize first what optoelectronics technology is and what impact it is making on various technological endeavors. Then we can see that the processing of our data into knowledge and making the knowledge readily available are the major challenges we have today. Business successes are influenced by the efficiency of information handling. Social development is likewise dependent. It is obvious that the previous statements must be interpreted in view of the definitions introduced earlier.

25.3 WHAT IS OPTOELECTRONICS TECHNOLOGY?

Advances in fabrication techniques for electronic material and in analytic tools and computational power have opened up an era in which electrons can be closely approximated to particles with specific energies, and photons, to coherent electromagnetic waves. This has greatly enhanced the prospect of being able to employ electrons and photons, respectively and jointly, to our best advantage. It is against this background that optoelectronics technology is developing into an ever-broadening and increasingly important technology base.

Vacuum tube technology, in which photon-excited free electrons and free electron-excited photons are used for optoelectronic conversion, continues to play an important role. However, the present trend is to bring about electron–photon interaction within a solid material. In principle, such an arrangement permits the creation of more robust, compact, reliable, and efficient devices. At the same time, the guidance of photons along low-loss dielectric structures adds another dimension to the use of electrons and photons. It is now possible to envisage the processing and transportation of both electrons and photons, with or without conversion, to achieve desired performance. Optoelectronics technology includes the generation, detection, and guidance of light and the integration of these actions. There is now the real prospect of achieving a number of significant advances in optoelectronic components with subpicosecond response times. These components include.

Single-frequency light sources
Detectors with sensitivity near the quantum limit
Transmission materials with practically zero loss and almost infinite bandwidth
Electrooptic and nonlinear optical materials with large coefficients

The benefits that can be accrued from these advances are likely to be in improved cost and performance for all system products that influence equipment for communication, control, and computer applications. More specifically, the impact will be felt in information transmission systems, storage systems, signal-processing applications, and sensor technology. The new technology will increasingly affect information networks, service, and general commerce.

25.4 TECHNOLOGICAL AREAS OF IMPACT

Information transmission, storage, and processing have been developing somewhat independently, while sensor development is a totally unrelated area of technology. The impact of optoelectronics technology in these areas is highly significant. Not only will the performance of such systems be improved but their use will be merged more closely.

In the long term, the implementation of information networks is likely to change radically as system capability makes broadband flexible networks realizable and as the demand for new services replaces traditional communication activities. These, in turn, will result in a redistribution of general commercial activities.

25.4.1 Information Transmission

Currently, information-transmission systems handle mainly the transmission and distribution of telephone and data traffic. Television and video signals are distributed separately, usually in a broadcast mode. The introduction of optical fiber transmission systems and optoelectronics technology has already demonstrated significant cost reduction for the trunk connections. This technology will soon prove to be cost-effective for local area networks (LANs) and subscriber-distribution applications. The near-term quest is to utilize optical fiber systems along with the existing copper-wire network. The use of both old and new technologies will permit an infrastructure of optical fiber systems with a much larger potential capacity for carrying information to be built up gradually. Narrowband signals for voice and data will be handled in an integrated fashion in the Integrated Service Digital Network (ISDN) environment. Efforts are also under way to include videophone service in a new broadband network. The evolution from a narrowband into a broadband network is, however, a complex issue involving considerations of both system cost and service provisions. An optical fiber transmission system is essential to this development since an increase of nearly a thousand-fold in bandwidth is required. Such an increase in bandwidth cannot be handled without optical fibers. However, the service provisions and eventual network configurations are open-ended challenges. The current situation calls initially for mass-produceable, low-cost, optical-to-electronic and electronic-to-optical conversion units. Optoelectronics technology is certainly progressing in the right direction.

25.4.2 Information Storage

Up to now, information storage has been associated with computer systems. Large amounts of data are stored in random access memories (RAM) as well as in slower but larger storage media such as magnetic tapes and disks. With the inception of computer networking, there is an increasing need to transport stored data. Optoelectronics technology is beginning to make an impact in this area with the introduction of laser optical disks. Furthermore, optical interconnections for electronic circuits are reducing electrical interference and signal differential-delay problems.

In the ISDN environment, machine-to-machine communication is seen as another service aspect of the network. This will make information storage a part of the communication system requirement. High-speed store-and-forward procedures for data will undoubtedly be demanded.

25.4.3 Information Processing

Information processing is needed for a variety of reasons. Speech and visual information must be transformed into a convenient electric form for ease of transmission and storage. Data may have to be transformed for the same reasons. Data may also have to be manipulated in computation and coding. Speech and visual information may have to be manipulated to extract features or to reduce redundant information. Speech and picture synthesis and recognition are examples of information-processing actions; coding and encryption are other examples. These latter two activities are needed to improve data safety. Obviously, information processing takes a great variety of forms and requires different ways of handling. However, greater signal-processing speed can usually be used advantageously.

Optoelectronics has played crucial roles in signal processing, particularly with visual and pictorial information. Photocopiers, video cameras, and optical radar are a few examples of devices that use optoelectronics. Recent signal-processing activities include laser bar code scanners, holographic interference deformation measurement, and laser printing. Also, there are a great variety of applications of optical wave-front transformations to achieve Fourier transformation, correlation, and convolution of signals.

25.4.4 Sensor Technology

A relatively minor although widespread use of optoelectronic devices is in intrusion-detection systems that sense the interruption of an optical beam. Alarm systems work with a collimated optical beam generated by a light-emitting diode (LED) and a collimating lens directed at a photodetector that is placed in line of sight with the beam but at a remote location. The interruption of the beam triggers the alarm. Elevators use such systems to initiate the closing of doors. Some escalators and automatic water-flushing systems use beam interruption as a command to start or stop. Another sensing system that uses optoelectronic detection is light-level sensors. A photodetector is used to sense light levels for camera systems or to switch street lights on automatically.

The introduction of lasers and optical fibers, however, has catapulted optoelectronics into the sensor field in a major way. The coherent laser has been recognized as a range-and-rate sensor. Lidar, optical radar, is capable of measuring distances very accurately. The laser has even been proposed as a means of establishing standards for length and time. Furthermore, an oscillating ring laser is a rotational-rate sensor with no moving parts that can be made with resolutions equal to, or exceeding, the resolution of mechanical gyroscopes. The Doppler shift effect, combined with light scattering, can also be used in flow-velocity sensors. The fiber waveguide, in conjunction with lasers and detectors, has very good feasibility as a universal sensor for physical parameters such as temperature, pressure, strain, and rates.

Fiber-waveguide sensors work by registering the change in the propagation characteristic of the fiber caused by a change in the specific physical parameter to be measured. For example, a fiber can be used as a temperature sensor if a temperature change causes a specific change in some parameter of the propagation of light through it. One way to do this is to sense the phase of a part of an optical signal traveling through a length of fiber. This phase signal is compared with that of another part of the same signal traveling through a reference fiber. The sensing fiber is exposed to the temperature change. As a result, the length of the sensing fiber changes due to thermal expansion, and the phase of the signal through that fiber is altered. An optical detection arrangement can readily be made, for example, in the form of an interferometer, which can be used to detect the phase change. Since the phase change is dependent on fiber length, extremely high sensitivity can be achieved by the use of long lengths of low-loss fiber. The same arrangement can be used for pressure sensing. In fact, the first proposed sensor application was as an acoustic pressure sensor in a sonar system.

Fibers in coil form have already been demonstrated as rotational-rate sensors. Fibers can also be used as a tether to moving vehicles so that other sensing equipment carried on the vehicle can be linked to ground-based or to other moving-platform-based equipment to execute sensing and navigation. These types of applications suggest that optoelectronic sensing is evolving. Not only are a broad range of improved and novel sensors available, but the signals they generate can be used conveniently as direct inputs to signal-processing and transmission equipment.

Fiber sensors are not without disadvantages. Many practical problems exist in the mechanical design and in the separation of wanted signals from unwanted signals.

25.5 MERGING OF TECHNOLOGIES

Unmistakably, the trend is toward the merging of various technologies that previously were concerned with separate application areas. In many ways this comes as no surprise. In the course of technological development in our society we have repeatedly reassessed separate developments into a more coherent whole. The transmission, storage, and processing of information are being integrated into a single area under the impetus of providing integrated service. If sensors are considered another form of service, then they are a part of the same system environment.

Indeed, the development of robotics will accelerate this trend. Robots are merely computer-controlled machine tools that use many sensors. As such, robots must eventually be incorporated into the information network for efficient and versatile use.

25.6 INFORMATION NETWORKS

Humans and machines already form an intimately and mutually dependent network. Many organizations store much of their operational data on computers. Payroll departments use computers to gather and analyze the information needed to prepare employees' paychecks. Banks cannot operate without computers to distribute daily interest and balance clients' accounts. Without such machines, many of the transactions could not be completed within the available time. Similarly, the turnaround of goods in stores and factories would be much slower. The available variety of competitive goods would be much smaller and possibly would offer much less value to the customers.

Consider the case of a cheese vendor who sells a line of gourmet cheeses. He has determined through his experience with customers that soft French cheeses constitute a competitive product line. His variety of these cheeses attracts customers away from supermarkets whose offerings are very limited. His selection of exotic cheeses, together with his knowledge of cheese, improve his customer relations. By carrying appealing accessories, he can make large-margin sales through impulse buying prompted by his shop's ambience. In operating his enterprise, he uses the telephone to order stock and to check deliveries. He has a personal computer to track his inventory and to keep his accounts. He receives advertisements from suppliers. He monitors the promotional activities of his competitors on radio and TV and in newspapers and magazines. His information network is not integrated, nor is his use of information quantitatively assessed.

The vendor could easily be set on a downward spiral if a shrewd competitor appeared on the scene who was determined to wrest business away from him. This competitor could develop and use a more effective information network. Instead of using the PC only to keep accounts, he could use it to do better sales analysis. Programs are available for PCs to allow the competitor to minimize inventory cost, to accelerate turnover, and to plan sales campaigns. The competitor could link his PC to a network of cheese vendors to receive statistical data on popularity trends and new products. He could also link his PC to a data bank to obtain local population data. Armed with these data, the competitor could organize his sales promotions and design special programs to lure the speciality customers from other vendors by offering comparable or superior knowledge and helpful information. The competitor might choose to branch off into wine sales rather than offering cheese accessories, since wine is also a consumable and tends not to produce a saturated market.

Of course, the outcome of such a hypothetical rivalry is not certain, and return on investment is not guaranteed. However, the integrated network concept is likely to give the second merchant the competitive edge to succeed against the original vendor who opts for narrow specialization.

Perhaps it is the promise of a ubiquitous communication capability that integrated service systems offer that is the reason that the communication networks of the world are being developed toward that end. ISDN of Europe and INS of Japan are strong efforts to establish a network that will first handle voice and data and then extend to include a broader range of services. These additional services will eventually include broadband services based on video signals. The development of local area networks (LANs) is also extremely strong. These LANs cater to the local needs of a group of customers. A LAN can be set up within an office complex at a geographically confined location, within a university complex, or at a single company with many geographically dispersed locations. These LANs usually interconnect computers, data stores, facsimile and telex equipment, and telephones, and may include video conferencing.

Another network development is in television distribution and multiservice broadband distribution. In the United States, TV distribution comes in several forms. The three principal networks—NBC, CBS, and ABC—and a number of local independent stations broadcast over free space and distribute their nationwide programs via satellite or trunk cables. The cable TV companies distribute the TV programs via dedicated cable both from network stations and from a few specialized stations that broadcast only to the cable TV distributors via satellite. These specialized stations generate all-news or all-sports programs or only movies. Cable TV subscribers pay for the basic cable service and usually pay additional fees to receive certain special stations such as one or more movie channels. Most cable TV distributors offer at least 26 channels. The third type of TV distribution provides pay-as-you-view services via cable TV or special satellites. This type of TV distribution network is spreading in many countries at different rates.

At the same time, a number of countries have built small-scale, field-trial systems of broadband multiservice networks. Canada, Japan, the United Kingdom, France, and Germany each has one or more such experimental networks. These systems usually use optical fibers as the transmission medium and distribute one to three TV stations simultaneously, using a switching arrangement, into a subscriber's home. At the same time, the system provides FM radio reception and data and telephone channels. Some or all of the households are equipped with videophones. This type of effort is undoubtedly promoted by the recognition that a broadband distribution network to all households could increase per capita productivity, promote expansion and redistribution of trade, and change the social structure. All of these are the expected results of improved and more effective usage of information.

Momentum for building the communication network of the future is increasing. Many ambitious nationwide and continent-wide systems are being planned. It is interesting to note that the current communication network based on copper conductors provides telephone services to around 1 billion households using 10 billion km of wires. Studies indicate that the per capita GNP is directly proportional to the number of subscribers per 100 households. While it may be an overstatement to say that the wealth of a nation can be due entirely to the quality of the communication network, the two certainly seem to go hand in hand. It is also interesting to point out that the telephone requires one-thousandth of the bandwidth needed for videophone. The future network, therefore, must handle information rates at least a thousand times larger when the use of videophone reaches the level of use of today's telephone. At present, the largest bandwidth system needed for telephones is around 1.7 Gb/s. This means that the bandwidth needed in the future will be in the range of 1700 Gb/s. This is a staggering frequency range, and one that possibly can be reached only by using optoelectronic techniques.

25.7 SERVICES

Telephone, data links, facsimile, and videotext are some familiar services available for our communication needs. These services support our trade, industry, and individual activities. Television and radio are two other important services that provide entertainment and promote community links and trade through advertising. Physical transportation and printed material are also vital to our trade, industry, and individual and community needs. When broadband services can be provided over a new network, the possible types of services and their use will undoubtedly evolve as the result of perceived needs and competitive forces.

Whenever information in any of its various forms helps to increase the competitive edge in business, specialized information services immediately spring up. These services provide distillation and analysis of pertinent data for special markets. Such information services are themselves subject to competitive pressure, and only the ones that do the best job of synthesizing data will survive. These services assist the growth of per capita productivity in the manufacturing industry and foster profitable trading of manufactured products. Furthermore, improved information allows the consideration of factors such as worldwide resource limits so that worldwide commerce can be expanded without generating any types of crises.

An example will illustrate the argument just presented. Company A is large enough to set up a facility for analyzing control data and for collecting the analyses and disseminating the results to its units to facilitate rigorous product planning. Company B is too small to afford such a facility, and yet it must compete. The choice is simple; Company B must cooperate with other companies in the same situation to have some enterprising, qualified people set up a data analysis facility on a commercial basis to help each of them. The assumption here is that information is too vital to be ignored in the planning stage for any product that must be competitive. Therefore, Company C, a service organization, emerges. This process is likely to be mutually reinforcing, with the result that a hierarchy and network of services will be established.

Videotext is a service that has been evolving over recent years. Initially it was offered as readily accessible news in text form for TV viewers. On-demand news, timetables for various daily events, weather, and other similar items are available to the viewers at any time. Later, more ambitious schemes such as Presstel in the U.K., Bildschirmtext in Germany, Captain in Japan, Antiope in France, and Teletext in Canada were offered. In the United States, legal search and financial data are offered to paying customers via special data links. So far, the penetration of these services is not widespread in the general domestic area; however, special interest groups have been well satisfied with the services that are offered.

Financial institutious have found up-to-date financial data of immense benefit for improving their trading postures. Using a legal-search facility, as opposed to manual-search efforts, lawyers can obtain the kinds of substantive evidence and corroborative precedents to help them win cases. When a group of travel agencies started using data on travel and vacation facilities available through a travel data service, both the quality and quantity of their offerings improved. The travel agencies could satisfy the individual requirements of their customers more rapidly and could provide better matches. As a matter of fact, the fortunes of major and minor airlines are strongly dependent on data-handling facilities. Recently, some airlines offered computerized preboarding seat selection for any future flight. This resulted in better control on overbooking and in a reduction in airline staff at ticket counters and in gate areas. Customer satisfaction increased, and revenues started to climb.

These examples suggest strongly that information services are vital in our daily life as well as for industry and commerce. However, it is evident that service development is lagging behind technological progress. This is due in part to the inertia of our socioeconomic system, but it is also due in part to the lack of ready availability of the necessary technological tools. Nevertheless, the conjunction of technology and the need for information-related devices is imminent.

The existence of a broadband network capable of high-quality video transmission to every home and very high speed data interconnections will allow the development of services not as yet envisaged. The pay-as-you-view movie service offered by cable TV operators is making substantial gains in acceptance despite early subscriber resistance. The change came about when film distributors found that video cassettes are not popular sales items while cassette rental is highly popular. Hence, when new technological advances allowed the cable TV operators to offer pay-as-you-view films, cassette renters found it more convenient to order the film by phone rather than to make the trip to a rental place.

An ambitious scheme is under way to transform TV advertising. Advertising has traditionally been approached on an antagonistic basis. The crux of successful advertising is to catch the attention of the potential customer and lure him or her into becoming a customer. The latest scheme is to package commercials in such a way that they have entertainment value in their own right. Potential customers are then tempted to watch commercials voluntarily as part of the fun. This trend has two really significant consequences. First, advertising must be packaged differently. Initially it may be very entertaining, but eventually it probably will contain substantive discussions of values and comparisons with competition. Eventually advertising will become interactive with the viewer. The second consequence is even more far-reaching. Successful advertising must win the potential customer's confidence and become a trusted friend. The would-be customer discovers the fact that he requires information to understand his needs and to help him determine how these needs should be satisfied. This implies a major shift in allocation of the resource. Manufacturers and service providers would invest substantially in improving the information network on which their trading activity must take place. The cost per subscriber for the construction of the hardware for the broadband distribution network can be substantially higher since it can be repaid very handsomely through serving the information vendors. The investment that has been directed toward advertising aimed at consumers will be directed toward the information network that will provide a data bank as well as serving consumers' needs.

25.8 COMMERCIAL ACTIVITIES

"Business as usual" is still true even with the changes in the competitive environment generated by an overabundance of product choices created by a host of available technologies. It is still risky and uncertain, but full of opportunities.

The overall volume of commercial activities increases. Apparently more of this increase occurs in the service section as information transmission, storage, and processing facilities allow the collection, sorting and analysis, and dissemination of information to be carried out to meet the needs of the activities. Products will have greater variety, better distribution, increased value, better reliability, and more planned obsolescence. Most important, products will be lower in cost. The net result of these factors is that more wealth will be generated for more people. Of course, this statement applies at any time to describe our civilization. However, in our present era of the "information society," the emphasis is on limiting the proliferation of redundant and trivial data; gathering and distilling needed data into useful form; and broadcasting data in the most informative way for users. All of these are being done with a view to making better products more cheaply and to meeting real human needs.

Today's environment is different from that of the earlier industrial age. The era of long product cycles and product differentiation by price and features is all but over. With the abundance of technology and the corresponding abundance in products, "product cycle" will be meaningful only in terms of a sequence of products meeting a dynamically evolving need. Price and product differentiation become secondary to identifying needs for the target customer and developing the customer's perception of need for the product. Obviously, information is the key to success. Optoelectronics, with its ability to improve information-related technology, will play an increasingly dominant role in shaping this environment.

25.9 ANALYSIS OF THE SCALE OF IMPACT

Our excursion into a discussion of the nature of optoelectronics technology and the actions in the areas where optoelectronics technology will make an impact allows an analysis of the scale of the impact to be meaningfully presented.

Optoelectronics introduced the following advantages:

A transmission medium with nearly zero loss, infinite bandwidth, and zero cost.

An operational speed envisaged to be one thousand times the current speed of 500 Mb/s

An integrated signal transmission, storage, and processing system

Sensors integrated into the information system

The technical areas where the impact of optoelectronics will be greater are

Information transmission
Information storage
Information processing
Sensors

In turn, optoelectronics will penetrate into the information network, services, and commerce.

For this discussion the scale of impact will be analyzed only for the technical areas.

25.9.1 Information Transmission

With the prospect of single-frequency lasers and dispersion-compensated fibers operating in the $1.2-1.6$-μm wavelength region, transmission systems based on silica fiber are expected to show steadily improving performance levels. Repeater spans will increase to several hundred kilometers. Transmitter speed will reach 16 Gb/s. Receiver sensitivity will approach 3 dB from the quantum limit. Many wavelengths, at 20 nm down to several gigahertz, can be simultaneously carried by a single fiber in a bidirectional operating mode. For special repeaterless applications in the transoceanic environment, super-low-loss fibers will eventually be available. Hence, the means for establishing an extremely broadband and low-cost transmission system is available. These developments will open the way to the establishment of a totally transparent broadband network.

25.9.2 Information Storage and Processing

The versatility of the interaction of electromagnetic waves with matter has not been explored in any depth. Even so, optical storage disks with huge capacities are already available. Holographic techniques promise to increase storage density by utilizing the volume rather than merely the surface areas of the storage medium. Integrated optical techniques for stable, portable, physically convenient optical phase detection opens the way for powerful signal processors.

Even without integrated optics, optical signal processors provide a convenient means to achieve Fourier transform correlation and convolution. With new optically active materials such as electrooptic photorefractive and magnetooptic materials, sequential, parallel, and systolic signal-processing arrangements are expected. Reconfiguration is a real possibility waiting for development. Reconfiguration by holographic masks for interconnections on VLSI circuits has already been suggested. This technique will allow a new approach to solving interconnection problems, as well as offering a possibility to improve reliability and to provide firmware with on-chip programming. Lastly, since optoelectronics has the capability for high operational speeds, present approaches to signal processing are likely to be reexamined with a view to achieving a better trade-off of bandwidth versus ease of implementation and versatility. For example, in a speech-recognition system, the analysis of the signal could be extended, within the short time that is available, so that better features can be extracted. At the same time, the same technique could be combined with a coding scheme to produce compressed speech, say at 9.6 kb/s, with better fidelity.

25.9.3 Sensors

Optoelectronics opens the possibility of introducing a new variety of sensors for physical parameters that will not only extend versatility and sensitivity, but also conveniently interface with optoelectronic signal processors, sorters, and transmission systems. These sensors represent a set of key components for creating humanless factories and robots capable of working cooperatively with each other. This prediction is based on presently demonstrated sensitivities of sensors using coherent signals and fibers that are beyond those of all conventional sensors. Additionally, these sensors are inert and have remote-sensing capabilities.

25.10 SEMIQUANTITATIVE ESTIMATION OF RETURN ON INVESTMENT

It is an interesting challenge to estimate the probable return on investment for optoelectronics technology. We have a technology that has the potential for impact on an incredibly wide cross section of other technologies and, in turn, an impact on the entire basis of commercial activities. On an optimistic but not realistic global basis, it is possible to state that the return on investment by mankind in this technology may approach infinity. Based on current commercial practices, a quantitative basis for estimations could be as follows.

On the investment side:

Specific basic research applied to optoelectronics technology
Specific development effort toward key components
Related development of products that are subsystems or systems built
directly around a key optoelectronic component

On the return side, the corresponding results are:

Applied research stimulated by the basic research
Component sales
Subsystem and system sales

Each company can work out the return on investment for each area of its own activities in which it has chosen to participate. More correctly, however, a semiquantitative return on investment should have one additional component of investment:

Planning, organizing, and establishing the new markets and their new products

This investment should be made consciously and deliberately. If each activity in the optoelectronics area made such an analysis of return on investment, the total return on investment globally would be a measure of the total growth of the sum of the GNPs for all nations. In other words, optoelectronics is needed to allow the growth of our world economy that is currently directly dependent on how well we manage the utilization of information.

25.11 SOME REFLECTIONS

Do we have an abundance of information, or do we really have an abundance of data? After all the discussions of the impact of optoelectronics, it is not really clear. However, it can be deduced from the argument that we need to advance optoelectronics technology rapidly so that we can construct more powerful equipment for the transmission, storage, and processing of data. These equipments are going to be very popular with users, since much collation, sorting, and analysis work must be done to reduce a vast amount of data into useful information and knowledge. The information and knowledge can then be used to aid our business activities. In fact, data are being generated at an exponential growth rate since the interaction and cross-fertilization of the contributions of individual workers have been limited by the inherent limitations in our communication skill and facilities.

It is not unusual to hear someone say, "I invented this device in the U.S. and two days later another guy in Europe announced that he did the same thing" or "I must work hard on this concept; I'd like to reduce it to a practice that is patentable before someone else does it." These illustrate the fact that not only are we not communicating efficiently, but also we are deliberately avoiding communication for fear of generating competition. Hence, data are growing exponentially, since they are generated by individual efforts. Much of the data generated are redundant, or at least similar. On the other hand, conferences, seminars, teaching and public relations activities, and particularly TV broadcasts are effective in promoting communication. As a result, even in the parts of the world most remote from civilization, people are no longer staggered by airplanes and radios. Hence, the exponential growth of data is not strictly taking place.

At the same time, an idea can be extremely influential when applied at the right time. The Chinese have an ancient proverb that illustrates this fact:

"A spark can start a fire which consumes the entire land." Reduction of data to information is a highly complex issue. The addition of powerful machines merely helps us initially to do mechanical and rule-based analysis. We will struggle to find ways and means to capture the essence of our experience, our flexibility of manipulating and interpreting data, and many other mental functions, and transfer these to ever more powerful machines.

By extending our thought processes to machines, it is conceivable that an interdependent man–machine age will come. In some minor ways it has already arrived. Without computers we are already incapable of doing many of our daily tasks such as banking, not to mention such astonishing feats as landing on the moon. Without the car, many of us could not reach our office.

In this context, the impact of optoelectronics technology on our information society can be dimly understood.

Index

Heterick Memorial Library
Ohio Northern University

DUE	RETURNED	DUE	RETURNED
1.		13.	
2.		14.	
3.		15.	
4.		16.	
5.		17.	
6.		18.	
7.		19.	
8.		20.	
9.		21.	
10.		22.	
11.		23.	
12.		24.